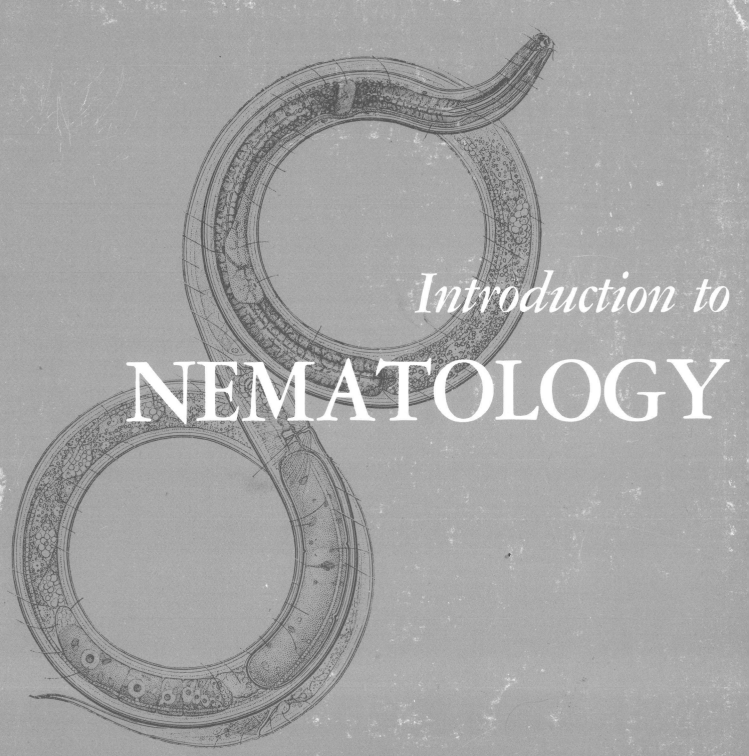

Introduction to

NEMATOLOGY

By the late **B. G. Chitwood** and **M. B. Chitwood**

UNIVERSITY PARK PRESS **BALTIMORE · LONDON · TOKYO**

INTRODUCTION
to
NEMATOLOGY

B. G. CHITWOOD and M. B. CHITWOOD

National Center for Primate Biology
University of California, Davis

INTRODUCTION
to
NEMATOLOGY

with contributions by

R. O. Christenson, L. Jacobs, F. G. Wallace, A. C. Walton, J. R. Christie, A. C. Chandler, J. E. Alicata, J. E. Ackert, J. H. Whitlock, T. von Brand, and T. L. Jahn

University Park Press
Baltimore • London • Tokyo

REPRINTED IN MEMORY OF
BENJAMIN G. CHITWOOD, 1907-1972

University Park Press
International Publishers in Science and Medicine
Chamber of Commerce Building
Baltimore, Maryland 21202

Library of Congress Cataloging in Publication Data
Chitwood, Benjamin Goodwin, 1907-1972.
Introduction to nematology
Includes bibliographies.
1. Nematoda. I. Chitwood, May Belle Hutson,
1908- joint author. II. Title.
(DNLM: 1. Nematoda. QX203 C543i)
QL391.N4C3623 1974 595'.182 74-19081
ISBN 0-8391-0697-1

CONTENTS

PREFACE

The *Introduction to Nematology* is herein presented under one cover for the first time. Section I, Parts One (1937), Two (1938), and Three (1941), with some revision, were reprinted as a single volume in 1950. Section II, Parts One and Two, covering postembryonic development and life histories, were not included in the revision, this being their first reprinting. I regret that time does not permit a thorough revision incorporating the many excellent life-cycle studies, especially those of the Spirurida.

Despite the abundance of data published in the meantime, the basic concepts of morphology and classification are as valid now as when first presented by B. G. Chitwood in 1937. Chapters in Section II, Part Two that were clearly outdated, by some of the more recent work pertaining to histochemistry and other aspects, have been omitted. The Volume, as herein presented, remains the most comprehensive treatise on the relationships, classification, and characteristics of all nematodes, and I sincerely believe that it will be very useful to students and scholars of Nematology around the world.

M. B. Chitwood

AN INTRODUCTION TO NEMATOLOGY

"Die Natur geht ihren Gang, und was uns als Ausnahme erscheint, ist in der Regel."

CHAPTER I

INTRODUCTION

B. G. CHITWOOD

Nematology is one of the younger sciences in that it has only recently begun to receive the attention due a subject of such wide academic and economic interest. Nematology, so named by the late Dr. N. A. Cobb, one of its most distinguished exponents, is by definition the study of what are commonly called thread worms or roundworms, or what Cobb called nemas. The mode of attack on the subject in this text is by a comparison of the organisms known as nematodes (or nemas) with one another and with other groups. Similar structures, similar life habits, or similar gross appearances whether they happen to occur in closely related or widely separated groups, are brought out. The exceptional diversity of nematodes makes nematology a fertile field for the study of analogy, convergence, homology, and divergence.

Nematodes have habitats more varied than have any other group of animals save the arthropods, and even this group is foreshadowed by the immense numbers of individual nematodes. Yet, of this massive assembly that live and thrive over the whole earth, few are even heard of by others than nematologists and even these are rarely seen.

Since the word *nema* is not yet in common use, the terms *worm* or *nematode* are usually applied. These words both have an immediate false or inadequate connotation which is soon dissipated if the hearer is shown one of those "beautiful little beasts." The grace of movement of some lowly soil-inhabiting forms finds little equal among other living organisms, being comparable to the gliding of snakes, which Solomon noted as one of the four mysteries of life. The complex patterns of their body markings and of the head and other parts might well be used in designing ladies' dresses. None of their grace and beauty is suggested by a name that carries the stigma "worm."

Now let us ask, where do nematodes live and how? Why are they so seldom seen by persons in general? The common forms which live in soil and water are usually too small to be seen by ordinary methods of visual examination; others, particularly some marine and parasitic species, are easily within the range of vision. Some are but 1/125 inch long, while others grow to relatively enormous dimensions. One species living in the kidneys of dogs and other mammals may be a yard long, the length of a fair sized snake, and attain the thickness of the little finger; this nematode is the "scarlet scourge," the kidney worm. Where do nematodes live? Almost anywhere. Nematodes antedated man in their conquest of the whole earth. Think of a place where any living organisms might survive, then go and look, and you will probably find nematodes.

As Cobb (1915) so aptly put it, "They occur in arid deserts and at the bottom of lakes and rivers, in the waters of hot springs and in the polar seas where the temperature is constantly below the freezing point of fresh water. They were thawed out alive from Antarctic ice in the Far South by members of the Shackleton Expedition. They occur at enormous depths in Alpine lakes and in the ocean. As parasites of fishes they traverse the seas; as parasites of birds they float across continents and over high mountain ranges." Man, without wings, flies in aeroplanes. Nematodes, without wings, fly in birds, bats, bees, flies, or fleas, or just catch on as these go by and sail with them. Few nemas have anything resembling feet, but here again they need not exert themselves in walking for representatives of the whole animal kingdom act as their common carriers, and even the winds may on occasion stoop to lift them and take them to their destination.

Think a moment. By what means may animals obtain their sustenance? The lowliest is the scavenger that lives upon the offal, decaying stuff, the putrid remains of other life. Of such beasts, there is among the nematodes a multitude that follow after other things, purifying, cleansing, bringing the dead back

to life in another form. That which dies must first be reduced into its lowest terms before it is available as food for other things. The buzzard has been honored through the ages and is still protected by law in many states. Yet does the buzzard do more to relieve the earth of waste than the unceasing efforts of these millions of small *worms?* They do not need our protection, for they are not seen by ordinary man. Soil nematodes are of chief importance in the destruction of dead plant material but they also take part in the latter part of decomposition of animal matter. Their interrelationships with other organisms, both dead and alive, might easily be so great that if tomorrow they all disappeared, a few weeks hence the foul odor of death might pervade the whole earth as the balance of life was destroyed.

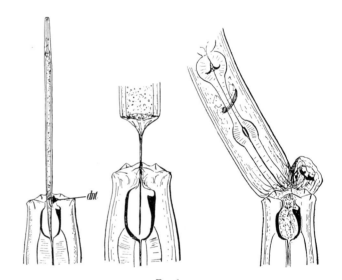

FIG. 1.
Mononchus feeding on *Rhabditis.* After Steiner & Heinly, 1922, J. Wash. Acad. Sc.

Some animals are herbivores, living entirely upon vegetable food. Of such forms we know many in the Nematoda. Some live on the outside of the plant, drawing its life juices by means of a minute hypodermic needle, the stylet; others may enter and wander through the tissues, causing much destruction; while others upon entering remain sedentary and through their fluids stimulate the formation of "nectaries," the cells of which supply their nourishment.

Among the nematodes there are many carnivores that prey upon the other microscopic animals living in their environment. Of these some might be compared to snakes that swallow their prey whole, while others, merely suck the living juices and cast away the shell as do the spiders. Others are equipped with massive jaws that rip and tear the bodies of their victims. Some live upon the smaller beasts, some upon their herbivorous cousins, while others live upon forms larger than themselves.

Man has divided nematodes into two groups, the free-living and the parasitic. These groupings are not always recognized by the animals themselves. The "host" can be only a habitat, and as such may or may not be suitable. We find among the nematodes every gradient from a free, unfettered life to a highly specialized parasitism. One form can not live upon or in a host; another can live upon a host at intervals but is not

1

forced to do so; another must pass a short part of its life within a host; and finally, there are others which can live only in a host throughout their lives.

How do they reproduce? Most nematodes are bisexual, the female producing ova which are retained within her uterus until fertilized by spermatozoa transferred to her by the male at copulation. After this a shell is formed and then the egg may be deposited, or it may be retained for a period within the mother, sometimes even until it hatches; thus some may be said to be oviparous while others are viviparous.[1]

Some forms may change from oviparous to viviparous during the life of the individual, being oviparous during early life and becoming viviparous with age. From the typical bisexual condition, we find every variant to the hermaphrodite. In many species males are rare and not entirely necessary, while in others males are entirely unknown. In such forms hermaphroditism, rather than parthenogenetism, appears to be the rule. Spermatozoa and ova may develop in the same gonad, a phenomenon known as syngonism.

The life of nematodes is usually a simple story; hatching from the egg as a small organism similar to the adult except for sexual characters, it grows and passes through a series of stages marked by ecdyses. Many parasites live a part of their life free in soil and change slightly in appearance as they become ready to enter their host and develop to sexual maturity. Some, however, may change considerably in appearance, and some must pass through one or even two intermediate hosts before reaching their final destination, and a few can use an indefinite series of intermediate hosts by reencysting as larvae infective for a final host. But far more spectacular than these are the few species in which there exist two alternate generations, one generation living free in soil while the other is parasitic. Such alternative generations of a single species may differ from one another in their morphology and habits more than members of larger groups which we term superfamilies.

How long may these lowly forms live? The time for completion of development from fertilized egg to sexually mature adult may be but three days in some saprophagous forms, while in other forms, such as the Guinea worm, a year may be necessary, and in some cases development may be delayed several years. We know little of their longevity as adults; some probably cannot survive over 3 weeks to a month while others may live many years. The greatest ages are probably attained by forms in immature resistant stages. Species such as *Anguina tritici*, the "eelworm" of wheat, may survive over 27 years in a dormant stage, according to Baker (Referred to by Needham in 1775).

Nematodes are known most widely, of course, as parasites of plants and animals, in which they are found in a great diversity of locations. In plants we find them attacking the root, bulb, stem, leaf, or flower, while in animals we find them in eye, mouth, tongue, digestive tract, lungs, liver, body cavity, muscle or joint, and, in fact, nearly any location seems to be suitable for some species of nematodes. Yet, though they are common as parasites, they are even more common as free inhabitants of soil and water.

Cobb (1915, p. 459) once estimated that 3,000,000,000 nematodes lived on a single acre of soil, the great majority in the top 3 inches. The beach sands often appear to be teeming with life, and of this life nematodes make up no small part. Cobb (1929) estimated the number of nematodes in the upper 20 mm. of marine beach sand as 1,500,000,000 per acre. The wheat gall formed from the kernel by *Anguina tritici* may contain from 11,575 to 18,051 specimens in moderate sized galls as determined by Byars (1920), and cases have been recorded with as many as 90,000 in a single gall, according to Cobb. In animals the number of specimens usually does not attain such great figures, but the larger size of the organisms makes records scarcely believable; the oxyurids found in the caecum of some iguanas filled a vessel apparently larger in size than the organ from which they were taken; 5,544 pinworms were passed by a man after one treatment; a small pup had a thousand ascarids. Rare instances are known in which human beings have suffered from more than a hundred Guinea worms, the entire skin being covered by a network of living worms. These cases are, of course, exceptional, but many similar records could be quoted.

Nematodes have great economic importance for man because of the large numbers and variety of nematodes that attack him. Among the parasites of man, hookworms and ascarids are two nematodes of major importance. The damage they may do can hardly be estimated, since we cannot accurately state the value of human life and health. However, though deaths due to such forms are not now relatively large, the loss of efficiency and ability to work due to the hookworm alone has caused the Rockefeller Foundation to spend over five million dollars in efforts to eradicate or control this parasite in the southern part of the United States. The campaign against the hookworm, begun by Dr. C. W. Stiles in 1903, and continued by local public health groups, the U. S. Public Health Service, and the Rockefeller Foundation, has undoubtedly been a great factor in the improvement of general sanitary conditions in the South. Despite the great advances which have been made in the past, the control of the hookworm still remains one of the major problems of medical parasitology.

A number of other nematode parasites of man, while less commonly known, are also of importance. *Trichinella spiralis* is probably the most dangerous single species, having in the past caused hundreds of deaths due to infestations acquired through the eating of uncooked or inadequately cooked pork. Although it is generally known that meats are rendered safe for human consumption if properly cooked, the knowledge is not universal, and either from lack of knowledge or failure to use it, apparently 10 per cent of the people in the United States acquire infestations with *Trichinella*, probably mostly non-clinical or sub-clinical.

Though man is said to be subject to parasitism by 32 species of nematodes, death due to nematode infestation is relatively rare and generally indicates a mass infestation. Among the parasitic diseases of man, dracontiasis, filariasis, elephantiasis, and onchocerciasis, which are caused by nematodes are of insignificant importance or unknown in North America. The common hookworm and ascarid, which do occur on this continent, cause emaciation and loss of vitality, and the hookworm often causes anaemia. The common pinworm, *Enterobius vermicularis*, known to many people as the cause of intense itching as it migrates from the rectum to the perineum in the vicinity of the anus, is probably the most cosmopolitan nematode of man for it has been found in all races and creeds, rich and poor, clean and filthy.

The nematodes parasitizing domesticated animals are too large in number to enumerate in this text. It is sufficient to say that the following numbers of species are known from these hosts: Dog, 36; cats, 33; cattle, 51; sheep, 63; horses, 69; and swine, 33. It has been estimated that the parasites of domesticated animals cause an annual loss of $500,000,000 in the United States. Of the various species involved, the large roundworm of swine, *Ascaris lumbricoides* v. *suis*, the swine kidney worm *Stephanurus dentatus*, swine lungworms, horse ascarids and strongyles, sheep lungworms and trichostrongyles, and poultry ascarids, cecum worms and gapeworms are probably the most destructive forms.

The nematodes parasitizing plants have, for the most part, been inadequately investigated. The root-knot nematodes are the best known parasites of plants and were until recently, considered a single species (*Heterodera marioni*) with a world wide distribution and a host list of some 1800 plants. Inconsistent results in experimental work led to re-study of the organisms and today we realize that they constitute a genus, *Meloidogyne* with several species. Members of this genus parasitize practically all types of commercial crops as well as weeds, trees and grasses. They are one of the major factors in crop production today, levying a tax of about 10% on all crops in the United States. Soil fumigation, crop rotation and selection of resistant varieties are the modern routine control procedures.

The genus *Heterodera* contains a number of more host specialized organisms which are cyst formers and hence particularly difficult to combat. Among these, *Heterodera schachtii*, the sugar beet nematode, *Heterodera rostochiensis*, the golden nematode of potatoes, and *Heterodera goettingiana*, the pea nematode, are economically important species which were apparently introduced into the United States from Europe. The distribution of *Heterodera schachtii* is today practically identical with that of established commercial plantings in this country while *H. goettingiana* is known in widely scattered locations from coast to coast. Thus far, *H. rostochiensis* appears to be confined to a limited area in New York State.

The genus *Ditylenchus*, or stem and bulb eelworms, includes such economically important pests as *Ditylenchus putrefasci-*

---

[1] The term oviparous is herein used to designate forms in which the young are passed before hatching from the vitelline membrane and without regard to the presence or absence of shell or the stage of development. Viviparous, on the other hand, signifies that the form gives rise to living young no longer covered by a vitelline membrane, even though a shell was formed within the mother.

ens of onions and *Ditylenchus destructor* of potatoes, as well as numerous other less conspicuous root parasites.

The eelworm, *Anguina tritici* is practically world wide in distribution due to its movement in seeds. In recent years meadow nematodes, *Pratylenchus* spp., ring nematodes *Criconemoides* spp., *Paratylenchus,* sting nematodes of the genus *Belonolaimus* and many other root parasites have been found to be deleterious to plant growth.

The genus *Aphelenchoides* includes many extremely destructive species attacking leaves and buds of plants. Among these *Aphelenchoides ritzema-bosi* of chrysanthemums, *A. fragarie* and *A. besseyi* of strawberries, and *A. oleisistus* of ferns and lilies are the outstanding species.

The place of nematodes in the development of general zoology is seldom recognized. Work on nematodes has contributed chiefly to the subject of cytology, this being largely due to the ease with which such studies may be made in nematodes. Such figures as Bütschli, Boveri, Hertwig, zur Strassen, E. van Beneden, and Goldschmidt used nematodes upon which to make the studies for which they are justly famous.

Bütschli (1875) is recorded as the first observer to witness formation of polar bodies by subdivision of the nucleus of the ovum, and he also gave full details regarding the character of the achromatic figure. E. van Beneden (1883) is credited with the original discovery of the separation of halves of each of the chromosomes from the 2 parents, and also for the discovery of the mechanism of mendelian heredity, the reduction division by which homologous chromosomes are separated. Regarding this work, E. B. Wilson states: "The cytological discoveries of this period reached their climax in the splendid researches of Edouard van Beneden on the history of the nuclei during fertilization of the egg of the nematode *Ascaris megalocephala,* which demonstrated that the chromosomes of the offspring are derived in equal numbers from the nuclei of the two conjugating germ-cells, and hence equally from the two parents. This fundamental discovery opened remarkable new possibilities for the detailed analysis of the nuclear organization and the cytological study of heredity and development . . . ."

Boveri (1893) followed the foregoing workers with the discovery that there is a cytological basis for the theory of Nussbaum and Weismann that there is a continuity of germ plasm, and that the soma may be regarded as a by-product without influence upon heredity. It was Boveri's work upon *Ascaris megalocephala* (= *Parascaris equorum*) which definitely established the significance of differentiation of animal cells, for he found that in the early cleavages there was a definite difference in the potentialities of the various cells, and that those cells which later formed the reproductive cells were set aside at the fourth cleavage. Other work in the same line by Boveri (1899, 1909), zur Strassen (1896, 1903, 1906), Martini (1903), Müller (1903) and Pai (1928) as well as several others, have contributed greatly to the study of cell lineage. Such work has led inevitably to the study of cytology and anatomy from the standpoint of cell constancy and has had results in the classic works of Goldschmidt (1903-1911) and Martini (1916).

The subject of nematology has a history of its own quite as ancient as any other field of zoology. This history follows in general the political and social history of Europe. Among the ancients, Moses, Hippocrates, Agatharchides, Aristotle, Pliny, Celsus and Vegetius may be cited. As would be expected, the oldest known forms are parasites of man. Preceding the European phase, however, there are some records left from the early Egyptian physicians. The "Papyrus Ebers," a manuscript which has been dated 1553-1550 B. C., was obtained by a German egyptologist, Ebers, in 1872. This manuscript contains passages indicating a knowledge of the large round worm, *Ascaris lumbricoides,* and the Guinea worm, *Dracunculus medinensis.* A knowledge of the disease caused by the hookworm, *Ancylostoma duodenale* is also claimed to be evident, but this may be due to confusion with schistosomiasis, a disease caused by trematodes. Any discussion of the ancients for nematology would be judged incomplete if it did not cite those much discussed verses of the Bible in which Moses (ca. 1250 B. C.) is supposed to have referred to the Guinea worm as the fiery serpent of the Israelites. This passage, Numbers 21: 6-9, reads as follows:

6. "And the Lord sent fiery serpents among the people, and they bit the people; and much people of Israel died."

7. "Therefore, the people came to Moses, and said, 'We have sinned, for we have spoken against the Lord, and against thee; pray unto the Lord that he take away the serpents from us.' And Moses prayed for the people."

8. "And the Lord said unto Moses, 'Make thee a fiery serpent, and set it upon a pole, and it shall come to pass, that everyone that it bitten, when he looketh upon it shall live.' "

9. "And Moses made a serpent of brass, and put it upon a pole, and it came to pass, that if a serpent has bitten any man, when he beheld the serpent of brass, he lived."

Whether or not this passage actually refers to the Guinea worm must even today be regarded as questionable. Küchenmeister (1857) studied the subject extensively and makes a convincing case in his comprehensive discussion. It has been suggested that the passage might be read ". . . . inflammatory serpents among the people and they layed upon the people" and also that the setting up of the brass image was of actual significance in impressing the people of the necessity of extracting the worm entire, for the chief damage of Guinea worm is due to incomplete extraction.

However true or false may be the reference to the fiery serpent, there can be no doubt concerning the passage which Plutarch attributes to Agatharchides of Cnidus (181-146 B. C.), teacher of Ptolmaeus Alexander, for herein we find the origin of the very name of the Guinea worm. The passage may be quoted as follows: "That the people taken ill on the Red Sea suffered from many strange and unheard of attacks, amongst others, worms, upon them, which gnawed away their legs and arms, and when touched, retracted themselves up in the muscles and there gave rise to the most unsupportable pains; but their evil has only been found then, neither before nor since amongst any other people." (The Greek word here used for worm is *dracontia micra,* literally "little snake," from which the generic name *Dracunculus* was derived. It has been suggested that this is actually a more detailed account of the plague visited upon the Israelites in the form of fiery serpents, in support of which there is the statement that the Children of Israel had only recently passed through the region of the Red Sea (from Hor towards Oboth on the way from the Sea of Suph) and that the period of elapsed time necessary to traverse this distance was between two and 12 months. This indeed approximates the developmental period of the Guinea worm, which is about 10 to 12 months.

There was a period between 450 B. C. and 200 A. D. which may be regarded as the time in which general knowledge of the people, and more particularly of the physicians of the day, was reduced to writing. Hippocrates (ca. 430 B. C.), in his "Aphorisms" evidenced some knowledge of the existence of worms, for he states: "To persons somewhat older, affections

FIG. 2.

Extraction of the Guinea worm (*Dracunculus medinensis*). After Velschius, 1674.

of the tonsils, incurvation of the spine, at the vertebra next to the occiput, asthma, calculus, roundworms, ascarids, acrochordon, ..." The roundworms of which he speaks were probably *Enterobius vermicularis*, for in another passage he mentions their occurrence in the vagina of women and of similar worms in horses.

Aristotle (384-322 B. C.), who has often been called the father of zoology, casually mentions nematodes on numerous occasions in his "Historia Animalium." Some of these references are as follows: ". . . . and some excrement yet within the living animal, like helminthes or intestinal worms. And of these intestinal worms there are three species: One named the flat worm, another the round worm, and the third the ascarid. These intestinal worms do not in any case propagate their kind."

"Gnats grow from ascarids; and ascarids are engendered in the slime of wells, or in other places where there is a deposit left by the draining off of water."

"Dogs when they suffer from worms eat standing corn."

In order to understand these references, one must realize the general opinions which Aristotle, as well as many scientists of later days, held. The theory of spontaneous generation, which is as old as science itself, was firmly believed by Aristotle. He saw no reason to doubt that lower forms of life, such as worms, might easily originate in mud and slime, for in such places life is certainly abundant and earthworms may be easily confused with ascarids.

Celsus (53 B. C.-7 A. D.) distinguished between roundworms (nematodes) and flatworms (cestodes), but added little more to our knowledge. Columella (ca. 100 A. D.) is said to have first mentioned an ascarid (*Neoascaris vitulorum*) from a calf, and Vegetius (ca. 400 A. D.) to have first mentioned the horse ascarid (*Parascaris equorum*). Galen (130-200 A. D.) adds somewhat to our knowledge, but he must be regarded chiefly as a collector and as a commentator of the works of previous authors. Among other things he noted worms in the mouth and flesh of the European red mullet, *Trigla* sp., and these were found also in place of the young, possibly referring to a philometrid. He mentions the Guinea worm but evinces some doubt as to its animality, though Pliny (27-79 A. D.) had called it a worm. This same doubt, apparently, was carried down through the ages and remnants of it persisted until the latter part of the 18th century.

The next several hundred years may well be considered the dark ages of nematology, as of all science. Little was done beyond the cataloguing of some previous Greek, Roman, and Arabic knowledge in the field of medicine. During this period, Paulus Aegineta (ca. 600 A. D.) is notable through the size of his work which was purely one of compilation. Following this writer, there is a nearly blank period of about 400 years, terminated by Avicenna (980-1037) who is notable through his text on medicine in which he refers incidentally to ascarids and dracunculids. An early glimmer of awakening is seen in the observations of Albertus Magnus (1200-1280) whose comments upon nematodes from falcons appears to be the first record of nematodes from birds of any type.

The reawakening of scientific investigation apparently began in the 16th century, and the period from this time until the latter part of the 18th century may be regarded as the medieval period of our subject. Caesalpinus (1519-1630) is quoted by Redi as the discoverer of *Dioctophyma renale* in the kidney of a dog. Vinegia (1547) is quoted as having described a filarid, as well as two species of intestinal nematodes, from a falcon. Aldrovandus (1602) is said to have first used the term Vermes as a group which he placed within the larger group Insecta. Redi (1684) recorded previous reports of worms in vertebrates, and added numerous new records from hosts as diverse as polecat, lion and fish. Tyson (1683) apparently made the first early attempts to study nemic anatomy; he described the lips, digestive tract, anus and eggs of *Ascaris lumbricoides*. The use of the microscope contributed greatly to advance in our knowledge, which thereafter becomes rapid. Borellus (1656) discovered the first free-living nematode, the vinegar eel; the same form was also studied by Robert Hooke (1667); the latter author also figures an "eel in paste," *Panagrellus redivivus*. Other early microscopists also showed an interest in "eelworms," among them being Joblot (1716, 1718, 1754), Baker (1742, 1744), who exploded a popular myth, saying: "Some people erroneously assert that the sharpness of vinegar is owing to nothing but the striking of these creatures upon the tongue and palate with their acute tails," others, later, were Ledermüller (1763), Leeuwenhoek (1719-1722) and Spallanzani (1769, 1787). Spallanzani described a free-living nematode, saying "it was more beautiful than the eel found in vinegar, but smaller." Spallanzani quotes Valisnieri (1710) as having claimed that vinegar eels metamorphose "en moucherons." Needham (1745) described and figured for the first time a plant-parasitic nematode, *Anguina tritici*, saying that blighted wheat was dissected and numerous minute fibers were observed. In order better to study them he placed them in water, upon which they came to life. This peculiar phenomenon is one of the outstanding features of the wheat eelworm. O F. Müller (1786) later described several species of free-living nematodes from fresh water.

In the field of parasitic nematodes this same period was characterized by reports and observations of additional species, among which were those of Andry (1750), Fröhlich (1789-1802), Leske (1779-1784), Unzer (1751), Werner (1782), Göze (1782), Zeder (1800), and Rudolphi (1809-1819). To Zeder we owe the vernacular grouping of the parasitic worms into five classes, namely Roundworms, hooked worms, suckered worms, flat worms and bladder worms. These groups received their latin names from Rudolphi.

As we have seen, the origin of intestinal worms was the subject of much philosophical debate from the time of the ancients down to the latter part of the 18th Century. It was at this time (1780) that the Copenhagen Academy of Science announced a prize for the best essay on "Ob der Saamen der Intestinalwurmer: als Bandwürmer, der Faden-oder Drathwürmer, (*Gordius*), der Spulwürmer, der Egelwürmer (*Fasciola*) u.s.w. den Thieren angeboren sei oder von aussen hineinkommen welches durch Erfahrung und andre Gründe zu beweisen und im letztern Fall Mittel dagegen vorzuschlagen." There were three competitors, Göze, Bloch and Werner, the winner being Bloch. Both Bloch and Göze upheld the view that intestinal worms were inborn, while Werner took the opposite view. This appears to be one of those interesting cases in which good presentation triumphs over fact. Bloch (1788, p. 83) stated that he believed that intestinal worms are inborn in the animals and that it is their true destination to live there; also, that they constitute a unique class of the animal kingdom. The points which he presented are too many to enumerate but they included the occurrence of worms in infants and in the body cavity and muscles of animals, and finally the fact that species of animals living in the same place, eating the same food, and drinking the same water have different types of worms. Modern nematology now recognizes that Bloch was not entirely wrong. Worms (ex ascarids, hookworms) can be inborn if the dam is infected at the proper stage of pregnancy.

In the period between 1820 and 1870, there was an uptrend in the character of work done. We find in this period the pioneers Bojanus (1817, 1821) and Cloquet (1824) both of whose work on the anatomy of ascarids, though crude from our present point of view, was exact for that period. Both of these workers made free-hand sections and careful dissections, thereby adding much to the general knowledge. They were followed by Owen (1835) who discovered *Trichinella spiralis* in human muscle. Leidy (1846) found it in swine muscles. Herbst (1851-52) and Virchow (1859) followed with work on the transmission of this worm from rats to pigs. Later, Leuckart (1860) and Zenker (1860) made important contributions to our knowledge of this important parasite. It was during this same period that knowledge of the life history of other nematodes began to increase. Most noteworthy of these were the discoveries by Leuckart (1865a-b) and Metchnikoff (1865, 1866) of the alternation of generations in the life history of *Rhabdias*, and the development of *Camallanus lacustris* in the intermediate host, *Cyclops*. Priority in both instances is claimed by Metchnikoff, though Leuckart, who first published the observations, stated that Metchnikoff merely followed Leuckart's directions. At Leuckart's suggestion for an investigation of *Dracunculus*, based on the similarity of *Camallanus* and *Dracunculus* larvae, Fedtschenko undertook the work which later (1871) resulted in his discovery that *Dracunculus medinensis* also utilizes *Cyclops* as an intermediate host.

While various others contributed to the knowledge of life histories, the observations of Davaine (1858) and Leuckart (1866) seem to be most worthy of recording. Davaine's studies and experiments with the eggs of *Ascaris lumbricoides* and *Trichuris trichiura* led him to the conclusion that segmentation and development to the vermiform stage in the egg without hatching indicates that the egg must then be introduced into the final host for completion of development, a rule subsequently found to have many exceptions. Leuckart contributed a classification of the types of life cycle known in nematodes, listing *Oxyuris vermicularis*, *Heterakis gallinarum* and *Trichuris ovis* as forms reintroduced into the final or primary host in the egg stage; *Protospirura muris* (intermediate host, *Tenebrio*), *Camallanus lacustris* (intermediate host, *Cyclops*), *Ollulanus tricuspis*, *Trichinella spiralis* (rats and pigs both act as intermediate and final hosts) ; and *Raphidascaris acus* (both intermediate and final host, fish), as forms in which the larvae develop in a second or intermediate host before being capable of developing to the adult stage in the final host; *Uncinaria stenocephala* as a form which hatches outside of the body, passes through a rhabditoid stage and is then infective for the final host (per os) ; and finally *Rhabdias bufonis* as a form with an alternation of generations, one parasitic and one free living.

In the field of taxonomy, Dujardin (1845), Diesing (1850-1851), and Schneider (1866) added many species and gathered together the information in a more usable form. Bastian (1865) widened the field of nematology through his descriptions, in a single paper, of 100 new species of free-living nematodes. Following this work interest in the field received a decided impetus.[2]

It is difficult to set any particular date as the beginning of recent nematology for, indeed, its origin was gradual. The great span of years during which such men as Leuckart worked really place them in more than one period. For convenience we may fix 1870 as an approximate date, because following this time there was a market uptrend in the amount and average quality of the work. In the field of free-living nematodes the beginning of publication by Bütschli and de Man very definitely sets the time, while Leuckart's extensive treatise on human helminthology (1876) serves as a landmark in the work on parasitic nematodes.

The greater part of our knowledge concerning life history, anatomy, and systematics is due to recent efforts. Thus we find the experimental evidence of the direct infection of man by *Ascaris* eggs done by Grassi (1887, 1888), Lutz (1887, 1888), Calandruccio (1890) and Epstein (1892), after experimental failure by earlier workers. Calandruccio (1890; see also Grassi, 1887) presented experimental evidence of a similar life cycle for *Trichuris* in man, and Grassi (1888) did the same in the case of an ascarid in dogs. While none of these discoveries were remarkable in the light of previous predictions by Davaine, Leuckart, and others, they led the way to actual demonstrations of life cycles.

Similarly, the life history of *Ancylostoma duodenale* as worked out by Looss (1896-1911) was foreshadowed by the work of Leuckart (1866) on *Uncinaria* and incomplete studies by other workers on *Ancylostoma*. However, the discovery by Looss of skin penetration as a mode of entry was a definite addition to knowledge, and the excellence and completeness of his life history study must stand unparalleled as yet.

Manson's (1878) discovery that mosquitoes act as intermediate hosts in the transmission of *Wuchereria bancrofti* is perhaps the most famous work occurring during this period, its fame hardly justified from the zoologist's standpoint since the life history of this form is essentially similar to that of *Camallanus*, but nevertheless interesting since it was the first case involving a biting insect as intermediate host.

Later workers have revealed the life histories of countless forms which cannot be enumerated here. The discovery of intermediate hosts (insects) for *Physocephalus* and *Ascarops* by Seurat (1913) ; the discovery of the migration of *Ascaris lumbricoides* through the lungs by Stewart (1916) ; and the addition of a new type of intermediate host (earthworms) for lungworms by Hobmaier (1929) are cited as perhaps the most interesting of recent discoveries.

In the general subject of anatomy, Looss' monograph on *Ancylostoma duodenale*, Martini's on *Oxyuris equi*, and the several papers by Goldschmidt (1903-1911), Rauther (1906-1918) and de Man (1886) are probably the most outstanding.

During recent years there has been an increasing number of species and genera, and these genera have necessitated expansion into more families and superfamilies. Among the recent taxonomists de Man, Cobb, Micoletzky, Filipjev, Steiner, Goodey, Allgén, Stekhoven, Kreis, de Coninck, and Thorne account for most of the genera of free living and plant parasitic nematodes. The nematode parasites of arthropods have been studied by Steiner, G. W. Müller, Christie, Basir and Bovien. Prominent taxonomists concerned with nematode parasites of vertebrates have been Looss, Railliet, Travassos, Walton, Skrjabin, Baylis, Maplestone, Cram, Seurat, Ortlepp, Mönnig, Lent, Freitas, and Caballero.

---

[2] For further information concerning the early history of nematology, consult Schneider (1866) and Pagenstecher (1879).

Bibliography

AEGINETA, PAULUS (ca. 600 A. D.) 1844.—The seven books of Paulus Aegineta translated from the Greek; by Francis Adams, v. 1. 683 pp. London.

AGATHARCHIDES (181-146 B. C.). See Plutarch.

ALBERTUS MAGNUS (1193-1280) 1495.—De animalibus libri vigintisex novissime impressi. Venezia.

ALDROVANDUS, U. 1602.—De animalibus insectis libri septem, cum singulorum iconibus ad vivum expressis. 767 pp. (pp. 765, 720-721 cited).

ANDRY, N. 1701.—De la Generation des vers dans le corps de l'homme . . . etc. 317 pp. Amsterdam.

ARISTOTLE (384-322 B. C.) 1910.—Historia animalium. Translated by D'Arcy Wentworth Thompson.

AVICENNA (980-1036) 1564.—Libri in re medica omnes, qui hactenus ad nos pervenere. Id est libri canonis quinque etc.). Venetiis. (Lib. IV, Sect. iii, tract ii, cap. 21, pp. 460-461.)

BAKER, HENRY, 1742.—The microscope made easy. London. (p. 79, pl. 7, fig. 10 cited).
    1743.—Idem. (pp. 81-83 cited).

BASTIAN, H. C. 1865.—Monograph on the Anguillulidae, or free nematoids, marine, land and freshwater; with descriptions of 100 new species. Tr. Linn. Soc. Lond., v. 25 (2) : 73-184, pls. 9-13, figs. 1-248.

BENEDEN, E. VAN. 1883.—Recherches sur la maturation de l'oeuf, la fécondation et la division cellulaire. Gand & Leipzig. 424 pp., pls. 3, 10-19 bis.

BLOCH, M. E. 1788.—Traité de la generation des vers des intestinis et des vermifuges. Strasbourg. 127 pp., 10 pls.

BOJANUS, L. H. 1817.—Bemerkungen aus dem Gebiete der vergleichenden Anatomie. Russ. Samml. Naturf. u. Heilk. Riga. u. Leipzig, v. 2(4) : 523-552.
    1821.—Enthelminthica. Isis (Oken) v. 1(2) : 162-190, pls. 2-3.

BORELLUS, P. 1653.—Historiarum, et observationum medicophysicarum, centuria prima, etc. Castris. 240 pp.
    1656.—Ibid . . . centuriae quattuor. 384 pp. Parisiis.

BOVERI, T. 1893.—Ueber die Entstehung des Gegensatzes zwischen den Geschlectszellen und den somatischen Zellen bei *Ascaris megalocephala*, nebst Bemerkungen zur Entwicklungsgeschichte der Nematoden. Sitzungsb. Gesellsch. Morph. u. Physiol. in München (1892), v. 8 (2-3): 114-125, figs. 1-5.
    1899.—Die Entwickelung von *Ascaris megalocephala* mit besonderer Rücksicht auf die Kernverhältnisse. Festschrift. v. Carl v. Kupffer, Jena. pp. 383-430, figs. 1-6, pls. 40-45, figs. 1-45.
    1909.—Die Blastomerenkerne von *Ascaris megalocephala* und die Theorie der Chromosomenindividualität. Arch. Zellforsch., v. 3: 181-268, pls. 7-9.

BRYAN, C. P. 1931.—The papyrus Ebers. 167 pp. New York.

BÜTSCHLI, O. 1875.—Vorläufige Mittheilung über Untersuchungen betreffend die ersten Entwickelungsvorgänge im befruchehen Ei von Nematoden und Schnecken. Ztschr. Wiss. Zool. v. 25:201-213.

BYARS, L. P. 1920.—The nematode disease of wheat caused by *Tylenchus tritici*. U. S. Dept. Agric. Bull. No. 842, pp. 1-40, figs. 1-6.

CALANDRUCCIO, S. 1890.—Animali parasiti dell 'uomo in Sicilia. Atti Accad. Gioenia Sc. Nat. in Catonia. 1889-1890, An. 66, 4. s., v. 2:95-135.

CELSUS, A. C. (53 B. C.-7 A. D.) 1657.—De medicina libri octo, ex recognitione Joh. Antonidae von Linden D. & Prof. Med. Pract. Ord. 558 pp. (Lib. 4, cap. 17 cited).
    1876.—Traité de médicine de A. C. Celse. Précédée d'une preface par Paul Broca. Paris. 797 pp.

CLOQUET, J. 1824.—Anatomie des vers intestinaux ascaride lumbricoide et échinorhynque géant. 130 pp. Paris.

COBB, N. A. 1915.—Nematodes and their relationships. U. S. Dept. Agric. Yearbook for 1914, pp. 457-490.

DAVAINE, C. J. 1857.—Recherches sur le développement et la propagation du trichocéphale de l'homme et de l'ascaride lombricoide. Compt. Rend. Acad. Nat. Sc. Paris, v. 46(25) : 1217-1219.
    1859.—Idem. J. physiol. del'homme. Paris, v. 2: 295-300.

DIESING, K. M. 1850.—Systema Helminthum. v. 1, 679 pp. Vindobonae.

    1851.—Idem., v. 2, 588 pp., Vindobonae.

DUJARDIN, F. 1845.—Histoire naturelle des helminthes ou vers intestinaux. 654 pp., Paris.

EPSTEIN, A. 1892.—Ueber die Uebertragung des menschlichen Spulwurms (*Ascaris lumbricoides*). Verhandl. d. Versamml. d. Gesellsch. f. Kinderheilk. deutsche Naturf. Aerzte, Wiesb. (1891) v. 9:1-16.

FEDTSCHENKO, B. A. 1871.—On the formation and increase of *Filaria medinensis* L. (Russ.). Inviest, Imp. Obsh. Liub., Estestvozn., Antrop., Moskva, v. 8(1):71-78.

FROELICH, J. A. VON. 1789.—Beschreibungen einiger neuen Eingeweidewürmer. Naturforsch. v. 24: 101-162, pl. 4, figs. 1-31.

    1791.—Beyträge zur Naturgeschichte der Eingeweidewürmer. Ibid. v. 25: 52-113, pl. 3, figs. 1-17.

    1802.—Idem., v. 29:5-96, pls. 1-2.

GALEN, C. (130-200) 1552.—De simplicum medicamentorum facultatibus libre xi. Lugdoni. 725 pp.

GOEZE, J. A. E. 1782.—Versuch einer Naturgeschichte der Eingeweidewürmer thierischer Körper. 471 pp., 44 pls., Blankenberg.

GOLDSCHMIDT, R. 1903.—Histologische Untersuchungen an Nematoden. Zool. Jahrb. Abt. Anat., v. 18(1): 1-57, figs. A-D, pls. 1-5.

    1908.—Das Nervensystem von *Ascaris lumbricoides und megalocephala*. I. Ztschr. Wiss. Zool. v. 90: 73-136, figs. A-W, pls. 1-4.

    1909.—Idem. II. Ibid. v. 92(2): 306-357, figs. 1-21, pls. 1-3.

    1910.—Idem. III. Festschrift Hertwig, v. 2: 256-354, figs. 1-29, pls. 17-23.

GRASSI, G. B. 1887.—*Trichocephalus* und *Ascaris*-Entwicklung. Preliminarnote. Centralbl. Bakt. u. Parasitenk. 1 J., v. 1(5): 131-132.

    1888.—Weiteres zur Frage der *Ascaris*-Entwickelung. Ibid., 2. J. v. 3(24): 748-749 dated April 4.

    1888.—Beiträge zur Kenntnis des Entwicklungscyclus von fünf Parasiten des Hundes. Ibid., 2 J., v. 4(20): 609-621.

HERBST, G. 1851.—Beobachtungen über *Trichina spiralis*, in Betreff der Uebertragung der Eingeweidewürmer. Nach. Gesellsch. K. Wiss. Göttingen. (19): 260-264.

    1852.—Ibid., (12) 183-204.

HIPPOCRATES (460-375 B.C.) 1849.—Works of Hippocrates, Translated by F. Adams. London, "Aphorisms," 3.26.

HOBMAIER, M. 1929.—Ueber die Entwicklung der Lungenwürmer des Schweines *Metastrongylus elongatus* und *Choerostrongylus brevivaginatus*. Eesti Loomaarstlik Ringvaade, v. 5(3):67-71.

HOOKE, R. 1667.—Micrographia: etc. London. (pp. 216-217, pl. 25, fig. 3 cited).

    1754.—Micrographia Restaurata or copper plates of Dr. Hooke's wonderful discoveries by the microscope. London. 65 pp., 33 pls. (pp. 45-46, pl. 23, fig. 3 cited).

JOBLOT, L. 1716.—Description de plusieurs nouveaux microscopes sur grande multitude d'insectes que vaissent dans les liqueurs, etc. (Not seen.) Cited by Joblot, 1754.

    1718.—Descriptions et usages de plusieurs nouveaux microscopes (etc.). 78 pp. Paris (Pt. 2, pp. 2-12, 27-30, pl. 1, pl. 9, figs. e-f. Cited in text).

    1754.—Observations d'histoire naturelle (etc.). 2 v. in 1. 38 pp., 124 pp., 78 pp., 27 pp. Paris (1, p. 2 T. 1 aceti, cited in text.)

KUECHENMEISTER, G. 1857.—On animal and vegetable parasites of the human body, a manual of their natural history, diagnosis, and treatment. Translated by Edwin-Lankester, v. 1: 1-452, 8 pls.

LEDERMUELLER, M. F. 1762.—Nachlese seiner mikroskopischen Gemüths-und Augen-Ergötzung (etc.) 94 pp., 50 pls. Nürnberg.

    1763.—Ibid. 202 pp., 101 pls., Nürnberg.

LEEUWENHOEK, O. 1722.—Opera omnia seu arcana naturae (etc.) 258 pp., Lugduni Batavorum.

    1798.—The select works of Leeuwenhoek. Samuel Hoole, London, pp. 126-128.

LEIDY, J. 1846.—Entozoon in the superficial part of the extensor muscles of the thigh of a hog. Proc. Acad. Nat. Sc., Phila., v. 3(5):107-108.

LESKE, N. G. 1779.—Anfangsgründe der Naturgeschichte. Leipzig. (Not verified).

    1784.—Idem. Leipzig. (Not verified).

LEUCKART, R. 1860—Untersuchungen über *Trichina spiralis*. 57 pp., 2 pls. Leipzig & Heidelberg.

    1865.—Zur Entwickelungsgeschichte des *Ascaris nigrovenosa*. Zugleich eine Erwiederung gegen Herrn Candidat Mecznikow Arch. Anat. Physiol. u. Wiss. Med. pp. 641-658.

    1865.—Helminthologische Experimentaluntersuchungen. Vierte Reihe. Nachr. v. d. k. Gesellsch. d. Wiss. u. d. Georg. Aug. Univ. Göttingen. (8):219-232.

    1866.—Untersuchungen über *Trichina spiralis*. 120 pp., 2 pls. Leipzig & Heidelberg.

    1876.—Die menschlichen Parasiten. v. 2., 3. Lief., 882 pp. Leipzig & Heidelberg.

LOOSS, A. 1896.—Notizen zur Helminthologie Egyptens I. Centralbl. Bakt. Parasitenk., 1 Abt., v. 20(24-25):863-870.

    1879.—Zur Lebengeschichte des *Ankylostoma duodenale*. Centralbl. Bakt., etc. 1 Abt. v. 24:483-488.

    1901.—Ueber das Eindringen der Ankylostoma Larven in die menschliche Haut. Ibid., v. 29(18):733-739, 1 pl., figs. 1-3.

    1905.—The anatomy and life history of *Agchylostoma duodenale* Dub. A monograph. Rec. Egypt. Govt. School Med., v. 3: 1-158, pls. 1-9 figs. 1-100, pl. 10, photos. 1-6.

    1911.—Idem., Part II. Ibid. v. 4:159-613, pls. 11-19, figs. 101-208, photographs 1-41.

LUTZ, A. 1888.—Weiteres zur Frage der Uebertragung des menschlichen Spulwurms. Centralbl. Bakt. u. Parasitenk. 2. J., v. 3(9):265-268.

    1888.—Zur Frage der Uebertragung des menschlichen Spulwurms. Weitere Mittheilungen. Ibid., 2. J., v. 3(14):425-428.

MAN, J. G. DE. 1886.—Anatomische Untersuchungen über freilebende Norsee-Nematoden. 82 pp., 13 pls.

MANSON, P. 1878.—On the development of *Filaria sanguinis hominis*, and on the mosquito considered as a nurse. J. Linn. Soc. Lond., Zool. (75), v. 14:304-311.

MARTINI, E. 1903.—Ueber Furchung und Gastrulation bei *Cucullanus elegans* Zed. Ztschr. Wiss. Zool. v. 74(4):501-556, pls. 26-28.

    1916.—Die Anatomie der *Oxyuris curvula*. Ztschr. Wiss. Zool. v. 116: 137-534, pls. 6-20.

METCHNIKOFF, I. I. 1865.—Ueber die Entwickelung von *Ascaris nigrovenosa*. Arch. Anat., Physiol. u. Wiss. Med., pp. 409-420, pl. 10, figs. 1-11A.

    1866.—Entgegen auf die Erwiderung des Herrn Prof. Leuckart in Giesen, in Betreff der Frage über Nematodenentwicklung. 23 pp. Göttingen.

MUELLER, H. 1903.—Beitrag zur Embryonalentwickelung der *Ascaris megalocephala*. Zoologica, Stuttgart, Heft 41, v. 17, Lief. 3-4: 1-30, figs. 1-12, pls. 1-5.

MUELLER, O. F. 1786.—Animalcula Infusoria fluviatilia et marina. (etc.) 367 pp., 50 pls., Hauniae.

NEEDHAM, F. 1745.—An account of some new microscopical discoveries. London. 126 pp., 6 pls. (pp. 85-89, pl. 5, figs. 6-7 cited).

    1775.—Lettre ecrite à l'auteur de ce recueil: Obs. Phys. Hist. Nat. v. Cinquème: 226-228.

OWEN, R. 1835.—Description of a microscopic entozoon infesting the muscles of the human body. Tr. Zool. Soc. Lond. v. 1:315-324, pl. 41, figs. 1-9.

PAGENSTECHER, H. 1879.—In Bronn's Klassen u. Ordnung des Their-Reichs., v. 4:1-252.

PAI, S. 1928.—Die Phasen der Lebenscyclus der *Anguillula aceti* Ehrbg. u. die ihre experimental morphologisches Beeinflusung. Ztschr. Wiss. Zool. v. 131: 293-344.

PLUTARCH.—Symposiacon (9th question, 8th book cited).

RAUTHER, M. 1906.—Beiträge zur Kenntnis von *Mermis albicans* v. Sieb. Habilitationsschrift. 76 pp., 3 pls., 26 figs. Jena. Also, Zool. Jahrb., Abt. Anat., 1906.

    1907.—Ueber den Bau des Oesophagus und die Lokalisation der Nierenfunktion bei freilebenden Nematoden. Zool. Jahrb. Abt. M., v. 23: 703-740, pl. 38, figs. 1-9.

    1918.—Mitteilungen zur Nematodenkunde. Zool. Jahrb. Abt. Anat., v. 40:441-513, pls. 20-24, figs. 1-40.

REDI, F. 1684.—Osservazioni . . . intorno agli animali viventi che si trovano negli animali viventi. 253 pp., 26 pls., Firenze.

Rudolphi, C. A. 1809.—Entozoorum sive vermium intestinalium historia naturalis. v. 2 (1), 457 pp., Amstelaedami.

1819.—Entozoorum synopsis cui accedunt mantesia duplex et indices locupletissimi, 811 pp. Berolini.

Schneider, A. 1866.—Monographie der Nematoden. 357 pp., 122 figs., 28 pls. 343 figs. Berlin.

Seurat, L. G. 1913.—Sur l'évolution du *Physocephalus sexalatus* (Molin). Compt. rend. Soc. Biol. v. 79:517-520, figs. 1-4.

1916.—Contributiones a l'étude des formes larvaires des nématodes parasites héteroxenes. Bull. Sc. France et Belgique. 7 s., v. 49(4):297-377.

Spallanzani, L. 1769.—Nouvelles récherches sur les découvertes microscopiques, etc. 2 pts., 298 pp., 293 pp. Londres & Paris.

1787.—Opuscules de Physique animal et végétale. v. 1.

Stewart, F. H. 1916.—On the life history of *Ascaris lumbricoides.* Brit. Med. J., v. 2:6-7, 486-488, 753-754.

Strassen, O. zur. 1896.—Embryonalentwickelung der *Ascaris megalocephala.* Arch. Entwicklungsmech. v. 3:27-105, 133-190.

1903.—Geschichte der T-Riesen von *Ascaris megalocephala.* Teil I. Zool., v. 17, Heft 40(1):1-37, pls. 1-5.

1906.—Die Geschichte der T-Riesen von *Ascaris megalocephala* als Grundlage zu einer Entwickelungsmechanic dieser Spezies. Zool. v. 17, Heft 40(2):39-342.

Thorne, G. & Giddings, L. A. 1922.—The sugar-beet nematode in the Western States. U. S. Dept. Agric. Farmers' Bull. No. 1248, 16 pp.

Tyson, E. 1683.—*Lumbricus teres,* or some anatomical observations on the round worm bred in human bodies. Phil. Tr., Lond. (146), v. 13:154-161, 1 pl., figs. 1-4.

Unzer, J. A. 1751.—Beobachtungen von den breiten Würmern (Vermes, Cucurbitini) Hamb. Mag., v. 8(3):312-315.

Vegetius (ca. 400 A. D.). 1781.—Artis veterinariae sive mulomedicinae libri quatuor. 37 pp. Mannheim.

Virchow, R. L. 1859.—Recherches sur le développement du *Trichina spiralis.* Compt. Rend. Acad. Sc., Paris, v. 49 (19):660-662.

Werner, P. C. V. 1782.—Vermium intestinalium. 144 pp., pls. 1-7; in 28 pp., pls. 8-9.

Wilson, E. B. 1925.—The cell in development and heredity. New York. 1232 pp.

Zenker, F. A. von 1860.—Ueber die Trichinen-Krankheit des Menschen. Arch. Path. Anat. v. 18, n. F., v. 8(5-6): 561-572.

CHAPTER II

GENERAL STRUCTURE OF NEMATODES

B. G. CHITWOOD

Having briefly considered the history of general nematology, the question naturally arises, "What is a nematode?" In answering this question it is necessary first to designate a nematode as a triploblastic, bilaterally symmetric, unsegmented, non-coelomate animal of the Series Scolecida Huxley, 1865 (or Vermes Amera Bütschli, 1910). Within this group Hyman (1940) recognizes two subdivisions or subseries, namely Acoelomata, including the Platyhelminthes and Nemertea and the Pseudocoelomata, including the Rotatoria, Gastrotricha, Kinorhyncha (or Echinodera), Nematoda, Nematomorpha, Acanthocephala, and Entoprocta (Bryozoa). The latter group corresponds to the Aschelminthes Grobben, 1908. The single character is the presence of a body cavity either without complete mesodermal lining or with such a lining formed by migratory mesenchymatous cells. Such groupment of the scolecidan phyla leave certain questionable points since the Acanthocephala show many similarities to the Platyhelminthes and the Entoprocta presents similarities to the Ectoprocta which are considered schizocoelous.

The Phylum Nemathelminthes or threadworms was formerly used as a group to include the Nematoda, Acanthocephala, and Gordiacea (or Nematomorpha). The Acanthocephala or spiny headed worms have recently been placed as a separate phylum associated with the Cestoda, a class of flat worms of the Phylum Platyhelminthes. The Nematomorpha or horse hair worms pass through a stage in their development which is grossly very similar to the Echinodera. They also differ from the Nematoda in having an apparently unsymmetrical esophagus and in having gonads opening at the posterior end of the body through a cloaca in both sexes. The Nematomorpha is a highly specialized group and was placed in the Nematoda only on the assumption that they evolved from such forms; this view is hardly tenable since at the present time all of our evidence indicates that the gonads opened into the cloaca of both sexes in primitive forms, while in the female nematode they open separately on the ventral side of the body. For this reason the Nematomorpha is considered a separate phylum. The Rotifera or wheel animalcules, like the Nematomorpha, differ from nematodes in that the female reproductive system empties into the cloaca and in addition are characterized by ciliation of the intestinal epithelium while in nematodes the intestinal epithelium is lined with a degenerate form of cilia, not known to be vibratile, the *bacillary layer.* Gastrotrichs and echinoders are interesting little creatures presenting many similarities to nematodes. They differ from nematodes in that they are either parthenogenetic or hermaphroditic but never definitely bisexual, while nematodes usually have two distinct sexes, though hermaphroditic and parthenogenetic females are known to occur; cilia also are characteristic of the former groups and occur in the excretory systems in both groups, and sometimes on the body surface of the Gastrotrichs.

The Nematoda is an exceedingly variable group and there is scarcely a positive statement that could be made regarding their anatomy which would apply to all forms. For this reason the reader should indulge the authors' many qualifications of their statements and likewise consider that any statement not qualified probably should be.

GROSS MORPHOLOGY

Nematodes are more or less cylindrical, sometimes fusiform, rarely sac-like, pear shaped or otherwise modified, particularly in the adult female. There is usually an oral opening and the opening is usually surrounded by lips bearing sensory organs, though the lips are absent in several groups. The mouth is terminal, at the anterior end of the body; it is followed by a mouth cavity, or stoma, an esophagus, intestine, and a rectum terminating in a ventral terminal, or subterminal anus, or cloacal opening. The body is covered with cuticle; internally the body is not metamerically segmented. There are usually no external appendages, but appendages do occur in rare forms. The body wall is composed of a hypodermis or epithelium which is situated beneath the cuticle, and a single layer of longitudinal muscles. Sexes are usually separate, the male reproductive system opening directly into the rectum forming a cloaca, while the reproductive system of the female has a separate opening, the ventrally situated vulva. Excretory and nervous systems are present but there is no circulatory system.

The general character of each of the organs may be described as follows:

External covering. Nematodes are covered by a non-cellular layer termed the cuticle. This layer is often spoken of as chitinous but it differs chemically from the material present in the chitinous exoskeleton of arthropods, consequently the term cuticle is adopted for the external covering of nematodes, the term being used in a loose sense and without chemical significance.

The cuticle is usually marked by regularly arranged transverse grooves or striae, deeper striae sometimes occur at intervals in which case they are spoken of as annulations and the intervals between them as annules. So far as known, this annulation is superficial and involves only the cuticle. Longitudinal striae are sometimes present also and a few forms are known which have oblique or cross-hatched cuticular markings. In addition to these markings the cuticle may be longitudinally thickened forming expansions which are termed *alae.* These alae may be lateral or sublateral and extend the entire length of the body; they may be confined to the anterior end of the body, the "neck region," in which case they are termed *cervical alae;* or they may be confined to the posterior part of the body being then termed *caudal alae.* The latter are found on the tail of the male sex in several groups and apparently aid in the attachment of the male to the female during copulation. Caudal alae are sometimes modified into forms very characteristic of particular groups and may be termed *bursae*.

Nematodes, like arthropods, pass through a series of stages in development; these stages are marked by molts, a second cuticle being formed beneath the old, and the earlier shed at each molt. There are usually four molts in the development of a nematode, the stage following the fourth molt being the fifth or adult stage. During the development of some nematodes the first one or two cuticula may be retained instead of being cast off and such a stage is characterized by an unusual ability to withstand adverse environmental conditions. Such larvae are said to be ensheathed.

Unlike the arthropods, the "larvae" of most nematodes are very similar in body form to the adult, differing only in size

and development of the reproductive system. From this point of view it would be more proper to refer to the developmental stages of nematodes as "nymphs," but unfortunately the term larva is so deeply intrenched in the literature that it could hardly be suppressed at the present time.

Hypodermis. Nematodes do not have a true epithelium in that the outermost cellular layer is not in direct relation with the environment but is covered by a cuticle and for this reason it is termed the hypodermis. This layer takes various forms but it usually consists of a delicate protoplasmic tube internally thickened in four longitudinal lines, these thickenings being termed the dorsal, lateral, and ventral *chords*. Usually the regions between the chords, *interchordal areas*, contain no nuclei the nuclei being confined to the chords. Distinct cell walls may be present or the entire hypodermis may be syncytial. Sometimes the hypodermis is slightly thickened in four longitudinal lines between the customary chords. When these

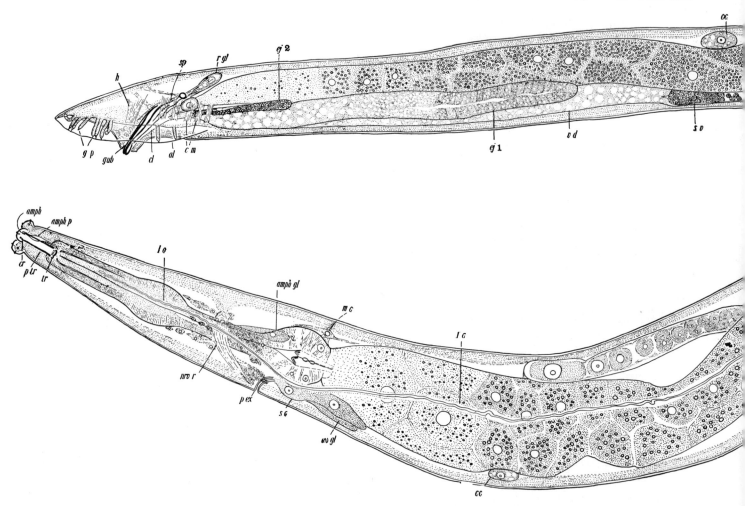

Fig. 3.
Rhabditis strongyloides. Male and female showing general anatomy.

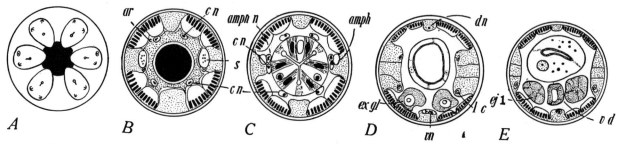

Fig. 4.

Rhabditis strongyloides. Cross sections. Diagrammatic. A—Head. B—Stomatal region showing cephalic nerves, arcade, and sensilla.

C—Esophageal region (precorpus) shows amphidial glands. D—Postbulbar region showing subventral excretory cells and lateral canals. E—Posterior part of male showing vas deferens and large ejaculatory glands.

8

thickenings are conspicuous they are often termed chords but this term should more properly be limited to instances in which the submedian thickenings contain nuclei of the hypodermis.

Musculature. The musculature of the body may be divided into two general types, the somatic musculature and the specialized muscles. The somatic musculature is the general muscular layer of the body wall and is composed of a single layer of more or less spindle shaped cells attached to the hypodermis throughout their lengths. These groups of cells lying parallel are separated by the longitudinal thickenings of the hypodermis into four primary muscle fields, the dorso- and ventro-submedian. The less conspicuous submedian thickenings of the hypodermis further subdivide them into eight fields, the subdorsal, dorsolateral, ventrolateral, and subventral fields. The muscle cells in each field usually act as a unit during contrac-

apparently of the same origin as the somatic musculature but limited to some particular part of the body such as labial muscles, somato-esophageal muscles, somato-intestinal muscles, rectal muscles and copulatory muscles.

Digestive tract. Lips. In free living nematodes there are usually six lips, two subdorsal, two lateral, and two subventral; these surrounding the oral opening. In some forms the lips are quite large while in other forms they are very inconspicuous. Sometimes lips are absent and often the two dorsal, the subventral and lateral lips on each side become fused, forming three lips. In still other cases two lateral lips or labial structures may be present. When lips are absent or rudimentary, the oral opening is usually surrounded by a delicate circum-oral membrane which is sometimes thickened to form a circum-oral elevation. In some specialized forms structures such as leaf crown, probolae, or hooks may take the place of lips.

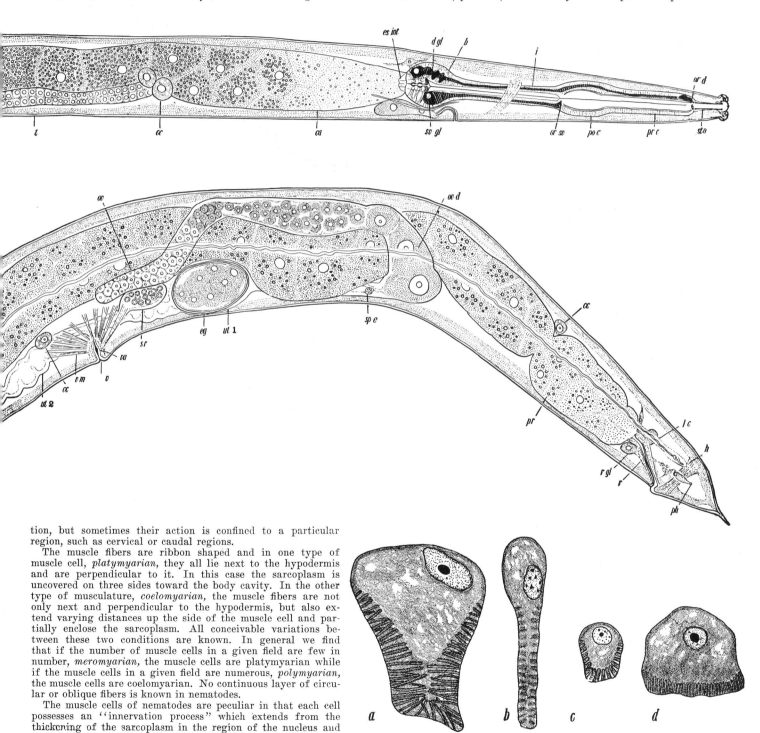

tion, but sometimes their action is confined to a particular region, such as cervical or caudal regions.

The muscle fibers are ribbon shaped and in one type of muscle cell, *platymyarian*, they all lie next to the hypodermis and are perpendicular to it. In this case the sarcoplasm is uncovered on three sides toward the body cavity. In the other type of musculature, *coelomyarian*, the muscle fibers are not only next and perpendicular to the hypodermis, but also extend varying distances up the side of the muscle cell and partially enclose the sarcoplasm. All conceivable variations between these two conditions are known. In general we find that if the number of muscle cells in a given field are few in number, *meromyarian*, the muscle cells are platymyarian while if the muscle cells in a given field are numerous, *polymyarian*, the muscle cells are coelomyarian. No continuous layer of circular or oblique fibers is known in nematodes.

The muscle cells of nematodes are peculiar in that each cell possesses an "innervation process" which extends from the thickening of the sarcoplasm in the region of the nucleus and communicates with the nerve ring or with a dorsal, ventral, or submedian nerve.

Specialized muscles. Under this heading we list muscles

Fig. 5.

Cross sections of muscle cells. a—*Heterakis gallinarum.* c—*Metoncholaimus pristiuris.* b—*Trichuris ovis.* d—*Oesophagostomum dentatum.* After Chitwood, 1931, Ztschr. Morph.

STOMA. This term is used to designate that portion of the digestive tract between the oral opening and the definitely tri-radiate beginning of the esophagus. It is lined with cuticle, usually covered by a fibrous tissue and surrounded by a mass of cells known as "arcadial tissue"; sometimes the stoma is surrounded by muscular tissue due to an anterior growth of the esophagus. Under the heading *stoma* we include the mouth cavity or buccal cavity together with its walls, the various forms taken by the organ being commonly spoken of under the names buccal capsule, vestibule, pharynx, etc. The stoma may be cylindrical, prismoidal, subglobular, and many other forms, it may contain teeth, denticles, or rasps; it may be either large and well developed or indistinguishable. It may be transformed into a protrusible spear in which case the structure is termed a stomatostyle. A tooth may be transformed into a stylet in which case this structure may be termed an onchiostyle.

Basically the stoma is divisible into three parts; a lip cavity, cheilostom; an elongated tube, protostom; and a short valve at the junction with the esophagus, telostom. The protostom is subdivisible into three regions, namely, prostom, mesostom, and metastom. The latter subdivisions of the protostom are indicated by more or less distinct joints of the cuticular lining. The walls of the various parts of the stoma are named rhabdions, each rhabdion deriving its name from the corresponding stomatal region (cheilorhabdion, protorhabdion, prorhabdion, etc.). Further details on this subject are given by Steiner (1933) and Chitwood and Wehr (1934).

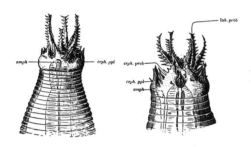

FIG. 6.
Acrobeles crossatus. Cephalic region showing probolae. After Steiner, 1929, Ztschr. Morph.

ESOPHAGUS. The esophagus is a muscular structure comparable to the pharynx of trematodes, gastrotrichs and echinoders. It is lined with cuticle and has a triradiate lumen, one ray directed ventrad while the other two are directed subdorsad. The rays of the lumen or esophageal radii may converge distally or be rounded distally forming marginal "tubes." The gross appearance is highly varied but these variations are centered around a few general forms. The esophagus may consist of corpus, isthmus and bulb, the first part being cylindrical and of moderate width, the second narrow and cylindrical, and the third pyriform. In some instances the corpus is distinctly subdivided into a cylindrical anterior part (procorpus) and an oval posterior part (metacorpus). The isthmus is sometimes indicated only by a slight constriction. The posterior swelling (bulbar region) may or may not possess well developed valves. In other forms the esophagus is clavate, cylindrical, or composed of a short, narrow anterior, muscular part and a long, wide posterior grandular part. In rare instances the esophagus may become highly degenerate showing little evidence either of its triradiate character or of muscle tissue.

There are usually three esophageal glands situated within the wall of the esophagus, one dorsal and two subventral, the dorsal gland opening near the anterior end of the esophagus and the subventrals somewhat further posteriad, but in some instances all three of the esophageal glands open into the stoma and in other instances all three may open far back in the esophagus. The esophageal glands may be serially reduplicated and due to degeneration of muscular tissue, may project outside the normal esophageal contour forming a stichosome, the individual cells being termed stichocytes. Sometimes the glands are multinucleate but more often they are uninucleate.

ESOPHAGO-INTESTINAL VALVE. The esophagus is connected with the intestine through a short structure termed the esophago-intestinal valve which often projects some distance into the lumen of the intestine. Like the esophagus it is lined with cuticle. This structure may be triradiate, dorsoventrally or laterally elongated.

INTESTINE. The intestine or mesenteron is composed of a single layer of epithelial cells. The internal surface of these cells is usually but not always covered by a layer of minute rod-like structures, the *Bacillary layer*. The intestine is often covered by irregular muscle fibers and a connective tissue sheath.

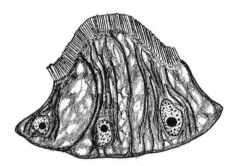

FIG. 7.
Cephalobellus papilliger. Intestinal epithelium. After Chitwood, 1931. Ztschr. Morph.

Usually the intestine is a straight tube but sometimes outpocketings or *ceca* occur near the anterior end of the intestine. More rarely definite twists or loops occur.

During postembryonic development of some parasitic species (the Mermithoidea) the cell walls of the intestinal cells disappear as does the lumen, the resulting syncytium being known as a fat body or trophosome while in others (the Strongylina) little or no cell division takes place but nuclear division proceeds resulting in the formation of multinucleate cells.

Toward its extremities, the intestine is often somewhat modified in character and sometimes the regions are more or less clearly set off from the remainder, the anterior part being termed the cardiac region and the posterior part, the prerectal region.

RECTUM OR CLOACA. The hind gut is lined with cuticle as is the esophagus, it is not triradiate in character but is more bilateral, and dorsoventrally flattened. At the anterior end the hind gut connects with the intestine through the intestino-rectal sphincter or valve which is formed of intestinal tissue. Usually there are one dorsal and two subventral rectal glands which open into the rectum while in the male the vas deferens also opens into the rectum from the ventral side forming a cloaca. In both sexes the hind gut usually empties posteriad on the ventral surface of the body. Various developments of the dorsal wall of the cloaca give rise to accessory sexual structures, such as the spicules and gubernaculum of the male.

Nervous system. The nervous system is composed of a circum-esophageal commissure, nerve ring, with which several ganglia and a number of longitudinal nerves are associated. There are six nerves which are directed anteriad from the nerve ring, each of which has three chief distal branches. These branches innervate sensory organs of the anterior extremity which may be either tactoreceptors (sensory papillae or setae) or chemoreceptors. The chemoreceptors of the anterior extremity are the amphids, two in number, lateral, or dorsolateral in position. Externally they may be pore-like, spiral, circular or of numerous modified forms. Internally the amphids have a pouch containing a group of nerve endings, the sensilla, connected with the amphidial nerve. The pouch itself, is a dilation of the amphidial gland, the body of which is situated in the lateral chord posterior to the nerve ring. There are a dorsal, a ventral, four submedian, and one, two, or three pairs of lateral nerves situated in the hypodermis posterior to the nerve ring. A variable number of commissures are present between the longitudinal nerves. Neither the dorsal nor the submedian nerves contain ganglia; the lateral nerves contain a few ganglia; the ventral nerve is in reality a chain of ganglia. "Innervation processes" extend from the somatic muscle cells to the dorsal, ventral, and submedian nerves. Scattered tactile sensory organs may be present either generally over the surface of the body or they may be confined to the posterior extremity of the male (genital papillae). One pair of tactoreceptors (the deirids or cervical papillae), situated near the nerve ring is fairly constant in many large groups. In all cases they are innervated by the lateral nerve. Paired postanal lateral sensory organs similar to the amphids are present in many nematodes, these structures being termed phasmids.

10

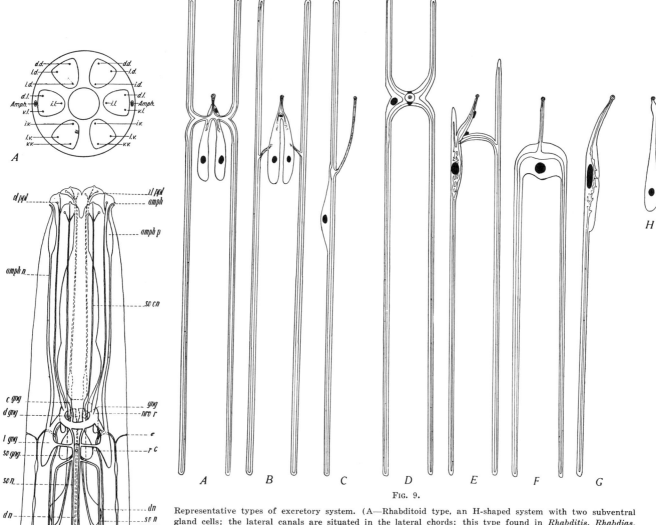

FIG. 9.

Representative types of excretory system. (A—Rhabditoid type, an H-shaped system with two subventral gland cells; the lateral canals are situated in the lateral chords; this type found in *Rhabditis*, *Rhabdias*, and the Strongylina in general. B—Variant of A, found in *Oesophagostomum*. C—Tylenchoid type, an asymmetric system with the lateral canals and gland cell confined to one chord. D—Oxyuroid type, an H-shaped system without subventral glands, with a greatly shortened terminal duct usually in form of a reservoir. E—Ascaroid type, a shortened H-type nearly ∩ in form; no excretory reservoir is apparent.

F—Cephaloboid type of excretory system, ∩ in shape; members of the Spirurida have an essentially similar system except the terminal duct is not so well developed. G—Anisakid type, a reduced form related to E in which one lateral canal completely disappeared. [Arrangement according to J. F. Mueller, 1927]. H—Single ventral cell type, present in Chromadorina, Monhysterina [except Plectidae] and Enoploidea.)

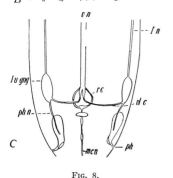

FIG. 8.

Diagrammatic representation of nervous system. A—Pattern of cephalic sensory organs of hypothetical primitive nematode. B—Anterior part of nervous system of *Rhabditis terricola*. C—Posterior part of nervous system of *Rhabditis terricola*. A & B after Chitwood & Wehr, 1934, Ztschr, Parasit.

Excretory system. The excretory system is perhaps, the most varied system in nematode anatomy and in some large groups it appears to be entirely absent. In some nematodes there are two lateral canals situated in the lateral chords and connected with each other, anteriorly and ventrally, by an excretory sinus. The canals may extend anterior to the sinus (H-shaped system) or may not (inverted U-shaped system. Sometimes in the former case there are two subventral excretory glands situated in the body cavity, and sometimes the canals are confined to only one lateral chord. The excretory sinus is connected with the ventrally situated excretory pore by a cuticularly lined reservoir or by an elongated terminal duct similarly lined. This type of system may be composed of numerous cells. In other nematodes the excretory system consists of a single excretory gland cell situated in the body cavity and without canalicular connections. Such a cell may be greatly elongated having an intracellular duct but in this case the duct is not lined with cuticle, there is no cuticularly lined terminal duct or reservoir. It connects with the excretory pore directly and at the connection there may be a dilation or ampulla.

11

Reproductive system. The male reproductive system consists of one or two tubular testes which empty into the vas deferens which in turn unites posteriorly with the rectum forming a cloaca. The vas deferens may or may not have a particular region termed the seminal vesicle. The female reproductive system consists of one or two, rarely larger numbers, of tubular ovaries connected with a uterus or uteri which terminate in the vaginal opening on the ventral surface of the body at the vulva. In parasitic nematodes the vagina and/or uterus is usually more highly developed with more prominent musculature than in free-living forms.

Nematodes develop usually, from fertilized eggs, the spermatozoa being furnished by the male at copulation, but in some nematodes spermatozoa develop in the same gonad as the ova in worms having secondary sexual characters of the females, and in still other cases parthenogenesis takes place.

Body cavity. The body cavity of nematodes is completely or partially lined by a delicate connective tissue layer of mesenchymatous origin.

Circulatory and *respiratory systems* are not known, the movement of the fluids of the body cavity apparently serving these purposes.

AN OUTLINE CLASSIFICATION OF THE NEMATODA

CLASS PHASMIDIA

Phasmids present, caudal glands and hypodermal glands absent; terminal excretory duct cuticularly lined, lateral canals present; cephalic sensory organs papilloid (very rarely setose); external amphids pore-like, usually small and labial in position (some diplogasterids exceptional); somatic papillae usually absent (exception *Labidurus*); somatic setae absent; deirids (cervical papillae) usually present; caudal alae or cuticular bursa commonly present; male with paired genital papillae, sometimes with one medio-preanal papilloid supplement; pseudocoelomocytes few, usually 4-6; subventral esophageal glands never opening at or near anterior end of esophagus. Inhabit soil (very rarely aquatic), plants and animals.

ORDER RHABDITIDA

Esophagus fundamentally composed of corpus, isthmus and bulbar region, best differentiated in larval stages; often clavate to cylindrical in adult stage; lips usually 3 or 6; esophageal glands uni- or binucleate; intestinal cells uninucleate or polynucleate; mero- or polymyarian. Parasitic representatives may or may not require intermediate host.

Suborder Rhabditina

Stylet absent; lips 6, 3, 2, or 0; corona radiata absent; female reproductive system relatively simple, vagina usually transverse and usually not heavily muscled; terminal excretory duct tubular and elongated, system varied but lateral canals paired; esophagus may or may not be terminated by a valved bulb; intestinal cells uni-, bi- or tetranucleate; intestinal lumen well developed; rectal glands present; caudal alae (so-called bursa) if present, containing papillae rather than "rays". (Saprophages in soil (rarely water), parasites of annelids, molluscs, arthropods, or vertebrates, or associated with plants.)

Superfamily Rhabditoidea Travassos, 1920

Stoma commonly distinct, lips usually distinct; esophagus consisting basically of corpus (pro-and metacorpus), isthmus and valved or nonvalved swelling; females with 2 or 1 ovary; males usually with two spicules and gubernaculum; *phasmids pore-like.* (Saprophagous or parasites of annelids, molluscs, arthropods or vertebrates, often associated with plants).

Rhabditidae Oerley, 1880.—3 or 6 lips or 2 pairs of oral hooks; stoma cylindrical, walls rigid, terminated by glottoid apparatus; esophagus usually with distinct procorpus and metacorpus, isthmus and valved bulb; female with 2 or 1 ovary, if 1, vulva preanal; male with caudal alae, variously developed, usually 9 pairs of genital papillae and phasmids in pattern. Monogenetic. (Parasites of molluscs, annelids, crustacea, insects, amphibia, saprophagous).

Rhabditinae Micoletzky, 1922.—Lips not replaced by hooks; cuticle not highly ornamented. [*Rhabditis, Cruznema, Rhabditella, Rhabditoides, Parasitorhabditis, Brevibucca*].

Diploscapterinae Micoletzky, 1922.—Lips replaced by hooks; cuticle not highly ornamented [*Diploscapter*].

Bunonematinae Micoletzky, 1922.—Lips not replaced by hooks; cuticle highly ornamented. [*Bunonema, Bogdanowia, Craspedonema, Rhodolaimus*].

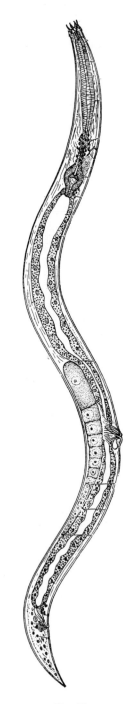

FIG. 10.
Acrobeles elaboratus
(Cephalobidae). After
Thorne, 1925, Tr.
Am. Mic. Soc.

Cylindrocorporidae Goodey, 1939.—3 or 6 lips, stoma cylindrical, walls rigid, extremely elongate; telorhabdions platelike; esophagus with pro- and metacorpus fused and short, isthmus, and pyriform glandular swelling; female with 2 or 1 ovary; male with genital papillae in pattern, 5 to 9 pairs. Monogenetic. (Saprozoic or parasites of gut of amphibians, reptiles, and mammals). [*Cylindrocorpus, Goodeyus, Myctolaimus, Longibucca*].

Rhabdiasidae Railliet, 1916.—Free-living generation as in Rhabditinae (2 ovaries); parasitic generation with reduced or capsuliform stoma with thick walls; esophagus clavate; females hermaphroditic. Heterogenetic. (Free-living generation saprophagous, parasitic generation in lungs of amphibians and reptiles). [*Rhabdias, Entomelas, Acanthorhabdias*].

Angiostomatidae (Blanchard, 1895).—Lips 3 indistinct, stoma thick walled, infundibuliform or capsuliform; esophagus with corpus indistinctly subdivided isthmus and muscular pyriform bulb with reduced valve; female amphidelphic, male with caudal alae and patterned genital papillae. Monogenetic. (Parasites in gut of snails, amphibia, and ? *Erinaceus*). [*Angiostoma, ? Spiruroides*].

Steinernematidae Chitw. and Chitw., 1937.—Lips indistinct, stoma reduced, vestibuliform, esophagus with simple corpus, indistinct isthmus and muscular reduced bulb; female amphidelphic; male without caudal alae, genital papillae in paired linear preanal series. Monogenetic. (Parasites of insects) [*Steinernema, Neoaplectana*].

Diplogasteridae Steiner, 1929.—Lips usually weak or absent; stoma extremely variable, usually with 1 or more retrorse teeth; esophagus with distinct procorpus and metacorpus (latter usually a very powerful suction bulb) isthmus and pyriform glandular region (latter with no valves but containing distinct muscles). Female usually with two ovaries; male with 9 pairs of genital papillae and phasmids; caudal alae narrow, absent, or very rarely of moderate width; spicules setose in wide arc; gubernaculum commonly hooked (External or internal consorts of insects, free in soil, and associated with plants).

Diplogasteroidinae Filipjev & Stekhoven, 1941.—Stoma cylindroid to prismoidal or collapsed; labial rugae absent or very weak; esophagus elongate. [*Rhabditolaimus, Rhabditidoides, Diplogasteroides, Neodiplogaster ? Demaniella*].

Cephalobiinae Filipjev, 1934.—Stoma consisting of 2 well developed, distinctly sclerotized tandem parts, teeth if present, in posterior part of stoma; labial rugae weak or absent; esophagus elongate. [*Cephalobium, Acrostichus, Loxolaimus, Butlerius, Odontopharynx*].

Pristionchinae n. subf.—Stoma very short and wide, retrorse teeth at base present or absent; labial rugae absent or weak; esophagus relatively short and thick. [*Pristionchus, Peronilaimus Lycolaimus*].

Diplogasterinae Micoletzky, 1922.—Prostom wide, heavily sclerotized, containing large retrorse teeth, mesostom collapsed or very weakly sclerotized, labial rugae very prominent; esophagus elongate or short. [*Diplogaster, Mononchoides*].

Tylopharynginae Filipjev, 1934.—Stoma narrow with basal enlargments simulating a stylet. [*Tylopharynx, Tylenchodon*].

Strongyloididae Chitw. & McIntosh, 1934.—Free-living generation with 2 lateral lips, stoma reduced, surrounded by esophageal tissue, esophagus with procorpus, metacorpus, isthmus and valved bulb; female with 2 ovaries; male without caudal alae, medio ventral preanal organ and patterned genital papillae. Parasitic generation with greatly elongate esophagus, Heterogenetic. (Free-living generation saprophagous, parasitic generation in gut of all vertebrates except fish). [*Strongyloides, Parastrongyloides*].

Cephalobidae Chitw. & McIntosh, 1934.—3, 6 or 4 lips, or 3 or 6 probolae, or 6 odontia and 6 cirri; stoma with jointed rhabdions or collapsed or both; esophagus with corpus (pro- and metacorpus sometime distinct, isthmus and valved bulb or glandular swelling. Female usually with 1 ovary and vulva just behind middle of body, rarely amphidelphic (*Alloionematinae*). Male without caudal alae or rarely inconspicuous ones; genital papillae in pattern, preanal organ present or absent. Monogenetic or heterogenetic with generations similar. (Saprophagous, or consorts of plants and land dwelling invertebrates).

Panagrolaiminae Thorne, 1937.—Stoma not completely surrounded by esophageal tissue; esophagus terminated by valved bulb; female with 1 ovary not doubly flexed posterior to vulva. [*Panagrolaimus, Macrolaimus, Tricephalobus, Panagrodontus, Plectonchus, Procephalobus, Neocephalobus, Turbatrix, Panagrellus, Panagrobelus, Chambersiella*].

Alloionematinae Chitw. & McIntosh, 1934.—Stoma not completely surrounded by esophageal tissue; esophagus terminated by valved bulb; female with 2 ovaries. [*Alloionema, Cheilobus, Rhabditophanes*].

Cephalobinae Filipjev, 1934.—Stoma collapsed, surrounded by esophageal tissue; esophageal bulb valved; 1 ovary doubly flexed posterior to vulva. [*Cephalobus, Eucephalobus, Acrobeles, Acrobeloides, Zeldia, Chiloplacus, Placodira, Cervidellus, Stegella*].

Daubayliinae Chitw. & Chitw., 1934.—Stoma collapsed; posterior swelling of esophagus non-muscular, not valved; 1 ovary not doubly flexed. [*Daubaylia*].

Superfamily Drilonematoidea n. superf.

Stoma greatly reduced, lips rudimentary or (?) absent; esophagus of adult consisting of elongate precorpus, metacorpus, isthmus and pyriform glandular region (never a valved bulb) to clavate short esophagus, not subdivisible; females with 1 (? always) ovary. Males with 2 or 0 spicules, gubernaculum present or absent. Phasmids usually (? always) large pocket-like. (Parasites in body cavity of earthworms).

Drilonematidae (Pierantoni, 1916).—Cephalic hooks absent.

Drilonematinae n. subf.—Corpus of esophagus not enlarged. [*Drilonema, (? Mesonema, Opistonema), Pelodira*].

Pharyngonematinae n. subfam.—Corpus of esophagus enlarged. [*Pharyngonema*].

Ungellidae n. fam.—Two cephalic hooks present [*Ungella, Synoecnema, Onchonema*].

Y

Appendix

Scolecophilidae Baylis, 1943.—Cephalic hooks absent, phasmids not observed, esophagus with short corpus and pyriform glandular region. Female with 1 ovary, vulva near base of esophagus. Male with 2 spicules; [*Scolecophilus, Scolecophiloides*].

Creagrocercidae Baylis, 1943.—This family was based on a single genus and species, *Creagrocercus barbatus*. The description states the organism has circular amphids and caudal glands. If this is true the family should be placed in the Order Chromadorida of the Aphasmidia. However, most of us have at some time mistaken a greatly enlarged ventral chord in the postanal region for caudal glands. Hence, because of the over-all similarity of the organism to other drilonematoids, we feel the description needs verification.

Suborder Tylenchina

Stylet present, esophagus fundamentally composed of procorpus, metacorpus, isthmus and glandular region; intestinal cells multinucleate, lumen minute; rectal glands absent; lateral excretory canal in only one chord; meromyarian; vaginal and uterine musculature usually not highly developed. (Nematodes of this group feed on liquid living cell contents as parasites of plants, carnivores or internal parasites of invertebrates).

Superfamily Tylenchoidea Chitw. & Chitw, 1937

Dorsal esophageal gland orifice in procorpus.

Tylenchidae Filipjev, 1934.—Head without internal sclerotization; stylet moderately developed to delicate; esophagus consisting of pro- and metacorpus; isthmus and glandular region; metacorpus not distinctly set off, usually elongate, not spheroid, with or without internal sclerotization, musculature present but weak; cuticle usually thin, striation faint, longitudinal incisures or ridges usually present; female with 1 or 2 ovaries; male with caudal alae adanal or subterminal. (Usually parasites of plants, some carnivorous, some consorts or invertebrates).

Neotylenchinae Thorne, 1941.—Head framework octagonal; metacorpus not bulbar or clearly ovoid; glands may or may not lie free but base of esophagus without stem-like extension; female with 1 ovary; male with caudal alae terminal or subterminal.

Nothotylenchinae Thorne, 1941.—Head framework hexagonal; metacorpus not bulbar, sometimes fusiform but without clear internal sclerotization; esophagus clearly set off from intestine, glands enclosed; females with 1 ovary; male with caudal alae adanal.

Paurodontinae Thorne, 1941.—Head framework hexagonal; metacorpus not bulbar, esophagus terminated by stem-like extension; females with 1 ovary; male with caudal alae adanal.

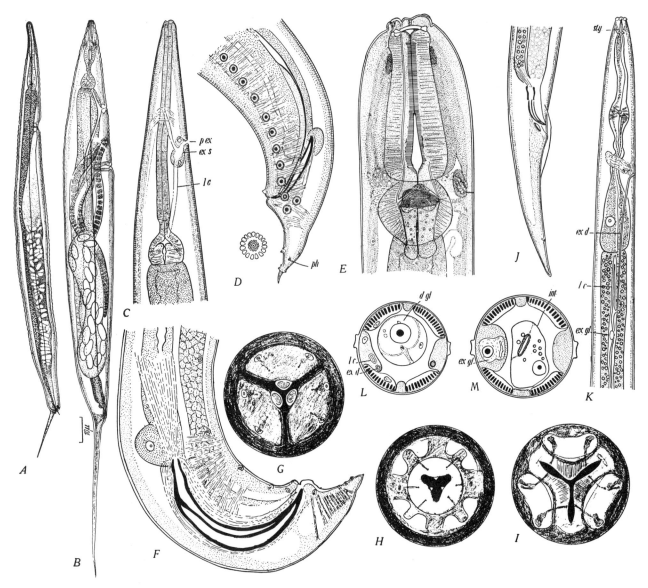

Fig. 11.

A-B—*Hammerschmidtiella diesingi* [Thelastomatidae] (A—Male; B—Female). C-D—*Cosmocercoides dukae* [Cosmocercidae] (C.—Esophageal region; D—Male tail). E-F—*Rhigonema infectum* [Rhigonematidae] (E—Esophageal region; F—Male tail). G—*Enterobius vermicularis*, [Oxyuridae] (Head). H—*Binema binema* [Thelastomatidae] (Head). I—*Spironoura affine* [Kathlaniidae] (Head). J-M—*Ditylenchus Sp.* [Tylenchidae] (J—Male tail; K—Esophageal region; L—Cross section at bulbar region of esophagus; M—Cross section at level of excretory cell). A-B after Chitwood, 1932, Ztschr. Parasit.; G-I—after Chitwood & Wehr, 1934, Ztschr. Parasit.

Tylenchinae Filipjev, 1934.—Head framework hexagonal; metacorpus fusiform with internal sclerotization; esophagus terminated by normal glandular region; females with 1 or 2 ovaries; males with adanal to terminal caudal alae. [Includes *Tylenchus, Psilenchus, Halenchus, Tetylenchus, Ditylenchus, Anguina, ? Chitinotylenchus, Tylenchulus*].

This group is rather heterogenous and includes genera extremely similar to those in the Neotylenchinae and Nothotylenchinae. A regrouping will be necessary.

Heteroderidae Thorne, 1949.—Head internally sclerotized; stylet well developed but shaft not greatly elongated; esophagus consisting of well defined procorpus and metacorpus, isthmus and glandular region; metacorpus usually spheroid, well set off with very well developed musculature and internal sclerotization; cuticle usually coarsely striated, usually with lateral ridges or incisures (except adult females of Heteroderinae); female with 1 or 2 ovaries; male with caudal alae adanal, terminal or absent. (Usually parasites of plant roots).

Heteroderinae Filipjev, 1934.—Esophageal glands free, on ventral side of body; females pyriform to lemon-shaped; males eel-shaped, tail short, bluntly rounded, without caudal alae; females with 2 ovaries, vulva terminal. (Sedentary parasites of plant roots).

Hoplolaiminae Filipjev, 1934.—Esophageal glands enclosed or free; males and females eel-shaped; tail of female usually short and somewhat rounded; male with caudal alae adanal or terminal; females with 2 ovaries, vulva near middle of body. (Vagrant parasites of plant roots, internal or partially external). [Includes *Hoplolaimus, Helicotylenchus, Rotylenchus, Tylenchorynchus, Radopholus,* and *Pratylenchus*].

Nacobbinae n. subf.—Esophageal glands enclosed or free on dorsal side of intestine; females saccate, vulva postequatorial or subterminal (not terminal), with 2 ovaries (*Rotylenchulus*) or 1 ovary (*Nacobbus*); males eel-shaped, caudal alae subterminal. (Ectoparasites of plant roots, females sedentary).

Criconematidae Thorne, 1949.—Stylet shaft greatly elongated; cuticle coarsely striated, annulated or scaly; metacorpus greatly enlarged, elongated. (Parasites of plant roots, usually external).

Criconematinae Taylor, 1936.—Cuticle coarsely annulated, scaly or spined, striae not regularly interrupted by lateral incisures or ridges; females very short and thick. [*Criconema, Criconemoides, Hemicycliophora*].

Paratylenchinae Thorne, 1949.—Cuticle finely annulated, Striae interrupted laterally; females minute, elongate (*Paratylenchus*) or short and thick (*Cacopaurus*). Males with weak alae.

14

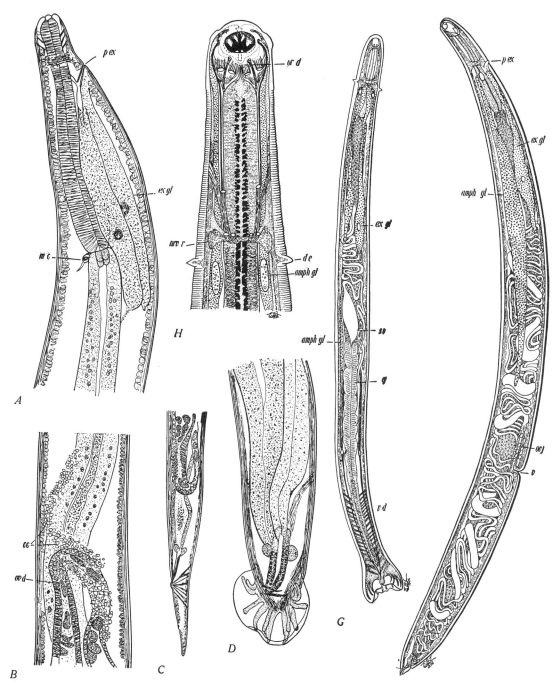

A-D—*Dictyocaulus viviparus* [Metastrongylidae] (A—Esophageal region; B—Postesophageal region showing coelomocyte; C—Tail of female; D—Tail of male). G-I—*Ancyclostoma duodenale* [Ancylostomatidae] (G—Male ven-

FIG. 12.

tral view; H—Cephalic region dorsal view; I—Female lateral view). A-D, originals by Mr. A. G. Dinaburg; G-I, after Looss, 1905, Rec. Egypt Govt. School Med.

Dolichodorinae n. subf.—Cuticle coarsely striated, interrupted laterally by 3 or more lateral incisures; both sexes of moderate size, very elongate, eel-shaped; females with 2 ovaries, male with caudal alae terminal (*Dolichodorus*) or adanal (*Belonolaimus*).

Allantonematidae Chitw. & Chitw., 1937.—Head of gravid female not set off, extremely degenerate reproductive sac, stylet extremely delicate, procorpus and metacorpus rarely distinguishable (*Howardula*), no evidence of esophageal musculature, gravid females completely transformed into reproductive sacs, often viviparous, 1 ovary; intestine without bacillary layer so far as known; male with adanal or subterminal caudal alae; phasmids large and pocketlike where known. (Parasites of hemocoel of insects).

Allantonematinae (Pereira, 1932).—Female uterus not prolapsed. [*Allantonema, Tylenchinema, Howardula, Parasitylenchus, Aphelenchulus, Chondronema, Scatonema, Fergusobia, Heterotylenchus, Bradynema*].

Sphaerulariinae Filipjev, 1929.—Female uterus prolapsed. [*Sphaerularia, Tripius*].

Superfamily *Aphelenchoidea* [*Fuchs, 1937*]

Dorsal esophageal gland orifice in metacorpus.

Aphelenchidae Steiner, 1949.—Characters of superfamily.
Aphelenchinae Stekh. & Teunissen, 1938.—Idem. [*Aphelenchus, Aphelenchoides, Metaphelenchus, Paraphelenchus, Schistonchus, Seinura, Bursaphelenchus, Laimaphelenchus, Parasitaphelenchus, Anomyctus, Crytaphelenchus, Ektaphelenchus*].

Appendix to *Tylenchina*

Myenchidae Pereira, 1932.—Excretory pore sucker-like. Parasites of leeches and amphibians. Includes [*Myenchus and Myoryctes*].

15

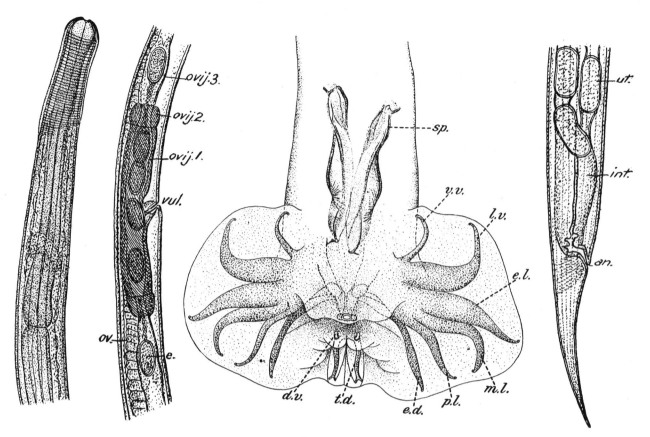

Fig. 13.
Cooperia curticei [Trichostronylidae]. After Ransom, 1911, U. S. D. A. Bull. 127.

Suborder Strongylina

Stylet absent; lips 3, 6, or corona radiata; stoma varied, well developed to rudimentary but not collapsed, and surrounded by esophageal tissue in form of vestibule; esophagus of larva consisting of pro- and meta-corpus, isthmus and bulbar region; adult with clavate type esophagus; somatic musculature mero- or polymyarian; excretory system with paired lateral canals and paired subventral glands; intestine with few polynucleate cells; rectal glands present; female with transverse or short vagina vera and single or double vagina uterina often heavily muscled; reproductive system usually complex; male with caudal alae usually containing well developed muscles forming the true or strongylatid bursa known only in this group; paired genital papillae in bursal rays with primary pattern as follows: prebursal papillae, (ventroventral, lateroventral), (externolateral, mediolateral, posterolateral), externodorsal, dorsal (triple, median one probably phasmid); spicules 2, equal. (Parasites of vertebrates in adult stage; early larval stages saprophagous or parasites of annelids or molluscs).

Superfamily Strongyloidea [*Weinland, 1858*]

Oral opening often surrounded by corona radiata; stoma usually well developed, subglobular, rather thick, without longitudinal ridges, body as a whole relatively thick; musculature meromyarian; esophagus of hatched larva usually as in genus *Rhabditis;* male with well developed bursa, rays not fused. (Adults usually inhabit gut, kidney or respiratory tract of reptiles, birds, and mammals, larval stages usually saprophagous, rarely parasites of earthworms).

Diaphanocephalidae Travassos, 1919.—Corona radiata absent; two lateral jaws; oral opening not guarded by cutting edges or teeth. (Habitat.—Gut of reptiles and amphibians; larval stages saprophagous).

Ancylostomatidae (Looss, 1905).—Corona radiata absent; lateral jaws absent; oral opening guarded by teeth or cutting plates. (Habitat.—Gut of mammals, larval stages saprophagous).

Ancylostomatinae Lane, 1923.—Oral opening guarded by 1 or more pairs of teeth.

Uncinariinae Stiles, 1903.—Oral opening guarded by cutting edges.

Strongylidae Baird, 1853.—Oral opening surrounded by corona radiata, without teeth or cutting edges, wall of stoma not composed of 6 longitudinal units. (Parasitic in reptiles, birds and mammals).

Strongylinae Railliet, 1885.—Stoma usually large, subglobular or elongated (infundibuliform); ventral cervical groove absent; duct of dorsal gland prolonged into stoma causing asymmetric development. (Parasitic in gut of equines, elephants, ostriches, rodents and varanids; larval stages saprophagous).

Cyathostominae Nicoll, 1927.—Stoma variable from small to large; duct of dorsal esophageal gland seldom projecting anteriad in wall of stoma; cervical groove absent.— In gut of equines, elephants, rhinoceros, marsupials, reptiles, rarely pigs and man; larval stages saprophagous.

Oesophagostominae Railliet, 1915.—Stoma usually short and subcylindrical, rarely large and subglobular; ventral cervical groove present. (Habitat.—In gut of mammals).

Cloacinidae Trav., 1919.—Oral opening surrounded by 6 large lips; stoma subcylindrical; corona radiata absent; teeth, cutting edges and jaws absent. (Habitat.—Marsupials).

Syngamidae Leiper, 1912.—Oral opening more or less hexagonal, without corona radiata or distinct lips (stomatal margin sometimes striated); stoma moderate to large, usually with 6 longitudinal thickenings. (Habitat.—Respiratory, kidney or gut of birds and mammals; larvae with or without intermediate host).

Syngaminae Baylis & Daubney, 1926.—Male with externolateral ray separate from other laterals; vulva anterior. (Habitat.—Respiratory system of birds and mammals).

Stephanurinae Railliet, Henry & Bauche, 1911.—Male with externolaterals arising from common trunk with other laterals; female with vulva posterior. (Habitat.—Renal tissues of pigs).

Deletrocephalinae Railliet, 1916.—Male with externolateral ray arising separately from other laterals; female with vulva median to posterior. (Habitat.—Gut of birds and rodents).

16

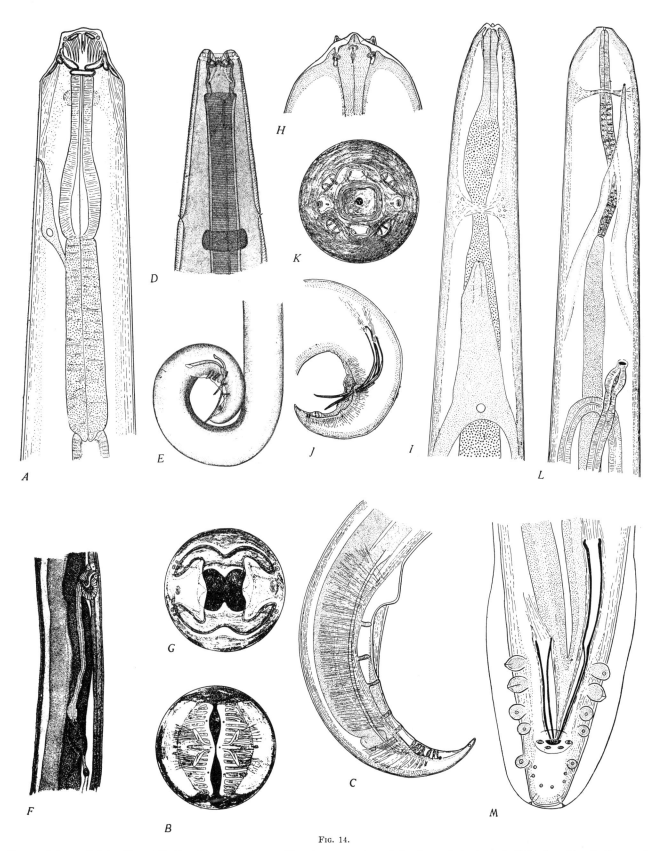

Fig. 14.

Representatives of Spirurida. A-C—*Camallanus americanus* [Camallanidae] (A—Esophageal region; B—Male tail; C—Head). D-G—*Habronema microstoma* [Spiruridae] (D—Head, median view; E—Male tail; F—Vulvar region; G—Head, en face). H-J—*Dracunculus dahomensis* [Dracunculidae] (H—Head, lateral view; I—Esophageal region of female, ventral view; J—Tail of male). K—*Dracunculus medinensis* [Dracunculidae] (Head of female). L-M—*Dirofilaria immitis* [Dipetalonematidae] (L—Esophageal region of female; M—Tail of male). C, after Chitwood & Wehr, 1934, Ztschr. Parasit.; D-F, after Ransom, 1913.

17

Superfamily *Trichostrongyloidea* Cram, 1927

Oral opening surrounded by 3 or 6 inconspicuous lips or lips absent; corona radiata absent; cuticle usually inflated in cephalic region; cuticle usually thick with numerous longitudinal ridges causing organisms to be somewhat "wiry"; musculature meromyarian or polymyarian; esophagus of hatched larva may or may not have valve of *Rhabditis;* male with well developed dorsal rays (dorsal ray sometimes reduced). (Adults usually inhabit digestive tract of vertebrates except fish, rarely found in lungs; larval stages usually saprophagous, sometimes molt in egg shell).

Travassos (1937) has monographed this group placing all members in the family Trichostrongylidae. He lists 14 subfamilies as follows: Trichostrongylinae, Ollulaninae, Strongylacanthinae, Mecistocirrinae, Spinostrongylinae, Amidostominae, Nematodirinae, Graphidiinae, Ornithostrongylinae, Heligmosominae, Viannaiinae, and Oswaldoneminae.

Metastrongyloidea Cram, 1927

Oral opening surrounded by 6 lips or lips rudimentary; corona radiata absent; stoma reduced, rudimentary, or absent; thin (rarely thick bodied); polymyarian; longitudinal cuticular ridges absent, cuticle moderately thin (ex. *Crenosoma*); bursa often reduced, rays somethat fused; first stage larva without valved bulb; tail usually with asymmetric development (except *Dictyocaulus*); intermediate host usually required. (Habitat.—Usually in lungs of mammals).

Metastrongylidae Leiper, 1909.—With the characters of the superfamily. (Recent work by Dougherty, 1949 and Gerichter, 1949 has left the superfamily Metastrongyloidea in an unsettled state and subdivision into families is impossible. The following classification of subfamilies is supplied as a working tool until such time as a thorough investigation has been made).

Skrjabingylinae Skrjabin, 1933; females amphidelphic, paired sphincters well developed; males with dorsal ray reduced, lateral rays well developed, gubernaculum simple or absent. [*Skrjabingylus, Crenosoma, Dictyocaulus, Otostrongylus, Troglostrongylus, Bronchostrongylus*].

Protostrongylinae Kamensky, 1905.—Worms elongate, females prodelphic, sphincter weak or absent; male with lateral rays rather well developed, dorsal ray reduced, gubernaculum complex. [*Protostrongylus, Spiculocaulus, Cystocaulus, Orthostrongylus, Neostrongylus, Leptostrongylus, Pneumostrongylus, Elaphostrongylus, Varestrongylus, Muellerius*].

Pseudaliinae Railliet & Henry, 1909.—Lips weak; female prodelphic with well developed sphincter at junction of vagina vera and vagina uterina; male with bursa greatly reduced with considerable fusion of rays but lateral ray stalks distinct, gubernaculum simple. [*Pseudalius, Pharurus, Stenurus, Halocercus.*].

Metastrongylinae Railliet & Henry, 1909.— Female prodelphic without sphincter at junction of vagina vera and vagina uterina or sphincter weak; males with well developed bursa containing stalked rays, gubernaculum simple. [*Metastrongylus, Angiostrongylus, Aelurostrongylus, Gurltia, Neometastrongylus, and ? Heterostrongylus*].

Filaroidinae Skrjabin, 1933.—Female prodelphic, without sphincter or with weak sphincter (except *Anafilaroides*); males with bursa greatly reduced, rays not stalked, often papilloid; gubernaculum simple. [*Filaroides, Metathelazia, Parafilaroides, Anafilaroides*].

Suborder Ascaridina

Stylet absent; lips 3, 2, or 0; corona radiata absent; adults with all gradations from rhabditoid to cylindrical esophagus; female reproductive system usually complex, vagina usually heavily muscled; terminal excretory duct usually short or vesicular, system usually with 2 lateral canals, subventral excretory glands always absent; intestinal cells typically uninucleate, moderate to large in number; rectal glands present; caudal alae, if present, with papillae, spicules 2, 1, or 0. Adult stage parasitic in gut of terrestrial arthropods, molluscs, and vertebrates; life cycle generally direct).

Superfamily *Oxyuroidea* Railliet, 1905

Ventrolateral cephalic papillae absent, 4 double or 8 simple papillae of external circle, internal circle of 6 papillae usually very minute; deirids ? absent; stoma cylindroid or short, not surrounded by esophageal tissue; esophagus terminated by valved bulb (few exceptions): terminal excretory duct short usually vesicular; meromyarian, life history direct.

Oxyuridae Cobbold, 1884.—External circle of 4 double cephalic papillae; corpus of esophagus not extremely short; male with 1 or 0 spicules. (Adults usually in hind gut of vertebrates, chiefly amphibians, reptiles and mammals).

Oxyurinae Hall, 1916.—Male without postanal cuticular projection.

Pharyngodoninae Trav., 1920.—Male with postanal cuticular projection.

Thelastomatidae (Trav., 1929).—External circle of 8 cephalic papillae or labiopapillae (This matter needs investigation); corpus of esophagus not extremely short; male with 1 or 0 spicule. (Adults in gut of terrestrial arthropods).

Rhigonematidae (Artigas, 1930).—External circle of 4 large cephalic papillae; corpus of esophagus extremely short and thick; male with 2 spicules. (Adults in gut of millepeds).

Rhigonematinae (Artigas, 1930).—One dorsal and two subventral lips.

Icthyocephalinae Artigas, 1930.—Two lateral jaws.

Atractidae Trav., 1920.—External circle of 4 double papillae; corpus of esophagus not extremely short; male with 2 or 0 spicules. (Adults in terrestrial arthropods and vertebrates except birds).

Atractinae Railliet, 1917.—Excretory vesicle sucker-like, entire corpus cylindrical; usually viviparous. (Intestine of fish, amphibians, reptiles, equines, elephants, hippopotamus and rhinoceros).

Ransomnematinae (Trav., 1929).—Excretory vesicle not sucker-like; entire corpus cylindrical or fusiform; oviparous. (Intestine of terrestrial arthropods).

Labidurinae York & Maplestone, 1926.—Excretory vesicle not sucker-like, procorpus fusiform, metacorpus cylindrical; viviparous. (Intestine of tortoises).

Superfamily *Ascaridoidea* (*Railliet & Henry*, 1915)

Ventrolateral cephalic papillae well developed and 4 large double submedian papillae of external circle; stoma usually surrounded by esophageal tissue and collapsed (except Subulurinae); terminal excretory duct usually short; meromyarian or polymyarian; male with 2 spicules; life history direct or indirect.

Cosmocercidae Trav., 1925.—Stoma surrounded by esophageal tissue forming a vestibule; esophagus consisting of cylindrical corpus, elongate isthmus and valved bulb containing uninucleate gland cells; oviparous or viviparous. (Intestine of molluscs, amphibians and reptiles).

Kathlaniidae (Trav., 1918).—Stoma surrounded by esophageal tissue forming a vestibule, esophagus consisting of an elongate cylindrical corpus, subspheroid isthmus, and valved bulb containing uninucleate esophageal glands; male with or without preanal sucker; meromyarian; oviparous. (Intestine of amphibians, reptiles and marsupial mammals).

Kathlaniinae Lane, 1914.—Subventral lips without pinnate cuticular leaves.

Cissophyllinae Yorke & Maplestone, 1926.—Subventral lips with pinnate cuticular leaves.

Heterakidae Railliet & Henry, 1914.—Stoma short and wide or collapsed forming a vestibule; esophagus consisting of clavate corpus, a short but not spheroid, isthmus, and valved bulb with subventral binucleate esophageal glands (esophagus rarely cylindrical); male usually with preanal sucker; meromyarian or polymyarian; life cycle direct or indirect.

Oniscicolinae Trav., 1929.—Lips distinct; vestibule present; sucker of male with sclerotized rim; meromyarian. (Body cavity of crustacean). This organism strangely similar to the genus *Spinicauda*.

Heterakinae Railliet & Henry, 1912.—Lips distinct, vestibule present; sucker with sclerotized rim; polymyarian. (Intestine of all vertebrate groups except fish).

Ascaridiinae Trav., 1919.—Lips well developed, vestibule not distinct; esophagus cylindrical; sucker with sclerotized rim; polymyarian. (Intestine of birds and reptiles).

Subulurinae Trav., 1914.—Lips represented only by apical lobes; stoma distinct, short and not in form of vestibule; sucker usually without sclerotized rim; polymyarian. (Intestine of birds and mammals).

Quimperiinae (Gendre, 1928).—Lips reduced to apical lobes; stoma rudimentary or in form of vestibule; esophagus with cylindrical corpus followed by shorter cylindrical glandular bulb; sucker present or absent; meromyarian. (Intestine of fish). There seems to be some doubt as to whether this subfamily should be placed with the Heterakidae or the Cucullanidae.

Ascarididae Blanchard, 1896.—Lips well developed; interlabia present or absent; distinct stoma present or absent (except *Crossophorus*); esophagus cylindrical or terminated by non-valved cylindroid to short bulbar region containing uninucleate esophageal glands; intestinal cecae commonly present; preanal sucker absent; polymyarian; oviparous; life cycle direct or indirect. (Intestine of all groups of vertebrates).

Ascaridinae Lane, 1923.—Esophagus plain, i.e., without set off bulbar region or diverticulum.

Anisakinae Railliet & Henry, 1912.—Esophagus terminated by set off bulbar region, ventriculus or esophageal diverticulum.

ORDER SPIRURIDA

Stylet absent; esophagus fundamentally composed of 2 parts, both cylindroid, never returning to corpus, isthmus and bulbar region even in first stage larva. Oral opening surrounded by 6 weak apical lip lobes, cuticular circumoral elevation, or paired lateral pseudolabia, ventrolateral cephalic papillae absent; subventral paired excretory glands absent, lateral canals in both chords; intestinal cells numerous, usually uninucleate; polymyarian. (Parasites of vertebrates in adult stage, larvae requiring intermediate host, usually arthropod).

Suborder Camallanina

Esophageal glands uninucleate (ex. *Philonema*); larva without cephalic hook, phasmids of larva large, pocket-like. (Intermediate host, copepods).

Superfamily Camallanoidea Trav., 1920

Internal circle of cephalic papillae minute, external circle partially fused; stoma usually well developed; oviparous or viviparous; larva with pocket-like phasmids. (Intestinal parasites of fish, amphibia and reptiles).

Camallanidae Railliet & Henry, 1917.—4 well developed and 4 rudimentary papillae of external circle; stoma not surrounded by esophageal tissue; usually two lateral jaws; esophagus composed of 2 tandem parts; male with 2 spicules.

Cucullanidae Cobbold, 1864.—4 large double papillae of external circle; stoma well developed, surrounded by esophageal tissue forming two lateral jaws; esophagus clavate; male with 2 spicules.

Anguillicolidae Yamaguti, 1935.—Cephalic structures not described; stoma distinct but walls not heavily sclerotized; esophagus clavate; male without spicules; female amphidelphic. (Swim bladder of *Anguilla*). Yamaguti labels structures caudal glands that are most certainly rectal glands.

Superfamily Dracunculoidea Cameron, 1934

Internal circle of cephalic papillae well developed, external circle of 8 well developed papillae; stoma rudimentary; vulva in mid-region; viviparous. (Intermediate host copepods).

Dracunculidae Leiper, 1912.—With internal cephalic sclerotization; esophagus with swollen anterior end, short cylindrical muscular part, glandular enlargement, constriction and greatly elongate glandular region.

Dracunculinae (Stiles, 1907).— Circumoral elevation present. (Tissue parasites of reptiles and mammals).

Avioserpentsinae Wehr & Chitwood, 1934.—Circumoral elevation absent. (Tissue parasites of birds).

Philometridae Baylis & Daubney, 1926.—Circumoral elevation and internal cephalic sclerotization absent.

Philometrinae Yamaguti, 1935.—Esophagus with swollen anterior end; larva with dorsal tooth. (Tissue parasites of fish).

Micropleurinae Baylis & Daubney, 1926.—Esophagus not swollen at anterior end; larva without dorsal tooth. (Tissue parasites of crocodiles and alligators).

Suborder Spirurina

Esophageal glands multinucleate; larva commonly with cephalic hook and phasmids pore-like. Intermediate hosts various-rarely copepods.

Superfamily Spiruroidea Raillet & Henry, 1915

Stoma distinctly developed and/or 2 lateral pseudolabia; vulva usually near middle or posterior part of body.

Thelaziidae Railliet, 1916.—Pseudolabia absent (except *Physocephalus*).

Thelaziinae Baylis & Daubney, 1926.—Plaques absent; spines or hooks absent, protorhabdions not rugose; caudal alae absent.

Spirocercinae Chitw. & Wehr, 1932.—Plaques absent; spines or hooks absent; protorhabdions not rugose; caudal alae present.

Ascaropsinae Alicata & McIntosh, 1933.—Plaques absent; spines or hooks absent; protorhabdions rugose; caudal alae present.

Gongylonematinae (Hall, 1916).—Plaques present; spines or hooks absent; protorhabdions not rugose; caudal alae present.

Rictulariinae Hall, 1913.—Plaques absent; spines or hooks present; protorhabdions not rugose; caudal alae present.

Spiruridae Oerley, 1885.—Pseudolabia well developed and lobed; papillae usually posterior to pseudolabia; interlabia present or absent.

Spirurinae Railliet, 1916.—Interlabia absent; sexes not dimorphic; female without caudal hook.

Habronematinae (Chitw. & Wehr, 1932).—Interlabia present; sexes not dimorphic; female without caudal hook.

Tetramerinae Railliet, 1915.—Interlabia present; sexes dimorphic; female without caudal hook.

Hedrurinae Railliet, 1916.—Interlabia present; sexes not dimorphic; female with caudal hook.

Acuariidae Seurat, 1913.—Pseudolabia well developed, not lobed; 4 completely fused double papillae; interlabia absent; cephalic ornamentation in form of cordons, collarettes or appendages.

Acuariinae Railliet, Henry & Sissoff, 1912.—Cordons present.

Schistorophinae Trav., 1918.—Hood or appendages present.

Seuratiinae Chitw. & Wehr, 1932.—Denticulate cephalic collarette present.

Gnathostomatidae (Railliet, 1895).—Pseudolabia well developed, lobed; 8 partially fused double papillae· or pseudolabia.

Gnathostomatinae (Baylis & Lane, 1920).—Cephalic bulb present; pseudolabia without fleshy growths.

Spiroxyinae.—Baylis & Lane, 1920.—Cephalic bulb absent; pseudolabia without fleshy growths.

Ancyracanthinae.—Yorke & Maplestone, 1926.—Cephalic bulb absent; elongated fleshy growths extending from sublateral lobes of pseudolabia.

Physalopteridae Leiper, 1908.—Pseudolabia well developed, not lobed; 4 completely fused double papillae or pseudolabia; interlabia absent.

Superfamily [Filarioidea, Weinland, 1858]

Oral opening circular or dorsoventrally elongated, circumoral elevation present or absent; cephalic papillae consisting of an internal circle of 0, 2, or 4 and an external circle of 8; lips, pseudolabia and jaws absent; stoma usually rudimentary or reduced; vulva usually near anterior end of body; oviparous or viviparous. (Tissue parasites of vertebrates except fish).

Filariidae Claus, 1885.—Oral opening not surrounded by crown of spines, circumoral elevation present or absent; stoma rudimentary; esophagus usually divided externally into short, narrow, anterior part and long, wide posterior part; oviparous or viviparous (if oviparous, eggs with thick shells); first stage larva usually short and stout, its anterior end bearing several rows of spines and larva with conically attenuated tail, or bluntly rounded tail with single row of spines.

Filariinae Stiles, 1907.—Oral opening circular, special cephalic structures absent; spicules unequal and dissimilar; caudal alae reduced or absent; vulva near oral opening; oviparous.

Aproctinae Yorke & Maplestone, 1926.—Oral opening circular; specialized cephalic structures absent; spicules subequal; caudal alae present or absent; vulva not near region of oral opening; oviparous; first stage larva with conoid tail.

Dicheilonematinae Wehr, 1935.—Oral opening usually dorsoventrally elongated; circumoral elevation, pseudonchia, and/or epaulettes present; spicules unequal or dissimilar; caudal alae present (except *Setaria*); oviparous (except *Setaria*).

Diplotriaeninae Skrjabin, 1916.—Oral opening dorsoventrally elongated; paired lateral trident-like structures; circumoral specializations absent; oviparous; caudal alae absent; first stage larva with bluntly rounded tail.

Tetracheilonematinae Wehr, 1935.— Oral opening surrounded by 4 conoid pseudonchia; tridents absent; spicules equal; caudal alae absent; oviparous.

Dipetalonematidae Wehr, 1935.—Amphids pore-like, head without spinate collarette; eggs without chitinous shell, covered only by vitelline membrane; larva microfilarioid (i.e., incompletely differentiated); head of adult without circumoral elevation, tridents, or pseudonchia; stoma rudimentary or stomatorhabdions distinct; vulva anterior.

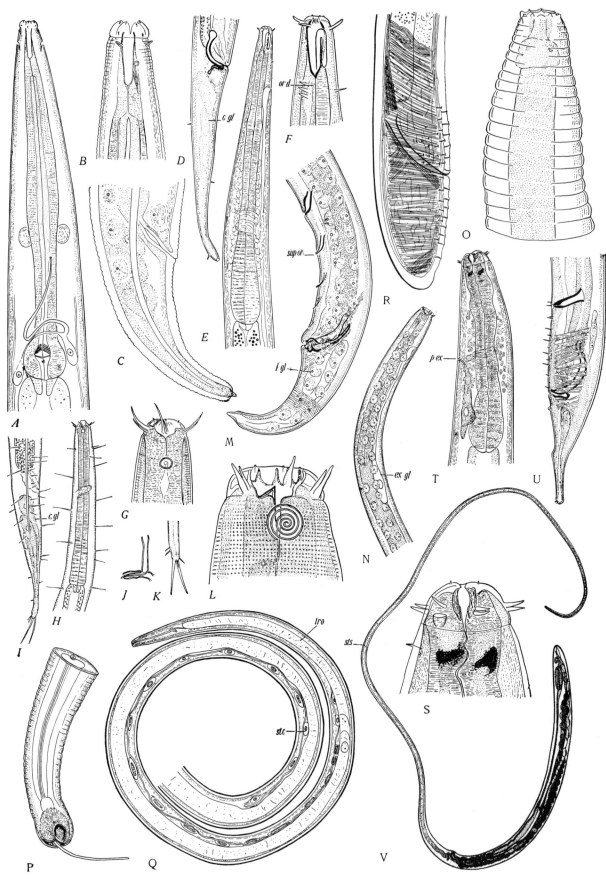

FIG. 15.

Dipetalonematinae Wehr, 1935.—Caudal alae narrow or absent; body not swollen at excretory sinus.

Dirofilariinae Wehr, 1935.—Caudal alae well developed; body not swollen at excretory sinus.

Onchocercinae Leiper, 1911.—Caudal alae narrow or absent; body swollen at excretory sinus.

Desmidocercidae Cram, 1927.—Amphids pore-like; head without spinate collarette; eggs with vitelline membrane; larva differentiated; spicules unequal or subequal; stoma cylindrical; rhabdions not distinct; vulva pre- or post-equatorial.

Stephanofilariidae Wehr, 1935.—Amphids massive, head with spinate collarette; eggs with vitelline membrane; larva differentiated; spicules unequal; stoma rudimentary; vulva anterior.

CLASS APHASMIDIA

Phasmids absent, caudal glands and hypodermal glands usually present (Caudal glands absent in all members of Dorylaimina and Dioctophymatina); terminal excretory duct not lined with cuticle (except some Plectinae), lateral canals absent; cephalic sensory organs setose to papilloid, external amphids circular, spiral, shepherd's crook, pocket-like or sometimes pore like, usually postlabial in position; somatic papillae usually (? always) present; deirids (specialized cervical papillae) absent; somatic setae commonly present; caudal alae or bursa absent (except *Anoplostoma* and *Oncholaimellus*); male usually with single ventral series of preanal supplements, sometimes with double row of supplements, sometimes with paired genital papillae; pseudocoelomocytes usually more than 6, sometimes very numerous; subventral esophageal glands may or may not open at anterior end of esophagus; rectal gland usually absent. Primarily aquatic, with some terrestrial and some inhabitants of invertebrates and vertebrates.

ORDER CHROMADORIDA

Amphids spiral, circular, vesiculate or other forms derivable from spiral; caudal glands practically always present; subventral esophageal glands never opening near anterior end of esophagus, never duplicated, never multinucleate; esophagus fundamentally divisible into 3 regions, corpus, isthmus and bulbar region, though sometimes grossly cylindrical or with cylindroid bulbar region, but not greatly elongated. (Freeliving in soil, fresh water or marine, sometimes commensal in gills or similar places but never parasitic).

Suborder Monhysterina

Esophago-intestinal valve dorsoventrally flattened or circular, moderately large to very large, never triradiate or vertically flattened; stoma if well developed, unarmed or containing 1 or 3 small teeth or 6 outwardly acting teeth and stoma may or may not be surrounded by esophageal tissue; ovaries outstretched or reflexed; papilloid or tuboid supplementary organs present or absent.

Superfamily Plectoidea Chitw., 1937

Amphids "plectoid" or 1-2 spiral; ovaries reflexed; ends of esophageal radii tuboid.

Plectidae Oerley, 1880.—Bulbar region of esophagus muscular. (Fresh water or marine).

Plectinae Mic., 1922.—Characters as of family.

Camacolaimidae Stek. & de Coninck, 1933.—Bulbar region of esophagus glandular.

Camacolaiminae Micol., 1924.—Supplementary organs papilloid; amphids unispiral, anterior to cephalic setae; ocelli present or absent. (Marine).

Aphanolaiminae Chitw., 1935.—Supplementary organs tuboid; amphids unispiral or circular, usually posterior to cephalic setae; ocelli absent. (Fresh water or marine).

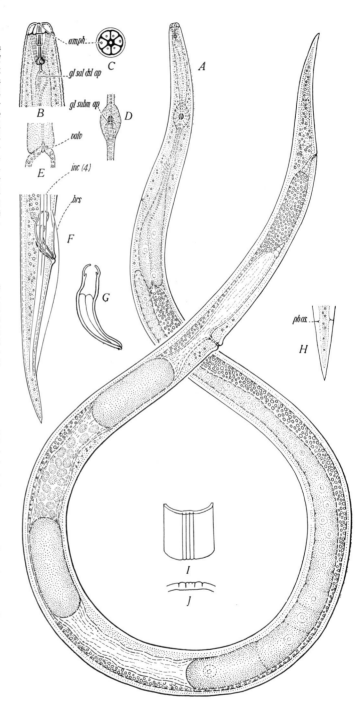

FIG. 15.

Representatives of Chromadorida and Enoplida. A-C—*Plectus parietinus* [Plectidae] (A—Esophageal region, ventral view; B—Stomatal region, lateral view; C—Female tail). D-F—*Axonolaimus spinosus* [Axonolaimidae] (D—Male tail; E—Esophageal region, lateral view; F—Lateral view). G-K—*Theristus setosus* [Monhysteridae] (G—Head, lateral view; H—Esophageal region; I—Female tail). (J—Spicules and gubernaculum; K—Tip of female tail). L-N—*Acanthonchus viviparous* [Cyatholaimidae] (L—Head, lateral view; M—Male tail; N—Esophageal region). O-P—*Eustrongylides perpapillatus* [Dioctophymatidae] (O—Head; P—Male tail). Q-R—*Agamermis decaudata* v. *paraguayensis* [Mermithidae] (Q—Esophageal region; R—Male tail). S-U—*Enoplus communis* v. *meridionalis* [Enoplidae] (S—Head; T—Esophageal region; U—Male tail). V—*Trichuris trichiura* [Trichuridae] (Female). O-P, after Jaegerskioeld, 1909, Nova Acta, ... Upsaliensis; Q-R, after Steiner, 1924, Centralbl. Bakt.; S-U, after Chitwood, 1936, Tr. Am. Mic. Soc.; V, after Urioste, 1923.

Fig. 16.

Ditylenchus dipsaci. A—Adult female; B—Head of female; C—en face; D—Metacorpus; E—Base of esophagus; F—Male tail; G—Spicule; H—Tip of female tail, ventral view; I & J views of lateral sector of cuticle. After Thorne, 1945, Proc. Helm. Soc. Wash. v. 12 (2): 27-34.

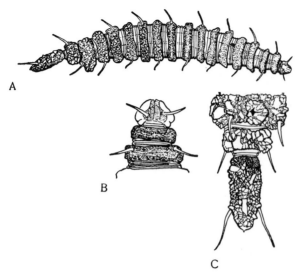

FIG. 17.

Desmoscolex americanus [Desmoscolecidae] (A—Male, lateral view; B—Head, ventral view; C—Tail, ventral view). After Chitwood, 1936, Proc. Helminth. Soc. Wash.

Appendix to Plectoidea.—Bastianiidae Stek. & Teunissen, 1938.—Cephalic setae of external circle 10, stoma very long and narrow, surrounded by esophageal tissue, or stoma rudimentary; esophagus cylindrical, greatly elongate, internal structure not investigated; amphids circular to unispiral; female with reflexed ovaries; male with numerous papilloid supplements, short, straight spicules. This group shows similarities to the Tripylidae and Alaimidae of the Tripyloidea. It seems possible that they may account for the origin of the Tripyloidea.

Superfamily Axonolaimoidea Chitw., 1937

Amphids spiral or variants, ovaries outstretched (rarely reflexed); ends of esophageal radii tuboid.

Axonolaimidae Stek & de Coninck, 1933.—Cuticle very minutely punctate or punctations apparently absent; amphids not multispiral; ovaries outstretched. (Marine).

Axonolaiminae Micol., 1924.—Stoma conoid, stomatorhabdions thick; amphids circular, open unispire or shepherd's crook. (Marine).

Diplopeltinae Rauther, 1930.—Stoma inconspicuous; amphids situated on cuticular plaques. (Marine).

Campylaiminae Chitw., 1937.—Stoma cylindroid or inconspicuous; amphids much elongated, hook-like; plaque absent. (Marine).

Cylindrolaiminae Micol., 1922.—Stoma cylindrical, sometimes short, stomatorhabdions thin; amphids unispiral or circular, plaque absent. (Fresh water or marine).

Comesomatidae (Stek. & de Coninck, 1933).—Cuticle coarsely punctate; amphids multispiral; ovaries sometimes reflexed. (Marine).

Superfamily Monhysteroidea Stek. & de Coninck, 1933

Amphids circular; ovaries outstretched; ends of esophageal radii convergent.

Monhysteridae Oerley, 1880.—Stoma not styletiform; radial muscles of esophagus diffuse, without cuticular attachment points. (Marine or fresh water).

Linhomoeidae Filip., 1929.—Stoma not styletiform; radial muscles of esophagus in 6 bands, cuticular attachment points often present. (Marine or fresh water).

Linhomoeinae Filip., 1929.—Stoma small; esophago-intestinal valve elongate.

Sphaerolaiminae Filip., 1929.—Stoma massive, globoid; esophago-intestinal valve short.

Siphonolaimidae Chitw., 1937.—Stoma styletiform; radial muscles of esophagus concentrated in 6 bands. (Marine).

Suborder Chromadorina

Esophago-intestinal valve triradiate or vertically flattened, usually very short; stoma if well developed containing a large dorsal tooth, 3 jaws or 6 inwardly acting teeth; stoma surrounded by esophageal tissue; ovaries reflexed; cup-like or tuboid supplementary organs present or absent.

Superfamily Chromadoroidea Stek. & Coninck, 1933

Cuticle striated, punctate; helmet absent; amphids spiral to kidney shaped; ambullatory setae absent.

Chromadoridae Filip., 1917.—Amphids kidney shaped to spiral; esophago-intestinal valve extremely small; cuticular punctation coarse.

Microlaimidae de Coninck & Stek., 1933.—Amphids 1-2 spiral; esophago-intestinal valve elongate; cuticular punctations minute; gubernaculum simple.

Cyatholaimidae de Coninck & Stek., 1933.—Amphids unispiral to multispiral; esophago-intestinal valve well developed, triradiate; gubernaculum often complicated; cuticular punctations coarse.

Cyatholaiminae Micol., 1922.—Stoma shallow or mesostom collapsed, without ribs to base and without jaws.

Choanolaiminae Filip., 1934.—Stoma deep, with ribs or jaws.

Tripyloididae Chitw., 1937.—Amphids 1-2 spiral; esophago-intestinal valve well developed, triradiate, gubernaculum complicated; cuticle with minute punctation, striation fine.

Superfamily Desmodoroidea Steiner, 1927

Cuticle annulated, not punctate; helmet often present; amphids spiral, shepherd's crook, circular or slit-like; ovaries reflexed; ambullatory setae present or absent; glandular setae present or absent.

Desmodoridae Micol., 1924.—Body not epsilonoid, ambullatory bristles absent; glandular setae absent.

Desmodorinae Micol., 1924.—Amphids spiral; helmet present; dorsal tooth usually present; cuticle not tiled.

Richtersiinae Cobb, 1933.—Same as Desmodorinae except helmet absent.

Stilbonematinae Chitw., 1936.—Helmet present or absent; amphids minute, slit-like, dorsal tooth rudimentary or absent; otherwise as in Desmodorinae.

Ceramonematinae (Cobb, 1933).—Amphids spiral to shepherd's crook; helmet present; dorsal tooth absent; cuticle tiled.

Monoposthiinae Filip., 1934.—Amphids circular; helmet more or less distinct; dorsal tooth present; cuticle with longitudinal ridges.

Epsilonematidae Steiner, 1927.—Body epsilonoid; ambullatory bristles present; glandular setae absent.

Draconematidae Steiner, 1930.—Body not epsilonoid; tubular glandular setae present.

Superfamily Desmoscolecoidea Stekh., 1935

Cuticle coarsely annulated or without distinct annules; helmet absent; amphids vesiculate; ovaries reflexed; tubular glandular setae present.

Desmoscolecidae Southern, 1914.—Cuticle coarsely annulated, not covered by hairy coat.

Greeffiellidae (Filip., 1929).—Cuticle not coarsely annulated, covered by hairy coat.

ORDER ENOPLIDA

Amphids pocket-like to pore-like or tuboid; caudal glands present or absent; subventral esophageal glands often opening through teeth or at anterior end of esophagus, sometimes duplicate, outside walls of esophagus, sometimes multinucleate; esophagus cylindrical or conoid, sometimes in form of narrow anterior and wide posterior glandular regions, often greatly elongated; setae present or absent. (Habitat.—Marine, fresh water, soil, arthropods and vertebrates).

Suborder Enoplina

Head usually bearing 6 plus 10, 10, or 6 plus 4 setae (setae absent in Mononchidae, Alaimidae, and a few scattered genera of other families); amphids usually pocket-like; typical stylet absent; esophageal gland orifices commonly in stomatal region, glands uninucleate; esophagus grossly cylindrical, conoid or multibulbar; intestine functional; somato-intestinal muscles not in 4 rows; female reproductive system relatively simple, vagina transverse; male with 2 (very rarely 1 or 0) spicules, testes typically paired; muscular caudal sucker absent; polymyarian or meromyarian; hypodermal glands commonly well developed; caudal glands usually present; ventral excretory cell usually present. (Habitat.—Primarily aquatic, marine or fresh water, sometimes moist soil).

Superfamily Enoploidea Stekh. & de Coninck, 1933

Cuticle of head duplicate; subventral esophageal gland orifices near anterior end of esophagus or through teeth; male supplementary organs 0, 1, or 2. (Habitat.—Marine or brackish water).

Enoplidae Baird, 1853.—Stomatorhabdions feebly sclerotized, stomatal region surrounded by esophageal tissue.

Enoplinae Micol., 1922.—Cephalic region set off by groove; esophagus cylindrical; amphids not elongated; 10 cephalic setae in external circle; male with 1 tuboid supplement.

Leptosomatinae Micol., 1922.—Cephalic region not set off by groove, esophagus cylindrical to conoid but never vesiculate, muscles of posterior region well developed; 10 cephalic setae of external circle; male with 1 or 2 tuboid supplements.

Phanodermatinae Filip., 1927.—Cephalic region not set off by groove; esophagus conoid, posterior part crenate in outline, vesiculate, muscles reduced; 10 cephalic setae of external circle; male with 1 or 2 tuboid supplements.

Oxystomininae (Micol., 1924).—Cephalic region not set off by groove; esophagus conoid, with smooth outline, muscles rudimentary; 6 plus 4 setae of external circle; amphids sometimes tubular; supplementary organs absent.

Oncholaimidae Baylis & Daubney, 1926.—Stomatorhabdions heavily sclerotized, only posterior part of stoma surrounded by esophageal tissue.

Oncholaiminae Micol., 1922.—Esophagus cylindroid, never crenate or multibulbar; supplementary organ absent or pedunculate.

Eurystominae (Filip., 1934).—Esophagus conoid, crenate or multibulbar; 1 or 2 cup-like supplements.

Enchelidiinae (Micol., 1924).—Esophagus conoid, sometimes crenate or multibulbar; supplements papilloid.

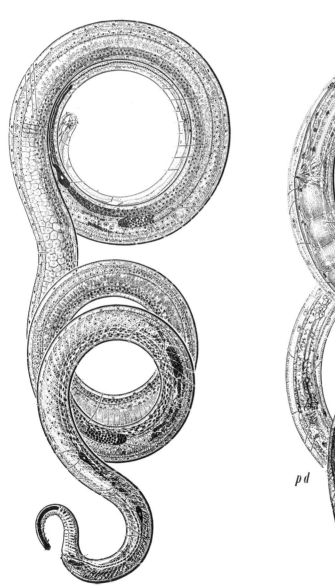
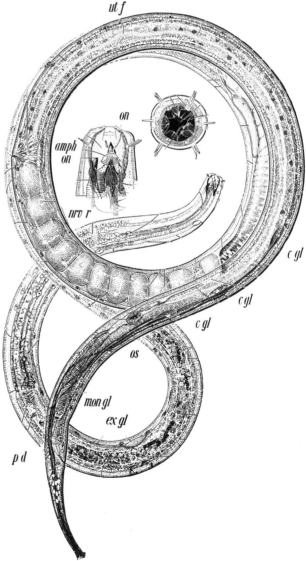

FIG. 18.

Metoncholaimus pristiuris [Oncholaimidae] (Male and female. After Cobb, 1932, J. Wash. Acad. Sc.)

23

Superfamily Tripyloidea Chitw., 1937

Cuticle of head not duplicate; subventral esophageal gland orifices anterior or posterior to nerve ring; esophago-intestinal valve usually large and thick; male supplementary organs usually 3 or more. (Habitat.—Fresh or brackish water and moist soil).

Tripylidae Oerly, 1880.—Dorsal and 2 subventral esophageal gland orifices in stomatal region; stomatal walls not well sclerotized; esophago-intestinal valve bulb-like.

Mononchidae Chitw., 1937.—All esophageal gland orifices posterior to nerve ring; stomatal walls strongly sclerotized; esophago-intestinal valve not bulb-like.

of narrow cylindrical anterior part and wide cylindrical posterior part or glands free in body cavity as 1 or 2 reduplicate series; somato-intestinal muscles in 4 rows; reproductive system varies according to superfamily; muscular caudal sucker absent; caudal glands absent; excretory system absent or very poorly developed. (Habitat.—Soil (or fresh water), parasites of insects or vertebrates).

Superfamily Dorylaimoidea Thorne, 1934

Amphids pocket-like; esophagus two part cylindrical; esophageal glands not free in body cavity; intestine not grown anterior to base of esophagus; male with 2 spicules and usually 2 testes; female with transverse vagina; eggs not operculate. (Habitat.—Soil, aquatic, many carnivorous, some possibly parasitize plant roots, some diatomivorous).

Dorylaimidae de Man, 1876.—Amphid apertures obscure, slit-like; esophagus with posterior third or more enlarged, not surrounded by spiral muscles; testes two; ventromedial row and paired adanal supplements present; polymyarian.

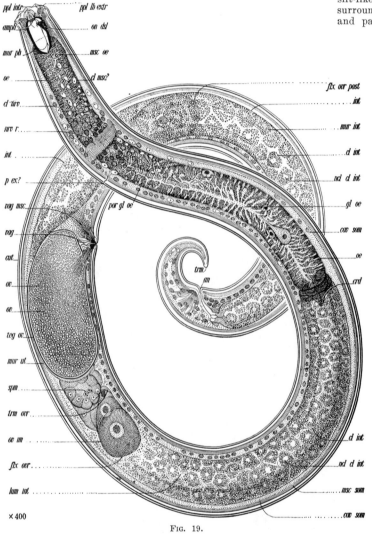

×400

Fig. 19.

Mononchus papillatus [Mononchidae] (Female). After Cobb, 1917. Soil Science.

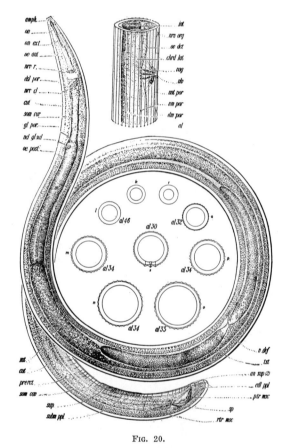

Fig. 20.

Dorylaimus stagnalis [Dorylaimidae]. (Male, full length with cross section outlines and vulvar region of female). After Thorne & Swanger, 1936, Capita Zoologica.

Alaimidae Micol., 1922.—Dorsal gland orifice near anterior end of esophagus; stoma rudimentary; esophago-intestinal valve not large.

Ironidae de Man, 1876.—Dorsal and subventral gland orifices into or near base of stoma; stoma well developed, cylindrical; esophago-intestinal valve not bulb-like.

Ironinae Micol., 1922.—3, 4, or 6 teeth at anterior end of stoma.

Cryptonchinae Chitw., 1937.—O teeth at anterior end of stoma.

Suborder Dorylaimina

Setae absent; stylet present; esophageal gland orifices posterior to nerve ring, glands uninucleate, esophagus consisting

Dorylaiminae Filipjev, 1928.—Stylet axial; not greatly attenuated and without basal extensions; stomatal walls not heavily sclerotized.

Nygolaiminae Thorne, 1935.—Stylet mural; not attenuated; stomatal walls not heavily sclerotized.

Actinolaiminae Thorne, 1939.—Stylet axial; not attenuated or modified; stoma with heavily sclerotized walls.

Longidorinae Thorne, 1935.—Stylet axial, greatly attenuated, often with marked extensions; stomatal walls not heavily sclerotized.

Tylencholaiminae Filip., 1934.—Stylet axial; not greatly attenuated, with basal extensions, rod-like or knob-like; stomatal walls not heavily sclerotized.

Leptonchidae Thorne, 1935.—Amphid apertures obscure, slit-like, esophagus with short basal swelling; not surrounded by spiral muscles; prerectum present; testes two; adanal supplements present: meromyarian.

Diphtherophoridae Thorne, 1935.—Amphid apertures ellipsoidal and conspicuous; esophagus with pyriform or elongate basal bulb; not surrounded by spiral muscles; testes single; adanal supplements absent; prerectum absent; meromyarian.

Belondiridae Thorne, 1939.—Amphid apertures obscure, slit-like; enlarged portion of esophagus surrounded by sheath of spiral muscles; prerectum present; testes two; adanal supplements present; polymyarian.

Superfamily Mermithoidea Wülker, 1924

Amphids modified, pocket-like to pore-like; esophageal glands free in body cavity forming reduplicate series termed a stichosome; intestine extending anterior to base of esophagus, usually without lumen; male with 1 or 2 spicules and usually 2 testes; female with highly developed reproductive system, usually two ovaries and tubular vagina; eggs modified but not operculate. (Parasites of land and fresh water arthropods in larval stages).

Mermithidae Braun, 1883.—Esophagus not dorylaimoid in larval stage.

Tetradonematidae Cobb, 1919.—Esophagus dorylaimoid in larval stage.

Superfamily Trichuroidea Raillet, 1916

Amphids pore-like; esophageal glands free in body cavity, forming stichosome; intestine not extending anterior to base of esophagus, with well developed lumen; male with 1 or 0 spicule and 1 testis; female with elongate tubular vagina and 1 ovary; eggs typically operculate. (Parasites of vertebrates in adult stage; life history direct or indirect).

Trichuridae Railliet, 1916.—Oviparous, male with 1 spicule; stichosome 1 cell row.

Trichurinae Ransom, 1911.—Male not degenerate; posterior part of body distinctly wider than anterior part.

Capillariinae Railliet, 1915.—Male not degenerate; posterior part of body not distinctly wider than anterior part.

Trichosomoidinae Hall, 1916.—Male degenerate, parasitic in uterus of female; posterior part of body not much wider than anterior part.

Trichinellidae Ward, 1907.—Viviparous; male without spicule; stichosome 1 cell row.

Cystoopsidae Skrjabin, 1923.—Oviparous; male with 1 spicule; stichosome in 2 cell rows.

Suborder Dioctophymatina

Setae absent; stylet absent at least in adult stage; dorsal and subventral esophageal gland orifices at anterior end of esophagus, glands highly polynucleate, inside wall of cylindrical esophagus; somato-intestinal muscles in 4 rows; reproductive system single in both sexes, highly developed, male with 1 spicule and caudal sucker; caudal glands absent; ventral excretory cell apparently absent; eggs operculate. (Parasites of vertebrates, life history not substantiated) .

Dioctophymatidae (Railliet, 1915).—Muscular cephalic sucker absent.

Dioctophymatinae (Cast. & Chalmers, 1910).—Vulva in anterior part of body. (Adults parasites of mammals).

Eustrongylidinae Chitw. & Chitw., 1937.—Vulva in posterior part of body. (Adults parasites of birds).

Soboliphymatidae Petrov, 1930.—Muscular cephalic sucker present. (Adults parasites of carnivores and fish).

Bibliography

ALLEN, M. W. 1940.—*Anomyctus xenurus*, a new genus and species of Tylenchoidea (Nematoda). Proc. Helm. Soc. Wash., v. 7(2):96-98.

ALLGEN, C. A. 1933.—Freilebende Nematoden aus den Trondhjemsfjord. Capita Zool., v. 4 (2):1-162.

1934.—Die Arten und die Systematische Stellung der Phanodermatinae, einer Unterfamilie der Enoplidae. Capita Zool. v.4 (4): 1-36.

BASIR, M. A. 1942.—Nematodes parasitic in *Gryllotalpa*. Rec. Ind. Mus. v.44 (1): 95-106.

BASTIAN, H. C. 1865.—Monograph on the Anguillulidae, or free nematodes, marine, land, and fresh water; with descriptions of 100 new species. Tr. Linn. Soc. London, v.25 (2): 73-184.

1866.—On the anatomy and physiology of the nematoids, parasitic and free; with observations on their zoological position and affinities to the echinoderms. Phil. Tr. Roy. Soc. London, v.156: 545-638.

BAYLIS, H. A. 1920-1923.—On the classification of the Ascaridae. Parts 1-3. Parasit. v.12: 253-264; v.12: 411-426; v.15:223-232.

1929.—A manual of helminthology, medical and veterinary. 303 pp., London.

1936.—Nematoda, v.1 (Ascaroidea and Strongyloidea). The Fauna of British India including Ceylon and Burma. 408 pp., London.

1939.—Nematoda. v.2 (Filarioidea, Dioctophymoidea, and Trichinelloidea). Ibid., 274 pp., London.

1943.—Some nematode parasites of earthworms from the Indo-Malay region. Parasit., v.35 (3): 112-127.

BAYLIS, H. A. & DAUBNEY, R. 1926.—A synopsis of the families and genera of Nematoda. London, Brit. Museum, 277 pp.

BAYLIS, H. A. & LANE, C. 1920.—A revision of the nematode family Gnathostomidae. Proc. Zool. Soc. London (3): 245-310.

BRUMPT, E. 1927.—Précis de parasitologie, 4 ed., 1452 pp., Paris.

BUETSCHLI, O. 1873.—Beiträge zur Kenntniss der freilebenden Nematoden. Nova Acta Acad. Nat. Curios., v.36(5): 1-124.

CALVENTE, I. G. 1948.—Revision del género *Pharyngodon* y descripcion especies nuevas. Rev.Iberica Parasit., v.8(4): 367-410.

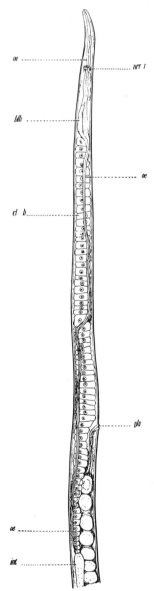

FIG. 21.

Trichinella spiralis [Trichinellidae] (Female, esophageal region). After Chitwood, 1930, J. Parasit.

CAMERON, T. W. M. 1934.—The internal parasites of domestic animals. A manual for veterinary surgeons. 292 pp., London.

CHANDLER, A. C. 1949.—An introduction to parasitology. New York, 756 pp.

CHITWOOD, B. G. 1937.—A revised classification of the Nematoda. Papers in helminthology, 30 year Jubileum K. I. Skrjabin, Moscou, pp., 69-80.
————— 1949.—''Root-knot nematodes'' Part I. A revision of genus *Meloidogyne* Goeldi, 1887. Proc. Helm. Soc. Wash., v.16 (2): 90-104.

CHITWOOD, B. G. & WEHR, E. E. 1934.—The value of cephalic structures as characters in nematode classification, with special reference to the superfamily Spiruroidea. Ztschr. Parasit., v.7(3): 273-335.

COBB, N. A. 1898.—Extract from MS report on the parasites of stock. Misc. Publ. No. 215, Dept. Agric. N. S. Wales. 62 pp.
————— 1914-1934.—Contributions to a science of nematology. Parts 1-26, 490 pp. [For sale by Frieda Cobb Blanchard, Univ. Mich., Ann Arbor, Mich.].
————— 1918.—Free-living nematodes. Chapter 15, pp. 459-505, in Ward and Whipple's Fresh Water Biology.

CRAM, E. B. 1927.—Bird parasites of the nematode suborders Strongylata, Ascaridata, and Spirurata. U. S. Nat. Mus. Bull. No. 140.

DIESING, K. M. 1850-1851.—Systema helminthum v.1-2. Vindobonae.

DITLEVSEN, H. 1926.—Free-living nematodes. The Danish Ingolf Expedition. Copenhagen.

DOUGHERTY, E. 1949.—The phylogeny of the nematode family Metastrongylidae Leiper, [1909]: a correlation of host and symbiote evolution. Parasit. v.39 (3/4): 222-234.

DUJARDIN, F. 1845.—Histoire naturelle des helminthes. Paris.

FANTHAM, H. B.; STEPHANS, J. W. W.; AND THEOBALD, F. V. 1916.—The animal parasites of man. 900 pp., London.

FAUST, E. C. 1949.—Human helminthology. 3rd ed., 744 pp., Lea & Febiger, Philadelphia.

FILIPJEV, I. N. 1927.—Les nématodes libres des mers septentrionales appartenant à la famille des Enoplidae. Arch. Naturg. (1925) 913: 1-216.
————— 1934.—The classification of the free-living nematodes and their relation to the parasitic nematodes. Smithson. Misc. Coll. v.89(6), pp. 1-63.
————— 1934.—Nematody Vrednye i poleznye v sel' skomkhoziaisjve, Moskva. 238 pp.

FILIPJEV, I. N. & STEKHOVEN, J. H. 1941.—A manual of agricultural helminthology. Leiden, 878 pp.

FRANKLIN, M. T. 1940.—On the specific status of the so-called biological strains of *Heterodera schachtii* Schmidt. J. Helminth. v.18(4): 193-208.

FREITAS, J. F. T. & ALMEIDA, J. L. DE 1935.—O genero *Capillaria* Zeder, 1800 (Nematoda-Trichuroidea) e as capillarioses nas aves domesticas. Rev. Dept. Nac. Prod. Animal Brasil, v.2(4-6): 310-363.

FUCHS, G. 1930.—Neue an Borken- und Rüsselkäfer gebunde Nematoden, halbparasitische und Wohnungseinermieter. Zool. Jahrb. v.59: 505-646.
————— 1937.—Neue parasitische und halbparasitische Nematoden bei Börkenkäfern und einige andere Nematoden. Zool. Jahrb. v.70 (5/6): 291-380.

GEDOELST, L. 1911.—Synopsis de parasitologie de l'homme et des animaux domestiques. 332 pp., Lierre & Bruxelles.

GERICHTER, C. 1949.—Studies on the nematodes parasitic in the lungs of Felidae in Palestine. Parasit. v.39 (3/4): 251-262.

GOFFART, H. 1930.—Die Aphelenchen der Kulturpflanzen. Monog. Pflanzenschutz No. 4, 105 pp., Berlin.

GOODEY, T. 1933.—Plant parasitic nematodes and the diseases they cause. Methuen & Co., London, 306 pp.
————— 1941.—On the morphology of *Mermithonema entomophilum* n.g., n.sp., a nematode parasite of the fly *Sepsis cynipsea* L. J. Helminth. v.19 (3/4):105-114.
————— 1943.—On the systematic relationships of the vinegar eelworm, *Turbatrix aceti* and its congeners with a description of a new species. J. Helminth. v.21 (1): 1-9.
————— 1945.—A note on the subfamily Turbatricinae and the genus *Turbator* Goodey, 1943. Ibid., v.21(2/3): 69-70.

————— 1946.—*Domorganus macronephriticus* n.g., n.sp., a new cylindrolaimid free-living soil nematode. Ibid., v.21 (4): 175-180.

HALL, M. C. 1916.—Nematode parasites of mammals of the orders Rodentia, Lagomorpha, and Hyracoidea. Proc. U. S. Nat. Mus. 1: 1-258.

HYMAN, L. 1940.—The invertebrates: Protozoa through Ctenophora. McGraw-Hill, New York & London.

JAEGERSKIOELD, L. A. 1909.—Freilebende Süsswassernematoden. Süsswasserfauna Deutschlands v.15:1-46.
————— 1909.—Zur Kenntnis der Nematoden Gattungen *Eustrongylides* u. *Hystrichis*. Nova Acta Reg. Soc. Sc. Upsaliensis, 4 s., v.2:1-48.

JANICKI, C. & RASIN, K. 1930.—Bemerkungen über *Cystoöpsis acipenseri* des Wolga-Sterlets, sowie über die Entwicklung dieses Nematoden im Zwischenwirt. Ztschr. Wiss. Zool. v.136: 1-37.

JUNGE, W. 1938.—Systematik und Variabilität der pflanzenparasitischen Aphelenchen, sowie deren Verbreitung an verschiedenen Wirtspflanzen. Ztschr. Parasit. v.10 (5): 559-607.

KREIS, H. A. 1934.—Oncholaiminae Filipjev, 1916, eine Monographische Studie. Capita Zool. v.4 (5): 1-270.

LAPAGE, G. 1937.—Nematodes parasitic in animals. Methuen & Co., London, 163 pp.

LINFORD, M. B. & OLIVEIRA, J. M. 1940.—*Rotylenchulus reniformis*, nov.gen., n.sp., a nematode parasite of roots. Proc. Helm. Soc. Wash. v.7 (1): 35-42.

LINSTOW, O. VON 1878-1889.—Compendium der Helminthologie. Hanover, 382 & 151 pp.
————— 1909.—Parasitische Nematoden. Süsswasserfauna Deutschlands v. 15: 47-81.

LOOS, C. A. 1948.—Notes on free-living and plant parasitic nematodes of Ceylon, 3. Ceylon J. Sc. v.23 (3): 119-124.

LOOSS, A. 1902.—The Sclerostomidae of horses and donkeys in Egypt. Rec. Egypt. Govt. School Med., v.1: 25-139.

LUCASIAK, J. 1930.—Badania anatomiczne i rozwojowe nad *Dioctophyme renale* (Goeze, 1782). Arch. Biol. de la Soc. d. Sc. et Lett. d. Varsovie, v.3: 1-100.

MAN, J. G. DE 1884.—Die frei in der reinen Erde und im süssen Wasser lebende Nematoden. Leiden, 206 pp.

MICOLETZKY, H. 1922.—Die freilebenden Erd-Nematoden. Arch. Naturg. v.87, Abt. A, (8-9): 1-650.
————— 1930.—Freilebende marine Nematoden von Sunda-Inseln. I. Enoplidae. Vidensk. Medd. Dansk. naturh. Foren, v. 87: 243-339.

MOENNIG, H. O. 1934.—Veterinary helminthology and entomology. The diseases of domesticated animals caused by helminth and arthropod parasites. 402 pp., London & Baltimore.

MOLIN, R. 1858.—Versuch einer Monographie der Filarien. Sitz. k. Akad. Wiss., Wien, v.28: 365-461.

MUELLER, G. W. 1931.—Ueber Mermithiden. Ztschr. Morph. u. Oekol. v.24:82-147.

NEVEU-LEMAIRE, M. 1936.—Traité d'helminthologie médicale et vétérinaire. 1514 pp., Paris.

OERLEY, L. 1880.—Monographie der Anguilluliden. Budapest, 165 pp.
————— 1886.—Die Rhabditiden und ihre medicinische Bedeutung.. Berlin, 84 pp.

ORTLEPP, R. J. 1923.—The nematode genus *Physaloptera* Rud. Proc. Zool. Soc. London (1922): 999-1107.

PEREIRA, C. 1932.—*Myenchus botelhoi* n.sp., curiosa nematoide parasite de *Limnobdella brasiliensis* Pinto (Hirudinea). Thesis, Fac. Med., São Paulo, 32 pp.

PETROW, A. W. 1930.—Zur Characteristik des Nematoden aus Kamtschatkaer Zobeln. *Soboliphyme baturini* nov. gen., nov. spec. Zool. Anz. v.86:265-271.

RAUTHER, M. 1930.—Vierte Klasse des Cladus Nemathelminthes. Nematodes, Nematoidea-Fadenwürmer. Handb. d. Zool. (Kükenthal u. Krumbach) v.2, 8 Lief., 4. Teil, Bogen 23-32, pp., 249-402.

REITER, M. 1928.—Zur Systematik und Oekologie der zweigeschlechtlichen Rhabditiden. Arb. Zool. Inst. Univ. Innsbruch v.3 (4): 93-184.

SANDGROUND, J. H. 1934.—On the validity of the various species of the genus *Onchocerca* Diesing. In Strong et al. Contr. Dept. Trop. Med. & Inst. Trop. Biol. & Med. No. 6: 135-172.

SCHEPOTTEFF, A. 1908.—Die Chaetosomatiden. Zool, Jahrb. v.26: 401-414.

SCHNEIDER, WILHELM 1939.—Würmer oder Vermes II. Fadenwürmer oder Nematoden. Die Tierwelt Deutschlands u. der angrenzenden Meeresteile. Jena, 260 pp.

SEURAT, L. G. 1920.—Histoire naturelle des nématodes de la Beribérie. Première Partie. Morphologie, Développement, ethologie et affinités des nématodes. 221 pp., Alger.

SKRJABIN, K. I. & SHIKHOBALOVA, N. P. 1948.—[Filariae of animals and man. Russian text]. 608 pp., Moskva.

SPREHN, C. E. W. 1932.—Lehrbuch der Helminthologie. 998 pp., Berlin.

SKRJABIN, K. I. ET AL. 1934.—Veterinary parasitology and invasion diseases of domestic animals. 600 pp., Moskva. [In Russian].

STEINER, G. 1919.—Untersuchungen über den allgemeinen Bauplan des Nematodenkörpers. Diss., 96 pp.
    1932.—Die arktischen Mermithiden, Gordioiden und Nectonemathoiden. Fauna Arctica v.60:161-174.
    1933.—The nematode Cylindrogaster longistoma (Stefanski) Goodey and its relationship. J. Parasit. v.20 (1): 66-68.
    1937.—Opuscula miscellanea nematologica V. Tylenchorhynchus claytoni n.sp., an apparently rare nemic parasite of the tobacco plant. Proc. Helm. Soc. Wash. v.4 (1): 33-36.
    1945.—Helicotylenchus a new genus of plant parasitic nematodes and its relationship to Rotylenchus Filipjev. Proc. Helm. Soc. Wash. v.12(2): 34-38.

STEKHOVEN, J. H. S. 1937-1939.—Nematodes and Nematomorpha. Bronn's Klassen v.4, Abt. 2, Buch 3, Liefs. 5-6.

STEKHOVEN, J. H. S., ADAM, W. & CONINCK, L. DE 1931-1933.—The free-living marine nematodes of the Belgian Coast. Mem. Mus. Roy. Hist. Nat. Belg. No. s 49 & 58.

TAYLOR, A. L. 1936.—The genera and species of the Criconematinae, a subfamily of the Anguillulinidae (Nematoda). Tr. Am. Micr. Soc. v.55 (4): 391-421.

THEILER, G. 1924.—The strongylid and other nematodes parasitic in the intestinal tract of South African Equines. 9th & 10th Reports of the Dir. Vet. Ed. & Res., pp. 601-773.

THORNE, G. 1937.—A revision of the nematode family Cephalobidae Chitwood and Chitwood, 1934. Proc. Helm. Soc. Wash. v.4(1): 1-16.
    1939.—A monograph of the nematodes of the superfamily Dorylaimoidea. Capita Zool. v.8 (5): 1-261.
    1941.—Some nematodes of the family Tylenchidae which do not possess a valvular median esophageal bulb. Great Basin Naturalist. v.2 (2): 37-85.
    1943.—Cacopaurus pestis nov.gen., nov.spec. (Nematoda: Criconematinae), a destructive parasite of the Walnut, Juglans regia Linn. Proc. Helm. Soc. Wash. v.10 (2): 78-83.
    1949.—On the classification of the Tylenchida, new order (Nematoda, Phasmidia). Ibid., v.16 (2): 37-73.

THORNE, G. & ALLEN, M. W. 1944.—Nacobbus dorsalis, nov. gen., nov.spec. (Nematoda: Tylenchidae) producing galls on the roots of alfileria, Erodium cicutarium (L.) Ibid., v.11 (1): 28-31.

THORNE, G. & SWANGER, H. H. 1936.—A monograph of the nematode genera Dorylaimus Dujardin, Aporcelaimus n.g., Dorylaimoides n.g., and Pungentus n.g. Capita Zool. v. 6 (4): 1-156.

TOERNQUIST, N. 1931.—Die Nematodenfamilien Cucullanidae und Camallanidae. Göteborgs Kungl. Vetensk. och Vitterh.-Samh. Handl. (5), 441 pp.

TRAVASSOS, L. 1937.—Revisão da familia Trichostrongylidae Leiper. Monog. Inst. Oswaldo Cruz No. 1. 511 pp.

WEHR, E. E. 1935.—A revised classification of the nematode superfamily Filarioidea. Proc. Helm. Soc. Wash. v.2: 84-88.

WUELKER, G. & STEKHOVEN, J. H. S. 1933.—Nematoda. Die Tierwelt der Nord- u. Ostsee. Teil 5, a-c.

YAMAGUTI, S. 1935.—Studies on the helminth fauna of Japan. Part 9, Nematodes of fishes 1. Japan. J. Zool. v.6 (2):337-386; Part 10, Amphibia nematodes. Ibid., v.6 (2): 387-392; Part 11, Reptilian nematodes. Ibid., v.6 (2): 393-402; Part 12, Avian nematodes Ibid., v.6 (2): 403-431; Part 13, Mammalian nematodes, Ibid., v.6 (2): 433-457.
    1941.—Idem., Nematodes of fishes II. Ibid., v.9(3): 343-395; Amphibian nematodes II. Ibid., 409-439; Avian nematodes, Ibid., 441-480.

YOKAGAWA, S. & MORISHITA, K. 1933.—Handbook of human parasitology v.2, 603 pp., Tokyo.

YORKE, W. & MAPLESTONE, P. A. 1926.—The nematode parasites of vertebrates. Phila., 536 pp.

ZEDER, J. G. H. 1800.—Erster Nachtrag zur Naturgeschichte der Eingeweidewürmer, mit Zufässen und Anmerkungen herausgegeben. Leipzig.

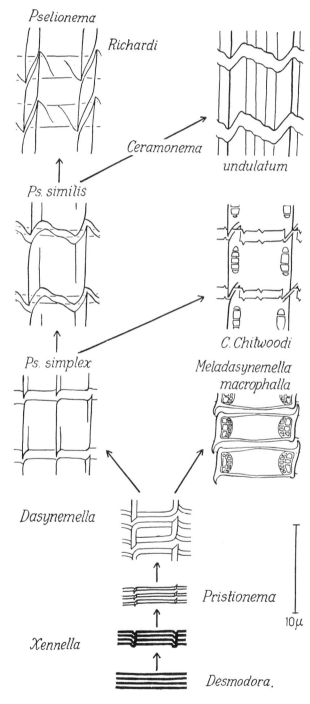

FIG. 22.

Cuticular modifications in the Desmodoridae.
To the left. Evolutionary series in the annulation of the cuticle from the plain annules of *Desmodora* through the modified annules of *Xennella*, *Pristionema*, and *Dasynemella* to the elaborate armor of *Pselionema*, *Ceramonema* and *Meladasynemella*.

27

CHAPTER III

THE EXTERNAL CUTICLE AND HYPODERMIS
A. THE CUTICLE

B. G. CHITWOOD and M. B. CHITWOOD

The cuticle of nematodes is composed of a more or less distinctly layered elastic sheath covering the body and extending inwardly at mouth, anus and vulva. The part of the cuticle covering the exterior of the body differs in some respects from that lining parts of the digestive tract and vagina. In order to differentiate between the types of cuticle the term *external cuticle* is applied to the exoskeletal covering, while the terms esophageal, rectal, cloacal or vaginal cuticle, respectively, are applied to that part of the cuticle lining these organs.

The external cuticle is intimately connected with, and undoubtedly is a product of the hypodermis. Whether it is an extracellular deposit or an intracellular formation is not certain. It would be difficult to prove whether the protoplasm of hypodermal cells does or does not extend into the cuticle; in most instances it does not appear to do so.

A. GENERAL MORPHOLOGY

There are several types of gross cuticular markings, namely, transverse, longitudinal and oblique markings, inflation, spination and other special modifications. All of these are due to the modifications of the various cuticular layers which are treated under the heading of "Cuticular layering." In addition to these markings there are appendages or superficial organs such as setae, supplementary organs, etc., which involve both the cuticle and hypodermis.

Transverse marking. In the living organism the cuticle appears, upon casual observation, to be a clear, refractive, often apparently homogenous substance. Its superficial markings are

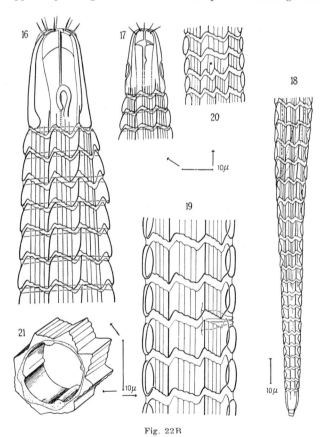

Fig. 22B

in the nature of grooves or ridges. Transverse grooves are striae and the distance between two striae is the *interstrial region.* Such transverse striae are of very common occurrence in the Nematoda, being, in fact, more often present than absent. It is only in such forms as *Enoplus* (Enoploidea), *Mononchus* (Tripyloidea), and *Dorylaimus* (Dorylaimoidea) and *Mermis* (Mermithoidea) that transverse striation appears to be totally absent and even in some forms related to these, transverse striation does occur.

Deep striae very commonly occur at intervals and are known as annulations, the distances between them being termed annules. Both striation and annulation commonly occur in nematodes, particularly in forms parasitic in vertebrates, such as ascarids (Fig. 23G), oxyurids, filaroids and spiruroids. Annulation is less common among free-living nematodes, being confined for the most part to marine groups such as desmodorids, desmocolecids (Fig. 17) and epsilonematids. Striation and annulation are often accompanied by additional minute markings, each of which may be treated separately. Sometimes these markings are distributed among the genera of a particular family and are of general significance but in other cases the lack of uniformity in their occurrence indicates that they have little phylogenetic significance; this apparent lack of significance may, perhaps, be due to our inadequate knowledge of a sufficient number of species. In members of the Phasmidia particularly, both annulation and striation are present in the same form. In *Ascaris lumbricoides* (Fig. 23G) deep striae are found at intervals, these deep striae involving the entire corticle layer as well as the greater part of the matrix layer of the cuticle.

PUNCTATION is a frequent type of marking, and appears as minute rounded areas of the cuticle which in a given species are arranged in a definite pattern. This type of marking is particularly well developed in *Rhabditis lambdiensis.* When studied under moderate magnification the cuticle of this species shows distinct transverse striae. With the highest magnifications, however, each interstrial area is resolvable into two transverse rows of punctations, (Fig. 23K) situated within longitudinal ridges. Such cuticular markings are common in the superfamily Rhabditoidea, particularly in the families Rhabditidae and Diplogasteridae. Punctations are not easily observed in many species in these families but their presence has been demonstrated in so many instances that it seems reasonable to assume that they are present in all species of these groups as well as in species of related groups. The same type of marking is found in larval strongyloids as in rhabditids and diplogasterids though they have not been demonstrated in the adults. The presence of superficially distinct punctations does not appear to be harmonious with thick cuticula and deep striation. In the remaining forms of the Phasmidia, the Ascaridina and Tylenchoidea, there is a tendency toward deep striation and it is in these groups that punctations are particularly uncommon.

In the Chromadoridae the punctations may occur in several patterns. They may consist of a single transverse row of round or elongated markings for each interstrial region; these markings being interrupted laterally, the sublateral rows of punctations being larger than the others (ex. *Spilophorella paradoxa,* Fig. 23, P. Z.). As a further variation, the "punctations" may appear as a corrugated band as in *Chromodorella* sp. (Fig. 23T-Y). De Man (1886) found the punctations of *Euchromadora vulgaris* are in the form of a basketwork and are composed of rows of elongated hexagonal pieces connected

---

Fig. 23 Continued

lateral break; FFF—A triply split annule; GGG—Annules with lateral flexible region; HHH—"Cut" annule. III-QQQ—Various modifications of the edges of annules in the Epsilonematidae showing means by which additional rigidity is obtained. RRR—*Longistriata hasalli* (showing cephalic inflation; SSS—*Gongylonema pulchrum* (Inflations in form of plaques); TTT—*Onchocerca gutterosa* (Thickened transverse ridges). OO-QQ, After Filipjev, 1934, Nematody. RR-QQQ, After Steiner, 1930, Deutsche Sudpolar-Expedition.

To the right. *Ceramonema undulatum;* 16—Anterior end of female, left side; 17—Same, dorsal view; 18—Tail of female, left lateral view; 19—Vulvar region, right lateral view; 20—Ventral view at excretory pore; 21—Schematic cross section. All figs. after de Coninck, 1942, Bull. Mus. Roy. Hist. Nat. Belg., v. 18 (22): 1-17.

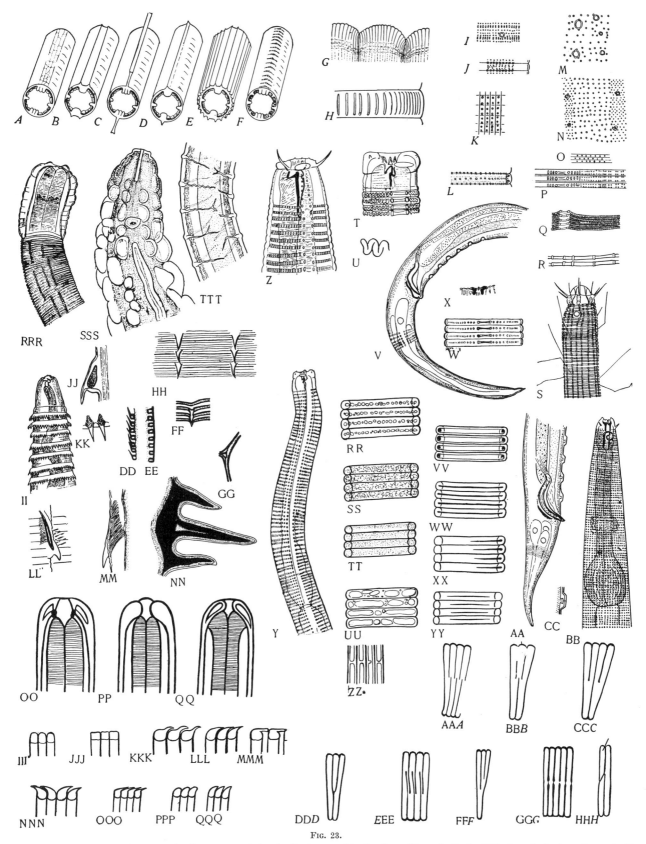

Fig. 23.

Cuticular Marking. A-F—Diagrams showing types of longitudinal markings (A—Alae absent, simple form; B—Sublateral alae in an oxyurid; C—Wide double lateral alae in an oxyurid; D—Narrow lateral alae; E—Longitudinal ridges in a Trichostrongyle; F—Chromadorid type of marking). G—*Ascaris lumbricoides* [Ascaridae]. (Cuticle at side of body. H—Undescribed oncholaim [Oncholaimidae] (Showing longitudinal ridges broken by striae). I-J—*Synonchiella truncata* [Cyatholaimidae] (I—Mid-region; J—Cervical region). K—*Rhabditis lambdiensis* [Rhabditidae]. L—*Pomponema*-like form [Cyatholaimidae]. M-O—*Paracanthonchus* sp. [Cyatholaimidae] (M—Cephalic region, lateral; N—Mid-region, lateral; O—Mid-region, median). P—*Spilophorella paradoxa* [Chromadoridae] (lateral). Q—*Pomponema mirabile* [Cyatholaimidae] (lateral). R-S—*Xyala striata* [Monhysteridae]. T-Y —*Chromadorella* sp. [Chromadoridae]. Z—*Spilophorella paradoxa*. AA-CC—*Ethmolaimus revaliensis* [Chromadoridae]. DD-GG—*Monopos-*

thia hexalata [Desmodoridae] (DD—Longitudinal section, at ridge; EE—Between ridges; FF—Tangential section through ridge; GG—Cross section). HH—*Seuratum* sp. [Cucullanidae]. II-KK—*Spinitectus* sp. [Spiruridae]. LL-MM—*Rictularia coloradiensis* [Thelaziidae] (spines). NN—*Physocephalus sexalatus* [Thelaziidae] (Cross section of cervical ala). OO-QQ—Cuticle at head (OO—Monhysteridae; PP—Camacolaimidae; QQ—Enoplidae). RR-YY—Diagrams of types of annules in Epsilonematidae. (RR—Vacuolar type; SS—Spongy type; TT—Cement-like; UU—Vesiculate; VV—Wide lumen type; WW—Narrow lumen; XX—Solid except dorso-submedial region; YY—Solid with lateral cavity). ZZ-HHH—Diagrams of lateral modifications of annules in Epsilonematidae. (ZZ—With lateral cross beam joints; AAA—Annules broken laterally and united; BBB—Two complete and one broken annule; CCC—Two dorsal split annules; DDD—Two complete and one partial annule; EEE—Annules united through a long

29

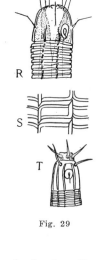

Fig. 29

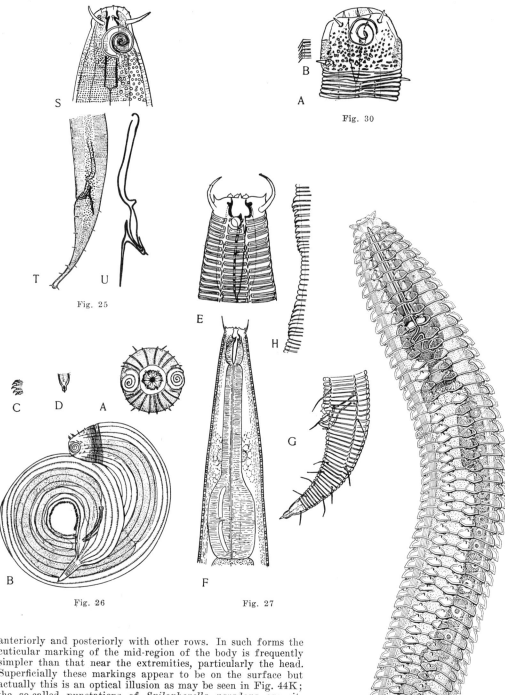

Fig. 25

Fig. 30

Fig. 26   Fig. 27

FIG. 24. *Criconema octangulare* (Criconematidae). After Cobb, 1915, U. S. D. A. Yearbook for 1914.

anteriorly and posteriorly with other rows. In such forms the cuticular marking of the mid-region of the body is frequently simpler than that near the extremities, particularly the head. Superficially these markings appear to be on the surface but actually this is an optical illusion as may be seen in Fig. 44K; the so-called punctations of *Spilophorella paradoxa* are situated between the cortical and basal cuticular layers.

In the Cyatholaimidae punctations are similar to those in the Chromadoridae, but may assume additional patterns. In *Pomponema mirabile* (Fig. 23Q) practically the same cuticular marking is found as in *Spilophorella paradoxa* except that in deep focus additional rows of punctations appear in the striae as well as between them, while in a related Pomponema-like form (Fig. 23L) the same arrangement is present except that no lateral interruption occurs and the punctations at the striae appear to be double. *Synonchiella truncata* presents further modifications in that three rows of round punctations occur between each pair of striations. Each striation is indicated in deep focus by longitudinally elongated punctations in the mid-region of the body (Fig. 23I) and by double rows of transversely elongated punctations in the cervical region (Fig. 23J). *Paracanthonchus* does not have deep or distinct striae but alternating rows of small punctations (Fig. 23O). In addition, however, larger and more distinctly placed punctations are present in the region of the lateral chords (Fig. 23N). In many cyatholaims, including *Synonchiella* and *Paracanthonchus*, larger "rimmed" spots occur laterally. These are the orifices of the sublateral glands or gland setae. Punctations in cyatholaims, as well as in comesomatids, are underneath the surface as in *Spilophorella* though this fact may be difficult to believe if one relies on totomount specimens instead of sections.

EVOLUTIONARY TRENDS. The evolution of transverse marking from striae to apparent segmentation is an interesting characteristic of some groups. The stylet bearing phasmidians, such as members of the Tylenchoidea, show a trend in this direction. Species of the genera *Aphelenchoides* and *Ditylenchus* have indistinct striae; *Aphelenchus avenae* and *Tylenchus costatus* with stouter proportions have somewhat deeper striae and closely approach that occurrence in the genus *Hoplolaimus* in which these markings reach the depth of annules. From these forms the trend is carried further through the genus *Hemicycliophora* to the curious genus *Criconema* (Fig. 24) in which there is little gross resemblance to ordinary nematodes. In *Criconema*, according to Taylor (1936), the annulations are very deep and the edge of each annule projects over the following annule, first as a series of scales, later as a circle of spines. Spination occurs in many groups of nematodes, and will be dealt with later.

Modification of cuticular striae becomes most amusing in the Heteroderinae. Females of *Meloidogyne* have rather normal striae in the mid-region of the body but in the perineal region the striae become highly convoluted forming whorls, arches, etc., which are practically as individual as finger prints. However, the basic pattern is specific. In the genus *Heterodera* the cuticular marking becomes even more involved. The females of the species living on shadscale have rather plain striae with minor transverse interruptions while the females *Heterodera schachtii* have highly folded, transverse marking. Internal layers further complicate the picture in *Heterodera* so that we may appear to have "punctation" present or absent, and if present the "punctations" may be fine, coarse, regular or irregular. Upon such characters, with excellent microscopy, species may readily be identified.

The chromadoroids, as previously mentioned, have transverse cuticular markings somewhat similar to those of rhabditoids. The desmodoroids, a related group, differ from chromadoroids in that definite annulation is the rule. Progressive thickenings of the cuticle, development of annulation and rigidity of the body seem to be the evolutionary trend. Within the genus *Euchromadora* (Chromadoroidea) evidences foreshadowing such development are present. In *Euchromadora vulgaris* the dorsal and ventral areas of the body surface appear as a series of plates broken by striae; in each of these plates there is a transverse row of longitudinal markings which project into the interstrial region; at the base of the esophageal region the median rows of plates are divided. The remainder of the body surface is covered by elongate hexagonal rods. Edges of "annules" project slightly anterior in the anterior end of the body and posterior in the posterior end of the body. Transverse striations extend almost to both extremities in *E. archaica*, a character correlated with the moderate transverse marking.

With cuticular thickening and more extreme transverse striation, as in the desmodoroids, minute patterns such as punctations are apparently lost, and the thickened cuticle is not annulated near the extremities but forms a "helmet" at the anterior and a caudal "cane" at the posterior extremity. In such forms as *Acanthopharynx japonicus* Steiner and Hoeppli (1926) described over-edging of the annules to a moderate degree, and further modification "coupled annules." In this type of reinforcement adjoining annules are often inserted into one another giving added rigidity and at the same time providing for elasticity. The tendency toward rigidity apparently reaches a maximum in the armored ceramonematids. In *Pselionema* and *Ceramonema* the annules zigzag in form so that they overlap considerably (Fig. 22). In such cases there are six or eight longitudinal joints in the cuticle corresponding to longitudinal ridges or alae but due to the extremely deep striation of the annules they are subsidiary. *Dasynemella* and *Dasynemoides* (Fig. 29) form connecting links between the wiry ceramonematids and the more serpentine desmodorids.

The plate-like form of the cuticle of the Monoposthiinae, another desmodorid group, is likewise due to deep striation coincident with longitudinal ridges. In this case the body may be definitely 6, 8, 10, 12 or more sided (Fig. 27), and according to the body region the ridges project into corresponding grooves of the preceding or succeeding annule (Fig. 27E), thus giving the appearance of spines. It may also be noted that the cortical and basal layers of the cuticle are separated by a cavity (Fig. 50G); this seems to be the rule in the Desmodoroidea.

In the Epsilonematide according to Steiner (1930, 1931) the large annules show a distinct cavity which may have either wide or narrow lumen, or the cavity may contain vacuoles, granules or spongy material (Fig. 23RR-YY). This material, usually absent in desmodoroids, doubtless corresponds to the matrix layer of other nematodes.

The annules of desmodoroids might be characterized as "inflated annules" but the degree of inflation is moderate. In

the desmoscolecids inflated annules are the most spectacular in form being separated by narrow, low annules which cause the worms to appear as if surrounded by a series of balloon tires; in other forms foreign material adheres to the large inflated annules giving to them a bizarre appearance (Fig. 17).[1]

Monhysterids are often said to have a "smooth" cuticle, but as a rule, transverse striation is visible and in at least one example, *Xyala striata*, deep striation and longitudinal ridges join to give to the worm an armored appearance very similar to that observed in desmodorids (Fig. 23R-S).

In a few genera, such as *Hamatospiculum* and *Onchocerca* of the Filarioidea, transverse cuticular ridges occur. These ridges, like alae, involve only the cortical and matrix layers of the cuticle (Fig. 44F).

Longitudinal markings. Longitudinal markings may take the form of ridges or alae, or they may be merely the result of interruptions in the transverse markings.

LONGITUDINAL RIDGES are raised areas which extend the length of the body and are present on the submedian as well as on the lateral surfaces (Fig. 23E). Such ridges are particularly common in the diplogasterids, corresponding apparently to rows of punctations. In *Rhabditis lambdiensis* (Fig. 50A) a cross section of the worm shows that the entire surface bears small ridges which in surface view appear as transverse rows of elongated rods (previously mentioned as containing punctations). On the ventral surface of the bursa these same rods or ridges appear as rounded elevations. Similar longitudinal ridges are commonly present and are compared to the ventral surface of the bursa of *Physaloptera* (Fig. 33F), *Spirocerca* and other spiruroids, as well as strongyloids. Trichostrongyloids, however, usually have more or less pronounced (unbroken) longitudinal ridges throughout the body length. *Oswaldocruzia* and *Longistriata* are favorable examples for such study, and in these forms it has been found that the longitudinal ridges involve only the two external cuticular layers (Fig. 35). Similar ridges, though inconspicuous, are sometimes present in representatives of the Aphasmidia such as *Dorylaimus* (Fig. 20) and oncholaimids. The appearance of such ridges may be altered by the presence or absence of transverse striae which cause the ridges to be broken into transverse rows of rods, or they may retain their identity in the various forms but in either case their nature is always the same. Longitudinal ridges also occur in such forms as *Monoposthia*, *Ceramonema*, etc.; these cases have already been discussed since they are subservient to the annular formations.

ALAE are usually lateral or sublateral cuticular thickenings or projections and may be divided into three general types as follows: *Longitudinal alae*, extending the length of the body; *cervical alae*, confined to the anterior part of the body; and *caudal arae*, confined to the posterior part of the body and to the male sex.

Longitudinal alae are usually sublateral, four in number (Fig. 23B), seldom very wide, and occur in both sexes. Sublateral alae are commonly present in members of the Rhabditina and sometimes in the Ascaridina, while true lateral alae are more common in the Ascaridina and Chromadorina. In the Oxyuroidea the alae may be extremely wide, in some instances equal to the diameter of the body. In *Parathelandros anolis* narrow sublateral alae are found in the female and wide lateral alae in the male (Fig. 35I, J); in this case the lateral alae are distally bifid indicating that they are fused sublateral alae. Sometimes both lateral and sublateral alae are said to be present in free-living forms (ex. *Cephalobus similis*). Wide lateral alae occur in a few scattered examples of the Strongylina (*Pharaurus alatus*) and Spirurina (*Foleyella* spp.) but are most common in the Oxyuridae and Thelastomatidae (Ascaridina). In the latter groups wide alae (*Parathelandros anolis*) are most characteristic of larvae and males rather than of females. The alae are used as fins for swimming, the movement of the body being undulant.

Among the members of the Tylenchina, longitudinal cuticular thickenings may appear as one solid ridge covering the width of the lateral chords or this ridge may be subdivided by incisures so that 2, 4, 6, 8 or 10 lines are observed in superficial examination. (Compare *Aphelenchoides*, *Ditylenchus*, *Aphelenchus*). In the subfamily Heteroderinae there is considerable sexual dimorphism in this regard. Larvae show 3 lateral ridges (4 incisures) as do males while females show no evidence of such structures except in the single species *Meloidogyne javanica*.

Fig. 25. *Dorylaimopsis metatypicus* (Comesomatidae). After Chitwood, 1936, Proc. Helminth. Soc. Wash.

Fig. 26. *Richtersia beauforti* (Desmodoridae). After Chitwood, 1936, Ibid.

Fig. 27. *Monoposthia hexalatha.* (H-Preanal region of male). After Chitwood, 1936. Ibid.

Fig. 29. R-S—*Dasynemella phalangida.* T—*Dasynemoides setosum.* (Desmodoridae). After Chitwood, 1936, Ibid.

Fig. 30. *Desmodorella cephalata.* (Desmodoridae). After Chitwood, 1936, Ibid.

[1] For further information on annulation and its significance see Steiner and Hoeppli (1926).

Cervical alae. These structures are modified, usually rather wide, lateral alae and are known only in parasitic nematodes, occurring in the Strongylina, Ascaridina (Fig. 31), and Spirurina. Usually they are simple, sometimes bifid, and very rarely distinctly trifid (*Physocephalus sexalatus* Fig. 23NN). They are formed from the cortical and matrix layers of the cuticle and usually contain a hard lateral skeletal support (44 I,O).

Caudal alae. Alae confined to the caudal region of the body are limited to the males and apparently serve as clasping organs during copulation. Apparently they are not homologous to lateral alae since in genera of the Oxyuroidea and Metastrongyloidea in which both occur, there is a distinct division between the two types of alae (*Pharurus alatus*, Fig. 33K). Dujardin (1845) introduced the term bursa or bell as a descriptive rather than as an anatomic term; he applied it to the caudal expansions of various strongylatids and *Dioctophyma*. The term *bursa* has since been adopted by workers as an anatomic term for the caudal alae in *Rhabditis* as well as in other representatives of the Rhabditina and to caudal alae in the Strongylina. It has not, however, been applied in the Ascaridina or Spirurina although there is no particular reason for the distinction. Whatever their form, caudal alae consist only of the cortical and matrix layers of the cuticle, that is, the oblique fiber layers have no part in the formation of such structures (Fig. 35C). It is rather peculiar that while caudal alae are characteristic of certain large groups of nematodes, such as the Strongylina, they may or may not be present in closely related genera in other groups. Caudal alae are found generally distributed throughout the Phasmidia and are absent in the Aphasmidia except in the genera *Oncholaimellus* and *Anoplostoma* (Fig. 33G-H). The form of caudal alae is variable and may best be discussed under the various groups in which they occur.

Genital papillae are situated within the caudal alae and are sometimes said to support them; this is hardly reconcilable with their function as tactile organs. Muscles accompanying the genital papillae as in the "rays" of strongylatids might be said more correctly to support the alae but even this is a dubious distinction, the fact being that alae support papillae or rays since the semi-transparent cuticle is a relatively hard substance while the structures embedded in it are soft.

Within the Rhabditina caudal alae may be present in either of the two superfamilies, Rhabditoidea and Tylenchoidea, and there is no simple means by which they may be differentiated. In general, the caudal alae of rhabditoids contain at least seven or more pairs of genital·papillae while the caudal alae of tylenchoids contain three or fewer pairs of genital papillae. Aside from this difference the variations may be discussed together. Caudal alae are absent in many diplogasterid genera, *Longibucca*, *Aphelenchoides*, *Parapholenchus*, *Criconema*, and in cephalobids, while they are characteristic of rhabditids, *Ditylenchus*, *Hexatylus*, *Angiostoma*, rhabdiasids, and cylindrogasterids. In several of the families there is a series of gradations from large wide alae to very narrow alae. This seems to indicate parallel evolution but the problem is too complex in the Rhabditina to be treated from the viewpoint of evolution and consequently will be treated from the standpoint of general appearance.

Schneider (1866) introduced a terminology for forms having caudal alae which is of some aid in description. To nematodes having caudal alae which do not meet posterior to the tail of the body proper he applied the term *leptoderan* while to forms having caudal alae meeting posterior to the tail he applied the term *peloderan*. A great variation in degree of development of caudal alae (Fig. 33) occurs in leptoderan forms. There are some species in which the tail is long and filiform that possess simple narrow caudal alae extending from some distance anterior to the cloacal orifice to some distance posterior to it (*Rhabditis longispina*), while in other forms the caudal alae are double, but otherwise similar (*Rhabditella*, *Rhabditoides*). Wide caudal alae more often occur in forms with a short tail (*Rhabditis lucanii*, *R. aspera*) than in forms with a long tail (*Rhabditis gracilicauda*). Narrow double caudal alae do occur in some short-tailed rhabditids (*R. oxycerca*). In peloderan forms with wide simple alae every conceivable degree of loss of tail is found, the extreme occurring in some rhabditids (*R. lambdiensis*, *R. strongyloides*) and some cephalobids, diplogasterids, cylindrogasterids, tylenchids, etc. A further development of the caudal alae in which they meet anteriorly forming a complete oval occurs in a few rhabditid species (*R. elegans and R. coarctata*). For the peloderan type of caudal alae the term bursa is applied and is used in a loose sense for all forms of caudal alae. For the type of bursa occurring in rhabditids we shall apply the term rhabditoid bursa.

The type of caudal alae occurring in strongylatids appears to be a direct development from the wide caudal alae found in rhabditids. The *strongylatid bursa* has undergone two modifications from the rhabditoid type, first there is a lobing, three chief lobes being formed, a dorsal and two lateral; second, the caudal papillae are accompanied by muscular tissue forming "rays." The dorsal lobe may be compared to the posterior part of the rhabditoid bursa since it contains the terminal part of the body (tail) and three pairs of genital papillae (Figs. 33I-J), which together with their accompanying muscles are termed the dorsal rays. Further discussion of the rays will be taken up under the nervous system. The bursae of the strongylatids are usually regarded as terminal but actually they are lateral as in rhabditids, the difference in appearance being due to the shortening of the tail and posterior projection of the region around the cloacal orifice forming a cloacal prominence (genital cone) as first pointed out by Looss (1905). One group of strongylatids, the Metastrongyloidea, differs from the others in that the bursa is shortened, narrow, the rays short, and the dorsal lobe and its rays always greatly reduced in size (Fig. 33K).

The Ascaridina also shows a variation of caudal alae. In the Oxyuroidea the caudal alae are usually narrow, confined to a limited region immediately surrounding the cloacal orifice, and since the tail commonly extends posterior to the alae they are considered leptoderan forms. Some representatives have no caudal alae whatever as in *Rhigonema* (Fig. 11F) (Rhigonematidae), *Cephalobellus* and *Blatticola* (Thelastomatidae), *Atractis* (Atractidae) and *Thelandros* (Oxyuridae). In the Ascaridoidea a much greater variation of caudal alae is found; in some families the caudal alae are entirely absent as in most genera of the Cosmocercidae (Fig. 11D), some heterakids and kathliniids while in others they are well developed but as they do not meet posterior to the anus they are, therefore, leptoderan (some heterakids, Fig. 33OP). In the Ascarididae the caudal alae are usually short and narrow, or absent. In none of these forms are the genital papillae accompanied by large muscular developments forming rays.

FIG. 31

Toxocara canis. [Ascaridae]. (Ventral view showing cervical alae).

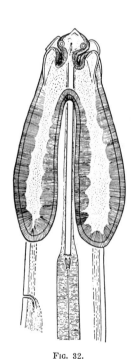

FIG. 32.

Cosmocephalus obvelatus [Acuariidae] (Showing cordons; note deirid is distally bifurcate).

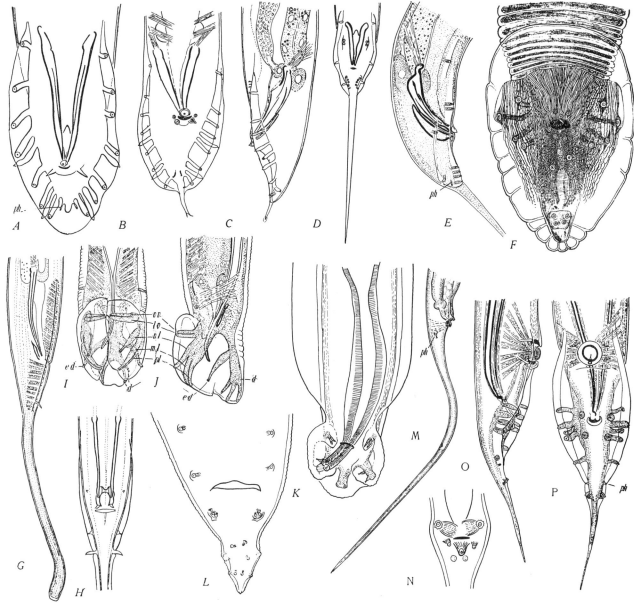

Fig. 33

Tails of males. A—*Rhabditis caussaneli.* [Rhabditidae]. B-C — *Rhabditis aspera.* D-E—*Rhabditella axei* [Rhabditidae]. F— *Physaloptera turgida* [Physalopteridae]. G-H—*Anoplostoma viviparum* [Oncholaimidae]. I-J—*Oesophagostomum longicaudatum* [Strongylidae]. K—*Pharurus alatus* [Pseudaliidae]. M-N—*Leidynema nocalum* [Thelastomatidae]. O-P—*Heterakis gallinarum* [Heterakidae]. D, After Reiter, 1928, Arb. Zool. Inst. Innsbruck; G-H, After de Man, 1907, Mem. Soc. Zool. France; M-N, After Chitwood & Chitwood, 1933, Philip. J. Sc.

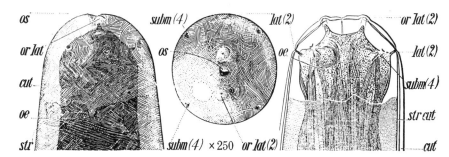

Fig. 34. *Mermis subnigrescens* (Mermithidae). (Side, en face, and median views of head). After Cobb, 1926, J. Parasit.

The Spirurida contains variants (Figs. 14 and 33F) similar to the Ascaridina except that there are no forms which parallel the long-tailed oxyurids (Fig. 33M-N). Primarily this group appears to be peloderan, with forms having wide caudal alae (rhabditoid bursa) occurring in numerous families (Filariidae, Dipetalonematidae, Spiruridae, Acuariidae, Physalopteridae, and Thelaziidae), but forms with narrow caudal alae not quite meeting posterior to the caudal extremity occur in some families (Dipetalonematidae, Spiruridae, Camallanidae, Gnathosto-matidae); caudal alae are absent in some forms (Thelaziidae, Spiruridae, Cucullanidae, Filariidae, Dipetalonematidae). The most outstanding modification of caudal alae occurring in this group is to be found in the asymmetrical development of some representatives of the Acuariidae. As in the Ascaridina the bursa, if present, is rhabitoid rather than strongyloid, since the genital papillae are not accompanied by muscular tissue.

Oblique Markings. Oblique fibers in the cuticle may be very conspicuous in some nematodes, the cross-hatching of two lay-

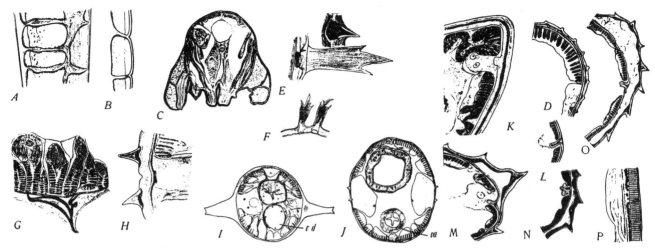

FIG. 35.

A-B—*Prooncholaimus* sp. [Oncholaimidae]. Showing vesiculate area in cuticle (A—lateral view; B—median view). C—*Heterakis gallinarum* (Section postanal showing caudal alae and genital papillae). D, K-P—trichostrongyl cuticle. (D—*Oswaldocruzia* sp. K—*Haemonchus contortus;* L-M & P—*Nippostrongylus muris;* N—*Longistriata noviberiae;* O—*L. musculi*). E-F—*Gnathostoma* [Gnathostomatidae]. (Various types of spines). G—*Rictularia coloradiensis* (Cross section through spine). H—*Hystrignathus rigidus* [Thelastomatidae] (Cross section through spine). I-J—*Parathelandros anolis* [Thelastomatidae] (I-Male; J-Female.) D & K-F, after Lucker, 1936, Parasit., I-J, after Chitwood, 1934, Smithson. Misc. Coll.

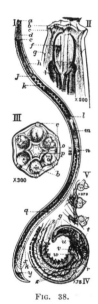

FIG. 38.

Mononchus major.
After Cobb, 1917,
Soil Science.

FIG. 36.

Heth dimorphum [Atractidae]. After M. B. Chitwood, 1935, Proc. Helminth. Soc. Wash.

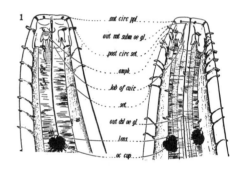

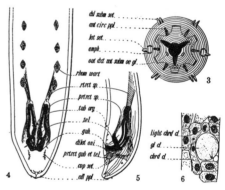

FIG. 37.

Deontostoma californicum [Enoplidae]. After Steiner & Albin, 1933, J. Wash. Acad. Sc.

ers of parallel fibers having the appearance of being on the surface (Fig. 34). Actually the fibers are always beneath the cortical layer and they are not visible in totomount preparations of most nematodes.

The current use of ''presence'' or ''absence'' of oblique fibers as a generic character in the Mermithidae is inaccurate. The position, form, and degree of development may well be of taxonomic value.

Inflation. When the cuticle is swollen in a blister-like manner this condition may be termed inflation. An outstanding example of this cuticular inflation may be seen in *Gongylonema.* In this form the cuticle of the cephalic and cervical regions is inflated in more or less longitudinal rows of blisters or *plaques*, the general appearance simulating deep annulations divided longitudinally. In the genus *Prooncholaimus* (Fig. 35A-B) of the Enoploidea, inflation is as well marked as in *Gongylonema* but here there are definite lateral and median rows of vesicles. Cuticular inflation also occurs in oxyurids, but in these forms it is confined to a limited lateral region just posterior to the lips, as in *Enterobius vermicularis*, these areas being termed lateral cephalic inflations. In most instances it is obvious that they do not correspond to annules in any way, for the cuticle of the inflation is definitely striated. Very similar but complete cephalic inflation is characteristic of numerous genera of the Trichostrongyloidea, i. e., *Longistriata*, *Nematodirus*, *Cooperia*, etc.; the size and extent of inflation in these forms are quite variable.

Inflation of the cuticle may be generalized over the whole body surface instead of restricted to certain areas (*Rhabdias bufonis*). Placing specimens of such a form in a hypertonic solution causes shrinkage and almost complete disappearance of the inflation. Conversely, one must be on guard against producing inflation artificially in parasitic nematodes by placing them in hypotonic solution or in reagents such as dilute acetic acid. Normal inflation of the caudal alae is common in spirurids such as *Ascarophis*. Inflation is caused by the separation of the cortical layer from the remaining layers of the cuticle and is generally caused by liquefication of the matrix layer.

Spination. The development of spines on the body surface is a common tendency and is usually considered in zoology as an evidence of senescense. In nematodes spination occurs in

34

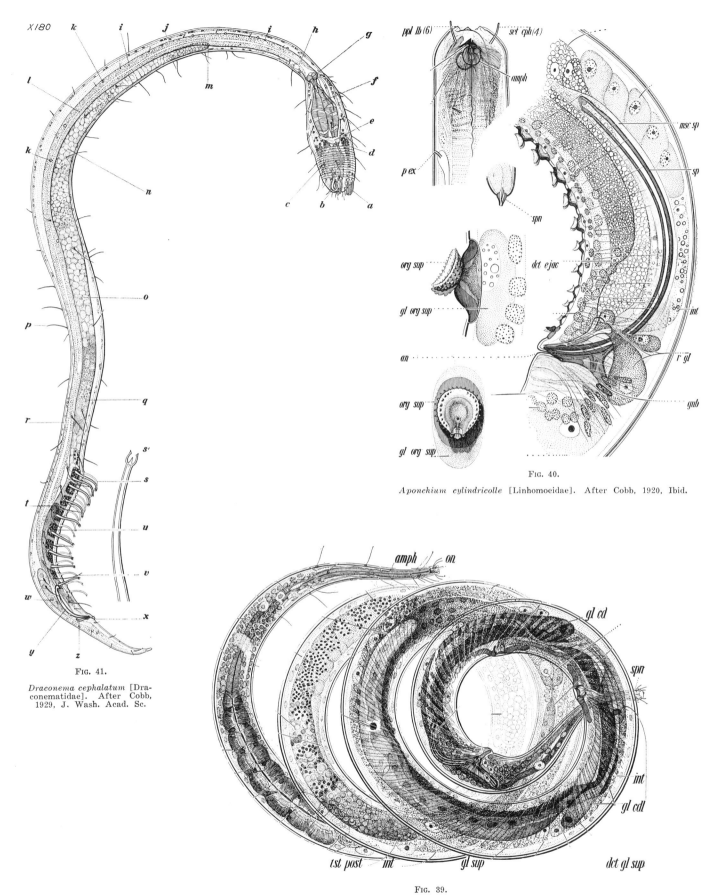

Fig. 41.

Draconema cephalatum [Dra-
conematidae]. After Cobb,
1929, J. Wash. Acad. Sc.

Fig. 40.

Aponchium cylindricolle [Linhomoeidae]. After Cobb, 1920, Ibid.

Fig. 39.

Bolbella tenuidens [Oncholaimidae]. After Cobb, 1920, Contrib. Sc. Nemat. IX.

various groups with little regard for apparent relationships and the results are sometimes a little grotesque. We find spination occurring more generally in the Phasmidia than in the Aphasmidia, and in parasites rather than in free-living forms. Thus, we find in the Metastrongylidae a genus, *Crenosoma*, in which the posterior edges of the annules may form spinate collarettes. The origin in this case is obviously through posterior elongation of the annular rings with indentations of their margins. A practically identical condition occurs in *Spinitectus* (Spiruridae) (Fig. 23II-KK) and here the same origin is indicated; in *Seuratum* (Cucullanidae) a similar condition is found but the origin is not so obvious as in *Crenosoma* and *Spinitectus*. The annules of the cuticle of *Ascarophis* sometimes project posteriorly and have a serrate appearance but their edges are not broken as in *Spinitectus*.

In the Ascaridina we find the Oxyuroidea affords many examples for the study of spination; the genera, *Carnoya*, *Hystrignathus*, *Lepidonema* and *Auchenacantha*, are the outstanding forms in this respect. *Carnoya vitiensis* serves well to illustrate the tendency toward spination. Like *Spinitectus* the spines correspond to annules and are limited to the cervical region, but unlike *Spinitectus* they originate from the anterior part of the annule. There are 12 rows in the female and only two in the male. Gilson (1898) demonstrated that by contraction of the cervical region the spines of the female may be directed anteriorly. In the genera *Rondonema* and *Angra* of the Ransomnematinae, lateral rows of spines and flanges, respectively, are present in the cervical regions of the females but are absent in the males; in the genus *Heth* (Fig. 36) there is a single fringed collar and two or more lateral spines; and in the genera *Lepidonema* and *Hystrignathus* spines are only present in the cervical region and then only in the adult females.

Returning to the spiruroids, posteriorly directed hooks are formed on the surface of *Gnathostoma* (Fig. 35E-F), these being particularly developed in the anterior part of the body and in transverse rows. In *Rictularia* and the males of some species of *Tetrameres* a pair of ventral rows of massive combs and spines sometimes extend to the anal region (Fig. 23LL-MM and 35G). In the Phasmidia the spines usually, if not always, consist of cortical and matrix layers of the cuticle and contain a dense supporting core which stains intensely with the aniline blue in Mallory's collagen stain; in this respect they are similar in structure to the cervical alae.

In some forms such as *Criconema*, *Gnathostoma*, *Spinitectus* and *Rictularia* the spines may be of real service in locomotion. In moving through tissues undulation of the body alone would be sufficient to force a spined nematode through the tissue since the spines are posteriorly directed. In the Aphasmidia, on the contrary, spination, when present, appears to be of service as reinforcement at joints of the large annules, as, for example, in *Monoposthia* (Fig. 27), rather than of service in locomotion.

Special modifications. In isolated groups modifications of the cuticle occur which may be of numerous types but usually have to do with copulation and are confined to the male sex. These modifications may take the form of suckers, medioventral preanal organs, inverse punctations surrounding genital papillae (rosettes) and cuticular internal supports around genital papillae (plectanes).

Suckers (Figs. 15P and 33O-P) are of two general types, both being dependent upon hypodermis and musculature for their development. A single preanal sucker may be present in heterakids and camallanoids (*Cucullanus*) in which case it is oval or rounded, sometimes, but not always, having a distinct rim. A series of stages in sucker development occurs in kathlaniids indicating that there is first a concentration of copulatory muscles in this area which later becomes a sucker through modification of the cuticle. The terminal genital sucker (erroneously termed bursa though not homologous with that structure) of the Dioctophymatoidea is an entirely different structure formed through development of copulatory muscles but without modification of the cuticle itself.

Rosette formation occurs around some but not all genital papillae of some cosmocercids (Fig. 11D) and sometimes thelastomatids (Fig. 33N), while plectanes are additional cuticular structures also occurring in some cosmocercids, apparently acting as supports of the papillae; similar structures occur in the Leptosomatinae (*Deontostoma californicum* Fig. 37).

A single medioventral preanal organ sometimes occurs in rhabditoids, and spiruroids but these structures may not all be homologous, since they appear to be simple papillae (Fig. 33F) in some instances (*Physaloptera*) and more like the supple-

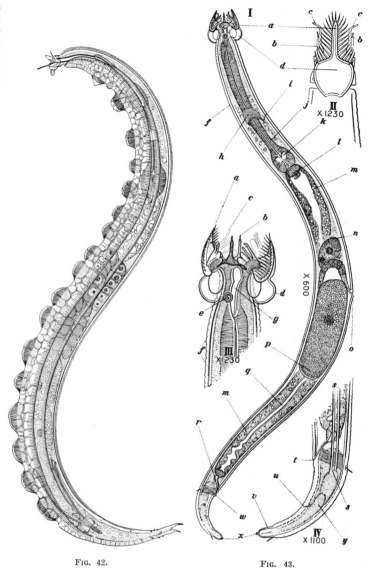

FIG. 42.

Bunonema inequale [Rhabditidae].
After Cobb, 1915, U. S. D. A.
Yearbook for 1914.

FIG. 43.

Wilsonema cephalatum [Plectidae].
After Cobb, 1915, Ibid.

mentary organs of the Aphasmidia (*Strongyloides*) in other instances.

Series of preanal, ventral, supplementary organs (Figs. 15, 21, 23 and 46H-N) occur very commonly in the Aphasmidia and these structures appear to be related to genital papillae (i. e., tactile organs), but in many instances large glandular and cuticular development indicates that they have been transformed into copulatory organs. Thus in the chromadorids and desmodorids the medioventral row of preanal papillae may be supported by internal cup-shaped cuticular thickenings while in other forms (such as *Camacolaimus*) the ventral row of papillae is unmodified. The preanal ventral "papillae" are sometimes surrounded by slightly raised areas of cuticle (*Mononchus*, *Aulolaimus*, *Dorylaimus*) and in extreme instances these extend to the esophageal region (*Prismatolaimus intermedius*, *Bastiania exilis*, *Tripyla lata* and *Anonchus mirabilis*) while in other instances there is either a single raised papilloid organ (*Cylindrolaimus melancholia*, *Fiacra brevispinosa*) or a ventral row of depressions confined to the esophageal region of the male (*Deontolaimus papillatus*). Further modifications take the form of ventral preanal conoid or tubular cuticular, apparently protrusible, structures. The tubes which are not very distinct in mononchs (Fig. 38) have been observed to connect with glands or sensory organs; in some instances the papillae are contained in a vesiculate swelling bearing denticles (*Mononchus*, *Trilobus*). Two to several tuboid protrusible preanal organs are known in many nematodes (*Plectus granu-*

losus, *Anonchus mirabilis*, *Aphanolaimus atentus*) and in some of these forms, such as *Anonchus*, the anterior organs are cup-like while the remainder are tuboid. Another variant of the tube type of supplementary organ is that occurring in forms such as *Enoplus communis*, in which there is only one preanal glandular tube; from such forms variants occur such as shortening and heavy cuticularization of the tube (*Thoracostoma*).

One oncholaim, *Oncholaimium appendiculatum*, possesses a large digitiform preanal organ connected with a gland. The glandular connection of these organs reaches extreme development in forms having several preanal organs (*Bolbella* and *Eurystomina*); each gland may be 1/6 to 1/10 of the length of the body but, nevertheless, unicellular (Fig. 39). In *Aponchium* we find the preanal organs to be small external cups bearing denticles (Fig. 40), while in *Polysigma* there are two rows or tuboid structures (total about 7ß) extending nearly the entire length of the body.

As regards the function of these supplementary organs, it is possible that they represent developments of sensory papillae which have taken on the functions of clasping and adhesion through development of cuticular structures, known at least in some instances to be connected with special muscles, and through the development of adhesive glands. It may be noted that these structures occur chiefly in the Aphasmidia, in forms with caudal glands (i. e., usually aquatic) and without caudal alae.

Tubiform setae. Hollow ambulatory setae or "legs," connected with glands, occur in the families Draconematidae and Desmoscolecidae. In the Draconematidae the setae are confined to the posterior surface and are in the form of two or four ventral rows of "legs." Glandular setae in the Desmoscolecidae are likewise paired but they occur both on the dorsal and ventral surfaces of the body and are not confined to the posterior part of the body. A pseudometamerism (Fig. 17) is produced through the occurrence of only one pair of setae on an annule (the glands will be discussed under the heading Hypodermal glands). In the Greeffiellidae the glandular setae are somewhat scattered over the surface of the body. According to Cobb (1929) the setae of *Draconema* are hollow, each connected internally or proximally with a three-celled gland, while distally they are cup-shaped with a small internal spine (Fig. 41). Presumably pressure on the spine releases the flow of secreted matter.

Bristle setae. Ambulatory setae apparently connected with neither glands nor· sensory organs and presumably acting entirely as organs of traction occur chiefly in two groups, the Greeffiellidae, in which they form a hairy coat over the surface of the body and the Epsilonematidae, in which they occur in groups near the permanently curved regions of the body. Tiny longitudinal rows of bristles corresponding to the spines of the longitudinal ridges also occur in desmodorids (Figs. 26 and 30).

Helmets. (Figs. 28-30, 37.) The labial region of nematodes is commonly set off from the remainder of the body by a groove but the cuticle of the cephalic region is usually not modified to any great extent. In certain of the more heavily armored groups, however, striation ceases some distance posterior to the labial region, the cuticle being very strongly thickened forming a structure known as a helmet. In such forms the cuticle of the labial region itself is not thickened but remains pliable. *Ceramonema*, *Acanthopharynx* and *Desmodorella* are examples of helmet bearing nematodes. The helmet of *Ceramonema* is relatively simple, that of *Acanthopharyngoides* is sutured, and that of *Desmodorella* is internally etched.

In such enoplids as *Deontostoma* there is an internal cephalic capsule forming an endoskeletal helmet, while in *Enoplus* (Fig. 15S) there is a distinct subcephalic groove and the cephalic cuticle is apparently duplicate. Filipjev (1922, 1934, a, b) regards this cuticular duplication as characteristic of the Enoplidae (equivalent to the Enoploidea in this text) (Fig. 2300-QQ).

Other specializations. In addition to the cuticular structures already discussed there are many instances of special developments, such as the "crests" and "tubercles" or *Bunonema* (Fig. 42); the cuticular growth uniting the cephalic setae of *Wilsonema* (Fig. 43); the cordons (ribbon-shaped, paired bands) of acuariids (Fig. 32); cephalic collars and many other structures. Cephalic prominences connected with the labial region form a great part of these specializations and may well be considered in connection with a discussion of the lips.

B CUTICULAR LAYERING

The first to call attention to the many layered structure of the cuticle of *Ascaris lumbricoides* was Von Siebold (1848) whose observations have been enlarged upon and added to by Czermak (1852), Bastian (1866), van Bömmel (1894, 1895), and Goldschmidt (1905). Other ascarids have been studied by Toldt (1899-1912), by Glaue (1910), and by K. C. Schneider (1902). A comparable study of *Oxyuris equi* was made by Martini (1912, 1916) but representatives of other phasmidian groups have not been previously examined from the standpoint of cuticular layering. De Man (1886) demonstrated for the first time, cuticular layering in an aphasmidian, *Enoplus communis*, and later (1904) illustrated similar structural details in *Mononchus gerlachei*, and *Thoracostoma* spp., while Rauther (1906) showed a similar formation of the cuticle of *Mermis albicans*. Investigations by the present writers indicate that in all probability the cuticle of all nematodes is actually made up of several layers and that thickness is the only marked difference between the forms in which structural formation has been observed in the past (Ascaridina, Enoplina, and Dorylaimina) and the remaining, forms. Since *Ascaris lumbricoides* and *Parascaris equorum* have been most thoroughly studied and do not differ notably from each other the structure of the cuticle in these forms will be discussed in detail.

The cuticle of *Ascaris lumbricoides* is divisible into 9 distinct layers as follows (Fig. 44A-C): (1) An external cortical layer; (2) an internal cortical layer; (3) a fibrillar layer; (4) a matrix layer (so called "homogenous layer"); (5) a boundary layer (Bandschichte of van Bömmel); (6, 7 and 8) external middle and internal fiber layers; and (9) a basal lamella.

The external cortical layer is itself subdivisible into 2 parts of which the outer (a) is much denser than the inner part (b). This layer is thinner and less dense at each transverse striation where the fibrillar layer extends through the cortical layer to the delicate membranous covering.

The internal cortical layer does not appear to be an actual entity since it is continuous with and often quite similar in consistency to the matrix layer from which it is grossly separated by the fibrillar layer. It has the appearance of a solid mass of spongy matrix depending upon the part of the body studied and the technic followed.

The fibrillar layer consists of a condensation of this spongy matrix (or mass of fibrils) which forms a closely woven network between the internal cortical layer and the matrix layer, strands of condensed material (fibrils, gallery fibers, etc.) extending into the external cortical layer (1) and the boundary layer (5).

The matrix layer (4), commonly said to be homogenous, usually has the appearance of a finely alveolar or spongy mass.

The boundary layer (5) is distinct only in some instances and probably should be interpreted as a condensation layer.

The three fiber layers (6, 7, and 8) are oblique, ribbon-like, possibly spiral, layers of very dense connective tissue. The external and internal fiber layers (6 and 8) are parallel and opposite in direction to the middle fiber layer.

The basal lamella is a thin layer which in cross section appears to be striated; in tangential section it appears to have fine longitudinal striations (i. e., alternating light and dark bands).

The above description is in substance the same as that given by van Bömmel. Considerable discussion has centered around the nature of the fibrillar layer and its ramifications. Toldt, working on *Parascaris equorum*, considered the fibrillar layer, its connection with the cortical layer, and its ramifications in the matrix as a system of "feeding channels." Goldschmidt, on the other hand, upheld the view that these structures are supporting. Peptic-hydrochloric acid digestion definitely proves Goldschmidt's contention to be correct. When the matrix and fiber layers have been dissolved the "feeding channels" remain attached to the cortical layer and have the same glassy refractive appearance.

The cuticle of *Parascaris equorum*, *Toxocara canis*, *Toxascaris leonina* and other ascarids has the same fundamental structure. Glaue (1910) found that the external cortical layer may contain dense transverse bands, one being present between each two striae in *Toxocara canis* (Fig. 44H), two in *Toxocara cati*, and none in *Toxascaris leonina*.

The cuticle of *Contracaecum spiculigerum* (Fig. 44P) differs considerably from that of *Ascaris* in having an external and internal cortical layer (1, 2), an external matrix layer (3), two layers of coarse oblique fibers (4-5), an internal matrix

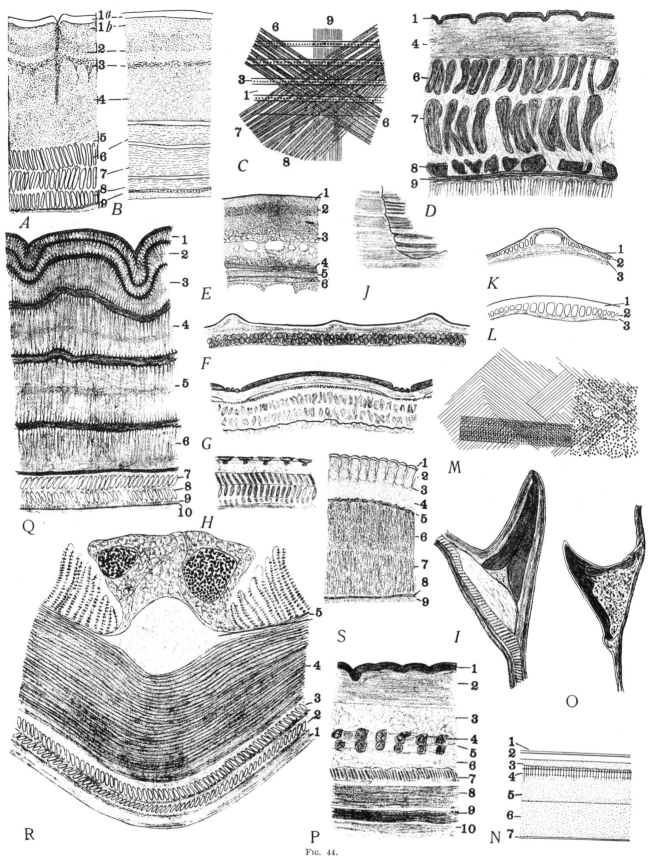

F_{IG.} 44.

Cuticular layering. A-C—*Ascaris lumbricoides* (A—Longitudinal section; B—Cross section) C—Surface view after removal of hypodermis. 1a—Outer part of external cortical layer. 1b—Inner part of external cortical layer. 2—Internal cortical layer. 3—Fibril layer. 4—Matrix layer. 5—Boundary zone. 6—External fiber layer. 7—Median fiber layer. 8—Internal fiber layer. 9—Basal lamella. D—*Dirofilaria immitis.* (Londitudinal section). 1—External cortical layer. 4—Matrix layer. 6-8 Fiber layers. 9—Basal lamella. E—*Physaloptera turgida,* (Cross section). 1—External cortical layer. 2—Internal cortical layer. 3—Matrix. 4—External fiber layer. 5—Median fiber layer. 6—Internal fiber. F—*Onchocerca gutterosa.* (Longitudinal section showing oblique thickenings do not involve fiber layers). G—*Oxyuris equi.* (Longitudinal section). H—*Toxocara canis.* (Longitudinal section). I—*Toxocara canis.* (Cross section through cervical ala). J—*Oesophagostomum dentatum* (Tangential section). K—*Spilophorella paradoxa.* (Cross section showing "alae," punctations, etc. 1—Cortical layer; 2—Hollow matrix containing fibril layer columns; 3—Basal layer). L—*Dorylaimopsis metatypicus.* (Cross section in lateral region, layers as in K). M-N—*Enoplus communis* (M—Surface view [punctations are more scattered in lateral areas]; N—Optical longitudinal section).

Figure 44C is incorrectly labelled and shaded (Layer 6 should be 7 and vice versa).

layer (6), three layers of oblique fibers (7, 8, 9) and a basal lamella (10). Martini found (1912, 1916) that the cuticle of *Oxyuris equi* (Fig. 44G) consists of a cortical layer underneath which a fibril layer and two layers of oblique fibers are embedded in a homogenous matrix.

The three oblique fiber layers and basal lamella are present in both *Physaloptera* and *Dirofilaria*. In *Dirofilaria immitis* the dense external cortical layer and but slightly less dense internal cortical layer rest directly upon the external layer of oblique fibers (Fig. 44D) while in *Physaloptera turgida* the distance between the external cortical layer and external fiber layer is greatly increased and consists of a spongy matrix (Fig. 44E). The latter is not dense externally and gradually becomes less dense and large vacuoles are present at a depth of about 2/3 of its thickness; thereafter, proceeding internally, it becomes more dense. In some physalopterids as well as in some other nematodes the cuticle appears to be separated from the hypodermis. This may be explained more easily if one assumes that the spongy matrix breaks down as appears probable from observations on *P. turgida*.

Franklin (1940) called attention to the thickened layered cuticle of *Heterodera* spp. In these organisms the cortical layer takes several folded pattern forms. Beneath this layer we see "punctations" and beneath these are the fibrous layers. However, the layering of the cuticle does not explain the remarkable persistence of *Heterodera* cysts in soil. Ellenby (1946) submitted evidence that the cuticle is hardened by the action of polyphenol oxidases.

In the Trichostrongyloidea (Fig. 35D and K-F) Lucker (1936) found the cuticle to be composed of several layers, 3-4 layers being distinct. In such forms as *Haemonchus contortus* there is a distinct, thin external cortical layer and wide basal layer. No distinct matrix layer is present nor are oblique fiber layers observable. The latter, together with the basal lamella probably correspond to the wide basal layer of this form. The basal layer is inconspicuously separable into two parts, an internal narrow layer (? basal lamella) and an external wide layer (oblique fiber layers or ? matrix layer); laterally this layer is distinctly thickened and forms the internal lateral ridges which are supporting or skeletal structures. In *Longistriata musculi* and *Nippostrongylus muris* the cortical layers are definitely separated from the basal layer. Between the cortical layers and the basal layer there is a fluid material which may precipitate upon fixation. The basal layer is relatively much thinner but it is otherwise apparently identical in structure. Upon removal of the cortical layer longitudinal rows of inverse punctations on the basal layer were observed in *N. muris;* the same may be observed as small elevations in longitudinal section.

The cuticle of *Oesophagostomum dentatum*, a member of the Strongyloidea, shows a slightly different type of cuticular layering in having a layer of sub-surface transverse fibrous bands (Fig. 44J). The cuticle of *Strongylus* (Fig. 44S) exhibits the chief cuticular layers seen in *Ascaris*. The cortical layer in longitudinal sections shows very marked transverse bands of condensation; the internal lateral ridges have the same marked affinity for fuchsin in van Gieson's stain) as do the oblique fiber layers and they are definitely attached to the internal layer of oblique fibers.

De Man (1886) found the cuticle of *Enoplus communis* (Fig. 44M-N) to consist of a thin cortical layer (1), an external layer of oblique fibers (2), an internal layer of oblique fibers (3), a layer of minute punctations from which "tubes" (4) extend into an external matrix layer (5) beneath which lies an internal matrix layer (6), and the basal lamella (7). *Thoracostoma, Dorylaimus, Mononchus* and other members of the Enoplida commonly show distinct multi-layered cuticular structure but unfortunately the details have not been worked out in these forms.

Rauther (1906) described the cuticle of *Hexamermis albi-*

cans with exactitude and the present writers have found the cuticle of *Mermis subnigrescens* (Fig. 44R) to be practically identical in structure. The cortical layer (1) of this form is subdivisible as in *Ascaris* but each division is apparently homogenous; as in *Enoplus* the cortical layer is followed by two layers of oblique fibers (2-3) bordered externally, medially and internally by areas of dense matrix substance giving somewhat the appearance of the cuticle of *Contracaecum*. Internal to these layers there is a thick multilayered matrix (4) followed by a delicate basal lamella (5) which is greatly thickened laterally. (The latter layer is not described by Rauther.)

In the cuticle of *Dioctophyma renale* (Fig. 44Q) we find a peculiar modification of layering. There are three oblique fiber layers (7-9) situated internally next to the basal lamella (10) as in *Ascaris*, but here the similarity ceases for there are six similar concentric external layers which, except for their position, appear to have their closest counterparts in the matrix layer of *Mermis* and the cuticle of *Thoracostoma*. No distinct cortical layer is present. Each "matrix layer" appears as an individual entity bounded externally and internally by a region of condensation and divided medially by a very minute hyaline line. Surface views show transverse dark "lines," presumably comparable to the "processes" seen in longitudinal section.

In the Chromadorida the cuticle usually appears to consist of two dense layers, a cortical layer (1) and a basal layer (3) separated by a more or less distinct matrix layer (2). In *Aphanolaimus* (Fig. 50J) the matrix layer is scarcely distinguishable from the basal layer. In *Spilophorella paradoxa* (Fig. 44K), *Dorylaimopsis* (Fig. 44L), *Halichoanolaimus robustus* and similar forms the matrix layer appears in cross section as a cavity traversed by the fibril layer; the "tubes" of the fibril layer cause the appearance of punctations when viewed in toto. In *Monoposthia hexalata* the cortical and basal layers are separated by a distinct cavity which contains a pair of lateral cuticular ridges (Fig. 50G). The cuticle of *Desmoscolex americanus* (Fig. 50C) likewise consists of two gross layers separated by a cavity but in this case the cavity is considerably wider at the annules than between them. In addition, the cuticle of this form bears an agglomeration of foreign material which can be removed mechanically.

From the above data it may be concluded that all nematodes have a layered cuticle and that the type of form of the layers is essentially quite similar in the various groups. The cuticle of such forms as chromadorids probably contains oblique fibers but they have not, as yet, been observed; presumably the fiber layers are represented by the basal layer. Punctation is a manifestation of the fibril layer. It has been shown (Fig. 44C) that in at least one nematode (*Ascaris lumbricoides*) in which punctation is not ordinarily visible, the effect may be obtained by proper preparation. Regarding the nature of the cuticle, studies of the layers seems to indicate that at the time of their formation they were actually a part of living cells. On purely morphological grounds one may hypothesize that the layers are each formed in turn as protoplasmic condensations in the exterior hypodermis. The entire cuticle seems to be somewhat in the nature of a connective tissue exoskeleton. Mallory's, van Gieson's, and Weigert's stains are all useful in the differentiation of cuticular layers.

There seems to be no sound evidence to support the view that the cuticle is a secretion. The very complexity of its structure is difficult to reconcile with such a view. The recent exposition by G. W. Müller (1936) regarding the formation of the cuticle as a secretion seems to be based on fundamental misinformation as to the structures involved.

C CHEMICAL CONSTITUTION

There has been much confusion in the past regarding the chemical nature of the cuticle of nematodes. Lassaigne (1843) demonstrated the solubility of the cuticle of *Ascaris* in potassium hydroxide and pointed out that on this basis it could not be called chitin. In spite of the fact that Odier (1823) had previously given the name chitin to the highly insoluble substance composing the exoskeleton of insects, Grube (1850) called the cuticle of ascarids and earthworms by this name and its erroneous use as regards the exoskeleton of nematodes has persisted up to the present time.

Flury (1912) identified the cuticle of *Ascaris* with keratin; Magath (1919) termed the cuticle of *Ascaris* and *Camallanus* cornein; and Mueller (1929), working with *Ascaris*, concluded that it is formed by two substances neither of which may be identified with any known chemical compound.

(1—Cortical layer; 2-3—Fiber layers; 4—Fibril layer, 5-6—Matrix layers; 7—Basal lamella). O—*Heterakis gallinarum*. (Cross section through cervical ala). P—*Contracaecum spiculigerum*. (Longitudinal section. 1—External cortical layer; 2—Internal cortical layer; 3—External matrix; 4—External coarse fiber layer; 5—Internal coarse fiber layer; 6—Internal matrix; 7—External fine fiber layer; 8—Median fine fiber layer; 9—Internal fine fiber layer; 10—Basal lamella). Q—*Dioctophyma renale*. (Longitudinal section. 1-6—Wide matrix layers; 7-9—Fiber layers; 10—Basal lamella). R—*Mermis subnigrescens*. (Cross section. 1—Cortical layer; 2-3—Fiber layers; 4—Matrix layer; 5—Basal lamella). S—*Strongylus equinus*. (Longitudinal section. 1—External cortical layer; 2—internal cortical layer; 3—Fibril layer; 4—Matrix layer. 5—External boundary layer; 6-7—Fiber layers; 8—Basal lamella; 9—Hypodermis). G, after Martini, 1916, Ztschr. Wirs. Zool.; N-O, after de Man, 1886, Nordsee Nematoden.

Some workers have used the term cutin in an effort to avoid terms of chemical significance. This usage is unfortunate since cutin is a compound related to cellulose and found only in plants. The use of the term corncin was likewise unfortunate since it was applied by Reichard to the horny substance of corals.

It is not proper to say that the cuticle of nematodes is composed of any one substance. As Sukataschoff (1899) demonstrated, the cortical layer differs from the matrix and fiber layers by b ing resistant to peptic digestion. This observation was confirmed by Reichard (1902) who, with Mueller (1929), added the further information that the matrix and fiber layers are more soluble in standard solvents than is the cortical layer. The observations of Sukataschoff, Reichard and Mueller indicated rather definitely the presence of at least two substances in the cuticle.

Flury (1912) based his observations on cuticle collected by digesting fresh ascarids in artificial gastric juice and since we know that the matrix and fiber layers are digested under such conditions we might assume that the "cuticle" with which he worked was actually the cortical layer. However, he obtained 1.0 to 1.5 gms. of "cuticle" from 100 gms. of fresh worms. This amount is many times greater than the amount of cortical layer one would expect to obtain under such conditions.

Chitwood (1936) found the cuticle of *Ascaris lumbricoides* (prepared by dissection) to be composed of at least five distinct substances, these substances being albumins, a glucoprotein (mucoid), a fibroid, a collagen, and a keratin.

The albumins or water soluble proteins comprise 25 per cent of the dry weight of the cuticle; there are probably two such proteins since coagulation was found to take place at 50 to 60° C. and filtered, the filtrate coagulated at 75 to 85° C. The glucoprotein was obtained by extraction with lime water but no quantitative estimation was made. Neither the albumins nor the glucoprotein were identified with morphologic parts of the cuticle.

The fibroid, named matricin, comprised about 35 per cent of the dry weight of the cuticle. Like elastin and reticulin it was found to be digested by both artificial gastric juice and some preparations of pancreatic extracts. Like other fibroids, it gives strong xanthoproteic, aldehyde, and mercuric nitrate reactions indicating the presence of tryptophane and tyrosine. Unlike other fibroids, it also contains considerable unoxidized sulphur as indicated by the formation of lead sulphide when heated with lead acetate in sodium hydroxide. This substance corresponds to the matrix layer.

The collagen, named ascarocollagen, comprises about 29 per cent of the dry weight of the cuticle. Like the collagen of vertebrates it is digested by artificial gastric juice but shows considerable resistance to digestion by pancreatic extracts. Ordinarily it is very difficult to hydrolize in hot water and due to the prolonged heating necessary it is not possible to obtain a gelatin. However, if the cuticle has been previously treated with running water, lime water and pancreatic extracts, the collagen is readily hydrolized to a gelatin in hot water. The gelatin so obtained, when dissolved in water to make a 4 per cent solution and then cooled to 10° C., sets to a hard jell. This gelatin agrees with other collagens in the ordinary protein color reactions but differs in the presence of unoxidized sulphur; for this substance the term ascarogelatin was proposed. The non-hydrolized form, ascaro-collagen, corresponds to the fiber layers, the fibril layer and the internal cortical layer although it is possible that the last two layers differ slightly in their chemical composition from the fiber layers since they are much less readily soluble in such reagents as sodium hydroxide and acetic acid.

The keratin comprises about 2.2 per cent of the dry weight of the cuticle. Like other keratins it is relatively insoluble in standard protein solvents, but is soluble upon boiling in 5 per cent sodium hydroxide or in 10 per cent hydrochloric acid; it is also resistant to peptic digestion and fairly resistant to pancreatic digestion (effected by only one of four extracts). It gives intense xanthoproteic, aldehyde, mercuric nitrate, and sulphide reactions. It corresponds to the external cortical layer.

It should be understood that the above discussion refers specifically to the cuticle of the external surface of the body and does not have any bearing on the nature of the stomatal esophageal, rectal, or vaginal linings, the spicules or the egg shell. The hard refractive part of the egg covering (true shell) of *Ascaris* has been definitely identified as a form of chitin by many workers, the lining of the stoma and esophagus of *Strongylus edentatus* was incorrectly identified by Imminck (1924) as chitin; and the spicules appear to be either a mucoid, a keratin, or a combination of the two.

D PERMEABILITY

The permeability of the cuticle of nematodes varies greatly with the normal habitat and mode of life of the species. Generally intravitam stains are used to determine this factor. Methylene blue, neutral red, and the various carmines have been favored for such work. In general, marine nematodes such as *Oncholaimus* appear to have a more permeable cuticle than terricolous forms, such as *Rhabditis*, or inhabitants of the digestive tract, such as oxyurids. In rhabditids intravitam stains were found by Chitwood (1930) to enter much more readily at the normal body openings and at thin places in the cuticle, such as the stoma, anus, vulva, excretory pore, amphids, phasmids, and in the case of methylene blue, at the papillae, than elsewhere. Christie (unpublished) has found that in preparasitic and adult specimens of *Agamermis decaudata* the cuticle is relatively impermeable to intra-vitam stains while in the parasitic stage, normally bathed in the fluids of the body cavity of its host the cuticle affords very little resistance to stains. The writers have found the cuticle of first stage larvae of rhabditoids and strongyloids comparatively permeable to stains but relatively impermeable in the later stages of these forms. This fact is particularly emphasized in the study of *Strongyloides* infected dogs. When gentian violet is administered daily this drug penetrates and kills all newly hatched larvae. However, it has no effect on adult females in the intestine of the host.

Chitwood (1938) found that the cuticle of *Ditylenchus* acts as a semi-permeable membrane. The permeability of this membrane is greatly affected by temperature. The semipermeability is destroyed by washing in acetone. Enzymes act differently on cut and uncut specimens, before and after washing in acetone. Use of sodium hypochlorite served to destroy all tissues but left the remains of the organisms apparently held together by a mere halo, possibly a monomolecular layer. This halo disappeared in acetone. These facts lead to the conclusion that the hard parts of the cuticle do not constitute the major protection from the environment. The major protection seemed to be a thermolabile membrane with solubilities of fat solvents but not reactive to fat stains.

Ellenby (1946) working on the cysts of *Heterodera rostochiensis* was unable to identify sterols in the cuticle. He was able to identify polyphenols but could not locate them as a structure. Females of this organism are first white and soft bodied. At this time they sink in water. Later they turn yellow, then red-brown. At such time they float. Ellenby found the color change to be an oxidative phenomenon. Unfortunately his work was not accompanied by morphologic studies and does not appear to be a full explanation of the situation. Fat solvents change the permeability of the cyst and the only effective soil fumigants are soluble in, dissolve, or, are dissolved by fat solvents. Whether or not the material is wax-like or merely functions in this manner has yet to be determined.

Filipjev (1934) interprets the greater permeability of the cuticle of marine nematodes as an evidence of primitivity. The extremely common occurrence of hypodermal gland orifices in the cuticle of marine nematodes makes it difficult in these forms to be certain of the actual means of ingress of the stain.

Mueller (1929) found that the body wall of *Ascaris* is a semi-permeable membrane. He found in an experiment in which a dialysis cell was formed from the body of *Ascaris* that urea and potassium iodide would pass through the wall while glucose would not.

Trim (1949) has found that the cuticle of *Ascaris lumbricoides* var. *suis* acts as a semi-permeable membrane, governed by lipids. The outermost layer of the cuticle is probably the main barrier to penetration.

In conclusion, it must be admitted that actually little is known about the permeability of the cuticle of nematodes. In all probability it corresponds in some measure with the habitat, but to just what extent absorption or excretion may take place directly through the cuticle is not known. The subject should provide a fertile field of investigation.

Bibliography
(Cuticle: Morphology)

BASTIAN, H. C. 1866.—On the anatomy and physiology of the nematoids, parasitic and free; with observations on their zoological position and affinities to the echinoderms. Phil. Trans. Roy. Soc. London, v. 156: 545-638.

BOEMMEL, A. VAN. 1894.—Ueber Cuticular-Bildungen bei einigen Nematoden. Thesis (Würzburg). pp. 191-212. Wiesbaden.
    1895.—Ibid., Arb. Zool.-zoot. inst. Würzburg, v. 10: 189-211, figs. 1-17, pl. 11.

CHITWOOD, B. G. 1936.—Some marine nematodes from North Carolina. Proc. Helm. Soc. Wash., v. 3(1): 1-16, figs. 1-4.

CHITWOOD, B. G. 1949.—"Root-knot nematodes" Part I. A revision of the genus *Meloidogyne* Goeldi, 1887. Ibid., v. 16 (2):90-104.

COBB, N. A. 1913.—*Draconema;* a remarkable genus of marine free-living nematodes. J. Wash. Acad. Sc., v. 3(5): 145-149, 1 fig.
    1915.—The asymmetry of the nematode *Bunonema inequale*, n. sp., Contrib. Sc. Nemat., (3): 101-112, figs. 1-2.
    1920.—One hundred new nemas. Contrib. Sc. Nemat. (9):217-343, figs. 1-118c.
    1929.—The ambulatory tubes and other features of the nema *Draconema cephalatum*. J. Wash. Acad. Sc., v. 19(12): 255-260, figs. 1-2.

CONINCK, L. DE. 1942.—Sur quelques espèces nouvelles de nematodes libres (Ceramonematinae Cobb, 1933), avec quelques remarques de systematique. Bull. Mus. Roy. Hist. Nat. Belg., v. 18 (22):1-37.

CZERMAK, J. 1852.—Ueber den Bau und das optische Verhalten der Haut von *Ascaris lumbricoides*. Sitz. K. Akad. Wiss. Wien. Math.-Naturw. Kl. v. 9(4): 755-762.

DUJARDIN, F. 1845.—Histoire naturelle des helminthes ou vers intestinaux. Paris. 654 pp., 12 pls.

FILIPJEV, I. N. 1934a.—The classification of the free-living nematodes and their relation to the parasitic nematodes. Smithson. Misc. Coll. (3216), v. 89(6): 1-63, pls. 1-8, figs. 1-70.
    1934b.—Nematody vrednye i poleznye v. sel'skom khoziaistve Moskva. 439 pp., 333 figs. (Russian).

FRANKLIN, M. T. 1940.—On the specific status of the so-called biological strains of *Heterodera schachtii* Schmidt. J. Helminth. v. 18(4):193-208.

GILSON, G. 1898.—Note sur un nématode nouveau des Iles Fiji: *Carnoya vitiensis* Gilson, nov. gen. La Cellule, v. 14(2): 333-369, 1 pl. Liège and Louvain.

GLAUE, H. 1910.—Beiträge zur systematik der Nematoden. Zool. Anz., v. 35: 744-759, figs. 1-5.
    1910.—Beiträge zu einer Monographie der Nematodenspecies *Ascaris felis* und *Ascaris canis*. Ztschr. Wiss. Zool. v. 95: 551-593, figs. 1-26.

GOLDSCHMIDT, R. 1905.—Über die Cuticula von *Ascaris*. Zool. Anz., v. 28: 259-266, figs. 1-9.

IMMINK, B. D. C. M. 1924.—On the microscopical anatomy of the digestive system of *Strongylus edentatus* Looss. Arch. Anat., Hist. & Embryol., v. 3 (4-6): 281-326, figs. 1-46.

LUCKER, J. T. 1936.—Preparasitic moults of *Nippostrongylus muris*, with remarks on the structure of the cuticula of *Trichostrongyles*. Parasitol., v. 28(2): 161-171, figs. 1-13.

MAN, J. G. DE. 1886.—Anatomische Untersuchungen über freilebende Nordsee-Nematoden. 82 pp., 13 pls. Leipzig.
    1904.—Nématodes libres. Résultats du voyage du S. Y. Belgica. Expéd. Antarct. Belg. Anvers. 55 pp., 11 pls.

MARTINI, E. 1912.—Bemerkungen über den Bau der Oxyuren. Zool. Anz., v. 39: 49-53, figs. 1-2.
    1916.—Die Anatomie der *Oxyuris curvula*. Ztschr. Wiss. Zool. v. 116:137-534, figs. 1-121, pls. 6-20.

MUELLER, G. W. 1936.—Die Haut der Nematoden. 26 pp., 15 figs., Greifswald.

RAUTHER, M. 1906.—Beiträge zur Kenntnis von *Mermis albicans* v. Sieb. Zool. Jahrb., Abt. Anat., v. 23(1): 1-76, pls. 1-3, figs. 1-26.

SCHEPOTIEFF, A. 1908.—Die Desmoscoleciden. Ztschr. Wiss. Zool., v. 90: 181-204, pls. 8-10.

SCHNEIDER, A. 1866.—Monographie der Nematoden. 357 pp., 122 figs., 28 pls. Berlin.

SCHNEIDER, K. C. 1902.—Lehrbuch der vergleichenden histologie der Tiere. 988 pp., 691 figs., Jena.

SIEBOLD, C. TH. VON. 1848.—Lehrbuch der vergleichenden Anatomie der wirbellosen Tiere. Berlin.

STEINER, G. 1930.—Die Nematoden. I. Teil. Deutsche Südpolar-Expedition 1901-1903. v. 20, Zool., v. 12: 170-215, figs. 1-19, pls. 25-41.
    1931.—Ibid., II Teil. Ibid., 307-433, figs. 20-108, pls. 48-57.

STEINER, G. and HOEPPLI, R. 1926.—Studies on the exoskeleton of some Japanese marine nemas. Arch. Schiffs. ü. Tropenhyg., v. 30: 547-576, figs. A-Q, pls. 1-2, figs. 1-6.

TAYLOR, A. L. 1936.—The genera and species of the Criconematinae. A subfamily of the Anguillulinidae (Nematoda). Tr. Am. Micr. Soc., v. 55(4): 391-421, figs. 1-63.

THORNE, G. 1945.—*Ditylenchus destructor*, n.sp., the potato rot nematode, and *Ditylenchus dipsaci* (Kühn) Filipjev, 1936, the teasel nematode (Nematoda: Tylenchidae). Proc. Helm. Soc. Wash. v. 12 (2): 27-34.

TOLDT, K. 1899.—Über den feineren Bau der Cuticula von *Ascaris megalocephala* Cloquet. Arb. Zool. Inst. Wien., v. 11: 289-326, Text figs. A-B. pl., figs. 1-9.
    1904.—Die Saftbahnen in der Cuticula von *Ascaris megalocephala* Cloquet. Zool. Anz., v. 27: 728-730.
    1905.—Über die differenzierungen in der Cuticula von *Ascaris megalocephala* Cloquet. Zool. Anz., v. 28: 539-542, figs. 1-3.
    1912.—Bemerkungen zur neuerlichen Diskussion über den Bau der Cuticula von *Ascaris megalocephala*. Zool. Anz., v. 39(15-16): 495-497.

Bibliography
(Cuticle: Chemical Constitution)

CHITWOOD, B. G. 1936.—Observations on the chemical nature of the cuticle of *Ascaris lumbricoides* var. *suis*. Proc. Helm. Soc. Wash., v. 3(2): 39-49.

CHITWOOD, B. G. 1938.—Further studies on nemic skeletoids and their significance in the chemical control of nemic pests. Proc. Helm. Soc. Wash., v. 5 (2): 68-75.

ELLENBY, C. 1946.—Nature of the cyst wall of the potato-root eelworm, *Heterodera rostochiensis* Wollenweber, and its permeability to water. Nature v. 157: 302.

FLURY, F. 1912.—Zur Chemie und Toxicologie der Ascariden. Arch. Exper. Path., v. 67: 275-392.

GRUBE, A. E. 1850.—Die Familien der Anneliden. Arch. Naturg., v. 16: 249-364.

LASSAIGNE, M. 1843.—Sur le tissu tégumentaire des insectes de différentes ordres. Compt. rend. Acad. Sc., Paris., v. 16: 1087-1089.

MAGATH, T. B. 1919.—*Camallanus americanus* nov. spec. Tr. Am. Micr. Soc., v. 38(2): 49-170, figs. A-Q, pls. 7-16, figs. 1-134.
    1928.—The cuticula of nematodes. Science, v. 67: 194-195.

MUELLER, J. F. 1927.—The cuticula of the Nemathelminthes. J. Parasit., v. 14: 131.
    1929.—Studies on the microscopical anatomy and physiology of *Ascaris lumbricoides* and *Ascaris megalocephala*. Ztschr. Zellforsch., v. 8(3):361-403, pls. 9-18.

ODIER, M. A. 1823.—Mémoires sur la composition chimique des parties carnées des insectes. Mém. Soc. Hist. Nat., v. 1: 29-42.

REICHARD, A. 1902.—Über Cuticular—und Gerüst-Substanzen bei wirbellosen Tieren. Diss. Frankfurt. 46 pp.

SUKATSCHOFF, B. 1899.—Über den feineren Bau einiger Cuticulae und Spongienfasern. Ztschr. Wiss. Zool., v. 66: 377-406, pls. 24-26.

Bibliography
(Cuticle: Permeability)

CHITWOOD, B. G. 1930.—Studies on some physiological functions and morphological characters of *Rhabditis* (Rhabditidae, Nematodes). J. Morph. & Physiol., v. 49(1): 251-275, figs. A-H, pls. 1-3, figs. 1-24.

FILIPJEV, I. N. 1934.—The classification of the free-living nematodes and their relation to the parasitic nematodes. Smithson. Misc. Coll., (3216) v.89(6): 1-63, pls. 1-8, figs. 1-70.

MUELLER, J. F. 1929.—Studies on the microscopical anatomy and physiology of *Ascaris lumbricoides* and *Ascaris megalocephala*. Ztschr. Zellforsch., v. 8(3): 361-403, pls. 9-18, figs. 1-80.

TRIM, A. R. 1949.—The kinetics of the penetration of some representative anthelmintics and related compounds into *Ascaris lumbricoides* var. *suis*. Parasitol., v. 39 (3-4): 281-290.

B. THE HYPODERMIS

A. GENERAL MORPHOLOGY

Our knowledge of the nature of the hypodermis of nematodes is due chiefly to the work of Martini (1903-1909) but many others have made contributions to this subject. As previously noted, the hypodermis usually consists of a thin layer beneath the cuticle and has longitudinal thickenings protruding internally between the sectors of the longitudinal muscles in the form of bands. These bands, known as the chords, generally contain the nuclei of the hypodermis. In the anterior and posterior parts of the body the arrangement of nuclei is modified to some extent from the four chord formation. The chords contain the greater part of the nerve cells. They carry the longitudinal nerves of the body and also, in some groups, the lateral excretory canals.

Chordal and Interchordal regions. Among the simplest nematodes there are four chords, one dorsal, two lateral, and one ventral. In the anterior part of the body the dorsal and lateral chords each contain one row of nuclei, while the ventral chord contains two rows of nuclei (Fig. 45A). In the remainder of the body, except in the tail, the dorsal chord is without nuclei, the lateral chords contain three rows of nuclei (a dorsolateral or sublateral, a lateral, and a ventrolateral or sublateral), while the ventral chord contains two rows of nuclei. In such forms, as Retzius (1906) first observed, cell walls are present which separate the nuclei and extend into and subdivide the interchordal region of the hypodermis. The chords may therefore be considered as representing the cell bodies of the epithelial cells forming the hypodermis.

The simplest nematodes have a very small number of hypodermal cells for no nuclear or cell division takes place after the larva hatches; in more complex forms the number of cells and/or nuclei is increased by division after hatching. There is a natural tendency in nematodes toward increase in size and complexity which in free-living forms takes the form of increase in cell number while in parasitic forms the same tendency is apparent but with increase in food supply and without the size limitations of a free-living environment the increase in size may be out of proportion to the increase in cell number. This naturally results in gigantism. Having a relatively limited number of cells, gigantic forms would have an unbalanced nucleo-cytoplasmic ratio if there were no more nuclei. Additional nuclear divisions without cell division results in syncytium formation. The greater the number of nuclei, the greater the loss of constancy in arrangement and number, the final end product of such a line of development being absolute absence of constancy, nuclear nests appearing and nuclei distributed in the interchordal hypodermis. Series such as this may occur in any group without regard to phylogenetic relationship.

The simplest nematodes, such as *Rhabditis* and *Rhabdias* (Rhabditoidea), have very few hypodermal cells, probably no more than are present in the larva at the time of hatching. In these forms the nuclei of the lateral and sublateral rows are about 12 to 14 in number, the nuclei of the lateral row being of about the same size or smaller than those of the sublateral rows (Fig. 45E).

From this simple arrangement there are numerous variants in increased complexity. Further nuclear division, cell division, and syncytium formation, sometimes with nuclear budding are known to occur. There is a general tendency in both the Phasmidia and Aphasmidia toward increased complexity, but the courses followed are not identical. The several variants in the two subclasses will be discussed separately.

In forms slightly more complex than *Rhabditis* and *Rhabdias* the nuclei of the sublateral rows in the lateral chords undergo division and forms may be found having only 12 to 14 large nuclei in each lateral row and two to four times as many nuclei (Fig. 45F) in each sublateral row, *Hammerschmidtiella diesingi*, *Cephalobellus papilliger*, *Enterobius vermicularis* (Oxyuroidea) and *Oswaldocruzia auricularis* (Trichostrongyloidea). In such forms, walls separate the lateral and sublateral rows but only contours have been observed between the nuclei of a particular row; no walls or contours have been observed in the interchordal areas. The nuclei of the lateral cell rows in these examples are usually much larger than those of the sublateral rows; the nuclei of the ventral cell rows and those of the anterior part of the dorsal chord are similar in size to those of the lateral cell row. Forms occur which are slightly more advanced than the two types mentioned above, the relative number of nuclei in the lateral and sublateral rows being about 1:2, but the nuclei of the lateral row are slightly smaller than those of the sublateral rows (Fig. 45G). In such cases the nuclei of the lateral rows have apparently undergone one additional division while those of the sublateral rows may have undergone two divisions (*Amidostomum nodularis* [Trichostrongyloidea] and *Crenosoma striatus* [Metastrongyloidea]). The relative size of the nuclei remains about the same in other forms having increased number of nuclei in the sublateral rows (*Dictyocaulus filaria* [Metastrongyloidea]).

A further stage in increased complexity from that discussed above is one in which the nuclei of the lateral rows are larger than those in the sublateral rows. There are many more nuclei in each of these rows than are present in the embryo (Fig. 45I). The walls between the lateral and sublateral rows of nuclei may remain distinct (*Pseudalius tumidus* [Metastrongyloidea]) or the walls may actually disappear (*Cosmocerca ornata, Heterakis gallinarum* [Ascaridoidea]). (Fig. 45H.)

In other forms of still greater complexity there are many more nuclei in the lateral rows than in the embryo, these nuclei being smaller than those of the sublateral rows, though the nuclei of the latter rows are more numerous (*Camallanus lacustris* [Camallanoidea], *Contracaecum clavatum* [Ascaridoidea]) (Fig. 45J). Further nuclear division in related forms results in irregularity of the nuclei of the sublateral rows (*Raphidascaris acus* [Ascaridoidea]), while increased division of the nuclei of the sublateral rows causes greater and greater irregularity and the relative size of the band of tissue containing the lateral nuclei diminishes (*Setaria equina* [Filarioidea]) (Fig. 45L). In the latter type nuclear "nests" are present in the sublateral rows, these apparently indicating centers of division of the nuclei of the syncytium. In these nests the nuclei are smaller while nuclei which are progressively further from the nests are larger, possibly indicating that they are older and have grown in size. In support of this view that interchordal nuclei have migrated from the chordal region Martini found that the nuclei of the sublateral rows are in clear "nests" in young specimens, 3 cm, long, of *Toxocara mystax* and no nuclei are present in the interchordal areas, while in older specimens, 17 cm. long, the nests have broken up and some nuclei have migrated to the interchordal areas (Fig. 45N). *Parascaris equorum* and *Ascaris lumbricoides* have such a hypodermis, there being a narrow lateral row containing moderately large nuclei, wide sublateral syncytia containing numerous small nuclei, and similar small nuclei scattered throughout the interchordal hypodermis and the dorsal as well as the ventral chords. All members of the Spirurida thus far investigated have syncytial chords with relatively large numbers of nuclei.

Returning to simple forms showing no increase in number of nuclei of the lateral rows, that is, with only 12 to 14 nuclei in each cell row it will be found that syncytium formation in the sublateral rows has occurred even in such forms. Multiplication of nuclei in the sublateral rows without migration into the interchordal region has occurred in some species (*Strongylus equinus* and *Oesophagostomum dentatum* of the Strongyloidea, and *Passalurus ambiguus* of the Oxyuroidea). A parallel to the series of types found in the Ascaridoidea seems also to be present in the Oxyuroidea, for nuclear nests and interchordal nuclei occur in *Oxyuris equi* although the nuclei in the lateral cell row remain as in the simplest forms (Fig. 45M).

These variations from the generalized type have all been among the parasitic nematodes. Variations do occur, however, in free-living groups but these variations are of a somewhat different nature than those occurring in parasitic nematodes; increase in number of cells in common but syncytium formation does not occur. Forms such as *Plectus* (Plectoidea) and *Tripyla* (Tripyloidea) are similar to *Rhabdias* except that further division of cells in all four chords has occurred. Three rows of cells are usually present in the anterior part of each of the four chords of enoploids and dorylaimoids, while more posteriad (Fig. 45C) there is a single row in the ventral chord, two in the lateral, four in the sublateral and one in the dorsal (*Paroncholaimus zernovi*). In those aphasmidian forms so far studied there has been a definite increase in the number of hypodermal cells over that number present in *Rhabditis*. In *Thoracostoma* four to five cell rows have been observed in each lateral chord, while in *Siphonolaimus* Zur Strassen (1904) illustrates two cell rows in the dorsal chord, three or more cell rows in the lateral chords, two cell rows in the ventral chord and 1 to 2 cell rows in each of the four submedian chords. It

should be noted, however, that these illustrations are of the esophageal region. One notable and apparently general distinction between phasmidian and aphasmidian forms is that the lateral cell rows do not differ from the sublateral cell rows as much in size or consistency in the aphasmidian.

Low submedian chords are of very common occurrence in the free-living representatives of the Aphasmidia. They do not, apparently, indicate close relationship since they occur in such diverse forms as *Desmoscolex* (Fig. 50C) *Halanonchus*, *Monhystera*, *Terschellingia* and *Sabatieria*. Nuclei have not been observed in the submedian chords of any of these forms.

Parasitic nematodes which are apparently related to aphasmidian free-living nematodes show a characteristic difference from, though parallel development to, the phasmidian parasites. Mermithids have been found to have eight chords, one

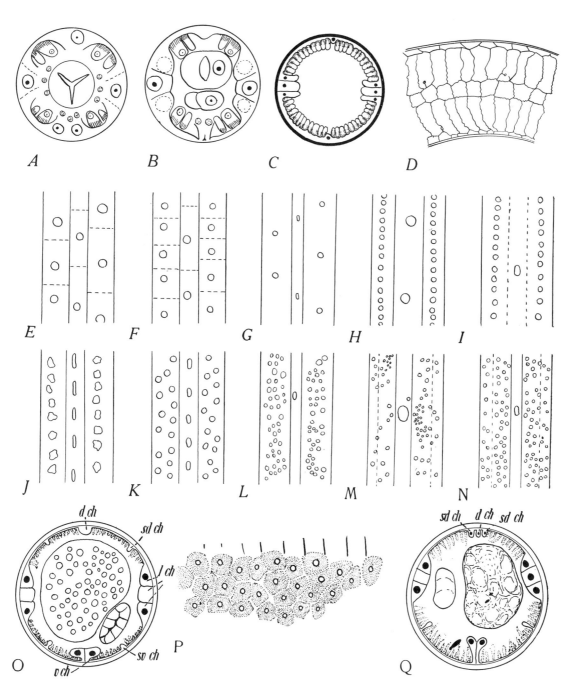

Fig. 45

Chords. A—Diagram of chords in larval nematode, esophageal region. B—Diagram of same in intestinal region. C—Diagram of chords in *Paroncholaimus*. D—*Oncholaimus vulgaris*. E—*Rhabditis*. F—*Oswaldocruzia*. G—*Amidostomum*. H—*Heterakis*. I—*Pseudalius*. J—*Contracaecum*. K—*Raphidascaris*. L—*Setaria*. M—*Oxyuris*. N—*Toxocara*. O—*Eu-* *mermis*. P—*Trichuris ovis* (Surface view of bacillary region). Q—*Mesomermis*. A-B, after Martini, 1909, Ztschr. Wiss. Zool. C, after Filipjev, 1924, Zool. Anz. D, from Rauther, 1930, Handb. Zool., after Retzius; E-N based on Martini 1909, loc. cit.; O & Q based on Steiner, 1929, Zool. Jahrb.

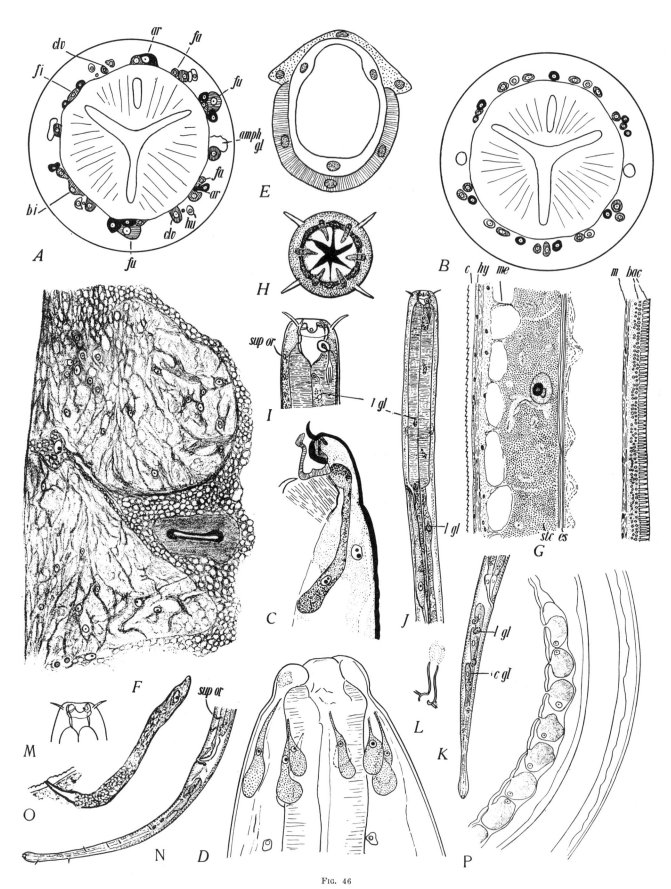

Fig. 46

Specialized hypodermal structures. A—*Oxyuris equi.* (Diagram of cells of anterior end. Similar cells are shaded alike). B—*Ascaris lumbricoides.* (Diagram of cells of anterior end, types of cells shaded as *Ascaris lumbricoides.* (Lateral chord). C—*Oxyuris equi* (Longinal section showing dorsal arcade cell; D—Totomount preparation showing arcade cells). E—*Strongylus edentatus.* (Arcade cells ["ligamentum esophageale"] which surround anterior part of stoma). F—*Ascaris lumbricoides.* (Lateral chords). G—*Trichuris suis.* (Longitudinal section showing bacillary layer). H-N—*Anonchus mirabilis* [Plectidae]. (Note supplementary organs [s] near head [I], preanal N, detail L and lateral hypodermal glands [l. gl.] and caudal glands [c. gl.]). O—*Plectus granulosus.* (Supplementary organ showing gland). P—*Axonolaimus parapinosus* (Supplementary organ glands). A-D, after Martini, 1926, Arch. Schiffs. u. Tropenhyg.; modified; E, after Immink, 1924, Arch. Anat.; G, after Rauther, 1918, Zool. Jahrb., modified; P. after de Coninck & Stekhoven, 1933, Mem. Mus. Roy. Hist. Nat. Belg.

dorsal, one ventral, two lateral and four submedian (Fig. 45O). Anterior to the nerve ring the lateral, dorsal and ventral chords each have three or more cell rows. Posterior to this region the dorsal chord contains a single cell row, the laterals, three rows of cells, and the ventral, two rows of cells; the submedian chords are mere hypodermal thickenings without nuclei (*Hexamermis albicans*, fide Rauther, 1906). Nuclei have been observed in the submedian chords in *Mermis nigrescens* by Linstow (1892) and in *Agamermis decaudata* by Steiner (1924); in some mermithids the dorsosubmedian chords have fused with the dorsal chord (Fig. 45Q), the result being a dorsal chord formed from three cell rows, two lateral chords of three cell rows and a ventral chord of two cell rows. Syncytium formation appears to have proceeded to a moderate extent in some mermithids, particularly posterior to the nerve ring.

Dioctophymatoids present many similarities to mermithids but development has proceeded further. In these forms there are not only eight chords, all containing nuclei, but also numerous inter-chordal nuclei (Fig. 50P). Near the anterior end of the body all four of the principal chords have three rows of nuclei, but posterior to the nerve ring rows of nuclei are difficult or impossible to distinguish. Nuclei are concentrated to some extent in the chords, often two to three in each submedian dorsal and ventral chord but additional thickenings of the hypodermis containing nuclei also occur between the chords.

In trichuroids only four chords are present; anterior to the nerve ring they are similar in most respects to those of other nematodes. Posterior to the nerve ring one or both lateral chords become unusually large and contain many cells; this is due to the presence of a bacillary band (Figs. 45P & 50N) which contains a mass of lateral hypodermal glands. Throughout the body, except at the anterior end, nuclei are present in the interchordal areas of the hypodermis, the lateral chords may be extremely low or even covered by somatic musculature and the median chords inconspicuous. Since the bacillary band is a special modification of the hypodermis it will be discussed later.

Cephalic modifications of the hypodermis. There is a completely separate organization of cells of the head of a nematode, but unfortunately it is not known to what form of tissue each belongs and what their function may be. Only a few nematodes have been studied in this regard. Goldschmidt (1903) and Hoeppli (1925) have described the cells as found in ascaroids; Martini (1916, 1926) described them in *Oxyuris equi* and *Tachygonetria microstoma* and Chitwood and Chitwood (1933) described them in *Cephalobellus papilliger*. The arcade consists of a circular band of tissue which usually surrounds the esophagus at the base of the lips (Fig. 4B). From it nine cells, one dorsal, two pairs of subdorsal, one pair of ventrolateral, and one pair of ventral cells, project posteriad into the body cavity (Fig. 46A-D). From the arcade, in ascarids, there are three anterior projections, one into each lip (lobus impar). In addition to these, certain other cells, two subdorsal and two ventrolateral *fiber cells* and six *clavate cells*, two dorsal submedian, two lateral, and two ventrosubmedian, contribute to the formation of the lips. Other cells are present which appear to be more intimately connected with the nervous system or connective tissue. There is a remarkable constancy of these structures in the Ascaridina, but only isolated bits of information regarding these are available in other groups. Imminck (1924) described nine cells (Fig. 46E) surrounding the stoma of *Strongylus edentatus* which he termed the ligamentum cephalo-esophageal. They probably should be considered homologues of the arcade cells. Rauther (1906) has described fusion of the submedian and median chords at the extreme anterior end of the body in *Hexamermis albicans;* the mass so formed contains numerous nuclei which conceivably might correspond to those of the arcade. Though there is but one genus, *Siphonolaimus*, in the Aphasmidia for which arcade cells have been described, it seems unlikely that the cephalic organization of the aphasmidians should differ materially from that of the phasmidians. Arcade cells are often pigmented and sometimes do have the appearance of glands in the Phasmidia as evidenced by the fact that Thomas (1930) described them as glands in *Rhigonema.* Filipjev (1924) reported that the epithelium of the anterior end of *Paroncholaimus zernovi* is formed of six cells situated in the lips, and Kreis (1934) described and figured a multitude of cells situated in the labial region of *Adoncholaimus meridionalis.* The writers, however, find absolutely *no nuclei* (and hence no complete cells) within the first *two stomatal lengths* of the anterior extremity of *Metoncholaimus pristiuris*, and since it is well known that the histological anatomy of closely related nematodes shows little

or no variation, we cannot regard as probable the observations concerning *Paroncholaimus* and *Adoncholaimus.*

Caudal modifications. In the postanal region of the body the chords are generally modified, particularly toward the end of the tail, and also to some extent even in the rectal region of the short-tailed forms. These changes are connected with the association of the ventral chord and anus, the disappearance of the dorsal chord, and the presence of copulatory muscles in the male. As Looss (1905) stated, the ventral chord passes to the sides of the posterior part of the rectum and is continuous with the swollen cells on the dorsal surface of the rectum, which he termed the *postanal pulvillus.* In some forms such as *Rhabditis* (Fig. 49B) the postanal pulvillus is very large while in other forms such as *Ascaris* this structure is quite inconspicuous. The pulvillus continues posteriorly as the typical ventral chord. The dorsal chord stops short of the posterior extremity even in short tailed forms; often, as in *Rhabditis*, it completely disappears some distance posterior to the anus. In such cases the lateral chords usually attain a dorsolateral position. Transverse copulatory muscles complicate the appearance of the chords in the male for they are often inserted in the upper half of the lateral chords (*Hexamermis albicans, Rhabditis strongyloides*). In addition to the lateral chords Rauther (1906) described multicellular subventral chords in *Hexamermis albicans.*

Hypodermal glands. There are glands connected with the amphids and phasmids but since these organs are considered chiefly sensory the glands connected with them will be described in connection with the nervous system. The excretory system is probably a hypodermal gland but this system will also be treated separately. Aside from these glands hypodermal glands are not known in the Phasmidia, being confined to the Aphasmidia. Such glands are of three types, namely, (1) lateral hypodermal glands (2) caudal glands and (3) supplementary glands.

(1) *Lateral hypodermal glands.* Very commonly there are two sublateral rows of unicellular glands situated in the lateral chords and opening by short ducts through pores in the cuticle in both the Enoplida and Chromadorida; these glands are not known to occur in other orders of the Nematoda. Such glands are known to occur in many groups of these orders but apparently they may be either present or absent in closely related genera. (Possibly this is due to faulty observation, the glands being minute and inconspicuous in forms in which they have not been observed.) In the Enoplina they are known to occur in such forms as *Thoracostoma, Deontostoma, Cylicolaimus, Oxystomina, Oncholaimus* and *Metoncholaimus.* In the first four genera these glands are large and conspicuous in whole specimens but in the last two genera the glands and pores are minute. These were first observed in *Oncholaimus vulgaris* by Stewart (1906).

In *Trichuris* and related forms there may be one or two socalled *bacillary bands* situated laterally (not medially as is generally stated) in the esophageal region. The lateral surface of these bands in superficial view (Fig. 45P) appears to be made up of small hexagons containing circles; the hexagons represent the ends of elongated columnar cells projecting into the cuticle and the circles are pores. Internally (Figs. 46GB, 50N) these elongated cells are situated in the lateral chords. Jägersköld (1901) and Rauther (1918) considered the bacillary bands to be hypodermal glands, while G. W. Müller (1929) considered them to be feeding pores by which food might pass in through the cuticle and hypodermis into connective tissue and thence to the stichosome or cell body. The writers cannot concur in this viewpoint since it is based on the erroneous hypothesis that trichuroids do not draw in food through the esophagus. On either side of the bacillary band there may be vesicular cuticular swellings the significance of which is not known. Possibly they are connected with glands.

In the Chromadorida, hypodermal glands (Fig. 50L) are known in the Plectidae, Camacolaimidae, Comesomatidae, Cyatholaimidae, Desmoscolecidae, Greeffielliidae and scattered genera of other families such as *Terschellingia* of the Linhomoeidae. Sometimes, as in *Anonchus mirabilis* (Figs. 46 H-N & 50K), the hypodermal glands are pigmented and sometimes, as in *Desmoscolex americanus* (Fig. 50C), they are connected with setae. In this particular instance the glands are notable in that they are situated between the cortical and basal layers of the cuticle in the sublateral regions (connected, however, with the lateral chords). The presence of sublateral glands in such forms as *Oncholaimus* and *Terschellingia* indicates that a more critical study will show such glands to be present in many genera in which they are now said to be absent. Stekhoven and Teunissen (1938) studied the partial constancy of the hypodermal glands of *Anaplectus granulosus* finding increase in the number of these structures during growth.

(2) *Caudal glands.* Caudal glands are glands situated in the posterior part of the body or tail and, emptying subterminally or terminally through a pore, the *spinneret*. These glands are three in number, unicellular, and usually quite elongate. They are known to occur only in the suborders Chromadorina, Monhysterina, and Enoplina. In all instances in which a critical study has been made three caudal glands have been found but instances are not uncommon in which authors illustrate two, four or five such structures. The position and form of such glands conform to the form of the body, being long, short, parallel or tandem according to the available space. Each gland consists of an enlarged cell body containing the nucleus and an elongated duct. The ducts from the three units unite posteriorly to form a common duct which is terminated by an ampulla. The tail is commonly enlarged at the ampulla and bears a fine terminal tube, the spinneret. According to Cobb (1915, 1917) the spinneret works on the principle of a needle valve (Fig. 47); it consists of a small cone the apex of which projects into the tube pore, this cone being connected with retractor muscles which upon contraction permits the fluid to flow around the cone and out the tube pore. Caudal glands secret a substance which hardens in water to form a thread which serves as a means of attachment in much the same manner as the threads secreted by spiders (Fig. 48).

(3) *Supplementary glands.* Glands connected with ventral preanal supplementary organs in the male have previously been mentioned. Such glands are unicellular, sometimes moderately small as in *Anaplectus granulosus* (Fig. 46O), sometimes tremendously elongated as Cobb (1920) observed in *Bolbella* (Fig. 39). They lie chiefly in the body cavity and may develop from the ventral chord. There is no direct proof that they secrete an adhesive substance but the general supposition is that they do. Similar glands have been described in representatives of most groups possessing well developed, cuticularized, tuboid supplementary organs such as enoplids, eurystominids and plectids. De Coninck and Stekhoven (1933) have recently discovered unicellular glands within the minute papilloid supplementary organs of *Axonolaimus* (Fig. 46P).

B. DETAILED MORPHOLOGY

Framework.—The cells or syncytium forming the hypodermis are rather highly organized and have, at least in some cases, a very delicate framework. As Apathy (1893, 1894) first showed in *Ascaris*, the interchordal areas contain minute supporting fibrillae which form a network continuous with the fibrillar network of the adjacent muscle cells. These fibrillae were supposed by Apathy to be motor processes innervating the muscles. Goldschmidt (1909) has shown, however, that the processes are supporting rather than innervating. These processes pass, not only laterally and ventrally into the chords, but also radially into the cuticle. Martini (1916) found that the chords in *Oxyuris* contain longitudinal, circular and radial supporting fibrils. The radial fibrils are often branched toward the internal surface of the chords while the circular fibrils extend into the interchordal areas of the hypodermis.

Stored food.—Von Kemnitz (1912) and Martini (1916) working on *Ascaris* and *Oxyuris* respectively, found that the hypodermis is the chief place of glycogen storage in the nematode body. The general vacuolate appearance of the cells of the chord prepared by ordinary technic is doubtless due to the solubility of glycogen and fats. The sublateral areas of the lateral chords and the hypodermis of *Ascaris* contain most of the glycogen, while the lateral areas of the lateral chords in the vicinity of the excretory duct and the median chords are relatively poor in glycogen. The massive lateral cells of *Oxyuris*, however, are nearly all glycogen and for this reason the nucleus and surrounding cytoplasmic strands have a stellate appearance. This is probably the explanation for the stellate appearance of the lateral cell row in *Cephalobellus* (Fig. 50D). In addition to glycogen, von Kemnitz also found fat droplets in abundance in the interchordal and chordal areas of the hypodermis.

Granules.—The hypodermal glands usually contain numerous small "granules" which have a strong affinity for basic dyes. This same staining affinity is shared by the caudal glands as well as by the excretory glands. Additional larger pigment granules are also present in the sublateral glands of some forms such as *Anonchus* (Fig. 50K); these granules are brown to black in the living organism but usually become blue with hemotoxylin just as do the small granules.

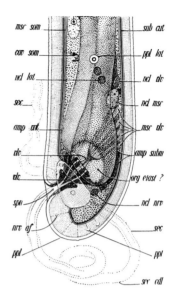

FIG. 47

Mononchulus ventralis Mononchidae. (Tail showing details of spinneret. Species atypical; usually spinneret is terminal). After Cobb, 1918, Contrib. Sc. Nemat. VII.

FIG. 48

Mononch tail showing caudal glands, spinneret, and secretion. After Cobb, 1917, Soil Science.

Bibliography (Hypodermis)

APATHY, S. 1893.—Über die Muskelfasern von *Ascaris*, etc. Ztschr. Wiss. Mikro. v. 10: 36-73, 319-361.
 1894.—Das leitende Element in den Muskelfasern von *Ascaris*. Arch. Mikro. Anat., v. 43: 886-911.

CHITWOOD, B. G. and CHITWOOD, M. B. 1933.—The histological anatomy of *Cephalobellus papilliger* Cobb, 1920. Ztschr. Zellforsch., v. 19(2): 309-355, figs. 1-34.

COBB, N. A. 1915.—The mechanism of the spinneret of a free-living nematode. J. Parasit., v. 2: 95.
 1917.—The mononchs (*Mononchus* Bastian, 1866). A genus of free-living predatory nematodes. Contrib. Sc. Nemat. (6). 129-184, figs. 1-72. Soil Science, v. 3(5): 431-486.

CONINCK, L. A. DE, & STEKHOVEN, J. H. SCHUURMANS. 1935.— The free living marine nemas of the Belgian Coast II. Mém. Mus. Roy. Hist. Nat. Belg. (58): 1-163, figs. 1-163.

DADAY, E. VON. 1909.—Beiträge zur Kenntnis der Süsswassern lebenden Mermithiden. Math. Naturw. Ber. aus. Ungarn, v. 27 [not seen].

EBERTH, C. J. 1863.—Untersuchungen über Nematoden. 77 pp., 9 pls., Leipzig.

FILIPJEV, I. N. 1924.—Über das Zellmosiak in der Epidermis von *Paroncholaimus zernovi*. Zool. Anz., v. 61(11/12): 268-277.

GOLDSCHMIDT, R. 1903.—Histologische Untersuchungen an Nematoden. I. Die Sinnesorgane von *Ascaris lumbricoides* L. and *A. megalocephala Cloqu*. Zool. Jahrb. Abt. Anat., v. 18(1): 1-57, pls. 1-5, figs. 1-40.

1906.—Mitteilungen zur Histologie von *Ascaris*. Zool. Anz., v. 29: 719-737, figs. 1-13.

1909.—Das Skellet der Muskelzelle von *Ascaris*. Arch. Zellforsch., v. 4(1): 81-119, pls. 6-9, figs. 1-19.

HAMANN, O. 1892.—Die Nemathelminthen. Jena. v. 2, 120 pp., 11 pls.

HOEPPLI, R. 1925.—Über das Vorderende der Ascariden. Ztschr. Zellforsch. v. 2(1): 1-68, figs. 1-27, pl, figs. 1-12.

IRWIN-SMITH, V. A. 1918.—On the Chaetosomatidae, with descriptions of new species, and a new genus from the coast of New South Wales. Proc. Linn. Soc. N. S. Wales (168), v. 42(4): 757-814, figs. 1-59, pls. 44-50.

JAEGERSKIOELD, L. A. 1901.—Weitere Beiträge zur Kenntnis der Nematoden. K. Svenska Vetenskaps-Akad. Handlinger. v. 35(2): 1-80, pls. 1-6.

KEMNITZ, G. VON. 1912.—Die Morphologie des Stoffwechsels bei *Ascaris lumbricoides*. Arch. Zellforsch., v. 7(4): 463-603.

KREIS, H. A. 1934.—Oncholaiminae Filipjev, 1916. Eine monographische Studie. Capita Zool., v. 4(5): 1-270, figs. 1-135.

LINSTOW, O. VON. 1892.—Ueber *Mermis nigrescens* Duj. Arch. Mikro. Anat., v. 40: 498-512, pls. 28-29, figs. 1-14.

1899.—Nematoden aus der Berliner Zoologischen Sammlung. Mitt. Zool. Samml. Mus. Naturk. Berlin, v. 1(2): 3-28, pls. 1-6, figs. 1-78.

LOOSS, A. 1905.—The anatomy and life history of *Agchylostoma duodenale*. Rec. Egypt. Govt. Sch. Med., v. 3: 1-158, pls. 1-9, figs. 1-100, pl. 10, photos. 1-6.

MAN, J. G. DE. 1904. Nématodes libres. Résultats du voyage du S. Y. Belgica. Expéd. Antarct. Belg. Anvers. 55 pp., 11 pls.

MARTINI, E. 1906.—Über Subcuticula und Seitenfelder einiger Nematoden. I. Ztschr. Wiss. Zool., v. 81(4): 699-766, pls. 31-33.

1907.—Idem. II. Ibid., v. 86(1): 1-54, pls. 1-3.

1908.—Idem. III. Ibid., v. 91(2): 191-235.

1909.—Idem. IV. Ibid., v. 93(4): 535-624, pls. 25-26.

1916.—Die Anatomie der *Oxyuris curvula*. Ztschr. Wiss. Zool., v. 116: 137-534, figs. 1-121, pls. 6-20.

1926.—Zur Anatomie des Vorderendes von *Oxyuris robusta*. Arch. Schiffs. u. Tropenhyg., v. 30: 491-503, figs. 1-9.

MÜLLER, G. W. 1929.—Die Ernährung einiger Trichuroideen. Ztschr. Morph., v. 15(1/2): 192-212, figs. 1-15.

RAUTHER, M. 1906.—Beiträge zur Kenntnis von *Mermis albicans* v. Sieb. Zool. Jahrb., Abt. Anat., v. 23 (1): 1-76, pls. 1-3, figs. 1-26.

1918.—Mitteilungen zur Nematodenkunde. Ibid., v. 40: 441-514, figs. A-P, pls. 20-24, figs. 1-40.

RETZIUS, G. 1906.—Zur Kenntnis der Hautschicht der Nematoden. Biol. Untersuch. n. F., 13: 101-106, figs. a-f.

SCHNEIDER, K. C. 1902.—Lehrbuch der vergleichenden Histologie der Tiere. Jena. 988 pp., 691 figs.

STEKHOVEN, J. H. & TEUNISSEN, R. J. H. 1938.—Nematodes libres terrestris. Exploration du Parc National Albert. Fasc. 22:1-299.

STEINER, G. 1924.—Beiträge zur Kenntnis der Mermithiden. 2 Teil. Mermithiden aus Paraguay in der Sammlung des Zoologischen Museum zu Berlin. Centralbl. Bakt., v. 62: 90-110, figs. 1-54.

1929.—On a collection of mermithids from the basin of the Volga River. Zool. Jahrb., Abt. Syst., v. 57: 303-328, figs. 1-42.

STEINER, G. and ALBIN, F. M. 1933.—On the morphology of *Deontostoma californicum* n. sp. (Leptosomatinae, Nematodes). J. Wash. Acad. Sc., v. 23(1): 25-30, figs. 1-7.

STEWART, F. H. 1906.—The anatomy of *Oncholaimus vulgaris* Bast., with notes on two parasitic nematodes. Quart. J. Micr. Sc., Lond., n. s., (197) v. 50(1): 101-150, figs. 1-9, pls. 7-9, figs. 1-40.

STRASSEN, O. ZUR. 1904.—*Anthraconema*, eine neue Gattung freilebender Nematoden. Zool. Jahrb., Suppl. 7, Festschrift Weismann: 301-346, figs. A-J, pls. 15-16, figs. 1-9.

THOMAS, L. J. 1930.—*Rhigonema nigella* spec., nov., a nematode and its plant commensal, *Enterobrus* sp. ? from the milleped. J. Parasit., v. 17 (1): 30-34, pls. 3-4, figs. 1-18.

TOERNQUIST, N. 1931.—Die nematoden familien Cucullanidae und Camallanidae, nebst weiteren Beiträgen zur Kenntnis der Anatomie und Histologie der Nematoden. Göteborgs K. Vetensk-o. Vitterhets Samh. Handl., 5 f., s. B, v. 2 (3), 441 pp., pls. 1-17.

TUERK, F. 1903.—Über einige im Golfe von Neapel frei lebende Nematoden. Thesis. Leipzig. 67 pp., pls. 10-11.

CHAPTER IV

SOMATIC MUSCULATURE, CONNECTIVE TISSUE, BODY CAVITY AND ORGANS OF BODY CAVITY

B. G. CHITWOOD and M. B. CHITWOOD

A. SOMATIC MUSCULATURE

1. GENERAL STRUCTURE

Unspecialized musculature. Early investigations of the somatic musculature of nematodes were initiated chiefly by the work of Schneider who first clearly differentiated between forms taken by the musculature of the body wall; the descriptive nomenclature used today is that proposed by Schneider. In 1860 he introduced the terms *platymyarian* and *coelomyarian;* which apply to the form taken by the individual muscle cell (for definition see p. 7), and in 1866 he introduced the terms *meromyarian, polymyarian,* and *holomyarian* which apply to the number of muscle cells between chords, that is, in a sector. Holomyarian nematodes were supposed to have only two or no chords, but Bütschli (1873) showed that at least four chords were present in the anterior end of trichurids and mermithids (classified by Schneider as holomyarian) even though only two chords may be present in other regions (*Trichuris,* Fig. 50N). Both nomenclatures, *platymyarian* vs. *coelomyarian* and *meromyarian* vs. *polymyar-*

ian, are useful since they are not entirely synonymous. However, it is true that meromyarian forms are usually platymyarian and polymyarian forms are usually coelomyarian. Between extremes numerous transitional forms occur which necessitates modifications of terminology. There is no hard and fast rule for separating nematodes into groups on the basis of type musculature. Nevertheless, Martini (1903, 1906, 1909) has shown that polymyarian nematodes are meromyarian and platymyarian in the first larval stage and that such conditions as polymyarity and coelomyarity are results of later development. This seems to rather indicate that the primitive type of somatic musculature in nematodes was platymyarian and meromyarian.

Simple meromyarian nematodes, such as *Rhabditis, Oxyuris* and *Strongylus,* have a very few muscle cells, there being in cross-section from two to five in each sector; these muscle cells project very little into the body cavity. The total number of somatic cells has been determined in two species, *Oxyuris equi* and *Strongylus* sp. In the former, Martini (1907, 1908, 1916) found a total of 65 muscle cells, eight in each laterodorsal row, nine in each dorsolateral and lateroventral row, seven in the right ventrolateral row and six in the left ventrolateral row. Similarly, Martini (1908) found a total of 87 somatic muscle cells in *Strongylus,* 11 in each row except the left lateroventral in which there are ten. In such forms there are usually only

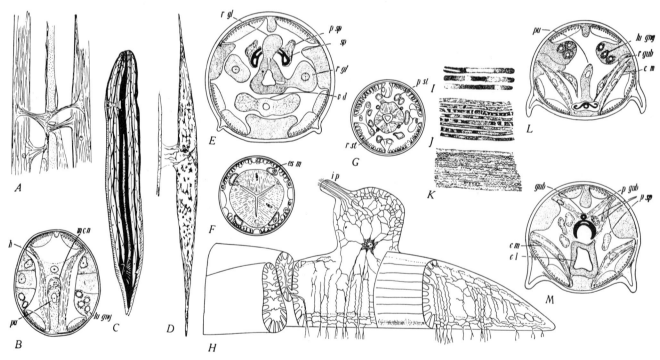

FIG. 49

Musculature. A—*Oxyuris equi* (Muscle cells connected with median nerve by innervation processes). B—Postanal region of female *Rhabditis strongyloides* showing depressor ani. (H-shaped muscle cell and pulvillus. C—*Oxyuris equi* dissected showing somatic musculature. D—*Oxyuris equi* (Muscle cell dissected out). E—*Rhabditis strongyloides* (Male. Section near posterior end of intestine showing rectal glands, retractor muscles and spicules [Distally fused in this species]). F—*Metoncholaimus pristiuris.* (Section near base of esophagus showing somato-esophageal muscles). G—*Dorylaimus obtusicaudatus.* (Section near base of stylet showing protractor and retractor muscles). H—Diagram of muscle cell in *Ascaris lumbricoides* showing fibrillar network. I-K—Longitudinal sections through contractile zone of muscle cell of *Ascaris lumbricoides* (I—Iron hematoxyin; J—Bielchowsky technic; K—Ikeda technic, showing longitudinal supporting fibrils. Dense zones indicate contraction). L-M—*Rhabditis strongyloides* male. (Showing spicular, gubernacular, and copulatory muscles. L—Near cloacal orifice; M—Further anterior). A & D, after Martini, 1916, Ztschr. Wiss. Zool.; C. from Rauther, 1930. Handb. Zool.. after Martini, modified; H. after Mueller, 1929, Ztschr. Zellf., modified; I-K, after Roskin, 1925, Ztschr. Zellf.

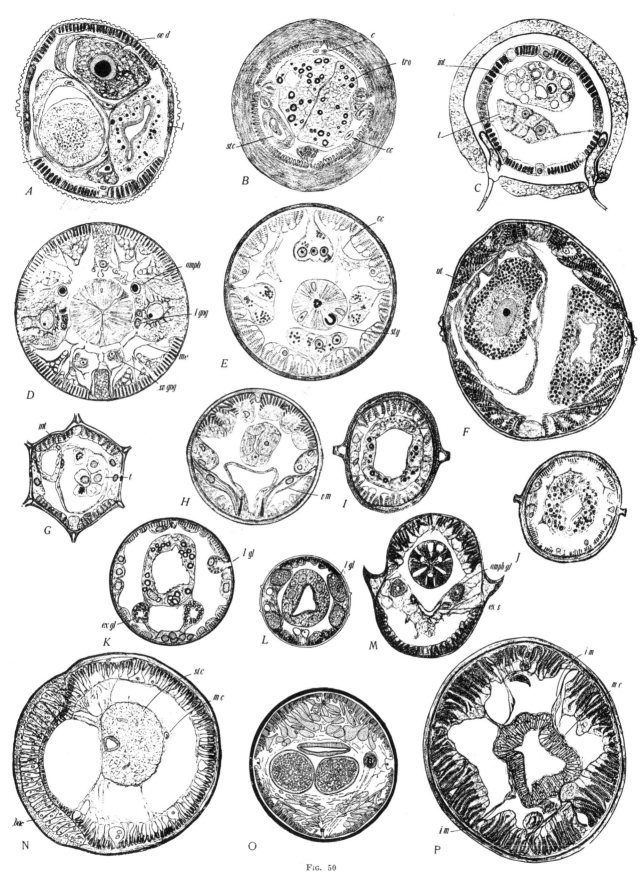

FIG. 50

Cross sections of various types of nematodes. A—*Rhabditis lambdiensis.* (Showing long ridges and chords). B—*Hexamermis albicans.* (Showing trophosome, stichocyte and coelomocytes). C—*Desmoscolex americanus.* (Showing glands of setae and cuticle inflated). D—*Cephalobellus papilliger.* E—*Dorylaimus obtusicaudatus.* F—*Theristus setosus.* G—*Monoposthia hexalata.* H—*Ethmolaimus revaliensis* (At site of vulva). I—*Spilophorella paradoxa.* J—*Aphanolaimus* sp. K—*Anonchus mirabilis.* L—*Plectus granulosus.* M—*Heterakis gallinarum.* N—*Trichuris ovis.* O—*Ascaris lumbricoides.* P—*Eustrongylides ignotus.* B, after Rauther, 1906. Zool. Jahrb., modified; D, after Chitwood & Chitwood, 1933, Ztschr. Zellf.; O, after Rauther, 1930, Handb. Zool.

four to eight muscle cells at the anterior end of the body, these cells being blunt at the ends, unlike the spindle-shaped cells further posteriad (Fig. 49C).

The characteristic number of cells in each sector of meromyarian nematodes is two to four, but occasionally a large number of cells may be present as in *Oesophagostomum dentatum*. Strictly speaking, meromyarian nematodes have only two to four muscle cells to a sector, the musculature being nearly always platymyarian; however, in some species, as in *Rhabdias bufonis* the muscle fibers may extend partially up the sides of the muscle cells. Such forms show faint indications of coelomyarity. With very slight increase in number of muscle cells this indication may become apparent (*Oesophagostomum*) and when the range of cells in a given sector reaches the number of six the muscle cells are usually of a type that might be termed shallow coelomyarian or transitional (*Plectus granulosus* Fig. 50L). Since there are usually fewer muscle cells in the anterior and posterior ends of the body, coelomyarity is less apparent in these regions. In descriptions, it is customary to note the condition and number of muscle cells near the middle of the body and not at the ends.

Meromyarian platymyarian nematodes, according to Chitwood and Chitwood, 1934, occur in several of the large groups. The Rhabditina consists entirely of such forms.

One superfamily of the Strongylina, the Strongyloidea, is characteristically meromyarian, while another, the Trichostrongyloidea, contains representatives of both types and still another, the Metastrongyloidea, is exclusively polymyarian. In the Trichostrongyloidea such forms as *Ostertagia* are typically platymyarian and meromyarian while other forms such as *Haemonchus* are peculiar in showing on cross-section four sublateral large platymyarian cells and 40-48 submedian small coelomyarian cells in the mid-region of the body. The latter form is typically platymyarian and meromyarian in the esophageal region. In the Ascaridina there is a similar diversity, members of the Oxyuroidea being meromyarian while members of the Ascaridoidea may be meromyarian (Cosmocercidae, Kathlaniidae), or polymyarian (Ascarididae), with one family (Heterakidae) polymyarian but showing indications of meromyarity in the cephalic region. The Spirurina and Dioctophymatina are the only suborders in which the representative members are exclusively polymyarian. In the Monhysterina forms such as *Anonchus*, *Camacolaimus* and *Aphanolaimus* are meromyarian and platymyarian, while other forms such as *Tripylium* are meromyarian and slightly coelomyarian, and still other forms such as *Axonolaimus*, *Theristus*, and *Siphonolaimus* are polymyarian and coelomyarian. Most members of the Chromadorina are somewhat transitional in having about six to eight muscle cells in a sector, these muscle cells being only slightly coelomyarian (*Spilophorella*) but some forms are extremely coelomyarian (Cyatholaimidae).

In the Enoplina there are a few meromyarian, platymyarian forms such as *Tripyla* and *Cryptonchus*, but the majority of forms are distinctly polymyarian and vary from low coelomyarian (*Metoncholaimus*) to high coelomyarian (*Enoplus*, *Thoracostoma*). In the Dorylaimina there is likewise a small number or residual group of meromyarian forms such as *Diphtherophora* but the great majority are polymyarian and coelomyarian (*Dorylaimus*, *Mermis*, *Trichuris*).

From this review it is obvious that the form or number of muscle cells is not in itself an indication of close relationship. Nevertheless, it is not entirely without evolutionary significance. As previously pointed out by the writers (1934) evolution from meromyarian to polymyarian musculature must have taken place in most of the large groups and meromyarity, therefore, can be regarded as indicative of the degree of primitivity. This view is in direct opposition to the view expressed by Filipjev (1934) who maintains that meromyarity is a larval characteristic retained only in abnormal (i. e., paedogenic) forms.

As each organ is discussed we shall see the tendency toward growth and cell multiplication repeated. Presumably those forms with a small number of cells or nuclei have not undergone as much evolution as forms with a large number of cells or nuclei.

Somato-esophageal muscles (Fig. 51). Under this heading a variety of muscles must be classified although they are of diverse form and origin. De Man (1886) first described somato-esophageal muscles extending from the body wall to the esophagus in *Enoplus* and *Oncholaimus*, while Türk (1903) found similar muscles in *Thoracostoma*. In *Enoplus* eight muscle bands are applied to the surface of the posterior end of the esophagus. Tracing anteriorly each of the four median bands fuse with the four sublateral bands and gradually the

bands leave the surface of the esophagus and are attached to the body wall near the submedian lines. Each muscle is composed of a single muscle cell the nucleus of which is situated in the part next to the body wall; that part of the cell which is next to the body wall is coelomyarian while the remainder is circomyarian.[1] In *Metoncholaimus pristiuris* there are likewise four somato-esophageal muscle cells and they are arranged in the same manner as those referred to above except that each muscle is not divided posteriorly and the nuclei (Fig. 49F) are situated in the four sarcoplasmic swellings at the base of the esophagus. Similar somato-esophageal muscles have been observed in *Halichoanolaimus* and *Eurystomina* but these have not been carefully studied.

In one of the stylet bearing nematodes, *Siphonolaimus*, another type of somato-esophageal muscle is present. In this peculiar form four muscles extend anteriorly from the body wall (where the nuclei are situated) just anterior to the nerve ring to the external surface of the esophagus where they form a muscular sheath over the short corpus. Anterior to the corpus six bands of muscle fibers, apparently continuous with the fibers covering the corpus, form an hexagonal cylinder around the stomatostyl and these fibers apparently terminate at the cuticle of the anterior extremity. In *Dorylaimus*, another stylet bearing form, there are 12 muscles (Fig. 49G) connected with the anterior end of the esophagus, and of these eight extend anteriorly from the external surface of the esophagus to the body wall and four extend posteriorly.

Looss (1905) described somato-esophageal muscles in *Ancylostoma duodenale*, and in this form, as also in *Oesophagostomum*, *Strongylus* and related genera, there are eight muscles connecting the esophagus and body wall. Anteriorly these muscles, four submedian and four sublateral, originate at the body wall near the base of the stoma and they extend posteriorly through the body cavity to the surface of the esophagus. Near the nerve ring each muscle is split, one portion passing posteriorly on the inside of the nerve ring and the other portion on the outside, the two portions uniting posterior to the nerve ring. These muscles are platymyarian at the body wall and on the surface of the esophagus, but between the two places they are deep coelomyarian to circomyarian; their nuclei are situated a short distance anterior to the nerve ring.

In *Camallanus*, as noted by Magath (1919) and Törnquist (1931), there are four massive modified sublateral somatic muscles attached anteriorly to the jaws, but in this form they rest against the body wall throughout their length. Törnquist (1931) also described four somato-esophageal muscles in *Cucullanus*. However, in this form the muscles extend from the outer surface of the jaws posteriorly to the esophagus where processes similar to those in *Strongylus*, make a loop through the nerve ring.

Muscles of such diverse arrangement must obviously have varying functions. In *Cucullanus* and *Camallanus* the muscles control the opening of the jaws, while in *Siphonolaimus* and *Dorylaimus* they must act as retractors and protractors of the stylet. In *Strongylus*, *Oncholaimus* and similar forms contraction of these muscles would probably force the esophagus forward, but whether or not this would aid in either dilating the oral opening or pressing the teeth against the animals' food is not known.

Somato-intestinal muscles. These are muscles which extend from the body wall to the intestine; they have received little attention, but are, nevertheless, of considerable significance since they play an important role in the movement of the intestinal centents. One suborder, The Dioctophymatina, is characterized by the presence of four submedian longitudinal rows of somato-intestinal muscles; these muscles are transverse, extending from the body wall to the intestine in a nearly perpendicular manner (Fig. 50P). During the life of the worm one may observe peristaltic movement of the gut contents as a result of the action of these muscles.

Thus far, no other type of nematode is known to possess such a series of somato-intestinal muscles. Martini (1916) described two groups of four muscles each in *Oxyuris equi*, one group near the middle of the intestinal length and the other near the posterior end of the intestine. These muscles extend obliquely through the body cavity, the fibers breaking up and extending longitudinally over the surface of the intestine. The writers have observed similar muscles in *Blatticola blattae* (Fig. 52), one subdorsal and one subventral pair extending anteriorly from the intestine to the body wall near the level

---

[1] The term circomyarian is introduced here as a descriptive term for muscle cells in which the fibroplasm completely surrounds the sarcoplasm. The distal parts of the coelomyarian muscle cells in *Ascaris* are sometimes circomyarian (Fig. 49H) but this term is chiefly applicable in nematodes to specialized muscles.

of the vulva, and another group extending from the intestine posteriorly to the body wall near the intestino-rectal valve. The two groups of muscles act in opposition to each other in *Blatticola blattae.*

In the genera *Enoplus* and *Metoncholaimus* one pair of subventral somato-intestinal muscles extends from the body wall posteriorly to the intestine to just anterior to the intestino-rectal valve.

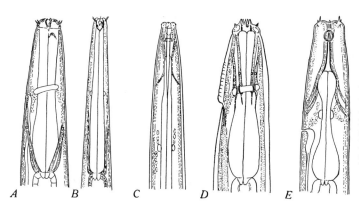

FIG. 51

Diagrams of somato-esophageal muscles. A—*Enoplus communis.* B—*Metoncholaimus pristiuris.* C—*Dorylaimus obtusicaudatus* D—*Oesophagostomum dentatum.* E—*Siphonolaimus conicus.*

Depressor ani. This H-shaped muscle is very characteristic in appearance (Figs. 49B & 52) and seems to be generally present in widely separated species. It has been described by Voltzenlogel (1902) in *Ascaris lumbricoides,* Looss (1905) in *Ancylostoma duodenale,* Martini (1916) in *Oxyuris equi* and by numerous subsequent authors in other forms. The muscle is H-shaped, consisting of two vertical groups of fibers between the dorsal wall of the cloaca or rectum, the posterior lip of the anus and the dorsolateral or subdorsal side of the body. Between the two groups of fibers there is a horizontal band of sarcoplasm containing a single nucleus. This muscle serves to dilate the rectum and elevate the posterior lip of the anus thus permitting defecation.

Dilator ani. Muscles have in some instances been described which extend from the ventral body wall to the posterior surface of the anus or cloacal orifice. Presumably their contraction would open the anus. Paired ventral preanal muscles connected with the anterior lip of the anus have also been described.

Copulatory muscles. Transverse muscles extending from the dorsal or ventral sides of the lateral chords to the subventral sides of the body are present in most male nematodes (Figs. 3 & 49L-M). Their number and extent vary extremely; in some forms, as in ascarids, there may be as many as 40-50 pairs of such muscles while in oxyurids there may be only three to four pairs. These muscles are usually confined to a region just anterior to the cloacal orifice, less commonly they may also extend posterior to the cloacal orifice (*Thoracostoma* fide Türk, 1903). Each muscle has a sarcoplasmic thickening, usually on its median face, in which the nucleus lies. Contraction of such muscles causes the ventral curvature of the posterior part of the body, which is so common in fixed material.

Bursal muscles. Bursal muscles may be considered as a kind of copulatory muscle but it is difficult at the present time to give a description of such structures. Looss (1905) lists the following specialized muscles (in addition to the ordinary copulatory muscles) as present in the male tail of *Ancylostoma duodenale* (1) *Musculus costae dorsalis.*—This muscle arises mediodorsally in the dorsal ray; it extends anteriorly and becomes first trifurcate then the median branches again split making four parts, the lateral parts pass anterior and are inserted dorsolaterally, next to the lateral chords; a single nucleus is present, lying in the sarcoplasm connecting the four branches of the muscle. (2) *Musculus costerum lateralium internis.*—This muscle arises as paired submedian muscles at the body wall anterior to the intestino-rectal valve, the two fiber groups being connected by a sarcoplasmic process, each branch extending laterally, entering the root of the lateral ray and then becoming branched, one branch entering the dorsal

and the other branch entering the ventral ray. Presumably the function of this muscle is to bend the bursa inwards. (3) *Musculus bursae basalis.*—This muscle arises from the ventral side of the bursa and extends dorsally to the root of the dorsal ray. (4) *Musculus costae lateralis externus posterior.*—This muscle arises anteriorly near the dorsal side of the lateral chords; it extends posteriorly to the base of the lateral rays where it becomes trifurcate. (5) *Musculus costae lateralis externus anterior.*—These muscles arise anterior to (4) and extend posteriorly to the base of the ventral rays. Both these muscles and (4) probably cause dilation or opening of the bursa.

Martini (1916) found eight special copulatory or bursal muscles of the male, one mediodorsal and seven ventral in *Oxyuris equi.* Presumably the dorsal muscle causes straightening of the tail and the ventrals cause curving.

Spicular muscles. Spicular muscles are of the following types: (1) *Retractor spiculi.*—Paired muscles extending anteriorly from the head of the spicule or spicules to the hypodermis in the region of the lateral chords (Fig. 49E), each muscle containing a nucleus in the sarcoplasm; they are circomyarian in *Ancylostoma,* according to Looss (1905), and platymyarian in *Oxyuris,* according to Martini (1916). (2) *Protractor spiculi.*—Paired muscles extending posteriorly and ventrally from the spicules to the ventral side of the body (Fig. 49L-M).

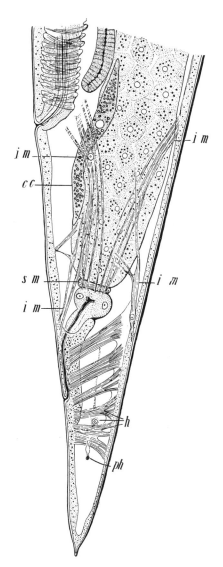

FIG. 52

Blatticola blattae. Posterior region of female showing somato-intestinal muscles, dorso-rectal muscle (Depressor ani) and one coelomocyte.

Gubernacular muscles. The muscles of the gubernaculum (Fig. 49L-M) may be of three types as follows: (1) *Retractor gubernaculi.*—These muscles extend from the gubernaculum to the dorsal wall of the body. (2) *Protractor gubernaculi.*—These are paired muscles extending from the ventral side of the body anteriorly to the gubernaculum. (3) *Seductor gubernaculi.*—Paired muscles extending from the lateral walls of the body to the gubernaculum.

Vulvar muscles. Muscles at the vulva may be of two types, the *dilator vulvae* and the *constrictor vulvae;* the latter type is little known. The *dilator vulvae* may consist of as few as eight bands of muscle fibers, four anterior and four posterior to the vulva; they are inserted ventrolaterally in the hypodermis (Fig. 50H). The bands may be much more numerous in some forms but are arranged in the same pattern. Looss (1905) found only one nucleus in the entire group of dilator fibers. In *Ethmolaimus* each of the eight bands of fibers has a separate nucleus. Circular fibers (the *constrictor vulvae*) have been described by Türk (1903) in *Thoracostoma.* It is possible that these are actually fibers of the vagina and not the vulva itself.

Other muscles. Dental or stomatal muscles have been described in some nematodes, these muscles extending from the teeth to the body wall. In other nematodes similar muscles attached to teeth are esophageal muscles. Muscles have also been mentioned in connection with the excretory pore, spinneret and male supplementary organs but they have never been adequately described.

2. FINER STRUCTURES OF THE MUSCLE CELL

Schneider (1866) first expressed the view that, unlike muscles in other animals, the muscle cells of nematodes have processes extending to the motor nerves rather than nerve processes extending to the muscles. Since this observation many papers have been written discussing this as well as other features of the somatic muscle cells. There has been general agreement regarding a few facts. The muscle cell, whether it be platymyarian, coelomyarian, or circomyarian, consists of (1) a sarcoplasmic part which contains the nucleus and from which there is a process extending to one of the median or submedian somatic nerves, the *innervation process* (Figs. 49A, D and H) (sometimes also secondary processes extending to adjacent muscle cells) and (2) a fasciate layer composed of (a) numerous ribbon-like "fibers" situated perpendicular to the cell surface and (b) sarcoplasm continuous with (1) and separating the individual "fibers." The fibers are often, if not generally, continuous from cell to cell so that in reality in some forms there are only eight functional, longitudinal muscles, each corresponding to a half sector (subdorsal, dorsolateral, ventrolateral, and subventral).

The detailed structure of the muscle cells of *Ascaris lumbricoides* and *Parascaris equorum* has been discussed extensively by Bütschli (1892), Apathy (1893, 1894), Rohde (1883, 1885, 1892), Goldschmidt (1904, 1909), Bilek (1909, 1910), Vejdovsky (1907), Cappe de Baillon (1911), Plenk (1924, 1925) and Roskin (1925), as well as by K. C. Schneider (1902) and Mueller (1929).

Briefly the views expressed have been as follows: Bütschli used the muscle cell to illustrate his alveolar or foam theory of protoplasm, each contractile element being described as formed of two rows of alveoli; there is no evidence to support this theory. Apathy observed fibrils extending from the hypodermis into the sarcoplasm of the muscle cell and forming an extensive system of neurofibrils between the contractile elements. He also conceived each "fiber" as being composed of a bundle of myofibrils. Schneider and Goldschmidt both differed with Apathy in his interpretation of the sarcoplasmic fibrils as being neuromotor in function; both authors considered these structures to be of a supporting nature. They agreed with Apathy, however, in thinking the contractile fibers contained myofibrils. Goldschmidt (1904, 1909) gave an elaborately detailed account of the fibrillar structure of the muscle cell, and like Schneider he considered the fibrils as supporting in nature. He observed both longitudinal and transverse fibrils between the fibers, these fibrils forming a network continuous with similar fibrils in the sarcoplasmic part of the cell (Fig. 49H). These "skeletal" fibrils were illustrated by Goldschmidt as radiating from the nucleus.

Cappe de Baillon showed for the first time that there are no myofibrils in the individual muscle fibers, but that these fibers constitute the actual contractile substance of the muscle cell; this observation appears to be correct in the light of more recent investigations. He was likewise correct in interpreting the transverse and longitudinal fibrils of the sarcoplasm as non-contractile or static fibers.

Plenk regarded the somatic musculature of *Ascaris* as striated muscle, designating the fibers as Q-stripes, the transverse fibrils as Z-stripes. As shown by Roskin, his observations are entirely incorrect and need no futher consideration here.

Roskin verified the observations of Cappe de Baillon in nearly all details. He showed by dissection that the fibril network is independent of the fibers as well as of the nucleus. He also found that the longitudinal fibrils (numbering seven to eight) between a given pair of fibers do not increase in diameter upon contraction but coil in a spiral (Fig. 49K). The fiber itself is composed of an elastic sheath and a homogeneous fluid material. Upon contraction numerous deeply staining "contraction swellings" are formed at intervals along the fiber (Fig. 49I-J).

Bibliography
(Musculature)

APÁTHY, S. 1893.—Ueber die Muskelfasern von *Ascaris*, nebst Bemerkungen über die von *Lumbricus* und *Hirudo*. Ztschr. Wiss. Mikr., v. 10(1): 36-73, pl. 3 (3): 310-361.

——— 1894.—Das leitende Element in den Muskelfasern von *Ascaris*. Arch. Mikr. Anat., v. 43(4):886-911, pl. 36.

BAILLON, P. CAPPE DE. 1911.—Étude sur les fibres musculaires d'*Ascaris*. I. Fibres pariétales. La Cellule, v. 27(1): 165-211, pls. 1-3, figs. 1-38.

BILEK, F. 1909.—Über die fibrillären Strukturen in den Muskel und Darmzellen der Ascariden. Ztschr. Wiss. Zool., v. 93:625-667, pls. 27-28, figs. 1-20.

——— 1910. Die Muskelzellen der grossen Ascaris-Arten. Anat. Anz., v. 37:67-78, figs. 1-10.

BUETSCHLI, O. 1873.—Giebt es Holomyarier? Ztschr. Wiss. Zool., v. 23(3):402-408, pl. 22, figs. 1-11.

——— 1892.—Ueber den feineren Bau der contractilen Substanz der Muskelzellen von *Ascaris*, nebst Bemerkungen über die Muskelzellen einiger anderer Wurmer. Festschr. Z. 70. Geburtst. R. Leuckarts pp. 328-336, pl. 34, figs. 1-9.

CHITWOOD, B. G. 1931.—A comparative histological study of certain nematodes. Ztschr. Morph., v. 23(1-2):237-284, figs. 1-23.

CHITWOOD, B. G., and CHITWOOD, M. B. 1934.—Somatic musculature in nematodes. Proc. Helm. Soc. Wash., v. 1(1): 9-10.

FILIPJEV, I. N. 1934.—The classification of the freeliving nematodes and their relation to the parasitic nematodes. Smithson. Misc. Coll. (3216), v. 89(6):63 pp., pls.

GOLDSCHMIDT, R. 1904a.—Der Chromidialapparat lebhaft funktionierender Gewebszellen. Biol. Centralbl., v. 24(7): 241-251, figs. 1-4.

——— 1904b.—Ibid. Histologische Untersuchungen am Nematoden. 2. Zool. Jahrb. Abt., Anat., v. 21(1):41-140, figs. A-Q, pls. 3-8, figs. 1-62.

——— 1909.—Das Skelett der Muskelzelle von *Ascaris* nebst Bemerkungen über den Chromidialapparat der Metazoenzelle. Arch. Zellforsch., v. 4(1):81-119, figs. a-c, pls. 6-9, figs. 1-19.

HARTMANN, M. 1925.—Allgemeine Biologie. I Teil. pp. 171-178. Jena.

LOOSS, A. 1905.—The anatomy and life history of *Agchylostoma duodenale*. Dub. A monograph. Rec. Egypt. Gov't School Med., v. 3: 1-158, pls. 1-9, figs. 1-100, pl. 10, photos 1-6.

MAGATH, T. B. 1919.—*Camallanus americanus* nov. spec. Trans. Am. Micr. Soc., v. 38(2):49-170, figs. A-Q, pls. 7-16, figs. 1-134.

MAN, J. G. DE. 1886.—Anatomische Untersuchungen über freilebende Nordsee-Nematoden. 82 pp., 13 pls.

MARTINI, E. 1903.—Ueber Furchung und Gastrulation bei *Cucullanus elegans*. Zed. Ztschr. Wiss. Zool., v. 74:501-556, illus., pls.

——— 1906.—Ueber Subcuticula und Seitenfelder einiger Nematoden. I. Ibid., v. 81:699-766, pls. 31-36.

——— 1907.—Ueber Konstanz histologischer Elemente bei erwachsenen Nematoden als Folge der determinierten Entwicklung. Sitzungsb. Naturf. Gesellsch. Rostock. J. 61 (8) :xxiii-xxvii.

——— 1908.—Die Konstanz histologischer Elemente bei Nematoden nach Abschluss der Entwickelungsperiode. Verhandl. Anat. Gesellsch., 22 J. v. 32:132-134.

1908.—Zur Anatomie der Gattung *Oxyuris* und zur Systematik der Nematoden. Zool. Anz., v. 32(19):551-559, 1 fig.

1909.—Ueber Subcuticula and Seitenfelder einiger Nematoden, IV, V. Ibid., v. 98:535-624, pls. 25-26.

MEULLER, J. F. 1929.—Studies on the microscopical anatomy and physiology of *Ascaris lumbricoides* and *Ascaris megalocephala*. Ztschr. Zellforsch., v. 8(3):361-403, pls. 9-13, figs. 1-80.

PLENK, H. 1924.—Nachweis von Querstreifung in der gesammten Muskulatur von *Ascaris*. (Not seen.)

1925.—Zur Histologie der Muskelfasern von *Ascaris, Lumbricus* und *Hirudo*. Verhandl Anat. Gesellsch. 34. Versamml, Wien, v. 60:273-275.

1926.—Beiträge zur Histologie der Muskelfasern von *Hirudo* und *Lumbricus*, nebst Berichtigung zu meinen Untersuchungen über den Bau der *Ascaris*—und Molluskenmuskelfasern. Ztschr. Mikr.-Anat. Forsch., v. 4:163-202, figs. 1-27.

ROHDE, E. 1883.—Beiträge zur Kenntniss der Nematoden. Diss. 26 pp. Breslau.

1885.—Beiträge zur Kenntniss der Anatomie der Nematoden. Zool. Beitr., Breslau, v.1(1): 11-32, pls. 2-6, figs. 1-35.

1892.—Gibt es Holomyarier? Sitz. K. Akad. Wiss. Berlin (35), 2:665-667.

1892.—Muskel und Nerf bei Nematoden. Sitzungsb. K. Akad. Wiss. Berlin, v. 28:515-526.

1893.—Muskel und Nerf. 1. *Ascaris*. 2. *Mermis* und *Amphioxus*. 3. *Gordius*. Zool. Beitr., Breslau, v. 3(1): 69-106, 6 pls., 161-192, 4 pls.

1894.—Apáthy als Reformator der Muskel-und Nervenlehre. Zool. Anz. (439), v. 17: 38-47, figs. 1-2.

ROSKIN, G. 1925.—Beiträge zur Kenntnis der glatten Muskelzellen. I. Die Muskelzelle von *Ascaris megalocephala*. Ztschr. Zellforsch., v. 2(5):766-782, figs. 1-9.

SCHNEIDER, A. 1860.—Ueber die Muskeln und Nerven der Nematoden. Arch., Physiol. u. Wiss. Med., pp. 224-242, pl. 5, figs. 1-12.

1863.—Neue Beiträge zur Anatomie und Morphologie der Nematoden. Ibid., pp. 1-25, pls. 1-2.

1864.—Ueber die Muskeln der Würmer und ihre Bedeutung für das System. Ibid., pp. 590-597.

1866.—Monographie der Nematoden. 357 pp., 28 pls., 343 figs. Berlin.

1869.—Noch ein Wort über Muskeln der Nematoden. Ztschr. Wiss. Zool., v. 19(2):284-286.

SCHNEIDER, K. C. 1902.—Lehrbuch der vergleichenden Histologie der Tiere, 988 pp. Jena.

TOERNQUIST, N. 1931.—Die Nematodenfamilien Cucullanidae und Camallanidae. Göteborgs K. Vetensk.—o.Vitterhets—Samh. Handl., 5. f., s. B, v. 2(3), 441 pp., pls. 1-17.

TUERK, F. 1903.—Ueber einige im Golfe von Neapel frei lebende Nematoden. Mitt. Zool. Stat. Neapel., v. 16:281-348, pls. 10-11.

VEJDOVSKY, F. 1907.—Neue Untersuchungen über die Reifung und Befruchtung. Königl. Gesellsch. Wiss. Prag.

VOLTZENLOGEL, E. 1902.—Untersuchungen über den anatomischen und histologischen Bau des Hinterendes von *Ascaris megalocephala* und *Ascaris lumbricoides*. Zool. Jahrb, Abt. Anat., v. 16:481-510, pls. 34-36, figs. 1-20.

B. BODY CAVITY, MEMBRANES AND COELOMOCYTES

The body cavity of nematodes has been compared to a segmentation cavity and to a coelome. As Rauther (1909) has pointed out, neither comparison is particularly apt. There is in all instances, some tissue surrounding at least part of the organs and separating them from the body cavity; the degree of development of this layer, termed "isolation tissue" by Goldschmidt (1906), varies both with the region of the body and the type of nematode studied. The tissue involved arises from mesenchymatous cells as do the other cells which lie free in the body cavity. Because of the mode of origin, the body cavity is here termed *pseudocoelome;* the membranes which cover and support the organs are termed *pseudocoelomic membranes* and *mesenteries;* the cells lying in the body cavity are termed *coelomocytes.*

Pseudocoelomic membranes and *mesenteries.*—These structures are best developed in ascarids, oxyurids, and other parasitic nematodes. In *Ascaris lumbricoides* we find a large cell situated dorsal to the esophagus between the levels of the nerve ring and the excretory sinus. From this cell a delicate spongy material extends anteriorly and posteriorly surrounding the esophagus and lining the body cavity. This pseudocoelomic membrane forms a delicate sheath over the internal surface of the muscle cells and extends between each pair of muscle cells to the hypodermis (Fig. 50D); it also covers the lateral and median chords. Mesenteries extend from the esophageal membrane to the membranes covering the musculature. The same material continues posteriorly covering the intestine and gonads. Thus far additional nuclei have not been observed in the posterior part of the body. Processes from stellate cells (see below) intermingle with the membrane and can scarcely be distinguished from it. Injection of the body cavity with *Escherichia coli* demonstrated very marked adhesiveness of the organisms to the membrane. An isolated large nucleus and cell body dorsal to the esophagus has been described in *Oxyuris equi, Cephalobellus papilliger, Camallanus* spp. and *Cucullanus* spp. The present writers can add that the occurrence of such a nucleus is practically universal in the Phasmidia and it is usually to be found dorsal to the base of the esophagus (*Rhabditis* [Fig. 3]). The same type of cell is also present in *Anonchus* and *Sabatieria* which are fairly representative of the aphasmidians of the order Chromadorida (indicating possibly, a general condition in this order), but in the Enoplida no such cell is known to occur.

In the Dioctophymatina and Trichuroidea membranes and mesenteries are relatively well developed, while in the remainder of the Enoplida they are poorly developed or possibly absent. The explanation of this observation can best be approached through comparison of the forms involved: In *Eustrongylides* (Fig. 50P) there are very thick membranes over the musculature and organs, these membranes being connected by a pair of discontinuous dorsolateral mesenteries. At regular intervals there is a pair of nuclei situated in the membranous covering of the intestine opposite the mesenteries. In *Trichuris* (Fig. 50N) we find in the esophageal region a similarly developed membranous layer which is often connected with the body wall by transverse mesenteries forming septa and causing a superficial appearance of metamerism. The mesenteries are not dorsolateral and not regular as in *Eustrongylides*. Regularly recurring pairs of lateral nuclei are present in the circum-esophageal membrane.

In *Hexamermis albicans* (Fig. 50B) Rauther has described two subdorsal rows of spheroid coelomocytes and in *Dorylaimus obtusicaudatus* the writers found two dorsolateral rows of similar coelomocytes (Fig. 50E). In neither instance have membranes been observed. Therefore, the obvious conclusion appears to be that coelomocytes and membrane cells are one and the same structure.

Martini (1916, 1925) and Hoeppli (1925) have described cells anterior to the nerve ring in oxyurids and ascarids which should properly be classified with the other types of coelomocytes and membrane cells. Three of these cells, the *fibril cells,* are adjacent to the esophagus and opposite the three esophageal radii, while three others, the *binding cells,* are also adjacent to the esophagus but are opposite the fibril cells, one being dorsal and the others ventrolateral (Fig. 46A-B).

Coelomocytes (Athrocytes, phagocytic cells, stellate cells, etc.).—Cells situated in the body cavity and apparently not a part of the general connective or isolation tissue are present in widely divergent groups and are probably universal. They have been described under various names indicating either their appearance or their function. Since they appear in all cases to be mesenchymatous in origin and therefore homologous, the term coelomocyte is preferred as a general term for all variations of such cells.

Bojanus (1821) first described four giant *stellate* cells in the anterior third of the body of *Parascaris equorum.* Similar cells were described in other ascarids by Jägerskiöld (1894, 1898, 1901) and Nassanov (1897-1900). Nassanov at first (1897) thought that the stellate cells were connected with the excretory system but this suggestion was later found to lack foundation. Shipley (1897) erroneously supposed that these cells were identical with the rectal glands. The subject was well summarized by Jägerskiöld (1898).

In general, it may be stated that there are two, four or six large cells situated in the body cavity of members of the Phasmidia. These cells may be ovoid, taeniaform, or highly branched (stellate); they may be ventral, lateral, or even dorsal in position.

In ascarids there are two or four always branched cells, there being four in such forms as *Ascaris lumbricoides* (Fig. 53A) and, according to Jägerskiöld, two in *Contracaecum osculatum*. Each cell is composed of a central mass in which there is a large nucleus (Fig. 53B) and numerous ray-like branches each bearing many small spherical "cellules" or "end organs." Nassanov was in doubt as to the significance of the terminal bodies, since with intravitam stains they often appear to contain nuclei. Nassanov found similar structures free in the body cavity and regarded them as leucocytes and compared the whole organ to a lymph gland. Metalnikoff (1923) pointed out that while distinct spherical bodies stained with methylene blue they were not nuclei. By killing and staining fresh material with acetic methyl green, and staining sections with hematoxylin the present writers definitely showed that the terminal bodies do not contain nuclei. However, they do contain numerous small globules which are dissolved by dehydration and clearing. These globules are neither glycogen nor fat. The globules are acid in reaction as shown by neutral red, neutral violet, brom-thymol blue and chlorphenol red. According to von Kemnitz (1912), urea and uric acid are not present. The cells show a very marked affinity for such stains as methylene blue and neutral red, extracting and concentrating the stain from the body fluid. Nassanov injected india ink, stains and *Bacillus anthracis* and found that such solid particles were phagocytized by the terminal swellings of the stellate cells. Metalnikoff (1923) repeated this work by injecting *Parascaris equorum* with tubercular bacilli, *Sarcina lutea* and *Micrococcus* from *Gulleria mellonella* and likewise found that such foreign materials were phagocytized. Mueller (1929) states that he found no evidence of actual phagocytic activity by any part of the stellate cells but only an agglomeration of injected materials on the surface of the terminal organs. Nassanov and Metalnikoff did not state whether they fixed and sectioned specimens after injections or examined them in toto. The present writers sectioned specimens of *Ascaris lumbricoides* fixed six hours after injection with india ink and *Escherichia coli*. In both cases the surfaces of the terminal organs were found to be covered with particles but though no evidence of actual phagocytosis of the particles of ink was observed (Fig. 53B) numerous terminal bodies were found containing bacilli (Fig. 53C).

Stefanski (1922) described four round cells as occurring in the body cavity of *Rhabditella axei* two of which were situated between the base of the esophagus and the anterior end of the ovary or testis, one at the blind end of the anterior gonad and one about ¾ of the body length from the anterior extremity; these cells contained large colorless vacuoles and were never found to respond to intravitam stains. The present writers have found the same condition to obtain with respect to these cells in *Rhabditis strongyloides* (Fig. 3).

Chitwood and Chitwood (1933) described two binucleated bodies in *Cephalobellus papilliger* which they called "X" bodies. They are rounded and unbranched, like the coelomocytes of *Rhabditis*. Four similar individual coelomocytes were observed in another thelastomatid, *Blatticola blattae*. In the male of this form, the first pair is situated near the anterior end of the testis and each cell is rounded (Fig. 53F) while the second pair (Figs. 52 and 53G) is near the second third of the body and each cell is slightly elongated. In the female the first pair is situated near the excretory pore and the second pair near the vulva and anus; all four cells are elongated, taeniaform (Fig. 53H). As in the case of *Ascaris lumbricoides*, the cells contain globules which give acid reactions with neutral red, neutral violet, brom-thymol blue, and chlorphenol red; they actively extract these dyes as well as methylene blue and crystal violet from the body fluid.

Nassanov (1900) described six "phagocytic organs" in *Strongylus* and Looss (1905) mentioned "strand-like organs" in the body cavity of *Ancylostoma duodenale*. These cells (Fig. 53D) are very similar to those of ascarids but contain refractive particles (Fig. 53E). Alicata (1935) found the branched cells of adult *Hyostrongylus rubidus* originate as small ovoid cells and that their positions indicate the sex of the larva (Figs. 53L-M).

In the Aphasmidia coelomocytes have been less commonly recognized. Schimewitsch (1899) first described such cells

in *Oncholaimus* sp. He found cells having strong affinity for methylene blue scattered in more or less distinct rows throughout the body cavity except in the vulvar region. It is uncertain, today, whether or not all the cells observed by Schimkewitsch were coelomocytes; some may have been hypodermal glands, but the one figured, ventral in position, is clearly recognizable as a coelomocyte. Jägerskiöld (1901) and Türk (1903) observed coelomocytes in various enoplids, such as *Enoplus*, *Thoracostoma*, and *Cylicolaimus*, and referred to them as "Fettzellen." Rauther (1909) described numerous branching "Fettzellen" which form a network over the intestine of *Enoplus communis* (Fig. 53J). As in the previous cases, these cells extract and concentrate methylene blue from the body fluid. The present writers have found, in *Metoncholaimus*

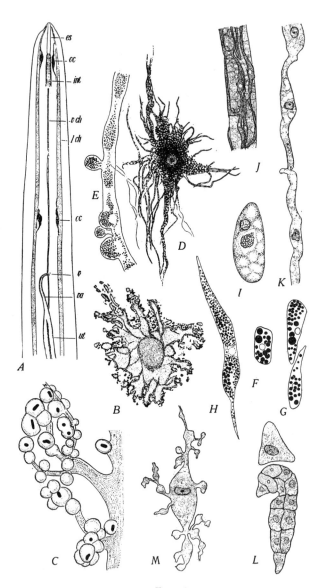

FIG. 53

Coelomocytes. A-C—*Ascaris lumbricoides*. (A—Dissected to show four stellate cells; B—Stellate cell of specimen injected with india ink; Later fixed and sectioned; C—Terminations of stellate cell of specimen injected with *Escherichia coli*, later fixed and sectioned. Note bacilli within terminal swellings). D-E—*Strongylus equinus*. (D—Stellate cell dissected out, black spots represent concentration of pigmented granules; E—Filament of stellate cell at greater magnification). F-H—*Blatticola blattae*. (F—Anterior coelomocyte of male showing vacuoles stained with neutral red; G—Posterior coelomocyte of male showing vacuoles stained with neutral red; H—Third coelomocyte of female showing vacuoles stained with methylene blue). I—*Metoncholaimus pristiuris*. (Coelomocyte as seen in longitudinal section, showing vacuole containing amorphus mass). J-K—*Enoplus communis*. (J—Intestine covered with anastomosing coelomocytes stained with methylene blue, K—Coelomocytes as seen in specimen, sectioned longitudinally). L-M—*Hyostrongylus rubidus*. (L—Male gonad and fourth coelomocyte in parasitic third stage larva; M—Coelomocyte in young adult female). J, after Rauther, 1909, Ergeb. u. Fortschr. Zool., modified; L-M, after Alicata, 1935, U. S. D. A., B. A. I. Tech. Bull.

pristiuris, a series of large circumscribed cells forming an irregular row near the ventral chord. Each cell (Fig. 53I) contains a moderately small nucleus, one or more large vacuoles, and several smaller vacuoles. The large vacuoles contain an amalgamation of clear, slightly yellowish, unstained bodies.

Meissner (1854, p. 230) described subspheroidal cells in the body cavity of *Hexamermis albicans,* stating that they were "Träger und Vermittler des Stoffwechsels" and contained fat droplets. Bugnion (1878) and Linstow (1900, p. 230) regarded them as blood corpuscles, but as Rauther (1906, p. 15) observed, they are fixed in position, being attached to the body wall by fine processes. They are highly vacuolate, rounded cells (Fig. 50B) concentrated for the most part in the region of the dorsal chord, but also present in the subventral regions especially near the posterior extremity.

As previously mentioned, Nassanov, in his several papers, referred to free migratory cells in the body cavity of ascarids and oxyurids; Bugnion (1878) and Linstow (1899) also refer to such cells, and Stefanski (1922) described several types of migratory cells in free-living nematodes. Such cells have not been mentioned by other workers. These observations may be explained in several ways. First, in such forms as ascarids the terminal swellings of stellate cells may be broken off. Second, fixed coelomocytes may be interpreted as free if the attachment processes are overlooked. Third, the appearance and disappearance of stained vacuoles in intravitam staining may give the impression of migratory cells. Fourth, the testis may be injured, thus setting free spermatozoa, or ameboid parasites may be present in the body cavity. Chitwood (1930) repeated Stefanski's work with negative results in *Rhabditis* and interpreted his observations as due to the progressive and regressive nature of the intravitam stains. Menjo (1934) found that there are no constant formed bodies in the body cavity of *Ascaris lumbricoides.*

As regards the possible functions of coelomocytes in nematodes, there are several possible interpretations. They may be considered as excretory cells which store the waste products in insoluble form, as phagocytic organs, as "blood purifiers," or as gland cells. The storage of excretory products is generally characterized by a salt formation with insoluble deposits such as crystals or amorphous pigmented masses. There is no evidence of products of such nature in the coelomocytes of ascarids, oxyurids, mermithids, or rhabditids, but granular pigmented bodies are present in the coelomocytes of *Strongylus* (Fig. 53E) and similar non-pigmented masses are present in the cells of *Metoncholaimus* (Fig. 53I). There is no evidence of the phagocytosis in any forms except ascarids and strongylids.

Evidence as to "blood purification," in the sense of actually absorbing and concentrating soluble materials, is unmistakable in ascarids, thelastomatids, and oncholaimids. Whether or not they store or digest soluble materials in normal life cannot be stated. Possible activity as gland cells in other senses, is likewise uncertain. The view that they might have the ability to break down glycogen very rapidly and thus contribute to the general metabolism was advanced by Kemnitz (1912), since after injecting ascarids with glycogen, he found none in the stellate cells.

In conclusion, it seems probable that since the coelomocytes of all nematodes are homologous, they have similar, if not identical, functions; the differences noted may easily be of degree rather than character. The writers would place them under the general heading of *fixed athrocytes,* as defined by Burian (1913). They are absorptive or phagocytic cells, comparable to the fixed histiocytes of vertebrates. As the body fluid flows by them it is assumed that they purify it in some manner.

Bibliography

(Connective Tissue, Coelomocytes, etc.)

ALICATA, J. E. 1935 [1936].—Early developmental stages of nematodes occurring in swine. U. S. Dept. Agric. Tech. Bull. No. 489, 96 pp., 30 figs.

BOJANUS, L. H. 1821.—Enthelminthica. Isis (Oken), Jena, v. 1(2):184-190.

BUGNION, E. 1878.—Note sur les globules sanguins du *Mermis aquatilis* Duj., etc. Act. Soc. Helv. Suisse, v. 60:247-255.

BURIAN, R. 1913.—Die Exkretion. Handb. vergl. Physiol., v. 2(2):257-443.

CHITWOOD, B. G. 1930.—Studies on some physiological functions and morphological characters of *Rhabditis* (Rhabditidae, Nematodes). J. Morph. & Physiol., v. 49(1):255-260, figs., pls.

CHITWOOD, B. G., and CHITWOOD, M. B. 1933.—The histological anatomy of *Cephalobellus papilliger* Cobb, 1920. Ztschr. Zellforsh., v. 19(2): 309-355, figs. 33-34.

FLURY, F. 1912.—Zur Chemie und Toxikologie der Ascariden. Arch. Exper. Path., v. 67:275-392.

GOLDSCHMIDT, R. 1906.—Mitteilungen zur Histologie von *Ascaris.* Zool. Anz., v. 29:719-737, figs. 1-13.

GOLOWIN, E. P. 1901.—Beobachtungen an Nematoden. 1. Phagocytäre Organe. 149 pp. 3 pls. Kasan (In Russian).

HOEPPLI, R. 1925.—Über das Vorderende der Ascariden. Ztschr. Zellforschr, v. 2(1):1-68.

JAEGERSKIOELD, L. A. 1894.—Beiträge zur Kenntnis der Nematoden. Zool. Jahrb., Abt. Anat., v. 7(3):449-532, pls. 24-28.

1898.—Über die büschelförmigen Organe bei der Ascaris Arten. Centralbl. Bakt., I Abt., v. 24(20): 737-741, 785-793, figs. 1-6.

1901.—Weitere Beiträge zur Kenntnis der Nematoden. K. Svenska Vetensk.-Akad. Handl. Stokholm, v. 35(2):1-80, figs. 1-8, pls. 1-6.

KEMNITZ, G. v. 1912.—Die Morphologie des Stoffwechsels bei *Ascaris lumbricoides* Arch. Zellforsch., v. 7:498-499, figs., pls.

KREIS, H. A. 1934.— Oncholaiminae. Filipjev, 1916. Eine Monographische Studie. Capita Zoologica., v. 4(5):1-270, figs. 1-135.

LINSTOW, O. VON. 1899.—Bericht über die wiss. Leistungen in der Naturgeschichte der Helminthen im Jahre 1893. Arch. Naturg., 60J., v. 2(3):230.

LOOSS, A. 1905.—The anatomy and life history of *Agchylostoma duodenale.* Dub. A monograph. Rec. Egypt. Govt. School Med., v. 3:158 pp., pls. 1-9, figs. 1-100, pl. 10, photos 1-6.

MARTINI, E. 1916.—Die Anatomie der *Oxyuris curvula.* Ztschr. Wiss. Zool., v. 116:137-534, figs. 1-121, pls. 6-20, figs. 1-269.

1926.—Zur Anatomie des Vorderendes von *Oxyuris robusta.* Arch. Schiffs.-u. Tropen-hyg., v. 30:491-503, figs. 1-9.

MEISSNER, G. 1853.—Beiträge zur Anatomie und Physiologie von *Mermis albicans.* Ztschr. Wiss. Zool., v. 5(⅔):207-284, pls. 11-15, figs. 1-55.

MENJO, J. 1934.—On microscopic ingredients of the coelomic fluid of *Ascaris.* Jap. with English summary. Keio Ig., v. 14(1):13-20. Not available. Abstract in Jap. J. Zool. v. 6(3):(58).

METALNIKOFF, S. I. 1897.—Sur les organes excreteurs de l'*Ascaris megalocephala.* Bull. Acad. Imp. Sc., St. Pétersb., 5 s., v. 7(5):473-480, figs. 1-7.

1923.—Les quatre phagocytes d'*Ascaris megalocephala* et leur rôle dans l'immunité. Ann. Inst. Pasteur, v. 37: 680-685, figs. 1-4.

MUELLER, J. F. 1929.—Studies on the microscopical anatomy and physiology of *Ascaris lumbricoides* and *Ascaris megalocephala.* Ztschr. Zellforsch., v. 8(3):395-396.

NASSONOV, N. 1897.—Sur les organes du systeme excreteur des Ascarides et des oxyurides. Zool. Anz., v. 20(533): 202-205, figs. 1-3.

1897.—Über Spengel's "Bemerkungen, etc.," in No. 536 des "Zoologischen Anzeiger's." Zool. Anz., v. 20 (543):412-415.

1897.—Sur les glandes lymphatiques des Ascarides. Zool. Anz., v. 20(548):524-530.

1898.—Sur les organe "terminaux" des cellules excréteurs de Mr. Hamann chez les ascarides. Zool. Anz., v. 21(550):48-50.

1898.—Sur les organes phagocytaires chez *Strongylus armatus.* Zool. Anz., v. 21(560):360-363, 1 fig.

1898.—Sur les organes phagocytaires des ascarides. Arch. Parasitol., v. 1(1):170-179, figs. 1-5.

1900.—Zur Kenntnis der phagocytären Organe bei den parasitischen Nematoden. Arch. Mikro. Anat., v. 55(4): 488-513, pls. 25-28.

RAUTHER, M. 1906.——Beiträge zur Kenntnis von *Mermis albicans* v. Sieb. Diss. Jena. 76 pp. pls. 1-3, figs. 1-25.
 1909.—Morphologie und Verwandtschaftsbeziehungen der Nematoden. Ergeb. u. Fortschr. zool., v. 1(3):491-596, figs. 1-21.

SCHIMKEWITSCH, W. 1899.—Ueber besondere Zellen in der Leibeshöhle der Nematoden. Biol. Centralbl., v. 19:407-410, figs. 1-2.

SHIPLEY, A. E. 1897.—Note on the excretory cells of Ascaridae. Zool. Anz., v. 20(541):342.

SPENGEL, J. W. 1897.—Bemerkungen zum Aufsatz von N. Nassonow uber die Excretionsorgane der Ascariden in No. 533 des "Zoologischen Anzeiger's" Zool. Anz., v. 20(536):245-248.

 1897.—Noch ein Wort über die Excretionszellen der Ascariden. Zool. Anz., v. 20(544):427-430.

STEFANSKI, W. 1922.—Excrétion chez les nématodes libres. Arch. Nauk. iBol. Towarz Nauk. Warszaw., v. 1(6):1-33, figs. 1-39.

STEWART, F. H. 1906.—The anatomy of *Oncholaimus vulgaris* Bast., with notes on two parasitic nematodes. Quart. J. Micro. Sc., Lond., n. s., (197), v. 50(1):101-150, figs. 1-9, pls. 7-9, figs. 1-40.

TUERK, F. 1903.—Über einige im Golfe von Neapel frei lebende Nematoden. Diss. Leipzig. Mitt. Zool. Stat. Neapel., v. 16. Reprint pp. 1-67, pls. 10-11.

CHAPTER V

CEPHALIC STRUCTURES AND STOMA

B. G. CHITWOOD

The various structures which go to make up the cephalic region of nematodes cannot be classified in a single category. Yet their study is naturally interlocked both in practical and developmental anatomy. Under the general heading "Cephalic structures" we shall discuss lips, pseudolabia, cephalic papillae, cephalic setae, amphids, probolae, collarettes, cordons and labial dentition while under the heading "Stoma" that part of the digestive tract between the oral opening and the anterior end of the esophagus will be considered. Of necessity, a discussion of cephalic structures must include parts of the nervous system, external cuticle and sometimes stomatal developments. Likewise, a discussion of the stoma overlaps to some extent both with the cephalic structures and the esophagus.

1. CEPHALIC STRUCTURES

Cephalic structures have been used, to a limited extent, as taxonomic characters since the appearance of Schneider's monograph (1866) which included *en face*, as well as lateral and medial views of the anterior extremity of many of the large nematodes. Such studies were extended by de Man (1886-1907) investigating free-living nemas and von Drasche (1883) working with parasitic nemas of Diesing and Molin's collections. Certain generalities came to be accepted as a result of the observations of Schneider and von Drasche. These were as follows: (1) That ascarids and heterakids have three lips, one dorsal and two subventral; (2) That spiruroids have two lateral "lips" and (3) That parasitic nemas generally have four submedian and two lateral cephalic papillae. The first two of these points are for the most part acceptable to us today but the third is no longer tenable. In parasitic nemas Looss (1902) introduced the use of cephalic structures in strongylid taxonomy causing them to be considered an integral part of generic and specific descriptions in this group but, apparently due to lack of interest or inadequacy of parasitological technic, little advance was made beyond Schneider, von Drasche, and Looss until very recently. In free-living nemas somewhat more steady progress has been made, partially attributable to the smaller size which makes critical study convenient and partially due to more widespread technical training. Though numerous workers have contributed to our knowledge of free living nemas, the chief impetus has come from the work of Cobb and Steiner. The glycerin jelly technic (see Cobb, 1920, and Chitwood and Wehr, 1934) developed in the laboratory under Cobb was introduced to the various visitors and associates; this technic made the study of nemas from en face practical.

Knowledge of the basic anatomy of the anterior end is due to the contributions of Goldschmidt (1903), Martini (1916), and Höeppli (1925). The confusion of two types of sensory organs, tactile structures (papillae) and chemoreceptors (amphids) in parasitic nemas caused much misunderstanding though Goldschmidt recognized the difference between the dorsal lateral organ (amphid) of *Ascaris lumbricoides* and the other sensory organs. The same differences both in the terminal sensilla and the internal nervous connection were brought out by Zur Strassen (1904), Looss (1905), and Martini (1916) in *Siphonolaimus* spp., *Ancylostoma duodenale* and *Oxyuris equi*, respectively, Zur Strassen even went so far as to state definitely that the dorsal lateral organ of *Ascaris* is the same as the circular lateral organ of *Siphonolaimus*. Other workers entirely ignored these observations until Cobb (1913) renamed the lateral organs *amphids* defining them as paired cephalic structures of specialized (unknown) function. It remained for the same author later (1923, 1924, 1928) to establish the general existence of pore like amphids in parasitic nemas through observation and constant reiteration that they are not "lateral papillae." Since then information has gradually accumulated showing their universal presence in the Nematoda. We shall not go into their internal anatomy at the present time since they are connected with the nervous system. It will be sufficient to note that each amphid is essentially a lateral or dorsolateral organ connected internally with the *lateroventral* commissure and with a gland (Fig. 3). Near the external orifice there is a dilation of the gland duct (amphidial pouch) in which nerve fibers terminate (the sensilla) (Fig. 8); the pouch is connected with the exterior either directly by an amphidial tube and pore or it opens into a pocket, circle or spiral external cuticular modification. In this part, only the external manifestation of the amphid (i. e., pore, pocket, spiral, etc.) will be considered.

In 1918 Filipjev introduced the external form of the amphids of free-living nemas as a family and subfamily character, later (1929, 1934 a, b) raising its use to suborders.

In the meantime information regarding the number and arrangement of cephalic sensory organs in both parasitic and free-living nemas was accumulating and Chitwood and Wehr (1932, 1934) brought out papers on the evolution and basic plan of cephalic structures with special reference to parasites while Stekhoven and de Coninck (1933), de Coninck (1935) and Stekhoven (1937) brought out similar papers with special reference to free living nemas. Differences in findings are for the most part matters of interpretation due to opposed schools of thought; the differences being in basic philosophy as to the evolution of nemas and not in the organisms themselves. The one school, represented by Filipjev, Stekhoven, and de Coninck regard polymyarian nemas as primitive and meromyarian nemas as neotenic while the other school, represented by Looss, Steiner and the writers, consider meromyarian nemas as primitive and polymyarian nemas as more highly evolved. The consequences are that each group sees the Nematoda from a separate point of vantage.

The basic plan of the anterior end appears to be six lips, two subdorsal, two lateral, and two subventral. On the summit of each lip there is a papilla, these six papillae constituting the internal circle and being known as internodorsals (id), internolaterals (il), and internoventrals (iv); situated more posteriorly on each of the submedial lips there are two papillae while on each of the lateral lips there is one papilla; these papillae constitute the external circle and have been named according to their position (Fig. 8a): dorsodorsals (dd), laterodorsals (ld), ventrolaterals (vl) (or externolaterals, el), lateroventrals (lv) and ventroventrals (vv). All of the members of the external circle are seldom exactly the same size or at exactly the same level. Stekhoven and de Coninck (1933) would therefore speak of them as constituting two circles and in some forms this in indeed the case. However, the papillae of the external circle are not always segregated in the same pattern. Thus the ventrolateral papillae tend to agree with the dorsodorsal and ventroventral papillae in their relative development in the Aphasmidia while they tend to agree with the laterodorsal and lateroventral papillae in the Rhabditina, Strongylina and Ascaridoidea. For that reason we regard the external circle as subdivisible into two papillary groups.

Fusion and reduction of cephalic papillae commonly modify the apparent cephalic arrangement but one can practically always recognize remnants of the original papillae and all cases may be explained in terms of the diagram presented (Fig. 8A).

As pointed out by Chitwood (1932) and the writers (1933) the cephalic papillary nerves are hexaradiately symmetrical and one would expect a hexaradiate symmetry to be basic for the papillae. Therefore, the external circle should consist of 12 papillae instead of 10. However, no rudiments of a dorsolateral pair are known except in some species of the Monhysteridae and Linhomoeidae. If these forms were the more primitive, one would expect to find rudiments of the aforementioned papillae in other groups and this is not the case.

The bilaterally symmetrical amphids are separately innervated and cannot be considered a part of the cephalic papillary symmetry. Unlike the papillary nerves, the amphidial nerves enter the nerve ring indirectly, through a commissure and their original position probably was posterior to the labial region as indicated by embryonic rhabditids and adult aphasmidians. Likewise, the amphidial orifice was probably larger and a bit like the plectoid amphid, if one is to interpret on the basis of embryonic rhabditids. As pointed out by the writers (1933) one cannot assume any existing form to represent the protonematode but if one combines characters of the genera *Rhabditis* and *Plectus* a common denominator of all nematodes is found. One cannot interpret aphasmidians entirely in terms of *Rhabditis* nor phasmidians in terms of *Plectus*, but the converse is moderately natural. Thus the amphids and papillae are basically labial in position in phasmidians while the amphids in aphasmidians are basically postlabial (a more primitive arrangement) and some of the papillae may be postlabial in position (a less primitive arrangement).

Regarding the basic number of lips, there are two choices. One may assume primitive triradiate symmetry in accordance with the symmetry of the esophagus as did Baylis and Daubney (1926) or a hexaradiate symmetry in accordance with the papillary arrangement as did the writers (1933). Since the lips are not formed from the esophageal primordium but from the cells of the anterior end (clavate cells of the papillary nerves and arcade lobes) the esophagus has nothing to do with them. The clavate cells are hexaradiately symmetrical while the arcade is bilaterally symmetrical with a gross triradiate and an actual 9-radiate symmetry (Fig. 46B). The actual 9-radiate symmetry is subdivisible into a triradiate and a hexaradiate formation rather than into three triradiate systems. We shall assume a hexaradiate symmetry as basic. Taking either extreme, six distinct lips or three distinct lips one finds repetitive series of transitions from the one to the other in the large groups. The lips themselves are subdivisible into two portions, the *apical part*, bearing the internal circle of papillae and the *basal part*, bearing the external circle of papillae and amphids in the Phasmidia and at least one subdivision of the external circle in the Aphasmidia. In some instances the two parts of a lip may be represented by separate lobes as in *Spironoura affine* and *Parascaris equorum* (Fig. 57) while in other instances, *Oxyuris equi*, *Metoncholaimus pristiurus* (Figs. 57 and 63) only the apical lobes may persist.

Original lips may totally disappear and be replaced by newly formed structures such as the pseudolabia of spiruroids, the probolae of cephalobids and the pseudonchia of filariids.

In the study of cephalic structures the student should be quite critical. It is not uncommon for two workers examining the same species to find greater differences than one worker would find examining representatives of two families. Lateral and medial views are often quite helpful but without an en face view they may be meaningless. It is due to this fact that the majority of older descriptions of the cephalic structures of parasitic nemas must be considered valueless. In examining en face views great care should be taken in focusing the microscope, and oil immersion is essential even in the study of the largest species.

A. PHASMIDIA

Phasmidians are similar to one another in the possession of simple external amphids, usually pore like and labial in position. The cephalic sensory organs are nearly always papilliform and in the most extreme cases are no more than setose papillae. In cases of reduction or fusion of the external circle it is always the dorsodorsals and ventroventrals that tend to disappear. The physiognomy of the various members of the class will be dealt with systematically.

RHABDITINA. Members of the suborder Rhabditina (Figs. 54-55) characteristically have six lips but three-lipped forms are quite common and many genera have no lips. The cephalic papillary arrangement is likewise diverse.

Rhabditidae.—Both six and three-lipped forms are common in the Rhabditidae and one finds every conceivable variation between the two. In *Rhabditis terricola* there are six large separate lips, an internal circle of six well developed papillae and an external circle of ten well developed papillae; the latter are not absolutely equal in form or size, the dorsodorsals and ventroventrals being similar as are the laterodorsals and lateroventrals while the ventrolaterals are more or less intermediate between the others. *Rhabditis lucanii* has three basally bilobed lips and the same number and arrangement of papillae but in this instance the ventrolaterals are small and like the dorsodorsals and ventroventrals. Other rhabditids vary between these types some having discrete, some confluent lips; in some the labial region is set off from the remainder of the body while in others this is not the case. In all instances the amphids are dorsolateral and labial in position and pore-like in character.

In *Diploscapter coronata* the lips have been entirely transformed into a pair of medial, outwardly acting, distally bifurcate fossores and a pair of lateral lamellae. Neither papillae nor amphids have been satisfactorily studied.

Cylindrocorpids have six discrete lips which may or may not be set off from the remainder of the body; there is an internal circle of six reduced papillae. In *Longibucca vivipara* and *L. lasiura* there is an external circle of four well developed papillae (laterodorsals and lateroventrals) and in the latter ventrolaterals are also present but reduced. In *Cylindrocorpus longistoma* there are six well developed (ld., vd., and lv.) and four reduced (dd. and vv.) papillae. The amphids are dorsolateral and labial in all forms.

Diplogasterids usually have no lips (*Diplogaster fictor*, *Mononchoides americanus* (Syn. *Diplogaster americanus*)) but instead they have a thin circum-oral membrane supported by longitudinal rugae which project anteriorly from the stoma. The number and degree of development of the rugae differs in the various species. In other genera inconspicuous remnants of six lips may be present (*Pristionchus*, *Rhabditidoides*, *Odontopharynx*). Among the forms thus far studied, *Pristionchus aerivora* is the only one known to show the full component of papillae (16); in this instance the dorsodorsal and ventroventral papillae are reduced while the others are well developed. In all other members of this subfamily studied the internal circle and the externomedial papillae are rudimentary or apparently absent. The remaining six papillae are often conically setose. Another peculiarity of the group is that the amphids vary from pore like and labial in position to slit or transversely elliptical (with similarities to *Plectus*).

The family Cephalobidae contains forms with six lips (*Panagrolaimus subelongatus*), three lips (*Cephalobus persegnis*, *Tricephalobus longicaudatus*) three simple (*Acrobeloides bütschlii*) or bifid labial probolae and sometimes six cephalic probolae (*Acrobeles complexus*). Labial probolae are cuticular labial structures which apparently replace the lips. They may be recognized by the fact that they are separated from the papillae bearing labial rudiments by a groove; cephalic probolae are known to occur only coincidentally with labial probolae, project anteriorly and often bear the cephalic papillae at their bases (*Acrobeles bodenheimeri*). Lips, on the contrary, bear papillae close to their apices. Another representative of this odd group (*Chambersiella rodens*) possesses six inwardly acting *odontia* internal to which there are six cirri (? cephalic probolae). When lips are present the full component of cephalic papillae (16) are often observable but in other instances they become difficult to recognize.

Of the remaining families of the Rhabditoidea the Steinernematidae (*Neoaplectana glaseri*) have a rounded oral opening,
no lips, and the full component of well developed papillae and
the Angiostomatidae (*Angiostoma plethodontis*) have three inconspicuous lips and the full component of papillae while the
Drilonematidae are devoid of lips and have only the external
circle of papillae, eight in *Dicelis nira* and four in *Ungella
secta*. Drilonematids may also have large hook-like teeth projecting from the stoma (*Ungella secta*).

Representatives of the Tylenchoidea have not been sufficiently
studied to make many statements regarding their cephalic
characters. As a rule the labial region is distinct, the lips
inconspicuous, six in number, and bearing four papillae and
the amphids. In a few forms such as *Anguina tritici* and
Rotylenchus similis an internal circle of six papillae have been
observed, these papillae being internal to the lips. In some
tylenchids such as *Hoplolaimus bradys* the labial region is
striated and supported by a sclerotized framework. In at
least one form (*Neotylenchus abulbosus*) there is a medial pair
of supplementary lips between the original six lips.

STRONGYLINA. *Strongyloidea.*—The majority of representatives of the superfamily Strongyloidea have no lips or they
are greatly reduced or rudimentary. In all instances the
amphids are dorsolateral or lateral and labial in position.

In the family Strongylidae lips are absent and in their place
one finds the *external corona radiata* or leaf crown (Fig. 56).
As we interpret these structures, they represent the divided
apical lobes of the original lips; they may number from six
to 40 or more. The labial region is generally set off from the
remainder of the body by a groove but it is not divided longitudinally as when large lips are present. Within this family
the internal circle is usually rudimentary or apparently absent
but it may be represented by reduced internolaterals (*Strongylus equinus*). The external circle is represented by four large,
often setose, submedian papillae each of which apparently consists of two original papillae that have fused (dd.—ld. and
vv.—lv.); ventrolaterals are rudimentary.

The closely related family Syngamidae supplies interesting
examples for comparison with the Strongylidae. In *Stephanurus dentatus* and *Syngamus trachea* the oral opening is sub-

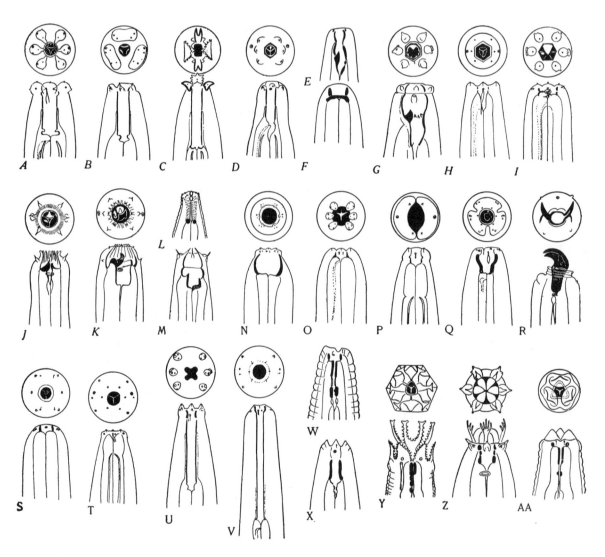

FIG. 54.

Cephalic regions in the Rhabditoidea. A-C—Rhabditidae. D-
M—Diplogasteridae. N-O—Rhabdiasidae. P—Strongyloididae.
Q—Angiostomatidae. R-S—Drilonematidae. T—Steinernematidae. U-V—Cylindrocorporidae W-AA—Cephalobidae. A—
Rhabditis terricola; B—*Rhabditis aspera*; C—*Diploscapter coronata*; D—*Rhabditidoides* sp.; E—*Acrostichus toledoi*; F—*Lycolaimus iheringi*; G—*Odontopharynx longicaudata*; H—*Alloionema appendiculatum* v. *dubia*; I—*Pristionchus aerivora*; J—
Diplogaster fictor; K—*Mononchoides americanus*; L—*Tylopharynx striata*; M—*Butlerius butleri*; N—*Entomelas entomelas*; O—*Rhabdias eustreptos*; P—*Strongyloides ransomi*; Q—
Angiostoma plethodontis; R—*Ungella secta*; S—*Dicelis nira*;
T—*Neoaplectana glaseri*; U—*Cylindrocorpus*; V—*Longibucca
lasiura*; W—*Cephalobus persegnis*; X—*Panagrolaimus sube-
longatus*; Y—*Acrobeles complexus*; Z—*Chambersiella rodens*;
AA—*Acrobeloides bütschlii*. E-F—After Rahm. 1929, Arch.
Inst. Biol. v. 2. G—After de Man, 1912, Zool. Jahrb. Abt. Syst.,
v. 33(6). H—After Chitwood and McIntosh, 1934, Proc. Helm.
Soc. Wash. v. 1(2). I—After Steiner, 1934, Proc. Helm. Soc.
Wash., v. 1(2). L-M—After Goodey, 1929, J. Helminth., v.
7(1). P—After Alicata, 1935, U. S. D. A. Tech. Bull. 489.
R—After Cobb, 1928, J. Wash. Acad. Sc., v. 18(7). S—After
Chitwood and Lucker, 1934, Proc. Helm. Soc. Wash., v. 1(2).
T—After Steiner, 1929, J. Wash. Acad. Sc. v. 19(19). U—
After Steiner, 1933, J. Parasit, v. 20(1). V—After McIntosh
and Chitwood, 1934, Parasit. v. 26(1). W-X—After Thorne,
1937, Proc. Helm. Soc. Wash. v. 4(1). Y and AA—After
Thorne, 1925, Tr. Am. Micr. Soc., v. 44(4). Z—After Cobb,
1920, Contrib. Sc. Nemat. 9. Remainder original.

hexagonal or one might say there are six rudimentary lips while in *Deletrocephalus demidiatus* there are six distinct lobes which might equally well be termed an external corona radiata or rudimentary labial lobes. In all three forms the internal circle is reduced but, nevertheless, distinct and the external circle consists of ten papillae. There is a distinct tendency toward fusion of papillae with coincident reduction in size of the dorsodorsals, ventrolaterals, and ventroventrals the median pairs of the external circle being nearly completely fused in *Stephanurus dentatus*, partially fused in *Syngamus trachea* and separate but approaching in pairs in *Deletrocephalus demidiatus*. As in the Strongylidae the labial region is usually set off by a groove but unlike the Strongylidae, the medial papillae of the external circle are never in the form of duplex setose papillae (Fig. 56).

The family Ancylostomatidae is characterized by the absence of both lips and a corona radiata; instead the oral opening is modified to the function of prostomatal teeth or cutting edges. As exemplified by *Necator americanus* the full component of papillae are represented, all of them being reduced with the exception of the laterodorsals and lateroventrals and one finds the medial pairs of the external circle closely approximate as in the Syngamidae. The labial region is not set off by a groove as in the previously mentioned families.

The family Diaphanocephalidae is characterized by a dorsoventrally elongate oral opening and is without lips, leaf crown, prostomatal teeth or cutting edges. The full component of papillae is represented, there being an internal circle of six reduced papillae and an external circle of four incompletely fused submedial and two simple ventrolateral (*Kalicephalus* sp.).

The family Cloacinidae is particularly noteworthy because of the presence of six massive lips, the laterals somewhat lower than the submedians. In *Zoniolaimus setifera* the internal circle is represented by reduced internolaterals and the external circle by four (? duplex) conoid papillae.

Trichostrongyloidea. Representatives of this superfamily often have a distinct cephalic inflation or cuticular helmets of numerous specialized forms which are used as generic characters. They are always devoid of a leaf crown and seldom show rudiments of either six or three lips. The oral opening may be of diverse form but is nearly always surrounded by an inconspicuous circumoral membrane. Representatives studied by the writers have an internal circle of six reduced papillae and an external circle of 10 simple papillae (medials

approaching in pairs), six medials partially or completely fused) or four (ventrolaterals apparently absent).

Metastrongyloidea. Members of this superfamily have neither the corona radiata of the Strongyloidea nor the cephalic inflation of the Trichostrongyloidea. Lips, if present, are much reduced except in *Metastrongylus* which has six massive lips, the largest of which are lateral. The oral opening is usually rounded and the labial rudiments, if present (*Filariopsis arator, Stenurus* sp.) set somewhat far back from the mouth and bear only the internal circle of papillae. The same tendency of papillary reduction and fusion observed in the Strongyloidea and Trichostrongyloidea follows also in this superfamily, medials of the external circle being smaller as are also the ventrolaterals (*Filariopsis, Stenurus, Dictyocaulus, Metastrongylus*).

ASCARIDINA. Members of the Ascaridina usually have three lips, one dorsal and two subventral (Fig. 57). While the internal circle of papillae is always reduced or rudimentary the two superfamilies differ as regards the external circle. In the Oxyuroidea the ventrolateral papillae are always rudimentary or absent while in the Ascaridoidea these papillae are well developed.

Oxyuroidea. In this superfamily the Thelastomatidae appear to be most primitive as regards cephalic papillae, for the external circle consists of eight quite separate papillae practically equal in size (*Leidynema appendiculatum, Protellina floridana, Aorurus philippinensis*). However, in this family lips are usually absent, there being a delicate circumoral membrane. In a few exceptional forms three reduced lips may be preserved (*Fontonema brachygaster*) and sometimes a lobing of the circumoral membrane may give the appearance of six reduced lips, two medial, four submedial (*Aorurus philippinensis*).

The family Oxyuridae appears to be a direct development of the Thelastomatidae in other structural characters but the fact that most members of this family retain three distinct lips (*Enterobius vermicularis*) and one form (*Oxyuris equi*) preserves the rudiments of six lips, seems to indicate that they must have originated rather early in thelastomatid phylogeny. Unlike thelastomatids, oxyurids have only four well developed papillae of the external circle. We interpret these as compound papillae formed by reduction of the mediomedials (dd. and vv.) and their complete fusion with the lateromedials (ld. and lv.). Secondary labial changes produce forms with two lateral lips by disappearance of the dorsal lip (*Macracis monhystera*) and others with four lips by division of the dorsal lip

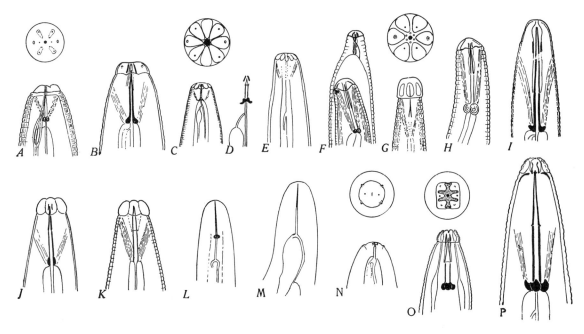

FIG. 55.

Cephalic region in the Tylenchoidae. A-K and O-P—Tylenchidae. M-N—Allantonematidae. A—*Anguina tritici*; B—*Ditylenchus dipsaci*; C-E—*Neotylenchus abulbosus* (C-D, female; E, male); F-H—*Rotylenchus similis* (F, female at last molt; G. upper, head of adult female and lower, adult male; H, adult female); I—*Paratylenchus macrophallos*; J—*Aphelenchoides parietinus*; K—*Aphelenchus avenae*; L—*Heterotylenchus aberrans* (adult free-living female); M—*Allantonema mirabile* (adult free-living female); N—*Chondronema passali*;

O—*Hoplolaimus bradys*; P—*Rotylenchus robustus*. A—After Steiner, 1925, Phytopath., v. 15(9). C-D—After Steiner, 1931, J. Wash. Acad. Sc., v. 21(21). E—After Steiner and Buhrer, 1932, J. Wash. Acad. Sc., v. 22(16). F-H—After Steiner and Buhrer, 1933, Ztschr. Parasitenk., v. 5(2). L-M—After Bovien, 1937, Some types of association between nematodes and insects. N—After Christie and Chitwood, 1931, J. Wash. Acad. Sc., v. 21(15). O—After Steiner and Le Hew, 1933, Zool. Anz. v. 101(9-10). Remainder original.

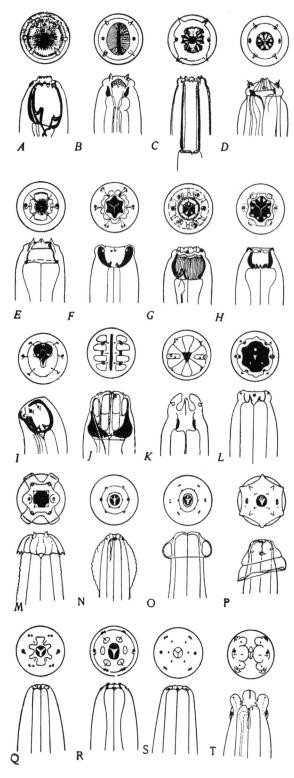

FIG. 56.

Cephalic regions in the Strongylina. A-E—Strongylidae. F-H—Syngamidae. I—Ancylostomatidae. J—Diaphanocephalidae. K—Cloacinidae. L-P—Trichostrongylidae. Q-R—Pseudaliidae. S-T—Metastrongylidae. A—*Strongylus equinus*; B—*Murshidia falcifera*; C—*Cylindropharynx rhodesiensis*; D—*Oesophagostomum dentatum*; E—*Cyclicocyclus insigne*. F—*Syngamus trachea*; G—*Deletrocephalus demidiatus*; H—*Stephanurus dentatus*; I—*Necator americanus*; J—*Kalicephalus* sp.; K—*Zoniolaimus setifera*; L—*Amidostomum cygni*; M—*Epomidiostomum uncinatum*; N—*Allintoshius nycticeius*; O—*Cheiropteronema globocephala*; P—*Tricholeiperia pearsei*; Q—*Filariopsis arator*; R—*Stenurus minor*; S—*Dictyocaulus filaria*; T—*Metastrongylus elongatus*. B—After Witenburg, 1925, Parasit. v. 17(3). C—After Yorke & Maplestone, 1926, Nematode parasites of vertebrates. L—After Wehr, 1933, J. Wash. Acad. Sc., v. 23(18): 391-396. M—After Wetzel, 1931, Proc. U. S. Nat. Mus. (2864) v. 78(21): 1-10. Q—After Wehr, 1935, J. Wash. Acad. Sc., v. 25(9). Remainder original.

(*Aspiculuris tetraptera*). Perhaps the strangest case of labial modification occurs between the closely related genera *Wellcomia* and *Syphacia*. In the latter genus there are the usual two subventral and one dorsal lip while in the former genus there is one ventral and two subdorsal. This absolute reversal in symmetry of the lips is not accompanied by reversal in other organs; the dorsodorsal papillae of *Wellcomia* are on the subdorsal lips, each of which has two compound papillae and an amphid in the usual symmetry; the ventral lip has no papillae.

The family Rhigonematidae is like the Oxyuridae in number of cephalic papillae and the subfamily Rhigonematinae contains forms with the common three lip symmetry (*Rhigonema infectum*). However, the other subfamily, Icthyocephalinae, presents a startling modification of symmetry. The head is divided forming two jaws and contrary to general opinion and to all other nematodes, the jaws are dorsal and ventral instead of lateral. The four compound papillae and lateral pore like amphids retain their normal positions not being modified by the change in symmetry.

Members of the Atractidae are the most diversified in cephalic characters of the whole suborder Ascaridina. Many of these forms are highly specialized and yet one must concede them a very ancient position in the Oxyuroidea very close to the Thelastomatidae. Like the oxyurids and rhigonematids they have only four compound papillae in the external circle and in this respect the thelastomatids should be more primitive. Six, three, and two lipped forms all occur in the Atractidae. The genus *Atractis* has six well developed lips not unlike *Rhabditis terricola*, while *Crossocephalus* has three lips like oxyurids. *Pulchrocephala* retains the three lips but has in addition cuticular projections from the labial region which may take innumerable forms but each element is grossly similar to an insect wing. *Heth*, on the other hand, has two lateral lips with corrugated edges and is provided with a spinate cephalic collarette, while in *Labidurus gulosus* the dorsal lip is replaced by a tuft like appendage. In the ransomnematids (*Heth*, *Pulchrocephala* etc.) the highly specialized or ornamental cephalic structures are confined to the female and do not make their appearance until the last molt.

Ascaridoidea. Members of the Ascaridoidea generally have three large conspicuous lips; the ventrolateral papillae and the other members of the external circle are all well developed. Throughout the entire group the medial pairs of the external circle are incompletely fused. One cannot assume ascaridoids, having the full component of papillae, arose from oxyuroids but neither can one assume the reverse for the entirely separate median pairs of thelastomatids could hardly have arisen from any known ascaridoid.

In cephalic papillary arrangement ascaridoids show practically no diversity but in labial developments diversity is marked.

Members of the Cosmocercidae (*Cosmocercoides dukae*), Heterakinae (*Heterakis gallinarum*) and Ascarididae (*Ascaris lumbricoides*) all have three large lips. In addition to the lips there may be posteriorly directed cuticular cordons (*Aspidodera* and *Heterocheilus*, Heterakinae and Anisakinae resp.) and between the lips there may be interlabia (*Porrocaecum* and *Parascaris*, Anisakinae and Ascaridinae resp.). The lips, themselves, may bear denticles on their internal surfaces (*Porrocaecum*, *Ascaris*); the apical lobes may be separated from the basal lobes by grooves (*Parascaris*); and the labial pulp may assume diverse forms which are considered specific (*Polydelphis quadricornis* and *P. boddaerti*). Members of the Subulurinae differ from other ascaridoids in that the lips are reduced to apical lobes bearing only the internal circle of papillae and there may be three (*Subulura distans*) six, or more apparent lobes. The grossly twelve lobed oral opening of *Aulonocephalus peramelis* is interpreted as having rudiments of six lips (the apical lobes) separated by six interlabia. Within the Kathlaniidae all manner of labial multiplicity is known, the genus *Spectatus* being characterized as having six lips, *Spironoura* as having three lips, *Kathlania* with about 16 labial divisions and *Cissophylus* with a bilaterally symmetric head. Of these only *Spironoura* and *Cissophylus* have been carefully studied. In *Spironoura* the lips are essentially ascaroid with the apical lobes separated from the basal by grooves. In *Cissophylus roseus* the dorsal lip is reduced, and transformed into a three pronged odontium while the subventral lips are massive and dentate.

CAMALLANINA. The suborder Camallanina (Fig. 58) differs from the Rhabditina, Strongylina and Ascaridoidea in that ventrolateral papillae are entirely unknown. Well developed lips are never present but rudiments of lips or lateral jaws may occur.

Camallanoidea. Most of the members of this superfamily have no lips but instead two lateral jaws. However, the genera *Omei* and *Haplonema* (Cucullanidae) as described by Hsü (1933) and the genus *Procamallanus* as described by Li (1935) preserve a less specialized condition. In *Omeia* six labial rudiments (apical lobes) are present, in *Haplonema* lips are absent and in *Procamallanus* the oral opening is hexagonal. The internal circle is in all instances reduced in size and the external circle represented by four papillae (duplex in Cucullanidae and simple in Camallanidae).

Dracunculoidea. Dracunculoids are devoid of both lips and jaws, the rounded oral opening being surrounded by a very thin circumoral membrane external to which there may (*Dracunculus, Avoisperpens*) or may not (*Philometra, Micropleura*) be a cuticularized circumoral elevation. The internal circle is well developed (a more primitive condition than in the Camallanoidea) and so also are the eight members of the external circle. In *Micropleura* and *Philometra* all of the papillae remain distinct while in *Dracunculus* and *Avioserpens* the medians of the external circle are partially fused. In *Dracunculus* the internodorsals and internoventrals fuse in the development of the female while the male retains the generalized condition. Members of the Dracunculidae also have a thickened cuticular helmet which projects anteriorly forming the circumoral elevation and posteriorly so as to surround the anterior end of the esophagus.

SPIRURINA. Like the Camallanina, this suborder contains no forms with ventrolateral cephalic papillae and true lips, if present, are represented only by rudimentary apical lobes. The first superfamily Spiruroidea shows a marked tendency toward the formation of false lips, *pseudolabia*, developed from the prorhabdions of the stoma while the second superfamily, Filarioidea, is characterized by the absence of both lips and pseudolabia. In their place there may be various types of labial structures. Within the entire suborder the internal circle of papillae is reduced, rudimentary or apparently absent.

Spiruroidea. The majority of spiruroids (Fig. 58) have two lateral pseudolabia but there is one exceptional group, the Thelaziidae. This group is apparently the most primitive of the superfamily and within it the development of pseudolabia is reproduced. The Thelaziinae, Spirocercinae, and Ascaropsinae contain forms with a rounded to hexagonal oral opening, the hexagonal form apparently corresponding to rudimentary apical lobes of six original lips. The internal circle of papillae is slightly reduced in all forms except *Physocephalus* in which it is rudimentary. The externodorsals and externoventrals are distinctly separate from the laterodorsals and lateroventrals in all members of the family but they are near the

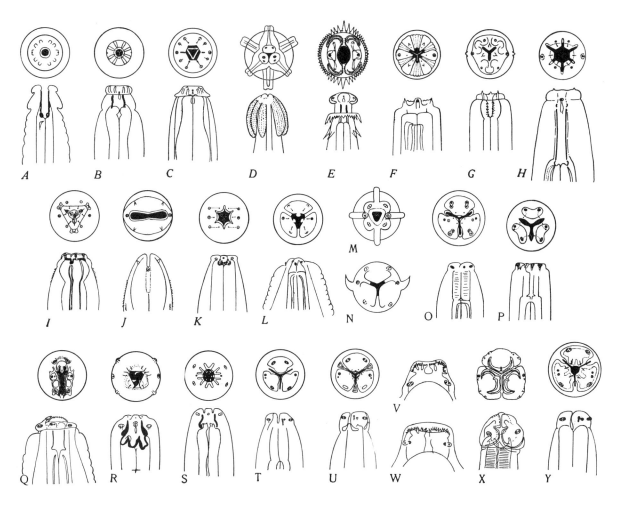

FIG. 57.

Cephalic regions in the Ascaridina. A-C—Thelastomatidae. D-H—Atractidae. I-J—Rhigonematidae. K-N—Oxyuridae. O—Cosmocercidae. P-Q—Kathlaniidae. R-T—Heterakidae. U-Y—Ascarididae. A—*Leidynema cranifera;* B—*Protrellina floridana;* C—*Aorurus philippinensis;* D—*Pulchrocephala* sp.; E—*Heth dimorphum;* F—*Atractis* sp.; G—*Crossocephalus viviparus;* H—*Probstmayria vivipara;* I—*Rhigonema infectum;* J—*Icthyocephalus* sp.; K—*Oxyuris equi;* L—*Enterobius vermicularis;* M—*Aspiculuris tetraptera;* N—*Dermatoxys veligera;* O—*Cosmocercoides dukae;* P—*Spironoura affine;* Q—*Cissophylus roseus;* R—*Subu-* lura distans; S—*Aulonocephalus peramelis;* T—*Heterakis gallinarum;* U—*Porrocaecum cheni;* V—*Polydelphis quadricornis;* W—*Polydelphis boddaerti;* X—*Parascaris equorum;* Y—*Ascaris lumbricoides.* A-B—After Chitwood, 1932, Ztschr. Parasit., v. 5(1); C—After Chitwood & Chitwood, 1934, Philipp. J. Sc., v. 52(4). S—After Baylis, 1930, Ann. & Mag. Nat. Hist. s. 10, v. 5. U—After Hsü, 1933, J. Parasit. v. 19(4). V-W—After Baylis, 1921, Parasit. v. 12(4). X—After Yorke & Maplestone, 1926, Nematode parasites of vertebrates. Remainder original.

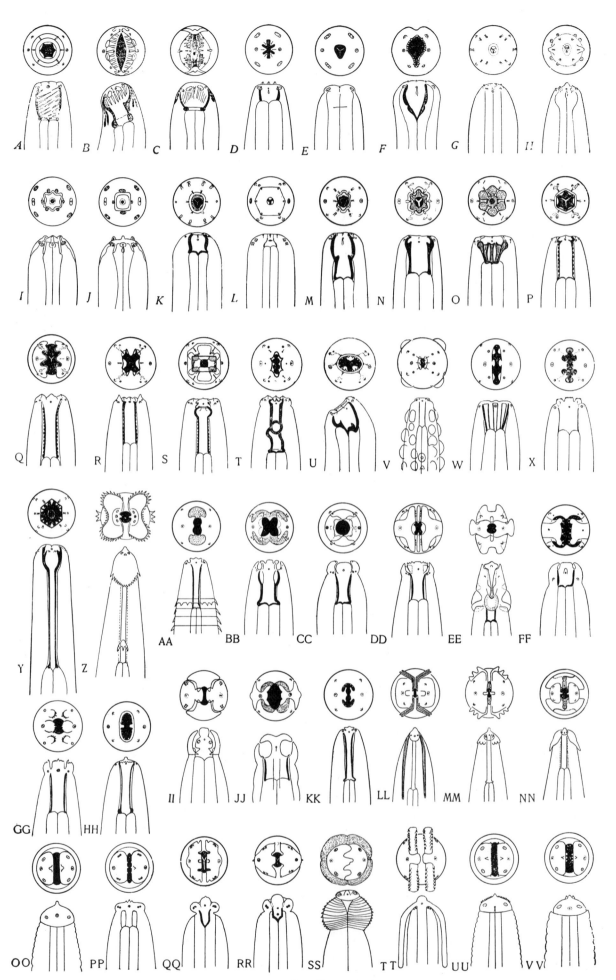

FIG. 58

62

size of the laterodorsals and lateroventrals in some forms, (*Thelazia californiensis, Pseudofilaria pertenue, Ascarops strongylina*) and are reduced or rudimentary in other forms (*Cylicospirura subaequalis, Spirocerca lupi*). In many of the species one notes six cuticular projections of the prostom (*Spirocerca lupi, Ascarops strongylina*). As shown by Chitwood and Wehr (1934) the third stage larva of *Physocephalus sexalatus* has both the six rudimentary labial lobes and six internal cuticular projections of adult *Ascarops*. The circumoral membrane (labial lobes) disappear in the adult and the original internal cuticular projections assume the form of paired trilobed lips which are termed pseudolabia. It is notable that these projections bear the internal circle of papillae but the papillae are rudimentary, not merely reduced as in other thelaziids. It is on the basis of these observations that the heads of other spiruroids are interpreted.

The family Spiruridae apparently contains the next most primitive representatives of the Spiruroidea. In these forms the pseudolabia are usually trilobed and with the exception of *Hedruris* they do not bear the external circle of papillae. In the Habronematinae (*Habronema*) and Hedrurinae (*Hedruris*) the median pairs of the external circle are very close together or partially fused, there being an accompanying reduction in the size of the dorsodorsals and ventroventrals while in the Tetramerinae (*Tetrameres americana*) and Spirurinae (*Protospirura* spp., *Mastophorus* spp.) there are four compound papillae due to more or less complete fusion. Labial structures are highly varied in this group and very valuable as generic and specific characters. Paired medial interlabia are present except in the Spirurinae and their shape, relative size and complexity make very useful taxonomic characters (*Tetrameres, Hedruris, Seurocyrnea*, etc). The pseudolabia are diverse in size, gross appearance and sometimes they have characteristic dentition. (*Mastophorus* vs. *Protospirura*). *Tetrameres americana* is anomalous in that the female has neither pseudolabia nor interlabia, a sexual dimorphism coinciding with the degeneration of the female to the form of a reproductive sac.

The family Acuariidae is interpreted as being most closely related to the Spirurinae of the family Spiruridae. As in the latter subfamily, there are only four well developed papillae, these being apparently the completely fused dorsodorsal-laterodorsals and ventroventral-lateroventrals. Unlike spirurids, the pseudolabia of acuariids are not trilobed and they bear the four papillae. Projecting posteriorly from the pseudolabia acuariids always have some type of cuticular ornamentation these ornaments taking the form of cordons in the Acuariinae, a spined cephalic collarette in the Seuratiinae, and four variously formed appendages in the Schistorophinae.

Passing now to the Gnathostomatinae we find that the pseudolabia have assumed a more massive size, become fleshy, but retained their lobed character. There are four double papillae in all forms, the degree of fusion varying in the different genera. Of this family the Spiroxyinae is undoubtedly the

most primitive for it contains such forms as *Spiroxys contorta* which superficially resemble *Protospirura* and *Mastophorus* of the Spirurinae. The subfamily Spiroxyinae differs from the Spirurinae in that the pseudolabia are massive, and bear the external circle of papillae in the former subfamily while they are inconspicuous and the papillae situated posterior to them in the latter subfamily. Gnathostomatids such as *Tanqua* have similar papillae but the pseudolabia are more irregularly lobed and posterior to them there is a large cephalic bulb formed by the anterior expansion of four internal posteriorly extending closed sacs, the ballonets (*Tanqua*). Neither the Ancyracanthinae nor the Spiroxyinae have a cephalic bulb but some authors have recorded ballonets in *Ancyracanthus*. The cephalic bulb apparently functions as a holdfast, being collapsed when the anterior end is inserted into the mucosa and thereafter being inflated. Spines or retrose annulation are provided to aid in this function. *Ancyracanthus* with four posteriorly directed cephalic appendages resembles *Schistorophus* of the Acuariidae but the incomplete fusion of the cephalic papillae and the lobed fleshy pseudolabia seem to definitely place it in the Gnathostomatidae.

The Physalopteridae apparently represent the final conclusion of evolutionary tendencies in the Spiruroidea. Here we find paired, massive fleshy, unlobed pseudolabia bearing both amphids and four completely fused compound papillae (laterodorsal-dorsodorsal and lateroventral-ventroventral). The various species of this family have been divided into genera by Schulz (1927) on the basis of their dentition. The genus *Physaloptera* is characterized by the presence of four teeth on the internolateral face of each pseudolabium, an internal group of three, two being sublateral and one lateral, and a single externolateral tooth. *Thubunaea* is similar but the teeth on one side are always rudimentary. *Abbreviata*, has the same lateral teeth but instead of sublaterals the entire margin of each pseudolabium is dentate and there are four double submedial teeth. Whether or not these teeth correspond to the original pseudolabial lobes is problematical but their development as labial structures seem to place them clearly in the category of *odontia* which statement also applies to the teeth of Protospirura, Mastophorus, Odontospirura, etc. of the family Spiruridae.

Filarioidea. Many filarioids have neither lips, pseudolabia, nor any other types of labial structures (Fig. 59). Such forms (*Dirofilaria, Dipetalonema, Elaeophora*) are placed in the family Dipetalonematidae. They are characterized by the absence of any structure which might be conceived to be of aid in feeding or penetration of tissue. The oral opening is rounded and bordered by a very delicate circumoral membrane. The majority of the remaining forms have some type of cephalic armature, such an armature sometimes taking the form of a circum-oral elevation which may bear lateral (*Dicheilonema*) or other anterior tooth like projections (pseudonchia), sometime taking the form of a sclerotized* helmet (*Squamofilaria*), sometimes having both circumoral elevation and helmet (*Dicheilonema*), and sometimes taking the form of lateral sclerotized tridents (*Diplotriaena*). These forms are all included in the family Filariidae. The remaining two families, Stephanofilariidae and Desmidocercidae each contain but one genus and may later be more closely associated with the other two families. The former family has one or two circles of cephalic spines while the later is devoid of external armature but possesses two internal cuticular projections of the prostom which may be homologous to the pseudolabia of spiruroids. In the number of cephalic papillae the superfamily Filarioidea is a remarkably constant group. There are always eight subequal large papillae which tend to take the form of two circles. However, we interpret these papillae as representing a subdivided external circle and the dorsodorsal and ventroventrals are usually anterior (*Dicheilonema, Dirofilaria*) to the laterodorsals and lateroventrals. The internal circle is apparently absent except in a few genera where it is represented by reduced internolaterals (*Dipetalonema, Litomosa*, and *Desmidocerca*). Many writers would interpret the four anterior papillae of filarioids as the internal circle but we cannot do this because in our comparative studies it is notable that throughout the entire Phasmidia there is a tendency toward reduction in papillary size. This tendency affects the internal circle first, and thereafter the externomedians. Furthermore, the rearrangement of the external circle into two circlets as a tendency

FIG. 58.

Cephalic regions in the Camallanoidea, Dracunculoidea and Spiruroidea. A–C—Camallanidae. D–F—Cucullanidae; G–H—Philometridae. I–J—Dracunculidae. K–V—Thelaziidae. W–KK (except Z)—Spiruridae. LL–NN & Z—Acuariidae; OO–PP & UU–VV—Physalopteridae. QQ–TT—Gnathostomatidae. A—*Procamallanus fulvidraconis*; B—*Camallanus sweeti*; C—*Camallanus microcephalus*; D—*Omeia hoepplii*; E—*Haplonema sinensis*; F—*Cucullanus truttae*; G—*Micropleura vivipara*; H—*Philometra rubra*; I—*Dracunculus medinensis*, male; J—*Dracunculus medinensis*, female; K—*Thelazia californiensis*; L—*Pseudofilaria pertenue*; M—*Oxyspirura mansoni*; N—*Spirocerca lupi*; O—*Cylicospirura subaequalis*; P—*Ascarops strongylina*; Q—*Physocephalus sexalatus*; R—*Simondsia paradoxa*; S—*Leiuris leptocephalus*; T—*Streptopharagus armatus*; U—*Rictularia coloradiensis*; V—*Gongylonema pulchrum*; W—*Spirura rytipleurites*; X—*Protospirura numidica*; Y—*Rhabdochona kidderi*; Z—*Stegophorus stellae-polaris*. AA—*Spinitectus carolini*; BB—*Habronema microstoma*; CC—*Draschia megastoma*; DD—*Seurocyrnea uncinipenis*; EE—*Parabronema indicum*; FF—*Odontospirura cetiopenis*; GG—*Mastophorus muris*; HH—*Cystidicola stigmatura*; II—*Hedruris* sp. JJ—*Tetrameres americana* (male); KK—*Ascarophis harwoodi*. LL—*Acuaria anthuris*; MM—*Yseria coronata*; NN—*Schistorophus cucullatus*; OO—*Skrjabinoptera phrynosoma*; PP—*Thubunaea leiolopismae*; QQ—*Spiroxys contorta*; RR—*Hartertia gallinarum*; SS—*Tanqua tiara*; TT—*Ancyracanthus pinnatifidus*; UU—*Abbreviata mordens*; VV—*Physaloptera maxillaris*. A—After Li, 1935, J. Parasit., v. 21(2). B—After Moorthy, 1937, J. Parasit., v. 23(3). D–E—After Hsü, 1933, Parasit., v. 24(4). L—After Sandground, 1935, Rev. Zool. Bot. Afr. v. 27(2). EE—After Baylis, 1921, Parasit., v. 13(1). FF—After Wehr, 1933, Proc. U. S. Nat. Mus., (2958) v. 87(17). Z. MM, NN—After Wehr, 1934. J. Wash. Acad. Sc., v. 24(8). Remaining figures based upon Chitwood & Wehr, 1934, Ztschr. Parasit. v. 7(3) and unpublished observations.

---

*The term sclerotized is used in the remainder of this text to indicate hardening without signifying the chemical composition. Eventually the chemistry of specialized cuticular structures will be discussed. Though many nemic structures superficially resemble chitin, this substance has been demonstrated only in the egg shell.

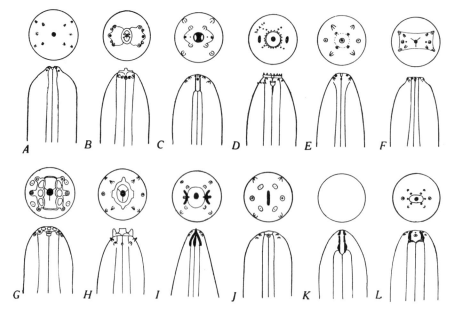

FIG. 59.

Cephalic regions in the Filarioidea. A, E-F, J-L—Dipetalonematidae. B, G-I—Filariidae. C—Desmidocercidae. D—Stephanofilariidae. A—*Dirofilaria immitis*; B—*Dicheilonema rheae*; C—*Desmidocerca numidica*; D—*Stephanofilaria stilesi*; E—*Elaeophora schneideri*; F—*Dipetalonema gracile*; G—*Squamofilaria thoracis*; H—*Setaria equina*; I—*Diplotriaena* sp.; J—*Cardianema cistudinis*; K—*Litomosoides hamletti*; L—*Litomosa americana*. E—After Wehr & Dikmans, 1935, Zool. Anz. v. 110(7-8). G—After Tubangui, 1934, Philipp. J. Sc., v. 55(2). L—After McIntosh & McIntosh, 1935, Proc. Helm. Soc. Wash. v. 2(1).

in the order Spirurida is noticeable in *Micropleura* (Dracunculoidea) and *Pseudofilaria* (Spiruroidea) in both of which there is the full component of papillae.

B. APHASMIDIA

Aphasmidians have externally modified amphids in all except the parasitic forms and even in these the modification persists in many of the Mermithoidea. Filipjev (1918, 1929, 1934) used the morphology of the amphids as one of the prime characters of major groups. More recently Stekhoven and de Coninck (1933) have reaffirmed such usage with modification. Of the many amphidial variants, there are three primary types in the Aphasmidia, these being the *spiral*, *circular*, and *cyathiform* (pocket-like). One easily recognizes transitions from *unispire* to *dispire* and *multispire*, and other series from unispire through *question mark*, to *shepherds crook*; these latter may be termed modified spiral amphids. Transition between unispire and circular is also an obvious step often indicated by a persistent break in the circle. Stekhoven and de Coninck (1933) further derive the *reniform* (transversely elongate) amphids of *Chromadora* from the circular amphids of *Microlaimus*. Interrelationship of these amphidial types seems scarcely questionable and this fact is used as a basis for the order Chromadorida.

The cyathiform type of amphid, characteristic of the order Enoplida, seems at first glance to be of an entirely different formation but as we shall see later, it also appears to have been derived from something close to the unispire.

Cephalic sensory organs in the Aphasmidia universally have one point in common, the lateral papillae of the external circle are externolateral rather than ventrolateral in position. This is correlated with the postlabial position of the amphids and is, perhaps, more primitive than the rhabditoid (and general phasmidian) arrangement. The size of cephalic setae presents another interesting field for observation; the external circle is always larger than the internal circle and whether the external circle is subdivided (*Plectus, Laimella*) or not (*Paracanthonchus, Anticoma, Theristus*) the elements are always of two sizes. If there are four large setae these are laterodorsal and lateroventral, but if there are six they are dorsodorsal, ventroventral and externolateral. The components of the internal circle are often papilla like and have been overlooked by many observers; the smaller members of the external circle may also be overlooked. It may be stated, however, that with

scarcely an exception the full component of cephalic papillae is present. Supplementary cephalic papillae or setae also occur and are very apt to cause confusion in the nomenclature. Two or more pairs of sublateral setae next to the amphid (paramphidial setae) are of the most common occurrence such being quite common in the Axonolaimoidea and Monhysteroidea. The only case of apparently true duplication of cephalic setae occurs in some monhysteroids (*Theristus*) in which the externolaterals are double. Mergence of somatic with cephalic setae (or papillae) is an unusual but not uncommon phenomenon. In such cases (*Eustrongylides, Mononchus, Metachromadora*) the sublateral or submedial somatic setae extend to the head region and become confused with the external circle of cephalic setae. Such added setae may become so numerous as to completely obscure the normal symmetry (*Steineria*). Fusion of cephalic sensory organs, so common in the Phasmidia, seems to be non-existent in the Aphasmidia.

Labial structures are entirely too diverse in this class for one to make satisfactory general statements. Both six and three-lipped forms occur in the two orders but six lips are definitely preponderant.

MONHYSTERINA. In cephalic sensory organs the suborder Monhysterina (Fig. 60) exhibits no real distinguishing character from Chromadorina, one can speak only of tendencies. The lips may be well developed, entire (*Plectus*), they may be represented only by the apical lobe (*Axonolaimus*) or they may be absent (*Sphaerolaimus*). In no instance are lips obviously replaced by cheilostomatal or prostomatal rugae.

Plectoidea. The Plectoidea is undoubtedly one of the most interesting groups of the entire Aphasmidia for it contains the potentialities of every structural diversity of the subclass. In tactile organs the great majority of the forms are uniform having an internal circle of six papillae and a subdivided external circle of six papillae (dd., vv. and el.) and four setae (ld. and lv.). Paramphidial setae are unknown in the group. The diversity of amphidial form in the group provides clues to the relationships of the whole Aphasmidia. In *Anonchus mirabilis* and *Plectus rhizophilus* one sees the typical unispire amphid, a double contour structure, each edge being the side of the groove. In *Aphanolaimus aquaticus* and *Camacolaimus prytherchi* there are nearly closed, unispire amphids of single contour but if one observes these en face, one sees that the central protuberance is present. These amphids are also unispire grooves. The broken circle (single contour) amphid of *Leptolaimus maximus* has not been studied en face but other species of the genus are known to have unispire amphids. *Anaplectus granulosus* (*Plectus granulosus*) is the final summation of all others for it combines features of the circular, unispire and cyathiform amphids. One might even term it a "universal amphid." At the surface it is transversely elliptical, but internally it is both unispire and cyathiform. One might argue that the plectoid amphid came to its formation through submergence of a spiral amphid or that the spiral amphid developed from the plectoid through emergence. In either case, this type must be considered a common denominator of the Aphasmidia.

De Coninck (1935) has recently placed the family Bastianiidae in approximate relation with plectoids on the basis of amphidial and male supplementary organ characters. In this group he places *Prismatolaimus, Tripyla, Trilobus* and *Bastiania*. Gross similarity in esophagi support his view but the writers cannot accept it. The three former genera seem best placed in the Enoplina (Tripylidae). De Coninck notes variation in *Bastiania parexilis* from unispire to transverse (cyathiform). The writers may add that in *Bastiania exilis* a single specimen had one unispire and one broken circle amphid. *Bastiania* and *Odontolaimus* are odd plectoids in having 10 cephalic setae, the external circle being partially subdivided. The oddity goes even further in that the six anterior setae (dd., el., vv.) are slightly longer than the four posterior (ld. and lv.). Thus we find the 6-6-4 symmetry of Plectoids remains but the size relationships in the two external subdivisions are reversed. The genera *Bastiania* and *Odontolaimus* seem best placed in the family Bastianiidae as an appendix to the Plectoidea.

Axonolaimoidea. In this superfamily one finds the same base symmetry as in the Plectoidea, namely an internal circle of six papillae, a subdivided external circle of six papillae or short setae (dd., el. vv.) and four long setae (ld. and lv.). In addition paramphidial setae are apt to be found just posterior to the cephalic setae. Such setae may be four (two pairs) in number, preamphidial (*Sabatieria longicaudata*) or postamphidial (*Laimella quadrisetosa*) or they may be eight in number (four pairs) (*Odontophora angustilaima, Axonolaimus subsimilis, A. odontophoroides*).

The family Comesomatidae is rather uniform in having multispiral amphids but the Axonolaimidae, are even more diverse than the Plectidae. *Araeolaimus* (*Araeolaimoides*) *zosterae Axonolaimus subsimilis* and *Odontophora* have rather distinct double contour unispire amphids, *Araeolaimus cylindrolaimus* has broken circle to single contour amphid and *Cylindrolaimus communis* a circular amphid. The amphids show numerous gradations in elongation from the unispire type in *Axonolaimus subsimilis* to the inverted U of *Axonolaimus spinosus* and shepherds crook, *Pseudolella granulifera*. As in the Plectoidea, each external amphid is a spiral groove, posterior closure of which leaves a central elevation. *Aegialoalaimus elegans* represents the sole instance of multispire amphids in the Axonolaimidae. However, the Diplopeltinae introduce still another variant, the presence of lateral shields accompanying the amphids (*Didelta*). In this group one may trace a repetition of the unispire—question mark—circular amphid development.

Monhysteroidea. Monhysteroids have great diversity in cephalic symmetry. The most common arrangement is an internal circle of six papillae and an undivided external circle of 10 or 12 setae. Stekhoven and de Coninck (1933) characterized the group as typically hexaradiate. The one unifying characteristic is that the amphids are nearly invariably circular.

The family Linhomoeidae includes some forms such as *Metalinhomoeus typicus* and *Desmolaimus zeelandicus* with a subdivided external circle of six papillae and four setae (ld. and lv.) or 6-6-4 symmetry; others such as *Paralinhomoeus lepturus, Monhysterium transitans* and *Halinema spinosum* with an undivided external circle of six short and four long setae or 6-(6-4) symmetry; and still others such as *Linhomoeus elongatus* with an undivided external circle of four short and six long setae or 6-(4-6) symmetry. There usually are, in addition, eight paramphidial setae. The genus *Sphaerolaimus* presents an array of setae that has not as yet been satisfactorily interpreted. One finds the normal internal circle of six papillae followed by a circle of six short setae (? dd., el. and vv.) followed by a circle of 16 setae, in eight pairs, two pairs sublateral, two submedial; in addition there are two pairs of preamphidial setae and somatic setae arranged in eight longitudinal rows (submedial and sublateral). One might, provisionally, assume that four papillae of the external circle (ld. and lv.) have been overlooked and somatic setae have

added to cephalic setae. On this basis the symmetry may be characterized as 6-10 (4?-6) + 16 (8-8) + 4 (2-2).

Members of the Monhysteridae sometimes have but six elongate setae (6-4-6) such as *Cytolaimium obtusicaudatum* but more commonly there are 10 or 12 setae in the external circle. *Halanonchus macramphidum* and *Theristus setosus* represent the more typical arrangement with six papillae in the internal circle, six short and six long setae in the external circle. Of the latter circle the longest six are the dorsodorsal, ventroventral and dorsolateral. Duplication of the externolateral results in a large dorsolateral and a small ventrolateral. Such symmetry may be characterized as 6-12 (6-6). Paired sublateral pre-amphidial setae are also present. *Scaptrella cincta* has a peculiar combination of axonolaimoid and monhysterid symmetry. The presence of 6 short setae in the internal circle and 12 setae in the external circle is characteristically monhysterid but the external circle consists of setae in three sizes; four are very long (ld. and lv.), 6 are moderately long (dd., vv., and dl.) and two are short (vl.). Therefore, the external circle might be characterized (2-6-4). *Omicronema litorium* and *Steineria* sp. represent types with increased cephalic setae. In the former instance three circles are described, an internal circle of six setae, an intermediate circle of four sublateral setae and an external circle of 18 setae in six groups of three, an arrangement which may be noted 6-4-18. It seems possible that a restudy of *Omicronema* will show it to be a sphaerolaim. *Steineria*, on the other hand, is typically monhysterid in character having an internal circle of six papillae and an external circle of 10 or 12 setae (according to the species). In addition to the cephalic setae, numerous somatic setae are grouped anteriorly in eight longitudinal rows, four submedial and four sublateral.

Siphonolaimus has an internal circle of six papillae, an external circle of 10 setae of which the four sublaterals (ld. and lv.) are the largest; this 6-10 (6+4) symmetry is more like *Paralinhomoeus* than any other forms discussed.

The circular amphid with central fleck is often mentioned as a characteristic of linhomoeids but the central fleck is neither confined to that group nor obvious in all members. As in plectoids, a central elevation represents the inner side of the amphidial groove, be it circular or spiral. The relative height between grooves determines the gross "presence" or "absence" of a central fleck.

Paramphidial setae of some type are nearly always present in monhysteroids; most commonly these are four in number, sublateral preamphidial in position. In addition to which there may be four sublateral postamphidial or four submedial preamphidial setae.

Six separate or three more or less lobed lips may be present in monhysteroids but if so, they are generally reduced, the labial elevation bearing faint longitudinal ridges; which are developments of the cheilorhabdions.

CHROMADORINA. Members of this suborder show all of the diverse symmetry arrangements (Fig. 61) of cephalic setae and nearly all of the amphidial forms present in monhysterins. True lips are seldom apparent but very highly developed cheilostomatal rugae usually take their place.

Chromadoroidea. The family Microlaimidae is characterized by unispire, postlabial amphids, an internal circle of six papillae and a subdivided external circle of six papillae or short setae and four long setae, 6-6-4 symmetry. True lips are rudimentary or absent, and 12 more or less projectable cheilostom rugae are present. Chromadorids differ from microlaims only in that the amphids are moved anteriad, usually to the level of the cephalic setae, and vary from unispire (*Odontonema*) to reniform (*Prochromadora*). The Cyatholaimidae, and Tripyloididae, differ from microlaimids and chromadorids in having an undivided external circle. In all forms studied by the writers there are four small and six large (dd., el., and vv.) papillae or setae in the external circle. De Coninck (1935) characterizes cyatholaimids as having two circles of six papillae and an external circle of ten setae. Since he specifies no form as exemplifying this condition we must judge by our own observations. In some cyatholaimids particularly members of the Choanolaiminae such as *Halichoanolaimus robustus*, the fiber trunk of each of the papillae of the internal circle shows in optical cross section much like a papilla but then it bends nearly at right angles over the stomatal cavity before reaching the true sensory terminus. Thus, by optical illusion one may see more papillae than exist. Rudimentary lips and heavy stomatal rugosities (usually 12) are conspicuous features of most members of the Cyatholaimidae (*Gammonema, Halichoanolaimus, Paracanthonchus* and *Pomponema*) while three well developed lips occur in members of the Tripyloidi-

65

dae. Both circular and unispire amphids are known in these families but multispire are the rule.

Desmodoroidea. Like microlaimids and chromadorids members of this superfamily have an internal circle of six papillae and a subdivided external circle of six papillae or short setae and four long setae (laterodorsal and lateroventral). *Monoposthia hexalata* and *Spirina parasitifera* are typical examples. Somatic setae are quite apt to become nearly cephalic in position as in *Metachromadora onyxoides* and *Croconema mammillatum* but by careful study one can usually segregate the two types. Helmet formation is often a conspicuous desmodoroid feature but this has been previously mentioned with the cuticle. Amphids in desmodoroids are primarily spiral, the unispire prevailing (*Epsilonema, Spirina*) but closed unispire (*Metachromadora onyxoides*), circular (*Monoposthia hexalata*), multispire (*Richtersia beauforti*) and elongate or shepherd's crook amphids (*Ceramonema*) also occur. True lips are rudimentary, sometimes indicated by six rudiments and stomatal rugae often replace them but these also disappear with reduction in stoma.

Desmoscolecoidea. The amphids of desmoscolecoids are a characteristic feature, usually being described as vesiculate. They are rather bubble like but internally one may distinguish evidences of a unispire character. Due to their small size the head of only one species, *Desmoscolex americanus,* has been studied en face. In this form there are six minute lips each bearing at least one (?two) papillae. Four large cephalic setae are known in all forms. Thus we may presume a possible desmodoroid relationship.

ENOPLINA. Members of the suborder Enoplina may have three, six or no lips, cephalic papillae or cephalic setae, but they are all similar in having cyathiform amphids.

Tripyloidea. Included in this group (Fig. 62) are many forms, which show relationships to other groups. The Monochidae, with six lips, an internal circle of six papillae and an external circle of ten papillae, seem clearly related to dorylaimoids. The family Tripylidae includes closely related forms with diverse symmetry. The genus *Tripyla* is characterized by three lips, an internal circle of six papillae and a subdivided external circle of six papillae or short setae and four papillae or long setae (ld. and lv.). *Trilobus longus* and *Prismatolaimus intermedius* have six small lips (apical lobes), an internal circle of six papillae and an external circle of 10 setae, four being short and six long (ld., el., and vv.). De Coninck places these forms in the Bastianiidae close to *Plectus* but we cannot agree with this placement on the basis of esophageal characters. Undoubtedly the Tripyloidea is the most primitive group of the order Enoplida, and hence most closely related to the Plectoidea but it is customary to place primitive groups with the forms that they gave rise to rather than with other primitive groups.

The family Ironidae includes forms which may (*Ironus, Ironella*) or may not (*Cryptonchus*) have cephalic setae. If such are present the external circle is subdivided with six anterior papillae or setae and four posterior setae (ld. and lv.). Lips may be moderately distinct but are usually rudimentary or absent.

Enoploidea. With the exception of the Oxystomininae the Enoploidea (Fig. 63) hold quite closely to a six-ten symmetry with an undivided external circle of which there are four small and six large setae (dd., el., and vv.). In the single exceptional group, the external circle is subdivided, with six large anterior and four large posterior setae (*Halalaimus caroliniensis* and *Oxystomina alpha*). Lips are seldom well developed in enoploids but one finds three massive apical lobes in *Enoploides* and *Enoplolaimus*, three small apical lobes in *Enoplus, Anticoma* and *Anoplostoma*. Six small apical lobes are general in the Oncholaiminae but even such lobes are not apparent in forms such as *Bolbella tenuidens* and *Enchelidium pauli*.

In most enoploids the amphids are typically cyathiform but in the Oxystomininae they are longitudinally elongate (*Oxystomina alpha, Halalaimus caroliniensis*). De Coninck (1936) considers the oxystomins related to the genus *Araeolaimus* and derives the amphids by elongation. This point requires more critical study. If the oxystomin amphid is a modified spiral, then it must be an open groove while if it is an elongate pocket, it is open only at its anterior end. Thus far such information has not been presented.

DORYLAIMINA. Like the Enoplina, the suborder Dorylaimina (Fig. 64) is characterized by cyathiform amphids but in parasitic members of this group the amphids may become externally pore-like. The cephalic papillae of dorylaimins, like those of monochids and phasmidians, never take the form of setae.

Dorylaimoidea. Members of this group for the most part, have six rather well developed lips; sometimes the lips are

set off from the remainder of the body as in mononchs and rhabditids. The full component of cephalic papillae are recognizable, there being an internal circle of six and an external circle of 10. Members of the latter circle are usually of two slightly different sizes and at two different levels on the lips. The large cyathiform amphids are situated just posterior to the labial region. Thorne (1935) described the amphids of diptherophorids as crescentic with reminiscences of spiral features.

Mermithoidea. Functional lips do not occur in mermithoids, the original lips being represented only by fiber tracts of the six papillary groups. These fiber tracts are (*Mesomermis bursata, Hexamermis albicans*) usually referred to as "papillae" in the literature. Careful study shows each of the four submedian "papillae" end in three tactile sensory organs (true papillae) while each of the laterals end in two; this corresponds exactly to the normal papillary symmetry with 16 papillae. The amphids may be associated with the lateral lip rudiment (*Hexamermis albicans*) or they may be quite separate from it (*Mesomermis bursata*). One finds all manner of amphidial types from large cyathiform as in *Mesomermis* to externally minute pore-like structures as in some *Hexamermis*. In addition, instances are known in which the amphids are joined dorsally by a fiber tract, a condition particularly common in the genus *Paramermis*. A shift of the oral opening ventrally is not uncommon in mermithoids (*Eumermis, Limnomermis*) and this migration is not accompanied by a shift of cephalic papillae, though the amphids may assume a more dorsal position.

Trichuroidea. Cephalic papillae and amphids have been studied in only one member of this superfamily, *Trichuris suis.* In this instance lips are absent, six papillae and externally pore like amphids were observed. The position of the lateral papillae (presumably of the external circle) in a true lateral position seems distinctly to indicate aphasmidian relationships.

DIOCTOPHYMATINA. Dioctophymatoids are well known for their cephalic symmetry or lack of it. Lips are absent in all members of the group. The oral opening in the Dioctophymatidae is surrounded by a cuticular circumoral membrane while in the Soboliphymatidae the body extends anteriad beyond the true oral opening forming an oral sucker. Presence of modified somatic muscle tissue and body cavity in this sucker is distinct evidence that the true oral opening should be regarded as the base of the sucker cavity; presumably, the cephalic papillae are on its internal surface.

"Cephalic papillae" are so numerous in members of the Dioctophymatidae (*Eustrongylides ignotus, Dioctophyma renale*) as to completely confuse one first observing them. The total number and arrangement varies within the species but certain papillae remain constant (Fig. 64). These latter are the true cephalic papillae; the others are considered as somatic papillae extending anteriorly from the lateral areas. There is an internal circle of six papillae and an external circle of four small (ld. and lv.) and six large papillae (dd., el. and vv.). This symmetry is the same as that to be found in a large part of the Dorylaimoidea and Enoploidea. It is in definite opposition to that found in phasmidians such as the Spiruroidea to which these nemas ars sometimes compared. The amphids of dioctophymatids are posterior to the externolateral papillae, sometimes appearing to be narrow, cyathiform, sometimes pore-like.

2. STOMA

The structure of the stoma has generally been used as a taxonomic character since the beginning of nematology. Its wide use for the classification of groups of all ranks makes a thorough consideration of its evolution necessary. That its morphology has great value can scarcely be doubted by any worker in the field but the weight that may be given to its gross form seems dubious. Since the method of feeding and character of the food itself is limited by the stoma and its armature, this organ is probably more directly influenced by environment than any other. A radical change of feeding habits of closely related forms would require change in stomatal morphology if forms are to survive. Likewise it is not at all inconceivable that quite unrelated nemas coming to feed in the same manner might eventually become grossly similar as regards stomatal morphology. For the above reasons one must consider stomatal morphology very closely.

Only one author, Cobb (1919), has attempted to use stomatal morphology as a major character. This author made the "presence" or "absence" of a stoma ("pharynx," buccal capsule etc.) the basis of dividing the "Nemates" into two classes, Laimia and Alaimia. This classification based upon a single character totally without correlation with any other organ has nothing in common with other classifications. Though their viewpoints may be diametrically opposed, the classifications of other writers all have considerable in common and they all are at complete variance with that of Cobb. The ''presence'' or ''absence'' of a stoma seems to be a rather bad point for classification purposes since all nematodes have something corresponding to the stomatal region.

Filipjev (1934) considered enoploids such as *Leptosomatum* which have no definite clear cut stoma as the more primitive while the writers regard forms such as *Rhabditis* and *Plectus* with distinct elongated stomata as primitive. Steiner (1933) proposed a nomenclature for the parts of the stoma which we shall follow as far as possible indicating homologous regions and structures and the apparent evolutionary trends and variations as we interpret them.

The cylindrical stoma of *Rhabditis* may be divided into three primary divisions (1) *cheilostom*—lip cavity, (2) *protostom*—cylindrical part of stoma and (3) *telostom*—end cavity, which in this form is often termed the glottoid apparatus. The corresponding walls of the stoma are termed cheilorhabdions, protorhabdions and telorhabdions, respectively. The walls of the protostom, i. e., protorhabdions, in *Rhabditis* are less distinctly subdivisible into three parts (a) prorhabdions, (b) mesorhabdions, (c) metarhabdions the corresponding regions being termed prostom, mesostom and metastom, respectively. In rhabditids the division between mesorhabdions and metarhabdions is scarcely visible in many species but manifests itself more prominently in related forms. The first subdivision, i. e., prorhabdions vs. meso-metarhabdions is quite obvious and has long been recognized as the ''mouth collar or mouth cuff.'' The basic parts of the stoma are apparently innate in the manner of their deposition and may best be observed in molting specimens or in specimens treated with reagents such as weak acids and alkalis. On the basis of studies made in this manner it would seem that the appearance of jointed protorhabdions in representatives of diverse groups is not convergence in the strict sense but rather manifestation of an innate characteristic which is ordinarily masked.

The stoma may take diverse shapes due to modifications of its parts and many different types of armature are developed according to the particular group involved. One of the most common types of armaments are teeth. Historically the term *onchia* has been most commonly applied to such structures though by derivation (*onkos* = hook, the barb of an arrow) its use does not seem apt. More recently the term *odontia* (*odous* = a tooth) has been applied and seems the more proper but in such case it should not be confused with the limited definition given by Cobb (1919). This author defined *odontia* as teeth arising by modification of the labial region while the term *onchia* was used by him to denote teeth arising more posteriad. Because of the common co-existence of both onchia and odontia, (in the Cobbian sense) one has need for two terms and the writers feel that though the first term is inapt, it is nevertheless worthy of preservation. We therefore retain Cobb's definitions.

Numerous other words are commonly used in a descriptive manner in specialized groups. Thus a region of the stoma or the margin of lips may be described as *dentate* (having teeth) or *denticulate* (having small teeth); basal onchia in the Strongyloidea are described as lancets; and the term *fossores* is used for outwardly acting teeth at the anterior extremity (often odontia). In some instances, as in the tylenchids, the stoma is transformed into a protrusible spear, termed a *stomatostyl,* while in others (dorylaimids) the stoma is to a greater or lesser extent filled by a large tooth, in which case the spear or stylet is termed an onchiostyl. Other specializations will be discussed with the various groups.

Convergence of stomatal formation accounts for the origin of stylets in four separate groups: the Tylenchoidea, of the Phasmidia, the Siphonolaimidae (Monhysteroidea), some representatives (*Anguinoides*) of the Camacolaimidae (Plectoidea) and the entire Dorylaimina. Similar convergence accounts for paired jaws formed essentially by the stomata rather than the lips in the Kalicephalidae (Strongyloidea), Icthyocephalidae (Oxyuroidea) and Camallanidae and Cucullanidae (Camallanoidea). One must be very hesitant in concluding relationships based upon such characters.

In describing the stoma it is the common practice to speak of certain parts or regions as being chitinized. As will be seen later, there is no real evidence that either the stoma or the denticular structures are actually chitin and we shall use the noncommittal term *sclerotized* for hardened refractive regions.

A. PHASMIDIA

In each of the large groups of the Phasmidia some forms exist that possess a cylindrical stoma very similar to that of *Rhabditis.*

RHABDITINA. This suborder (Fig. 54) is divided chiefly on the base of the stoma into two superfamilies, the Rhabditoidea in which the stoma is not transformed into a stylet and the Tylenchoidea in which such a transformation has taken place.

In the Rhabditoidea there are two families, Rhabditidae and Rhabdiasidae, in which the stoma is of a generalized structure consisting of cheilostom, protostom and telostom, the protostom being cylindrical, not surrounded by strong esophageal tissue. The parts are all well sclerotized and divisions of the protostom are not distinct. It is interesting that in the parasitic generation rhabdiasids have a relatively short (*Rhabdias*) or subglobular (*Entomelas*) stoma with well sclerotized walls showing no indications of cheilostom or telostom. This transformation takes place in the development of the individual after it enters the host.

The family Cylindrogasteridae (*Longibucca, Cylindrogaster*) is probably the next most closely related group and herein we again find a cylindrical stoma, distinct cheilorhabdions and telorhabdions (in form of small plates) and a greatly elongated protostom subdivisible only into pro- and meso-metastoms. Cephalobids, differ considerably in stomatal appearance, there being a more or less cylindrical stoma in *Panagrolaimus* and a collapsed stoma in *Acrobeloides* and *Cephalobus.* In all cases the stomatorhabdions are rather separate, giving the impression of a segmented stoma due to areas lacking in sclerotization; the extent of ''degeneration'' in stomatorhabdions is apparently correlated with the amount of esophageal musculature surrounding the stoma. With complete collapse of the stoma (*Daubaylia*) there is an entire absence of sclerotization and the base of the original stomatal region (telostom) is indicated only by a break in the esophageal musculature. The consequent ''stomatal region of the esophagus'' is termed a *vestibule.* Such a vestibule is all that remains in the related family Steinernematidae (*Neoaplectana*).

The Diplogasteridae is a highly variable group containing forms which link it with the Rhabditidae, Cephalobidae, Strongyloididae and Tylenchidae. Several series of genera are known in the Diplogasterinae. *Rhabditidoides* has a cylindrical protostom sclerotized as in *Rhabditis* but the cheilorhabdions are non-sclerotized and the telorhabdions asymmetrically developed; the closely related genus *Acrostichus* has distinct cheilorhabdions but the prostom and mesostom each form a distinct cavity, followed by a modified metastom containing a large dorsal tooth. *Odontopharynx* and *Butlerius* may be considered further members of this ''double stoma'' series. *Neodiplogaster* may be considered as a side branch of such a series originating from a form not unlike *Rhabditidoides* in which collapse of the stoma was followed by loss of sclerotization of stomatorhabdions with consequent convergence with cephalobids and steinernematids in the resulting vestibule. A second line of evolution seems to be indicated by a series from *Rhabditidoides* to *Mononchoides americanus, Diplogaster fictor, Pristionchus aerivora* and terminate with *Lycolaimus.* In this series there is collapse and non-sclerotization of the metarhabdions accompanied by shortening and thickening of the pro- and mesorhabdions, the development of a massive dorsal and a right subventral tooth and finally by complete amalgamation of these structures in *Lycolaimus.* The third and last line of evolution seems to antedate *Rhabditoides* in that the cheilorhabdions are preserved. *Tylopharynx* and *Tylenchodon* have a stoma that clearly simulates a stomatostyl, the stylet guide being formed by the cheilorhabdions, the anterior part of the stylet by prorhabdions, the shaft (or stylet) by meso-metarhabdions and basal knobs by telorhabdions. Such forms are retained in the Diplogasteridae since there is no proof that the ''stylets'' are protrusible. *Tylopharynx* and *Tylenchodon* form a definite link with the Tylenchidae. We have omitted remarking thus far on the presence of sclerotized rugae in the Diplogasterinae. As has been previously noted, members of this group have either no lips or much reduced lips. In many forms

---

*The root onch- seems to have been incorrectly derived from the Greek. However it is too deeply embedded in nematological literature to warrant changing. J. V. OWENS.

the prostom is longitudinally broken into numerous heavily sclerotized rugae, these rugae having much the same appearance as the internal leaf crown of strongylids. The degree of development of the rugae seems to be correlated with the size and degree of development of the dorsal tooth (*Pristionchus aerivora, Diplogaster fictor, Mononchoides americanus*). As to the origin and end of the series previously mentioned, there are some who would see the series in reverse, i. e., proceeding from short stoma, amalgamated forms such as *Lycolaimus* to cylindrical stoma forms such as *Rhabditidoides*. In reply we may state that study of young larvae show the series to be correctly oriented. Forms with a short or divided stoma in the adult stage have a more cylindrical, elongated, less divided stoma in the larval stage.

Returning to the other subfamily of the Diplogasteridae, the Alloionematinae we find less variation in stomatal development. In *Alloionema* and *Rhabditophanes* the cheilostom and prostom form the functional stoma and their rhabdions are well sclerotized but not always distinctly recognizable; the meso-metarhabdions may or may not be well sclerotized but they are always surrounded by esophageal tissue and the telorhabdions are rudimentary. *Strongyloides ransomi* (Strongyloididae) passes through this stage in its larval development preceding complete collapse of the mesostom with vestibule formation, *Seleneella* of the Alloionematinae apparently proceeds even further in this line of evolution with shortening and amalgamation of the cheilostom-prostom, the result being a nearly complete convergence with *Lycolaimus*.*

The remaining two families of the Rhabditoidea are insufficiently known for a general characterization of their stomata; we cite the figures of *Angiostoma plethodontis* (Angiostomatidae), *Dicelis nira* and *Ungella secta* (Drilonematidae) as representatives of the groups.

It has been previously noted that the Tylenchoidea (Fig. 55) have a stomatostyl homologous with the stoma of rhabditoids, probably having developed through some such forms as *Tylenchodon* or *Tylopharynx*. Such a stylet consists of four basic sclerotized parts (1) cheilorhabdions (stylet guide), (2) prorhabdions (conoid or insertable part of stylet), (3) meso-metarhabdions (stylet shaft) and (4) telorhabdions (stylet knobs). The stylet guide or cheilorhabdions are best developed in tylenchids with a sclerotized cephalic region such as *Hoplolaimus* and *Pratylenchus* but may be relatively distinct in less heavily sclerotized forms such as *Neotylenchus*. The basal knobs (telorhabdions) are primarily three in number but may be bilobed, particularly in cases where they are anteriorly bent such as *Hoplolaimus bradys*. In tylenchids the greatest diversity is in degree of development of the basal knobs and relative size of the stylet. A few forms such as *Aphelenchus avenae*, have no basal knobs and a very delicate stylet. Others such as *Aphelenchoides parietinus* have weak knobs and a delicate stylet while *Ditylenchus dipsaci* has both moderately developed. *Hoplolaimus bradys* represents a type with massive stylet while *Paratylenchus* and *Criconema* have a massive and greatly elongated stylet which may extend posteriorly into the metacorpus. A few instances are known in which there is sexual dimorphism in the degree of stylet development.

The entire family Allantonematidae is characterized by stylet degeneration, the cheilorhabdions and telorhabdions being most effected by this tendency. Stylet development in this family seems to be primarily dependent upon the life history, the stylet being best developed in very young larvae and degenerating or entirely disappearing after whatever stage has passed in which the organism enters its host animal. Thus Cobb (1928) showed that the stylet is vestigial in adult males while it is well developed in non-gravid females; the males and females copulate while free-living after which the male dies but the female re-enters its host. The stylet of the female also becomes degenerate after entrance into the host.*

STRONGYLINA. Stomata in the suborder Strongylina (Fig. 56) give one of the clearest cases exemplifying the *biogenetic* law that ontogeny recapitulates phylogeny. Though the structure in the adult stage is highly varied in the first stage larva it is always of a more or less rhabditoid form. In adult strongyloids the stomatorhabdions are always heavily sclerotized, the stoma well developed and capacious while in adult trichostrongyloids and metastrongyloids the stomatorhabdions are usually weakly sclerotized and the stoma reduced or rudimentary. Though the stoma becomes reduced or rudimentary in many representatives of the Strongylina, such reduction occurs through shortening rather than through overgrowth of esophageal tissue. Vestibule formation which occurs in so many other groups (Rhabditoidea, Ascaridoidea, Spiruroidea and Camallanoidea) is conspicuously absent in the Strongylina.

Strongyloidea. First stage strongyloids have a stoma identical with that of *Rhabditis*. This stoma gradually collapses in the second stage and in the third stage it may simulate a stomatostyl. Cobb (1923) described the third stage larva of *Necator americanus* as possessing a stylet but as shown by Stekhoven (1926) this was a false interpretation. The lumen of the stoma is merely partly closed and non-functional in this stage, the so called stylet not being protrusible. A similar appearance occurs in third stage *Rhabditis* larvae when they enter the resistant phase.

In the adult stage there are two genera of the Strongylidae with a rhabditoid stoma namely *Cylindropharynx* and *Pharyngostrongylus*. Interpreting on the basis of these two forms, the chief part of the stomatal wall, the so-called buccal capsule of strongyloids corresponds to the amalgamated protorhabdions of *Rhabditis*. The telorhabdions in some instances (*Cylindropharynx, Cylicocyclus*) may be represented by a transverse sclerotized basal plate but usually are not distinguishable. It has previously been noted that the external corona radiata seems to be homologous to the apical lobes of the original lips. The internal corona radiata appears to be a development of the cheilorhabdions (*Cylindropharynx, Strongylus*). Progressive shortening of the protorhabdions accounts for such forms as *Cylicocyclus, Murshidia* and *Oesophagostomum*. Thickening of the protorhabdions and dilation of the protostom account for *Strongylus* and its satellites. Teeth originating at the base of the stoma (lancets) are a common development of the Strongylidae, Syngamidae and Ancylostomatidae. Such onchia are considered products of the telorhabdions. In addition, one often notes a large dorsomedial tooth or a tube in the dorsal wall of the stoma; this tooth or tube is also thought to be a product of the telorhabdions though it may extend to the anterior end of the protostom (*Strongylus*) and is actually the duct of the dorsal esophageal gland. It is often termed the dorsal gutter.

Stomata in the Syngamidae are rather subglobular as in many of the Strongylidae but they differ in that the protorhabdions have six longitudinal thickenings, two medial and four sublateral (*Syngamus, Deletrocephalus, Stephanurus*). In the Diaphanocephalidae (*Kalicephalus*) the protorhabdions are split sagittally forming a pair of lateral jaws and they usually have four longitudinal thickenings on each side. In addition the telorhabdions form a thick, sclerotized basal plate.

In the Ancylostomatidae the stoma usually has a distinct dorsal bend and is asymmetrically developed as in *Strongylus* due to the large dorsal gutter. In addition there may be either teeth (*Ancylostoma*) or cutting edges developed from the cheilorhabdions on the ventral side.

Trichostrongyloids and metastrongyloids are both characterized by marked stomatal reduction. Usually there are no distinctly sclerotized stomatal structures. However in a few forms (*Amidostomum, Epomidiostomum-Trichostrongylidae* and *Stenurus-Pseudaliidae*) sclerotized protorhabdions persist to the adult stage. In the Trichostrongyloidea subventral lan-

---

*The Rhabditinae—Alloionematinae—Diplogasterinae complex seems to exhaust practically all of the possible combinations of stomatal, esophageal and bursal characters. Aside from tooth formation—absence of valved bulb and cephalic characters there is little or no correlation. One might easily prefer to arrange the narrow bursate cylindroid stoma, forms of the Rhabditidae (*Rhabditoides, Rhabditella*) and Diplogasterinae (*Rhabditidoides, Neodiplogaster*) together and the short stoma *Lycolaimus* with *Seleneella* in the Alloionematinae. However, on the basis of present evidence this would cause an unwarranted confusion and would be even more arbitrary than the present division.

*The anterior part of the stylet, as well as the cheilorhabdions are cast off with the exuvium at the last molt of *Ditylenchus*. It likewise differs in chemical nature. However, the extent of molting is no evidence of homology for the entire stomatal and esophageal lining is cast off at the last molt of *Camallanus, Ancylostoma* and *Agamermis* while apparently only the cheilorhabdions molt in *Rhabditis* and first stage *Ancylostoma* larvae. Recent observations indicate that the stylet is not developed in the first stage larvae of *Heterodera* and *Meloidogyne* spp. The first stage larvae molt within the egg shell, and this exuvium bears no evidence of the stylet. Whether or not this phenomenon is a general character of the suborder Tylenchina remains to be demonstrated.

cets are occasionally present (*Amidostomum*) and the dorsal esophageal gland usually empties into the stoma. In the latter instance its duct often takes the form of a sclerotized onchium which may be the only sclerotized part of the stoma (*Tricholeiperia, Haemonchus*). In metastrongyloids teeth of all forms are apparently absent (*Filariopsis, Stenurus, Dictyocaulus* and *Metastrongylus*). It is interesting to note that trichostrongyloids revert to their ancestral stomatal form, *Rhabditis*, in the first stage larva while metastrongyloids only partially revert in that stage for they have distinct cheilorhabdions and prorhabdions of rhabditoid form but though the mesostom and telostom are recognizable the rhabdions are non-sclerotized.

ASCARIDINA. Most ascaridins (Fig. 57) either have a rudimentary stoma or a weakly sclerotized vestibular region but some representatives of the more ancient families, Thelastomatidae and Atractidae preserve a cylindrical rhabditoid stoma. Thus *Leidynema cranifera* and *Probstmayria vivipara* both have stomata in which the various stomatorhabdions are distinctly sclerotized. In *Protrellina* and *Aorurus* of the Thelastomatidae one notes progressive shortening of the stomatorhabdions and loss of sclerotization. The telorhabdions may persist in the form of basal teeth or laminae or they may entirely disappear. The dorsal esophageal gland never empties into the stoma as in the Strongyloidea. In the Atractidae a cylindrical protostom (*Heth, Probstmayria*), weakly sclerotized stoma with collapsed mesostom (*Atractis*) and vestibule (*Crossocephalus*) are all known to occur. In the Oxyuridae the protostom is always greatly shortened, often feebly sclerotized (*Enterobius*) but the telorhabdions are commonly large and conspicuous laminae (*Oxyuris*). The family Rhigonematidae is characterized by a rudimentary stoma surrounded by esophageal tissue (vestibule). The stomatorhabdions are non-sclerotized, with the exception of the cheilorhabdions of *Rhigonema;* these latter take the form of three sclerotized dentate jaws, internal to the lips. In *Icthyocephalus* the stomatal region of the esophagus is horizontally split forming paired jaws.

The subfamily Subulurinae of the Ascaridoidea is the only group of that superfamily in which the stomatal region is not surrounded by esophageal tissue. Herein the short heavily sclerotized protostom is followed by a dentate telostom (*Subulura distans, Aulonocephalus peramelis*) which is strongly reminiscent of *Oxyuris* and *Enterobius*. The other families have weakly or non-sclerotized protorhabdions, the stoma collapsed and of subtriangular or triradiate form in cross section. In the Cosmocercidae, Kathlaniidae and Heterakinae the base of the vestibule (stomatal region of esophagus) is evidenced by a break in esophageal tissue at the original position of the telorhabdions. In the Ascarididae, with the exception of *Crossophorus* there is not the slightest evidence of the original stoma. The esophagus seems to extend uninterrupted to the base of the lips. The single exceptional genus gives the final proof that the anterior end of the esophagus of ascarids is homologous to the vestibule of cosmocercids and heterakids for in *Crossophorus* there is not only a distinct vestibule but the metastom is dilated and distinct telorhabdions are visible.

CAMALLANINA. The first superfamily Camallanoidea, is characterized by the presence of a well developed stoma in most forms and at least a distinct vestibule in the remainder while the second superfamily, Dracunculoidea, has a rudimentary stoma, the stomatorhabdions are non-sclerotized in all forms (Fig. 58). None of the adults of the Camallanoidea have that which might be termed a rhabditoid stoma but *Procamallanus* most closely approaches it. In this form the stoma is barrel-shaped, cheilorhabdions are not distinct, protorhabdions amalgamated and heavily sclerotized and followed by a transverse ring-shaped telorhabdion. Li (1935) has shown that as in other groups the larva more closely approaches the rhabditoid stoma than does the adult for in the first stage larva the stoma is much more narrow and cylindrical. In adults of other genera of the Camallanidae (*Camallanus sweeti, C. americanus*, etc.) the prostom is sagittally slit, forming two lateral jaws; longitudinal ridges of the internal wall of the protostom make their appearance and paired, sclerotized medial tridents are formed at the external surface of the stoma. Somatic muscles are attached to both the tridents and the exterior surfaces of the jaws.

In the Cucullanidae one observes, though not as completely, a repetition of the evolution in the Camallanidae. *Omeia* is the only form which retains the primitive, non-esophageal tissue surrounded stoma but even in this case there is little resemblance to the cylindrical stoma. *Haplonema* and *Seuratum*

appear to be products of some genus like *Omeia* in which the stoma collapsed, stomatorhabdions degenerated and were covered with esophageal tissue forming a vestibule. *Cucullanus* may be interpreted as more ancient than *Omeia* in distinct retention of pro- and meso-metarhabdions but less primitive in that the entire stoma is surrounded by esophageal tissue and sagittally divided forming lateral jaws. This is likewise one of the very few exceptional cases wherein esophageal tissue surrounding the stoma is not correlated with stomatorhabdion degeneration.

SPIRURINA. Most members of the Spiruroidea (Fig. 58) have a rather cylindrical stoma with strongly sclerotized protorhabdions, but distinct cheilorhabdions are unknown. The stomata of practically all forms are specialized to some extent and none can be regarded as prototypes of the superfamily. However, various members of the Thelaziidae (*Oxyspirura, Ascarops, Spirocerca* and *Rhabdochona*) indicate that a cylindrical protostom subdivisible into pro- and mesostomata and reduced telostom with plate-like telorhabdions were characteristic of the ancestor. One cannot but be struck by the similarity of the stomata of *Longibucca* and *Cylindrogaster* to spiruroids both in this respect and in the tendency toward bilaterality in cephalic structures. Nevertheless there is too wide a gap between the Cylindrogasteridae (Fig. 54) and Spirurida for one to assume relationships at the present time.

In a few forms of the Thelaziidae (*Thelazia, Pseudofilaria*) the protorhabdions are shortened and amalgamated but in the majority (*Oxyspirura, Ascarops*, etc.) the protostom is elongated and there are six onchia at the junction of protostom and mesostom. These onchia may take varied forms, sometimes rounded (*Ascarops*) and sometimes bi- or trifurcate (*Cylicospirura*); in still other instances they may be dentate (*Leiuris*) and sometimes opposed by medial plates (*Simondsia, Leiuris*). Spiral or transverse rugosities of the protostom are confined to the subfamily Ascaropsinae. Similar rugosities are known only in the genus *Pharyngostrongylus* of the Strongylidae. Basal onchia (lancets) are known only in *Rictularia*, presenting remarkable convergence with the Strongylidae. Careful consideration of the stomatal formation in thelaziids is essential to an understanding of the stomata of other spiruroids. The larva of *Physocephalus* has six prostomatal onchia and no lips like the adult *Ascarops*. During later development the circumoral membrane is apparently lost at the same time the prostom is everted. This results in the six prostomatal teeth coming to occupy an external position and they form the basis of two lateral trilobed pseudolabia. The writers interpret the stomata of other families of spiruroids as of the everted type. Later developments of the prorhabdions (pseudolabia) have been discussed with other labial structures.

Members of the Spiruridae tend to have a rather wide, cylindrical well sclerotized mesostom which may become laterally flattened in some genera (*Spirura, Protospirura*) but extreme development of pseudolabia and interlabia may obscure the stoma (*Tetrameres* male and *Hedruris*). Acuariids are rather uniform in the possession of a long narrow cylindrical mesostom.

Passing to the Gnathostomatidae and Physalopteridae we see the first and only tendency toward vestibule formation in the Spiruroidea. The genera *Spiroxys* and *Hartertia* (Gnathostomatidae) are the only representatives which retain a sclerotized mesostom and in these forms the stoma is much shortened and surrounded by esophageal tissue. *Thubunaea* (Physalopteridae) retains a laterally compressed vestibule while *Physaloptera, Abbreviata, Skrjabinoptera* (Physalopteridae), *Tanqua* and *Gnathostoma* (Gnathostomatidae) have completely rudimentary stomata, the esophageal tissue proceeding uninterrupted to the base of the pseudolabia. At the anterior end, in such forms, the esophageal lumen becomes dorsoventral before connecting with the labial bases.

The Filarioidea (Fig. 59) might be considered the "astomatous" twin of the Spiruroidea. A few genera, however, are known to have rather distinct, sclerotized stomata. It is interesting to note that such forms are not dissimilar to thelaziids. *Desmidocerca* is a striking counterpart for it not only has a cylindrical stoma but also a pair of lateral prostomatal onchia which may be homologous to those of the spiruroids. *Litomosa* has a short stoma practically identical with that of *Thelazia* while *Litomosoides* has a cylindrical stoma with separate distinct stomato-rhabdions. Even in such forms as *Dirofilaria immitis* one must assume a cylindrical stoma in the not too dim ancestry because such a stoma, although weakly sclerotized, is present in the third stage larva.

69

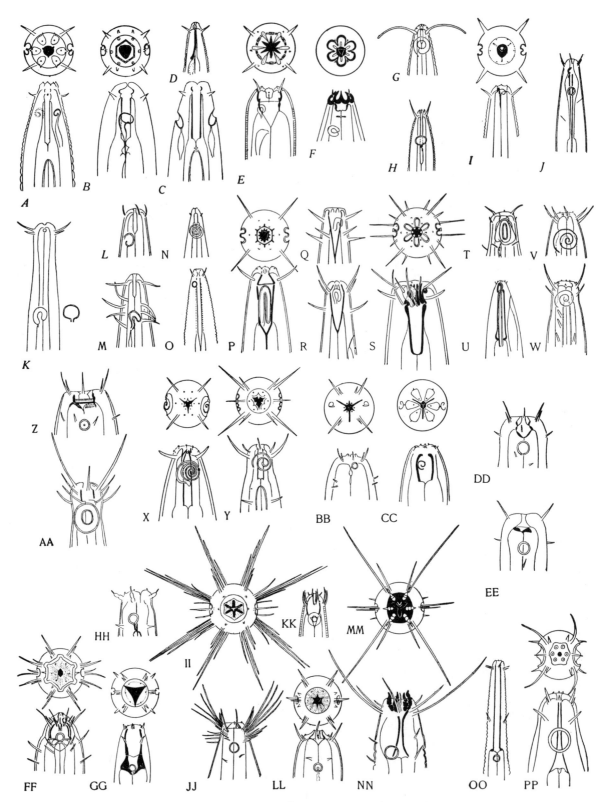

Fig. 60.

Cephalic regions in the Monhysterina. A-C, E-F, H—Plectidae. D, G, I—Camacolaimidae. J-K—Bastianiidae. L-U—Axonolaimidae. V-Y—Comesomatidae. Z-FF—Linhomoeidae. GG-OO—Monhysteridae. PP—Siphonolaimidae. A—*Plectus rhizophilus* (dorsal right); B—*Anaplectus granulosus*; C—*Anaplectus granulosus* (median view); D—*Anguinoides stylosum*; E—*Anonchus mirabilis*; F—*Teratocephalus cornutus*; G—*Aphanolaimus aquaticus*; H—*Leptolaimus maximus*; I—*Camacolaimus prytherchi*; J—*Odontolaimus chlorosus*; K—*Bastiania exilis*; L—*Araeolaimus cylindrolaimus*; M—*Araeolaimus zosterae*; N—*Aegialoalaimus elegans*; O—*Cylindrolaimus communis*; P—*Axonolaimus spinosus*; Q—*Axonolaimus odontophoroides*; R—*Axonolaimus subsimilis*; S—*Odontophora angustilaima?*; T—*Didelta maculata*; U—*Pseudolella granulifera* (dorsal right); V—*Sabatieria longicaudata*; W—*Comesoma minimum*; X—*Dorylaimopsis metatypicus*; Y—*Laimella quadrisetosa*; Z—*Linhomoeus elongatus*; AA—*Halinema spino-* sum; BB—*Monhysterium transitans*; CC—*Tripylium carcinicolum*; DD—*Paralinhomoeus lepturus*; EE—*Metalinhomoeus typicus*; FF—*Sphaerolaimus* sp.; GG—*Halanonchus macramphidum*; HH—*Cytolaimium obtusicaudatum*; II-JJ—*Steineria* sp.; KK—*Omicronema litorium*; LL—*Theristus setosus*; MM-NN—*Scaptrella cincta*; OO—*Rhynchonema cinctum*; PP—*Siphonolaimus* sp. D, H, L, Q, R, GG, HH—After Chitwood, 1936, Proc. Helm. Soc. Wash., v. 3 (1). F—After Cobb, 1914, Tr. Am. Micr. Soc., v. 33. I—After Chitwood, 1935, Proc. Helm. Soc. Wash., v. 2 (1). J & O—After de Man, 1884, Die frei in der reinen Erde—Nematoden. N—After Stekhoven, 1931, Ztschr. Morph. v. 20 (4). T, U, AA-CC, KK, OO—After Cobb, 1920, Contrib. Sc. Nemat. 9. W—After Chitwood, 1937, Proc. Helm. Soc. Wash., v. 4 (2). Z—After de Man, 1889, Mem. Soc. Zool. France. v. 2. DD-EE—After de Man, 1907, Ibid., v. 20. Remainder original.

B. APHASMIDIA

As in the Phasmidia, stomatal morphology in the Aphasmidia is of no value as an ordinal character. Forms with a rudimentary stoma occur in each large group and in many of the groups series extend from the cylindroid type, through various modifications ending in instances of convergence.

MONHYSTERINA. Members of the Monhysterina (Fig. 60) have one character in common and opposed to the related Chromadorina, namely, that the cheilorhabdions do not take the form of twelve sclerotized longitudinal ridges, (odontia), replacing lips. However, in axonolaimids and monhysterids one may sometimes note a longitudinal sclerotization which is apparently the homologue, or even predecessor of the chromadoroid type.

Plectoidea. Anaplectus granulosus provides us with the aphasmidian version of *Rhabditis* not only in esophagus and lips but also in the stoma. The cheilostom is hexangular, the protostom, subtriangular in cross section. Cheilorhabdions and protorhabdions are well sclerotized, telorhabdions only faintly sclerotized. *Anaplectus* is unusual in that the protostom has parallel walls; in most plectoids the walls converge posteriorly. In *Leptolaimus maximus* the stoma is extremely long and narrow, protorhabdions distinct while in some related forms the stoma collapses forming a greatly elongate vestibule. *Anonchus mirabilis* and *Teratocephalus cornutus* exemplify shortening and dilation of the stoma with distinct joints at junction of pro- and mesorhabdions. In *Teratocephalus* the protorhabdions are further modified taking the form of six inwardly acting teeth or odontia.

The family Camacolaimidae is characterized by a diminution in stomata in all forms. In *Aphanolaimus aquaticus* the stoma is minute, cylindrical, with practically non-sclerotized protorhabdions while in *Camacolaimus prytcherchi* only the dorsal stomatal wall is sclerotized and it projects anteriorly as an onchium. *Anguinoides stylosum* is a further example of the same tendency, in this instance the dorsal onchium is separate throughout its length terminating posteriorly in two knobs. *Anguinoides* is a striking parallel to *Ditylenchus dipsaci* of the Tylenchoidea but we must classify the spear as an onchiostyl in this instance.

Passing to the Bastianiidae, we may judge that the "astomatous" *Bastiania* arose from some such form as *Odontolaimus* which has a greatly elongate, narrow stoma.

Concluding our resumé of the Plectoidea we note that non-muscular esophageal tissue extends beyond the mesorhabdions in such members of the genera *Anaplectus* and *Plectus* as have been studied. Comparing with *Rhabditis*, we would consider this as a more advanced evolutionary development. Such a view is borne out by the somatic musculature of the forms studied. Since there are representatives in the Plectoidea (*Anonchus*), Axonolaimoidea (*Axonolaimus*) and Monhysteroidea (*Halanonchus*) in which esophageal tissue does not extend anteriad, we must conclude *Anaplectus* while primitive, does not fulfill all obligations of the Aphasmidian ancestor. Combining cephalic characters and general stomatal outline of *Anaplectus* with the more primitive stomatal and somatic muscle characteristics of *Anonchus* we may, perhaps, have the proper picture.

Axonolaimoidea. Primarily axonolaimoids have a cylindroid or conoid protostom. In the Axonolaiminae the protostom is conoid, the mesostom surrounded by esophageal tissue; the cheilostom is anteriorly conoid. Thus in *Axonolaimus* we have a close parallel with *Plectus.* Twelve weak longitudinal sclerotizations of the cheilostom are usually evident in *Axonolaimus* species. In *Odontophora angustilaima* these 12 sclerotizations are anteriorly fused forming six large outwardly acting odontia. As we shall see later in dealing with the Chromadorina, the 12 odontia replacing lips in that group probably originated in a form near *Axonolaimus.* The same tendency of the cheilorhabdions, with multiplication of elements may be seen in the Monhysteroidea.

Inconspicuous, weakly sclerotized, cylindroid stomata occur in the subfamily Cylindrolaiminae which includes forms in which esophageal tissue extends to the anterior end of the protostom (*Araeolaimus zosterae*) and forms in which this is not the case (*Cylindrolaimus communis, Aegialoalaimus elegans*). Cylindroid or collapsed stomata occur in representatives of the Campylaiminae (*Pseudolella granulifera*) and Diplopeltinae (*Didelta maculata*). In the former type the protorhabdions terminate anteriorly in three small teeth, a parallel to *Dorylaimopsis.*

The Comesomatidae have stomata of two general types. In the first the stoma is cylindroid, the protorhabdions are well developed and terminated anteriorly by three equal teeth (*Dorylaimopsis metatypicus* and *Laimella quadrisetosa*); the entire protostom is surrounded by esophageal tissue. In the second, the protostom is collapsed, the rhabdions are non-sclerotized, the esophageal tissue transforms the stomatal region to a vestibule (*Sabatieria longicaudata, Comesoma minimum*). In both instances the cheilorhabdions are short and do not converge anteriorly as in *Axonolaimus.*

Monhysteroidea. Stomatal diversity in this superfamily has thus far prevented adequate revision of the group into compact small units. In the majority of instances, when the stomatorhabdions are well sclerotized esophageal tissue does not surround them (except *Tripylium carcinicolum*). We may presume that when the stoma is rudimentary as in *Theristus* or *Cytolaimium,* it reached this condition through shortening rather than vestibule formation and collapse. A few forms with a large conspicuous stoma are retained in this group; such are *Rhynchonema cinctum* with an extremely long cylindrical stoma, *Halanonchus macramphidum, Omicronema litorium* and *Sphaerolaimus* sp, with wide, heavily sclerotized stomata. In *Sphaerolaimus* the cheilorhabdions consist of innumerable sclerotized rugae. These same rugae are retained though the protostom has disappeared in *Theristus setosus, Steineria* sp. and other typical monhysterids. In *Scaptrella cincta,* on the contrary, one finds the cheilorhabdions transformed into six outwardly acting odontia as in *Odontophora* of the Axonolaimoidea. Many linhomoeids (*Terschellingia pontica, Monhysterium transitans*) have no distinctly sclerotized rhabdions and in the remainder one notes degrees in shortening and reduction. Thus in *Linhomoeus elongatus* and *Halinema spinosum* the entire stoma is short and wide. In the former one notes subequal cheilorhabdions and prorhabdions with posteriorly converging mesorhabdions. *Paralinhomoeus lepturus* and *Metalinhomoeus typicus* seem to be further steps in stomatal reduction of this series.

Siphonolaims are a group apart, having the entire protostom transformed into a stomatostyl as in the Tylenchoidea but there is little or no resemblance in the organ itself.

In leaving the Monhysterina, one may note that prostomatal teeth, if present, are anterior, small and subequal. Sometimes a dorsal, anteriad pointing tooth is described in monhysterids. Such a structure is present at the base of the stoma in *Scaptrella,* it is non-sclerotized, and is probably an esophageal development through which the dorsal esophageal gland has its orifice. Heavily sclerotized dorsal teeth do not occur in this group, a distinct contrast with the group to follow.

CHROMADORINA. Muscular esophageal tissue always surrounds the protostom in members of this suborder (Fig. 61). Forms with a rudimentary stoma are numerous but they arise through collapse of the protostom rather than shortening. If the protorhabdions are well sclerotized the cheilorhabdions form odontia replacing the lips except in the Tripyloididae. Onchial development is usually apparent, taking the form of a large dorsal tooth opposed by smaller subventral ones.

Chromadoroidea. Of this superfamily the Microlaimidae appear to be most primitive from the standpoint of stomatal characters, including, as it does, forms with subcylindrical protostom and protorhabdions terminated anteriorly by onchia (*Ethmolaimus revaliensis*). In this instance the dorsal onchium is retrorse and the subventrals mere sclerotized oppositional thickenings. In *Statenia trichura* the same structure obtains except that the subventral onchia are also retrorse and but slightly smaller than the dorsal. In *Microlaimus dimorphus* and *Bolbolaimus cobbi* the dorsal onchia assume a mesostomatal position and are axially directed, the smaller subventral onchia being slightly posterior and oppositional in character. In all microlaimids the protorhabdions are moderately sclerotized and completely surrounded by muscular esophageal tissue; the base of the stomatal region is usually indicated by a groove or tissue differentiation. Cheilorhabdions take the form of 12 odontia replacing the lips in function but seldom protruding beyond them.

In the Chromadoridae examples of the same tendencies are evident, *Prochromadora oerleyi* corresponding to *Statenia* and *Odontonema guido-schneideri* to *Bolbolaimus.* In addition the stomatal region may not be set off posteriad (*Chromadora, Spilophorella paradoxa*); the prostomatal dorsal onchium in this case may be small or large, the protostom collapsed or apparent. Cheilorhabdions take the form of 12 odontia as in the Microlaimidae.

In the Cyatholaimidae there are two chief stomatal types. In the Cyatholaiminae the cheilorhabdions always take the form of 12 conspicuous odontia which may project anteriorly beyond the labial region (*Pomponema mirabile*); the prostom is wide, meso- and metastom narrow, with converging weakly sclerotized mesorhabdions. A large dorsal axial onchium at the junction of the pro- and mesostomata is usually present and may be opposed by smaller subventral teeth (*Acanthonchus viviparus*). Choanolaims differ in that axial mesostomatal teeth are inconspicuous or absent and the mesostom does not have converging walls. In all instances the mesorhabdions are divided into two or more elements to their base. In true choanolaims (*Halichoanolaimus dolichurus*, *Gammanema ferox*) the prostom is like an inverted cone and the mesostom short, prismoidal; cheilorhabdions fuse posteriorly, being continuous with six prostomatal and mesotomatal rugae. Sometimes, in addition, there are numerous denticles at the junctions of pro- and mesorhabdions (*Halichoanolaimus*). On the basis of stomatal characters we should, perhaps, recognize the subfamily Selachinematinae as a valid group of the Cyatholaimidae. While they are undoubtedly closely related to choanolaims they differ in that cheilorhabdions are feebly developed and the stomatal cavity greatly reduced. The protorhabdions are fused into three or two mandibles terminating anteriorly in several teeth. Of the diverse types, *Synonchiella truncata* seems the most generalized since it has three equal mandibles, one dorsal and two subventral; each of these mandibles is distally bifid and bears several hook-like onchia. *Cheironchus bulbosus* exemplifies a partial reduction of the dorsal mandible while *Selachinema ferox* represents complete reduction in the dorsal mandible and hypertrophy of the subventrals into lateral jaws.

The Tripyloididae seems misplaced in the Chromodoroidae, yet their relationship with the cyatholaims on the basis of other characters makes their position here obligatory. In these forms cheilorhabdions do not take the form of odontia. Instead there are three large lips. Protorhabdions may exhibit numerous joints, the protostom being on the whole inverted, wide conoidal in form (*Tripyloides vulgaris*, *Bathylaimus cobbi*).

Desmodoroidea. Whenever the protorhabdions are distinctly sclerotized and the protostom not collapsed, desmodoroids (*Metachromadora onyxoides*, *Monoposthia hexalata*, *Desmodora scaldensis*) exhibit all of the characteristics of the Chromadoridae, but quite often the stoma is completely atrophied, in which case protorhabdions cannot be detected (*Ceramonema reticulatum*, Fig. 28). Though axial teeth are present in *Spirina parasitifera* the protorhabdions are feebly developed and cheilorhabdions not apparent.

Desmoscolecoidea. So far as known, this group must be characterized as having a rudimentary stoma with no visible stomatorhabdions.

ENOPLINA. Members of the suborder Enoplina do not have a protrusible onchiostyle and the cheilorhabdions do not take the form of 12 odontia as in the Chromadorina. In a few instances they may form a transverse denticulate ridge. In such instances they do not form an armature of the lips as sometimes occurs in the Monhysterina.

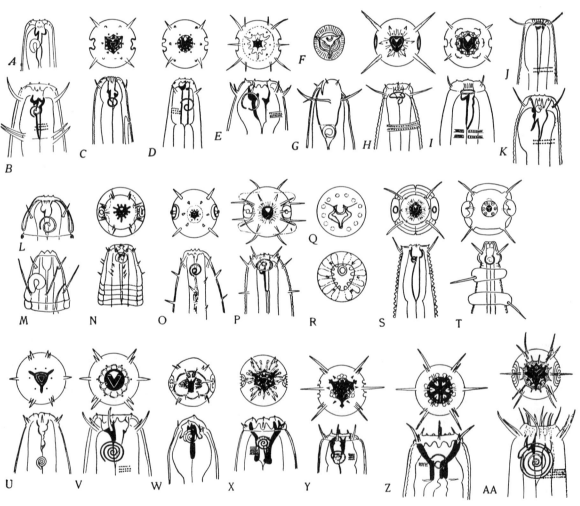

FIG. 61.

Cephalic regions in the Chromadorina. A-F—Microlaimidae. G & U—Tripyloididae. H-K—Chromadoridae. L, N-S—Desmodoridae. M—Epsilonematidae. T—Desmoscolecidae. V-AA—Cyatholaimidae. A—*Achromadora monohystera*; B—*Statenia trichura*; C—*Microlaimus dimorphus*; D—*Ethmolaimus revaliensis*; E—*Bolbolaimus cobbi*; F—*Bolbolaimus cobbi* (section through stomatal region); G—*Bathylaimus cobbi*. H—*Chromadora* sp.; I—*Spilophorella paradoxa*; J—*Prochromadora oerleyi*; K—*Odontonema guido-schneideri*; L—*Desmodora scaldensis*; M—*Epsilonematid* (schematic); N—*Croconema mammillatum*; O—*Spirina parasitifera*; P—*Metachromadora onyxoides*; Q—*Metachromadora onyxoides* (stomatal region); R—*Metachromadora onyxoides* (labial region); S—*Monoposthia hexalata*; T—*Desmoscolex americanus*; U—*Tripyloides vulgaris*; V—*Acanthonchus viviparus*; W—*Cheironchus bulbosa*; X—*Halichoanolaimus dolichurus*; Y—*Synonchiella truncata*; Z—*Gammanema ferox*; AA—*Pomponema mirabile*. G—After Filipjev, 1922, Act. Inst. Agron. Stauropol., v. 1 (16). J & K—After Filipjev, 1930, Arch. Hydrobiol, v. 20. L—After de Man, 1889, Mem. Soc. Zool. France, v. 2. M—After Steiner, 1930, Deutsche Sudpolar Expedition. N—After Steiner & Hoeppli, 1926, Arch. Schiffs.-u. Tropenhyg., v. 30. O—After Cobb, 1928, J. Wash. Acad. Sc. W—After Filipjev. 1918. Trav. Lab. Zool. Stat. Biol. Sebastopol, v. 2 (4). Remainder original.

Tripyloidea (Fig. 62). Herein are grouped four families, separable on stomatal characters, the Ironidae, with a much elongate narrow stoma and heavily sclerotized protorhabdions, the Alaimidae with a rudimentary stoma, the Mononchidae with a capacious stoma and very heavily sclerotized stomatorhabdions, and the Tripylidae with weakly sclerotized protorhabdions.

Mononchs usually have a subglobular stoma with a massive dorsal prostomatal onchium; the latter may be opposed by a variety of dental structures taking the form of transverse denticulate ridges, longitudinal ventral ridges or small onchia (*Mononchus gerlachei*). *Mononchus tunbridgensis* has all of the family characteristics except that the stoma is of moderate length and cylindroid. Such a form would be the presumptive ancestor of the family.

In the Tripylidae, onchia, if present, are basal. The weakly sclerotized protorhabdions may be many jointed (*Trilobus longus*), or not jointed (*Prismatolaimus intermedius*). The stoma may be subcylindrical (*Prismatolaimus*), conoid (*Trilobus*) or collapsed (*Tripyla*). Dorsal or subventral asymmetrically placed denticles are sometimes present.

In the Ironidae the long subcylindrical to prismoidal stoma is always surrounded by esophageal tissue and sometimes set off as a stomatal swelling (*Ironella prismatolaima*). Except in *Cryptonchus* the cheilorhabdions take the form of outwardly acting odontia (fossores). In *Ironus ignavus* the dorsal odontium is double and the subventrals simple while in *Ironella prismatolaima* the reverse is the case. *Syringolaimus smargidus* and *Dolicholaimus obtusus* have three double equal odontia.

Enoploidea (Fig. 63). The families Oncholaimidae and Enoplidae are separable through the fact that only the stomatal base is surrounded by esophageal tissue in the Oncholaimidae while the mesostom and often the prostom is surrounded by muscular esophageal tissue in the Enoplidae.

In the Oncholaiminae the stoma is typically capacious to subglobular, and armed with three unequal conoid mesosto-

matal onchia which act as orifices of the esophageal glands. One of the subventral onchia usually being the largest (*Metoncholaimus pristiuris*). Sometimes the onchia are multiple (*Polydontus marinus*) and sometimes there is but one, the large subventral (*Oncholaimellus clavodiscus*). A small residue of more primitive species with cylindroid stoma persist as in so many other groups (*Pseudopelagonema elegans, Anoplostoma viviparum*).

The subfamilies Eurystomininae and Enchelidiinae appear as slightly modified oncholaims. The stoma is more elongate, the stomatorhabdions are usually jointed and the onchia attenuated to a needle-like point. In these groups one subventral onchium is highly developed while the other two are minute if present (*Bolbella tenuidens*). In addition the cheilorhabdions or prorhabdions may bear minute denticles (*Eurystomina americana*). Sexual dimorphism in stomata makes its appearance in enchelidiids in a most surprising manner. The adult male has a completely rudimentary stoma while the female has a highly developed stoma (*Enchelidium pauli*).

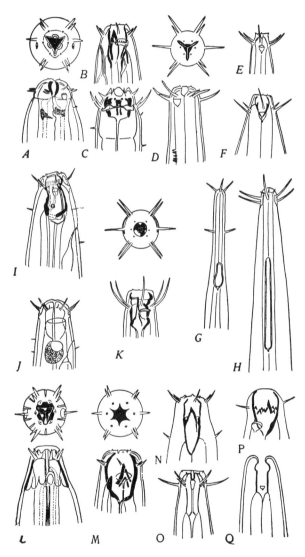

FIG. 63.

Cephalic regions in the Enoploidea. A-F & L—Enoplidae. G-K & M-Q—Oncholaimidae. A—*Enoplus communis*. B—*Eurystomina americana*. C—*Enoploides amphioxi*. D—*Anticoma litoris*. E—*Phanodermopsis longisetae*. F—*Rhabdodemania minima*. G—*Oxystomina alpha*. H—*Halalaimus caroliniensis*. I—*Enchelidium pauli*, v. *denticulatum*, female. J—*Enchelidium pauli*, v. *denticulatum*, male. K—*Bolbella tenuidens*. L—*Thoracostoma* (*Pseudocella*) sp. M—*Metoncholaimus pristuris*. N—*Oncholaimellus clavodiscus*. O—*Anoplostoma viviparum*. P—*Polydontus marinus*. Q—*Pseudopelagonema elegans*. B, D, E, F, H—After Chitwood, 1936, Tr. Amer. Micr. Soc., v. 55 (2). C—After Filipjev, 1918, Trav. Lab. Zool. Stat. iBol. Sebastopol, v. 2 (4). I, J—After Micoletzky, 1930, Vid. Medd. fra Dansk. Natur. Foren., v. 87. K—After Cobb, 1920, Contrib. Sc. Nemat. 9. N—After de Man, 1890, Mem. Soc. Zool. France, v. 3. O—After de Man, 1907, Mem. Soc. Zool. France, v. 20. P-Q—After Kreis, 1934, Capita Zool., v. 4 (5). Remainder original.

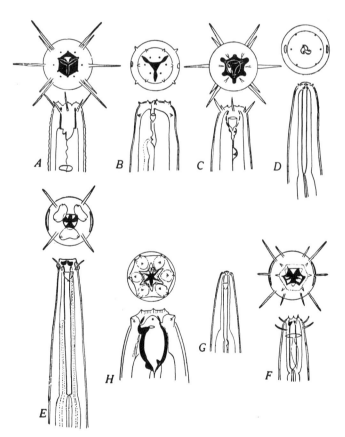

FIG. 62.

Cephalic regions in the Tripyloidea. A-C—Tripylidae. D-G—Ironidae. H—Mononchidae. A—*Prismatolaimus intermedius*. B—*Tripyla* sp. C—*Trilobus longus*. D—*Syringolaimus smargidus* (upper) and *S. brevicaudatus* (lower). E—*Ironus ignavus*. F—*Cryptonchus nudus*. G—*Ironella prismatolaima*. H—*Mononchus gerlachei*. D—After Cobb, 1928, J. Wash. Acad. Sc., v. 18 (9). H—After de Man, 1904, Expéd. Antarct. Belg. Remainder original.

In the Enoplidae stomata are definitely on the wane, being rudimentary in the Oxystomininae and Phanodermatinae, represented chiefly by three mandibles in Enoplinae and usually quite inconspicuous in the Leptosomatinae. In both, *Enoplus communis* and *Enoploides amphioxi*, the mandibles are the only sclerotized parts of the esophageal lining; these latter are anteriorly bifid and axially hooked. At their bases one finds the three esophageal gland orifices. In some enoplids it is said that small onchia corresponding to those of *Metoncholaimus* are present at the base of the mandibles.

In *Rhabdodemania minima* (Leptosomatinae) we have an example showing the maximum of stomatal development in the Enoplidae; the protosom is wide, the mesostom conoidal, the walls are moderately sclerotized. Three mesostomatal onchia are present. In other leptosomatids such as *Anticoma litoris* and *Pseudocella* sp. the stoma is completely collapsed and onchia, if present, are minute.

DORYLAIMINA. The suborder is characterized by the presence of a protrusible onchiostyl at least in the larval stages. It now seems that this structure persists to the adult stage in many forms in which its presence is not suspected. The group also differs from the Enoplina in that the esophageal glands never empty into the stoma or rudiment thereof.

Dorylaimoidea (Fig. 64). Onchiostyls are always well developed and a conspicuous feature of adult dorylaimoids. According to the observations of Thorne (1930, 1935) the onchiostyl originates as a subventral tooth such as one notes in *Nygolaimus brachyuris* or *Sectonema ventralis*. Such an onchiostyl is described as mural. Further development of a ventral groove finally results in a hollow cylinder through which food passes (*Dorylaimus stagnalis*, *Actinolaimus* sp.) in which case the stylet is *axial*. Evidence of its original formation is indicated by the dorsally oblique stylet aperture. It is of further interest to note that in such forms as *Dorylaimus* and *Actinolaimus* the stylet is formed as a cylinder in one of the subventral sectors of the esophageal wall; a new stylet moves up to its final position at each molt; while forming, the onchial cylinder has an open crack on the side *away from* the lumen. Most dorylaimids have little evidence of the original protorhabdions, these being best preserved in *Nygolaimus* and *Actinolaimus*. The latter is a most unusual member of the group having a radially striated cheilorhabdion and four massive, onchium-like, stylet guides in addition to heavily sclerotized pro- and mesorhabdions and *stylet guiding ring*. Since the latter structure is absent in such forms as *Nygolaimus* we may presume it to be the modified telorhabdions or metarhabdions. Its homology has not been determined. In dorylaimids the stylet may become tremendously elongated (*Trichodorus obtusus*, *Leptonchus granulosus*) and in addition it may be terminated in three flanges (*Tylencholaimus aequalis*) or knobs (*Xiphinema americanum*). In the latter instance we have a case of total convergence with tylenchoids in gross stylet morphology. The guiding ring and oblique stylet aperture provides us with a clue to its dorylaim origin. Stylets with flanges or knobs usually show a joint (*Xiphinema*); this joint indicates the extent of the true stylet. The anterior part is the only part formed in the esophageal wall; the posterior part with flanges or knobs, is formed in situ and may be considered a differentiated continuation of the esophageal lining.

The developmental origin and significance of the stylet in *Diphtherophora communis* and *D. perplexans* have not been solved as yet. These forms have a short massive stylet with a complicated dorsal arch at its anterior end and three large knobs at its posterior end.

Mermithoidea (Fig. 64). The first stage larva of such mermithoids as *Agamermis decaudata*, *Mermis subnigrescens* and *Hexamermis albicans* have an onchiostyl quite similar to that of *Dorylaimus* and the stoma is completely atrophied. In the adult, as a rule, one can distinguish no semblance of stylet or stoma, and the esophageal lining extends to the oral opening, but in a few forms a very minute stylet rudiment has been described.

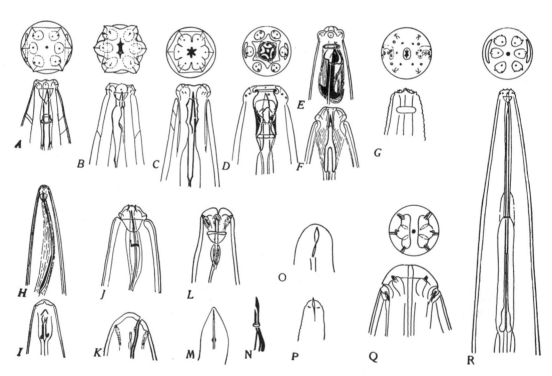

FIG. 64.

Cephalic regions in the Dorylaimina and Dioctophymatina. A–D, F & R—Dorylaimidae. E, H–I—Diphtherophoridae. J—Leptonchidae. G & M—Dioctophymatidae. K–L, N, Q—Mermithidae. O—Trichinellidae. P—Trichuridae. A—*Dorylaimus stagnalis*; B—*Sectonema ventralis*; C—*Nygolaimus brachyuris*; D—*Actinolaimus* sp.; E—*Diphtherophora perplexans* (dorsal on right side); F—*Tylencholaimus aequalis* (median view); G—*Eustrongylides ignotus*. H—*Trichodorus obtusus*; I—*Diphtherophora communis*; J—*Leptonchus granulosus*; K—*Eumermis behningi*; L—*Mesomermis bursata*; M—*Dioctophyma renale* (1st stage larva); N—*Mermis subnigrescens* (1st stage larva); O—*Trichinella spiralis* (1st stage larva); P—*Trichosomoides crassicauda* (1st stage larva); Q—*Hexamermis albicans*; R—*Xiphinema americanum*. A—After Thorne & Swanger, 1936, Capita Zool., v. 6 (4). B–C—After Thorne, 1930, J. Agric. Res., v. 41 (6). E & H—After Cobb, 1913, J. Wash. Acad. Sc., v. 3. F—After Cobb, 1918, U. S. D. A., B. P. I., Agric. Tech. Circ. 1. I—After de Man, 1884, Die frei in der reinen Erde . . . Nematoden. J—After Cobb, 1920, Contrib. Sc. Nemat. 9. K–L—After Steiner, 1929, Zool. Jahrb. Abt. Syst., v. 57. M—After Lukasiak, 1930, Arch. Biol. Soc., Sc. & Lettres Varsovie, v. 3 (3). N—After Cobb, 1926, J. Parasit., v. 13. O–P—After Feulleborn, 1923, Arch. Schiffs. & Tropenhyg., v. 27. Q—After Rauther, 1906, Zool. Jahrb. Abt. Anat., v. 23. Remainder original.

Trichuroidea (Fig. 64). Fülleborn (1923) showed that the first stage larvae of *Trichuris trichiura*, *Trichinella spiralis* and *Trichosomoides crassicauda* all have distinct stylets. It is customary to assume that all trace of stylet disappears in the adult stage but Li (1933) has found a distinct functional stylet in adult *Trichuris trichiura* and the writers have seen one in adult *Trichuris vulpis*. Such stylets are very difficult to observe in fixed material and should be studied intravitam, possibly with the addition of such stains as crystal violet or iodine—1 per cent sulphuric acid. Of all trichuroid stages observed by the writers, the neotenic male of *Trichosomoides crassicauda* has a stylet most easily studied.

DIOCTOPHYMATINA. In adult dioctophymatins the stoma is entirely rudimentary; no distinct cavity nor stomatorhabdions can be observed and the esophagus extends to the oral opening. Orifices of three large esophageal glands at the oral opening seem to indicate definite affinities with enoploids. However, Lukasiak (1930) described a protrusible stylet in the first stage larva of *Dioctophyma renale*. If this observation is verified it will lend additional weight to the trichuroid-dioctophymatoid relationship. Such a relationship was first proposed by Rauther (1918) on the basis of gonads and is supported by the similarity of eggs, spicule and protractor muscles of the spicule.

Bibliography

BASTIAN, H. C. 1865.—Monograph on the Anguillulidae, or free nematoids, marine, land, and fresh water; with descriptions of 100 new species. Tr. Linn. Soc. Lond., v. 25 (2): 73-184, pls. 9-13, figs. 1-248.

1866.—On the anatomy and physiology of the nematoids, parasitic and free; with observations on their zoological position and affinities to the echinoderms. Phil. Tr., Lond., v. 156: 545-638, pls. 22-28.

BAYLIS, H. A. 1921.—A new genus of nematodes parasitic in elephants. Parasit. v. 13 (1): 57-66, figs. 1-7.

1921.—On the classification of the Ascaridae. II. The polydelphis group; with some account of other ascarids parasitic in snakes. Parasit. v. 12 (4): 411-426, figs. 1-7.

1923.—Idem. III. A revision of the genus *Dujardinia* Gedöelst, with a description of a new genus of Anisakinae. Ibid., v. 15 (3): 223-232, figs. 1-8.

1930.—Some Heterakidae and Oxyuridae [Nematoda] from Queensland. Ann. & Mag. Nat. Hist. s. 10, v. 5: 354-366, figs. 1-10.

BAYLIS, H. A. and DAUBNEY, R. 1922.—Report on the parasitic nematodes in the collection of the Zoological Survey of India. Mem. Ind. Mus. v. 7 (4): 264-347, figs. 1-75.

1926.—A synopsis of the families and genera of Nematoda. London. Brit. Museum. 277 pp.

BUETSCHLI, O. 1873.—Beiträge zur Kenntniss der freilebenden Nematoden. Nova Acta Deutsch. Akad. Naturf. v. 36 (5): 1-144; pls. 17-27, figs. 1-69.

CHITWOOD, B. G. 1932a.—The basic plan of the nervous system of nematodes. J. Parasit., v. 19 (2): 167.

1932b.—A synopsis of the nematodes parasitic in insects of the family Blattidae. Ztschr. Parasit. v. 5 (1): 14-50; figs. 1-59.

CHITWOOD, B. G. and CHITWOOD, M. B. 1933.—The characters of a protonematode. J. Parasit., v. 20 (2): 130.

1934.—Nematodes parasitic in Philippine cockroaches. Philipp. J. Sc. v. 52 (4): 381-392; pls. 1-3, figs. 1-19.

CHITWOOD, B. G. and WEHR, E. E. 1932.—The value of head characters in nematode taxonomy and relationships. J. Parasit., v. 19 (2): 167-168.

1934.—The value of cephalic structures as characters in nematode classification, with special reference to the Spiruroidea. Ztschr. Parasit., v. 7 (3): 273-335, figs. 1-20, 1 plate.

CHRISTIE, J. R. 1924.—[The embryology of *Agamermis decaudata*.] J. Parasit., v. 11: 111-112, fig. 1.

1931.—Some nemic parasites (Oxyuridae) of Coleopterous larvae. J. Agric. Res., v. 42 (8): 463-482; figs. 1-14.

1936.—Life history of *Agamermis decaudata*, a nematode parasite of grasshoppers and other insects. J. Agric. Res., v. 52 (3): 161-198; figs. 1-20.

COBB, N. A. 1898.—Extract from Ms. report on the parasites of stock. Misc. Publ. No. 215, Dept. Agric., N. S. Wales, 62 pp., 129 figs.

1913.—[New terms for the lateral organs and ventral gland.] Science, N. S., v. 37 (952): 498.

1913.—New nematode genera found inhabiting fresh water and non-brackish soils. J. Wash. Acad. Sc., v. 3 (10): 432-444, 1 pl. attached.

1915.—*Selachinema*, a new nematode genus with remarkable mandibles. Contrib. Sc. Nemat. (4): 113-116.

1917.—Notes on nemas. Contrib. Sc. Nemat. (5): 117-128.

1917.—The Mononchs (*Mononchus* Bastian, 1866). A genus of free-living predatory nematodes. Contrib. Sc. Nemat. (6): 129-184, figs. 1-70. Also in Soil Sc., v. 3 (5): 431-486.

1918.—Estimating the nema population of the soil. U. S. D. A., B. P. I. Agric. Technology Circ. No. 1, 48 pp., 43 figs.

1919.—The orders and classes of nemas. Contrib. Sc. Nemat. (8): 214-216.

1920.—One hundred new nemas. Contrib. Sc. Nemat. (9): 217-343, figs. 1-118.

1920.—Microtechnique. Suggestions for methods and apparatus. Tr. Amer. Micr. Soc., v. 39 (4): 231-242, figs. 1-6.

1922.—A new species of *Nygolaimus*, an outstanding genus of the Dorylaimidae. J. Wash. Acad. Sc., v. 12 (18): 416-421; figs. 1-2.

1923.—The pharynx and alimentary canal of the hookworm larva-*Necator americanus*. J. Agric. Res., v. 25 (8): 359-361, plate 1. Ibid., (shortened), J. Parasit., v. 9: 244-245, fig. 5.

1923.—Amphids in parasitic nemas. J. Parasit., v. 9: 244.

1923.—Interesting features in the anatomy of nemas. J. Parasit., v. 9: 242-243; figs. 3-4.

1923.—Notes on *Paratylenchus*, a genus of nemas. Contrib. Sc. Nemat. (14): 367-370; 1 fig. Also in J. Wash. Acad. Sc., v. 13 (12): 254-257; 1 fig.

1923.—An emendation of *Hoplolaimus* Daday, 1905, nec auctores. J. Wash. Acad. Sc., v. 13 (10): 211-214; figs. 1-4. Contrib. Sc. Nemat. (13): 363-366; figs. 1-4.

1924.—The amphids of *Caconema* (nom. nov.) and of other nemas. J. Parasit., v. 11: 118-119; fig. N.

1924.—Food of rhabdites and their relatives with descriptions of two new rhabdites and a new rhabditoid genus. J. Parasit., v. 11: 116-117, figs. K and M.

1924.—Notes on the amphids of nemas. J. Parasit., v. 11: 110-111; figs. F-H.

1924.—Amphids on oxyurids. J. Parasit., v. 11: 108.

1925.—Amphidial structures in nemas. J. Parasit., v. 11: 222-223.

1926.—The species of *Mermis*. A group of very remarkable nemas infesting insects. J. Parasit., v. 13: 66-72; figs. 1-3.

1928.—The fossores of *Syringolaimus*. J. Parasit., v. 15: 69.

1928.—A new species of *Syringolaimus*. With a note on the fossorium of nemas. Contrib. Sc. Nemat. (19): 398-492; figs. 1-3. Also in J. Wash. Acad. Sc., v. 18 (9): 249-253; figs. 1-3.

1928.—*Ungella secta* n. gen., n. sp. A nemic parasite of the Burmese Oligochaete (earthworm) *Eutyphoeus rarus*. Contrib. Sc. Nemat. (16): 394-397; figs. 1-4. Also in J. Wash. Acad. Sc., v. 18 (7): 197-200; figs. 1-4.

1928.—*Howardula benigna*. A nemic parasite of the cucumber-beetle (*Diabrotica*). Contrib. Sc. Nemat. (10): 345-352; fig. 1-8.

1928.—Nemic Spermatogenesis. Contrib. Sc. Nemat. (16): 377-387; figs. 1-15.

1928.—*Spiroxys amydae* n. sp. J. Parasit., v. 15: 217-219; figs. 1-6.

1928.—The amphids of the nema *Physaloptera phrynosoma*. 2 figs. J. Parasit., v. 15: 70.

CONINCK, L. A. DE. 1935.—Contribution à la Connaissance des Nèmatodes libres du Congo Belge. I. Les nematodes libres des marais de la Nyamuamba (Rowenzori) et des sources chaudes du Mont Banze (Lac. Kivu). Rev. Zool. & Bot. Afric., v. 21 (2-3): 211-232, 249-326, figs. 1-80.

1936.—*Metaraeolaimoides oxystoma* n.g. n. sp. (Nematoda) en zijne afleiding van *Araeolaimoides* de Man, 1893 door Allometrie. (avec résumé en Français). Biol. Jaarb. 182-204; figs. 1-6 and A-C.

CONINCK, L. DE 1942.—De symmetrie-verhoudingen aan het Vooreinde der (vrijlevende) Nematoden. Naturwetensch. Tijd. v. 24 (2-4): 29-68.

CONINCK, L. A. DE and STEKHOVEN, J. H. S. 1933.—The free living marine nemas of the Belgian Coast II. With general remarks on the structure and the system of nemas. Mem. Mus. Roy. Hist. Nat. Belg. (58), 163 pp., 163 figs.

DRASCHE, R. VON, 1883.—Revision der in der Nematoden-Sammlung des k.k. zoologischen Hofcabinetes befinliche Original-Examplare Diesing's und Molin's. Verhandl. d.k.k. zool.-bot. Gesellsch. in Wien (1882), v. 32: 117-138, pls. 7-10.

FILIPJEV, I. N. 1918.—Free-living marine nematodes in the vicinity of Sevastopol. Biol. Stantsii Ross. Akad. Nauk. s. 2 (4): 1-352, 11 pls. Russian.
    1929.—Two new species of *Actinolaimus* from South Africa. Ann. & Mag. Nat. Hist. s. 10, v. 4: 433-439, figs. 1-2.
    1930.—Les nématodes libres de la baie de la Neva et de l'extrémité orientale du Golfe de Finlande. Arch. Hydrobiol., v. 20: 637-699, v. 21: 1-64; figs. 1-33.

FUELLEBORN, F. 1923.—Über den ''Mundstachel'' der Trichotracheliden-Larven und Bemerkungen über die jüngsten Stadien von *Trichocephalus trichiurus*. Arch. Schiffs. & Tropenhyg., v. 27: 421-425; pl. 11, figs. 1-18.

GOLDSCHMIDT, R. 1903.—Histologische Untersuchungen an Nematoden. I. Die Sinnesorgane von *Ascaris lumbricoides* L. und *A. megalocephala* Cloqu. Zool. Jahrb. Abt. Anat., v. 18 (1): 1-57; pls. 1-5, figs. 1-40.

HAGMEIER, A. 1912.—Beiträge zur Kenntnis der Mermithiden. I. Biologischen Notizen und systematische Beschreibung einiger alter und neuer Arter. Diss. Heidelberg. 92 pp., 5 pls. Also Zool. Jahrb., Abt. Syst., v. 32: 521-612.

HESSE, R. 1892.—Ueber das Nervensystem von *Ascaris megalocephala*. Ztschr. Wiss. Zool., v. 54 (3): 548-568; pls. 23-24, figs. 1-20. Also Diss. Halle, 23 pp., 2 pls.

HETHERINGTON, D. C. 1923.—Comparative studies on certain features of nematodes and their significance. Ill. Biol. Monog., v. 8 (2): 1-62; pls. 1-4, figs. 1-47.

HOEPPLI, R. 1925.—Über das Vorderende der Ascariden. Ztschr. Zellforsch., v. 2 (1): 1-68; figs. 1-27, 1 pl., figs. 1-12.

HOEPPLI, R., and HSUE, H. F. 1929.—Helminthologische Beiträge aus Fukien und Chekiang. II. Parasitische Nematoden aus Vogeln und einem Tummler. Arch. Schiffs. & Tropenhyg., v. 33 (1): 24-34; pls. 1-5, figs. 1-22.

HSUE, H. F. 1933.—On some parasitic nematodes collected in China. Parasit., v. 24 (4): 512-541; figs. 1-46.
    1933.—Remarks on some morphological characters of parasitic nematodes of man and dog with descriptions of a new *Goezia* species from Yangtze beaked sturgeon. Chinese Med. J., v. 47: 1289-1297; figs. 1-7.
    1933.—Some species of *Porrocaecum* (Nematoda) from birds in China. J. Parasit., v. 19 (4): 280-286; pls. 2-3, figs. 1-18.
    1935.—A study of some Strongyloidea and Spiruroidea from French Indo-China and of *Thelazia chungkingensis* Hsü, 1933, from China. Ztschr. Parasit., v. 7 (5): 579-600; figs. 1-31.

HSUE, H. F. and HOEPPLI, R. 1933.—On some parasitic nematodes collected in Amoy. Peking. Nat. Hist. Bull., v. 8: 155-168; figs. 1-15.

IMMINCK, B. D. C. M. 1924.—On the microscopical anatomy of the digestive system of *Strongylus edentatus* Looss. Arch. Anat., Hist. and Embryol., v. 3(4-6): 281-326, figs. 1-46.

JAEGERSKIOELD, L. A. 1897.—Uber den Oesphagus der Nematoden besonders bei *Strongylus armatus* Rud. und *Dochmius duodenalis* Dubini. Bihang K. Svenska Vetensk.-Akad. Handl., Stockholm, v. 23(5): 26 pp., 2 pls.

LI, H. C. 1933.—On the mouth-spear of *Trichocephalus trichiurus* and of a *Trichocephalus* sp. from monkey, *Macacus rhesus*. Chinese Med. J., v. 47: 1343-1346; pl. 1, figs. 1-2.
    1935.—The taxonomy and early development of *Procamallanus fulvidraconis* n. sp., J. Parasit., v. 21(2): 103-113; pls. 1-2, figs. 1-10.

LOOSS, A. 1902 (1901).—The Sclerostomidae of horses and donkeys in Egypt. Rec. Egypt. Govt. School Med. 25-139, pls. 1-3; figs. 1-172.

LUKASIAK, J. 1930.—Anatomische und Entwicklungs-geschichte Untersuchungen an *Dioctophyme renale*. (Goeze, 1782) [*Eustrongylus gigas* Rud.] Arch. Biol. Soc. Sc. and Lett. Varsovie, v. 3: 1-100; pls. 1-6, figs. 1-30.

MAGATH, T. B. 1919.—*Camallanus americanus* nov. spec. Tr. Am. Micr. Soc., v. 38(2): 49-170; figs. A-Q, pls. 7-16, figs. 1-134.

MAN, J. G. DE. 1884.—Die frei in der reinen Erde und im süssen Wasser lebenden Nematoden der niederländischen Fauna. 206 pp., 34 pls., 145 figs. Leiden.
    1886.—Anatomische Untersuchungen über freilebende Nordsee-Nematoden. 82 pp., 13 pls., 29 figs. Leipzig.
    1888.—Sur quelques nématodes libres de la mer du nord, nouveaux ou peu connus. Mém. Soc. Zool. France, v. 1(1): 1-51; pls. 1-4, figs. 1-20.
    1889.—Troisième note sur les nématodes libres de la mer du nord et de la manche. Mém. Soc. Zool. France, v. 2: 182-216; pls. 5-8, figs. 1-12.
    1890.—Quatrième note sur les nématodes libres de la mer du nord et de la manche. Mém. Soc. Zool. France, v. 3(2-3): 169-194; pls. 3-5, figs. 1-10.
    1904.—Nématodes Libres. Résultats Voyage du S. Y. Belgica. Expéd. Antarct. Belge. Anvers. 51 pp., 11 pls.
    1907.—Sur quelques espèces nouvelle ou peu connues de nématodes libres habitant les côtes de la Zélande. Mém. Soc. Zool. France, v. 20: 33-90; pls. 1-4, figs. 1-17.

MARCINOWSKI, K. 1909.—Parasitisch und semiparasitisch an Pflanzenlebende Nematoden. Arb. K. Biol. Anst. Land. u. Forstwirt. v. 7(1): 1-192, figs. 1-76, pl. 1.

MARTINI, E. 1916.— Die Anatomie der *Oxyuris curvula*. Ztschr. Wiss. Zool., v. 116: 137-534; figs. 1-121, pls. 6-20.

MICOLETZKY, H. 1914.—Freilebende Süsswasser-Nematoden der Ost-Alpen. Zool. Jahrb., Abt. Syst. v. 36(4-5): 331-546; pls. 9-19, figs. 1-36.
    1922.—Neue freilebende Nematoden aus Suez. Sitzungsb. Akad. Wiss. Wien. Math.-Naturw. Klasse. Abt. 1, v. 131(4-5): 77-103, figs. 1-13.
    1930.—Freilebende marine Nematoden von Sunda-Inseln. I. Enoplidae. Vidensk. Medd. Dansk. Naturh. Foren v. 87: 243-339, figs. 1-24.

ORTLEPP, R. J. 1932.—Some helminths from South African Chiroptera. 18. Rpt. Direct. Vet. Serv. and Animal Ind., Union S. Africa: 183-195, figs. 1-17.

RAHM, G. 1929.—Nematodes parasitas e semi-parasitas de diversas plantas culturaes do Brasil. Arch. Inst. Biol. v. 2: 67-136; pls. 13-23, figs. 1-145.

RAUTHER, M. 1918.—Mitteilungen zur Nematodenkunde. Zool. Jahrb., Abt. Anat., v. 40: 441-514; figs. A-P, pls. 20-24, figs. 1-40.

SANDGROUND, J. H. 1933.—Scientific results of an expedition to rain forest regions in Eastern Africa. Bull. Mus. Comp. Zool., v. 79(6): 341-366, figs. 1-22.
    1935.—A redescription of *Filaria pertenue* Rodhain, 1919 and the creation of a new genus, *Protofilaria*, for its reception. Rev. Zool. Bot. Afric., v. 27(2): 248-253, figs. 1-5.
    1935.—*Spirura michiganensis* n. sp. and *Rictularia halli* n. sp., two new parasitic nematodes from *Eutamias striatus lysteri* (Richardson). Tr. Am. Micr. Soc., v. 44(2): 155-166; pl. 28, figs. 1-9.

SCHNEIDER, A. 1866.—Monographie der Nematoden. 357 pp., 122 figs., 28 pls. 343 figs. Berlin.

SHULZ, R. ED. 1927.—Die Familie Physalopteridae Leiper, 1908, (Nematodes) und die Prinzipien ihrer Klassification. Samml. Helminth. Arbeit. Prof. K. I. Skrjabin gewidmet. Moskva: 287-312, 1 pl.

SKRJABIN, K. I. 1916.—Materialy pogelminto faunie Paraguaia (contributions a l'étude helminthologique du Paraguay) Zoologich. Vesnik, Petrograd, v. 1, pt. 4, pp. 705-735; 2 figs., pls. 24-25, figs. 1-27.

STEINER, G. 1916.—Freilebende Nematoden aus der Barentsee. Zool. Jahrb., Abt. Syst. v. 39(5-6): 511-676, pls. 16-36, figs. 1-46.
    1917.—Über die Verwandtschaftsverhältnisse und die Systematische Stellung der Mermithiden. Zool. Anz., v. 48(9): 263-267.
    1918.—Studien an Nematoden aus der Niederelbe. I Teil, Mermithiden. Mitt. Zool. Mus. Hamburg. (2 Beiheft. Jahrb. Hamburg Wiss. Anst.) v. 35: 75-100; figs. 1-13.
    1919.—Die von A. Monard gesammelten Nematoden, der Tiefenfauna des Neuenburgersees. Bull. Soc. Neuchâtel. Sc. Nat., v. 43: 1-104; figs. 1-18.
    1923.—Limicole Mermithiden aus dem Sarekgebirge und der Torne Lappmark. Naturwiss. Untersuch. Sarekgebirg. Schwed.—Lappland, v. 4(8): 805-827; figs. 1-29.

1923.—Beitrage zur Kenntnis der Mermithiden 2 Teil. Mermithiden aus Paraguay in der Sammlung des Zoologischen Museums zu Berlin. Pp. 90-110, figs. 1-34.

1924.—Some nemas from the alimentary tract of the Carolina tree frog (*Hyla carolinensis* Pennant). J. Parasit., v. 11: 1-32, figs. 1-65.

1924.—A remarkable new genus and species of mermithid worms from Jamaica. Proc. U. S. Nat. Mus. (2527), v. 65(14): 1-4, pls. 1-2.

1925.—The problem of host selection and host specialization of certain plant-infesting nemas and its application in the study of nemic pests. Phytopath., v. 15(9): 499-534; figs. 1-8.

1929.—*Neoaplectana glaseri*, n. g., n. sp., (Oxyuridae), a new nemic parasite of the Japanese beetle (*Popillia japonica* Newm.). J. Wash. Acad. Sc., v. 19(19): 436-440; fig. 1, A-I.

1929.—On the gross morphology of *Acrobeles* (*Acrobeles*) *crossatus* n. sp. (Rhabditidae, Nematodes) found in diseased bulbs of *Iris tingintana* Boiss. and Reut. with remarks on its ecology and life cycle. Ztschr. Morph., v. 15(4): 547-558, figs. 1-13.

1930.—The nemic fauna of the slime flux of the Carolina Poplar. J. Agric. Res., v. 41(6): 427-434; figs. 1-3.

1933.—The nematode *Cylindrogaster longistoma* (Stefanski) Goodey [sic], and its relationship. J. Parasit., v. 20(1): 66-68, fig. 1.

1935.—Opuscula miscellanea nematologia I. Proc. Helm. Soc., Wash., v. 2(1): 41-45; figs. 1-3.

1936.—Opuscula miscellanea nematologica IV. Proc. Helm. Soc., Wash., v. 3(2): 74-80; figs. 22-25.

STEINER, G., and ALBIN, F. M. 1933.—On the morphology of *Deontostoma californicum*. n. sp. (Leptosomatinae, Nematodes). J. Wash. Acad. Sc., v. 23(1): 25-30; figs. 1-7.

STEINER, G., and HOEPPLI, R. 1926.—Studies on the exoskeleton of some Japanese marine nemas. Arch. Schiffs. u Tropenhyg., v. 30: 547-576; figs. A-Q, plates 1-2.

STEKHOVEN, J. H. SCHUURMANS, 1926.—New facts concerning the larvae of *Anchylostoma caninum* and *Necator americanus*. Proc. 3rd. Pan-Pacific Science Congress. Tokyo. Pp. 2577-2580.

1937.—Parasitic Nematoda. Exploration du Parc National Albert, Mission G. F. de Witte (1933-1935) Fasc. 4: 1-40; figs. 1-116.

STEKHOVEN, J. H. S., and CONINCK, L. A. DE. 1933.—Morphologische Frage zur Systematik der freilebenden Nematoden. Verhandl. Deutsch. Zool. Gesellsch.: 138-143: figs. 1-2.

STRASSEN, O. ZUR. 1904.—*Anthraconema*, eine neue Gattung freilebender Nematoden Zool. Jahrb. Suppl. 7, Festschrift Weismann: 301-346; figs. A-J, pls. 15-16, figs. 1-9.

THORNE, G. 1925.—The genus *Acrobeles* von Linstow, 1877. Tr. Am. Micr. Soc., v. 44(4): 171-210; figs. 1-40.

1930.—Predacious nemas of the genus *Nygolaimus* and a new genus, *Sectonema*. J. Agric. Res., v. 41(6): 445-466; figs. 1-18.

1935.—Notes on free-living and plant-parasitic nematodes. II. Proc. Helm. Soc. Wash., v. 2(2): 96-98.

1937.—A revision of the nematode family Cephalobidae Chitwood and Chitwood, 1934. Proc. Helm. Soc. Wash., v. 4(1): 1-16 figs. 1-4.

THORNE, G., and SWANGER, H. H. 1936.—A monograph of the nematode genera *Dorylaimus* Dujardin, *Aporcelaimus* n. g., *Dorylaimoides* n. g., and *Pungentus* n. g. Capita Zool., v. 6(4): 1-223; pls. 1-31, figs. 1-188.

WALTON, A. C. 1936.—A new species of *Zanclophorus* from *Cryptobranchus alleganiensis*. Tr. Ill. State Acad. Sc., v. 28(2): 267-268, figs. 1-7.

WEDL, C. 1856.—Ueber die Mundwerkzeuge von Nematoden. Sitzungsb. Math. Naturw. Cl., v. 19(1): 33-68; pls. 1-3, figs. 1-3.

WEHR, E. E. 1933.—A new nematode from the Rhea. Proc. U. S. Nat. Mus. (2958) v. 82(17): 1-5; figs. 1-3.

1933.—Descriptions of two new parasitic nematodes from birds. J. Wash. Acad. Sc., v. 23(8): 391-396; figs. 1-8.

1934.—Descriptions of three bird nematodes, including a new genus and a new species. J. Wash. Acad. Sc., v. 24(8): 341-347; figs. 1-15.

1935.—A revised classification of the nematode superfamily Filarioidea. Proc. Helm. Soc. Wash., v. 2(2): 84-88.

1935.—A restudy of *Filariopsis arator* Chandler, 1931, with a discussion of the systematic position of the genus *Filariopsis* van Thiel 1926. J. Wash. Acad. Sc. v. 25(9): 415-418, figs. 1-7.

WU, H. W. 1934.—Notes on the parasitic nematodes from an Indian elephant, Sinensia, v. 5(5-6): 512-533; figs. 1-28.

CHAPTER VI

THE ESOPHAGUS INCLUDING THE ESOPHAGO-INTESTINAL VALVE*

B. G. CHITWOOD and M. B. CHITWOOD

1. GENERAL MORPHOLOGY

The esophagi of nematodes are extremely diverse in both gross anatomy and mode of function; this diversity is correlated with the widely differing feeding habits as well as with the phylogeny of this organ, the result being an assembly of numerous types with examples of convergence and of divergence. In addition features of no apparent present value to the organism are often preserved.

The esophagi of all nematodes have a few characteristics in common, for example their fundamental triradiateness, one ray of the lumen being directed ventrally (though torsion may cause a complete reversal in some regions), a cuticular lining; marginal nuclei and fibers connected with the rays of the lumen or *radii*, radial muscles and nuclei located in the regions between the esophageal radii; nerve cells of the esophago-sympathetic system; and glands, emptying into the lumen of the esophagus or into the stoma.

The esophagus in cross section has three sectors (Fig. 65), a dorsal sector and two subventral sectors. Cell walls are not distinguishable in the esophagus as a whole; it is, therefore, syncytial. However, the cell walls of nerve cells are usually distinct and the protoplasm of each of the esophageal glands usually retains its identity. The nuclei of the marginal and radial fibers are so placed as to leave no doubt that each belongs to a specific part of the fibers and therefore functions separately.

Through their contraction the radial muscles cause dilation of the esophageal lumen; the marginal fibers are apparently static rather than contractile and the marginal tissue and protoplasm, according to Martini (1908), correspond to the epithelium and form the cuticular lining of the esophagus. The esophagus is often covered externally by a semicuticular membrane which may possibly be a product of the marginal tissue. Longitudinal supporting fibers are usually absent in the esophageal wall. The esophageal glands apparently serve a digestive capacity, the secretion being either ejected through the mouth in case of extracorporeal digestion or passed into the intestine with the food and aiding in intracorporeal digestion. Other possible functions of the esophageal glands will be discussed later.

Nemic esophagi evidence many degrees of cell or nuclear constancy; Looss (1896) first discovered nuclear constancy with reference to the radial, marginal and gland nuclei in ascarids. Since this observation, representatives of the major groups have been studied by us with the result that we find many large groups characterized by a certain number of nuclei of a given type, arranged in a definite manner; loss of nuclear constancy is characteristic of certain groups, and evolution from forms with few cells or nuclei to forms with many cells or nuclei may be traced.

Other points of systematic value in the study of nemic esophagi are: The form of the lumen and esophageal lining, the position of the esophageal gland orifices, the form of the gland ducts and tubules, and finally, the form of the esophago-intestinal valve.

The esophageal lumen itself may be triradiate, the sides of the radii converging distally (Fig. 65B), or they may terminate in incomplete tubes (Fig. 65A). In either case the esophageal lining may have thickened *attachment points* at which the radial muscles are inserted (Fig. 65B); whenever the radii terminate in marginal tubes, or whenever attachment points are present, the radial muscles are said to be concentered (Fig. 65C) while in other instances they are usually dispersed (Fig. 65D).

The dorsal esophageal gland orifice is usually at or near the anterior end of the esophagus. In some cases it is situated on a dorsal tooth projected into the stoma (*Strongylus, Ancylostoma, Oncholaimus*) but in other cases the orifice is far removed from the anterior end (*Aphelenchoides, Prionchulus, Dorylaimus*). The subventral glands, on the contrary, usually have their orifices in the posterior part of the esophagus (metacorpus) and only in rare instances do they extend to the anterior end (Enoploidea, Tripyloidea, Dioctophymatoidea). As a general rule, the duct of an esophageal gland is situated in the middle of a sector and branches are given off into the protoplasm from each side of the central duct. In large forms it has been found that these branches terminate blindly in the form of tubules and the protoplasm is not in direct relation with the lumen. Only the terminal duct is lined with cuticle.

Because of numerous cases of convergence in gross form, the structural diversity will be discussed group by group uniting both the gross anatomy and the histology in order that evolutionary trends may be more readily observed.

A. CLASS PHASMIDIA

All phasmidians have one point in common with respect to the esophagus, namely, the subventral esophageal glands always have orifices far removed from the anterior extremity. On the basis of gross morphology they show no obvious separation into groups but on the basis of histology and developmental anatomy they are divisible into two groups, the orders Rhabditida and Spirurida. In the Rhabditida the esophagus shows evidences, in its larval stages at least, of being composed of two major parts, an elongate corpus and a short bulbar region, while in the Spirurida it is composed of a short muscular anterior part and a long wide glandular part. These are fundamental differences established in this histology of the organ.

RHABDITINA. The esophagus of members of the suborder Rhabditina consists of a cylindroid corpus (further divisible into procorpus and metacorpus*), a narrow isthmus and a pyriform or elongate bulbar region. In some forms, such as *Rhabditis* (Fig. 65), the bulbar region contains a well developed valve, while in others, such as *Diplogaster* (Fig. 76), no such valve is present. It will suffice for our purpose to describe the esophagus of *Rhabditis* and compare the other forms with it.

The esophageal lumen of *Rhabditis terricola* varies according to the region, the rays of the lumen of the procorpus terminating in well developed marginal "tubes" (Fig. 67a-b), those of the metacorpus in smaller "tubes" and those of the isthmus and bulbar region in acute angles. The lumen is modified at the valve and the lining thickened (Fig. 68c). The procorpus contains 18 nuclei: six radial nuclei (r_{1-6}) in one *set* (RI) near the base of the procorpus, one being on each side of each sector (Fig. 67a-b), and 12 nerve cell nuclei (n_{1-12}) arranged in three chains, one in the center of each sector. The radial muscles are concentered in six bands corresponding to the six radial nuclei. The metacorpus (Fig. 66 & 67c-d) contains 28 nuclei: three bilobed marginal nuclei (m_{1-3}) in one set (MI)[†], six radial nuclei (r_{7-12}) in one set (RII), and 19 nerve cell nuclei (n_{13-31}) forming three chains as in the procorpus. The isthmus is anucleate while the bulb contains 30 nuclei, as follows: six marginal (m_{4-6}, m_{7-9}), three of which are in the prevalar region (MII) and three in the postvalvar region (MIII); 12 radial (r_{13-18}, r_{19-21}, r_{22-24}), a set of six of which are in the prevalar region (RIII) and groups of three in the valvar (R IVa) and postvalvar regions (R IVb); nine nerve cell nuclei (n_{32-40}) and three esophageal gland nuclei (g_{1-3}). The esophago-intestinal valve has a simple triradiate

---

*Strictly speaking the esophago-intestinal valve is part of the esophagus. However, in conformity with common usage, the term esophagus as used herein excludes the esophago-intestinal valve unless otherwise specified.

*These terms are substituted for precorpus and postcorpus (p. 10) at the suggestion of Dr. Steiner.
†Not uncommonly the marginal nuclei may appear to be double, in forms with marginal tubes at the termination of esophageal radii; in such cases the lobes are designated m_{1a}, m_{1b}, etc.

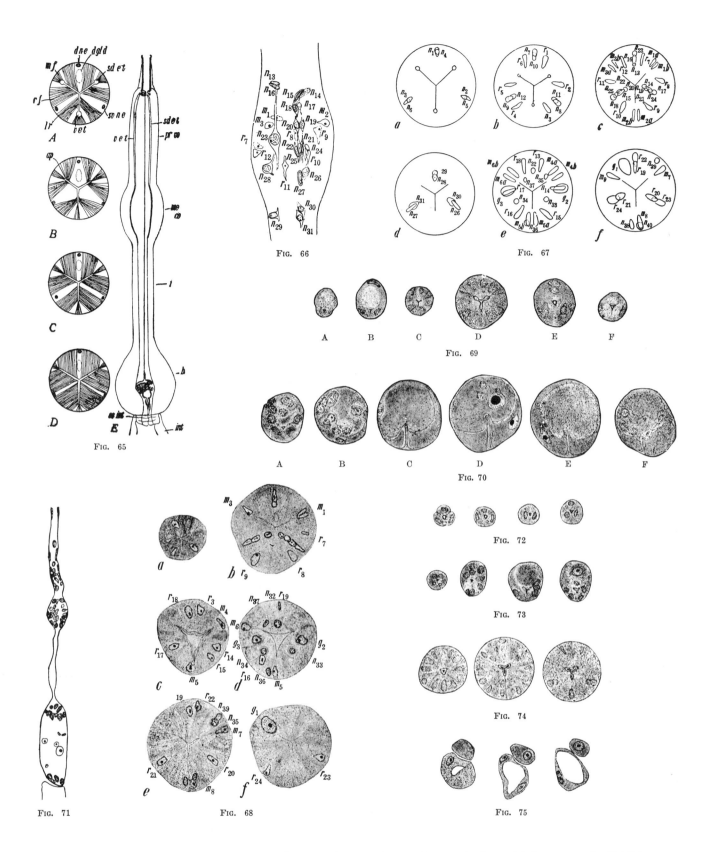

FIG. 65. Diagrams of esophagi. A-D—Cross sections showing various types of muscle arrangement. (A—Radii of esophageal lumen terminated by tubes, musculature concentered; ex. *Rhabditis, Leidynema, Heterakis* and first stage strongyles, *Plectus* and *Axonolaimus;* B—Sides of radii converging distally, musculature concentered, attachment points present; ex. *Oesophagostomum, Camallanus, Cucullanus, Sphaerolaimus* and *Metachromadora;* C—Sides of radii converging distally, musculature concentered, no attachment points, ex. *Ethmolaimus, Monoposthia, Sphilophorella, Acanthonchus;* D—Sides of radii parallel, musculature dispersed, ex. *Ascaris, Physaloptera, Theristus, Tripyla*). E—Diagram of rhabditid esophagus showing parts of esophagus and relative position of radial tubes of lumen. dne—dorsal esophageal nerve, sv ne subventral esophageal nerve, sd et subdorsal and v et ventral esophageal tubes, mf marginal fibers, lr radius of lumen, ap attachment point, pr co procorpus, me co metacorpus, i isthmus, b bulb. Original.

FIG. 66. Diagram showing nuclei of metacorpus in *Rhabditis.*

FIG. 67. Diagram showing groups of nuclei in esophagus of *Rhabditis terricola* as occurring in cross section. a-b—Procorpus; c-d—metacorpus; e-f—bulb.

FIG. 68. *Rhabditis terricola,* esophageal cross sections. a—procorpus; b—metacorpus; c-f—*Rhabditis lambdiensis* bulb.

FIG. 69. *Ditylenchus putrefasciens.* Sections through corpus.

FIG. 70. *Ditylenchus putrefasciens* (Sections through bulbar region).

FIG. 71. *Ditylenchus putrefasciens* (Reconstruction of esophagus with nuclei in position).

FIGS. 72-75. *Aphelenchus avenae.* 72—procorpus; 73—metacorpus; 74—bulbar region; 75—region of glandular appendage.

FIGS. 66-75—After Chitw. & Chitw. 1936, J. Wash. Acad. Sc., v. 26 (2); FIG. 65 original.

79

lumen; its wall consists of an internal layer of transverse fibrous tissue containing two nuclei and an external circular layer containing three nuclei.

The esophagus of *Rhabditis* functions as follows: Contraction of the radial muscles in the procorpus tends to triangulate the lumen thereby increasing its volume; thereafter contraction of the radial muscles of the metacorpus cause dilation of its lumen while relaxation of the muscles of the procorpus and dilation of the bulbar valve occur. In order to discuss the mechanism of the bulbar valve we shall label the parts; the esophageal lining has a series of three thickened regions. During rest the first piece (1) is convex anteriorly while the second and third pieces (2-3) nearly touch one another. Contraction causes a reversal of the position of *1*, point *a* (Fig. 97E-F) becoming nearly axial and point *b* moving from an axial position to a point formerly occupied by *a*. This is accomplished by contraction of the radial muscles of the prevalvar region (associated with r_{13-18}) and possibly of the valvar region (associated with r_{19-21}). Opposition to this movement, i. e., return to the position of rest, is accomplished by muscles of the valvar region (r_{19-21}). Movement of piece *1* to a position of dilation is followed by dilation of the lumen opposite pieces *2-3* in series (through contraction of the postvalvar radial muscles associated with r_{22-24}), opening of the esophago-intestinal valve and, finally, closure of this structure. There is no evidence that the marginal fibers are ever contractile. They appear to function entirely in the capacity of "fixed points" upon which the sectors are "hinged." In the corpus, relaxation proceeds slowly while contraction or dilation is spasmodic. Perhaps the tubular form of the esophageal radii contributes to the opposition of the radial muscles because of their elasticity. Muscle fibers definitely opposed by other muscle fibers are found only in the valvar region of the bulb. All muscle fibers of the esophagus are perpendicular or oblique to the esophageal axis; there are no circular or longitudinal fibers.

In *Rhabditis*, the dorsal esophageal gland orifice is at the anterior end of the procorpus (Fig. 76), the short cuticularly lined terminal duct is followed by a small ampulla and a long canal extending to the bulb where it becomes lost in the mass of dorsal esophageal gland protoplasm; the canal extends nearly to the base of the bulbar region where the gland nucleus is situated. The subventral gland orifices are situated at the base of the metacorpus and, likewise, each is provided with an ampulla and a canal leading posteriorly to the bulbar region. The mass of the subventral gland protoplasm is lateral in position (i. e., in the lateral part of the subventral sectors) and their nuclei are in the valvar region.

The nerve cells, previously mentioned, form the esophago-sympathetic system which consists of a nerve trunk in the center of each esophageal sector, these trunks being connected by three commissures, one at the base of the corpus, one in the prevalvar region and one in the postvalvar region. This system is connected with the central nervous system by means of a pair of nerves from the subventral trunks through the external surface of the procorpus.

Other members of the superfamily Rhabditoidea with a valved bulb apparently have the same structure as that described above. In *Diplogaster* (Fig. 76), and similar forms, in which the valve is absent but the radial muscles of the bulbar region not degenerate, slightly larger marginal "tubes" occur in the corpus and the radial muscles associated with r_{19-24} seem to act together instead of as two separate groups. *Rhabdias* and *Strongyloides* present peculiarities in that the esophagus has a well developed valve in the free-living generation and first stage larva of the parasitic generation, but this structure degenerates in the later development of the parasitic generation though there is no change in the number or arrangement of the esophageal nuclei.

' In the superfamily Tylenchoidea, actual degeneration of the radial muscles of the bulbar region takes place but the 30 nuclei of this region which correspond to those in *Rhabditis* still remain (Figs. 70-71). This degeneration of the musculature is correlated with increase in size of the esophageal glands which form practically the entire bulbar region in forms such as *Ditylenchus dipsaci* and may even extend beyond the base of the esophagus as one or more esophageal appendages as in *Aphelenchus avenae* (Figs. 75-76). In a few forms, such as *Aphelenchus* and *Aphelenchoides*, Cobb (1923) showed that the dorsal gland orifice is situated in the anterior part of the metacorpus instead of the anterior part of the procorpus as is generally the case. In many tylenchids the metacorpus is unusually large and acts as the chief "pump" of the esopha-

gus, while in the Allantonematidae the musculature of the entire esophagus appears degenerate.

STRONGYLINA.—Members of the suborder Strongylina have an esophagus which is grossly clavate to cylindroid but in some instances faint indications of procorpus, metacorpus, and bulbar region are observable in the adult. The first stage larvae of strongyloids and many trichostrongyloids have an esophagus identical with that of *Rhabditis;* during the second and third stages the esophagus becomes more elongate and the valves of the bulbar region disappear resulting in an esophagus reminiscent of that of *Diplogaster* except that the metacorpus is not enlarged or as distinctly set off. With progressive development the various regions become grossly obliterated to a greater or lesser degree. Metastrongyloids differ from strongyloids and most trichostrongyloids in that first stage larva does not have a valved bulb, but more closely corresponds in its morphology to the second and third stage larvae of the other two superfamilies.

The detailed structure of the esophagi of strongylins was studied by Jägerskiöld (1897), Looss (1905), Imminck (1921, 1924) and the writers (1934, 1935). It was found that the marginal, radial, and gland nuclei agree in number with those of *Rhabditis* but the prevalvar radial nuclei (r_{13-18}) are arranged in two groups of three nuclei each, near the center of the sectors. There is no essential difference in arrangment of the nerve cells though there is considerable variation as to the total number observed (from 29 in *Metastrongylus elongatus* to 44 in *Strongylus edentatum*). The triradiate esophago-intestinal valve seems to be composed of seven cells. The subventral esophageal gland nuclei are quite minute in members of the Strongyloidea and thus far gland orifices near the nerve ring have been observed but rarely. This may be correlated with the hypertrophy of the dorsal gland characteristic in such forms, and it may have a distinct bearing on the feeding habits.

The musculature is always "concentered" rather than "dispersed" in strongylins, and the lumen may have marginal tubes as in *Metastrongylus* or there may be a series of thickenings of the esophageal lining forming attachment points as in *Oesophagostomum dentatum* (Fig. 78) and *Ancylostoma duodenale*.

ASCARIDINA.—More work has been done on the esophagi of representatives of the suborder Ascaridina than of representatives of other groups. It was in *Parascaris equorum* that the orifices of esophageal glands were observed for the first time in any nematode by A. Schneider (1866). Subsequent work has been carried out by Jägerskiöld (1893, 1894), Hamann (1895), Looss (1896), Ehlers (1899), Jerke (1901), K. C. Schneider (1902), Goldschmidt (1904, 1909, 1910), Martini (1916, 1922), Kulmatycki (1918, 1922), Allgen (1921), Mueller (1929, 1931), Hsü (1929, 1933), Plenk (1924, 1925, 1926), Chitwood (1931), Chitwood and Hill (1931), Chitwood and Chitwood (1933, 1934, 1936), de Bruyn (1934), and Mackin (1936).

In ascaridins the esophagus varies in gross morphology more than in any other group. The members of the superfamily Oxyuroidea are less diverse in detailed anatomy than are the members of the superfamily Ascaridoidea. In the former group, the esophagus in the adult stage is basically rhabditoid, having, with a few questionable exceptions, a cylindroid corpus, a more or less distinct isthmus, and a valved bulb. In numerous forms, such as *Syphacia* and *Oxyuris*, the corpus may be clavate while in a few, such as *Ransomnema*, it may be fusiform. In some forms the metacorpus is enlarged and either cylindroid as in *Leidynema* or ovoid as in *Hammerschmidtiella;* in odd types the entire corpus is pyriform (*Aorurus*) and in still others the procorpus only is fusiform (*Ozolaimus*). The isthmus also is diverse in gross appearance, sometimes being greatly elongated as in *Atractis* or *Aorurus*, sometimes only moderate in size as in *Thelastoma*, sometimes recognizable only as a groove as in *Rhigonema* and sometimes subglobular as in *Labidurus*. Less diversity occurs in regard to the bulb; it is valved except in a few genera such as *Dermatoxys* and *Leiperenia* but the degree of development of the valves differs greatly between the typical rhabditoid form of *Thelastoma* and the much reduced type of *Oxyuris*.

One would presume that forms such as *Leidynema* (Fig. 76) and *Hammerschmidtiella* represent the primitive type since the metacorpus is enlarged as in *Rhabditis* and since careful study proves the corpus to be subdivided into pro- and metacorpus even in forms in which these parts otherwise are not grossly separable. However, primitivity apparently does not apply here since in the genera *Leidynema* and *Hammerschmidtiella* the esophagus does not reach this stage of development until adulthood of the female. The juvenile females, as well as the adult neotenic males, have a cylindrical corpus as in *Thelastoma*. Sexual dimorphism in the esophagus as well as cephalic structures, stomata, and cuticular ornamentation is one of the outstanding oddities of the Thelastomatidae and Ransomnematinae. Developmental modifications, without apparent bearing on relationship, have also been described in *Oxyuris equi* by Ihle and Van Oordt (1921) and Wetzel

(1930); in this species the fourth stage larva has a pseudostom formed by the dilation of the entire corpus (Fig. 97 O).

In their detailed anatomy, the oxyuroids show less diversity. The lumen of the corpus usually shows the marginal "tubes" as in *Rhabditis* and sometimes in addition (or sometimes instead of them) the esophageal lining has distinct cuticular thickenings or attachment points for the radial muscles. The three esophageal glands (some statements to the contrary) open as in *Rhabditis*, i. e., the dorsal near the base of the stoma, the subventrals near the base of the metacorpus. The

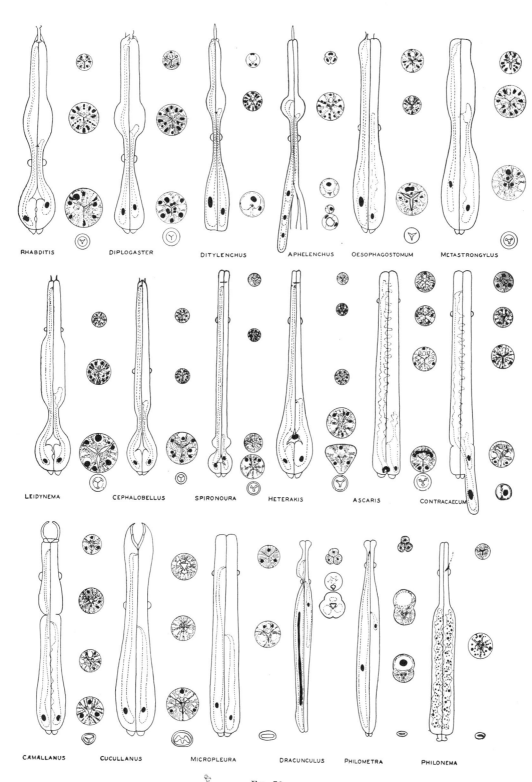

FIG. 76

Diagrams of esophagi, including representatives of the Rhabditoidea (*Rhabditis, Diplogaster*), Tylenchoidea (*Ditylenchus, Aphelenchus*), Strongyloidea (*Oesophagostomum*), Metastrongyloidea (*Metastrongylus*), Oxyuroidea (*Leidynema, Cephalobellus*), Ascaridoidea (*Spironoura* to *Contracaecum*), Camallanoidea (*Camallanus, Cucullanus*), Dracunculoidea (*Micropleura* to *Philonema*). Original.

number and arrangement of the nuclei corresponds totally or approximately to *Rhabditis* in all forms studied with one exception: Martini (1916) records three nuclei in each of the subventral esophageal glands of *Oxyuris equi* which fact might conceivably be due to either gigantism or misinterpretation. The esophago-intestinal valve is triradiate, more or less rhabditoid, and contains five to seven nuclei.

The Ascaridoidea includes some forms which have esophagi of the thelastomatid type such as *Cosmocercoides*. With the exception of the Ascarididae (in which no indication of a stomatal region is apparent) and the Subulurinae (in which there is a short wide stoma) the stomatal region in the Ascaridoidea is surrounded by esophageal tissue containing radial muscles. The modified stomatal region so formed is termed a *vestibule*. In some ascaridoids, the isthmus is obliterated, the bulb being

in direct continuity with the corpus such as *Heterakis*, while in still others (*Spironoura*), it may be ovoid to spheroid; a few forms have no valves, the bulbar region being elongated (*Quimperia*) and others have a cylindrical esophagus either with or without a distinctly set off bulbar region (*Ascaridia* and *Ascaris*, *Toxocara* and *Contracaecum*, etc.). It is in the latter group that we have the outstanding diversity of the superfamily. While the bulbar region is always distinct in the larva when removed from the egg it may totally disappear from the standpoint of gross examination during later stages in its development (*Ascaridia, Ascaris, Toxascaris*, etc.). In other forms it remains grossly unmodified but becomes relatively smaller due to the disproportionate increase in length of the corpus (*Toxocara*) and in some it becomes elongated forming a short cylindroid glandular region (*Anisakis*). Forms

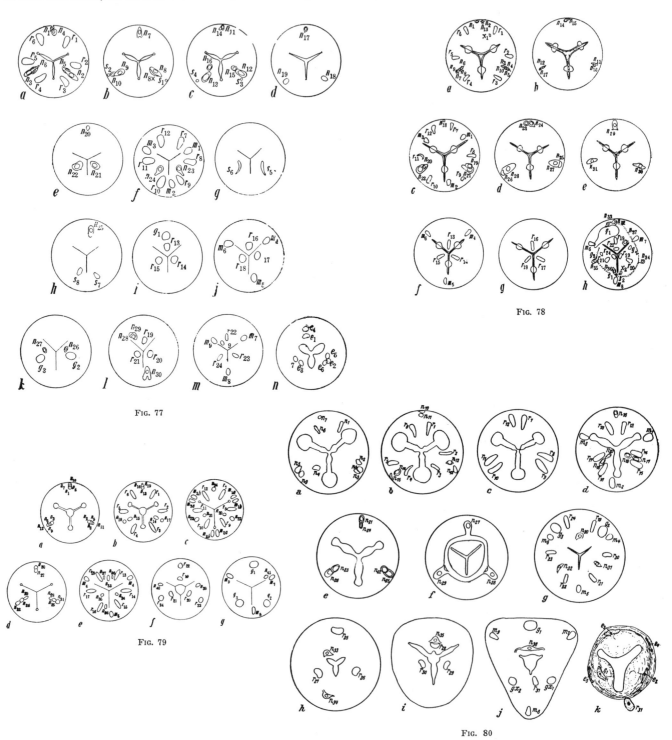

FIG. 77

FIG. 78

FIG. 79

FIG. 80

FIGS. 77-80

Diagrams showing distribution of nuclei in the esophagi of *Metastrongylus elongatus* (77), *Oesophagostomum dentatum* (78), *Leidynema appendiculatum* (79), and *Heterakis gallinarum* (80). After Chitw. & Chitw., J. Wash. Acad. Sc.

82

which retain a bulbar region often have one or more posteriorly directed bulbar appendages (*Contracaecum, Multicaecum*). The internal structure varies fully as much as the gross appearance and its variations will be dealt with separately.

Ascaridoids with a valved bulb invariably have marginal "tubes" in the region of the corpus and concentered radial fibers but the esophageal lining is without thickened attachment points in all forms.

Meromyarian forms such as cosmocercids and kathlaniids have approximately the same nuclei as are present in rhabditoids and oxyuroids (one group of three large marginals and two groups of six large radials in the corpus; two groups of three marginals, one group of six radial and two groups of three radials, and three gland nuclei in the bulb). Polymyarian forms, such as heterakids and ascaridids show evidence of additional cell division in the esophagus, for in heterakids there is an additional group of six radial nuclei in the procorpus making a total of 30 radials and the subventral esophageal glands are each binucleate. In ascarids there are 36 radials and 12 marginals, the additional nuclei (six radials and three marginals) being situated anterior to or opposite the orifice

sues. However, in such forms the subventral glands do not extend into the anterior part of the esophagus. According to Hsü, 1933, the dorsal gland nucleus completely surrounds the esophageal lumen in *Anisakis* and becomes ventrally situated in *Toxocara* and *Contracaecum*. Thus, though the ventrally situated bulbar appendage in *Contracaecum* represents the two subventral sectors of the esophagus and may or may not contain small lobes of the subventral esophageal glands, it is chiefly formed by the much enlarged dorsal gland (Fig. 76).

SPIRURIDA.—The essentially different form of the esophagus in the order Spirurida has been previously mentioned. The fact that the esophagus is always histologically divisible into a short muscular anterior part and a long glandular posterior part is not of course, noticeable in many forms with cylindroid or even clavate esophagi. However, the fundamental significance of this type of organization is evidenced by the absence of any stages in the life history showing reminiscences of the rhabditoid esophagus such as were mentioned for the Strongylina and Ascaridina.

CAMALLANINA.—The suborder Camallanina contains forms in which the esophageal glands are primarily uninucleate;

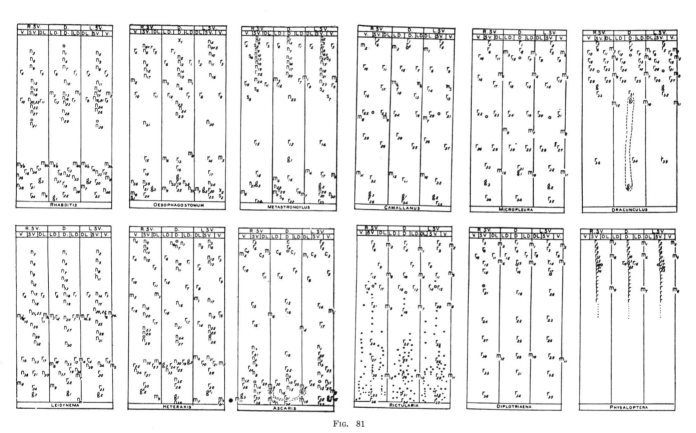

FIG. 81

Tables of esophageal nuclear distribution of Phasmidians. Original.

of the dorsal esophageal gland. The anterior radials (Fig. 81) in this case are located in the center of their sectors, in two groups of three, the remaining radial groups of the corpus are broken, i. e., each group of six is subdivided into two groups of three, one in each sector and these nuclei may or may not become centrally located according to the particular genus. Members of the Ascarididae have only three esophageal gland nuclei and for that reason *Ascaridia*, with binucleate subventral glands, though the esophagus is cylindrical, must be placed in the Heterakidae.

Whether or not the esophagus retains its gross separation into corpus and bulbar regions in the adult, the two regions are histologically separable, the base of the corpus being indicated by the basal commissure of the metacorpus. In ascaridids the esophageal glands undergo many peculiar modifications (see Hsü, 1929, 1933). The dorsal gland situated near the base of the bulbar region is bilobed in *Ascaris lumbricoides* the lobes are marginal and connected by a fine strand. In this form lobes of the dorsal gland extend into the subventral sectors of the corpus giving the appearance of fusion of tis-

secondarily a few forms with multinucleate glands appear to have arisen. There are two superfamilies, Camallanoidea and Dracunculoidea, the former being characterized by relatively well developed musculature usually having cuticular attachment points for the radial muscles and an esophago-intestinal valve that is either triradiate or shows reminiscences of this condition, while the latter group has enormously developed esophageal glands, relatively meager musculature in the glandular region, no attachment points and a dorso-ventrally flattened esophago-intestinal valve.

Esophagi of Camallanoids have been studied by Jägerskiöld (1909), Magath (1919), Törnquist (1931), and Hsü (1933). *Camallanus microcephalus* has a so-called "divided" esophagus (Figs. 76 & 98 E-F). Here the dorsal gland orifice is somewhat posterior to the base of the stoma, the subventral gland orifices at the anterior end of the posterior part of the esophagus. In the anterior part of the esophagus there are 18 radial nuclei and six marginal nuclei, the radials in four groups, two groups of three anterior to the dorsal gland orifice and two groups of six posterior to this level; the marginals are in two groups, one situated near the level of the dorsal gland orifice, the other between the third and fourth radial groups. In the posterior part of the esophagus there are likewise 18 radials in four groups and six marginals in two groups. Here the first group of six radials is situated near the anterior end of this part of the esophagus at the level of the subventral gland orifices, the second group of six radials somewhat anterior to the mid region of this part, and the third and fourth groups of three radials near the base of the esophagus. One marginal group is just posterior to the level of the subventral gland orifices while the other is near the level of the third group of radials. It would appear obvious from the nuclear distribution (Fig. 81) that the posterior or glandular region corresponds not only to the bulbar region of *Rhabditis* or *Ascaris* but also to the metacorpus of *Rhabditis* and part of the metacorpus region of *Ascaris*, since the bulbar region in these forms contains only six marginal and 12 radial nuclei while the posterior part of *Camallanus* contains 18 radial and six marginal nuclei. This view is supported by the fact that the commissure of the metacorpus is situated in the anterior part of the glandular region of the esophagus of *Camallanus*.

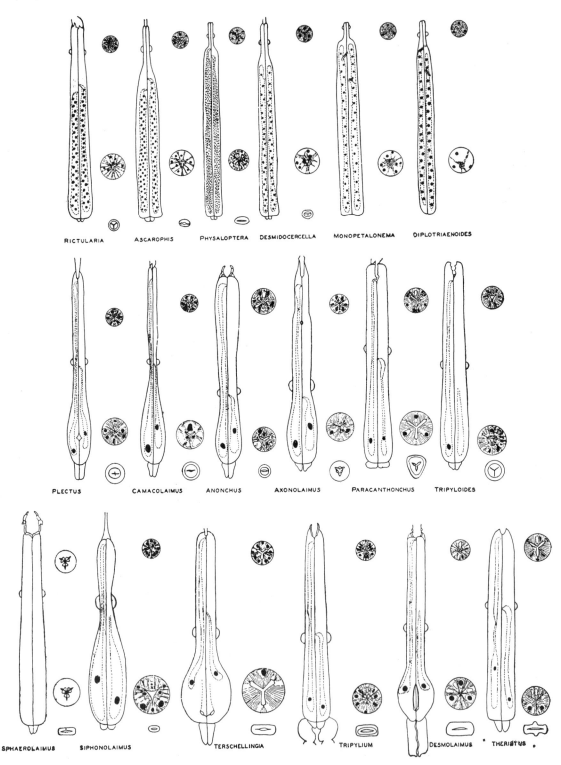

FIG. 82

Diagrams of esophagi; including representatives of the Spiruroidea (*Rictularia* to *Physaloptera*), Filarioidea (*Desmidocercella* to *Diplotriaenoides*), Plectoidea (*Plectus* to *Anonchus*), Axonolaimoidea (*Axonolaimus*), Chromadoroidea (*Paracanthonchus, Tripyloides*), Monhysteroidea (*Sphaerolaimus* to *Theristus*). Original.

In *Cucullanus* though the esophagus is clavate instead of being divided (Fig. 76) we observed essentially the same nuclear arrangement and gland orifice positions in the posterior swollen region. The nuclei correspond to those of the posterior part of the camallanid esophagus.

Esophagi of dracunculoids were studied by Jägerskiöld (1894), zur Strassen (1907), Mirza (1929), Hsü (1933), and Yamaguti (1935). In gross features the esophagus may be cylindrical as in *Micropleura*, clavate or fusiform as in *Philometra*, divided into a short narrow anterior muscular part and a long wide posterior glandular part as in *Philonema*, *Dracunculus* and *Avioserpens*. In *Dracunculus* and *Avioserpens* the glandular region is constricted in the latitude of the nerve ring (Fig. 76). In *Dracunculus*, *Avioserpens* and *Philometra* the anterior end of the esophagus takes the form of a subglobular swelling but does not do so in *Philonema* and *Micropleura*.

The radial and marginal nuclei in *Dracunculus*, *Avioserpens* and *Micropleura* follow the same arrangement as in *Camallanus*, R I and R VI being divided groups. All these nuclei are also in the same position relative to the level of the orifice of the esophageal glands. In *Dracunculus* and *Avioserpens* the region posterior to the constriction at the nerve ring corresponds to the posterior part of the esophagus of *Camallanus*; in the two former genera an additional peculiarity is observed; the dorsal esophageal gland and its nucleus are tremendously enlarged (Fig. 81). In *Philometra* the fourth and fifth groups of radial nuclei (R IV and R V) are also divided into two subgroups each, but the marginals and radials retain the same relative positions as in *Dracunculus*; however, in the latter the subventral glands (Fig. 76) are greatly reduced in size. Yamaguti (1935) described a genus, *Icthyofilaria*, with a posterior glandular appendage similar to that observed in *Contracaecum*. However a study of the histology of this structure has not been made.

Philonema represents the ultimate in esophageal gland development of the Dracunculoidea. This genus has the typical spiruroid-filarioid esophagus, a fact that opens the question whether or not it is correctly placed in the suborder Camallanina. All three esophageal gland orifices are located in the posterior, much enlarged glandular region of the esophagus and the esophageal glands are multinucleate. The radial nuclei of all six groups are arranged in triplets, i. e., all six groups are subdivided; since there are three groups of radials in the anterior muscular part of the esophagus one would judge this part to be homologous to the anterior portion of the esophagus of *Camallanus*; the position of the subventral gland orifices (Fig. 76) is in support of this view. The dorsal gland orifice is shifted considerably posteriad in this form. The esophageal glands each contain several hundred nuclei of varying sizes often arranged in apparent "constellations" such as one would expect from nuclear budding induced by gigantism. Perhaps this case is analogous rather than homologous to that of the multinucleate glands of spiruroids. In the spiruroids the many gland nuclei apparently arose through typical division because they are approximately equal in size. If *Philonema* did arise separately the gigantism of these glands in *Dracunculus* and *Philometra* might be correlated with the unequal nuclear divisions (amitosis?) in the gland of *Philonema*.

SPIRURINA.—There are two esophageal features common to the forms contained in the suborder Spirurina, namely, that the esophageal glands are always multinucleate and that the dorsal gland always opens in the glandular (posterior) part. Esophagi of members of this group are grossly cylindrical or divided into a short anterior muscular part and a wide posterior glandular part. Even in forms with a cylindrical esophagus these two parts are distinguishable on the basis of their consistency. The gross form of the esophagus appears to be of no phylogenetic significance since changes from distinctly "divided esophagi" to cylindrical esophagi occur sporadically within groups of closely related genera.

In all except one family the marginal and radial nuclei are of the same number as in the suborder Camallanina (totals of 12 and 36 respectively); in this one exceptional family, the Physalopteridae, there appears to be a non-limited number of radial nuclei (Fig 81); the radial nuclear sets usually are all subdivided, three in each group, but in some forms (Fig. 81) the second, third or fourth may be partially or not at all subdivided. Regardless of the gross apparent extent of the glandular region, in all forms the anterior muscular part contains only the first 12 radial nuclei and the posterior glandular part contains 24 radial nuclei (or more as in *Physaloptera*

maxillaris); the anterior part does not, therefore, correspond to the entire anterior part of the esophagus of *Camallanus*. Since the fifth and sixth groups of radial nuclei of *Camallanus* appear to correspond to the radial nuclei of the bulb of *Rhabditis* and the fifth and sixth groups of *Ascaris*, one would conclude that the remainder of the esophageal nuclei are homologous to those of the corpus; since the second group of marginal nuclei is typically anterior to or opposite the third group of radial nuclei, the third and fourth groups of radial nuclei of *Camallanus* are probably homologous to the second group of radial nuclei of *Rhabditis* which has divided in this form; hence the second group of marginal nuclei and third and fourth groups of radial nuclei are homologous to the metacorpus; if, likewise, the first group of radial nuclei of *Rhabditis* divided as also the first group of marginal nuclei of this form (the marginals migrating anteriorly) then the homologies indicate that whatever region contains the first and second group of radial nuclei in a spiruroid is the procorpus. Therefore, we may say with a reasonable degree of certainty that the anterior muscular region of these forms is the *procorpus* and the posterior glandular region is *metacorpus* and *bulbar region*.

The esophageal glands of spiruroids and filarioids have very many nuclei, varying from a minimum of about 30 each to a maximum that can scarcely be estimated in forms such as *Physaloptera*. The increase in nuclear number in these forms has apparently progressed with regularity in division since the nuclei are fairly equal in size.

Basir (1949) enumerated the nuclei of the esophagus of *Physaloptera varani* as follows: anterior muscular region with 2 sets of 3 marginal nuclei and 4 sets of 3 radial nuclei; posterior glandular region with unlimited number of radial nuclei arranged in sets of 3 or 6; only 5 nerve cells; esophageal gland nuclei extremely numerous.

The esophageal lumen may be simple-triradiate as in *Rictularia coloradiensis* (Fig. 82), somewhat dilated but convergent peripherally as in *Ascarophis* or contain marginal "tubes" as in *Desmidocercella numidica* and *Diplotriaenoides* (Fig. 82). Furthermore the esophageal lining may have thickened regions as in *Monopetalonema physalarum*, sometimes with distinct attachment points, or it may be unmodified. Thus far the possible phylogenetic significance of such structures has not been determined but modifications appear to be most pronounced in the forms found in body cavities or tissues, i. e., Filarioidea. The form of the esophago-intestinal valve changes from a distinctly triradiate shape such as is found in *Rictularia* to a dorsoventrally flattened structure such as is found in *Physaloptera*.

Further information on nuclear arrangement, esophageal lining and shape of lumen and esophago-intestinal valve, will result in substantial evidence bearing on the inter-relationship of members of the suborder Spirurina.

B. CLASS APHASMIDIA

Aphasmidians as a whole have no single esophageal feature in common. Like the subclass Phasmidia they are divisible into two major groups on the basis of the esophagus, the first order, Chromadorida, corresponding to the order Rhabditida, contains forms in which the esophagus is primarily divisible into corpus, and bulbar region, the second order, the Enoplida, contains forms in which this division is not grossly apparent.

CHROMADORIDA.—The esophagi of members of this order have three uninucleate esophageal glands; the dorsal gland orifice is situated at or near the base of the stomatal region (never, so far as is known, does the gland empty directly into the stoma through a tooth); the subventral gland orifices are at the base of the corpus. Representatives of this order have been studied by the writers (1936).

The suborder Monhysterina contains forms in which the esophago-intestinal valve is relatively well developed, often quite elongated, and dorso-ventrally flattened, or rounded rather than triradiate.

Of the whole Aphasmidia the representatives of the superfamily Plectoidea most closely approach the Rhabditoidea, for in this group forms such as *Plectus* have a rhabditoid esophagus with a well developed valve in the bulb, but unlike *Rhabditis* the pro- and metacorpus are seldom apparent on gross study. The lumen of the corpus terminates marginally in distinct tubes and the valve of the bulb works in a manner similar to that of *Rhabditis*. Other plectoids have an elongated bulbar region without valves and in some forms the corpus also may be quite elongate (*Leptolaimus*), while in others (*Anonchus*), the entire esophagus may be cylindrical (Fig. 82).

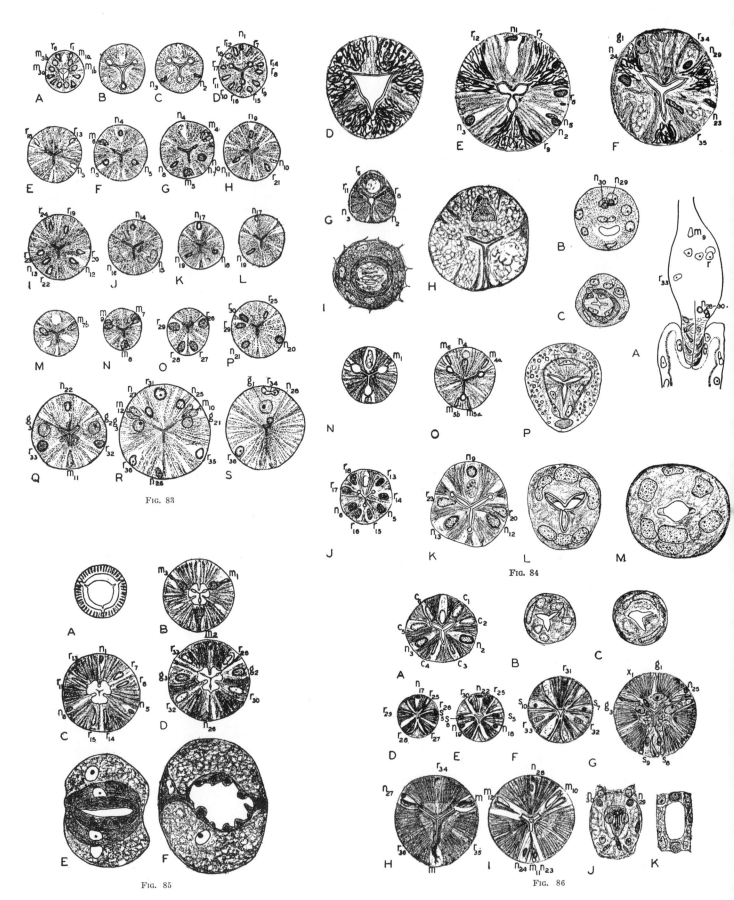

Fig. 83

Fig. 84

Fig. 85

Fig. 86

FIG. 83. Esophagus of *Anaplectus granulosus*. A–F—procorpus; F–L—metacorpus; M–S—bulb.

FIG. 84. A–C—*Anaplectus granulosus*. (A—longitudinal section through bulb and esophago-intestinal valve.) B–C—Cross sections, through esophago-intestinal valve (also includes very small part of bulb with n29-30). D–F—*Anonchus mirabilis*. (D—Corpus at base of stoma; E—corpus somewhat further posterior; F—bulbar region). G–I—*Camacolaimus prytherchi* (G—Corpus; H—base of bulbar region showing g1 and n28-30; I—esophago-intestinal valve). J—*Axonolaimus spinosus*, (corpus). K–M—*Sabatieria vulgaris* (K—Corpus; L–M—

esophago-intestinal valve). N–P—*Paracanthonchus* sp. (N—Anterior part of corpus; O—corpus somewhat more posterior; P—esophago-intestinal valve).

FIG. 85. *Tripylium carcinicolum* v. *calkinsi*. (A—Stomatal region; B–C—corpus; D—bulbar region; E–F—esophago-intestinal valve).

FIG. 86. A–C—*Microlaimus* sp. (A—Corpus; B–C—esophago-intestinal valve). D–K—*Chromadora* sp. (D–I—Serial section through bulbar region; J–K—esophago-intestinal valve).

All after Chitw. & Chitw., 1936, J. Wash. Acad. Sc., v. 26 (8).

The musculature of the bulbar region may be reduced so that this region forms an elongate glandular swelling (*Camacolaimus, Anguinoides, Aphanolaimus*) similar to that part of the esophagus of *Ditylenchus*, and in still other forms such as *Onchium ocellatum* the esophageal glands may project posteriorly beyond the base of the esophagus. It would appear, therefore, that not only in stylet and stomatal formation but also in esophageal formation the Plectoidea present a parallel series to the Rhabditina. In two points all plectoids are similar, the esophageal lumen of the corpus peripherally is terminated by marginal tubes and the esophago-intestinal valve is definitely dorso-ventral in symmetry (Figs. 83-84).

The corpus of all forms contains four groups of six radial nuclei (24) and two groups of three (or three double) marginal nuclei (total 6 or 12). The radial muscles of the corpus are more or less concentered but no forms are thus far known in which the lining is thickened forming attachment points. As in *Rhabditis* in the bulbar region there are 12 radial and 6 (or 12) marginal nuclei, the first set of 6 radials and the first set of marginals forming the prevalvar region; the succeeding radials are arranged in two groups of three which together with the second set of marginals form the postvalvar region. In *Anaplectus granulosus* (Fig. 83) and similar forms with unusually well developed "tubes" at the ends of the esophageal radii the marginal nuclei of each group are double, one member of each pair being on each side of

each tube, while in forms such as *Camacolaimus prytherchi* in which the tubes are minute (Fig. 84), no such doubling of marginal nuclei occurs.

Representatives of the superfamily Axonolaimoidea have esophagi like plectoids except that no representative of this group has a valved bulb and no forms are known in which the bulbar region is principally glandular through enlargement of esophageal glands at expense of muscular tissue (Figs. 82 &

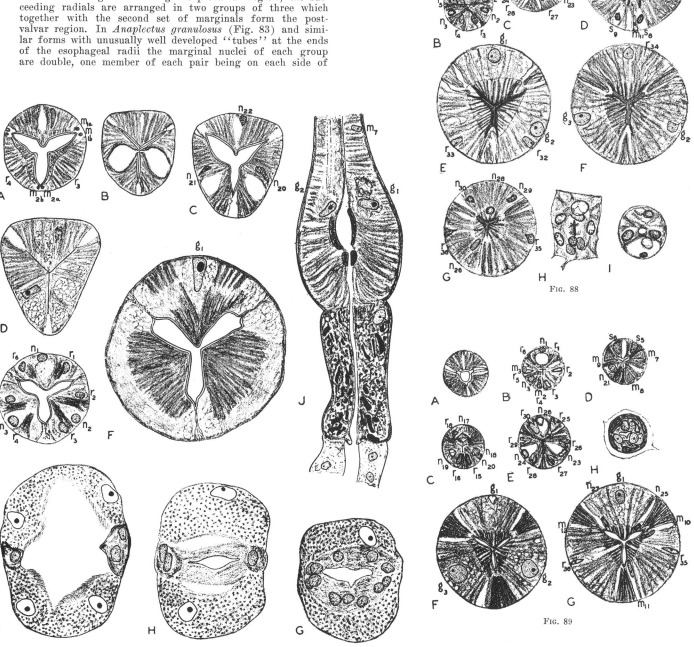

Fig. 87

Fig. 88

Fig. 89

FIG. 87. A-D—*Theristus setosus*. E-I—*Terschellingia pontica* (E—Corpus; F—bulbar region; G-I—serial sections through esophago-intestinal valve). J—*Desmolaimus zeelandicus* v. *americanus* (Longitudinal section through bulb, and esophago-intestinal valve).

FIG. 88. *Ethmolaimus revaliensis*. (A—Stomatal region; B—corpus; C-G—bulbar region; H-I—esophago-intestinal valve).

FIG. 89. *Monoposthia hexalata*. A-C—corpus (A—in stomatal region; B—just posterior to the orifice of dorsal gland; C—near base). D-G—bulbar region. H—esophago-intestinal valve.

All after Chitw. & Chitw., 1936, J. Wash. Acad. Sc., v. 26 (8).

84). The entire esophagus may be clavate (*Comesoma, Sabatieria*) or the corpus may remain grossly distinct (*Axonolaimus, Araeolaimus*); the bulbar region is always rather elongate, never in the form of a definite bulb. As a rule one may distinguish procorpus and metacorpus in totomount specimens through the change in the esophageal lining (Fig. 15E) as the size of the marginal tubes becomes reduced; no forms are known with thickened attachment points for the radial muscles although the muscles themselves are concentered.

The nuclei of the esophagus of axonolaimoids are essentially as in *Plectus* except that in some forms (*Sabatieria vulgaris*) the six posterior radial nuclei tend to assume a hexa-rather than a trisymmetry in their arrangement. The esophago-intestinal valve is as elongated as in plectoids and of slightly different form (Fig. 84 L-M); it contains about 10 to 12 nuclei.

Representatives of the superfamily Monhysteroidea are of three general types, linhomoeids, monhysterids and siphonolaims all characterized by a smaller number of radial nuclei than is found in plectoids and axonolaimoids. The Linhomoeidae contains forms with a clavate esophagus which may (*Desmolaimus*) or may not (*Linhomoeus*) be terminated by a well marked muscular swelling or bulb; the esophago-intestinal valve is usually definitely elongated but may be rather short (*Sphaerolaimus*). In these forms the radial muscles are concentered and this condition is usually accompanied by thickened cuticular attachment points on the lining (Figs. 82, 85, 87). The esophageal lumen while modified due to these attachment points is devoid of marginal tubes. Forms such as *Tripylium carcinicolum*, *Terschellingia pontica* and *Desmolaimus zeelandicus*, v. *americanus* have only 30 to 33 radial nuclei, the reduction or difference in number being in the metacorpus; the second group of marginals is also apparently lacking in *Terschellingia*. We find in these forms the same peculiar distribution of the radial nuclei of the posterior part of the bulb previously mentioned in the axonolaimoids (Fig. 90). The esophago-intestinal valve of linhomoeids is often associated with specially differentiated intestinal cells (*Tripylium*) and in some forms such as *Desmolaimus* this structure forms a separate and distinct organ which may be termed the ventricular column (Figs. 82, 87J). Monhysterids (Figs. 82, 87A-D) have a more cylindrical esophagus, though never a distinct linhomoeid bulb, and the ventricular column is never elongated; the posterior radial nuclei have the same odd type of arrangement as linhomoeids but the radial musculature is of a dispersed type throughout, the esophageal lining without thickenings, the lumen very simply triradiate and the esophago-intestinal valve less elongate but otherwise similar to that of *Terschellingia* (Fig. 87G-I); the latter structure contains 19 to 23 nuclei. Siphonolaims have a very short corpus, an elongate isthmus, a short glandular bulbar region and a short esophago-intestinal valve containing six nuclei. The nuclei of the bulbar region are as in linhomoeids and monhysterids but only three marginal and 18 radial nuclei are present in *Siphonolaimus conicus*. Monhysteroids are peculiar in having very minute marginal nuclei (Fig. 87A).

In the suborder Monhysterina one first encounters paired pigment spots in the subdorsal or dorsolateral regions of the procorpus in a few genera of the Camacolaimidae (*Onchium ocellatum*) and Axonolaimidae (*Araeolaimus*). These peculiar structures are situated in the dorsal parts of the subventral sectors (dorsolaterally) or the lateral parts of the dorsal sector (subdorsally); there is, in the first mentioned genus, a pair of acorn like pigment masses each provided with a hyaline lens. While these are termed ocelli, no one has thus far connected them with sensory nerves. In *Monhystera paludicola* similar ocelli are situated in the body cavity dorsal to the esophagus.

CHROMADORINA.—In the suborder Chromadorina the esophago-intestinal valve is usually short though sometimes elongate but never dorso-ventrally flattened; it always retains its triradiate character. None of the representatives of this group preserve semblance to rhabditoid or plectoid esophagi.

In the Chromadoroidea the esophagus consists essentially of a cylindrical corpus and a bulbar region (Fig. 91); the esophageal lumen is triradiate, the lining unmodified, so far as known, though the radial muscles are definitely concentered. Members of the Chromadoridae such as *Chromadora* and *Ethmolaimus revaliensis* have a very short esophago-intestinal valve containing 12 or 13 nuclei (Figs. 86, 88); 12 marginal and 36 radial nuclei are present in the esophagus, the posterior nuclei of the bulbar region being arranged in two typical groups of three. In *Microlaimus dentatum* the esophago-intestinal valve containing 11 nuclei is elongate reminding one of *Terschellingia*, but differs from that form in being tri-

radiate (Fig. 86 B-C). Chromadoroids show no tendencies toward diminution in nuclear numbers such as was noted in monhysteroids but, rather the opposite. Several "additional nuclei" make their first appearance in this group; the arrangement and position of these nuclei (s, c, and x) are characteristic of both chromadoroids and desmodoroids (Fig. 90).

Cyatholaims generally have a much elongated tripartite bulbar region or a cylindrical esophagus. In representatives of the former type (*Paracanthonchus coecus*) the lumen of the esophagus may be slightly enlarged marginally (Fig. 84 N-P). Choanolaims and tripyloids, on the contrary, have a typical triradiate lumen with long rays as in monhysterids rather than the short rays and minute lumen typical of the Chromadorina. All of these forms have a rather large and well developed but short esophago-intestinal valve.

Desmodoroids have the same general esophageal organization and the same general pattern of nuclear distribution (Fig. 89) as do chromadoroids but the esophageal lining may have thickened cuticular attachment points (*Metachromadora onyxoides*).

The so-called "multiple bulb" of chromadoroids and desmodoroids deserves special mention. In both superfamilies one may trace series of forms from a subspheroid bulb to a subcylindroid tripartite bulb thence to a cylindrical esophagus. The apparent sub-division of the bulbar region in forms such as *Monoposthia hexalata* and *Ethmolaimus revaliensis* (Fig. 91) into two parts is due to a particular arrangement of the musculature and glandular tissues and a break in the thickness of the esophageal lining at the points where one muscle ceases and another begins. This break indicates the separation of radial subgroups of muscles containing nuclei r_{31-33} and r_{34-36} respectively and is essential to the function of this type of bulb. The musculature containing the radial nuclei r_{25-30} is in the anterior part of the tripartite bulb and the corresponding muscles are a separate functional unit.

Pigment spots in the anterior part of the corpus are commonly found in members of the Chromadoridae and like those previously mentioned in the Camacolaimidae and Axonolaimidae they are situated in the subdorsal marginal regions of the esophagus but in this case the pigment is diffuse rather than concentrated and is not provided with a lens. It seems proper to designate these as mere *pigment spots* while reserving the term *ocelli* for concentrated pigment bodies accompanied by lenses. De Man (1889) described true ocelli situated dorsal to the esophagus in the body cavity of *Cyatholaimus demani* (Syn. *Cyatholaimus ocellatus* of de Man).

The esophagi of desmoscolecoids have not as yet been adequately studied. The esophagus is grossly rather cylindroid but narrow, the three esophageal glands projecting posteriorly indicating a reduction in the musculature of the bulbar region. Such forms (Fig. 17) have well developed, brilliantly colored pigment bodies dorsal to the base of the esophagus.

ENOPLIDA.—The esophagi of representatives of this order commonly (always?) have five or more uninucleate or multinucleate esophageal glands. The esophagus usually has an elongate muscular anterior part followed by an elongate glandular posterior part, such divisions resembling those of spiruroids and filarioids. The location of the esophageal gland orifices varies widely in the group.

ENOPLINA.—Esophagi of representatives of this sub-order have been studied by Marion (1870), de Man (1886, 1904), Jägerskiöld (1901), Türk (1903), Rauther (1907) and the writers (1937). In general we find two types of esophagi in this group: the first type in which the esophageal glands have orifices rather near this nuclei, that is, in the posterior glandular part of the esophagus (such forms are included in the superfamily Tripyloidea; and the second type, in which the subventral esophageal glands open anteriorly either near the base of the stomatal region or in subventral teeth (such forms may occur in either of the superfamilies Enoploidea or Tripyloidea).

In the Tripyloidea the esophagus either is cylindrical (*Prionchulus, Tripyla*) or consists of an elongated narrow corpus and a slightly wider elongate glandular region (*Alaimus*); in all such forms studied there are five uninucleate esophageal glands, one dorsal and four subventral (Fig. 91). Mononchs such as *Prionchulus muscorum* have concentered radial muscles and well developed cuticular attachment points (Fig. 92); the esophago-intestinal valve which is triradiate and quite massive (Fig. 92M), contains 22 nuclei in *P. muscorum*. A total of 36 radial and nine marginal nuclei have been observed, the radials (12 in procorpus, 12 in metacorpus and 12 in glandular region) are arranged in sets of six, indicating that even those of the posterior group act as a single unit rather than as two units

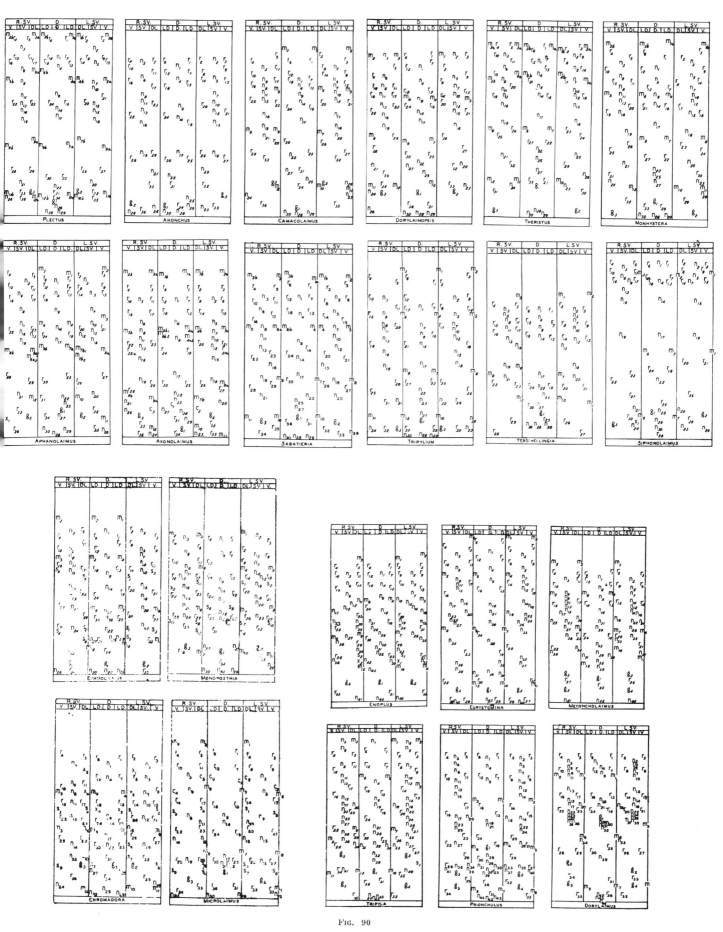

Fig. 90

Tables of esophageal nuclear distribution of Aphasmidians. Chitw. & Chitw., 1936, J. Wash. Acad. Sc., v. 26 (8) & 1937, v. 27 (12).

as in rhabditids, plectids and chromadorids. Nuclear distribution (Fig. 90) indicates quite definitely that the glandular region of the esophagus of *Prionchulus* as well as other tripyloids, enaploids and dorylaimoids is homologous to the bulbar region of the orders Rhabditida and Chromadorida rather than the glandular region of spiruroids since it does not contain the radial nuclei characteristic of the metacorpus (total number of radials 18 in spiruroids). The five esophageal gland nuclei are subsequal in size and each of the accompanying glands opens nearly directly into the esophageal lumen (Fig. 94D).

In *Tripyla papillata* and *Trilobus longus* the musculature is dispersed, no cuticular attachments are present and the esophago-intestinal valve is quite massive, consisting of an external part containing six or seven large nuclei and an internal part containing up to 100 nuclei (Fig. 94 E-L). This peculiar structure, sometimes termed a pseudo-bulb, or bulb, is commonly thought to be a part of the esophagus proper but this does not appear to be the case. It is a further development of the type of valve found in *Prionchulus*. The five esophageal glands are similar to those of the latter genus except that the

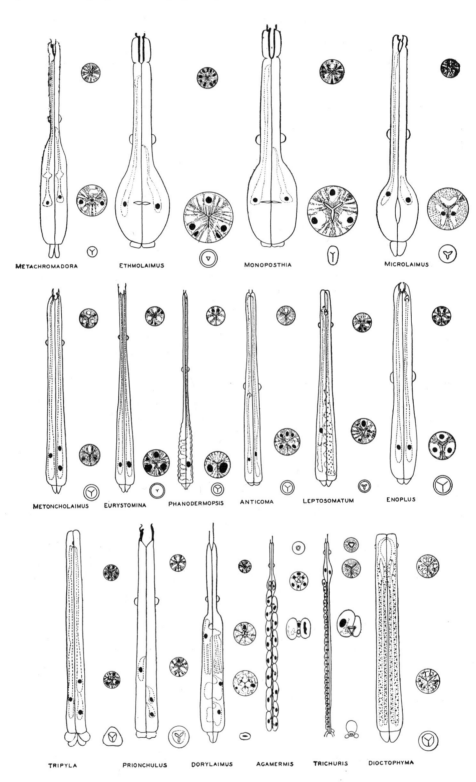

FIG. 91

Diagrams of esophagi including representatives of the Desmodoroidea (*Metachromadora, Monoposthia*), Chromadoroidea (*Ethmolaimus, Microlaimus*), Enoploidea (*Metoncholaimus* to *Enoplus*), Tripyloidea (*Tripyla, Prionchulus*), Dorylaimoidea, Trichuroidea and Dioctophymatoidea. Original.

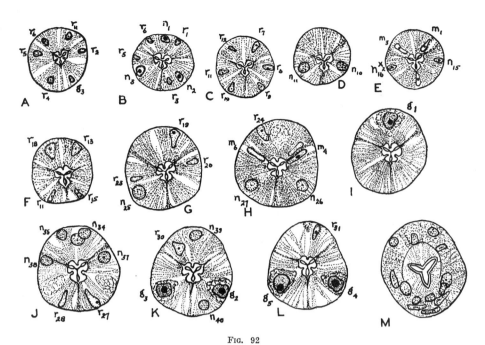

FIG. 92

Prionchulus muscorum, sections (a few left out)
in series, M—esophogo-intestinal valve. Original.

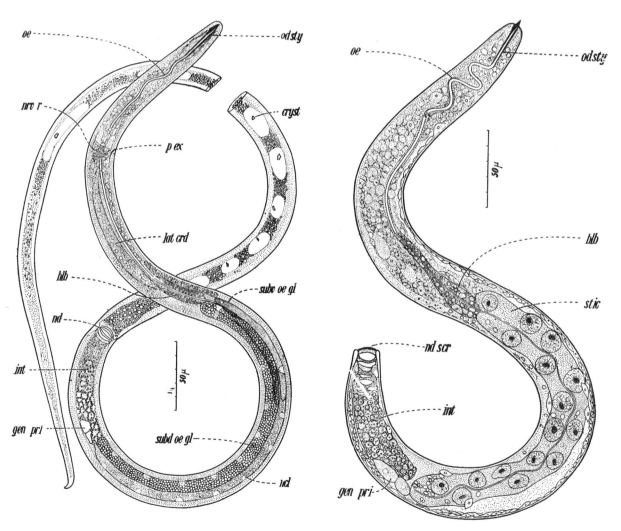

FIG. 93

Agamermis decaudata. A—Preparasitic larva. (Note primary esophageal glands, subd. oe. gl., subv. oe. gl.). B—Larva after 4 days in
host. (Note development of stichocytes.) After Christie. 1936, J.
Agric. Res., v. 52 (3).

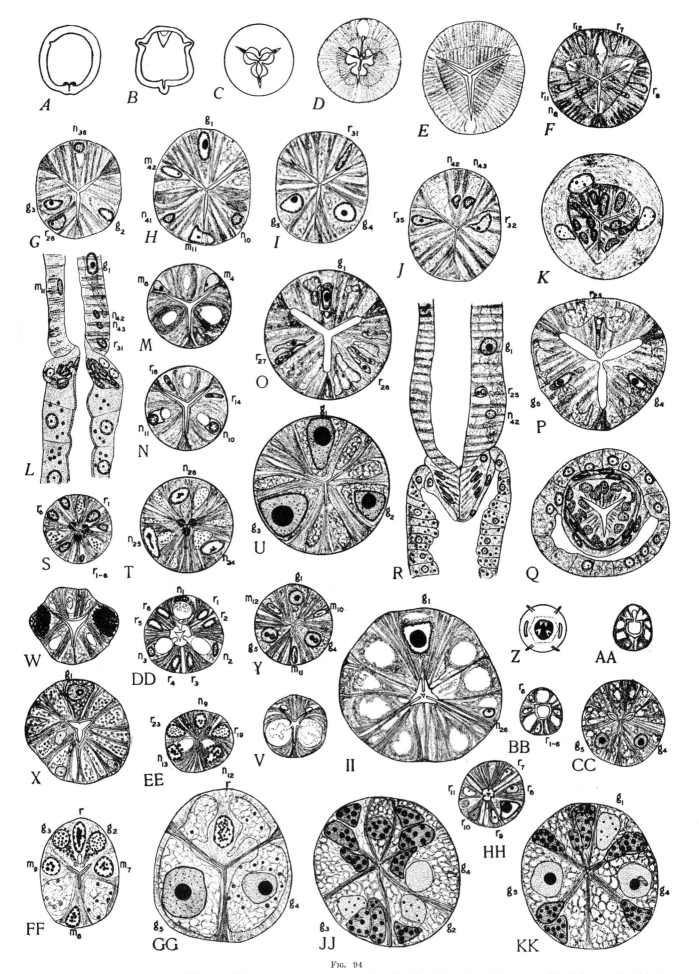

Fig. 94

Esophagi of members of the Enoplida. A-D—*Prionchulus muscorum* (A-B—Showing cuticular lining in the stomatal region; C—procorpus; D—bulbar region at level of dorsal gland orifice) E-L—*Tripyla papil-lata.* (E—Stomatal region; F—corpus; G-I—Bulbar region (glandular); K—esophago-intestinal valve or "pseudobulb"; L—Longitudinal through regions covered by H-K). M-R—*Metoncholaimus pristiuris*

92

dorsal and the first pair of subventral esophageal glands extend the base of the stomatal region where they open. *Alaimus* is similar to *Tripyla* except that the radial muscles are *con-centered;* the triradiate esophago-intestinal valve is quite small and inconspicuous, and the subventral glands do not extend beyond the enlarged glandular region.

In the superfamily Enoploidea the gross form of the esophagus varies considerably, there being types with a cylindrical esophagus, types in which the narrow corpus is followed by a gradually expanding elongate glandular region (conoid) and types in which the corpus is slightly narrower and set off internally from an elongate cylindrical bulbar region. Due to the distribution of radial muscles the conoid type of esophagus may be multibulbar (*Polygastrophora, Bolbella*), crenate in outline (*Phanodermopsis*) or smooth in contour (*Oxystomina, Eurystomina, Leptosomatum*). Among the latter, two types are distinguishable in regard to the muscular development. In *Oxystomina* and its relatives the musculature is poorly developed while in the remaining forms it is well developed.

In the Oncholaiminae the esophagus is of a cylindrical type, the esophageal lining simple, the lumen also simple and triradiate throughout; the esophago-intestinal valve is triradiate, moderately elongate and contains numerous nuclei. There are 12 marginal nuclei but only 27 radial nuclei in *Metoncholaimus pristiuris* and according to their distribution (Fig. 90), we may judge that compared with *Tripyla*, the third group of radial nuclei (of *Metoncholaimus*) represents a case of failure of cleavage of the last set of radials of the corpus of *Tripyla*. In the later form the final cleavage results in the third and fourth sets (RIII and RIV) or 12 nuclei while in *Metoncholaimus* only six nuclei are present. Likewise the three giant radial nuclei (Fig. 94 O-P) of the bulbar region also may represent a cleavage failure. In this form one dorsal and two pairs of subventral gland nuclei are to be found in the posterior part of the esophagus; separate orifices of the two glands in each sector have not been distinguished. The glandular protoplasm of the subventral sectors seems to be continuous and one concludes that in this case the subventral glands are binucleate; the ducts from all three glands extend anteriorly into the teeth of the stoma where each opens in a minute pore. Rauther (1907) recorded four esophageal glands in the subventral sectors of *Oncholaimus vulgaris*. He found separate pores for the hindmost pair situated somewhat anterior to the nerve ring and dorsal to the center of their respective sectors.

Eurystomina americana has a very similar esophagus to *Metoncholaimus* but its narrow part (corpus) has distinct cuticular attachment points and the musculature is concentered in the corpus (Fig. 94T) but not in the bulbar region; 30 radial nuclei are present, there being 18 in the corpus as in *Metoncholaimus*, but there are 12 in the glandular (bulbar) region arranged in two sets of six as in *Tripyla* and *Prionchulus*. The esophageal glands extend the entire length of the esophagus as in oncholaims but only three gigantic gland nuclei (Fig. 94U) have been observed, the right subventral being much the largest. Only eight nuclei have been seen in the esophago-intestinal valve, the anterior part of which is triradiate while the posterior part is dorsoventrally flattened.

Enoplus communis has a cylindrical esophagus more like that of oncholaims than *Eurystomina* but 33 radial nuclei are present, the corpus containing the full number of 24 nuclei as in *Tripyla*, while the bulbar region (glandular part) contains only nine as in *Metoncholaimus*, the last group of three indicating again a case of suppressed cleavage (Fig. 95A-AA). There are three large subequal esophageal gland nuclei in the bulbar region and the corresponding glands each has an orifice at the anterior end of the esophagus, at the base of the teeth. The short, well developed, triradiate esophago-intestinal valve of *Enoplus* contains 11-15 nuclei (number questionable). Rauther (1907) obtained the same total as the writer's (i.e., 109 nuclei which includes nerve cell nuclei and esophago-in-

testinal valve nuclei) but he differs in some cases as to the functions he attributes to specific ones. He also describes two lateral glands (the nuclei of which we designate x_{1-2} in Fig. 90 & 95S) near the mid region of the esophagus which open anteriorly near the level of the amphids in the marginal region. We agree that these probably are esophageal glands but were unable to distinguish their orifices.

The esophagus of *Phanodermopsis longisetae* is interesting because it is representative of forms with a crenate glandular esophagus which is commonly termed "cellular." The illusion of cells (Fig. 91) is created through the localization of extremely sparse musculature at intervals, separating swollen regions of nearly exclusive glandular tissue. Though the esophagus is conoid and crenate the nuclear distribution is similar to that found in *Enoplus* with the following exceptions: There are 36 radial nuclei, the most posterior group being subdivided into two sets of three nuclei; the two chief subventral gland nuclei are much enlarged as are the glands, while the dorsal gland nucleus is situated far forward and much reduced in size, and the small subsidiary (lateral glands of Rauther) subventral gland nuclei are situated near the dorsal gland nucleus (Fig. 94FF). The chief subventral glands (Fig. 94GG) empty into the stoma while the dorsal and subsidiary glands appear to have their orifices posterior to the nerve ring.

Leptosomatum elongatum, representative of still another type, has multinucleate subventral esophageal glands. The conoid esophagus is of smooth contour, somewhat vesiculate internally. The anterior part (corpus) contains the same nuclei found in *Phanodermopsis* while the posterior part contains such a large number of nuclei (about 23 in each chief subventral gland) that it would be difficult to designate the types accurately. The dorsal gland is uninucleate, the nucleus (Fig. 94X) much larger than other nuclei of the esophagus; the subventral gland orifices are at the anterior end of the esophagus while the dorsal gland orifice is somewhat further posterior. In the corpus the musculature is concentered (Fig. 94W) and the esophageal lining thick but without attachment points. According to Jägerskiöld (1901), Türk (1903) and Rauther (1907) the esophagi of the related genera *Thoracostoma* and *Cylicolaimus* (which are grossly cylindrical) have similar esophagi except that the esophageal lining has definite cuticular attachment points. Rauther was able to distinguish small subsidiary subventral (lateral) glands as in *Enoplus*. Other enoplids, such as *Anticoma* (Fig. 94V) and *Rhabdodemania*, with cylindrical esophagi have gland orifices in similar positions to those above described, with a simple i.e., enoploid esophageal lining.

In the whole order Enoplida pigment spots or ocelli occur only in the superfamily Enoploidea, families Enoplidae and Oncholaimidae, and are of sporadic appearance in these groups. In the Oncholaiminae such pigment is rather diffuse in the musculature of the corpus while in *Enoplus* it is concentrated in a pair of subdorsal spots in the marginal areas near the anterior end. Well developed "ocelli" have been described in leptosomatids, phanodermatids and enchelidiids.

Finally we come to the family Ironidae which in many ways appears to have closer affinities with the Mononchidae and Dorylaimidae than with other enoploids. *Ironus* (Fig. 94Z-CC) and *Ironella* have cylindrical esophagi with well developed cuticular attachment points, concentered radial muscles, and 5 subequal esophageal glands. Like *Enoplus, Oncholaimus,* and *Tripyla,* three of the glands have orifices into the stomatal region. *Cryptonchus nudus,* though otherwise very similar to *Ironus,* has its esophageal glands confined to the posterior part of the esophagus (Fig. 94DD & Y). It seems, therefore, to be intermediate between such forms as *Prionchulus* and *Ironus.*

DORYLAIMINA.—This suborder is perhaps more compact in fundamental esophageal organization than the suborder Enoplina though the gross morphology is certainly more diverse and includes more odd types (Fig. 91). Dorylaimins have one point in common with each other and with the family Mononchidae, namely that none of the esophageal glands extend to the stomatal region. The suborder contains superfamilies with esophagi of two types, one in which the glandular region is tremendously elongated, the subventral glands reduplicated and protruding from the esophageal contour, and the other in which the glandular region is either short or only moderately elongated, and in which only two pairs of subventral glands are present. This latter group includes the soil and aquatic species of the superfamily Dorylaimoidea while the former group includes the parasitic nemas of the superfamilies Mermithoidea and Trichuroidea.

Dorylaimoids have a cylindrical corpus followed by either an elongated cylindrical glandular region (Dorylaimidae) or a short pyriform glandular region (Diphtherophoridae, Leptonchidae) the parts sometimes separated in the latter instance

(M-N—Corpus; O-P—bulbar region; Q—esophago-intestinal valve; R—longitudinal sagittal through regions shown in O-Q dorsal to reader's right). S-U—*Eurystomina americana* (S—anterior part of corpus; T—posterior part of corpus; U—glandular region). V—*Anticoma litoris.* Near anterior end of esophagus, subventral gland orifice and subdorsal (dorsolateral) ducts. W-X—*Leptosomatum elongatum* v. *acephalatum.* (W—Region of ocelli, 5 ducts also visible; X—glandular region). Y-DD—*Cryptonchus nudus,* (Y—posterior end of glandular region; DD—corpus). Z-CC—*Ironus ignavus* (Z-head; AA—dorsal gland orifice (stomatal region); BB—beginning of corpus or end of stomatal region; CC—posterior part of glandular region (corpus proper is like fig. DD). EE-GG—*Phanodermopsis longisetae.* (EE—Posterior part of corpus; FF—anterior part of glandular region; GG—posterior part of glandular region). HH-II—*Dorylaimus obtusicaudata.* (HH—Corpus at level of stylet cell; II—glandular region at level of dorsal gland nucleus). JJ-KK—*Triplonchium* sp. (Glandular region showing five esophageal gland nuclei.)

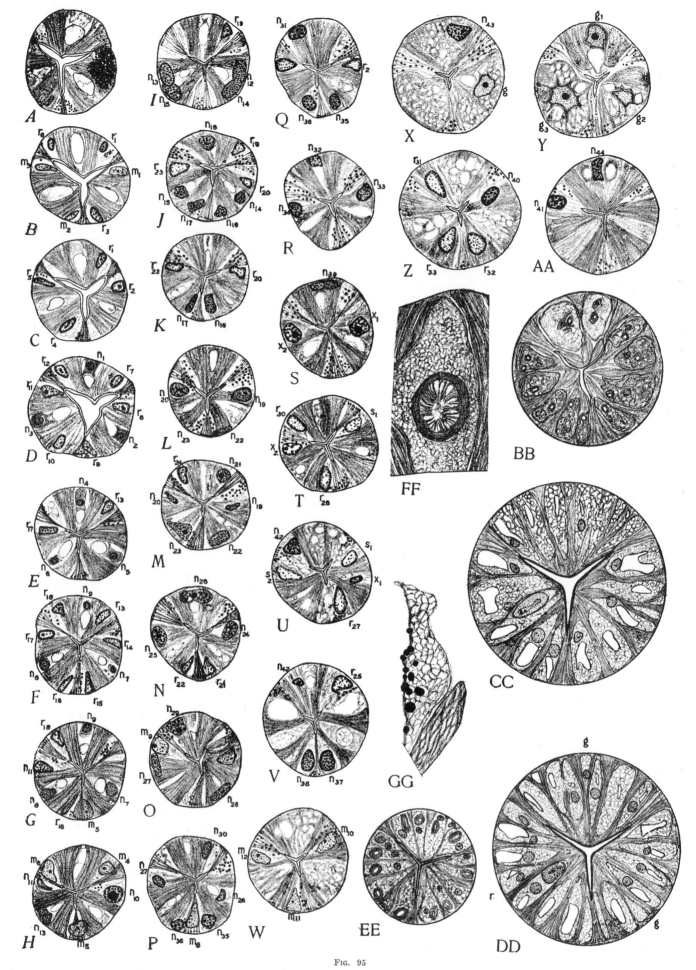

Fig. 95

A-AA—*Enoplus communis* v. *meridionalis.* (Serial sections with a few left out.) BB—*Dioctophyma renale.* CC-GG—*Soboliphyme baturini.* (CC—basal region of esophagus, radial nucleus also showing); DD— anterior to CC; EE—level of secondary tubules; FF—secondary tubule enlarged; GG—acidophylic granules in base of subventral gland tubules. Original.

by a more or less distinct isthmus; the esophageal lumen is subtriangular anteriorly, rapidly becoming minute and triradiate with marked cuticular thickenings of the esophageal lining (Fig. 94HH); the musculature is strongly concentered. The radial musculature is well developed throughout the esophagus in leptonchids and dorylaimids but rather degenerate in the glandular region of diphtherophorids.

Dorylaimus obtusicaudatus has an esophagus extremely similar to that found in *Prionchulus* for the anterior part (corpus) contains four sets of six radial nuclei (three sets in the narrow part, one at the junction of the anterior and posterior parts) and only one set of three marginals (Fig. 90); the posterior glandular (bulbar) region contains two sets of six radial and two sets of three marginal nuclei, the hindmost set of radials being somewhat subdivided but arranged in a manner indicating they act as a unit. Each of the five esophageal glands has its orifice near the level of the nucleus with a very short duct. The dorsal gland bifurcates, each branch continuing to divide dichotomously and the branches enter the subventral sectors, eventually coming to fill the entire nonmuscular part of the esophagus anterior to the subventral glands (Fig. 94II). These latter are found in two pairs considerably posterior to the dorsal gland; the first subventral gland on the right side is considerably smaller than the others. The esophago-intestinal valve of dorylaimids is very well developed, elongate, dorsoventrally flattened, and contains 27 nuclei.

There is in addition to the usual number of nuclei of the esophagus a large nucleus in the left subventral sector which, with its surrounding protoplasm, acts as a generative nucleus of the stylet (Fig. 94HH). As previously stated the stylet is formed in the procorpus and moves anteriorly to take its final position attached to the anterior end of the esophageal lining.

Leptonchus has an esophagus apparently identical internally with that of dorylaimids but only three large esophageal glands have been observed; presumably both of the first pair of subventral glands in this instance are small. *Triplonchium* of the Diphtherophoridae has very feebly developed, if any, thickened cuticular attachment points of the esophageal lining, and the dorsal and first pair of subventral gland nuclei (situated at the same level) is only slightly smaller than the posterior subventral pair (Fig. 94 JJ-KK).

The esophagi of the mermithoids have been studied by Rauther (1906), Hagmeier (1912), G. W. Müller (1931), Christie (1936) and the writers (1935, 1937) as well as many others who have added scattered bits of information. The peculiarity of the mermithoid esophagus is its relatively great length and tenuousness. In the adult stage, which does not feed, the greater part of the esophagus posterior to the nerve ring appears as a slender tube covered with a bit of protoplasm to which various types of cells are attached at intervals. The esophageal musculature appears to be entirely degenerate. One must turn to the larval stages of mermithids for an understanding of these structures.

The preparasitic larva has an esophagus consisting of a short anterior part terminated by a muscular swelling and followed by an elongate narrow posterior part with which glands are associated. There are three unicellular elongate esophageal glands extending posteriorly into the body cavity from the base of the anterior (non-glandular) portion, the dorsal being considerably larger than the subventrals; these *primary glands* (Fig. 93) become atrophied after the entrance of the larva into its host. In addition there is a double series of large cells (stichocytes) attached to and coextensive with the posterior part of the esophagus. These cells are secondary esophageal glands, probably representing a reduplication of the posterior pair of subventral esophageal glands of dorylaims and enoplids. In the parasitic stage these glands become much enlarged (Fig. 93) and later they too atrophy as the larva approaches adulthood. The lining of the esophagus of mermithids is rather clearly triradiate in the preparasitic larva, becoming subtriangular in later stages until finally the basic triradiate character is scarcely discernible (Fig. 96A-E). In *Agamermis decaudata* 48 large nuclei corresponding to the marginals and radials are present, most of them arranged with little indication of pattern, and there is even less indication of muscular fibers in the parasitic larvae (Fig. 96). There are 27 such nuclei anterior to the orifices of the primary glands indicating that this region corresponds to the corpus while there are 21 in the posterior part, indicating that it corresponds approximately to the bulbar region. These nuclei are situated within the wall of the esophagus proper. Each of the glands of both types has an orifice not distant from its nucleus. There is one peculiarity

of the esophagus which does not seem reconciled with the tremendous esophageal gland development in the parasitic stage, namely, that the lining disappears posterior to the orifice of the last stichocyte and the lumen does not connect with the degenerate intestine. The stichocytes are quite obviously functional and it is difficult to believe that fluid food is not drawn in through the esophagus but if so, we do not know its destination. The number of stichocytes varies from four to 16 or possibly more, according to the particular form.

The peculiar esophagus of trichuroids has long been a subject for study. Eberth (1860, 1863), Leuckart (1866), Rauther (1918), G. W. Müller (1929), Christenson (1935) and the writers (1929, 1935, 1937) have investigated various forms of this group. Ward (1917) proposed a separate suborder for trichuroids and mermithoids (Trichosyringata) based on the peculiarity of the esophagus. The esophagi of mermithoids and trichuroids are similar but not fundamentally different from other nematodes. Ward stated "A type of radically different character is the capillary esophagus . . . It consists of a row of cells, pierced throughout its entire length by a delicate tube." The cells of which he speaks are stichocytes or esophageal glands attached to but not "pierced by" the esophagus proper which is to a greater or lesser degree embedded in these cells. As in mermithids, the wall of the esophagus contains its own nuclei. Much discussion has centered around the significance and nature of the structures, but since the points now are clear, further discussion seems unnecessary.

The anterior part of the esophagus of *Trichuris ovis* is narrow and muscular, terminated by an elongated swelling; the lumen is triradiate, the lining thick but without attachment points. Within this region, besides nerve cells, one group of three marginal nuclei and two groups of six distinct radial nuclei are present. On this basis one might presume it to be the procorpus but within the terminal swelling a set of three large nuclei of esophageal glands is present (Fig. 91 & 96I). These glands, whose orifices are posterior to the nerve ring, doubtless correspond to the primary glands of mermithids. The posterior part of the esophagus is quite narrow and embedded in a single series of large cells (cell body "zellkorper" of authors). The narrow "tube" or esophagus proper is triradiate to hexaradiate and has its own wall containing radial nuclei as well as nerve cells occurring at intervals. Contrary to general supposition, this wall contains well developed radial muscle fibers (Fig. 96M). The large cells in which it is embedded are esophageal glands, each having a separate orifice reached or formed by a tube through the esophageal wall. This being the case, there is no fundamental difference from other nematodes in which the esophageal glands may come to lie outside the esophageal contour (*Contracaecum, Aphelenchus, Onchium*). Because of the fact that in larval trichuroids the stichocytes are more or less alternately paired and the orifices in the adult tend to alternate it seems reasonable to assume that the single row stichosome of *Trichuris* is a later evolutionary development from a double row of stichocytes such as is found in mermithoids. This view is supported by the illustrations of Janicki and Rosin (1930) of the esophageal region of *Cystoopsis* which shows two rows of stichocytes. The number of stichocytes seems to be variable to some extent within a given species of trichuroid.

At the end of the esophagus, unlike mermithoids, we find a direct connection of the esophagus with the functional intestine formed by a dorsoventrally flattened esophago-intestinal valve such as is present in dorylaimoids. Two large cells (Fig. 96O) described as glands with direct openings into the esophageal lumen, by Eberth (1860, 1863), are attached to the esophagus at its base. Neither Rauther (1918) nor the writers were able to distinguish any protoplasmic connection of these cells with the esophageal lining or any tubules in the cytoplasm of these cells. It now seems possible that these cells are enlarged mesenterial cells, homologous to the series of smaller cells supporting and covering the esophagus and stichosome.

DIOCTOPHYMATINA.—The esophagus of the suborder Dioctophymatina has received little attention since the first report of Schneider (1866) that the walls of the cylindrical esophagus of such forms contain numerous longitudinal "tubes." No additional information was added until the subject was recently reopened by the writers (1937). The esophageal lining is simple, the lumen triradiate and the well developed musculature dispersed in representatives of this group. There is no division, either grossly or internally, into anterior and posterior parts. The three massive esophageal glands have their orifices at the anterior end of the esophagus and begin branching dichotomously near the level of the nerve ring; glandular tissue is thereafter interspersed between

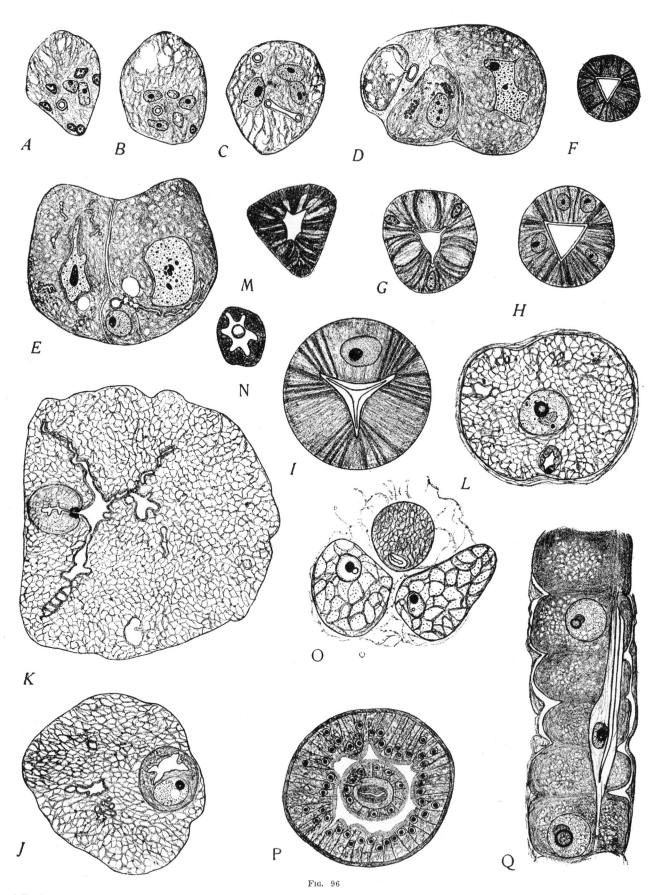

Fig. 96

A-E—*Agamermis decaudata*. (4 mm long parasitic larva.) (A-B—Anterior end of esophagus containing problematic nuclei, small, and first group of radial nuclei, larger; C—just anterior to nerve ring including r_{9-12}; D—at posterior level of primary glands showing most posterior gland orifice and one primary gland nucleus, the other gland stopped anterior to this section. Large cell to the right is the first stichocyte; E—mid-stichosome region showing esophageal tube, r type nucleus, two stichocytes and branched tubule). F-Q—*Trichuris ovis*. (F—Section through 3 anteriormost nuclei s_{1-3}; G—section at level of first marginal group; H—section at level of first radial group; I—section dorsal primary esophageal gland. J—stichocyte enclosing esophagus at level of radial nucleus; K—tubule orifice and branching; L—stichocyte surrounding esophagus in wall of which there is a small nucleus (? nerve); M—esophageal wall from stichosome surrounded region, stained with fuchsin to show musculature; N—Esophageal wall from stichosome surrounded region containing red blood corpuscle in lumen; O—End of esophagus with 2 accompanying cells (lower); P—Esophago-intestinal valve and intestine; Q—Longitudinal section showing relationship of stichocytes and radial muscle cells to esophageal lumen). Original.

the radial fibers to the base of the esophagus. It appears that there are 36 radial and nine marginal (or possibly 12) nuclei in the esophagus. The radials are arranged in subgroups of three, one near the center of each sector. The esophago-intestinal valve is triradiate.

The highly remarkable esophageal glands deserve special note. Each has a short *terminal duct* lined with cuticle, followed by a short, thick walled *primary tubule* which bifurcates into secondary tubules. In *Soboliphyme baturini* the secondary tubules appear to be lined with cilia (Fig. 95FF). Here and in *Eustrongylides* these secondary tubules branch dichotomously, time after time, throughout the length of the esophagus and come to nearly fill the non-muscular part of their sectors (Fig. 95CC-DD), but do not enter other sectors. The marginal tubules end blindly (Fig. 97C) and their position is taken by others formed by the branching of the more centrally located tubules. In *Dioctophyma renale* the same condition exists with the exception that the secondary tubules of the dorsal gland do not undergo further branching (Fig. 95BB) and tubules from the subventral glands take a marginal position in the dorsal sector (Fig. 97B). In all cases we find dense glandular protoplasm containing numerous gland nuclei (Fig. 95CC-DD) surrounding the tubules. There are literally hundreds of such nuclei. Whether the subventral glands correspond only to the anterior pair of subventral esophageal glands or to both anterior and posterior pairs of subventral glands of enoplids is uncertain.

The dioctophymatid esophagus resembles the enoplid esophagus in the position of the orifices of the esophageal glands and also resembles to some extent that of the leptosomatids in the multinucleate condition of the glands. However, the peculiar dichotomous branching of the tubules has its only parallel in the branched dorsal gland of *Dorylaimus*.

2. ESOPHAGO-SYMPATHETIC NERVOUS SYSTEM

The presence of nerve cells in the wall of the esophagus was first mentioned by Looss (1896) and this system was described briefly in *Ancylostoma duodenale* by that writer in 1905. Later (1910) Goldschmidt studied the system in *Ascaris lumbricoides* followed by Martini (1916) in *Oxyuris equi*, Imminck (1921, 1924) in *Strongylus edentatus*, de Bruyn (1934) in *Angusticaecum holopterum*, and various observations by the writers (1933, 1937) refer to free-living and parasitic forms.

In substance, this system consists of three longitudinal nerves, one situated near the center of each sector and extending from the base nearly to the anterior end of the esophagus (Fig. 97R). These nerves contain in their course a series of nerve cells and two or three commissures joining the longitudinal nerves. In most forms we find a commissure at the base of the corpus, another in the anterior part of the bulbar region and a third in the posterior part of the bulbar region. Nerve cell nuclei may usually be distinguished from the other types of nuclei though they vary considerably in size, but in smaller forms it is often not possible to identify all nuclei with certainty. In such cases one must place reliance purely upon considerations based on comparative anatomy. In the majority of instances no attempt has been made to trace the nerve fibers but the nerve cell pattern has been recorded and found to be of considerable value from the standpoint of comparative histology. In very large nematodes the nerve cells of the esophagus are often disproportionately small and may easily be overlooked. This seems a probable reason for the small number of such cells reported to occur in ascarids and spiruroids. In forms with multinucleate esophageal glands the nerve cells may be easily confused with gland nuclei. Of course, it is also possible that in such forms (always devoid of a valved bulb) there is no need for the complicated esophago-sympathetic system of smaller forms.

In *Spironoura affine* (Fig. 97A) there are seven cells in each nerve anterior to the nerve ring (Fig. 97P); two of these are glia cells (n_{1-3}, n_{19-21}), the remainder nerve cells (n_{4-18}). The nerve fibers give off lateral branches (Fig. 97P) into the radial muscle regions. No further nerve cells are present in the procorpus. The metacorpus contains three large nerve cells in the dorsal nerve ($n_{24, 29, 32}$) and five nerve cells in each subventral nerve ($n_{22-23, 25-31, 33-34}$). At the base of the metacorpus we find a well developed commissure between n_{29-31} and n_{32-34}. The bulbar region (Fig. 97Q) contains nine cells attributable to the nervous system of which at least two

(n_{41-42}) and possibly a third (n_{38}) are probably glia cells. There are two commissures in the bulbar region, an anterior and a posterior (Fig. 97A). The connection of the esophago-sympathetic with the central nervous system has not been observed for *Spironoura*. However, we presume it to be similar to that of *Ascaris* in which a process of each of the subventral nerves passes through the external wall of the esophagus near the level of the dorsal gland orifice (Fig. 97R). This process continues posteriorly on the outside of the esophagus connecting with a bipolar cell and through this cell with the nerve ring.

Goldschmidt recorded only 17 cells in this system in *Ascaris lumbricoides* in which Hsü recorded 18 cells; de Bruyn recorded 27 in *Angusticaecum holopterum* and Martini 20 in *Oxyuris equi*. All of these writers mention cells of dubious or unknown significance in the esophagus, particularly in the procorpus. Our own observations on *Ascaris lumbricoides* indicate that a set of six cells near the level of the dorsal gland orifice are homologous to n_{4-9} of *Spironoura*.

The peculiar distribution of nerve cells in representatives of the suborder Enoplina is worthy of particular note since it is very probably indicative of relationship. As may be seen from the diagrams (Fig. 90) n_{19-20} and n_{33-34} are usually marginal in position in the subventral sectors as are n_{30-31} in the dorsal sector. Together with n_{29} and n_{32} the latter nuclei form a quadrangle in forms such as *Tripyla papillata*, *Dorylaimus obtusicaudatus* and *Metoncholaimus pristiuris*.

3. FINER STRUCTURE OF THE ESOPHAGUS

Fibers.—Considerable discussion has taken place as to whether or not the marginal fibers are contractile. Hamann (1895), Rauther (1907), Allgen (1921), Plenk (1924, 1925, 1926) and Looss (1905) maintained that the marginals are contractile, while Loose (1896), Schneider (1902), Goldschmidt (1904) and Martini (1916, 1922) hold that they are supporting or skeletal structures and observations of the writers support this view. In the marginal region of *Ascaris* one finds two types of fibers, a first type, extending from the esophageal lining to the esophageal covering, a second type extending longitudinally in two more or less distinct rows, one on each side of the esophageal margin. This second type, the "fiber plates" of Goldschmidt (1904), is known only in ascarids.

The radial fibers extend more or less perpendicularly from the esophageal lining to the external covering of the esophagus but often run rather obliquely so that their contraction might easily shorten the esophagus to a moderate extent. K. C. Schneider (1902) characterized the radial fibers as striated muscle. This view was supported by Martini (1916) though other authors including Goldschmidt (1904) disagree, stating that the appearance of striation is due to a minute system of supporting fibrils. Convincing evidence of striation of the muscle is not yet established. It is therefore concluded that if such is present, it must be of a rather peculiar character hardly comparable to striated muscle in other organisms.

Finally the "fenestrated membrane" describe by Goldschmidt (1904), Kulmatycki (1918, 1922) and de Bruyn (1934) in ascarids shall be discussed. This is a longitudinal membrane between the external and the internal coverings of the esophagus beginning at some distance from the head, ending near the base of the esophagus. The writers find no indication of such a structure. However, Kulmatycki (1922) states that it is actually double and in places can be seen to be a distinct tube. He illustrates that which is obviously an esophageal gland tubule. Since the gland branches coincide with the location of the "fenestrated membrane," and since we can find no other structure fitting the description, the whole probably is a misinterpretation.

Ducts and Tubules.—The internal structure of the esophageal glands has been very little studied and a generalization from the meager information available would be presumptuous. In the past the glands have been considered as rather simple structures, most investigators regarding the ducts as in direct continuity with the gland protoplasm. Actually, this is never the case. The esophageal gland ducts are continuous with tubules of various types. When the gland orifice is some distance from the actual beginning of the gland substance, there is a long unbranched central duct. This duct

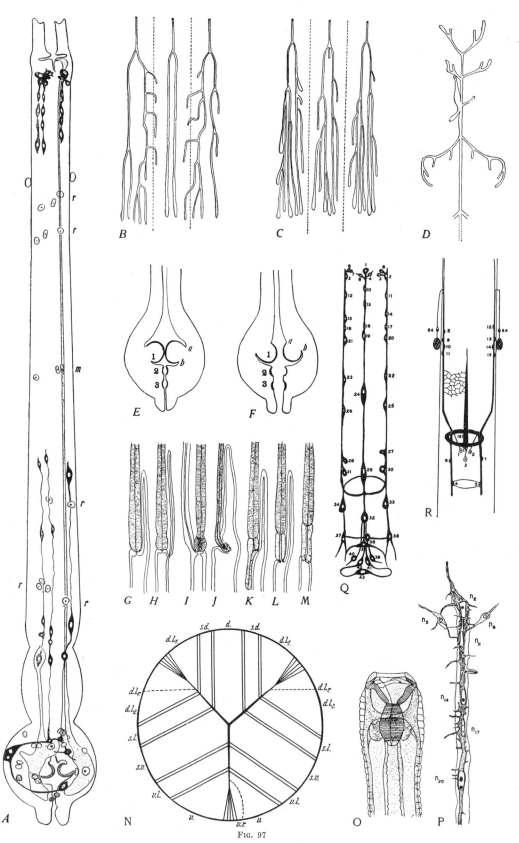

FIG. 97

A—Reconstruction of esophagus of *Spironoura* showing nuclei, glands and esophago-sympathetic. (Dorsal to reader's right.) B—*Dioctophyma renale*, esophageal gland tubule branching. C—*Soboliphyme baturini*, esophageal gland tubule. D—*Ascaris lumbricoides*, dorsal esophageal gland tubule. E-F—Rhabditoid bulb. (E—at rest; F—contracted). G-M—Ascaridid esophagi and intestines. (G—*Amplicaecum*; H—*Angusticaecum*; I—*Dujardinia halicornis*; J—*Dujardinia helecina*; K—*Contracaecum*; L—*Porrocaecum*; M—*Anisakis*. N—Diagram of esophageal symmetry. (d. dorsal area; dl₁ dorsolateral area of dorsal section; dl₂ dorsolateral area of subventral sector; dl_r dorsolateral (i. e., subdorsal) esophageal radius; sd subdorsal area; sl sublateral area; sv subventral area; vl ventrolateral area; vr ventral esophageal radius). O—*Oxyuris equi*, (Fourth stage larva). P-Q—*Spironoura affine*, esophago-sympathetic nervous system. (Q—Diagrammatic connection of cells, dorsal view, compare with A.; P—Detail of right subventral nerve trunk, lateral view). R—*Ascaris lumbricoides* (Esophago-sympathetic nervous system, diagrammatic dorsal view. D & N after Chitw. & Hill, 1932, Ztschr. Zellforsch, v. 14 (4). G-M, after Baylis, 1920, Parasit., v. 12 (3); O, after Ihle & Oordt, 1921, Proc. Sec. Sc. K. Akad Wetensch. Amsterdam.

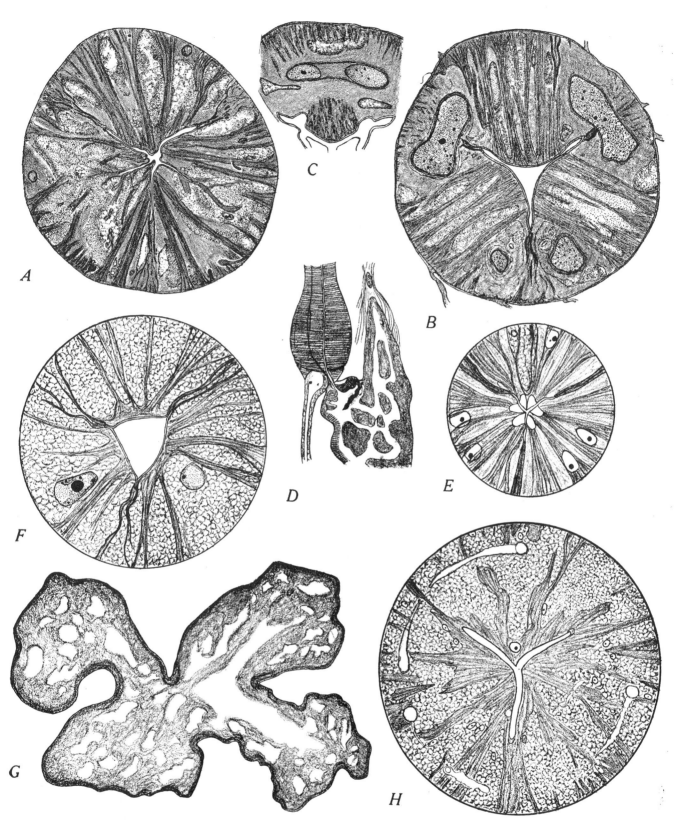

F.G. 98

A-C—*Ascaris lumbricoides*. (A—At level of subventral gland orifice; B-C—at level of esophageal gland nuclei [the dorsal gland nucleus is bilobed, lobes coming together in C]); D—*Goezia annulata* (Showing esophageal appendix left, and intestinal caecum, right, also connective tissue cell). E-F—*Camallanus microcephalus* (E—through corpus at level of first group of radial nuclei; F—glandular region at level of subventral gland nuclei). G—Terminal part of dorsal esophageal gland tubule of *Ascaris lumbricoides*. H—*Physaloptera maxillaris*—(Glandular region, note 1 radial nucleus central in dorsal sector, numerous gland nuclei, and duct branching. A-C & G, after Chitw. & Hill, 1932, Ztschr. Zellforsch., v. 14 (4). D, after Hamann, 1895, Die Nemathelminthen, v. 2.

may continue after reaching the glandular region and give off lateral paired side branches, "pinnate branching." The terminations of the branches may be "tubular" or "alveolar (acinus)."

Hsü (1933) speaks of "simple tubular" glands in *Philometra* and *Dracunculus houdemeri* stating that other nematodes have "branched tubular" glands. It is very easy to fall into error regarding such structures, particularly when the glands terminate in acini or branched acini. This is, in fact, the case in regard to the glands of *Dracunculus dahomensis*, the subventral ones being compound branched alveolar glands, while the dorsal is apparently of the simple branched alveolar type.

In forms such as *Ascaris lumbricoides* the dorsal gland is simple (pinnately branched); the subventrals are compound (palmately branched) and tubular. In *Physaloptera maxillaris* all three glands are simple (pinnately branched) and apparently tubular (Fig. 98H). However, one has difficulty in tracing the secondary and tertiary tubules; they might easily terminate in acini. The compound form of the dichotomously branching tubular glands of dioctophymoids is obvious, as are the simple branched tubular glands of *Trichuris ovis* and *Agamermis decaudata* (Fig. 96).

Alveolar glands are particularly difficult to study. Preservation of the tubules in satisfactory condition for study is actually a rarity and for answer to many of the problems one must look to living specimens. Quite often the lobulations of the gland protoplasm itself are mistaken for the tubules. Such lobulation is dependent upon muscle distribution and may have no bearing on the tubular system (the stichocytes of *Trichuris ovis* and *Agamermis decaudata* have branched tubules though the glands are not highly lobulated).

"CHROMIDIA."—Working with *Ascaris lumbricoides* Goldschmidt (1904) described bodies in the plasma of the muscles associated with the radial nuclei. They stained intensely with hematoxylin and he termed them chromidia. These bodies he conceived to be "vegetative" nuclei originating from the radial nuclei. Vejdovsky (1907) and Bilek (1909) considered them artefacts but Hirschler (1910, 1912) demonstrated their existence in the living cell. Neither Hirschler, von Kemnitz (1912) nor Kulmatycki (1922) found any evidence that they originated from the nucleus. These bodies are found only in the large ascarids and are most numerous in the vicinity of the radial nuclei, but they may also occur in the marginal areas or at considerable distance from the nuclei. They vary in number and appearance. When few are present they tend to take the form of coiled fibers, and when many are present the form of thick flecks; both forms may be seen in a single chromidium. Kulmatycki (1922) found such structures as well as a golgi body in cells of the spicule sheath, thereby eliminating the possibility that the chromidia are homologues of the golgi body; he named them ascaridochondria, relating them to mitochondria, chondrosomes, etc. "Chromidia" occur in marginal areas of *Anonchus mirabilis* (Fig. 84 D-F), even more spectacularly than in *Ascaris*. Their significance, today, is unknown.

OCELLI AND PIGMENT SPOTS.—The occurrence of pigment masses in the Aphasmidia has been mentioned. That such structures exist, has been known since the time of Bastian. Because of their general appearance they have been widely accepted as photoreceptors despite a total absence of evidence that they are connected with the nervous system. Brownish to red granules may be rather irregularly and generally distributed in the esophageal tissue of such forms as oncholaims. Similar pigmented granules may be slightly more concentrated in the subdorsal marginal areas of forms such as *Enoplus* and *Chromadora*. In these forms there is a definite pair of "pigment spots" but the pigment also extends posteriorly from the spots and may be present to a considerable extent in the subventral marginal regions of the esophagus. Rauther (1907) regarded the pigment granules of *Enoplus* and *Oncholaimus* as excretory granules and thought they were eliminated through the esophageal glands. We find no evidence of the "refractive granules" in the ducts of the esophageal glands and see no reason to assume they are excretion products. Schulz (1931b) upon finding the spots to be within the esophagus in *Enoplus* reaffirmed Rauther's interpretation and differentiated such bodies from true ocelli (with lenses) which he observed in *Leptosomatum*, *Thoracostoma* and *Parasymplocostoma*. In the latter type he described the lens as an invagination of the superficial cuticle of the body surrounded by pigment and connected with a special ocellus cell. He states that he does not believe the true ocellus is connected in any way with the esophagus. On the contrary, the writers have found the ocelli

of *Leptosomatum elongatum* to be completely enclosed within the wall of the esophagus; the lens seems to be formed from the external covering of the esophagus and no special cell is associated with the ocellus. We must conclude that even in this instance the ocellus is a part of the esophagus. If it is innervated, as one would presume, then the esophago-sympathetic nervous system must include also, the "optic nerve." In the Desmoscolecoidea, there is definite evidence that the pigment bodies are outside the esophagus; in this group the posterior part of the esophagus is degenerate and the esophageal glands outside the general contour. In *Monhystera paludicola* true ocelli are present; these are likewise situated outside the esophagus.

There remains one additional case of pigmentation in the cephalic region associated with photoperception. In the gravid female of *Mermis subnigrescens* Cobb (1926, 1929) described diffuse reddish pigment anterior to the nerve ring. Such pigment is absent from the head of young females and males, only being found in specimens ready to deposit eggs. Though the exact location of the pigment was not determined, the case is interesting since it supplied the only actual evidence of photoperception in a nematode, for egg laying only takes place in the light, ceasing in darkness. That heat is not the stimulus is indicated by the fact that such "egg laying" females will continue to lay eggs though placed on ice in a dish of water so long as the light continues.

Bibliography

ALLGEN, CARL. 1921.—Über die Natur und die Bedeutung der Fasersysteme im Oesophagus einiger Nematoden. Zool. Anz., v. 53(3/4): 76-87, no figs.

BASIR, M. A. 1949.—The histological anatomy of the esophagus of *Physaloptera varani* Parona, 1889. Tr. Am. Micr. Soc. v. 67 (4): 352-358.

BAYLIS, H. A. 1920.—On the classification of the Ascaridae. I. The systematic value of certain characters of the alimentary canal. Parasitol., v. 12(3): 254-264, figs. 1-6.

BILEK, F. 1909.—Ueber die fibrillären Strukturen in den Muskel und Darmzellen der Ascariden. Ztschr. Wiss. Zool., v. 93: 625-637, pls. 27-28, figs. 1-20.

BRUYN, W. M. DE. 1934.—Beiträge zur Kenntnis von *Angusticaecum holopterum* (Rud.) einem Nematoden aus *Testudo graeca* L. Diss. Amsterdam. 120 pp., 47 figs., 4 pls., 8 figs.

CHITWOOD, B. G. 1930.—The structure of the esophagus in the Trichuroidea. J. Parasit., v. 17: 35-42, pls. 5-6.
    1931.—A comparative histological study of certain nematodes. Ztschr. Morph., v. 23(1/2): 237-284, figs. 1-23.
    1935.—The nature of the "Cell body" of *Trichuris* and "Stichosome" of *Agamermis*. J. Parasit., v. 21(3): 225.
    1936.—The value of esophageal structures in nemic classification. J. Parasit., v. 22(6): 528.

CHITWOOD, B. G., and M. B. 1933.—The histological anatomy of *Cephalobellus papilliger* Cobb, 1920. Ztschr. Zellforsch., v. 19(2): 309-355, figs. 1-34.
    1934.—The histology of nemic esophagi. I. The esophagus of *Rhabdias eustreptos* (MacCallum, 1921). Ibid., v. 22(1): 29-37, figs. 1-4.
    1934.—Idem. II. The esophagus of *Heterakis gallinae*. Ibid., v. 22(1): 38-46, figs. 1-4.
    1934.—Idem. III. The esophagus of *Oesophagostomum dentatum*. (Rudolphi). J. Wash. Acad. Sc. v. 24(12): 557-562, figs. 1-3.
    1935.—Idem. IV. The esophagus of *Metastrongylus elongatus*. Ibid., v. 25(5): 230-237, figs. 1-4.
    1936.—Idem. V. The esophagus of *Rhabditis*, *Anguillulina* and *Aphelenchus*. Ibid., v. 26(2): 52-59, figs. 1-6.
    1936.—Idem. VI. The esophagus of members of the Chromadorida. Ibid., v. 26(8): 331-346, figs. 1-4.
    1936.—Idem. VII. The esophagus of *Leidynema appendiculatum* (Leidy, 1850). Ibid., v. 26(10): 414-419, figs. 1-4.
    1937.—Idem. VIII. The esophagi of representatives of the Enoplida. Ibid. v. 27 (12): 517-531, figs. 1-2.

CHITWOOD, B. G., & HILL, C. H. 1932.—A note on the esophageal glands of *Ascaris lumbricoides*. Ztschr. Zellforsch., v. 14(4): 605-615, figs. 1-17.

CHRISTENSON, REED O., 1935.—Studies on the morphology of the common lungworm, *Capillaria aerophila* (Creplin, 1839). Tr. Am. Micr. Soc., v. 54(2): 145-154, figs. 1-3, pl. 27.

CHRISTIE, J. R. 1936.—Life History of *Agamermis decaudata*, a nematode parasite of grasshoppers and other insects. J. Agric. Res., v. 52(3): 161-198, figs. 1-20.

COBB, N. A. 1926.—The species of *Mermis*. J. Parasit., v. 13: 66-72, figs. 1-3, pl. 2, fig. 2.
    1929.—The chromatropism of *Mermis subnigrescens*, a nemic parasite of grasshoppers. J. Wash. Acad. Sc., v. 19(8): 159-166, fig. 1.

EBERTH, C. J. 1859.—Beiträge zur Anatomie und Physiologie des *Trichocephalus dispar*. Ztschr. Wiss. Zool. 10 Band, 2 Heft.
    1860.—Beiträge zur Anatomie und Physiologie des *Trichocephalus dispar*. Ztschr. Wiss. Zool. v. 10; 233-258, pls. 17-18, figs. 1-24.
    1863.—Untersuchungen über Nematoden. Leipzig, 77 pp., 9 pls.

EHLERS, H. 1899.—Zur Kenntnis der Anatomie und Biologie von *Oxyuris curvula* Arch. Naturg., 65 J., v. 1 (1): 1-26 pls. 1-2, figs. 1-20.

GOLDSCHMIDT, R. 1904.—Der Chromidialapparat lebhaft funktionierender Gewebszellen. (Histologische Untersuchungen an Nematoden II). Zool. Jahrb. Abt. Anat., v. 21 (1): 41-140, figs. A-Q, pls. 3-6, figs. 1-62.
    1909.—Das Skelett der Muskelzelle von *Ascaris* etc. Arch. Zellf., v. 4 (1): 81-119, figs. A-C, pls. 6-9, figs. 1-19.
    1910.—Das Nervensystem von *Ascaris lumbricoides und megalocephala*. III. Teil. Festschrift R. Hertwig, v. 2: 255-354, figs. 1-29, pls. 17-23, figs. 1-125.

HAGMEIER, A. 1912.—Beiträge zur Kenntnis der Mermithiden. Diss. Heidelberg. 92 pp., 4 pls., 55 figs. Also Zool. Jahrb., v. 32 (6): 521-612, figs. a-g, pls. 17-21.

HAMANN, O. 1895.—Die Nemathelminthen, v. 2, pp. 1-120, pls. 1-9, Jena.

HEINE, P. 1900.—Beiträge zur Anatomie und Histologie der Trichocephalen insbesondere des *Trichocephalus affinis*. Centrabl. Bakt. I. Abt., v. 28: 779-787, 809-817, pls. 1-2, figs. 1-13.

HIRSCHLER, J. 1910.—Cytologische Untersuchungen an Ascariszellen. Bull. Internat. Acad. Sc. Cracovie., v. 78: 638-645.
    1912.—Über einige strittige Fragen den Ascariton cytologie. Verh. VIII. Internat. Zool. Kongress in Graz: 932-936.

HSÜ, H. F. 1929.—On the esophagus of *Ascaris lumbricoides*. Ztschr. Zellforsch., v. 9 (2): 313-326, figs. 1-10.
    1933.—On *Dracunculus houdemeri* n. sp. *Dracunculus globocephalus*, and *Dracunculus medienensis*. Ztschr. Parasit., v. 6 (1): 101-118, figs. 1-38.
    1933.—A study of the oesophageal glands of some species of Spiruroidea and Filarioidea. Ztschr. Parasit., v. 6 (3): 277-287, figs. 1-6.
    1933.—Study of the oesophageal glands of parasitic Nematoda, superfamily Ascaroidea. Chinese Med. J., v. 47: 1247-1288, ps. 1-10, figs. 1-53.
    1933.—The esophageal glands of nematodes. Lingnan Sc. J., v. 12 (Suppl.): 13-21.

HSÜ, H. F. and HOEPPLI, R. 1933.—Die Oesophagusdrüsen einer *Proleptus* sp. und von *Thelazia callipaeda* (Nematoda) Ztschr. Parasit., v. 6 (3): 273-276, figs. 1-3.

IHLE, I. E. W. and OORDT, G. J. VAN. 1921.—On the larval development of *Oxyuris equi* (Schrank). Proc. Sec. Sc. K. Akad. Wetensch. Amsterdam, v. 23: 603-612, figs. 1-6.

IMMINCK, B. D. C. M. 1921.—Bijdrage tot de kennis va den boun van den voordarm van *Sclerostomum edentatum* Looss. Diss. Leiden.
    1924.—On the microscopical anatomy of the digestive system of *Strongylus edentatus* Looss. Arch. Anat., v. 3 (4-6): 281-326, figs. 1-46.

JÄGERSKIÖLD, L. A. 1893.—Bidrag till Kännedomen om Nematoderna. Diss. Stockholm. 86 pp., 5 pls., figs. 1-43.
    1894.—Beiträge zur Kenntnis der Nematoden Zool. Jahrb. Abt. Anat., v. 7 (3): 449-532, pls. 24-28.
    1897.—Ueber den Oesophagus der Nematoden besonders bei *Strongylus armatus* Rud. und *Dochmius duodealis* Dubini. Bihang K. Svenska Vetensk. Akad. Handl. Stockholm, v. 23, Afd. 4 (5), 26 pp., 2 figs., 1 pl., figs. 1-6.
    1909.—Nematoden aus Aegypten und dem Sudan. Results Swedish Zool. Exped. to Egypt and White Nile, 1901. No. 25: 66 pp., 23 figs., 4 pls.

JANICKI, C. and RASIN, K. 1930.—Bemerkungen über *Cystoopsis acipenseri* des Wolga-Sterlets, sowie über die Entwicklung dieses Nematoden im Zwischenwirt. Ztschr. Wiss. Zool., v. 136 (1): 1-37, figs. 1-26.

KEMNITZ, G. VON. 1912.—Die Morphologie des Stoffwechsels bei *Ascaris lumbricoides*. Arch. Zellf., v. 7: 463-603, figs. 1-9, pls. 34-38.

KULMATYCKI, W. J. 1918.—Einige Bemerkungen über den Bau der Deckmuskelzellen im Oesophagus sowie dessen Funktion bei *Ascaris megalocephala*. Anat. Anz., v. 51 (1): 18-29, figs. 1-4.
    1922.—Bemerkungen über den Bau einiger zellen von *Ascaris megalocephala*, mit besonderer Berücksichtigung des sogenannten Chromidialapparates. Arch. Zellforsch., v. 16: 473-551, pls. 22-26, figs. 1-36.

LEUCKART, K. G. F. R. 1866.—Untersuchungen über *Trichina spiralis*. 120 pp., 7 figs., 2 pls. Leipzig & Heidelberg.
    1876.—Die Menschlichen Parasiten., v. 2; 513-882, 119 figs. Leipzig, v. 2: 368-369.

LOOSS, A. 1895.—*Strongylus subtilis* n. sp., ein bischer unbekannter Parasit des Menschen in Egypt. Centrabl. Bakt. Abt. I., v. 18 (6): 161-169, figs. 1-8.
    1896.—Ueber den Bau des Oesophagus bei einigen Ascariden. Centrabl. Bakt. I Abt., v. 19 (1): 5-13.
    1901.—The Sclerostomidae of horses and donkeys in Egypt. Rec. Egypt. Govt. Sch. Med., pp. 25-139, pls. 1-13, figs. 1-172.
    1905.—The anatomy and life history of *Agchylostoma duodenale*. Rec. Egypt. Govt. Schl. Med., v. 3: 1-158, pls. 1-9, figs. 1-100, pl. 10, photos 1-6.

MACKIN, J. G. 1936.—Studies on the morphology and life history of nematodes in the genus *Spironoura*. Univ. Ill. Bull., v. 33 (52), 64 pp., 69 figs.

MAGATH, T. B. 1919.—*Camallanus americanus* nov. spec. Tr. Amer. Micr. Soc., v. 38 (2): 49-170, figs. A-Q, pls. 7-16, figs. 1-134.

MAN, J. G. DE. 1886.—Anatomische Untersuchungen über freilebende Nordsee-Nematoden. 82 pp., 13 pls. Leipzig.
    1889.—Troisième note sur les nématodes libre de la mer du nord et de la manche. Mem. Soc. Zool. France, v. 2: 182-216, pls. 5-8, figs. 1-12e.
    1904.—Nématodes libres, Résultats du voyage du S. Y. Belgica. Expéd. Antarct. Belg. Anvers. 55 pp., 11 pls.

MARION, A. 1870.—Nematoides non parasites marins. Ann. Sc. Nat. Paris. Zool., 5. s., v. 13, art. 14, 100 pp., pls. 16-26.

MARTINI, E. 1916.—Die Anatomie der *Oxyuris curvula*. Ztschr. Wiss. Zool., v. 116: 137-534, figs. 1-121, pls. 6-20.
    1922.—Ueber die Fibrillensysteme im Pharynx der Nematoden. Zool. Anz., v. 54 (9/10): 193-198, 1 fig.

MIRZA, M. B. 1929.—Beiträge zur Kenntnis des Baues von *Dracunculus medinensis*. Velsch. Ztschr. Parasit., v. 2 (2): 129-156, figs. 1-33.

MÜLLER, G. W. 1929.—Die Ernährung einiger Trichuroideen. Ztschr. Morph., v. 15 (1/2): 192-212, figs. 1-15.
    1931.—Ueber Mermithiden. Ibid., v. 24 (1): 82-147, figs. 1-34.

MUELLER, J. F. 1931.—The esophageal glands of *Ascaris*. Ztschr. Zellforsch., v. 12 (3): 436-450, pls. 1-5, figs. 1-21.

PLENK, H. 1924.—Nachweis von Querstreifung in sämtlichen Muskelfasern von *Ascaris megalocephala*. Ztschr. Anat. & Entwickl. 1 Abt., v. 73: 358-388, figs. 1-29.
    1925.—Zur Histologie der Muskelfasern von *Ascaris Lumbricus* und *Hirudo*. Verh. der Anat. Ges. vom 21-24 April, 1925, 34 Vers. in Wien. Erg. heft zum 60 Bd. Anat. Anz. (1925-26): 273-275.
    1926.—Beiträge zur Histologie der Muskelfasern von *Hirudo* und *Lumbricus*, nebst Berichtigungen zu meinen Untersuchungen über den Bau der *Ascaris*—und Molluskenmuskelfasern. Ztschr. Mikrosc. Anat. Forsch., v. 4: 163-202, figs. 1-27.

RAUTHER, M. 1906.—Beiträge zur Kenntnis von *Mermis albicans* v. Sieb. Zool. Jahrb. Abt. Anat., v. 23 (1): 1-76, pls. 1-3, figs. 1-26.
    1907.—Ueber den Bau des Oesophagus und die Lokalisation der Nierenfunktion bei freilebenden Nematoden. Ibid., v. 23 (4): 703-738, pl. 38, figs. 1-9.
    1918.—Mitteilungen zur Nematodenkunde. Zool. Jahrb., Abt. Anat., v. 40: 441-514, pls. 20-24, figs. 1-40.

SCHNEIDER, A. 1866.—Monographie der Nematoden. 357 pp., 122 figs., 28 pls., 343 figs. Berlin.

SCHNEIDER, K. C. 1902.—Lehrbuch der vergleichenden Histologie der Tiere. 988 pp., 691 figs., Jena.

SCHULZ, E. 1931a.—Betrachtungen über Augen freilebender Nematoden. Zool. Anz., v. 95 (9/10): 241-244, figs. 1-3.
1931b.—Nachtrag zu der Arbeit: Betrachtungen über die Augen freilebender Nematoden. Zool. Anz., v. 96 (5/6): 159-160, fig. 1.

STADELMANN, H. 1891.—Ueber den anatomischen Bau des Strongylus convolutus Ostertag nebst einigen Bemerkungen zu seiner Biologie. Diss. 39 pp. Berlin.
1892.—Idem. Arch. Naturg., 58, J., v. 1 (2): 149-176. pl. 1-.

STEKHOVEN, J. H. SCHUURMANS and BOTMAN, P. J. 1932.—Zur Ernährungs-biologie von Proleptus obtusus Duj. und die von diesem Parasiten hervorgerufen reaktiven Aenderungen des Wirtsgewebes. Zeit. f. Parasitenk. (Z. F. W. Z. Ab. F.). v. 4 (2): 220-239.

STRASSEN, O. ZUR. 1907.—Filaria medinensis und Ichthyonema. Verhandl. Deutsch. Zool. Gesellsch. 17 J., 110-129, figs. 1-8.

TÖRNQUIST, N. 1931.—Die Nematodenfamilien Cucullanidae und Camallanidae—etc. Göteborgs K. Vetensk.-o. Vitterhets-Samh. Handl., 5 f., s. B, v. 2 (3), 441 pp., pls. 1-17.

TÜRK, F. 1903.—Ueber einige im Golfe von Neapel freilebenden Nematoden. Thesis. Leipzig. 67 pp., pls. 10-11. Also Mitth. Zool. Stat. Neapel, v. 16 (3): 281-348, pls. 10-11.

VEGLIA, F. 1916.—The anatomy and life history of the Haemonchus contortus (Rud.). 3rd & 4th Rpt. Vet. Res. Union S. Af. pp. 349-500, 28 pls., figs. 1-60, charts 1-18.

VEJDOVSKY, F. 1907.—Neue Untersuchungen über die Reifung und Befruchtung. Konigl. Böhm. Gesellsch. Wiss. Prag. (Not seen.)

WARD, H. B. 1917.—On the structure and classification of North American parasitic worms. J. Parasit., v. 4: 1-12.

YAMAGUTI, S. 1935.—Studies on the helminth fauna of Japan. Part 9. Nematodes of fishes, I. Jap. J. Zool., v. 6 (2): 337-386, figs. 1-65.

WETZEL, R. 1930.—On the biology of the fourth stage larva of Oxyuris equi (Schrank). J. Parasit., v. 17 (2): 95-97, pl. 12.

CHAPTER VII

THE INTESTINE OR MESENTERON

B. G. CHITWOOD and M. B. CHITWOOD

A. GENERAL MORPHOLOGY

The intestine of nematodes is a tube the wall of which is composed of epithelial cells. Its gross morphology does not vary markedly in different groups of nematodes. Usually it is a simple, more or less straight tube accommodating itself to the reproductive organs and space in the body cavity.

SUBDIVISIONS OF INTESTINE.—The intestine may be divided into three regions: the anterior part or ventricular region; mid-region or intestine proper; and the posterior part or prerectal region. The ventricular and prerectal regions commonly differ from the mid-region in the height of the cells and shape of the lumen. Usually, there is also some difference in the type of cell inclusions present in these regions. When a region is quite definitely differentiated from the remainder of the intestine it is herein termed either a ventriculus (anterior) or prerectum (posterior) while an adjectival usage is retained when the differentiation is not marked.

APPENDAGES.—Two types of cecae or diverticulae occur in the ventricular region, one directed anteriorly, the other posteriorly. The first type is by far the more common, occurring in various degrees of development in members of several groups in the Phasmidia. Only one free-living nematode is known with such a structure, namely Rhabditis cylindrica, and in this instance the cecum is very small, scarcely a third of the intestinal diameter in length. Likewise but one member of the Strongylina, Grammocephalus (Ancylostomatidae) has been described as possessing a short intestinal cecum and but a few representatives of the order Spirurida (Dacnitis spp. and Dichelyne spp.) have such structures. The cecae in these forms are quite small. Development of ventricular cecae is most common in the Ascaridoidea, sporadically occurring in such forms as Contracaecum, Angusticaecum and Amplicaecum (Ascarididae, Anisakinae). In the last mentioned forms the cecum may be very large (Fig. 97 G-M), extending far beyond the base of the esophagus, even to the nerve ring. A posteriorly directed cecum (Fig. 99D) is known to exist only in females of the genus Leidynema (Thelastomatidae).

No satisfactory explanation of the intestinal cecum development in nematodes has yet been made. The sporadic occurrence of this structure does not seem to be correlated with feeding habits. Phylogeny throws no light on the subject, for closely related forms may differ in this respect. Though the cecum is always a development of the ventricular region its cells do not differ cytologically from the remainder of this region, indicating no functional specialization (Fig. 98D).

LAYERS.—The intestinal wall consists of a single layer of epithelial cells which usually bear on their internal surface a bacillary layer (Stäbchensaum, bordeur en brousse), and some times a distinct subbacillary layer (Deckschicht) is apparent. The external surface of the cells may be quite naked, or it may be covered by a distinct basal lamella, a muscularis mucosae and a mesenterial membrane. One or more of these coverings may be present or they all may appear to be absent.

Sometimes the protoplasm of the epithelial cells is divisible into distinct zones. The ectoplasmic zone is a layer of dense cytoplasm bordering the sides of the cell; when definitely thickened on the side of the cell facing the lumen it is called a plasma cap (Fig. 103J3, Z3). The remainder of the cell is termed the endoplasm; it contains the nucleus, cell inclusions and sometimes other structures such as plasma strings (Fig. 103J5), basioplasm (Fig. 103J6), basal fibrillae, etc.

The Bacillary and Subbacillary Layers.—The bacillary layer consists of an internal border appearing to be made up of fine rods or "cilia" beneath which one often finds a subbacillary layer (Deckschicht) which stains with iron-hematoxylin. The first layer varies markedly in appearance, the bacilli sometimes being rather large and well separated, sometimes compact, sometimes fine and hair like. Under ordinary circumstances the bacillary layer has a compact appearance but the elements have been seen quite discretely in living specimens of Rhabditis strongyloides. In general, the bacillary and subbacillary layers are most highly developed (thickest) in representatives of the Strongylina (Fig. 102K) and impart a characteristic appearance which one is not likely to confuse with that of other nematode groups. Sometimes due to fixation the bacillary layer may have pulled away from the epithelium, giving the appearance of a peritrophic membrane. Like the bacillary layer of other organisms it is digested by proteolytic enzymes and is therefore non-chitinous. (The peritrophic membrane of arthropods is chitin).

Somewhat extended discussions of the significance of the bacillary layer have been given in the past. Since this layer is not peculiar to the Nematoda, but occurs in the intestine of various groups of worms as well as in arthropods and vertebrates, conclusions based on the study of nematodes alone could scarcely be considered valid. The several viewpoints expressed have been as follows: (1) The bacillary layer is a development of minute tubes which aid in resorption or excretion; (2) it is itself a secretion product of protective nature; (3) it is a layer of amalgamated or degenerate cilia. The second possibility seems least probable in the light of comparison by which one finds the layer very well developed in

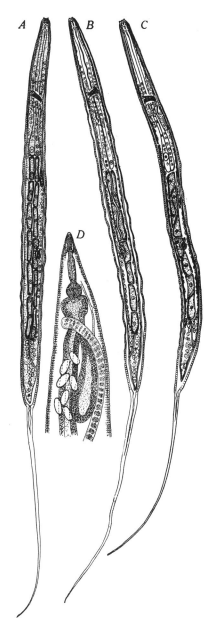

FIG. 99

A-C—Third stage larval strongyles showing oligocyty. (A—*Gyalocephalus capitatus*; B—*Cylicocercus goldi*; C—*Cylicocercus catinatus*). D—*Leidynema appendiculatum* (Adult female showing cecum). A-C, after Lucker, 1936, Proc. Helm. Soc. Wash., v. 3(1). D, after Chitwood, 1932, Ztschr. Parasitenk., v. 5(1).

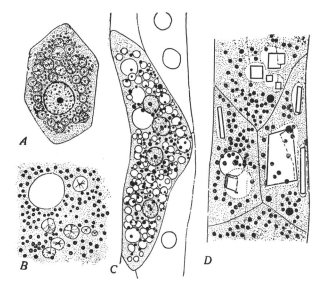

FIG. 100

Intestinal cell inclusions. A—*Rhabditis strongyloides* (Living intestinal cell, surface view; large radially striated rhabditis sphaerocrystals and small, soluble granules). B—*Theristus setosus* (Area of intestinal cell showing nucleus [not shaded], rhabditin sphaerocrystals, and olivaceous sphaeroids [small shaded]). C—*Ditylenchus dipsaci* (Cell to left with four shaded nuclei, numerous colorless fat droplets and small black [actually purple] protein globules [stain, crystal violet]). D—*Diploscapter coronata* (fat globules black, olivaceous sphaeroids shaded, nucleus colorless, crystals colorless, in vacuoles [osmic]). Original.

forms which hold to a liquid or semiliquid diet (*Rhabditis, Trichuris*) and is absent in some forms with a "solid" diet (*Metoncholaimus*). Quack (1913), following Bütschli's alveolar theory of protoplasm, held the bacilli of the bacillary and subbacillary layers to be rows of vacuoles and not actual entities, while Hetherington (1923) held the bacillary layer to be cilia and the subbacillary layer basal granules. Upon the basis of appearance and comparative morphology one must conclude that Hetherington's view is the more probable. The function is a problem for general zoology rather than nematology.

Protoplasmic zones.—The presence of protoplasmic zones, likewise, seems to be of no special significance in nematology since differentiation into ectoplasm and endoplasm is of widespread occurrence in cells of living animals. Certain authors, such as K. C. Schneider (1902), have seen fit to call the plasma

cap of *Ascaris* a "nutritorische zone." Quack has shown that in *Ascaris* such material is not confined to the periphery but extends into deeper parts of the cell as irregular masses (plasma strands) in *Ascaris* and may be so distributed as to form a mantle (Quack's fig. 21). Goldschmidt (1904) interpreted the strands as a "Chromidialapparat" but Hirschler (1910), von Kemnitz (1912), and Quack (1913) all have found this to be an error. Quack found that starved specimens of *Ascaris* showed no diminution in the plasma cap or plasma strands and hence eliminated the possibility that the material involved is absorbed albumen, concluding that it is differentiated functional cytoplasm (Compare Figs. 103J-M).

External coverings.—In large myriocytous (see p. 104) nematodes one often finds a homogeneous, slightly basophilic layer in immediate contact with the external cell surface, this layer being termed the *basal lamella* (Fig. 103J7, Z6). Apparently this layer is a differentiation acting as a supporting structure or it is a secretion product of the intestinal epithelium. It is not subdivided into areas corresponding to the cells and the ectoplasm is attached to it rather than continuous with it. Fibrillar strands of the ectoplasm reach its surface but do not appear to enter into it as one might expect if it were merely a differentiation of the outer cell surfaces. It acts more in the nature of a sheath and has affinity for collagen stains. At the present time there is no actual proof that the basal lamella is formed by the intestine. Though such a layer is plain in *Ascaris, Physaloptera, Tanqua, Trichuris* and *Dioctophyma*, in other forms it is generally not visible. The extent of its development is obviously not correlated with phylogenetic relationships but rather with cell numbers for in all of the above mentioned forms the intestine is myriocytous.

In most free-living nematodes one can discern no distinct mesenterial sheath over the intestine but in *Dorylaimus* as well as in the majority of parasitic nematodes an extremely thin membrane isolates the intestine from the body cavity and is termed the pseudocoelomic membrane. Beneath the membrane or mesentery, muscle fibers may be present but such fibers do not form a continuous layer and they are usually confined to the posterior part of the intestine. However, in unusual instances they may form a coarse mesh work (Fig. 103H) which in cross section gives the appearance of a separate muscle layer. Such muscle fibers are classified as specialized somatic muscles rather than as a *muscularis mucosae*.

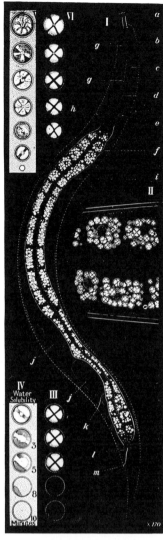

FIG. 101

Rhabditis monhystera. Rhabditin sphaerocrystals as seen in polarized light. After Cobb. 1914, J. Parasit. v 1(1).

B. MODIFICATIONS OF SUPERFICIAL APPEARANCE; FORM OF LUMEN

The superficial appearance of the intestine as observed in toto depends upon the total number of cells, the character of the cells and the character of cell inclusions. The shape of the intestinal lumen is likewise dependent on the number of cells, the form of the cells and whether or not they are equal or unequal in height.

CELL NUMBER.—Like the hypodermis and musculature, the intestine of various nematodes presents a series of stages in increased complexity; this series recapitulates to a greater or a lesser extent the ontogeny of the individual. As long ago as 1866, Schneider called attention to the fact that strongyles have an intestine composed of but a few cells, 18-20, there being only two in a given intestinal circumference. Maupas (1900) noted that rhabditids also have but few intestinal cells in the adult stage, 30 being recorded in *Rhabditis elegans* while 18 were counted in newly hatched larvae of this species. Similarly Pai (1928) found only 18 intestinal cells in adult *Turbatrix aceti* and the writers find 64 intestinal cells in adult *Rhabditis strongyloides* and 20 in first stage larvae. On the other extreme we have forms such as *Ascaris lumbricoides* with innumerable intestinal cells (about 1,000,000), forms such as *Heterakis gallinarum* with about 12,000 and intermediate forms such as *Prionchulus*, *Hystrignathus*, and *Metoncholaimus* with about 600, 400, and 5,000 respectively. In forms with 64 or

less intestinal cells the most notable and obvious feature is that the cells tend to be longitudinally elongate and rectangular (Figs. 99A-C, 100C, 102C). When the number is 64 (Fig. 100A) the characteristic hexagonal appearance is first noticeable in only a few of the cells but when the number reaches 128 all are hexagonal.

The picture becomes clearer when this information is examined in the light of embryonic development. Martini (1903) found that when 10 cleavages have occurred the definitive larva is formed—an organism with a theoretic number of 1,024 cells. However, there is a definite lag of cleavages in the endodermal stem cell since the cells of this line actually number 16 to 20 instead of 128 as would be expected if no lagging occurred. Comparing this information with facts concerning the somatic musculature one notes that the tenth cleavage has taken place in the mesodermal stem cell since 64 cells are present at hatching, this being the total number to be expected, as well as the typical number of cells in adult meromyarian nematodes. From this point of view, one might say that the course of regular cell division has not been fulfilled in a nematode with less than 128 intestinal cells and that cell division has only proceeded beyond "completion" when the intestinal cell number exceeds 128.

Thus, on the basis of the number of intestinal cells, one may classify nematodes into two groups, namely those which have not exceeded the "fore ordained" number and those which have exceeded this number. For the first condition we propose the term *oligocytous* while for the second condition the term *polycytous* may be used. However, there is a tremendous variation in the possible number of cells in the latter instance and for descriptive purposes a further division seems to be advantageous. Such a division is difficult but one finds a moderate correlation between the number and height of cells in a cross section and the total number of cells of the intestine.

Forms with less than 8,224 cells (16 cleavages) have more or less cuboidal epithelium with a maximum of 20-50 cells, usually of equal height, in a given circumference. Where raised areas occur in the lumen they are generally due to high individual cells. On the other hand forms with over 8,224 cells have 100 or more in a given circumference and definite plicae or villae are formed by groups of higher cells. The term *polycytous* is arbitrarily limited to forms with the former type of intestine (256-8,224 cells) while the term *myriocytous* is introduced for forms with the latter type of intestine (over 8,224 cells).

CELL CHARACTER.—It has previously been noted that cells in various regions of the intestine may differ in character; upon some occasions specialized cells may be scattered in the intestinal epithelium. Forms in which such cells are present may be termed *heterocytous* while forms in which the alternative is true may be termed *homocytous*.

CELL SIZE.—The intestinal lumen may be rounded, subpolygonal or quite irregular. A lumen of smooth contour occurs in oligocytous forms, though even in such forms it may become irregular due to folds or rugae, involving the entire epithelial wall. A subpolygonal lumen is characteristic of polycytous forms wherein each individual cell tends to cause a concavity in the outline of the lumen. Sometimes in *polycytous* forms, but more commonly in myriocytous forms, there are definite projections of cells into the lumen, such cells being taller than their neighbors; these groups of cells form villi or plicae (Fig. 103 E, F, I). This inequality in the height of cells in a given cross section may be termed *anisocyty* while the reverse would be *isocyty*.

NUCLEAR NUMBER.—As a rule intestinal cells in nematodes are uninucleate but exceptions to this rule are becoming more and more numerous with critical observation. The causes of polynucleation are not known; one can only interpret from scattered observation. Normal "Gigantism" of oligocytous forms appears as one of the factors. Tremendous increase in cell size such as occurs in *Strongylus equinus*, where one finds single intestinal cells 4 mm long by 500 microns wide, apparently increase the requirements of nuclear material to such an extent that uninucleate cell may be at a disadvantage. Given an equivalent amount of nucleoplasm many nuclei provide for more nuclear surface and a closer association of cytoplasm with nucleoplasm than could be obtained with a single nucleus. One might say that nemas inherently unable to continue cell cleavage compensate for this by undergoing nuclear division when natural forces no longer limit their size. Polynucleation is known to occur only in parasites. When characteristic of the entire intestine it is usually present in an entire natural group, but the exceptional sporadic cases (*Gnathostoma*, *Philometra*) are not explainable at present.

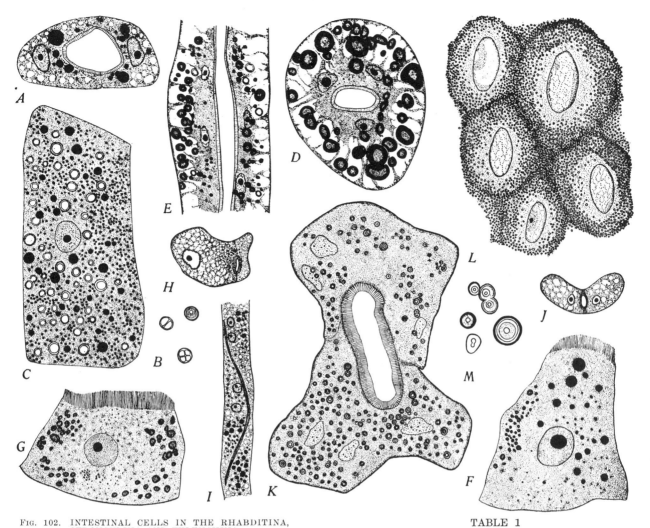

FIG. 102. INTESTINAL CELLS IN THE RHABDITINA, STRONGYLINA AND TYLENCHINA

A-C—*Rhabditis terricola* (A—Cross, C—horizontal section of intestine; B—partially dissolved sphaerocrystals in neutral violet); D-E—*Panagrolaimus subelongatus* (D—Cross, E—longitudinal section of intestine); F-G—*Rhabdias eustreptos.* (Cells seen in cross section, F—anterior, G—mid to posterior); H-J—*Ditylenchus dipsaci* (I—longitudinal section; H & J—cross sections); K—*Oesophagostomum dentatum* (Cross section); L-M—*Strongylus edentatus* (L—Surface view; M—isolated sphaerocrystals).

Increase in cell size in tylenchids can not be the cause of polynucleation since these forms are no longer than rhabditids. Quite obviously an insufficient amount of information has thus far been gathered to permit far reaching general conclusions.

Classifying nematodes according to the number of cells of the intestine, number of nuclei, specialization or lack of specialization, and equality or inequality of cells provides an interesting survey of the Nematoda. Examples of the known types are given in the accompanying table.

C. CELL INCLUSIONS

Under this heading are included all substances which are not a part of the active cytoplasm whether organic or inorganic, food reserves or waste products. Numerous types of stored food and waste products have been observed and in addition there is a residuum of non-classified material termed sphaeroids or granules. Food reserves are known to include glycogen, rhabditin, fats, and protein. Waste products are for the most part not classified chemically.

(1) RESERVE FOOD MATERIALS

Glycogen. This substance, when present, is in a liquid or semiliquid state, since it is water soluble. In fresh material it may be identified through its coloration with iodine-potassium-iodide solution or the Best's carmine technic as described by Lee (1928). Giovannola (1936) has recently employed the

TABLE 1

Genus	Cell No. (1)	Nuclear No. (2)	Cell Size (3)	Cell Character (4)
Rhabditis	+	+	+	+
Ditylenchus	+	+	+	+
Chondronema	−	+	+	+
Strongylus	+	−	+	+
Rhabdias	−	+	+	+
Hystrignathus	−	+	+	+
Spironoura	=	+	−	+
Ascaris	=	+	−	+
Physaloptera	=	+	−	+
Tanqua	=	−	−	+
Gnathostoma	=	−	+	+
Philometra	=	−	−	−
Prionchulus	−	+	+	−
Metoncholaimus	−	+	+	−
Monhystera	+	+	+	+
Anonchus	−	+	+	+
Plectus	−	+	+	+
Halanonchus	−	+	+	−
Dorylaimopsis	−	+	+	−
Synonchiella	−	+	+	−
Tripyla	±	+	+	+
Leptosomatum	−	+	+	−
Enoplus	−	+	+	+
Ironus	−	+	+	+
Dorylaimus	−	+	+	+
Leptonchus	? +	+	+	+
Agamermis	=	−	+	+
Trichuris	=	+	+	+
Dioctophyma	=	+	±	+

(1)	(2)	(3)	(4)
+ is oligocytous	+ is uninucleate	+ is isocytous	+ is homocytous
− is polycytous	− is polynucleate	− is anisocytous	− is heterocytous
= is myriocytous			

105

new Bauer (1933) technic for staining glycogen in the intestine of preparasitic larval *Ancylostoma caninum*, *Necator americanus*, and *Nippostrongylus muris*. It is always best to use a saliva enzyme control, because that which is removed by saliva is presumptively carbohydrate in nature. Busch (1905), von Kemnitz (1912) and Quack (1913) found glycogen to be the chief stored food in the intestinal epithelium of adult *Ascaris* and *Strongylus*. Giovannola reported glycogen to be the chief food reserve in the larvae of parasitic nematodes preceding and during rapid growth.

Rhabditin. This occurs as birefringent sphaerocrystals described by Maupas (1900), Cobb (1914) and Jacobs and Chitwood (1937) from the intestinal epithelium of *Rhabditis* spp. The sphaerocrystals are grey in color, bright spots in dark field illumination and bright spots with a central cross when observed between crossed Nichols of a polariscope (Figs. 100A-B & 101). They are slowly soluble in cold water, more rapidly on boiling; they are moderately soluble in 5% formalin, and in 10% acetic acid; rapidly soluble in dilute and concentrated hydrochloric, sulfuric, and nitric acids, in 50% formalin, and in sodium and ammonium hydroxides; they are insoluble in alcohol, glycerin, and xylol. When the intestines of specimens are mashed out under a cover slip and exposed to saliva or diastase of 37.5°C., these birefringents disappear from the intestine in one-half to one hour while approximately twice this time is necessary in water and in inactivated saliva controls. Iodine-potassium iodide has no effect. Presumably rhabditin is a carbohydrate but attempts to starve specimens and reduce the number of crystals were without effect. They disappear, however, when the larvae enter the encysted third stage (become ''dauer'' larvae).

Similar sphaerocrystals were described from the intestine of *Theristus setosus* by the writers (1938).

Fats and Fatty Acids. These substances are present as colorless globules imparting a grey opaque color to the organism. In dark field illumination they appear as bright circles, and between crossed Nichols they are not visible. Such material may be identified through its coloration with Sudan III, Scharlach R, Nile blue sulphate, osmic acid, and Flemming's Strong fixative. It is not dissolved by saliva, water, or hydrochloric acid and gives neither xanthoproteic nor ninhydrin reactions. It is, of course, soluble in alcohol, xylol and ether. Standard histological technic results in the appearance of large empty spaces or vacuoles wherever fats were present in the cell. Semipermanent mounts of small nematodes may be obtained by alcohol fixation, and evaporation of glycerin in Scharlach R or Nile blue sulphate according to the procedure of Goodey (1930). Pleasing temporary mounts can be made by placing living specimens in alcoholic solutions of Scharlach R. The most exact method is to cut the specimen, let the intestine flow out of the body and stain with Scharlach R, osmic acid or Flemming's fixative. If desirable they may be counterstained with haematoxylin. Permanent preparations may be made by sectioning osmicated specimens.

Von Kemnitz (1912) and Quack (1913) found fat globules in small amount in the intestine of adult *Ascaris* and *Strongylus*. Giovannola (1936) concluded that fat is the primary food reserve in larval parasitic nematodes in stages preceding a period of fasting. He further states that the quantity of fat globules is an index to the ''physiological age'' of preparasitic strongyloid larvae. More critical investigations along such lines would seem promising. Goodey (1930) identified fat globules in the intestine of representatives of the Tylenchidae, Rhabditidae, -Diplogasteridae, Cephalobidae, Plectidae, and Mononchidae. The writers have identified fat as the chief form of stored food in *Cephalobellus* and *Blatticola* (Thelastomatidae), *Chondronema* (Allantonematidae), *Spironoura* (Kathlaniidae), and various tylenchs, hoplolaims and criconematids and *Dorylaimus stagnalis*.

Stored Protein. Such substances occur as non-briefringent colorless globules similar to fat globules in transmitted light and dark field illumination. As described by Chitwood and Jacobs (1937), they are insoluble in water, alcohol, xylol, ether, and ½-saturated ammonium sulphate; dissolve in 10% acetic acid and in 5% KOH; are not affected by saliva; are pale yellow in Flemming's fixative; give positive xanthoproteic and ninhydrin reactions; and are digested by artificial gastric juice. The globules stain with gentian violet or haematoxylin; they also stain blue with Nile blue sulphate and orange with Scharlach R. These reactions apparently place them as complex proteins of a conjugated nature. The majority of globules of *Agamermis* are composed of this type of substance and not fats as is commonly supposed. Similar globules have been identified by the writers (1938) in the intestine of *Dity-*

lenchus dipsaci (Fig. 100C). It is quite possible that the colorless, insoluble (in alcohol-xylol), basophilic globules, present in *Rhabditis*, *Panagrolaimus*, *Aphelenchoides* and *Plectus* etc., are of the same nature. The fact that such globules stain with Nile blue sulphate in the same manner as fatty acids indicates that staining technics are not necessarily indicative of fat. Proteins may stain as do fats but they may be distinguished through their insolubility in fat solvents, digestion in artificial gastric juice, and positive xanthoproteic and ninhydrin reactions.

(2) WASTE PRODUCTS

Inorganic Sphaerocrystals. Reddish-brown, weakly bi-refringent sphaerocrystals occur in the intestine of many parasitic nematodes, including *Ascaris*, *Camallanus*, *Strongylus*, *Ancylostoma*, and *Trichuris*. These structures are similar in appearance to rhabditin both in transmitted light and between crossed Nichols, although in totomount preparations and sections they are not birefringent. The optical activity in this case can only be observed when the crystals are isolated. Unlike rhabditin, they are dark in dark field illumination and are insoluble in water, acetic acid, NaOH (all concentrations) and saliva. They are also insoluble in alcohol and xylol and are not affected by gastric or pancreatic enzymes. Askanazy (1896), Looss (1905) and Fauré-Fremiet (1912) regarded them as products of haemoglobin resorption, while Liévre (1934) was unable to establish the presence of haemoglobin in the intestine of *Ascaris lumbricoides* and *Parascaris equorum* by spectroscopic analyses. He demonstrated the presence of haemoglobin by this means in 75% of the specimens of *Toxocara canis* examined. Von Kemnitz (1912) identified them as zymogen granules and Quack (1913) identified them as gypsum ($CaSO_4.2H_2O$). Rogers (1940) found no evidence of gypsum in the intestine of *Strongylus* but he did find large quantities of zinc and sulfur together with traces of iron. The writers find that these crystals may be obtained relatively pure by boiling the intestine in 10% KOH and washing in a centrifuge. Crystals prepared in this manner are not charred by heating on a glass slide to the melting point of glass. They may be dissolved by heating in concentrated HCl and recrystallized. Such crystals are birefringent and obliquely extinct. However, a reddish-brown residue is left when the slide is dried. This residue stains blue in dilute HCl-potassium ferrocyanide, indicating the presence of iron; direct experiments on the sphaerocrystals produces the same result in partially dissolved (swollen) crystals. Some compound containing iron is evidently present as an adsorption within the sphaerocrystals. As evidence of the association of sphaerocrystals with a blood feeding mode of life, Törnquist (1930) pointed out that *Camalanus*, which is known to feed on blood, has them, while *Cucullanus*, which does not feed on blood, does not possess them. One might add that adult oxyurids and thelastomatids, as well as the first three larval stages of *Strongylus*, and *Camallanus* are also devoid of them. The evidence is entirely circumstantial. The occurrence of grossly similar insoluble sphaeroids in the intestine of *Theristus setosus* (see Olivaceous sphaeroids) and other free-living nematodes casts some doubt on the above interpretation since they also contain iron. Liévre (1934) interprets positive tests for iron in the intestine of *Ascaris* as due to substances obtained from animal and vegetable food, not haemoglobin.

Olivaceous sphaeroids. Reddish brown, apparently non-birefringent sphaeroids were observed by the writers (1938) in the intestinal cells of *Theristus setosus* (Fig. 100B). They have the following characteristics: Not blackened by osmic acid; not colored by Scharlach R, insoluble in alcohol; ninhydrin and xanthoproteic reactions negative; blue in neutral violet; blue in crystal violet, blue in nile-blue sulphate; not digested by artificial gastric juice or diastase; soluble in 10% HCl and 2% KOH but not in 10% acetic acid or 2% HCl; blue in potassium ferricyanide followed by 1% HCl. From these observations it seems that the sphaeroids must consist of an organic ferrous iron salt or a salt of a weak base ($Fe(OH)_2$) and a weak acid. Similar sphaeroids were also observed in *Dorylaimus stagnalis* and an unidentified oncholaimid and *Diploscapter coronata* (Fig. 100D).

Crystals. Stefanski (1916) and Cobb (1918) observed polyhedral colorless birefringent crystals in the intestine of *Ironus*. They are very similar in appearance (Fig. 105P) to triple phosphate, and, according to Stefanski, they are very soluble in acetic acid and potassium hydroxide; slightly soluble in hot water and insoluble in cold water, alcohol, ether, chloroform, and acetone; are not stained by iodine-potassium-iodide, but stain with eosin and fuchsin. Isolated colorless polyhedral crystals have also been observed in *Tripyla* (Fig. 107A).

Crystal aggregates in mermithids were observed by Meissner (1853), Rauther (1906), Hagmeier (1912) and Christie (1936). These are first seen in vacuoles of the post-nodal region of preparasitic larval *Agamermis decaudata*. Their number increases with age and in old adult specimens similar crystals have been observed in the body cavity. They are very similar in appearance (Fig. 107D) to uric acid and allantoin; are birefriengent, obliquely extinct; insoluble in water, alcohol, ether, 10% ammonium hydroxide, 10% acetic acid, 10% HCl, glycerin, and ½-saturated ammonium sulphate. Presumably they represent a nitrogenous product.

In *Diploscapter coronata* large quadrate tablets, colorless to yellowish brown, have been observed by the writers (Fig. 100D). Like olivaceous sphaeroids, they are non-birefringent, soluble in 2% sodium hydroxide and 10% hydrochloric acid but are insoluble in 1% HCl, alcohol and glycerin. They give a negative ninhydrin reaction and a positive potassium ferricyanide—1% HCl reaction. Therefore they appear to be a ferrous iron compound, probably organic in nature. It is possible that they may be crystals of the same substance composing the olivaceous sphaeroids.

(3) MISCELLANEOUS ''GRANULES''

In the majority of instances the cell contents of the nematode intestine have not been studied chemically. The term ''granule'' is, of course, chemically meaningless. Since at least four distinct substances are known to exist in a sphaeroidal state, namely, rhabditin, fat, protein and olivaceous sphaeroids each form must be considered with care. With living specimens, dark field illumination is sufficient to separate the globules of fat and protein from the sphaerocrystals of rhabditin and strongylin. In addition, the first two are colorless while the last three are yellowish brown to reddish brown. However, still other types may be discovered. Non-birefringent brownish or yellowish refractive ''granules'' are present in the intestinal epithelium of *Metoncholaimus*, *Siphonolaimus*, *Ironus*, and other forms. Such ''granules'' are strongly refractive and non-staining (? olivaceous sphaeroids). In addition one finds moderately refractive basophilic ''granules'' in the intestine of such forms as *Dorylaimopsis*, and *Plectus*, and special cells of the intestine of *Synonchiella*. Cobb (1922) has described birefringent sphaerocrystals in special cells of *Eurystomina* as ''marionellin'' and we find these to be relatively insoluble and basophilic. In other forms acidophilic ''granules'' have been observed. For morphological purposes the term *globule* will be restricted to those inclusions which are known to be non-birefringent and appear as bright circles in dark field illumination, i.e. fats and proteins. The term *sphaerocrystal* is restricted to inclusions known to be birefringent, while *sphaeroid* is applied to strongly refractive non-birefringent or apparently non-birefringent bodies, and *granule* is reserved for moderately or weakly refractive bodies of unknown optical activity. It will appear obvious that weakly birefringent substances such as strongylin may easily be classified as sphaeroids pending critical study.

The function of strongylin, sphaeroids and granules is for the most part unknown. Sphaerocrystals of strongylin were seen to be thrown out or ''excreted'' from the cells of *Ascaris* and *Strongylus*. The yellowish brown sphaeroids of *Rhabdias* and *Ironus* were also observed to be eliminated from the intestine (Fig. 105R).

(4) INTESTINAL PARASITES

Protozoan parasites are apt to occur in the intestinal cells as well as in other organs of nematodes and might easily be confused with cell inclusions or degenerating cells. Micoletzky (1922) described sporozoan parasites of the intestinal wall of *Dorylaimus carteri* and *Plectus cirratus*, and Kudo and Hetherington (1922) described a microsporidian named *Thelohania reniformis* from the intestinal epithelium of *Mastophorus muris*. The writers have encountered similar forms (Fig. 103C) in the intestine as well as the musculature, gonads, and chords of *Spironoura affine*. It is sufficient, for the present, to merely call attention to their existence. Many protozoan and fungous parasites of nematodes have been described and such information will be presented in a later part.

D. COMPARATIVE MORPHOLOGY

Up to the present time no consistent attempt has been made to record, much less present, specific information regarding the intestine in the various groups of the Nematoda. All workers recognize various impressions upon which they may have an ''intuition'' as to the group to which a nematode may belong. Whenever it is possible, in morphology, to reduce these sensory impressions to words, it invariably contributes to our understanding of relationships and to the transfer of knowledge from one worker to another. The present writers must of necessity deal in terms of examples. By giving a sufficient number of examples, it is hoped that a skeleton outline may be provided around which others can build a structure of some value.

Rhabditina. For the members of the Rhabditina we have the observations of Maupas (1900), Cobb (1914), Goodey (1930), and Giovannola (1936) as our only direct attacks on the problem, but numerous observations from the time of Bütschli, preserved chiefly in the form of drawings, serve as a foundation upon which we may build.

The Rhabditina may be characterized as oligocytous, homocytous, and isocytous with the exception of the Rhabdiasidae, Drilonematidae, and Allantonematidae, which appear to be wholly or in part polycytous. *Rhabditis* (Fig. 3, 100A, 101, 102A-C), *Turbatrix*, and *Diplogaster* retain the simple uninucleate condition in the intestinal epithelium. In these forms the cellular outlines are distinct and quite often emphasized by the absence of cell inclusions. The intestinal lumen tends to be flattened; the cells alternate, giving a zig-zag appearance in lateral view. Free-living stages of the Rhabdiasidae are also oligocytous, and in general quite similar to *Rhabditis*, while the parasitic adult is very definitely polycytous, the cells cuboidal (Fig. 102F-G). Ventricular and prerectal regions are almost always differentiated from the remainder of the intestine through absence of, or marked diminution of, cell inclusions. The intestine cells of many species of *Rhabditis* contain birefringent sphaerocrystals of rhabditin, but this substance has not been identified in any other members of the suborder. Sphaeroids, apparently non-birefringent or weakly birefringent, are very noticeable in the parasitic female of *Rhabdias* but these are of the insoluble type. Anteriorly the intestinal circumference consists of about 12 cells, very low, containing few or no inclusions. In the mid-region there are six to eight somewhat higher cells containing a moderate number of small sphaeroids, numerous larger basophilic globules and small basophilic granules (Fig. 102F), while posteriorly the sphaeroids are larger, more numerous, and the basophilic bodies minute and numerous (Fig. 102G). In these forms the sphaeroids are a deep red-brown in color.

Maupas (1900) noted that in some species of *Rhabditis* birefringents are absent. In such forms he found ''albuminofatty'' globules to be more prevalent.

Cephalobids are usually described as having an intestine composed of two series of cells (rows). Though the lumen is dorsoventrally flattened and often appears zig-zag, as in *Rhabditis*, no cell walls are distinguishable in sectioned *Panagrolaimus subelongatus* (Fig. 102E). The nuclei are near the lumen, two to six, and usually four, in a given circumference, with a total of about 210. Since disappearance of cell walls occurs in the fourth stage larva, at the same time as nuclear division (there are about 20 uninucleate cells in the third stage larva), it seems best to interpret this form as primarily oligocytous, secondarily polynucleate and/or syncytial. Perhaps polynucleate cells will eventually be distinguished as in tylenchids. The cephalobid intestine in section is characterised by its large faintly basophilic globules, scattered brownish shells, and large empty spaces which presumably were filled with fatty substances.

The Tylenchoidea present a picture in contrast to the Rhabditoidea when the intestine is considered. Although Debray and Maupas (1896) were able to distinguish 16 cells forming the intestine of *Ditylenchus dipsaci* in the fourth stage larva, distinct cells have seldom been observed in the adult members of the Tylenchidae. The intestine appears as an opaque mass of large globules, very beautiful in dark field illumination, but not visible in polarized light. This material is of a fatty character. Sections of *D. dipsaci* and *Aphelenchoides parietinus* fail to show clear evidence of cell walls (Figs. 102H-J) in the adult stage. In dissected specimens the large quadrinucleate cells are seen (Fig. 100C). The cytoplasm is highly vacuolate due to the removal of fats. Basophilic globules are also seen in sections, but these appear to be very erratic in disposition. In *D. dipsaci* the lumen is dorsoventrally flattened anteriorly and zig-zagged posteriorly (Fig. 102H-J); 56 nuclei were counted in one specimen. In *A. parietinus* the intestine is quite similar except that its sides in the mid-region tend to be more nearly equal and tend to surround the gonads. In this form a maximum of two nuclei has been observed in

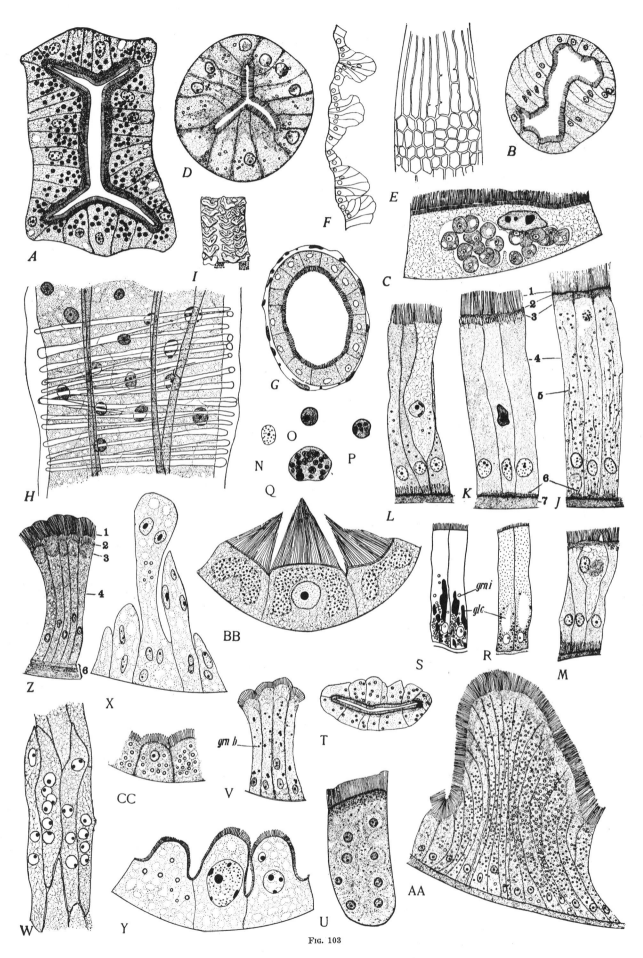

Fɪɢ. 103

108

one section and on one side of the intestine. It is notable that the lumen is relatively much smaller in tylenchids than in rhabditoids, and the bacillary layer relatively shorter and more compact. Apparently we have no increase in the number of intestinal cells, 16, between hatching and adulthood. In *Chondronema passali*, the only representative of the Allantonematidae studied, the intestine is definitely polycytous, there being four to six hexagonal cells in a circumference. The cells are filled with fat globules as in the tylenchids. Unlike the latter, however, no bacillary layer appears to line the round intestinal lumen. The lumen contains a glassy-appearing substance, possibly of protein nature, which is slightly basophilic and apparently represents partially digested body fluid from the host. Aside from the presence of a few birefringents (nature unknown) in the posterior end of the intestine of *Chondronema*, there is no known case in which either ventricular or prerectal regions differ from the mid-region in members of the Tylenchoidea.

Strongylina (Figs. 99 & 102). The Strongylina as a group appear to be oligocytous or low polycytous and the cells are polynucleate as noted in *Strongylus* by Schneider (1866), Looss (1901) and Quack (1913) and in *Ancylostoma* by Looss (1905). Other representatives of the same suborder (*Stephanurus, Oesophagostomum, Ostertagia, Longistriata, Metastrongylus* and *Dictyocaulus*) examined by the writers exhibit the same characteristics. The intestine consists of two rows each of 10 or more cells, each cell containing 10 to 500 nuclei. The smaller numbers, 10 to 20 nuclei, occur in members of the Trichostrongyloidea. In such forms as *Ostertagia* there are two rows of nuclei in each intestinal cell, four in an intestinal circumference while in *Trichostrongylus instabilis* Looss (1895) found each of the two cell rows of the intestine to contain a single row of nuclei. About 40 to 50 nuclei are present in an intestinal cell of *Oesophagostomum*, 500 in one of *Stongylus* while no reliable estimate has yet been made for *Metastrongylus*. The shape of the lumen is quite diverse, being dorsoventrally flattened in *Ostertagia*, irregular due to longitudinally folded walls in *Stephanurus*, or rounded to ovoid in *Oesophagostomum* and *Metastrongylus*. A thick, compact bacillary layer resting on a well developed subbacillary layer is characteristic of the Strongyloidea. Wetzel (1931) and Lucker (1935, 1936, 1938) have made use of the specific intestinal cell constancy of the third-stage larvae of horse strongyles as a means of differentiating the species in this stage. Thus Wetzel found two rows of eight long subtriangular cells in *Strongylus equinus*, two rows of ten cells in *S. edentatus*, and two rows of 16 cells in *S. vulgaris*. Lucker found a total of only eight cells in the intestine of *Cylicodontophorus ultrajectinus, Cylicocercus goldi,* * and *C. catinatus* larvae, of 12 in *Gyalocephalus capitatus* and of 16 in *Poteriostomum ratzii*. Similar identification of species of strongylin parasites of sheep has been shown practicable by Dikmans and Andrews (1933) on the basis of the intestinal cells of the third stage larvae.

Intestinal nuclei of strongylins are sphaeroid in such extremely divergent forms as *Ostertagia* and *Strongylus* while in *Oesophagostomum, Stephanurus, Ancylostoma* and *Kalicephalus* they are irregularly elongate or even tuboid. There is

usually a slight, though distinct, diminution in the number of insoluble sphaerocrystals both in the ventricular and prerectal regions of members of the Strongylina; in such regions the lumen is often slightly larger and the epithelium thinner than in the remainder of the intestine. Glycogen, stored in the endoplasm, appears to be the chief food reserve. Like tylenchoids, however, there is relatively little absorptive surface in the strongylin intestine.

Ascaridina (Fig. 103). The Ascaridina present a very different picture for this group is polycytous to myriocytous. Polynucleate cells are rare; they never constitute more than a small proportion of the intestinal cells in a species. Insoluble sphaerocrystals appear to be absent in the Oxyuridae, Thelastomatidae, and Kathlaniidae while they are present in members of the Rhigonematidae, Heterakidae, and Ascarididae. The smaller representatives of the Thelastomatidae (*Blatticola, Cephalobellus*) and Oxyuridae (*Macracis*) are polycytous and isocytous while larger oxyuroids such as *Oxyuris equi* and *Enterobius vermicularis* as well as rhigonematids, kathlaniids, heterakids, and ascaridids are myriocytous and anisocytous. Distinct ventricular enlargements are characteristic of the polycytous oxyuroids such as *Macracis, Blatticola,* and *Cephalobellus* though such may also occur in some myriocytous forms such as *Heterakis*. In addition sections of the intestine of polycytous oxyuroids (also *Spironoura*) show numerous large vacuolate areas which correspond to fat globules seen in the living specimens. Passing to the myriocytous forms we find the ventricular region less and less apparent with increased cell number. There is also a marked tendency toward anisocyty manifesting itself in *Rhigonema* by the formation of an oblong lumen; in *Spironoura* by an I-shaped lumen formed by two staggered longitudinal ridges (Mackin, 1936) and in *Heterakis* by a triangular lumen or three longitudinal ridges (Baker, 1936). Tuft or villus formation has been described as a further development of anisocyty in *Oxyuris equi* by Martini (1916); in this instance the tufts which are composed of numerous cells may give a hexagonal appearance when seen in toto (Fig. 103E). Jägerskiöld (1893, 1894) described groups of elongate cells especially developed in the anterior part of the intestine of *Contracaecum spiculigerum*, of *C. osculatum* and *Raphidascaris decipiens;* such cell groups reduce the lumen to a narrow, folded canal. Similar cell groups, according to Cobb (1888), take the form of V-shaped ridges (Fig. 103I) in *Anisakis simplex*. Anisocyty in *Ascaris*, on the other hand, is limited in such a manner that the small cells are lateral and the lumen, consequently, takes the form of a dorsoventrally flattened tube (Fig. 500). Glycogen constitutes the chief stored food in *Ascaris* according to Kemnitz (1912) and Quack (1913) while glycogen is absent in *Oxyuris* according to Martini (1916) and this is in agreement with the writers' findings that fatty substances are the energy reserves of oxyurids while appearing in negligible quantity in ascaridids.

Polynucleate cells with two to three nuclei have been observed upon rare occasions in *Spironoura* but they appear to be quite common in the lateral areas of the intestine of *Ascaris*. Ehrlich (1909) associated polynucleation in *Ascaris* with nuclear degeneration but Quack (1913) was unable to substantiate this view. Rather extensive studies of both nuclear and cytoplasmic "degeneration" have been made with this form by Ehrlich, Quack, and Guieysse-Pellissier (1909). Nuclear degeneration involves an enlargement of the nucleus, increased basophily and the formation of strongly refractive sphaeroids (Fig. 103N-Q) within the nucleus; these changes are followed by elimination of the cell, or a portion of the cell containing the nucleus, into the intestinal lumen. So-called cytoplasmic degeneration (Fig. 103M) involves the formation of an acidophilic mass, usually near the base of the cell, inclusion of normal cytoplasmic sphaerocrystals within the mass, movement of the whole toward the lumen and final elimination. Martini (1916) illustrated such elimination of "degenerate cells" in *Oxyuris*. The regularity of the occurrence of "degeneration" in *Ascaris* leads one to suspect that it is a normal physiologic process not necessarily retrogressive in nature. Assuming that insoluble sphaerocrystals are waste products, "cytoplasmic degeneration" might be considered a mode of excretion. In *Rhabdias* and *Strongylus* it has been previously noted that sphaerocrystals are normally eliminated in the faeces.

Spirurida (Fig. 103). For the order Spirurida there is a surprising dearth of recorded knowledge concerning the intestine. Jägerskiöld (1893, 1894), Magath (1919), Hetherington (1923) and Törnquist (1931) seem to have been the only authors who gave the intestine consideration. It is peculiar that in this group there appear to be as many instances

---

*Regarding *Cylicocercus goldi* Lucker (1938) states that the lumen passes between the first three cells "through the cytoplasm of the five posterior cells." He assures us that there are only 8 nuclei and 8 cells in the third stage larvae. It still seems possible that he may have overlooked something. Compare figs. 100 C, 102 H-J, and figs. 99 C.

FIG. 103. INTESTINE IN ASCARIDINA, SPIRURINA AND CAMALLANINA

A—*Rhigonema infecta*. B-C—*Spironoura affine*. (B—cross; C—longitudinal in prerectal region showing parasites). D—*Heterakis gallinarum*. E-F—*Oxyuris equi* (E—superficial; F—longitudinal). G-H—*Macracis monhystera* (G—cross; H—superficial). I—Anisakinid internal view showing ridges as seen in dissection. J-Q—*Ascaris lumbricoides* (J—normal; 1. bacillary layer; 2. subbacillary layer; 3. plasma cap; 4. endoplasm; 5. plasma string; 6. basioplasm; 7. basal lamella; K—binucleate cell, one nucleus degenerating; L—early cytoplasmic degeneration; M—late cytoplasmic degeneration; N—normal nucleus; O—early degenerating nucleus; P—mid degenerating nucleus; Q—late degenerating nucleus). R-S—*Toxocara canis* (R—intestinal cell with Cajal silver; S—with Iodine vapor. glc. glycogen, grn i insoluble granule (sphaerocrystal). T-U—*Gnathostoma spinigerum*. V—*Rictularia coloradiensis* (grn b basophilic granule). W-Y—*Philometra rubra* (W—surface view; X—ventricular region; Y—mid-region). Z—*Physaloptera retusa* (1. Bacillary layer; 2. subbacillary layer; 3. plasma cap; 4. endoplasm; 5. plasma string; 6. basal lamella). AA—*Tanqua tiara*. BB—*Ascarophis* (*Cystidicola harwoodi*). CC—*Camallanus americanus*. E-F, after Martini, 1916, Ztschr. Wiss. Zool. v. 116; I, after Cobb, 1888, Beitraege zur Anatomie und Ontogenie der Nematoden; R-S, after Argeseanu, 1934, Compt. Rend. Soc. Biol. v. 116; remainder original.

of marked dissimilarity of the intestine in closely related forms as there are instances of similarity. All of the members of this order appear to be myriocytous with the possible exceptions of *Gnathostoma* and *Philometra*; the latter have relatively few, large, polynucleate cells (Fig. 103U & W). Insoluble sphaerocrystals are present in *Philometra*, *Dracunculus*, *Micropleura*, *Camallanus*, *Gnathostoma*, and *Tanqua*, while they are absent in *Ascarophis* (*Metabronema*), *Cucullanus*, *Physaloptera* and *Rictularia*. Basophilic globules, probably of a protein nature, are present in *Rictularia*. Most of the representatives of this group have very tall, narrow intestinal cells and are anisocytous because the epithelium exhibits either

Sabatieria and *Dorylaimopsis*, of the Comesomatidae, are polycytous, having 256 to 500 intestinal cells, six to eight in a circumference, they also have a flattened lumen, a low bacillary layer and large basophilic globules. They differ from one another in that *Sabatieria* is apparently homocytous while *Dorylaimopsis* is heterocytous having scattered cells containing large acidophilic masses (Fig. 104G).

In the Axonolaimidae the intestine is approximately as in the Comesomatidae, there being around 256 cells, six in circumference, a rounded to flat lumen and reddish-brown, nonstaining, sphaeroids. Like *Sabatieria*, *Axonolaimus* appears circumference and reddish-brown non-staining granules. to be homocytous.

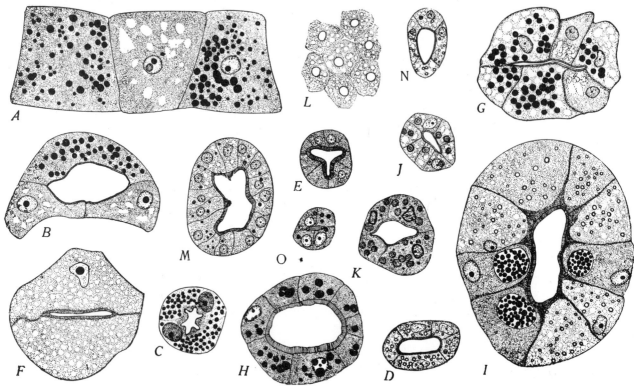

FIG. 104. INTESTINE IN CHROMADORIDA

A-B—*Synonchiella truncata*. C—*Monhystera cambari*. D—*Axonolaimus spinosus*. E—*Wilsonema bacillivorus*. F—*Tripylium carcinicolum* v. *calkinsi*. G—*Dorylaimopsis metatypicus*. H—*Anaplectus granu-* *losus*. I—*Halichoanolaimus robustus*. J—*Ethmolaimus revaliensis*. K—*Anonchus mirabilis*. L—*Chromadora* sp. M—*Halanonchus macramphidum*. N—*Chromadora* sp. O—*Chronogaster gracilis*. Original.

longitudinal ridges and valleys or villi. However, three forms are conspicuous exceptions to this rule, namely, *Camallanus*, *Ascarophis* and *Gnathostoma*. Diversity in height of cells and character of the bacillary layer and basal lamella are also conspicuous features of the group.

Chromadorida (Fig. 104). In the Chromadorida the only observations regarding the intestine have been of an incidental nature. We have records such as those of de Man (1884) in which species of the genus *Monhystera* are differentiated on the basis of their having a black or grey intestine and "two cell rows" or more than two cell rows. The rich red-brown to black pigmentation of the intestine of *Siphonolaimus* was recorded by zur Strassen (1904). The number of cells in an intestinal circumference was mentioned by Cobb (1920) in many forms of this group. Zur Strassen (1904) and Schepotieff (1908) were the only previous workers to study sections of forms of this order.

Members of the family Plectidae have relatively few intestinal cells, 120 to 930, the form with the smallest number, *Anonchus*, being oligocytous while the remaining forms studied, *Plectus*, *Chronogaster*, and *Wilsonema* are polycytous. *Anonchus* and *Chronogaster* have only four cells in a circumference, a very low bacillary layer, flat lumen, and large eosinophilic granules. *Wilsonema* has up to eight cells in a circumference, a lobed lumen, high bacillary layer and no granules and *Plectus* has up to 12 cells, a rounded lumen, high bacillary layer and basophilic globules (? protein).

In the family Camacolaimidae, *Aphanolaimus* has around 100 cells (oligocytous), a flat lumen, a high bacillary layer, four cells in a circumference and basophilic globules like *Plectus* while *Camacolaimus* has around 256 cells (low polycytous), a rounded lumen, low bacillary layer, six cells in a

In the Monhysteridae there are two quite different types of intestine. In the first, exemplified by *Monhystera* and *Theristus*, the lumen is multiradiate (Fig. 104C) though there are, respectively, 60 and 120 cells, two and four in circumference; the bacillary layer is low and compact and the intestinal inclusions are brownish or grey and basophilic. In the second, exemplified by *Halanonchus*, there are about 566 cells, six to 16 in a circumference; the bacillary layer is relatively higher, less compact, the lumen irregular, and the sphaeroids are brownish and non-staining.

Linhomoeids commonly have few intestinal cells in a circumference, usually two in the mid-region of the intestine, but the total number of cells varies considerably. *Tripylium* has 26 cells while *Terschellingia* and *Desmolaimus* exhibit around 128; the lumen is flat to rounded, the bacillary layer low, compact (not resolvable in *Tripylium* but so in the other two examples); the colorless globules in *Tripylium* are soluble in alcohol (therefore presumably fatty), while in *Terschellingia* and *Desmolaimus* the cell inclusions are sphaeroids, slightly brownish, and insoluble.

Unlike other monhysterids and linhomoeids, the siphonolaim intestine is deep brown to black in color, the pigmentation being due to refractive, insoluble sphaeroids. Zur Strassen (1904) found the intestine of *Siphonolaimus weismanni* to consist of 22 cells in circumference and to be composed of a total of 6,000 cells (estimation from statements in description); its lumen is rounded, the bacillary layer unusually high.

Members of the Chromadoridae, Microlaimidae and Desmodoridae fall within the lower limits of polycyty, varying within the narrow range of 128 to 256 cells. Seemingly all have a four cell circumference in the mid-region though there may be six to eight in a circumference in the ventricular region. Members of these families have a very low bacillary

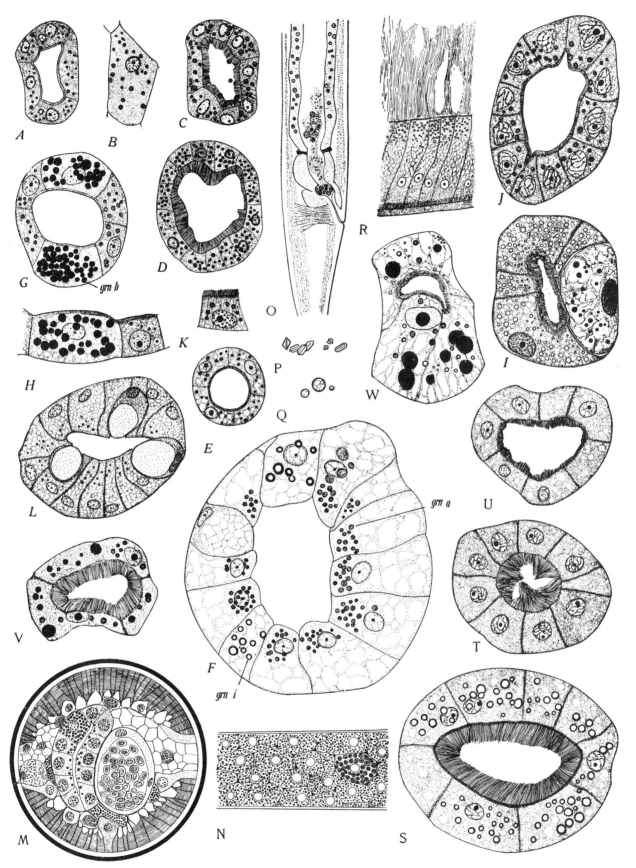

Fig. 105. INTESTINE IN ENOPLIDA

A-C—*Tripyla papillata* (A—posterior part; B—surface view; C—ventricular region). D-E—*Prionchulus muscorum* (D—ventricular region; E—mid-region). F—*Metoncholaimus pristiuris*. G-H—*Eurystomina americana*. I—*Phanodermopsis longisetae*. J-K—*Enoplus communis v. meridionalis* (J—mid-region; K—ventricular region). L—*Leptosomatum elongatum v. acephalatum*. M-N—*Thoracostoma strasseni*. O-Q—*Ironus tenuicaudatus* (O— note sphaeroids in rectum; P—isolated crystals; Q—isolated degenerate sphaeroid shells). R—*Dioctophyma renale*. S-U—*Dorylaimus stagnalis* (S—mid region; T—posterior part of prerectum; U—prerectum). V-W—*Leptonchus* sp. (V—ventricular region; W—mid-region). M-N, after Tuerk, 1903, Mitt. Zool. Stat. Neapel, v. 16; remainder original.

111

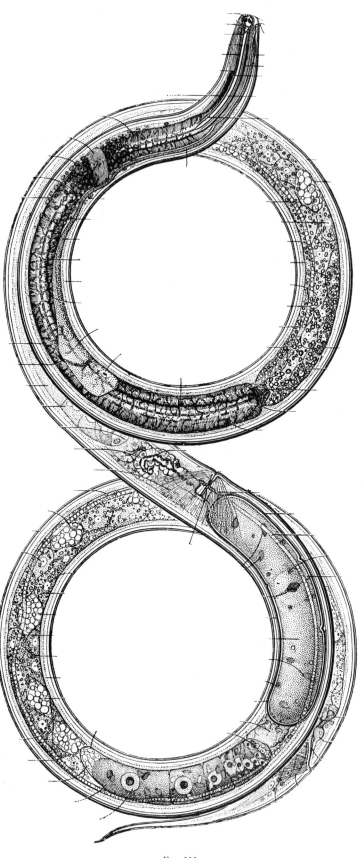

F G. 106

Ironus tenuicaudatus. (Note specialized intestinal cells containing large globules, also small polyhedral crystals in ordinary intestinal cells). After Cobb, 1918, Contrib. Sc. Nemat. 7.

layer and a rounded to subpolygonal lumen; a moderate number of somewhat basophilic sphaeroids is usually present and in addition, from the coarse vacuolate appearance of the cytoplasm, one might suspect a considerable amount of fatty substances. These three families appear to be homocytous and isocytous.

Members of the Cyatholaimidae, on the contrary, are more distinctly polycytous, varying in cell number from around 256 to 1,000 with from three to 12 cells in circumference; they appear to be uniformly heterocytous. In *Halichoanolaimus* the normal cells contain brown sphaeroids (often appearing in section as basophilic shells) while the scattered heterocytes are devoid of these bodies but contain instead, large vacuoles packed with basophilic globules. In *Synonchiella* the normal cells are vacuolate, without sphaeroids, and the heterocytes are dense, filled with basophilic globules (Fig. 104A). The similarity of the intestine of *Halichoanolaimus* to that of *Dorylaimopsis* is very striking.

Schepotieff (1908) described the intestine of *Desmoscolex* as consisting of few cells and as containing very large brownish globules which were insoluble in alcohol-xylol.

Enoplida (Figs. 105-107). The order Enoplida, containing both simple and complex free-living forms, as well as diverse types of parasites, shows extreme variation in the form of the intestine.

Within the Tripyloidea, Stefanski (1916) studied the cell inclusions of *Ironus* and Cobb (1917, 1918) described the intestine of *Ironus* and *Mononchus*. *Ironus* is polycytous and heterocytous (Fig. 106), the number of heterocytes apparently varying with the species. The ordinary cells contain yellowish non-staining sphaeroids which appear as shells with irregular contents in formalin-preserved material. These sphaeroids are sometimes eliminated through the anus (Fig. 105O). *Tripyla* (Fig. 105A-C) is homocytous and barely polycytous, having 136 to 150 intestinal cells, a low bacillary layer, acidophilic granules and scattered polygonal crystals. *Prionchulus* is likewise polycytous and homocytous (Fig. 105D-E); the species differ in having from 170 to 500 intestinal cells; anteriorly the cells in a circumference are more numerous, higher, and have a much more pronounced bacillary layer than in the mid-region but no definite ventriculus is present. Like *Ironus*, *Prionchulus* has acidophilic granules but crystals are absent.

In the Enoploidea Türk (1903), Jägerskiöld (1901), and de Man (1904) studied the intestine of *Thoracostoma* and *Cylicolaimus* and Rauther (1907) that of *Enoplus*; Cobb (1922, 1924a) investigated the intestine of *Eurystomina* and *Anticoma* and Chitwood (1931) that of *Metoncholaimus*. So far as known, all members of this group are markedly polycytous, isocytous, and have uninucleate cells. With the exception of *Enoplus* they are all heterocytous and even in this form cells are occasionally found which differ from their neighbors in the presence of large acidophilic bodies. Türk found the homocytes of *Thoracostoma* to contain greenish-brown granules. Specimens kept in clean white sand had a clear intestine free from such inclusions. He judged these inclusions to be resorption vacuoles of plant food. Occasional heterocytes he interpreted as fat cells (Fig. 105M-N). A bacillary layer appears to be totally absent in *Metoncholaimus*, *Thoracostoma* and *Cylicolaimus*. This layer is represented merely by a peripheral condensation in *Eurystomina* (Fig. 105G-H) and *Leptosomatum* (Fig. 105L) while it is moderately high and distinct in *Enoplus* and very high, especially in the ventricular region, of *Phanodermopsis*. The ordinary cells (homocytes) of *Phanodermopsis* contain yellowish non-staining sphaeroid shells (Fig. 105I) while the corresponding cells of *Leptosomatum* have a conspicuously vacuolate plasma (? fat vacuoles) and a few basophilic globules; the homocytes of the remaining forms contain acidophilic granules. Heterocytes in *Eurystomina* and *Phanodermopsis* are filled with basophilic globules while the heterocytes of *Leptosomatum* include a large amorphous acidophilic vacuole and those of *Metoncholaimus* may either be basophilic with a large vacuole or contain scattered large yellowish non-staining sphaeroids (Fig. 105F). Chitwood and Chitwood (1938) identified fats and ferrous iron salts as the chief cell inclusions of an oncholaimid.

The number of dorylaimoids of which the intestine has been studied is inadequate. Members of the Dorylaimidae all seem to be polycytous but the number of cells in an intestinal circumference varies from four to 20. Anteriorly the bacillary layer is highest in the ventricular region and posteriorly a conspicuous change is notable in this layer in the prerectum. Throughout ventricular and mid-regions the cells contain yellowish brown non-staining sphaeroids (appearing as shells in section) while these structures are absent in the prerectum.

The latter is set off as a distinct section of the intestine in the Dorylaimidae (Figs. 20-21) and Leptonchidae and may even be subdivided into two distinct units in *Actinolaimus*. *Leptonchus* resembles dorylaimids in general but differs in that the number of intestinal cells is smaller (?oligocytous) and the cells contain massive basophilic globules. Nothing is known concerning the intestine of the diphtherophorids aside from the fact that they have no prerectum.

The intestine of the Mermithoidea has been given more attention than that of other groups because, as was early recognized, the peculiar nature of the intestine constitutes one of the major characteristics of the group. Schneider (1860) first recognized that the solid mass of tissue which Meissner (1853) called the "Fettkörper" corresponds to the mesenteron of other nematodes. The work of Meissner (1853, 1856), Schneider (1860), Rauther (1906, 1909), Hagmeier (1912), Steiner

dromermis sp. the nuclei are relatively larger (Fig. 107E) and apparently less numerous than in the previously mentioned forms but unfortunately no exact information is available. Large vacuoles containing crystals or crystal aggregates have been observed in several mermithids in all stages from the preparasitic lavra to the senile adult; these crystals accumulate with age, becoming a conspicuous feature of specimens after reproduction has ceased. The increase in crystals and vacuoles coincides with diminution of intestinal globules. Rauther (1906) compared the crystals with uric acid but was unable to obtain an unmistakable murexide reaction. Concerning the globules of nutritive reserve, the following observations have been made on *Agamermis decaudata*: Sections of young parasitic larvae contain only a few basophilic globules in a rather dense cytoplasm (Fig. 107B); the larvae at emergence and the young adults are literally packed with such globules in a

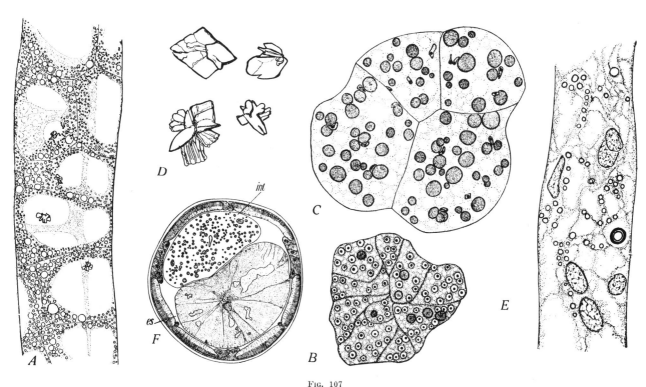

FIG. 107

Intestine of mermithoids. A-D—*Agamermis decaudata* (A—trophozome of senile male, dissected alive; note remnants of protoplasm near edge of cells and large vacuoles; B—Cross section of large larva (1 cm) from body cavity of grasshopper showing dense protoplasm, at time protein globules [shaded] begin to appear, nuclei circles with central nucleolus; C—Cross section of adult, non-senile showing numerous protein globules, note relatively minute nuclei; D—Crystals as seen in A). E—*Hydromermis* sp. (longitudinal section of adult). F—Cross section of entire *Mermis* sp. showing intestine with lumen. A-E, original; F, after Rauther 1909, Ergeb. u. Fortschr. Zool. v. 1(3).

(1933), and Christie (1936) makes it possible to characterize the mermithid intestine as an organ of food storage in which the larva, during the parasitic stage, stores the nutrient matter on which it draws throughout adult life and reproduction. In order to meet these requirements the intestine grows anterior to the base of the esophagus, regularly reaching the level of the nerve ring. Rauther (1909) observed a lumen (Fig. 107F) in the anterior part of the intestine of *Mermis* sp. and interpreted this part of the intestine as a caecum. Steiner (1933) stated that in some mermithids the trophosome has an axial cavity and a wall of polynucleate cells, while in others the axial cavity disappears but the polynucleate cellular condition persists, and in still others the cell walls disappear forming a syncytium. Rauther found that the intestine of *Hexamermis albicans* consists of two longitudinal rows of cells each containing 10 to 15 nuclei in the adult stage while the writers found the intestine of *Agamermis decaudata* (Figs. 107A-C) to be four to 10 cells in circumference and each cell to contain 22 to 25 nuclei. The total number of cells in these forms appears to fall within the upper limits of polycyty. In *Hy-*

vacuolate cytoplasm; the cells of specimens in the emerging larvae and adults are filled with colorless, oily appearing globules (whence the name fat body); according to Chitwood and Jacobs (1937) only a small proportion of these globules is fat, the great majority being protein.

The superfamily Trichuroidea is typically myriocytous, anisocytous, homocytous and the intestinal cells are uninucleate. Villus formation in *Trichuris* is uniform, the subpolygonal units (groups of cells) causing much the same appearance as large individual cells in surface view; from this standpoint there is distinct parallelism with *Oxyuris* (Fig. 103E). Actually each unit is composed of 50 to 100 tall narrow cells. The bacillary layer is quite high in the mid-region and the basal layer is unusually thick. Reddish-brown sphaerocrystals are present throughout the mid-region of the intestine.

The intestine of the dioctophymoids is much like that of the trichuroids, differing only in that the bacillary layer (Fig. 105R) may reach a height nearly equal to that of the cell proper. The sphaerocrystals are localized on the side toward the lumen and villi are not uniform.

113

Bibliography

ARGESEANU, S. 1934.—Les constituants de la cellule intestinale des ascarides. Compt. Rend. Soc. Biol. Paris, v. 116: 754-756, figs. 1-2.

ASKANAZY, M. 1896.—Der Peitschenwurm ein blutsaugender Parasit. Deutsches Arch. Klin. Med. v. 57(1-2): 104-117, pl. 2, figs. 1-9.

BAKER, A. D. 1936.—Studies on *Heterakis gallinarum* (Gmelin, 1790) Freeborn, 1923, a nematode parasite of fowls. Tr. Royal Canad. Inst. v. 20(2): 179-215, v. 21(1): 51-86, pls. 1-15, figs. 1-164.

BAUER, H. 1933.—Mikroskopisch-chemischer Nachweis von Glycogen und einigen anderen Polysacchariden. Ztschr. Mikr. Anat. Forsch. v. 33: 143-160.

BEST, F. 1906.—Ueber Kerminfärbung des Glycogens und der Kerne. Ztschr. Wiss. Mikr. v. 23: 319-322.

BILEK, F. 1909.—Ueber die fibrillären Strukturen in den Muskel und Darmzellen der Ascariden. Ztschr. Wiss. Zool. v. 93: 625-667, pls. 27-28, figs. 1-20.
    1910.—Noch ein Wort über der fibrillären Strukturen in den Darmzellen der Ascariden. Anat. Anz. v. 36: 17-25, figs.

BONNET, R. 1895.—Schlussleisten der Epithelzellen. Deutsche Med. Wochenschr. Ver. Berlin. p. 58.

BRAULT, A., & LEOPER, M. 1904.—La glycogène dans le developpement de certains parasites (cestodes et nematodes). J. Physiol. & Path. Gen. v. 6: 503-512.

BUSCH, F. W. C. M. 1905.—Sur la localisation du glycogéne chez quelques parasites intestineaux. Arch. Internat. Physiol. v. 3: 49-61, figs. 1-8.

CHITWOOD, B. G. 1931.—A comparative histological study of certain nematodes. Ztschr. Morph. v. 23(1-2): 237-284, figs. 1-23.

CHITWOOD, B. G., & CHITWOOD, M. B. 1933.—The histological anatomy of *Cephalobellus papilliger*, Cobb, 1920. Ztschr. Zellforsch. v. 19(2): 309-355, figs. 33-34.
    1938.—Further notes on intestinal cell inclusions in nemas. Proc. Helm. Soc. Wash. v. 5(1): 16-18.

CHITWOOD, B. G., & JACOBS, L. 1938.—Stored nutritive materials in the trophosome of the nematode, *Agamermis decaudata* (Mermithidae). J. Wash. Acad. Sc., v. 28(1): 12-13.

CHRISTIE, J. R. 1936.—Life history of *Agamermis decaudata*, a nematode parasite of grasshoppers and other insects. J. Agric. Res. v. 52(3): 161-198, figs. 1-20.

COBB, N. A. 1888.—Beiträge zur Anatomie und Ontogenie der Nematoden. Diss. Jena. 36 pp., 3 pls.
    1914.—Rhabditin. Contribution to a science of nematology. J. Parasit. v. 1(1): 40-41, 1 pl., figs. 1-6.
    1917.—The mononchs (*Mononchus* Bastian, 1866), a genus of free living predatory nematodes. Contrib. Sc. Nemat. (6): 129-184, 68 figs. Also in Soil Science v. 3: 431-486.
    1918.—Filter-bed nemas: Nematodes of the slow sand filter-beds of American cities. Contrib. Sc. Nemat. (7): 189-212, figs. 1-9.
    1920.—The use of the polariscope in determining the character of cell inclusions in nemas. J. Parasit. v. 6: 200.
    1920.—One hundred new nemas. Contrib. Sc. Nemat. (9): 217-343, 118 figs.
    1922.—*Marionella*. Contrib. Sc. Nemat. (11): 353-358. Also in J. Wash. Acad. Sc. v. 11(21): 504-509.
    1924a.—Minute birefringents in living cells. J. Parasit. v. 11: 102-104.
    1924b.—Specialization in the cells of the intestine of some nemas. J. Parasit. v. 11: 108-109.

DARRIBA, A. R. 1930.—Contribution al estudio del *Ganguleterakis spumosa* Med. Paises Cálidos v. 3(6): 481-513, figs. 1-28.

DEBRAY, F., & MAUPAS, E. 1896.—Le *Tylenchus devastatrix* Kühn et la maladie vermiculaire des fèves en Algérie. 55 pp., 1 pl., 17 figs. Alger.

DIKMANS, G., & ANDREWS, J. S. 1933.—A comparative morphological study of the infective larvae of the common nematodes parasitic in the alimentary tract of sheep. Tr. Am. Micr. Soc. v. 52(1): 1-25, pls. 1-6.

EHRLICH, R. 1909.—Die physiologische Degeneration der Epithelzellen des Ascaris Darmes. Arch. Zellforsch. v. 3: 81-123, pls. 2-4.

FAURE-FREMIET, E. 1913.—La cellule intestinale et la liquide cavitaire de l'*Ascaris megalocephala*. Compt. Rend. Soc. Biol. v. 74(11): 567-569.

GEHUCHTEN, A. VAN. 1893.—Contribution a l'étude du mécanisme de l'excrétion cellulaire. La Cellule. v. 9: 95-117, figs. 1-20.

GIOVANNOLA, A. 1936.—Energy and food reserves in the development of nematodes. J. Parasit. v. 22(2): 207-218, figs. 1-7.

GIROUD, A. 1922.—Note sur la tube digestif d'*Ascaris holoptera* (Rudolphi). Arch. Zool. Exper. & Gen. v. 61(1): notes & rev.: 17-20, figs. 1-2.
    1926.—Signification des bâtonnets basaux de certaines cellules, en particulier des cellules intestinales d'ascarides. Zool. Bericht. v. 10: 318 (Abstract of 1924, C. R. Ass. Anat., 19. Reunion, Strasbourg: 142-148.) Original not seen.
    1927.—La cellule intestinale des nématodes. Thèse d'agrégation. Not seen.

GOLDSCHMIDT, R. 1904a.—Der Chromidialapparat lebhaft funktionierender Gewebezellen. Biol. Centrlbl. v. 24(7): 241-251, figs. 1-4.
    1904b.—Idem. Zool. Jahrb., Abt. Anat. v. 21(1): 41-140, figs. A-Q, pls. 3-8, figs. 1-62.

GOODEY, T. 1930.—On the presence of fats in the intestinal wall of nematodes. J. Helminth. v. 8(2): 85-88.

GUERRINI, G. 1910.—Di alcuni fatti di secrezione studianti nell' epitelio intestinale dell, *Ascaris megalocephala*. Arch. Parasit. v. 14(2): 193-223, figs. A-D.

GUIEYSSE-PELLISSIER, A. 1909.—Étude de la division karyokinétique des cellules épithéliales de l'intestin d'*Ascaris megalocephala*. Compt. Rend. Assoc. Anat., v. 11: 82-91, figs. 1-4.

HAGMEIER, A. 1912.—Beiträge zur Kenntnis der Mermithiden. Diss. Heidelberg. 92 pp., 4 pls., 55 figs. Also in Zool. Jahrb., Abt. Syst. v. 32: 521-612.

HETHERINGTON, D. C. 1923.—Comparative studies on certain features of nematodes and their significance. Ill. Biol. Monog. v. 8(2): 1-62, pls. 1-4, figs. 1-47.

HIRSCHLER, J. 1910.—Cytologische Untersuchungen an Ascariden-Zellen. Bull. Internat. Akad. Sc. Cracovie. Math. & Nat., s. B. (7B): 638-645.

IMMINCK, B. D. C. M. 1924.—On the microscopical anatomy of the digestive system of *Strongylus edentatus* Looss. Arch. Anat. v. 3(4-6): 281-326, figs. 1-46.

JACOBS, L., & CHITWOOD, B. G. 1937.—A preliminary note on ''rhabditin'' sphaero-crystalloids. Proc. Herm. Soc. Wash. v. 4(2): 60.

JAEGERSKIOELD, L. A. 1893.—Bidrag till kännedomen om Nematoderna. Diss. 86 pp., 5 pls. Stockholm.
    1894.—Beiträge zur Kenntnis der Nematoden. Zool. Jahrb., Abt. Anat. v. 7(3): 449-532, pls. 24-28.
    1901.—Weitere Beiträge zur Kenntnis der Nematoden. K. Vetenskaps-Akad. Handl. v. 35(2): 1-80, figs. 1-6.

JANOWSKI, J. 1930.—Vacuome appareil de Golgi et mitochondries dans les cellules épithéliales de l'intestin moyen chez *Ascaris megalocephala*. Compt. Rend. Soc. Biol. v. 104: 1092-1093, figs. 1-3.

JOSEPH, H. 1903.—Beiträge zur Flimmerzellen und Centrosomerfrage. Arb. Zool. Inst. Univ. Wien. v. 14(1): 1-80, pls. 1-3, figs. 1-61.

KEMNITZ, G. VON. 1912.—Die Morphologie des Stoffwechsels bei *Ascaris lumbricoides*. Arch. Zellforsch. v. 7(4): 463-603, figs. A-J, pls. 34-38.

KUDO, R., & HETHERINGTON, D. C. 1922.—Notes on a microsporidian parasite of a nematode. J. Parasit. v. 8: 129-132, figs. 1-30.

KULMATYKI, W. J. 1922.—Bemerkungen über den Bau einiger Zellen von *Ascaris megalocephala* mit besonderer Berücksichtigung des sogenannten Chromidialapparates. Arch. Zellforsch, v. 16: 473-551, pls. 22-26, figs. 1-36.

LEE, BOLLES. 1928.—Microtomists' Vade Mecum. 9 ed. Blakiston, New York.

LIEVRE, H. 1934.—A propos de l'hématophagie des Ascaris. Compt. Rend. Soc. Biol. Paris, v. 116: 1079.

LOOSS, A. 1895.—*Strongylus subtilis*, n. sp., ein bischer unbekannter Parasit des Menschen in Egypten. Centrlbl. Bakt. b. 18(6): 161-169, figs. 1-8.

1901 (1902).—The Sclerostomidae of horses and donkeys in Egypt. Rec. Egypt. Govt. School. Med. :25-139, pls. 1-13, figs. 1-172.

1905.—The anatomy and life history of *Agchylostoma duodenale*. Rec. Egypt. Govt. School Med., v. 3: 1-58, pls. 1-9, figs. 1-100, photos 1-6.

LUCKER, J. T. 1934.—The morphology and development of the preparasitic larvae of *Poteriostomum ratzii*. J. Wash. Acad. Sc. v. 24(7): 302-310, figs. 1-12.

1935.—The morphology and development of the infective larvae of *Cylicodontophorus ultrajectinus* (Ihle). J. Parasit. v. 21(5): 381-385, figs. 1-3.

1936.—Comparative morphology and development of infective larvae of some horse strongyles. Proc. Helm. Soc. Wash. v. 3(1): 22-25, fig. 9.

1938.—Description and differentiation of infective larvae of three species of horse strongyles. Proc. Helm. Soc. Wash., v. 5(1): 1-5, figs. 1-2.

LUKJANOW, S. W. 1888.—Notizen über das Darmepithel bei *Ascaris mystax*. Arch. Mikr. Anat. v. 31: 293-302.

MACKIN, J. G. 1936.—Studies on the morphology and life history of nematodes in the genus *Spironoura*. Univ. Ill. Bull., v. 33(52), Ill. Biol. Monogr. v. 14(3): 1-64, pls. 1-6.

MAGATH, T. B. 1919.—*Camallanus americanus* nov. spec. Tr. Am. Micr. Soc. v. 38(2): 49-170, figs. A-Q, pls. 7-16, figs. 1-134.

MAN, J. G. DE. 1884.—Die frei in der reinen Erde und im süssen Wasser lebenden Nematoden der niederländischen Fauna. Leiden. 206 pp., 34 pls. 145 figs.

1904.—Nematodes libres. Résultats du Voyage du S. Y. Belgia. Anvers. 51 pp., 11 pls.

MARTINI, E. 1903.—Ueber Furchung und Gastrulation bei *Cucullanus elegans* Zed. Ztschr. Wiss. Zool. v. 74(4): 501-556, pls. 26-28.

1916.—Die Anatomie der *Oxyuris curvula*. Ztschr. Wiss. Zool. v. 116: 137-534, figs. 1-121, pls. 6-20.

MAUPAS, E. 1900.—Modes et formes de reproduction des nematodes. Arch. Zool. Expér. & Gén. 3. s., c. 8: 461-624, pls. 16-26.

MEISSNER, G. 1853.—Beiträge zur Anatomie und Physiologie von *Mermis albicans*. Ztschr. Wiss. Zool. v. 5(2-3); 207-284, pls. 11-15, figs. 1-55.

1856.—Beiträge zur Anatomie und Physiologie der Gordiaceen. Ztschr. Wiss. Zool. v. 7: 1-140, pls. 1-7.

MICOLETZKY, H. 1922.—Die freilebenden Erd-Nematoden. Arch. Naturg. v. 87 (1921) Abt. A. (8): 79-91, figs. G-P.

QUACK, M. 1913.—Ueber den feineren Bau der Mitteldarmzellen einiger Nematoden. Diss. Heidelberg. 50 pp., 3 pls. Also in Arch. Zellforsch. v. 11(1): 1-50, figs. a-L, pls. 1-3, figs. 1-36, 1-18.

RAUTHER, M. 1906.—Beiträge zur Kenntnis von *Mermis albicans* v. Sieb. Diss. Jena, 76 pp., pls. 1-3, figs. 1-25.

1907.—Ueber den Bau des Oesophagus und die Lokalisation der Nierenfunktion bei freilebenden Nematoden. Zool. Jahrb. Abt. Anat. v. 23(4): 703-740, pl. 38, figs. 1-9.

1909.—Morphologie und Verwandtschaftsbeziehungen der Nematoden. Ergeb. & Forstschr. Zool. v. 1(3): 491-596, figs. 1-21.

ROGERS, W. P. 1940.—The occurrence of zinc and other metals in the intestine of *Strongylus* spp. J. Helminth. v. 18 (2/3): 103-116.

ROMEIS, B. 1913.—Ueber Plastosomen und andere Zellstrukturen in den Uterus, Darm, und Muskelzellen von *Ascaris megalocephala*. Anat. Anz. v. 44 (11-12): 1-14, 1 pl., figs. 1-11.

SCHEPOTIEFF, A. 1908.—Die Desmoscoleciden. Ztschr. Wiss. Zool. v. 90: 181-204, pls. 8-10.

SCHNEIDER, A. 1860.—Bemerkungen über *Mermis*. Arch. Anat., Physiol. & Wiss. Med. pp. 243-252, pl. 6, figs. 13-18.

1866.—Monographie der Nematoden. 357 pp., 122 figs., 28 pls. Berlin.

SCHNEIDER, K. 1902.—Lehrbuch der vergleichenden histologie der Tiere. 988 pp., 691 figs. Jena.

STEFANSKI, W. 1916.—Die freilebenden Nematoden des Inn, ihre Verbreitung und Systematik. Zool. Anz. v. 46 (12-13): 363-385, figs. 1-4.

STEINER, G. 1933.—Some morphological and physiological characters of the mermithids in their relationship to parasitism. J. Parasit. v. 19 (3): 249-250.

STRASSEN, O. ZUR. 1904.—*Anthraconema*, eine neue Gattung freilebender Nematoden. Zool. Jahrb. Suppl. 7, Festschr. z. 70. Geburtst. A. Weismann, pp. 301-346, figs. A-J, pls. 15-16, figs. 1-9.

TAYLOR, A. L. 1936.—The genera and species of the Criconematinae, a subfamily of the Anguillulinidae (Nematoda). Tr. Am. Micr. Soc. v. 55(4): 391-421, figs. 1-63.

TOERNQUIST, N. 1931.—Die Nematodenfamilien Cucullanidae und Camallanidae. Göteborgs K. Vetensk.—o. Vitterhets—Samh. Handl., 5. f., s. B, v. 2 (3) 441 pp., pls. 1-17.

TUERK, F. 1903.—Ueber einige im Golf von Neapel frei lebende Nematoden Diss. Leipzig. Also in Mitth. a. d. Zool. Stat. zu Neapel v. 16: 281-348, pls. 10-11.

VIGNON, P. 1901.—Recherches de cytologie générale sur les épithéliums. Arch. Zool. Exper. & Gén. 3. s., v. 9: 371-715, figs. 1-6, pls. 15-25.

WETZEL, R. 1931.—On the differentiation of the third stage larva of *Strongylus equinus*, *S. edentatus*, and *S. vulgaris*. J. Parasit. v. 17(4): 235.

CHAPTER VIII

THE POSTERIOR GUT

(STRUCTURES OF THE PROCTODEUM)

B. G. CHITWOOD and M. B. CHITWOOD

The existence of a complete digestive tract terminated by an anus of separate sexes was discovered by Tyson (1683) in *Ascaris;* he probably observed the spicules also but was unable to interpret them correctly. Soon afterwards it was established that the intestine of the female connects by means of a valve or sphincter with the posterior gut (rectum) and thence with the outside through a ventrally situated anus. With one exception, the female reproductive system never connects with the rectum; in the genus *Rondonia* Travassos, 1920 (Atractidae) the vagina joins the rectum (Fig. 108L) to form a cloaca. This condition is approached in several other forms particularly *Aorurus agile* (= *A. subcloatus* Christie, 1931, Thelastomatidae) and *Eustrongylides tricolor* Sugimoto, 1931 (Dioctophymatidae) but an internal junction of vagina and cloaca exists only in *Rondonia.* In the male, the reproductive system always joins the rectum, forming a cloaca from the walls of which various copulatory structures develop. Since there is usually a definitely elongate tail, the anus or cloacal opening is ventral. In exceptional groups characterized by the absence of a tail, such as the Trichuroidea and Dioctophymatoidea and in scattered representatives of other groups the anus or cloacal opening may be terminal or subterminal. Male strongylins can hardly be placed in this category since the dorsal ray represents the tail and the *genital cone* is developmentally a ventral outgrowth of the cloacal lips. The intestino-rectal valve, cloaca, spicules, gubernaculum and telamon are all included under the general heading of posterior gut since, with the exception of the intestino-rectal valve, they are wholly formations of the proctodeum. The valve is quite diverse, formed sometimes chiefly, sometimes entirely from endodermal tissue, but in all instances it functions as a part of the rectum. Other structures are essentially modifications of the rectum and will be discussed from that standpoint.

A. RECTUM, INTESTINO-RECTAL VALVE AND RECTAL GLANDS

The rectum is a more or less flattened, subtriangular or irregular tube lined internally by a cuticular layer underneath which there is a layer of large epithelial cells, and covered externally by mesenterial and muscle tissue. Leuckart (1876) was under the impression that the cuticular layer of the rectum was continuous not only with external cuticle but also with the bacillary layer of the intestine. Voltzenlogel (1902) found Leuckart to be in error regarding the latter connection. The cuticular lining of the rectum ends slightly posterior to the junction of mesenteron and proctodeum leaving the rectum naked for a short distance. Though Voltzenlogel made this observation on *Ascaris* it was confirmed by Martini (1916) for *Oxyuris* and the writers for such diverse forms as *Metoncholaimus, Cephalobellus, Dioctophyma* and *Trichuris* (Oncholaimidae, Thelastomatidae, Dioctophymatidae, and Trichuridae, respectively). All investigators have found the rectal cuticle to be continuous with the external cuticle. It is known to be cast off at the molt with the remainder of the exuvium. Voltzenlogel and Martini both observed that the fiber layers and striation of the external cuticle cease at the inner side of the anal lips; farther inward the rectal lining consists of cortical, matrix and basal layers. However, there is considerable thickening of the first two layers in most parasitic nemas.

It is impractical to discuss the rectal epithelium without first considering the rectal glands since there has been much confusion in interpretation. Walter (1856) was supposed by Bastian (1866) to have first seen the large cells at the junction of the intestine and rectum in *Cosmocerca trispinosa* (*Oxyuris ornata*) and to have mislabeled them nerve cells. Actually Walter was entirely correct; the structures he illustrated were the paired preanal ventral ganglia. Shortly thereafter Claparede illustrated the cells now known as rectal glands in "*Ascaris commutata*" and "*A. mucronata*" labeling them

anal glands. Since that time similar structures have been reported from many parasitic nematodes. Eberth (1860, 1863) illustrated "anal glands" in *Heterakis vesicularis, Draschia megastoma,* and *Passalurus ambiguus;* Macalister (1865) mentioned them in *Atractis dactylura* and was the first to suggest that they might be homologues of the malpighian tubules of insects; Bastian (1865) described anal glands in *Anticoma* spp., *Linhomoeus, Halichoanolaimus,* and *Cyatholaimus;* Bütschli (1873) described anal glands in *Rhabditis aspera;* Leuckart (1876) mentioned six anal glands in *Ancylostoma;* de Man (1886) described various cells around the rectum in *Enoplus, Oncholaimus,* and *Anticoma* as anal glands; Hesse (1892) working on *Parascaris* interpreted the large cells as "Gewebepolster" cells; Augstein (1894) observed anal glands in *Dictyocaulus filaria;* Shipley (1894) described anal glands in *Toxascaris transfuga* but later (1897) presumed them to be identical with the giant "büschelförmige Organe" (Coelomocytes, Jägerskiöld (1893, 1894) described anal glands and a unicellular sphincter muscle in *Contracaecum clavatum;* Hamann (1895) gave a very good description of both rectal glands and rectal epithelium in *Goezia* (Anisakinae); Ehlers (1899) and Jerke (1901) mentioned rectal glands in *Oxyuris equi;* Looss (1901) described cells forming a "rectal ligament" in members of the Strongylidae considering the whole group of cells in this region as being non-glandular; Voltzenlogel (1902) gave an excellent description of the rectal glands, rectal epithelium, etc., of *Ascaris;* Looss (1905) denied the existence of rectal glands in *Ancylostoma* and interpreted these cells as part of a "rectal ligament" which view was concurred in by Imminck (1924) working on *Strongylus,* Törnquist (1931) working on *Cucullanus* and *Camallanus* and Mackin (1936) studying *Spironoura;* Martini (1916) published thorough descriptions of the rectal glands, epithelium and musculature of *Oxyuris;* finally Magath (1919) considered the rectal glands as sarcoplasm of the sphincter muscle.

FIG. 108

A-D—*Dioctophyma renale* (A—Cross section through posterior part of male showing spicular pouch, intestine and vas deferens; B—Cross section through cloaca and spicule near its entrance into cloaca; C—Cross section of spicular pouch showing cuticular lining, epithelium and protractor muscle; D—Cross section more posterior to A near junction of rectum and vas deferens). E-F—*Tripyla papillata* (E—Cross section of male in region of cloaca and spicular pouches; F—Longitudinal section in same region). G—*Parascaris equorum* (Longitudinal section through proximal end of spicule showing spicular cells). H—*Ascaris lumbricoides* (Longitudinal section through caudal region of male). I—*Parascaris equorum* (Longitudinal section through caudal region of male). J—*Goezia annulata* (Longitudinal section through caudal region of female). K—*Parascaris equorum* (Cross section of male anterior to intestino-rectal valve). L—*Rondonia rondoni* (Lateral view showing vagina opening into rectum). M-R—*Heterakis gallinarum* (All illustrations of male. M—At level of intestino-rectal valve; N—At level of rectal gland orifices; O—At level of preanal sucker; P—Reconstruction of cloacal region; Q—Longitudinal section at intestino-rectal valve showing sphincter muscle, double dorsal gland and secondary dorsal gland; R—Cross section considerably anterior to intestino-rectal valve showing paired subventral glands and their accompanying cells, secondary glands). S-V—*Leptosomatum elongatum* v. *acephalatum* (S—Intestino-rectal valve, inner cells as intestinal; T—Section following S; U—Rectum showing epithelial cells; V—Preadult male showing spicular primordia). W—*Dorylaimus stagnalis* (Cross section showing rectum of female). X-Z—*Metoncholaimus pristiuris* (X—Cross section of male at level of intestino-rectal valve; Y—Cross section of cloacal region of male showing spicules and gubernaculum; Z—Longitudinal section of female; there is no evidence of rectal glands or a break in the rectal cuticle such as one would expect in that case). AA—*Eurystomina americana* (Cross section of male at cloacal opening). BB-JJ—*Enoplus communis* v. *meridionalis* (BB—Longitudinal section of female in rectal region, as in Z, there is no evidence of rectal glands; CC—Cross section of preadult male showing primordia of spicules and gubernacular crura; DD-JJ—Serial sections through cloacal region of male, beginning postanal and going anteriad, some sections omitted between most anterior sections). G-I & K, after Voltzenlogel, 1902, Zool. Jahrb., Abt. Anat., v. 16; J. after Hamann, 1895, Die Nemathelminthen v. 2; L, after Baylis, 1936, Ann. & Mag. Nat. Hist. s. 10, v. 17; remainder original.

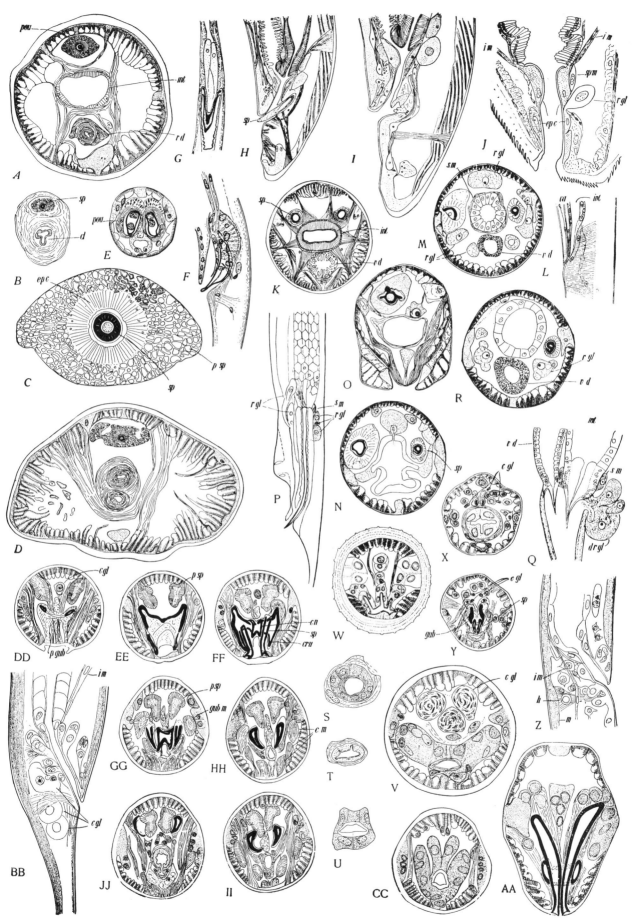

Fig. 108.

117

Controversy over the function of cells of the rectal region has confused the picture, especially since some workers deny the existence of functional glands opening into the rectum. Recently the writers (1930, 1931, 1933) observed the orifices of such glands into the rectum in *Rhabditis, Heterakis, Macracis, Cephalobellus,* and *Hystrignathus* (Rhabditidae, Heterakidae, Oxyuridae and Thelastomatidae respectively) and similar gland orifices were reported by Baker (1936) for *Heterakis.* It does not, however, necessarily follow that all of the structures in the past termed glands are homologous with the structures described by the above mentioned authors. As will be seen later, rectal glands are by no means a universal feature in nematodes. However, it is considered certain that the structure described by Jägerskiöld, Hamann, Voltzenlogel and Martini are rectal glands.

Hamann reports that the rectal epithelium of *Goezia* is composed of two pairs of cells, one pair anterior and one pair posterior to the rectal glands (Fig. 108J). In *Ascaris,* Voltzenlogel found four large epithelial cells forming an anterior circle (Fig. 108I) and additional cells posterior to them but the latter were not constant in position. In *Oxyuris* Martini describes the rectal epithelium as composed of seven cells, an anterior ring of three, (one dorsal and two subventral) and two pairs of cells arranged in tandem posterior to the first group (Fig. 109W-X). In females of *Goezia, Ascaris,* and *Oxyuris* there are three rectal glands projecting into the body cavity and having processes which penetrate the rectal epithelium. Voltzenlogel was the first to show that there is sexual dimorphism in the number of rectal glands; he reported six rectal glands for male ascarids; Martini later found the same number in males of *Oxyuris.*

Confusion in regard to the structures has been due to two factors; the rectal glands may be embedded in the lateral and dorsal chords or they may be associated with the vas deferens. Thus Mackin described the rectum of *Spironoura* as composed of 10 cells in the female, three forming a "rectal ligament," and 14 in the male, two in a dorsal "ligament," two in a "genital ligament," and one in each of two ligaments extending to the lateral chords. All of the cells designated "ligament cells" by Mackin are rectal glands; those of the "genital ligament" are embedded in the wall of the vas deferens but have separate orifices into the rectum (Fig. 110MM). The "small ejaculatory glands" described by Chitwood (1930, 1931) in *Rhabditis* (Fig. 3, ej 2) *Macracis* and *Heterakis* and by Baker (1936) in *Heterakis** are also rectal glands and correspond to the cells of the "genital ligament" described by Mackin.

From these observations, it appears that three rectal glands in the female and six in the male is the rule for members of the Rhabditoidea and Ascaridina and for at least some members of the Spiruroidea. However, rectal glands appear to be totally absent in the Tylenchoidea. The rectal ligament cells described by Looss and Imminck for strongyloids are in part rectal glands. Such glands occur in representatives of all suborders of the Phasmidia but appear to be absent in a few isolated types and groups such as *Dracunculus, Dirofilaria* and tylenchoids. Perhaps the absence of rectal glands in these forms will be explained in the future on physiological grounds when their function becomes known.

In the Aphasmidia no single case has thus far been definitely established of the existence of rectal glands. Though mentioned by Eberth (1863), Bastian (1865) and de Man (1886) it is notable that no later mention was made of such glands by de Man (1904) nor by Jägerskiöld (1901). The best substantiated record of such occurrence is in *Enoplus communis* as illustrated by de Man (1886). The writers have been able to identify the numerous cells shown by that author as attached to the rectum but all appear to be separated from the rectal lumen by the cuticular rectal lining (Fig. 108 BB).

This is in sharp contrast to the established cases such as *Spironoura* and *Heterakis* (Fig. 108N, 110 Y & MM) in which there is a distinct break in the cuticle at the level of each gland orifice. It is concluded that in *Enoplus communis* the cells in question are merely epithelial cells. Careful study of *Cylicolaimus* and *Thoracostoma* by Jägerskiöld (1901) and Türk (1903), of *Trichuris* by Rauther (1909, 1918), of *Metoncholaimus, Leptosomatum, Tripyla, Prionchulus, Dioctophyma, Soboliphyme, Aphanolaimus, Halichoanolaimus, Paracanthonchus* and *Dorylaimopsis* by the writers failed to reveal rectal glands in a single case. In *Anaplectus granulosus,* female totomount specimens appear to show three rectal glands but we have not been able to verify the point in sections. Pending further proof it is concluded that aphasmidians are usually without rectal glands. In the male of *Dorylaimus prolificus* there is a group of four pairs of cells near the anterior end of the prerectum which appear to be glandular; in living specimens one may trace a slender tube leading from each cell posteriad nearly to the intestino-rectal valve where each tube turns ventrad and disappears between the vas deferens and the intestine; the homologues of the above mentioned cells are not known.

The intestino-rectal valve is a very simple structure, consisting, in oligocytous forms such as *Rhabditis,* of the posterior parts of the prerectal intestinal cells surrounded by a sphincter muscle. In polycytous and myriocytous nematodes the intestinal cells in the valve region become much smaller, more numerous, and are often devoid of a bacillary layer. They may form a valve either by reflexure into the intestinal lumen or extension into the rectal lumen (Fig. 108Q & 109W).

The musculature controlling the intestino-rectal valve, rectum and anus has already been briefly discussed. The existence of a uninucleate sphincter muscle was first made known by Jägerskiöld (1893, 1894) and Gilson and Pantel (1894). Later workers have often confused other structures with the sphincter and described sphincter muscles with two, four or more nuclei (Magath, 1919, in *Camallanus,* and Chitwood, 1931, in *Macracis*). Reexamination of representatives of all groups of the Nematoda by the writers establishes the unicellular sphincter as universally present. It is a circular band of fibers containing a single nucleus which may be dorsal, ventral, or lateral in position. Its innervation process extends anteriorly to the dorsal nerve in *Ascaris* according to Voltzenlogel. This muscle closes the intestino-rectal valve preventing reentry of materials from the rectum into the intestine during defecation.

The depressor ani, an H-shaped muscle, is likewise unicellular and of universal occurrence. It is this muscle that elevates the dorsal wall of the rectum causing materials to be drawn into the rectal cavity; it then elevates the posterior lip of the anus thus permitting defecation. The rectum is devoid of circular muscles and for the most part defecation is accomplished by pressure. Subventral and subdorsal somato-intestinal muscles (Fig. 52) probably supply the pressure by dilating the prerectal lumen, thus drawing materials into that region from the mid-region of the intestine and forcing them into the rectum by relaxing at the same time that the rectal sphincter relaxes and the anterior part of the depressor ani contracts; by such means the rectal cavity is filled. Thereafter the rectal sphincter contracts, the anterior part of the depressor ani relaxes and the posterior part contracts; pressure on the walls of the distended rectum by the body fluid causes it to collapse and the waste products to be forced out.

B. CLOACA

The vas deferens enters the rectum from the ventral side in males of all groups with the sole exception of the Trichuroidea. In trichuroids Rauther (1909, 1918) found that the vas deferens enters the rectum dorsolaterally (Fig. 110 Z-DD). In phasmidians the junction of vas deferens with hind gut is nearly simultaneous with or immediately posterior to the intestino-rectal valve so that practically no rectum exists; the whole of the hind gut is then transformed into cloaca. In some aphasmidians, particularly enoploids, the vas deferens is apt to join the rectum somewhat more posteriad so that both a rectum and cloaca may coexist.

C. SPICULAR POUCH

With few exceptions, which will be discussed later, the spicules enter the cloaca from the dorsal side immediately anterior to the anus. They develop in a pair of cell masses, the spicular primordia, which develop as proliferations of the dorsal wall of the cloaca, first described correctly by Seurat

---

*In *Heterakis gallinarum* one finds six additional smaller cells in the body cavity, a dorsal pair, one situated on each side of and between the large pair of tandem dorsal rectal glands (Fig. 108, N, P), and two subventral pairs, one situated on each side of each of the large ventral rectal glands (Fig. 108 P, R). These cells seem to have ducts into the cells to which they are attached. They may conceivably be subsidiary rectal glands. Increase in rectal gland number from three to six and possibly to 12 seems to support the view of Macalister that they are homologues of the malpighian tubules of insects. Martini (1913) aptly indicated the unicellular rectal glands of nematodes as precursors of the groups of unicellular glands of *Macrobiotus* which in turn are undoubtedly forerunners of the multicellular tubular glands of other tardigrades. This last mentioned are, in turn, considered as identical with the malpighian tubules of insects. This view, though questioned by Seurat (1920) seems logical in view of the fact that multicellular tubular glands normally, in evolution, arise by reduplication of associated unicellular glands with common acini.

(1920) in *Falcaustra lambdiensis* (Fig. 110 OO). Schneider assumed the presence of a single primordium in nematodes with two spicules but this is incorrect. Previous to the formation of the spicules the primordia are without a lumen, as may be seen in *Enoplus* (Fig. 108CC), but later developmental phases indicate that the spicular primordia should be interpreted as instances of suppressed evagination followed by terminal invagination. The primordia become differentiated in such a manner that they form a *pouch* which contains the spicules. As shown by Voltzenlogel (1902) the pouch is lined with a cuticle continuous with that of the cloaca. It is covered by an epithelium which is also continuous with that of the cloaca. When two spicules are present, the paired spicular pouches always join before entering the cloaca. In parasitic nematodes this pouch and the protractor muscles of the spicules form an obvious spicular covering termed the sheath. In free-living nematodes the spicular pouch is often extremely delicate* and easily overlooked. Possibly for this reason Türk (1903) denied the existence of a pouch in *Thoracostoma* where it had been previously observed and illustrated by Jägerskiöld (1901). It has been seen in all free-living nematodes studied by the writers. The protractor muscles of the spicules form a complete longitudinal muscle layer on the surface of the pouch in most parasitic nematodes but in free-living nematodes these muscles are more commonly confined to the dorsolateral sides. The pouch and accompanying muscles are most conspicuous in *Tripyla* and *Triplonchium* of the free-living nematodes and give an appearance (Fig. 108F) which is peculiar to these genera and their relatives. In the Strongylina and Dioctophymatina the spicular pouch commonly joins the cloaca some distance anterior to the anus and in the Trichuroidea this tendency is carried to an extreme (Fig. 110 DD). In other groups of nematodes the spicules can scarcely ever be said to lie within the cloaca even for a short distance; actually they merely pass through a common cloaco-spicular orifice. Mueller (1925) and Chitwood and Chitwood (1933) found the spicules of *Proleptus* and *Cephalobellus* to have a separate aperture immediately posterior to the cloacal opening.

The spicular pouch should under no circumstances be confused with the so-called spicular sheath of *Trichuris*, which was named the *cirrus* by Rauther (1909). It is a unique occurrence in the Nematoda that in *Trichuris* and relatives the lining of the cloaca (cirrus) is itself evertible and often armed with teeth (Fig. 109L); this structure is capable of being inserted with the spicule into the vagina during copulation and undoubtedly serves as a true penis. It is in direct continuity with the external cuticle and when retracted there are three layers (Fig. 110 AA) of cloacal lining. The feces must pass through the cirrus for evacuation to take place.

D. SPICULES

Nematodes usually have two spicules, each spicule being essentially a tube covered by a sclerotized cuticle and containing a central protoplasmic core. Its cuticular covering is continuous with the cuticular lining of the spicular pouch and from that standpoint the spicules may be regarded as evaginations of the spicular pouch. The cuticle is often layered, the outer layer (Fig. 109 CC) being colorless and structureless while the inner layer or layers are tan to brownish and sometimes are composed of numerous prismoidal elements. In other cases the spicules may appear to be spongy or reticulated (Fig. 110 NN). The prismoidal elements Fig. 109 CC) should not be confused with the so called striation such as occurs in *Stenurus* (Fig. 33 K), *Protostrongylus* (Fig. 110 EE), etc. Such striation is due to the extension of the sclerotized ribs into a weak or nonsclerotized flange. The central protoplasmic core of the spicule may or may not contain the nuclei of the spicular epithelium. These cells and nuclei are often situated anterior to the proximal end of the spicules, surrounded by the retractor muscle. Their number is variable, four being recorded in *Ascaris* (Fig. 108 G & K). Ordinarily one finds each spicule provided with two retractor muscles and two protractor muscles. Both pairs of muscles are attached to the proximal end of the spicule. The retractors extend anteriad and toward the lateral chord where they may be attached to the body wall either subdorsally, dorsolaterally or ventrolaterally dependent upon the species involved. The protractors tend to surround the spicular sheath and may be inserted postanally to the body wall or to the dorsal side of the spicular pouch. In those groups in which a single spicule is present, the retractor muscles extend to both the right and the left body walls. In exceptional instances the protractors and retractors are composed of numerous cells which form a longitudinal muscle layer on the spicular pouch (Figs. 108 A, C & 110 DD) that is quite characteristic of the groups involved (Trichuroidea and Dioctophymatoidea). As a rule each spicule may be divided into three sections, the head (capitulum), shaft (calomus), and blade (lamina). The head is the modified portion of the proximal end; the shaft is the tube-like part between the head and the blade; the blade is the distal portion which is usually flanged (Fig. 110 DD-EE). There may be one or two flanges; if one, it is ventral as seen in cross section and if two, one is dorsal and one ventral (Fig. 110 MM). Though such differentiation of the spicule into regions is the rule, no such regions are apparent in *Dioctophyma* or *Trichuris*. Extensive diversity occurs in the spicular form.

Two is the basic number of spicules in nematodes and they originate from a double spicular primordia as stated by Seurat (1920). However a single spicule occurs in many groups of nematodes. It is a characteristic of the Trichuroidea, Dioctophymatoidea, Oxyuridae and Thelastomatidae and also occurs in isolated genera of the Desmodoridae (*Monoposthia*), and Mermithidae (*Hydromermis*). In other groups it is not uncommon for the spicules to be distally fused and stages with nearly complete fusion are known. Thus, in *Rhabditis strongyloides* (Fig. 109 EE, H-J), *R. terricola* (Rhabditidae) and *Nematodirus aspinosus* (Trichostrongylidae) the distal ends of the spicules are fused. In the subfamily Ransomnematinae of the Atractidae, there is a complete series of stages in fusion, including forms such as *Carnoya* with two entirely separate spicules, *Heth* and *Angra* with nearly completely fused spicules, and *Pulchrocephala* with no spicules. In the Mermithidae there is a similar series from *Mermis* with two separate spicules, to *Paramermis elegans* (Fig. 109 D-E) with partially fused spicules, and *Hydromermis* with a single spicule. Genera and species completely devoid of spicules are of sporadic occurrence but usually, if not always, such forms are confined to groups characterized by a tendency toward spicular fusion or presence of a single spicule. Thus we find *Trichinella* of the Trichuroidea, *Aspiculuris*, *Dermatoxys* and several but not all species of *Pharyngodon* (all representatives of the Oxyuridae), *Hystrignathus* of the Thelastomatidae and *Pulchrocephala* of the Atractidae characterized by the absence of spicules. There have been two divergent views on the interpretation of unpaired spicules. One group, led by Schneider (1866), maintains that it is a neotenic character resulting from a failure of division of an originally unpaired spicular primordium. As has been previously noted, this view is based on the erroneous assumption that in two-spicule nematodes the spicular primordium is originally single. Other authors have considered the single spicules as resulting from reduction by loss of one spicule. In view of the several series showing spicular fusion, this theory seems hardly tenable. As Cobb (1898) pointed out, the retractor muscles of the single-spicule forms go to both sides of the body wall, indicating the double character of the spicule.

Diversities in spicular morphology are too numerous to be covered completely. The general form is often quite diagnostic, being used as generic and specific characters throughout the Nematoda. Unfortunately the shapes vary so much and are sometimes so complicated that one must rely chiefly on illustrations. Cobb (1898) provided some descriptive terms which might well be more widely applied (Fig. 109 NN). As a rule the spicules of free-living nematodes are equal and similar. In parasitic nematodes, particularly the various groups of the Spiruroidea, Filarioidea, and some species of the Heterakidae, Cucullanidae, Camallanidae and Atractidae the spicules are unequal and dissimilar. Asymmetry in the form of the spicules is given varying significance in the different groups. Usually the left spicule is longer than the right but in some forms the converse is the case (*Heterakis gallinarum*). The blade of the longer spicule is generally alate (Fig. 109 T) and sometimes bears a distal hook. The shorter spicule is usually heavier and often terminates in a massive hook (Fig. 109 M, N, T).

While often specific, the form of the spicules sometimes varies and more than one specimen should always be studied. The writers found marked variation in spicular morphology in *Dirofilaria immitis* from an abnormal host (muskrat). In addition to the normal type, a specimen in which the left spicule blade was degenerate and another in which the left spicule was absent were found in a single muskrat. Such findings cause one to be somewhat dubious of the numerous species of filariids being described at the present time, and differentiated chiefly or wholly on diversity in spicular form.

---

*The pouch wall and cuticle were inadvertently omitted in Fig. 49 E, L, M. They are delicate but nevertheless present in this form.

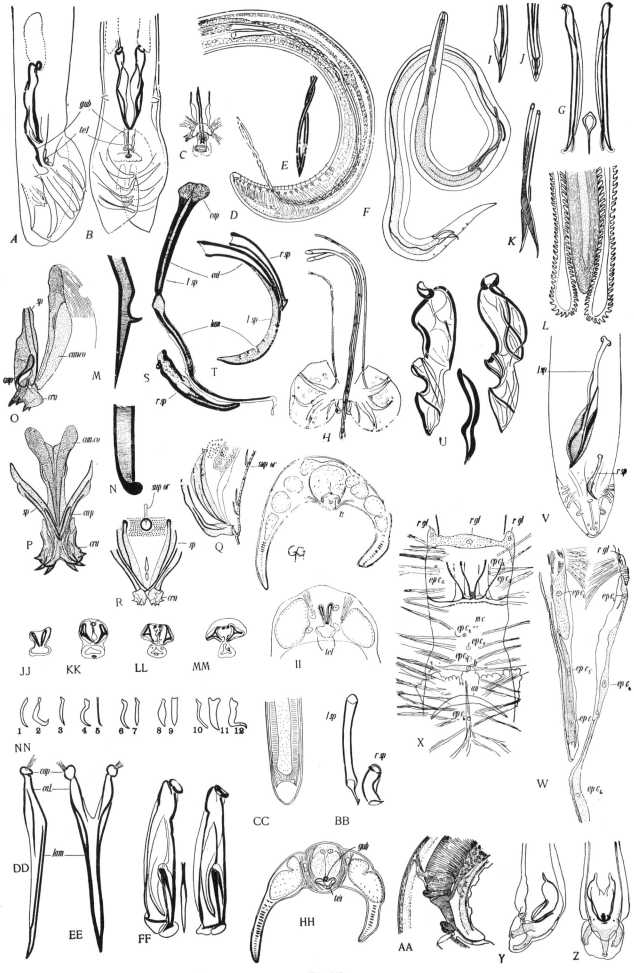

FIG. 109

120

Differences in spicular morphology of related species may be due to the degree of cephalation, length and differentiation of the shaft, or length and character of the blade. The blade may be highly twisted as in *Dicheilonema* (Fig. 109 V); twisting and irregularity of the blade is particularly characteristic of some genera of the Trichostrongylidae (Fig. B, U, FF). In addition the blade may be distally branched (Fig. 109 A, MM) or the flanges separated from the tube.

The spicules are seldom very long in proportion to the body in free-living nematodes, one of the most outstanding exceptions to this rule being *Metoncholaimus pristiuris* (Fig. 18) in which the spicules are one-twentieth of the body length (seven times as long as the anal body diameter). In parasitic nematodes both the absolute and relative spicule lengths may be many times greater. Thus, Wehr (1933) found the two equal spicules of *Odontospirura cetiopenis* (Spiruridae, Habronematinae) to attain lengths of 10-11 mm in specimens 15- to 17 mm. Ransom (1904) found the left spicule of *Gongylonema ingluvicola* to attain 17 to 19 mm in specimens 17 to 19 mm long (when retracted it twists and extends through only three-fourths of the body length); in the same form he found the right spicule to be only 0.1 mm long. Seurat (1913) found that the left spicule of *Tetrameres inermis* reaches 1.187 mm in length and the right only 0.075 mm in specimens of a total length of 2.125 mm (Fig. 109 F). Long spicules most commonly occur in representatives of the Spiruroidea, Filarioidea, Dioctophymatoidea, and Trichuroidea and in those forms where the vagina is long and tubular. One might attempt to correlate the length of the vagina with that of the spicules were it not that one finds related species with vaginae of about equal length and spicules relatively much smaller.

E. GUBERNACULUM AND TELAMON

The gubernaculum is by common definition a cuticular thickening (sclerotization) of the dorsal wall of the cloaca. This definition is at once misleading. The gubernaculum is formed from the wall of the spicular pouch. Proximally it is usually not in direct contact with the spicular cavity but is to be seen as one or more plates in the wall of the spicular pouch; distally it may come to be the dorsal wall of the pouch or it may project free into the lumen (Fig. 110 U). The cuticle of the gubernaculum, like that of the spicules, is in direct continuation with the pouch lining but unlike the spicules its sclerotized layers may extend beyond this covering internally. The gubernaculum is essentially a plate in the groove of which the spicules move. This condition is seen in many nematodes such as *Rhabditis* (Fig. 49 L), *Ancylostoma*, etc. In others, however, the medial part of the plate may protrude into the spicular pouch separating the spicules. The posterior part of the gubernaculum may then be termed the *corpus* and the anterior (medial) piece, the *cuneus*. Where the division occurs the spicules are usually alate. In *Spironoura* (Fig. 110 A-U) the gubernaculum divides near its proximal end giving off an

FIG. 109

A-C—*Hyostrongylus rubidus* (A—Lateral view of male tail; B—ventral view; C—Detail of telamon, specimen cleared in phenol). D-E—*Mermis elegans* (D—Lateral view of male tail; E—tip of spicules). F—*Tetrameres inermis* (Full length of male, showing greatly elongated left spicule). G—*Ostertagia circumcincta* (Spicules and gubernaculum); H-J—*Nematodirus aspinosus* (H—Ventral view of male tail, setaceous spicules distally fused; I & J—Lateral and ventral view of spicule tips). K—*Murshidia elephasi* (Spicules). L—*Trichuris vulpis* (Tip of cirrus and spicule). M-N—"*Habronema*" *seurati* (Tips of left and right spicules). O-P—*Cyatholaimus ocellatus* (Spicules and gubernaculum). Q-R—*Paracanthonchus caecus* (Spicules and gubernaculum). S—*Dirofilaria repens* (spicules). T—*Oswaldofilaria* sp. (Spicules). U—*Trichostrongylus probolurus* (Spicules and gubernaculum). V—*Dicheilonema horridum* (Ventral view of male tail, reversed sides). W-X—*Oxyuris equi* (W—Longitudinal section of female rectum; X—Dorsal view of female rectum), Y-Z—*Stenurus minor* (Lateral and ventral views of male tail). AA—*Proleptus obtusus* (Reconstruction of cloacal region of male). BB—*Setaria tundra* (Spicules). CC—*Trichuris vulpis* (Tip of spicule after exposure to Fairchild's trypsin). DD-EE—*Rhabditis strongyloides* (Spicules, distally fused). FF—*Ostertagia trifida* (Spicules and gubernaculum). GG-MM—*Hyostrongylus rubidus* (Serial sections through cloacal region of male showing telamon, gubernaculum, spicules, spicular pouch and cloaca). NN—Various spicular shapes (1, arcuate; 2, hamate; 3, arcuate distally; 4, falcate; 5, setaceous; 6, sigmoid; 7, linear; 8, fusiform; 9, elongate; 10, bent or boomerang shaped; 11, cuneiform; 12, L-shaped). D-E, after Hagmeier, 1912, Zool. Jahrb., Abt. Syst. v. 32; F, after Seurat, 1913, Bull. Soc. Nat. Afrique Nord Alger v. 5; G, U & FF, after Kalantarian, 1928, Trudy Gosudarstv. Inst. Eksper. Moskva, v. 5; H-J, after Rajewska, 1931, Ztschr. Infekt. v. 40; K, after Wu, 1934, Sinensia v. 5; M-N & V, after Skrjabin, 1917, Parasit. v. 9; O-R, after de Man, 1889, Mem. Soc. Zool. France, v. 2; W-X, after Martini, 1916, Ztschr. Wiss. Zool. v. 106; Y-Z, after Baylis and Daubney, 1925, Parasit. v. 17; AA, after J. F. Mueller, 1925, J. Parasit. v. 12; BB, after Rajewskaja, 1928, Die Setarien, etc.; NN, after Cobb, 1898, Misc. Publ. No. 215, Dept. Agric. N. S. Wales; remainder original.

anterior branch (cuneus) which comes to lie free in the spicular pouch; each dorsal spicular flange thus runs in a groove formed by the cuneus and corpus of the gubernaculum.

Hall (1921) proposed the term telamon for an "ornamental supporting structure" formed in the ventral and lateral walls of the cloaca of *Hyostrongylus rubidus* (Fig. 109 C). This structure differs from the gubernaculum and spicules in staining capacity and is more like the external cuticle. As may be seen from section (Fig. 109 GG-MM) Hall's interpretation as to the origin of the telamon is correct. It is an immovable sclerotized part of the cloacal wall which apparently serves to turn the spicules posteriorly when they are protruded from the spicular pouch into the cloaca; otherwise the spicules might break through the ventral cloacal wall. Ordinarily such a protective structure would be unnecessary since the spicular pouch orifice is immediately opposite the cloacal opening but in the Strongylina this is not the case.

Since Hall's original publication much confusion has resulted from the application of the terms gubernaculum and telamon, particularly in the Metastrongylidae. Cameron (1927) used the term telamon for lateral protrusible branches (crura) of the corpus which he described as "ornamental supporting structures near the cloacal aperture" (Fig. 110 EE-JJ). Gebauer (1932) applied the term telamon to a medial ventral sclerotization of the spicular pouch (Fig. 110 EE) (the structure now known as the *capitulum*) and the term gubernaculum to the unpaired and paired subdorsal sclerotizations of the spicular pouch (corpus and crura). The structures involved have been discussed by Shu'lts, Orlov and Kutass (1933), and by Dikmans (1935). A restudy of *Protostrongylus* indicates that the former authors correctly interpreted the entire complex as a gubernaculum. The most important differentiation between gubernaculum and telamon is that the *telamon is formed directly from the cloacal lining* while the *gubernaculum is formed from the spicular pouch*. Though the gubernaculum is primarily dorsal, it is also primarily medial, i. e., it may develop proximal to the union of the spicular pouches and, as in *Spironoura* it may be composed of two or more parts which are heavily sclerotized, these parts being joined by feebly sclerotized regions. Thus the medial (ventral) piece (Fig. 110 R-S) termed the telamon by Gebauer, is actually a part of the gubernaculum named the *capitulum* by Shu'lts, Orlov and Kutass: the two posteriorly directed pieces (crura) termed the telamon by Cameron and the gubernaculum by Gebauer are joined anteriorly to the capitulum by the unpaired piece, corpus (Fig. 110 S).

The confusion in terminology has led workers in other groups to misapply the term telamon. Thus Steiner and Albin (1933) termed the anterior parts of the gubernaculum (crura) of *Deontostoma californicum* (Fig. 37, 4-5) a telamon. In so far as the writers are aware, a true telamon does not exist outside the Strongylina.

The gubernaculum is often complex in free-living nematodes and the various parts are worthy of discussion. De Man (1886) described the gubernaculum of *Enoplus communis* (Figs. 110 A, C-D) as being composed of three parts: an unpaired medial piece (termed cuneus, projecting anteriorly into the cloacal cavity between the spicules) to which is joined posteriorly and laterally a less sclerotized structure (corpus) which in turn has two strongly sclerotized lateral pieces (crura). These lateral pieces project into the cloacal cavity on each side of the spicules, each of which moves in a separate groove between cuneus and crura, guided posteriorly by the corpus. In other aphasmidians many variations of the above described arrangement are known to occur. In *Acanthonchus viviparus* (Fig. 15 M), and *Paracanthonchus caecus* (Fig. 109 Q-R) the ends of the crura are dentate while in *Cyatholaimus elongatus* they are denticulate. There is a marked diversity in development of the parts in closely related forms. Thus de Man (1889) found the crura to be quite massive and apparently detached (cuneus and corpus absent) in *Paracanthonchus caecus* while in *Cyatholaimus ocellatus* he found the cuneus-corpus (Fig. 109 O-P) extremely large, the crura considerably smaller and in addition he found a small anterior ventral piece (capitulum). In the latter species the spicules are guided anteriorly, posteriorly, medially and laterally. In other forms such as *Syringolaimus striatocaudatus* (Fig. 111E) and *Anoplostoma viviparum*, de Man (1888, 1907) described well developed grooved crura accompanied by a weak corpus and in these cases each spicule moves in a groove of the corresponding crura. In *Theristus normandica* (Fig. 111 F-G) the gubernaculum appears to surround the spicules, and de Man (1890) found it to be scarcely differentiated into separate parts, but the medial

121

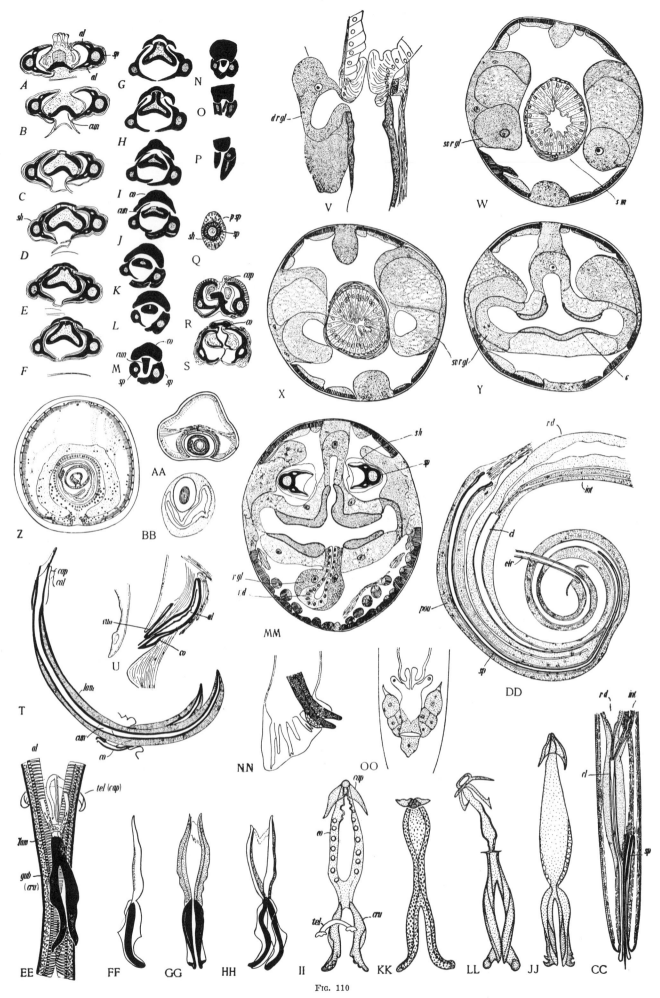

Fig. 110

region projects as a cuneus between the spicules and the lateral parts partially surround the spicules and serve as crura. In *Sphaerolaimus hirsutus* (Fig. 111 H-I) the same author found a gubernaculum much like that of *Theristus* but a capitulum in addition. Another type of variation, particularly characteristic of the Axonolaimidae and Linhomoeinae, though sometimes occurring in other groups, is due to the presence of two posterior prolongations of the corpus, termed the *apophyses*. In *Terschellingia longicaudata* the spicules are separated by a distinct cuneus and held in position laterally by flanges of the corpus, no distinct crura being present, while in *Metalinhomoeus typicus* (Fig. 111 J-K) capitulum, cuneus, crura, corpus, and apophyses may all be distinguished.

The musculature of the gubernaculum has been previously described but the origin and insertion of the various muscles is not limited as previously indicated. The gubernaculum often behaves as though hinged at the junction of the spicular pouch and cloaca. Muscles (protractor gubernaculi) attached to the proximal ends of the gubernacular corpus, crura or the distal ends of the apophyses extend posteriorly to the ventral body wall and their contraction moves the distal ends of the corpus, crura, and cuneus anteriorly, forcing the spicules outward, their tips pushed anteriorly (into the female). Muscles from the proximal ends of the crura in such an instance (*Enoplus*) extend laterally and ventrally to the body wall; their contraction moves the distal ends of the various gubernacular parts posteriorly withdrawing the spicules.

F. FUNCTION OF THE SPICULES

Though it was recognized by Schneider (1866, p. 244) that the spicules are never hollow and that spermatozoa do not ordinarily flow ''through them,'' the conception that they are hollow and act as true intromittent organs has somehow persisted. A locatory and excitatory function has been ascribed to the spicules by Schneider (1866), Bütschli (1872), Rauther (1909, 1918, 1930), Seurat (1920), Baylis (1929), Mueller (1930) and Chitwood & Chitwood (1933). On the other hand, Looss (1905) and zur Strassen (1907) considered that the two flanged spicules in *Ancylostoma* and *Philometra* come together in the form of a tube and that the sperm flows between them. The transmission of sperm down the groove or between the grooves of flanged spicules has been considered probable by Mueller (1925), Baylis (1929) and Rauther (1930). In cases where the spicule or spicules are devoid of flanges, their cross section being practically circular throughout (*Ascaris*, *Trichuris*, *Dioctophyma*) there seems to be no conceivable way by which they could ''conduct the sperm.'' Regarding this type of spicule, Mueller (1930) was of the opinion that they are withdrawn during the period of sperm movement for he found a copulating pair of *Ascaris* in which, upon section, the spicules were found not to be inserted. However, observations of copulating nematodes in which the spicules have been seen alternately inserted and withdrawn have been numerous and Mueller's finding seems hardly significant. The spicules take an active part in copulation and we have merely to define that part.

In forms with flanged spicules, do the flanges ever form a tube or groove by which the sperm pass to the female? Tube formation by two spicules in forms such as *Spironoura*, *Ancylostoma* and *Protostrongylus* is undeniable. This tube is formed by the orientation of the spicules by the gubernaculum (Fig. 110 A-T, 33K) while they are still in the spicular pouch. There seems to be no way by which the sperm could gain entrance to this closed tube after it reaches the cloaca.

The spicules might more plausibly play the role of sperm transmitters in forms where the spicules are dissimilar and one of them is flanged or where the spicules are distally fused with a median groove. Mueller (1925) described such an instance in *Proleptus* in which he noted that the blade of the left spicule had wide flanges forming a nearly complete tube open proximally as well as distally. He also found a cloaca-spicular canal connecting the upper part of the cloaca with the spicular pouch just anterior to the beginning of the spicular blade (Fig. 109 AA). Such a condition has been seen neither before nor since. The writers can more easily think of flanges as primarily for the purpose of increased rigidity and it is conceivable that the flanges aid in keeping an open passageway into the vaginal lumen.

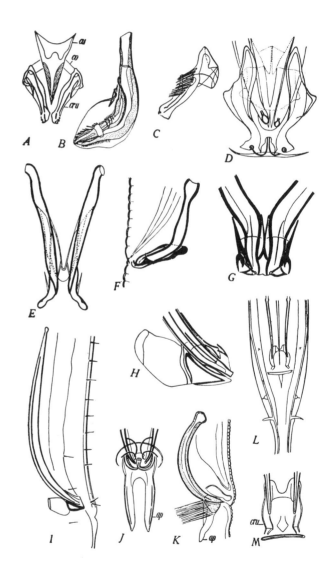

FIG. 110

A-P—*Spironoura affine* [Kathlaniidae] (Serial cross sections through spicules and gubernaculum). Q-S—*Protostrongylus rupricaprae* [Metastrongylidae] (Q—Cross section of shaft, i.e., calomus, of spicule; R—Cross sections of gubernaculum and blade, i.e. calomus, of spicule; R—T-Y—*Spironoura affine* (T—Spicules and gubernaculum cleared in 10 per cent NaOH); U—Gubernaculum and spicular ala in longitudinal section; V—Intestino-rectal valve and dorsal gland of female, longitudinal section; W-Y—Sections of female through intestino-rectal region showing dorsal and subventral rectal glands and their orifices. Z—*Trichuris suis* (Cross section of male near cloacal opening). AA—*T. trichiura* (Cross section near cloacal opening). BB—*T. suis* (Cross section of cloaca with spicular pouch about to emerge from its walls; according to Rauther more posteriorly the cloacal walls completely surround the spicule forming a double layer). CC—*T. suis* (Reconstruction of male tail; according to Rauther the internal lining of the cirrus is continuous with the cloacal lining anteriorly and is not the lining of the spicular pouch. The upper part of the cloaca has a double lining, the two layers fusing proximally). DD—*T. vulpis* (Tail of male cleared in phenol). EE—*Protostrongylus austriacus* (Gubernaculum and spicules labelled by Gebauer; correct terminology in parenthesis). FF-HH—*Protostrongylus rupricaprae* (Gubernaculum in various views). II—*P. kochi* (Gubernaculum); JJ—*P. raillieti* (Gubernaculum). KK—*P. leuckarti* (Gubernaculum). LL—*Cystocaulus nigrescens* (Gubernaculum). MM—*Spironoura affine* (Cross section of male at junction of rectum and vas deferens showing orifices of dorsal, ordinary subventral and secondary subventral rectal glands; cuticle is absent in region of gland orifice). NN—*Dictyocaulus filaria* (Lateral view of male tail showing spongy type of spicules). OO—*Falcaustra lambdiensis* (Ventral view of rectal region of preadult male showing spicular primordia). Z, BB & CC, after Rauther, 1918, Zool. Jahrb., Abt. Anat. v. 40. AA, after Rauther, 1909, Ergeb. u. Fortschr. Zool. v. 1; EE-HH, after Gebauer, 1932. Ztschr Parasitenk. v. 4; II-LL, after Schulz, Orlov & Kutass, 1933, Zool. Anz. v. 102; NN, after Yorke & Maplestone, 1926, Nematode Parasites of Vertebrates; OO, after Seurat, 1920. Hist Nat. Nemat. Berberie: remainder original.

FIG. 111

A-C—*Enoplus communis* (A—Gubernaculum, ventral view; B—spicule; C—gubernaculum). D—*E. brevis* (Gubernaculum and tips of spicules). E—*Syringolaimus striatocaudatus*. F-G—*Theristus normandica* (Spicules and gubernaculum, lateral view, F and ventral view, G). H-I—*Sphaerolaimus hirsutus* (H—Detail of gubernaculum as seen in I). J-K—*Metalinhomoeus typicus* (J—Ventral view; K—lateral view). L-M—*Anoplostoma viviparum* (L—Ventral view; M—detail of gubernaculum). A-D, after de Man, 1886, Nordsee Nematoden; E, after de Man, 1888, Mem. Soc. Zool. France, v. 1; F-G after de Man, 1890, Ibid., v. 3; H-M, after de Man, 1907, Ibid., v. 20.

The movement of the spicules back and forth during copulation would serve not only to keep the vulva and vagina open but also, at least to some extent, would actually propel the sperm into the female.* Branching of the spicular blade and twisted spicula formation such as occurs in trichostrongyles (Fig. 109 U, FF) would seem to be particularly adapted to such activity. In forms with markedly unequal and dissimilar spicules it is notable that the vagina is always quite long and tubular. Seurat (1920) advanced the view that in such species the short spicule opens the lips of the vulva and the proximal region of the vagina while the long spicule assures the progression of the sperm in the long vagina. It is true that the short spicule usually has a large hook which would be well adapted as a holdfast.

The gubernaculum is, of course, primarily a spicular guide and prevents the spicules from breaking through the wall of the spicular pouch and cloaca when exerted. In such forms as *Terschellingia, Enoplus* and *Spironoura* it may also act as a levator. Ordinarily the gubernaculum is not everted during copulation but in forms with denticulate or dentate crura such as *Paracanthonchus* and *Cyatholaimus* (Fig. 109 O-R) it would seem that they are adapted for gripping the vulvar lips and holding the vulva open.

---

*Seurat states that the spicules are absolutely immobile during copulation but in living free-living nematodes such as *Rhabditis* the writers have observed alternate withdrawal and insertion over extended periods.

Bibliography

ALICATA, J. E. 1935 (1936).—Early developmental stages of nematodes occurring in swine. U. S. Dept. Agric. Tech. Bull. No. 489, 96 pp., 30 figs.

AUGSTEIN, O. 1894.—*Strongylus filaria* R. Diss. Leipzig. 54 pp., 2 pls. Also in Arch. Naturg. 60 J., v. 1 (3): 255-304, pls. 13-14.

BAKER, A. D. 1936.—Studies on *Heterakis gallinae*, a nematode parasite of fowls. Tr. Roy. Canad. Inst., v. 20 (2): 179-215, v. 21 (1): 51-86; figs. A-G, pls. 1-15, figs. 1-164.

BASTIAN, H. C. 1865.—Monograph on the Anguillulidae, or free nematoids, marine, land, and fresh-water; with descriptions of 100 new species. Tr. Linn. Soc. Lond., v. 25 (2): 73-184, pls. 9-13, figs. 1-248.
　　1866.—On the anatomy and physiology of the nematoids, parasitic and free; with observations on their zoological position, and affinities to the echinoderms. Phil. Tr. Lond., v. 156: 545-638, pls. 22-28.

BAYLIS, H. A. 1929.—A manual of helminthology, medical and veterinary. 303 pp., 200 figs. London.
　　1936.—The nematode genus *Rondonia* Travassos. Ann. & Mag. Nat. Hist. s. 10, v. 17: 606-610, figs. 1-2.

BAYLIS, H. A. and DAUBNEY, R. 1925.—A revision of the lungworms of Cetacea. Parasit., v. 17 (2): 201-216, figs. 1-25.

BUETSCHLI, O. 1872.—Beobachtungen über mehrere Parasiten. Arch. Naturg. J. 38, v. 1 (2): 234-249, pls. 8-9.
　　1873.—Beiträge zur Kenntniss der freilebenden Nematoden. Nova Acta K. Leop.-Car. Akad. Naturf., Dresden, v. 36 (5): 124 pp., pls. 17-27, figs. 1-69.

CAMERON, T. W. M. 1927.—Studies on three new genera and some little-known species of the nematode family Protostrongylidae Leiper, 1926. J. Helminth., v. 5 (1): 1-24, figs. 1-14.

CHITWOOD, B. G. 1930.—Studies on some physiological functions and morphological characters of *Rhabditis* (Rhabditidae, Nematodes). J. Morph. & Physiol., v. 49 (1): 251-275, figs. A-H, pls. 1-3, figs. 1-24.
　　1931.—A comparative histological study of certain nematodes. Ztschr. Morph. & Oekol., v. 23 (1/): 237-284, figs. 1-23.

CHITWOOD, B. G. and CHITWOOD, M. B. 1933.—The histological anatomy of *Cephalobellus papilliger* Cobb, 1920. Ztschr. Zellforsch., v. 19 (2): 309-355, figs. 1-34.

CHRISTIE, J. R. 1931.—Some nemic parasites (Oxyuridae) of coleopterous larvae. J. Agric. Res., v. 42 (8): 463-482, figs. 1-14.

CLAPAREDE, E. 1859.—De la formation et de la fécondation des oeufs chez les vers nématodes. Mem. Soc. Phys. & Hist. Nat. Genève, v. 15 (1): 1-101, pls. 1-8.

COBB, N. A. 1898.—Extract from MS. report on the parasites of stock. Dept. Agric., Sydney, N. S. Wales, Misc. Publ. No. 215, 62 pp., 129 figs.

DIKMANS, G. 1935.—Two new lungworms, *Protostrongylus coburni* n. sp., and *Pneumostrongylus alpenae* n. sp., from the deer, *Odocoileus virginianus*, in Michigan. Tr. Am. Micr. Soc., v. 54 (2): 138-144, pls. 25-26.

EBERTH, C. J. 1860.—Zur Organisation von *Heterakis vesicularis*. Würzburg Naturw. Ztschr., v. 1: 41-60, pls. 2-4, figs. 1-29.
　　1863.—Untersuchungen über Nematoden. 77 pp., 9 pls. Leipzig.

EHLERS, H. 1899.—Zur Kenntnis der Anatomie und Biologie von *Oxyuris curvula* Rud. Diss. Marburg, 26 pp., 2 pls., 20 figs. Also in Arch. Naturg. 65 J., v. 1 (1): 1-26, pls. 1-2, figs. 1-20.

GEBAUER, O. 1932.—Zur Kenntnis der Parasitenfauna der Gemse. Ztschr. Parasit., v. 4 (2): 147-219, figs. 1-70.

GILSON, G. and PANTEL, J. 1894.—Sur quelques cellules musculaires de l'*Ascaris*. Anat. Anz., v. 9 (23): 724-727, figs. 1-2.

GLAUE, H. 1910a.—Beiträge zu einer Monographie der Nematodenspecies *Ascaris felis* und *Ascaris canis*. Ztschr. Wiss. Zool., v. 95 (4): 551-593, figs. 1-26.
　　1910b.—Beiträge zur Systematik der Nematoden. Zool. Anz., v. 35: 744-759, figs. 1-5.

HAGMEIER, A. 1912.—Beiträge zur Kenntnis der Mermithiden. Diss. Heidelberg. 92 pp., 5 pls., 55 figs. Also in Zool. Jahrb. Abt. Syst., v. 32 (6): 521-612, figs. a-g, pls. 17-21, figs. 1-55.

HALL, M. C. 1921.—Two new genera of nematodes with a note on a neglected nematode structure. Proc. U. S. Nat. Mus. v. 59(2386): 541-546, figs. 1-2.

HAMANN, 1895.—Die Nemathelminthen (2). Die Nematoden. 120 pp., 11 pls.

HESSE, R. 1892.—Ueber das Nervensystem von *Ascaris megalocephala*. Ztschr. Wiss. Zool. v. 54(3): 548-568, pls. 23-24, figs. 1-20.

HSÜ, H. F. 1933.—Some species of Porrocaecum (Nematoda) from birds in China. J. Parasit. v. 19(4): 280-285, pls. 2-3, figs. 1-18.
　　1933b.—On some parasitic nematodes collected in China. Parasit. v. 24(4): 512-541, figs. 1-46.

IMMINCK, B. D. C. M. 1924.—On the microscopical anatomy of the digestive system of *Strongylus edentatus* Looss. Arch. Anat., Hist. & Embryol., v. 3(4-6): 281-326, figs. 1-46.

JAEGERSKIOELD, L. A. 1893.—Bidrag till Kännedomen om Nematoderna. Diss. Stokholm. 86 pp., 5 pls., 43 figs.
　　1894.—Beiträge zur Kenntnis der Nematoden. Zool. Jahrb. Abt. Anat. v. 7(3): 449-532, pls. 24-28.
　　1901.—Weitere Beiträge zur Kenntnis der Nematoden. Kongl. Svenska Vetenskaps-Akad. Handl. v. 35 (2): 1-80, pls. 1-6.
　　1909.—Nematoden aus Aegypten und dem Sudan. Results of the Swedish Zoological Expedition to Egypt and the White Nile, 1901. 66 pp., 4 pls.

JERKE, M. 1901.—Zur Kenntnis der Oxyuren des Pferdes. Diss. Jena. 64 pp., 1 pl.

KALANTARIAN, E. V. 1928.—Zur Trichostrongyliden fauna der Schafe Armeniens. Trudy Gosudarstv. Inst. Eksper. Vet. Moskva, v. 5(2): 40-57, figs. 1-24 (Russian).

LEUCKART, R. 1868-1876.—Die Menschlichen Parasiten. Leipzig & Heidelberg. v. 2, 882 pp., 401 figs.

LOOSS, A. 1901.—The Sclerostomidae of horses and donkeys in Egypt. Rec. Egypt. Govt. School Med. pp. 25-139, pls. 1-13, figs. 1-172.
　　1905.—The anatomy and life history of *Agchylostoma duodenale* Dub. Rec. Egypt. Govt. School Med., v. 3: 1-158, pls. 1-9, figs. 1-100, pl. 10, photos 1-6.

MACALISTER, A. 1865.—On the presence of certain secreting organs in Nematoides. Ann. & Mag. Nat. Hist. 3. s. (91), v. 16(4): 45-48.

MACKIN, J. G. 1936.—Studies on the morphology and life history of nematodes in the genus *Spironoura*. Univ. Ill. Bull. v. 33(52), Ill. Biol. Monogr. v. 14(3): 64 pp., pls. 1-6, figs. 1-69.

MAGATH, T. B. 1919.—*Camallanus americanus* nov. spec. Tr. Am. Micr. Soc. v. 38(2): 49-170, figs. A-Q, pls. 7-16, figs. 1-134.

MAN, J. G. DE, 1886.—Anatomische Untersuchungen über freilebende Nordsee-Nematoden. 82 pp., 13 pls.

1888.—Sur quelques nématodes libres de la mer du nord, nouveaux ou peu connus. Mém. Soc. Zool. France v. 1: 1-51, pls. 1-3, figs. 1-20.

1889.—Troisième note sur les nématodes libres de la mer du nord et de la manche. Mém. Soc. Zool. France v. 2: 182-216, pls. 5-8, figs. 1-12.

1890.—Quatrième note sur les nématodes libres de la mer du norde et de la manche. Mém. Soc. Zool. France v. 3: 169-194, pls. 3-5, figs. 1-10.

1904.—Nématodes libres. Résultats du voyage du S. Y. Belgica. Expéd. Antarct. Belg. Anvers. 55 pp., 11 pls.

1907.—Sur quelques espèces nouvelles ou peu connues de nématodes libres habitant les côtes de la Zélande. Mém. Soc. Zool. France v. 20:33-90, pls. 1-4, figs. 1-17.

MARTINI, E. 1913.—Ueber die Stellung der Nematoden im System. Verhandl. Zool. Gesellsch. v. 23: 233-248.

1916.—Die Anatomie der *Oxyuris curvula*. Ztschr. Wiss. Zool. v. 116: 137-534, figs. 1-121, pls. 6-20.

MUELLER, J. F. 1925.—Some new features of nematode morphology in *Proleptus obtusus* Dujardin, J. Parasit. v. 12: 84-90, pl. 10, figs. 1-10.

1930.—The mechanism of copulation in the nematode *Ascaris lumbricoides*. Tr. Am. Micr. Soc. v. 49(1): 42-45, pl. 6, figs. 1-4.

RAJEWSKY, S. A. 1928.—Die Setarien und deren pathogenetische Bedeutung. Trudy Gosudarstv. Inst. Eksper. Vet. Moskva, v. 5(1): 53-108, figs. 1-34. German summary.

1931.—Zur Characteristik der Nematoden der Gattung *Nematodirus* Ransom, 1907. Versuch einer Monographischen Bearbeitung. Ztschr. Infektskrank, v. 40: 112-136, pls. 4-10, figs. 1-61.

RANSOM, B. H. 1904.—A new nematode (*Gongylonema ingluvicola*) parasitic in the crop of chickens. U. S. D. A. Bur. Anim. Ind., Circ. No. 64, 3 pp., 2 figs.

1911.—The nematodes parasitic in the alimentary tract of cattle, sheep, and other ruminants. U. S. D. A., Bur. Anim. Ind. Bull. No. 127, 132 pp. 152 figs.

RAUTHER, M. 1909.—Morphologie und Verwandtschaftsbeziehungen der Nematoden. Ergeb. & Fortschr. Zool. v. 1(3): 491-596, figs. 1-21.

1918.—Mitteilungen zur Nematodenkunde. Zool. Jahrb. Abt. Anat. v. 40: 441-514, figs. A-P, pls. 20-24, figs. 1-40.

1930.—Vierte Klasse des Cladus Nemathelminthes. Nematodes, Nematoidea = Fadenwürmer. Handb. Zool. (Kükenthal & Krumback) v. 2, 8 Lief., 4. Teil, Bogen 23-32, pp. 249-402, figs. 267-426.

SANDGROUND, J. H. 1933.—Report on the nematode parasites collected by the Kelley-Roosevelts expedition to Indo-China with descriptions of several new species. Ztschr. Parasitenk. v. 5(3/4): 542-583, figs. 1-33.

SCHNEIDER, A. 1866.—Monographie der Nematoden. 357 pp., 122 figs., 28 pls. Berlin.

SEURAT, L. G. 1913.—Observations sur le *Tropidocerca inermis* Linstow. Bull. Soc. Hist. Nat. Afrique du Nord, Alger, v. 5(8): 191-199, figs. 1-11.

1920.—Histoire naturelle des nématodes de la Bérberie. Première partie. Morphologie, développement, éthologie et affinités des nématodes. 221 pp., 34 figs. Alger.

SHIPLEY, A. E. 1894.—Notes on nematode parasites from the animals in the Zoological Gardens, London. Proc. Zool. Soc. Lond. (3): 531-533, pl. 35, figs. 1-6.

1897.—Note on the excretory cells of the Ascaridae. Zool. Anz. v. 20: 342.

SHUL'TS, R. ED., ORLOV, I. W., & KUTASS, A. J. 1933.—Zur Systematik der Subfamilie Synthetocaulinae Skrj. 1932 nebst Beschreibung einiger neuer Gattungen und Arten. Zool. Anz. v. 102(11-12): 303-310, figs. 1-10.

SKRJABIN, K. I. 1917.—Sur quelques nématodes des oiseaux de la Russie. Parasit. v. 9(4): 460-481, pls. 18-19, figs. 1-19.

STRASSEN, O. ZUR. 1907.—*Filaria medinensis* und *Ichthyonema*. Verhandl. Deutsch. Zool. Gesellsch. 17 J., 110-129, figs. 1-8.

SUGIMOTO, M. 1932.—On the parasitic nematode (*Eustrongylides tricolor* Sugimoto, 1931) in the proventriculus of the Formosoan domestic duck. J. Soc. Trop. Agric. v. 4: 103-116, figs. 1-4, pl. 2.

TOERNQUIST, N. 1931.—Die Nematodenfamilien Cucullanidae und Camallanidae nebst weiteren Beiträgen zur Kenntnis der Anatomie und Histologie der Nematoden. Götenborgs K. Vetensk.—o Vitterhets.—Samh. Handl., 5. f., s. B, v. 2(3), 441 pp., pls. 1-17.

TRAVASSOS, L. A. 1920.—Esboco de uma chave geral dos nematodes parasitos. Rev. Vet. & Zootech. Rio de Janeiro v. 10(2): 59-70, 1 table.

TUERK, F. 1903.—Ueber einige im Golfe von Neapel frei lebende Nematoden. Thesis Leipzig. 67 pp., pls. 10-11. Also in Mitt. Zool. Stat. Neapel, v. 16: 281-348, pls. 10-11.

TYSON, E. 1683.—*Lumbricus teres,* or some anatomical observations on the round worm bred in human bodies. Phil. Tr. Lond., (147), v. 13: 154-161, 1 pl., figs. 1-4.

VOLTZENLOGEL, E. 1902.—Untersuchungen über den anatomischen und histologischen Bau des Hinterendes von *Ascaris megalocephala* und *Ascaris lumbricoides*. Diss. Jena. 32 pp., 3 pls. Also in Zool. Jahrb., Abt. Anat. v. 16(3): 481-510, pls. 34-36.

WALTER, G. 1856.—Beiträge zur Anatomie und Physiologie von *Oxyuris ornata*. Ztschr. Wiss. Zool. v. 8(2): 163-201, pls. 5-6, figs. 1-28.

1858.—Fernere Beiträge zur Anatomie und Physiologie von *Oxyuris ornata*. Ztschr. Wiss. Zool. v. 9(4): 485-495, pl. 19, figs. 29-34.

WEHR, E. E. 1933.—A new nematode from the Rhea. (2958) Proc. U. S. N. M. v. 82(17): 1-5, figs. 1-3.

CHAPTER IX

THE EXCRETORY SYSTEM

M. B. CHITWOOD and B. G. CHITWOOD

Introduction

The structures which at present are termed "excretory" system have been the subject of many arguments into which little evidence has been introduced. An excretory function has not always been presumed for these structures and in some instances other structures have been termed excretory organs.

Bojanus (1817), studying *Parascaris equorum*, discovered a pair of lateral vessels contained within the lateral chords and anastomosing in a bridge beyond which they continued. Later (1821) the same author thought he saw a row of lateral stigmata in *Rhaphidascaris acus* and he finally concluded that the lateral canals were blood vessels and the "büschelförmigen Organe" with which they are sometimes associated were gills.

Cloquet (1824) likewise observed lateral vessels and their anastomosis anteriorly; it was his opinion also that the vessels were circulatory. Mehlis (1831) observed a gland opening near the head in *Contracaecum spiculigerum* and paired strand-like bodies (glands) in strongylids opening at the mouth. He presumed both to be salivary glands, but Schneider later interpreted these structures as excretory glands in both cases.

Shortly thereafter von Siebold in an appendix to a thesis by Bagge (1841) noted the existence of a ventral pore connected with paired lateral canals in *Oswaldocruzia* (*Strongylus auricularis*) and *Aplectana*. (*Ascaris acuminata*) but he did not, at that time, express any view to the function of these structures.

Blanchard (1847) injected specimens of *Ascaris lumbricoides* and upon seeing a large ovoid body in the left lateral chord (the gland nucleus, Fig. 112H) he decided that he had found the heart and he also maintained that there is a second pair of vessels just under the cuticle. On this basis he decided this system to be circulatory.

Leidy (1853) observed a "follicle communicating with the exterior and having its bottom connected by means of radiating bands to the external surface of the alimentary canal" in *Thelastoma attenuatum* and *Aorurus agile*. He was followed by Huxley (1856) and Wagener (1857) who recognized the full extent of the lateral vessels in *Oswaldocruzia filiformis* (*Strongylus auricularis*) and *Heterakis* sp., respectively. It was at this time that Davaine (1857) described a ventral tube and a lateral vessel (on one side only) in *Anguina tritici*.

Schneider (1858, 1860, 1863, 1866) was the first worker to draw the various isolated bits of information together, synthesize, examine critically and summarize. He first proved Blanchard's concept regarding *Ascaris lumbricoides* to be erroneous, showing that what was interpreted as the heart is a large nucleus in the vessel wall, that there is only one pair of vessels and they unite to open through the ventral pore. He likewise established the fact that the ventral pore and canals are joined in the oxyuroids and strongylins; he saw the labyrinthoid coils of the posterior termination of the vessels in *Alloionema appendiculatum;* and also observed the two strand-like bodies ("Subventral" or "cervical" glands) that are attached to the anastomosis or bridge in *Rhabditis strongyloides and Strongylus* spp. He further expressed the view that lateral vessels are present in all "mero" and "polymyarian" nemas and usually absent in "holomyarian" nemas and usually absent in "holomyarian" nemas (except *Anguina tritici*); and finally he (1866, p. 220) concluded that this system of vessels must be related to the excretion of chemical waste products as in the excretory system of all other worms.

Meanwhile Eberth (1860, 1863) erroneously described paired lateral vessels in *Heterakis gallinae,* with two anterior and two posterior lateral openings (amphids and phasmids respectively) and in *Passalurus* he apparently did not differentiate between vessel and chord. In reference to marine nematodes he was more accurate in describing a fine pore near the head and a clear tube proceeding posteriorly in the esophageal region of *Oncholaimus, Enchelidium,* and *Enoplus.*

Bastian (1866) verified many of Schneider's observations, proved the general existence of a "ventral" gland in marine and fresh water nemas and definitely showed that the "water vascular system" (vessels) in parasitic nemas and the ventral gland in free-living forms "are only modifications of one and the same structure." While drawing attention to a similar system in trematodes he noted that in no instance have vibratile cilia been observed in the canals of nematodes and that in neither nematodes nor trematodes is the system adapted to respiratory activity. He concluded that it must be excretory.

GENERAL MORPHOLOGY. The diversity of the nemic excretory system makes it a difficult system to interpret. The general concept of a unicellular system probably originated with Bastian's homologizing of the single ventral gland of marine nemas with the tubular system. Though neither Bastian nor Schneider emphasizes the point, only one nucleus (the sinus nucleus) was known in *Ascaris;* nevertheless they knew of two cells associated with the sinus in rhabditids and strongylins (Fig. 112 I-L). Cobb's description (1890) of the origin of the excretory system in the first stage larva of *Enterobius vermicularis* as an outgrowth of a single invaginated hypodermal cells has given much impetus to the primary single cell concept. Later (1925) the same author described the system in mature embryos of *Rhabditis icosiensis* as consisting of a single gland cell, terminal duct, and paired lateral canals in a ventral position, i.e., not in the lateral chords; according to his view the paired subventral glands are derived from the single cell by splitting and the canals are merely outgrowths. Study of the excretory system in young specimens is technically very difficult and open to considerable error due to the delicacy of the structures. As indicated by Cobb's own figures and verified in diverse instances by the writers the system in first stage larvae is much nearer to that of the adult than is commonly supposed. Actually we have been unable to establish with certainty any difference between larva and adult. In our opinion the common pore and cell illustration in larval nematodes represents only the obvious features. The fact that neither Cobb nor other adherents of the unicellular gland idea have accounted for or even recognized the existence of a terminal duct cell in addition to glands or sinus cells makes us most dubious of the entire concept.

The second concept of the system is based upon its identity with the protonephridial system of trematodes, rotifers, gastrotrichs, etc. This viewpoint was accepted by Bastian (1866) and Schneider (1866) without particular question. It likewise seemed reasonable to Bütschli (1876) and Martini (1916). It has fallen into disfavor recently because it cannot be accepted by those presuming the marine aphasmidian (having a single ventral gland cell) as primitive. These authors (Filipjev, Stekhoven) must assume the system to arise de novo in the Nematoda or assume that the remainder of the Animal Kingdom arose from nematodes. They have apparently chosen the first alternative though with odd consistency at the same time relating nematodes, rotifers, echinoderes, gastrotrichs and nematomorphs. Steiner (1919, 1920) brought out a series of diagrams hypothetically indicating the mode of evolution of the nemic excretory system from that of the rotifer. His diagrams give interesting points but overlook the essential complexity of the tubular system of both groups. We shall not go into such matters at the present. The most necessary evidence, critical embryological study, is lacking.

Golowin (1902) expressed the view that the tubular, cuticularly lined terminal duct cell is an invagination of the hypodermis which meets and fuses with the lateral canals and excretory sinus. This view was concurred in by Goldschmidt (1906) and has rather significant support. It accounts for the collecting tubes as a separate entity, not developed from a ventral gland; these tubes may be considered as derivatives of the basic protonephridial system without reorientation of the system; it accounts for the minimum two to three cell system in the Phasmidia as well as the ventral gland of marine forms without assuming either de-differentiation of the protonephridial system or de novo formation within the Nematoda. Properly speaking, the various nemic excretory systems have only one point in common, namely that they open through a ventral pore.

Von Linstow (1909) presented a classification of the Nematoda, based on the excretory system, which has considerable merit though it is in disrepute because the original author incorrectly placed many forms. The general outline may be summarized as follows:

I. Secernentes. Lateral canals emptying anteriorly through a ventral pore; chords narrow and high. [*Ascaris, Oxyuris, Oxysoma, Nematoxys, Heterakis, Strongylus, Cucullanus, Dacnitis, Spiroxys, Rictularia, Cheiracanthus (Gnathostoma), Tropidocerca, Ancyracanthus*].

II. Resorbentes. Lateral canals absent, excretory pore absent; chords 1/6 periphery; esophagus and gut often atrophied. Feeding by resorption through cuticle. [*Angistoma (Rhabdias), Eustrongylus (Dioctophyma), Hedruris, Dispharagus, Ichthyonema (Philometra), Filaria*].

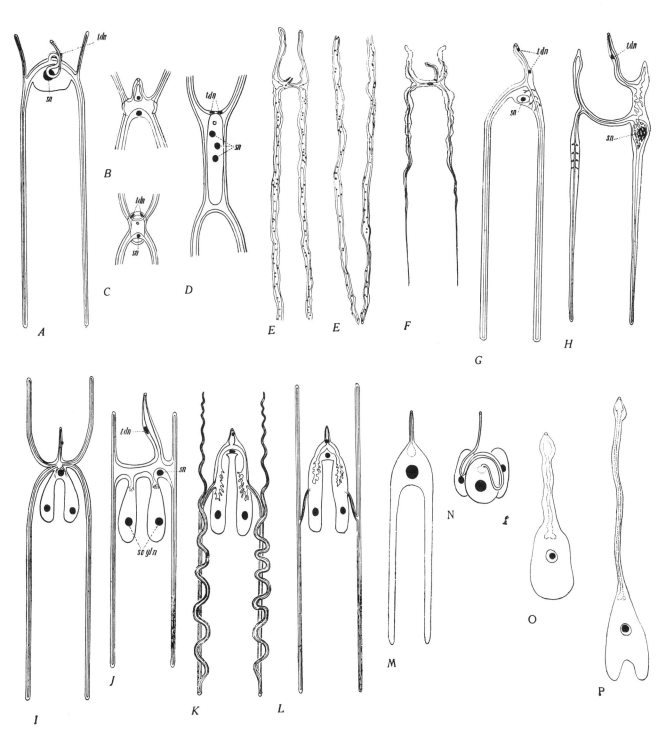

FIG. 112. DIAGRAMS OF NEMIC EXCRETORY SYSTEMS

A—*Rhabditis dolichura;* B—*Heterakis gallinarum;* C—*Macracis monhystera;* D—*Spironoura affine;* E—*Cucullanus heterochrous;* F—*Camallanus lacustris;* G—*Camallanus microcephalus;* H—*Ascaris lumbricoides;* I—*Rhabditis strongyloides;* J—*Metastrongylus elongatus;* K—*Strongylus equinus;* L—*Oesophagostomum dentatum;* M—*Anonchus mirabilis;* N—*Anaplectus granulosus;* O—*Chromadora quadrilinea;* P—*Phanodermopsis longisetae.* E & F, After Toernquist, 1931, Goeteborgs Kungl. Vetensk. Vitterhets—Samh. Handl. s. B., v. 2 (3); 1-441, figs. 1-13, pls. 1-17. Remainder original.

III. Pleuromyarii. Muscles extending over lateral lines; excretory system absent, male with one spicule. [*Hystrichus, Trichocephalus (Trichuris), Trichosoma (Trichosomoides)*].

IV. Adenophori. Small high lateral chords; ventral pore present, connected with a ventral cervical gland; two spicules; herein the free-living nematodes should be placed. (*Myenchus, Myoryctes, Rhabditis, Cephalobus*).

The Secernentes by definition correspond to the Phasmidia and the examples include representatives of four out of six of the major phasmidian groups, i.e., Strongylina, Ascaridina, Camallanina and Spirurina. The group Resorbentes was less fortunately constituted, since it was primarily based upon a supposed method of feeding which the author attempted to correlate with anatomy. On the basis of present day information all representatives except *Dioctophyma* would be placed in the Secernentes. The exceptional genus might be placed with the Pleuromyarii (on the basis of one spicule) or the Adenophori (on the basis of chords). From the standpoint of excretory system, or lack of it, the Pleuromyarii must be considered synonymous with the Adenophori. Steiner (1919) criticised von Linstow rather severely for having put forward such a classification because it was well known that *Rhabditis* and *Cephalobus* had lateral excretory canals. The evidence today shows that Linstow was in error as to all of the examples of the Adenophori which he listed. However, it should be recognized that Linstow was partially correct. Most free-living nemas differ from typical parasitic nemas in that they have no lateral excretory canals. Linstow had little choice of example since he was discussing the parasitic nemas and covered free-living forms only to the extent that they occurred as parasites.

It is interesting that we now find ourselves using some of the characters given by von Linstow in his groups Secernentes and Adenophori, though his examples were largely incorrect. If his definition of Secernentes is limited to the statement concerning the excretory system, then the group corresponds to the Phasmidia. Similar limitation of the diagnosis of the Adenophori to a statement concerning the excretory system would cause this group to correspond roughly to the Aphasmidia. It is of further interest to note that the presence of lateral canals is always evidenced by a cuticularly lined terminal duct or excretory vesicle; furthermore, the reverse is also true with only two exceptional instances (Plectidae and larval mermithids.) In all instances carefully investigated in recent years, the tubular excretory system occurring in the Phasmidia has been found to consist of not less than two cells, one or more forming the cuticularly lined terminal duct and one or more forming the lateral canals and glandular tissue. The Aphasmidian unicellular gland lacks the cuticular lining of the terminal duct and consists of a single gland cell with no canals. We may conceive of this system as the homologue not of the entire Phasmidia system as the homologue not of the entire Phasmidian system but only of the ectodermal part, i. e., the terminal duct cell.

A. Phasmidia

In the first group, there are four chief modifications of the excretory system (Fig. 9, p. 11), namely (1) the oxyuroid or simple H system (Oxyuroids, some ascaridoids, some scattered members of the Spirurina); (2) the rhabditoid system, a combination of the H type with two subventral glands (known in some rhabditids and strongylins); (3) the ascaridid inverted U-system, characteristic of ascaridids but occurring also in most members of the Spirurina and some free-living forms (*Panagrolaimus*); (4) the asymmetric system, known only in Anisakinae and Tylenchoidea.

Critical study indicates types (3) and (4) have appeared more than once in the general evolution and both types may be considered as ordinary variations of type (1). Type (2) is known only in one parasitic group, the Strongylina and in some representatives of one free-living group, the Rhabditidae. Normal progressive evolution would account for the origin of the rhabditoid system from the oxyuroid system.

1) SIMPLE H SYSTEM. This type, commonly spoken of as the oxyuroid system, is by no means limited to oxyuroids. There are several variants of the system, chiefly dependent upon the form of the terminal excretory duct.

    (a) Terminal duct long and tubular.
    (b) Terminal duct greatly shortened, but not vesicle-like.
    (c) Terminal duct elongate, vesiculate.
    (d) Terminal duct a very short vesicle.

(1a) Presumably, the elongate, cuticularly lined terminal duct is the more primitive type. In such a system one finds at least one, sometimes two nuclei attached to the terminal duct and one large sinus nucleus. Long anterior canals are known only in free-living nemas such as *Rhabditis dolichura* and *Cheilobus schneideri* (Fig. 112A). Excretory systems of such forms have been studied by Bütschli (1873), Jägerskiöld (1909), Magath (1919) and Törnquist (1931). Undoubtedly this type of system is more widespread in the Rhabditoidea than is known; it is very closely approached in ascaridids. Törnquist found representatives of both the Camallanidae and Cucullanidae with H type systems and both H and inverted U forms existing within the two genera *Procamallanus* and *Camallanus*, the H type being confined to camallanids of fish. There are really two quite distinct forms for this system. In *Camallanus* (Fig. 112 F-G) one finds two nuclei within the wall of the terminal duct, one nucleus near the pore and one near the anterior surface of the sinus. In this case the terminal duct connects with the sinus near the left lateral chord and there is a large sinus nucleus on the left side (Fig. 114 CC). The lateral canals begin as thick walled tubes, sometimes with indistinct, questionable, tributary tubules; thereafter the canals become smaller and they never have a thick lining nor have nuclei been observed in their walls. This, then, is a three nucleate or three celled system.

In *Cucullanus*, according to Törnquist, (1931) sometimes the terminal duct is dilated like a bladder near the pore (like b or d). There is no distinctive sinus nucleus but there are many similar nuclei within the walls of the posterior lateral canals which should probably be interpreted as multiples of the original sinus nucleus (Fig. 112 E). The walls of the lateral canals are thick throughout their length but are particularly massive in the cervical region and taper posteriad (Fig. 114 L). Distinct branching tubules penetrate the wall of each canal, these being best developed in the cervical region (Fig. 114 M); the tubules do not leave the canal tissue. Törnquist found such tubules arising more or less symmetrically in fours, two dorsolaterally and two ventrolaterally from the canal axis, while the writers have found six series of tubules more common, there being two lateral ones in addition to those previously mentioned.

The multinucleate character of the cucullanid excretory system was first noted by Jägerskiöld who considered the whole system as representing a polynucleate cell. Törnquist and the writers have confirmed Jägerskiöld in not being able to find cell walls. However, we may well question our interpretation of this case until further information is obtained relative to the occurrence of nuclei in lateral canals. Such findings are becoming common.

(1b) The shortened terminal duct type, as exemplified by *Heterakis gallinarum* (Fig. 112 B), is very similar to (1a) and probably should be considered a modification thereof. The observations of Wagener (1857), Eberth (1860) and Schneider (1860) have previously been mentioned. More recently Chitwood (1931) and Baker (1936) have restudied this form. The excretory orifice is guarded anteriorly by a lip cell (Fig. 113 RR) formed from the anterior side of the terminal duct. Thence internally the terminal duct is wide, irregular until it merges with the excretory sinus. On the lower left hand side a small nucleus is present in the terminal duct wall (Fig. 113 JJ). The only means of distinguishing sinus and duct is through the cuticular lining of the duct. There is another nucleus on the ventral side associated with the sinus and terminal duct but apparently exterior to their walls (Fig. 113 KK); one often sees fibrous tissue around the duct and near this nucleus, and it seems possible that the nucleus and fibers represent a sphincter muscle of the sinus and terminal duct. The sinus nucleus itself is situated medially and is not especially large in *Heterakis* (Fig. 113 LL). No additional nuclei have been observed in the lateral canals. The heterakid system, therefore, seems to consist of not less than three nuclei or cells, two belonging to the duct and one to the sinus and canal system.

(1c) The elongate, vesicular terminal duct is another modification of type 1a; it is confined, so far as known, to the Kathlaniidae occurring in such forms as *Spironoura affine* and *Cissophylus roseus*. Whether or not it is a character of the entire family is not known. As Mackin (1936) observed, the anterior and posterior canals unite with their mates and empty into a great elongate sinus which separates anterior and posterior canals (Fig. 112 D). This is distinctly different from other forms in which the two anterior and two posterior canals all come together at practically the same level. Apparently this sinus combines the functions of terminal duct and sinus since there is a delicate cuticular ventral lining connected with the excretory pore; thus we might say the dorsal side is sinus and the ventral side is terminal duct. The excretory pore is at the anterior end of the excretory region (Fig. 113 NN) and near this region two small subventral nuclei are present; these we attribute to the terminal duct. The dorsal side of the sinus contains three equal, rather large nuclei (Fig. 113 MM). The tissue in which they are situated is continuous with the tissue of the lateral canals and in the latter no nuclei have been observed. On the basis of present information the *Spironoura* system contains five nuclei, three belonging to the canal system (sinus) and two to the duct.

(1d) The typical oxyurid excretory system is a notable feature in most representatives of the Oxyuridae, Thelastomatidae, and Atractidae. Due to the clarity of the oxyurid body and to the rather large diameter of the lateral canals, this form has become synonymous in the literature with the H system even though it is but one variant.

Leidy (1853), Eberth (1863), Schneider (1866), Bütschli (1871), Cobb (1890), Martini (1916), Rauther (1918), Thapar (1925), Chitwood (1931) and Chitwood and Chitwood (1933) have studied the system as seen in *Thelastoma, Passalurus, Oxyuris* (sensu lato), *Leidynema* and *Hammerschmidtiella, Enterobius vermicularis, Oxyuris equi, Macracis monhystera* and *Thelandros alatus*, various reptilian oxyurids (Pharyngodoninae), *Hystrignathus* and *Macracis*, and *Cephalobellus* respectively. Subsequently extension of observations has shown the system to be practically universal in the families mentioned. The four lateral canals come together at about the same level forming, with the sinus, an X (Fig. 112 C). The sinus nucleus may be medial as in *Hystrignathus* and *Macracis* or on the left side as in *Cephalobellus;* it is not proportionately very large. Chitwood (1931) described regularly arranged nuclei in the lateral canals of *Macracis monhystera* (Fig. 113 FF-GG). In other forms no such nuclei appear to be present. Constancy and recognition of the sinus nucleus in *Macracis* argue against the concept that the canal nuclei are derivatives of the sinus nucleus.

The excretory vesicle, bladder or reservoir is a most interesting structure showing considerable variety in oxyuroids. As in types 1b and 1c the terminal duct is in open connection with the excretory sinus (Fig. 113 V, *Cephalobellus*), its cuticular lining forming the ventral part of the sinus wall. As observed by Rauther (1918) in *Macracis*, the terminal duct wall contains two nuclei or is composed or two cells, these cells may both be situated anterior to the excretory pore or one may be anterior and one posterior. The conspicuous rounded vesicle seen in totomounts is apparently not a storage structure (see figs. 113 BB-CC) but more on the order of a valve; i. e., the sinus cavity fuctions as the reservoir while the terminal duct is shortened. Sometimes (Fig. 113 Y-Z) there is a distinct circular sphincter muscle near the junction of sinus and duct. In representatives of the Atractidae the wall of the terminal duct is radially striated from the excretory pore giving the appearance of a sucker.

Martini (1916) saw structures which he considered might be rudimentary cilia near the blind ends of the excretory canals of *Oxyuris equi*. The writers have not observed such structures in the species studied.

(2) THE RHABDITOID SYSTEM (Fig. 112 I). The essential features of this system include (a) a terminal elongate cuticular duct, (b) an excretory sinus connected with paired lateral canals, and (c) paired subventral excretory glands also connected with the excretory sinus.

Mehlis (1831) supposedly observed the subventral glands (strand-like organs) in strongyles and thought they opened anteriorly as salivary glands. He may have confused the subventral excretory glands (cervical glands of Looss) with the amphidial glands (cephalic glands of Looss, fig. 12 G-I). In 1841 von Siebold observed the pore and vessels in *Oswaldocruzia (Strongylus auricularis)* and shortly thereafter Dubini (1843) associated the subventral glands of *Ancylostoma duodenale* with the esophagus and Bilharz (1852) observed the junction of subventral glands and excretory pore but did not see the lateral vessels. It was Schneider (1858-1866) who first observed the full gross morphology of the system by noting all three of the essential components (see above) and by joining them. However, he erred in one respect, by considering the gland cells as non-glandular attachments of the sinus (or bridge). Schneider's observations were made in reference to *Rhabditis strongyloides* (Fig. 112 I), *Cylicostomum tetracanthum, Strongylus armatus* and other strongyloids. Bütschli (1873) found *Rhabditis terricola* to have the same type of excretory system as that described by Schneider

for *R. strongyloides*. Leuckart (1876) did the same in regard to *Ancylostoma duodenale*. Since that time papers dealing with the excretory system as seen in rhabditids and strongylins have been published by Rzewuski (1887), *Metastrongylus elongatus;* Stadelmann (1891), *Ostertagia ostertagi;* Augstein (1894), *Dictyocaulus filaria;* Looss (1895). *Trichostrongylus colubriformis;* Poeppel (1897), *Strongylus vulgaris;* Looss (1905), *Ancylostoma duodenale;* Maupas (1916), *Rhabditis* spp.; Cobb (1925), *R. icosiensis;* Aubertot (1926), *R. pellio;* Stekhoven (1926), *Ancylostoma* and *Necator;* Chitwood (1930, 1931), *Rhabditis* spp., and *Oesophogostomum dentatum;* Eisma (1932), *Ancylostoma;* and Raven and Stekhoven, (1934), *Rhabditis* spp.*

According to our observations this system consists of not less than three, usually four cells in such diverse forms as *Rhabditis strongyloides, Oesophagostomum dentatum, Kalicephalus* sp. and *Metastrongylus elongatus*. The terminal duct is narrow, elongate and tubular in adult rhabditids as well as larval strongylins while it becomes folded and thick-walled (Fig. 113 K) in adult strongylins. Its nucleus lies just posterior to the end of the cuticular lining in these forms and corresponds to the carrying cell of the excretory vesicle described by Looss (1905) in *Ancylostoma*. It has been clearly identified in *Oesophagostomum, Metastrongylus* and *Strongylus*. More posteriad, the excretory canal widens into the transverse excretory sinus into which the paired subventral glands open. The excretory sinus contains a large conspicuous nucleus (Fig. 113 L) which may be medial (*Rhabditis strongyloides, R. coarctata, Oesophagostomum*) or left (*Metastrongylus*) in position. Thin-walled branching tubules extend posteriorly from the sinus into each of the subventral glands. The lateral canals may connect directly with the excretory sinus as in *Rhabditis* and *Metastrongylus* or they may connect indirectly by way of the subventral gland cells as in *Oesophagostomum, Ancylostoma* (vide Looss) and *Strongylus* (vide Schneider's description of double lateral canals). In the latter instance one finds two pairs of lateral canals in each chord extending nearly to the caudal extremity before they unite (Fig. 113 R). In other forms the lateral canals are tubular, delicate and usually extend nearly to the two extremities; often vacuoles may be seen in the wall of living specimens; each canal ends blindly at its terminus, often in a small ampulla. Thus far no one has identified nuclei in the lateral canals but this may be due to lack of critical study. In a complete series of cross sections of a male *Strongylus* the writers observed only one possible canal nucleus (Fig. 113 S) which was situated very close to the junction of the canals posteriorly; since it could not be verified in the corresponding canal in the other chord nor in other series, it may easily have been an artifact. Ordinarily the lateral canals in nemas are considered "outgrowths" of the cell which we have herein called the *sinus cell*. In the organization of *Oesophagostomum* and *Strongylus* such a relationship is seemingly precluded since the canals do not empty into the sinus directly, but through the subventral glands. One may conceive the *Rhabditis strongyloides* type (Fig. 112 I) to be most primitive, and transformation to the *Strongylus* type to have taken place by a posterior shifting of the canal union. In *Rhabditis* and *Strongylus* the sinus nucleus *plus the subventral gland nuclei* are therefore to be considered homologous to the sinus nucleus of oxyuroids and *Rhabditis dolichura*.

Maupas (1900) apparently first noted contraction in the lateral canals of *Rhabditis lucianii* and in 1916 the same author noted a contractile ampulla at the end of the terminal excretory duct in *R. terricola*. Similar observations were made concerning larvae of *R. pellio* by Aubertot (1926) and *Rhabditella axei* by Raven and Stekhoven (1934). The latter authors observed only one gland cell and presumed there would be two ampullae if there were that many gland cells. This is not necessarily the case. The ampulla corresponds apparently to the excretory sinus and may act as a central contractile bladder in which the various contents empty.

---

*As already noted, the "rhabditoid" excretory system does not occur in all rhabditoids, but is merely characteristic of a limited group centered around *Rhabditis strongyloides*. The majority of rhabditoids, include *Rhabditis dolichura* and other *Rhabditis sensu lato* have an H-shaped (oxyuroid) system. Raven and Stekhoven (1934) denied the existence of lateral canals in *Rhabditella axei*, but the writers have re-examined this species and can confirm the previous observation of Chitwood (1930) that the system in this form is of the "rhabditoid" type—an H system with distinct sinus nucleus and two subventral glands.

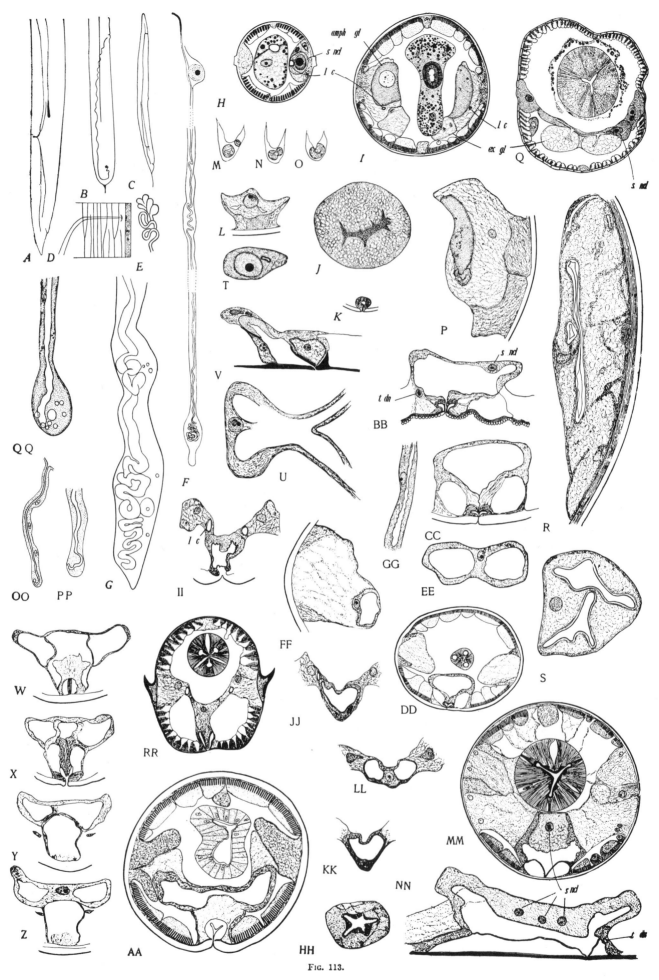

Plurality of lateral canals in the lateral chords has been recorded by Schneider (1866), Leuckart (1876), Poeppel (1897) and Looss (1902), the cases referring to strongyloids. Ultimately these canals have always been found to join and not to represent a real duplication or branching of the canals. However, in females of some species of rhabditids this is not the case. As first noted by Maupas (1916) there may be an auxiliary excretory system with paired openings dorso- or ventrolaterally posterior to the vulva. Maupas recorded such a system in *R. terricola*, *R. lucianii*, *R. axei*, *R. pellio*, *R. seurati*, *R. sergenti* and *Angiostoma limacis*. It was many years before the writers had the pleasure of seeing this system, and then only in *Rhabditis terricola*, *R. coarctata*, *R. cylindrica*, *Rhabditella axei* and *Rhabditoides paraelongata*. It does not seem to be present in *R. strongyloides*. Whether it may occur abnormally in that species or whether Maupas misidentified his species, we cannot say. The system as seen by us in *R. terricola* consists of two tubes, one in each lateral chord, connected by a ventrolateral pore near the vulva. The canals extend posteriorly to the caudal region, and are much coiled at the blind end (Fig. 113 F-G.) In *R. coarctata* a large nucleus seems to be associated with the terminal duct of the auxiliary system but it is not certain that the nucleus is in the duct wall. In *R. sergenti* Maupas found both an anterior and a posterior branch, each bifurcate (Fig. 113 A).

3. THE ASCARIDID OR INVERTED U-TYPE. Historically one of the systems most commonly studied and referred to, it is not confined to ascaridids but occurs in spiruroids and in at least one group of rhabditoids (Cephalobidae). As pointed out by Törnquist (1931) polyphyletic origin of the inverted U type from the H form occurs even within genera and this system cannot be regarded as separate and distinct from 1a. The essential parts are an elongate, cuticularly lined terminal duct, a sinus, and two posterior lateral canals. By custom *Ascaris lumbricoides* is retained in this classification though distinct rudiments of anterior canals are present (Fig. 112 H).

The system as described in *Ascaris lumbricoides* and *Parascaris equorum* by Goldschmidt (1906) consists of three cells, one nucleus in the wall of the terminal duct, another in the protoplasmic part of the "anterior bridge" and a third, the giant excretory nucleus in the left lateral chord. Mueller (1929) by means of injections showed Goldschmidt in error relative to the existence of both an anterior and posterior "bridge" (sinus). As found by Mueller and verified in sections by the writer, the terminal duct leads to the left lateral canal with which it connects very close to the anterior extremity of the canal; the right lateral canal extends further

FIG. 113.

A—*Rhabditis sergenti* (Female with both ordinary and auxiliary excretory systems shown in outline); C—*Rhabditis seurati* (Female excretory systems shown in outline); B, D, & E—*Rhabditis icosiensis* (B—Posterior part of female showing auxiliary excretory system; D—Terminal duct of auxiliary system; E—Blind end of auxiliary system, all drawings semi-diagrammatic); F—*Rhabditis coarctata* (Auxiliary canal, regions drawn camera lucida from living specimen; the large nucleus could not be seen in the specimen on the opposite side and it may not actually be within the duct wall); G—*Rhabditis terricola* (Blind end of auxiliary canal as seen in living specimen); H—*Ditylenchus putrefasciens* (Cross section showing sinus nucleus in lateral chord); I-P—*Oesophagostomum dentatum*, cross sections (I—Near base of esophagus showing subventral glands, lateral canals, and amphidial glands; J—Subventral gland at level of nucleus; K—Terminal duct at excretory pore; L—Sinus with its nucleus; M-O—Approach and fusion of lateral and transverse canals; P—Lateral chord with lateral canal in mid-region of body); Q—*Metastrongylus elongatus* (Cross section at level of sinus, with its nucleus and connection with subventral glands); R-S—*Strongylus equinus* (R—Cross section of lateral chord in posterior region of female showing two lateral canals, one in cross section, the other longitudinal; S—Cross section of lateral canals about to fuse in preanal region of male; note round body which stained like a nucleus); T—*Rictularia coloradiensis* (Terminal duct at level of nucleus); U-V—*Cephalobellus papilliger* (U—Excretory sinus; V—sagittal section through terminal duct); W-AA—*Hystrignathus rigidus* (W-Z—Serial sections through terminal duct and sinus; AA—Cross section of another speciment at same level as that shown in W); BB-GG—*Macracis monhystera* (BB—Longitudinal section through terminal duct; CC—Cross section through excretory sinus; DD—Cross section of entire specimen at level of CC; FF—Lateral canal in chord showing nucleus in its wall; GG—Same as FF, seen in longitudinal section); HH—*Cucullanus serratus* (Lateral canal showing bacillary layer or possible cilia); II-LL & RR—*Heterakis* (II—Terminal duct; JJ—Second terminal duct nucleus, at level of sinus; KK—Sinus with possible constrictor nucleus; LL—Sinus nucleus; RR—Terminal duct with first terminal duct nucleus); MM-NN—*Spironoura affine* (MM—Cross section at level of first sinus nucleus; NN—Sagittal section through sinus); OO-PP—*Goezia annulatus* (OO—Entire excretory system; PP—Blind end of system); QQ—*Panagrolaimus subelongatus* (Blind end of lateral canal). A-E, After Maupas, 1916. Compt. Rend. Soc. Biol. v. 79; I-J, After Chitwood, 1931, Ztschr. Morph. v. 23 (½); U-V, After Chitw. & Chitw., 1933, Zellforsch. v. 19 (2); HH—After Toernquist, 1931, Goeteborgs Kungl. Vetensk. o. Vitterhets. Handl. s. B., v. 2 (3); OO-PP, After Hamann, 1895, Die Nemathelminthen v. 2. Remainder original.

anterior than the left, indeed nearly to the level of the excretory pore. The excretory sinus is quite variable, its lumen branching and anastomosing (Fig. 114 A-B & E). The writers found a single nucleus (Fig. 114 F) definitely within the wall of the terminal duct about half way along its length. The gigantic sinus nucleus of *Ascaris* is moved to the left lateral chord and is surrounded by branching capillaries of the left lateral canal. These branchings become less marked posteriorly and finally fuse forming a typically simple canal. The canal may give off minute tubules (Fig. 114 C) as were previously noted in *Cucullanus* but no additional nuclei have thus far been observed in the canal walls. Goldschmidt (1906) thought the tubules extended into the adjacent areas of the chords and they were the actual secreting part or glandular structure. Subsequent work has disproved this hypothesis.

The camallanid inverted U system is essentially the same as the H system found in that group and requires no special comment. In filarioids, spiruroids and dracunculoids the inverted U system is apparent in totomounts used for taxonomic study (Fig. 14 I & L) but has not been subjected to critical analysis. The same is equally true of cephalobids such as *Panagrolaimus subelongatus*.

4. ASYMMETRIC SYSTEM. A single lateral canal confined to one (usually the left) lateral chord, is apparently a reduction phenomenon which arose twice; (a) in the Ascarididae and (b) in the Tylenchina.

(a) Shift of the sinus nucleus to the left lateral chord in *Ascaris* with submergence of the sinus to a minor role supplies a clue as to the origin of the anisakid system. Representatives of this group have been studied by Mehlis (1831), v. Siebold (1838), Schneider (1866), Cobb (1888), Hamann (1895), Jägerskiöld (1893, 1894, 1898) and Mueller (1927). In such forms the excretory pore is often situated far forward, even between the subventral lips (*Anisakis simplex*). The system as observed in *Contracaecum* and *Anisakis* consists of a terminal duct connected with a lateral canal in the left lateral chord, this canal being greatly enlarged and having numerous side branches (Fig. 114 O). As in *Ascaris*, branching is most noticeable at the level of the gigantic sinus nucleus. According to Jägerskiöld and Mueller the entire system consists of a single cell served by the one large nucleus (Fig. 114 H-J), but this seems improbable in the light of other observations which have shown that the terminal duct is a separate entity.

Hamann (1895) found an even greater reduction in the excretory system of *Goezia annulata*. Here the entire system is formed by a short tube (Fig. 113 OO-PP) which begins in the left lateral chord and ends between the subventral lips. This tube is composed of three cells in tandem, the lumen being intracellular.

4b. The tylenchoid system (Fig. 9C) was first noted by Davaine (1857) and later observed by Bütschli (1873), Strubell (1888), Debray and Maupas (1896), and Cobb (1914). Unlike the situation in the Anisakinae, there is no shift anterior of the excretory pore correlated with asymmetry. In these forms there is a very well developed, long terminal duct leading posteriorly to one lateral chord. This chord may be either right or left, usually right but the side varies not only with species and strains but also among individuals of a given population. The terminal duct apparently has one small nucleus in its wall and the lateral canal contains a gigantic sinus nucleus (Fig. 113 H). Tylenchoids also differ from Anisakinae in that the lateral canal extends anteriad to its junction with the terminal duct. Cobb (1914) described in this group the only nema thus far known with a posterior ventral excretory pore, *Tylenchulus semipenetrans*, and in this form there is sexual dimorphism, the excretory pore being near the middle of the body in the male.

There is no evidence of relationship between these two asymmetric systems any more than there is between the forms grouped under the inverted U system or the H system. Undoubtedly we would derive 4a from 3 (ascarid) and that from 1b just as 4b would be derived directly from 1a.

Aphasmidia

Thus far, only one type of non-canalicular excretory system is known, that being the simple ventral gland cell as first observed by Eberth (1863) in *Oncholaimus*, *Enoplus* and *Enchelidium*, (Enoploidea). Shortly thereafter Bastian (1866) described the same type of cell in *Cyatholaimus* (Chromadoroidea) and *Sphaerolaimus* (Monhysteroidea). Since then

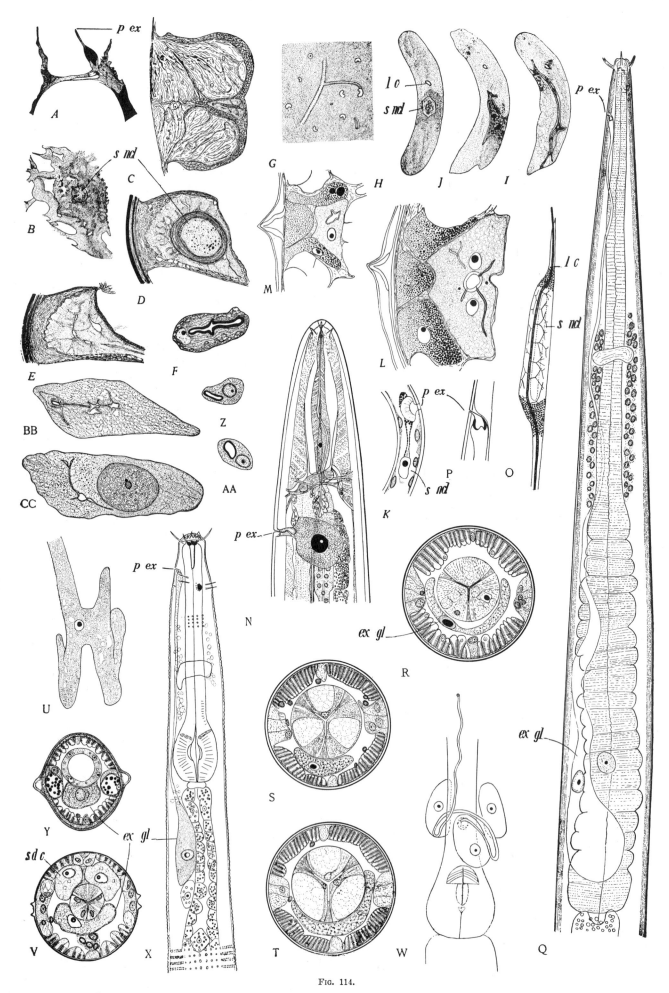

Fig. 114.

132

through the work of de Man, Cobb, Steiner and others, we have come to view this as the typical aphasmidian system. In it there is often a terminal ampulla (Fig. 112 O-P) and the excretory cell may have a greatly elongated neck (Fig. 114 Q) but it is never lined with a cuticle except at the terminus. It might be regarded as corresponding either to the sinus cell or the terminal duct cell. Because of the absence of lateral canals, which we associate with the presence of a sinus cell, we prefer the latter interpretation. According to this view the terminal duct cell of the Secernentes system is homologous to the ventral gland of the Adenophori. In chromadorids and desmodorids there are often one or more small elongate bodies posterior to the excretory cell which have been considered secondary or auxiliary excretory cells (see Steiner, 1916) but more recent study has shown them to be coelomocytes without apparent protoplasmic connection with the excretory system.

De Man (1886) found the ventral gland of *Enoplus communis* to be lobed in such a manner as to justify the term H-shaped (Fig. 114 U). Some more recent workers such as Wülker and Stekhoven (1933) have interpreted the structure found in *Enoplus*, as an early stage in the origin of the H-type excretory system. The writers regard the lobation in *Enoplus* a coincidence since it would not account for the existence of a cuticularly lined terminal duct. The typical ventral gland has been described in representatives of the Chromadoridae (Fig. 114 X), Microlaimidae, Cyatholaimidae, ? Desmodoridae, ? Epsilonematidae, ? Draconematidae, Camacolaimidae, Axonolaimidae, Comesomatidae, Monhysteridae, Linhomoeidae, Enoplidae, Oncholaimidae, and Ironidae. There is scarcely a group, however, in which a ventral gland has been found in all genera. Perhaps this is due to delicacy of the tissue which changes quickly on fixation. We have been unable to locate a ventral gland cell in *Ethmolaimus revaliensis*, *Monoposthia hexalata*, *Monhystera cambari* and *Theristus setosus* even after careful examination of serial sections.

Jägerskiöld (1901) associated the existence of sublateral hypodermal glands with degeneracy of the excretory system in aphasmidians in general and *Cylicolaimus magnus*, *Thoracostoma* spp. and trichuroids in particular. While it is true that hypodermal glands are very well developed in some such forms, they are no more so than in species with a distinct ventral gland such as *Acanthonchus viviparus* (Fig. 15 N). Furthermore, sublateral hypodermal glands are now being found in more and more groups in which they had been overlooked (linhomoeids, oncholaimids, dorylaimids, microlaimids, chromadorids).

Nevertheless, it seems to be an established fact that no excretory system is present in either the Trichuroidea or Dioctophymatoidea (Linstow's Pleuromyarii; Rauther's Holognia). In the related Dorylaimoidea ordinarily no excretory pore may be observed but a rudimentary pore has been mentioned by some authors and de Coninck (1931) illustrated a terminal excretory duct connected with a ventral gland cell in *Diphtherophora vanoyei*. A similar arrangement is indicated for the Mermithoidea in which an excretory pore and terminal excretory duct is commonly attributed to pre-parasitic and young parasitic mermithids (Fig. 93). In adult mermithids usually no pore is to be observed, but forms in which the structure has persisted have been described by Rau-

ther (1909), and Steiner (1919). Corti (1902) described in *Hydromermis rivicola* an excretory canal in one lateral chord but no excretory pore was observed and the structure was not illustrated. Hagmeier (1912) described and figured the ventral gland devoid of any terminal duct in adult *Mermis brevis*, and the writers observed a similar structure in *Hydromermis* sp.

The families Tripylidae and Mononchidae resemble dorylaimoids in that the excretory system is usually inconspicuous or overlooked. Cobb (1918) established the existence of an excretory pore in *Mononchulus ventralis* and the writers have seen such in sections of *Prionchulus muscorum*. As yet the internal connections of the excretory tube have not been determined.

The excretory system of plectids remains to be discussed. As in so many other characters, so here also, the plectids show greater resemblance to the rhabditoids than to any other phasmidians. The conception of the excretory system of plectids has been confused considerably by interpretations. The odd loop of the terminal duct (Fig. 112 N) gave the genus its name. At the level of this loop there are a large ventral gland cell and two dorsal cells (Fig. 114 W). We have previously (see Fig. 15 A) conceived the terminal duct as entering the two subdorsal cells but this seems to be wrong; it enters only the ventral cell. The only suggested function of the other two cells is athrocytic. It should be noted that *Anaplectus* and *Plectus* have no lateral canals though they have a well developed terminal duct; whether or not the duct itself has its own nucleus in addition to the gland nucleus we cannot say. *Anonchus mirabilis* provides another and even more interesting bit of evidence as to the evolution of the excretory system for in it one finds an inverted U-shaped system *in the body cavity* and not in the lateral chords; in this instance there is a distinct terminal duct, apparently with a separate nucleus, and a large sinus cell from which two apparently solid protoplasmic appendages extend posteriorly through the body cavity. For the time being, *Anaplectus* and *Plectus* might be considered aberrant offshoots while *Anonchus*, by slight transformation, could account equally well for the origin of the Adenophori system as for the Secernentes system. Thus, again we find the Plectidae holding the key to points in nemic evolution.

Bibliography

AUBERTOT, M. 1926.—Contractilité de l'appareil excréteur chez les larves du *Rhabditis pellio* (Schn.). Compt. Rend. Acad. Sc., Paris, v. 182 (2): 163-165, 2 figs.

AUGSTEIN, O. 1894.—*Strongylus filaria* R. Diss. (Leipzig). 54 pp., 2 pls., figs. 1-31, Berlin.

BAGGE, H. 1841.—Dissertatio inauguralis de evolutione strongyli auricularis et ascarids acuminatae vivaparorum. Diss. 16 pp., 1 pl., 30 figs. Erlangae.

BAKER, A. D. 1935-1936.—Studies of *Heterakis gallinae*. (Gmelin, 1790) Freeborn, 1923, a nematode parasite of domestic fowls. Tr. Roy. Canad. Inst., v. 20 (2): 179-215, figs. A-E; & v. 21 (1): 51-86, pls. 1-15, figs. 1-164.

BASIR, M. A. 1949.—The excretory system of *Physaloptera varani* Parona, 1889. Tr. Am. Micr. Soc. v. 68 (2): 118-122.

BASTIAN, H. C. 1866.—On the anatomy and physiology of the nematodes, parasitic and free; with observations on their zoological position and affinities to the echinoderms. Phil. Tr. Roy. Soc. Lond., v. 156: 545-638, pls. 22-28.

BERGH, R. S. 1885.—Die excretionsorgane der Würmer. Eine Uebersicht. Kosmos J. 9, v. 17 (2): 97-122; pl. 2, figs. 1-10.

BILHARZ, T. 1852.—Ein Beitrag zur Helminthographie humane, aus brieflichen Mittheilungen des Dr. Bilharz in Cairo, nebst Bemerkungen von Prof. C. Th. v. Siebold in Breslau. Ztschr. Wiss. Zool., v. 4 (1): 53-72, pl. 5, figs. 1-20.

BLANCHARD, E. 1847.—Recherches sur l'organisation des vers. Ann. Sc. Nat. Paris, Zool. 3 s., v. 7: 87-128: & v. 8, 119-149, 271-341, pls. 8-14.

BOJANUS, L. H. 1817.—Bemerkungen aus dem Gebiete der vergleichenden Anatomie. Russ. Samml. Naturw. & Heilk., Riga & Leipzig, v. 2 (4): 523-553.
        1821.—Enthelminthica. Isis, v. 1 (2): 162-190, pls. 21-22, figs. 1-29.

FIG. 114.

A-F—*Ascaris lumbricoides* (A—Analine blue-black injected specimen showing terminal duct, sinus and associated tubules; B—Details of same region showing sinus nucleus; C—Right lateral chord showing canal branches; D-E—Left lateral chord in sinus region, D—at level of sinus nucleus; E—anterior to that level; F—Terminal duct nucleus); G-J & O—*Anisakis simplex* (G—Detail of tubule branching and terminal acini; H-J—Cross sections through sinus cell at level of nucleus, H of a young specimen, I of an older specimen, O and J of a senescent specimen); K—*Loa loa* (Excretory pore and cell in microfilaria); L-M—*Cucullanus* sp. (L—Lateral chord and canal near sinus; M—same in posterior part of body); N—*Mermis brevis* (Lateral view of anterior end of adult, showing excretory cell); P-R—*Phanodermopsis longisetae* (P—Excretory pore and terminal ampulla; Q—Esophageal region; R—Cross section at level of excretory cell); S-U—*Enoplus communis* (S—Cross section at level of excretory nucleus; T—cross section posterior to S; U—Ventral view of excretory system and associated structures as seen in toto); V-W—*Anaplectus granulosus* (V—Cross section at level of ventral gland cell; W—Ventral view of excretory system and associated structures as seen in toto); X—*Chromadora quadrilinea* (Esophageal region); Y—*Spilophorella paradoxa* (Cross section at level of excretory cell); Z-CC—*Camallanus microcephalus* (Z—Terminal duct showing first nucleus; AA—Terminal duct showing second nucleus; BB—End of terminal duct connecting with sinus; CC—Sinus with its nucleus). A-B, After Mueller, 1929, Ztschr. Zellforsch. v. 8 (3); G-J & O, After Mueller, 1927, Ztschr. Zellforsch. v. 5 (4); K, After Fuelleborn, 1929, Handb. Path. Micro-organ. v. 6 (2): N. After Hagmeier, 1912, Zool. Jahrb. v. 32 (6); U, After de Man, 1886, Nordsee Nematoden. Remainder original.

BUETSCHLI, O. 1871.—Untersuchungen über die beiden Nematoden der *Periplaneta* (*Blatta*) *orientalis* L. Ztschr. Wiss. Zool., v. 21 (2): 252-292, pls. 21-22, figs. 1-29.

1873.—Beiträge zur Kenntniss der freilebenden Nematoden. Nova. Acta. K. Akad. Naturf., Dresden, v. 36 (5) 124 pp., pls. 17-27, 69 figs.

1876.—Untersuchungen über freilebende Nematoden und die Gattung *Chaetonotus*. Ztschr. Wiss. Zool., v. 26 (4): 363-413, pls. 23-26.

BURIAN, R. 1910.—Die Excretion. Winterstein's Handbuch Vergl. Physiol v. 2 (2): 257-304, figs. 1-14.

CHITWOOD, B. G. 1930.—Studies on some physiological functions and morphological characters of *Rhabditis*. J. Morph. & Physiol., v. 49 (1): 251-275, figs. A-H, pls. 1-3, figs. 1-24.

1931.—A comparative histological study of certain nematodes. Ztschr. Morph., v. 23 (1-2): 237-284, figs. 1-23.

CHITWOOD, B. G. and CHITWOOD, M. B. 1933.—The histological anatomy of *Cephalobellus papilliger* Cobb, 1920. Ztschr. Zellforsch., v. 19 (2): 309-355, figs. 1-34.

CLOQUET, J. G. 1824.—Anatomie des vers intestinaux ascaride, lumbricoide et échinorhynque géant. Mémoire courroné par l'Académie royal des sciences qui en avait mis le sujet au concours pour l'année 1818. 130 pp. Paris.

COBB, N. A. 1888.—Beiträge zur Anatomie und Ontogenie der Nematoden. Diss. 36 pp., 3 pls. 32 figs., Jena.

1890.—Oxyuris-larvae hatched in the human stomach under normal conditions. Proc. Linn. Soc. N. S. Wales, 2 s., v. 5 (1): 168-185, pl. 8, figs. 1-5.

1898.—Extract from MS. report on the parasites of stock. Agric. Gaz. N. S. Wales, v. 9 (3): 296-321, figs. 1-45: (4): 419-454, figs. 46-127. Also Miscellaneous Publ. No. 215, Dept. Agric. N. S. Wales, 62 pp. 129 figs.

1913.—New terms for the lateral organs and ventral gland. Science, n. s. v. 37 (952): 498.

1914.—Citrus-root nematode. J. Agric. Res., v. 2 (3): 217-230. Figs. 1-13.

1916.—Notes on new genera and species of nematodes. J. Parasit., v. 2 (4): 195-196.

1925.—*Rhabditis icosiensis* J. Parasit., v. 11 (4): 218-220, fig. A.

CONINCK, LUCIEN DE 1931.—Sur trois espèces nouvelles de nématodes libres trouvés en Belgique. Bull. Mus. Roy. Hist. Nat. Belg., v. 7 (11): 1-15, figs. 1-5.

CORTI, E. 1902.—Di un nuovo nematode parassita in larva de *Chironomus*. Reale. Ist. Lomb. Sc. e Lett. Rend. 2. s., v. 35 (2-3): 105-113.

DAVAINE, C. J. 1857.—Recherches sur l'Anguillule du blé niellé considérée au point de vue de l'histoire naturelle et de l'agriculture. Compt. Rend. Soc. Biol. (1856) 2. s., v. 3: 201-271, pls. 1-3.

DEBRAY, F. and MAUPAS, E. 1896.—Le *Tylenchus devastatrix* Kühn et la maladie vermiculaire des fèvesen Algérie. Algérie Agricola, 55 pp., 1 pl.

DUBINI, A. 1843.—Nuovo verme intestinale umano (*Agchylostoma duodenale*) constituente un sesto genera dei nematoidei dell'uomo. Ann. Uni. di Med. Milano, v. 106 (316): 5-13.

EBERTH, C. J. 1860.—Zur Organisation von *Heterakis vesicularis*. Würzburger. Naturwiss. Ztschr., v. 1: 41-60, pls. 2-4, figs. 1-29.

1863.—Untersuchungen über Nematoden. 77 pp., 9 pls. Leipzig.

EISMA, M. 1932.—De differentiatie van het derde stadium van de larven Ancylostomidae van mensch, hond und kat. Diss. (Leiden). 155 pp., figs. 1-99. Haarlem.

FUELLEBORN, F. 1929.—Filariosen des Menschen. Handb. Path. Mikroorg., v. 6 (28): 1043-1224, figs. 1-77, pls. 1-3, 31 figs.

GOLDSCHMIDT, R. 1906.—Mitteilungen zur Histologie von *Ascaris*. Zool. Anz., v. 29 (24): 719-737, figs. 1-13.

GOLOVIN, E. P. (1902).—[Observations on nematodes. II. Excretory apparatus] [Russian text]. Uchen. Zapiski Imp. Kazan. Univ., 120 pp., 4 pls.

GYORY, A. VON 1856.—Ueber *Oxyuris spirotheca* n. sp., 8 pp., 1 pl. Wien.

HAGMEIER, A. 1912.—Beiträge zur Kenntnis der Mermithiden. Diss. 92 pp., 7 text figs., 5 pls. Heidelberg.

HAMANN, O. 1895.—Die Nemathelminthen. 2. Heft., 120 pp., 11 pls. Jena.

HUXLEY, 1856.—Lectures on general natural history. Medical Times. (329) v. 12: 383-386, figs. 6-8.

JAEGERSKIOELD, L. A. 1894.—Beiträge zur Kenntnis der Nematoden. Zool. Jahrb. Abt. Anat., v. 7 (3): 449-532, pls. 24-28.

1898.—Ueber die büschelförmigen Organe bei den *Ascaris*-Arten. Centralbl. Bakt., I. Abt., v. 24 (20): 737-741; (21); 785-793, figs. 1-6.

1901.—Weitere Beiträge zur Kenntnis der Nematoden. Kongl. Svenska Vetensk.-Akad. Handl. v. 35 (2): 1-80, figs. 1-8, pls. 1-6.

1903.—Nematoden aus Aegypten und dem Sudan. Results Swedish Zool. Exped. to Egypt and White Nile 1901. pt. 3 (25), 66 pp., figs. 1-23, pls. 1-4.

JERKE, H. W. M. 1902.—Eine parasitische *Anguillula* des Pferdes. Arch. Wiss. Prakt. Thierheilk, v. 29 (1-2): 113-127, pl. 1, figs. 1-9.

JOSEPH, G. 1882.—Vorläufige Bemerkungen über Musculatur, Excretions-organe und peripherisches Nervensystem von *Ascaris megalocephala* und *lumbricoides*. Zool. Anz. (125), v. 5: 603-609.

LEIDY, J. 1853.—A flora and fauna within living animals. Smithsonian Contrib., v. 5 (2): 1-67, pls. 1-10.

LEUCKART, R. 1876.—Die menschlichen Parasiten und die von ihnen herrührenden Krankheiten, v. 2, 882 pp., 401 figs. Leipzig.

LINSTOW, O. VON 1905.—Neue Helminthen. Arch. Naturg., Berl., 71. J., v. 1 (3): 267-276, pl. 10, figs. 1-17.

1909.—II. Parasitische Nematoden. Süsswasserfauna Deutschlands (Brauer). Heft 15: 47-83, figs. 1-80.

LOOSS, A. 1895.—*Strongylus subtilis* n. sp., ein bisher unbekannter Parasit des Menschen in Egypten. Centralbl. Bakt. 1. Abt., v. 18 (6): 161-169, pl. 1, figs. 1-8.

1902. (1901).—The Sclerostomidae of horses and donkeys. Rec. Egypt. Govt. Sch. Med., pp. 25-139, pls. 1-13, figs. 1-172.

1905.—The anatomy and life history of *Agchylostoma duodenale* Dub. Rec. Egypt. Govt. Sch. Med. v. 3: 1-158, pls. 1-9, figs. 1-100, pl. photos. 1-6.

MACKIN, J. G. 1936.—Studies on the morphology and life history of nematodes in the genus *Spironoura*. Ill. Biol. Monogr., v. 14 (3): 1-64, figs. 1-2, pls. 1-6, figs. 1-69, Univ. Illinois Bull. v. 33 (52).

MAGATH, T. B. 1919.—*Camallanus americanus* nov. spec., a monograph on nematode species. Tr. Am. Micr. Soc., v. 38 (2): 49-170, Figs. A-Q, pls. 7-16, figs. 1-134.

MAN, J. G. DE, 1886.—Anatomische Untersuchungen über freilebende Nordsee-Nematoden, 82 pp., 13 pls. Leipzig.

MARTINI, E. 1916.—Die Anatomie der *Oxyuris curvula*. Ztschr. Wiss. Zool., v. 116: 137-534, figs. 1-121, pls. 6-20, figs. 1-269.

MAUPAS, E. 1900.—Modes et formes de reproduction des nématodes. Arch. Zool. Exper. & Gen., 3. s., v. 8: 463-624, pls. 16-26.

1916.—Nouveau *Rhabditis* d'Algérie. Compt. Rend. Soc. Biol., Paris. v. 79 (13): 607-613, figs. 1-5.

1919.—Essais d'hybridation chez les nèmatodes. Bull. Biol. France & Belg. v. 52 (4): 466-498, figs. 1-7.

MEHLIS, E. 1831.—Novae observationes de entozois. Isis (1): 68-99, pl. 2, figs. 1-18.

METALNIKOV, S. I. 1897.—[Ueber die Excretionsorgane von *Ascaris megalocephala*]. (Russian) Bull. Acad. Imp. Sc. St. Petersb., s. 5, v. 7: 473-480, figs. 1-7.

MIRZA, M. B. 1929.—Beiträge zur Kenntnis des Baues von *Dracunculus medinensis* Velsch. Ztschr. Parasit., v. 2 (2): 129-156, figs. 1-33.

MUELLER, J. F. 1927.—The excretory system of *Anisakis simplex*. Ztschr. Zellforsch., v. 5 (4): 495-504, figs. 1-4.

1929.—Studies on the microscopical anatomy and physiology of *Ascaris lumbricoides* and *Ascaris megalocephala*. Ztschr. Zellforsch., v. 8 (3): 361-403, pls. 9-18, figs. 1-30.

NASONOV, N. V. 1898.—Ueber die Anatomie und Biologie der Nematoden. Russian Varshavsk. Univ. Izviest, (5): 1-24; (6): 25-43, pls. 1-2, figs. 1-17.

1900.—Zur Kenntniss der phagocytären Organe bei den parasitischen Nematoden. Arch. Mikr. Anat., v. 55 (4): 488-513, pls. 26-28.

POEPPEL, E. 1897.—Untersuchungen über den Bau des *Strongylus armatus* s. *Sclerostomum equinum* (auctorum). Nebst einem Anhang über die Biologie desselben und das Aneurysma verminosum. Diss. 57 pp., 2 pls., 31 figs., Leipzig.

RAUTHER, M. 1907.—Ueber den Bau des Oesophagus und die Lokalisation der Nierenfunktion bei freilebenden Nematoden. Zool. Jahrb., Abt. Anat., v. 23 (4): 703-738, figs. A-G, pl. 38, figs. 1-9.
　　　1909.—Morphologie und Verwandschaftsbeziehungen der Nematoden. Ergeb. Fortschr. Zool. v. 1 (3): 491-596, figs. 1-21.

RAUTHER 1918.—Mitteilungen zur Nematodenkunde. Zool. Jahrb. Abt. Anat., v. 40 (4): 441-514, figs. A-P, pls. 20-24, figs. 1-40.
　　　1930.—Vierte Klasse des Cladus Nemathelminthes. Nematodes (Nematoidea-Fadenwürmer): Handb. Zool. Kükenthal & Krumbach v. 2, 8 Lief., Teil 4, 249-402, figs. 267-426.

RAVEN, B. and STEKHOVEN, J. H. S. 1934.—Zur Frage der Excretion bei den Rhabditiden. Zool. Anz., v. 106 (1-2): 17-20, fig. 1.

RZEWUSKI, R. B. E. VON 1887.—Untersuchungen über den anatomischen Bau von *Strongylus paradoxus*. Diss. 36 pp., 2 pls., 32 figs. Leipzig.

SCHEPOTIEFF, A. 1908.—Desmoscoleciden. Ztschr. Wiss. v. 90: 181-204, pls. 8-10.

SCHNEIDER, A. 1858.—Ueber die Seitenlinien und das Gefässystem der Nematoden. Arch. Anat., Physiol. & Wiss. Med., pp. 426-436, pl. 15, 9 figs.
　　　1860.—Ueber die Muskeln und Nerven der Nematoden. Arch. Anat. Physiol. & Wiss. Med., pp. 224-242, pl. 5, figs. 1-12.
　　　1863.—Neue Beiträge zur Anatomie und Morphologie der Nematoden. Arch. Anat., Physiol. & Wiss. Med., pp. 1-25, pls. 1-2.
　　　1865.—Ueber Haematozoen des Hundes. Arch. Anat., Physiol. & Wiss. Med., pp. 421-422.
　　　1866.—Monographie der Nematoden. 357 pp., 122 figs. 28 pls., 343 figs. Berlin.

SIEBOLD, C. TH. VON 1838.—Helminthologische Beiträge Vierter Beiträge über geschlechtslose Nematoideen. Arch. Naturg., 4. J., v. 1: 301-314.

STADELMAN, H. 1891.—Ueber den anatomischen Bau des *Strongylus convolutus* Ostertag. Diss. 39 pp. Berlin.

STEFANSKI, W. 1917.—Contribution a l'étude de l'excrétion chez les nématodes libres. Biol. Centralbl., v. 37 (6): 294-311, figs. 1-9.
　　　1922.—Excrétion chez les nématodes libres. Arch. Nauk. Biol. Towarz. Nauk. Warszaw., v. 1 (6): 1-33, figs. 1-39.

STEINER, G. 1916.—Freilebende Nematoden aus der Barentssee. Zool., Jahrb., Abt. Syst. v. 39 (5-6): 511-676, pls. 16-36, figs. 1-46.
　　　1919.—Untersuchungen über den allgemeinen Bauplan de Nematodenkörpers. 96 pp., figs. A-E. pls. 1-3, figs. 1-16, Jena. Also in Zool. Jahrb. (1921) Abt. Morph. v. 43.
　　　1920.—Betrachtungen zur Frage des Verwandschaftsverhältnisses der Rotatorien und Nematoden. Festschr. Zschokke, 16 pp., 15 figs. Basel.

STEKHOVEN, J. H. S. 1927.—The nemas *Anchylostoma* and *Necator*. Proc. Konigl. Acad. Wetensch. Amsterdam., v. 30 (1): 113-125, figs. A-C, pls. 1-3.

STRASSEN, O. ZUR 1907.—*Filaria medinensis* und *Ichthyonema*. Verhandl. Deutsche Zool. Gesellsch. : 110-129, figs. 1-8.

STRUBELL, A. 1888.—Untersuchungen über den Bau und die Entwicklung des Rübennematoden *Heterodera schachtii* Schmidt. Bibliotheca Zool. Orig. Abh. Gesammt. Zool., Heft. 2. 50 pp., 2 pls. 57 figs.

THAPAR, G. S. 1925.—Studies on the oxyurid parasites of reptiles. J. Helminth., v. 3 (2-4): 83-150, figs. 1-132.

TOERNQUIST, N. 1931.—Die Nematodenfamilien Cucullanidae und Camallanidae nebst weiteren Beiträgen zur Kenntnis der Anatomie und Histologie der Nematoden. Göteborgs K. Vetensk. -o Vitterhets- Samh. Handl., 5. f., s. B, v. 2 (3), 441 pp., pls. 1-17.

WAGENER, G. R. 1857.—Ueber *Dicyema* Kölliker. Arch. Anat. & Physiol. pp. 354-368, pls. 11-14.

WALTER, G. 1856.—Beiträge zur Anatomie und Physiologie von *Oxyuris ornata*. Ztschr. Wiss. Zool., v. 8 (2), pp. 163-201, pls. 5-6, figs. 1-28.

WUELKER, G. and STEKHOVEN, J. H. S. 1933.—Nematoda. Die Tierwelt Nord- und Ostsee. Teil 5a., 64 pp., 65 figs.

CHAPTER X

THE REPRODUCTIVE SYSTEM

B. G. CHITWOOD and M. B. CHITWOOD

Introduction

The reproductive system is more or less similar in both sexes of all nematodes, being composed of one or two (rarely multiple) tubular gonads, each gonad being comparable to a single testicular or ovarial tubule of arthropods or vertebrates.

Sexual dimorphism is, as a rule, limited to characters of the reproductive system such as the vulva and male copulatory apparatus (bursa, spicules, genital papillae) but it is manifested to a minor extent in the fact that females are nearly always larger than males. Among the parasitic species this tendency becomes most marked in *Trichosomoides crassicauda* in which the male is so small that it enters the female by way of the vulva and spends its life within the uterus (Fig. 115 D). Other marked but less spectacular cases are those of *Dracunculus medinensis* and *Philometra globiceps* in which the female becomes 18 and 33 times as large, respectively, as the male. In these cases, however, the female continues growth after copulation takes place. *Howardula benigna* (Allantonematidae) presents a similar case of copulation in a juvenile stage but afterwards the young female enters a new host where it undergoes the greater part of its development.

Sexual modifications of gross body form are rare and usually take a single course, the enlargement of the female so that she becomes a reproductive sac—*Heterodera* (Fig. 115 N), *Tetrameres* (Fig. 115 C), and *Tylenchulus*. The lobes of *Phlyctainophora* may only be the result of growth in a confined situation. The Allantonematidae (inhabiting the body cavity of insects) present more examples of reproductive sac formation than do all other nemic groups combined. In this group we find *Tylenchinema* (Fig. 115 J), *Chondronema*, *Scatonema* (Fig. 115 E), and *Allantonema* (Fig. 115 I) all of which show progressive stages of degeneration of the female to reproductive sac formation. *Tripius* (Fig. 115 K) and *Sphaerularia* (Fig. 115 A-B), of the same family, go much further in degeneration; after copulation of the precocious free-living adults the females enter the new host and growth of the reproductive system takes place at the expense of the remainder of the body. The uterus with the ovary enclosed is everted and continues to grow until it is many times as large as the female body.

Most of the remaining types of sexual dimorphism are attributed to failure of the male to attain complete development, such are the sexual differences in cephalic characters, esophagus, and cuticle of thelastomatids. Seurat (1920) considered the alae an spines of the male of *Tetrameres fissispina* as organs of propulsion an fixation necessary to its free life in the succenteric ventricle of its host and migration to the female, which lives sedentarily in the gastric glands of its host. Sexual dimorphism in free-living nemas is extremely rare, the most outstanding examples being members of the Enchelidiinae; in some of these forms there may be complete degeneration of stoma in the adult male (*Enchelidium pauli* Fig. 63 I-J). In male tylenchoids, the stylet is often not as well developed as that of the female, *Neotylenchus abulbosus* being an extreme example of this type of dimorphism. Sexual dimorphism in size of amphids, those of the male being the larger, has been described in such forms as *Trilobus gracilis* var. *homophysalidis*, *Ironus ignavus* and some mermithids.

Intersexes. Meissner (1853) first observed male secondary sexual characters (spicules and genital papillae) in normal female *Hexamermis albicans*. Such forms were erroneously regarded as hermaphrodites but this is a false impression as only one set (female) of reproductive organs is developed. Since the time of Meissner such intersexes have been described in *Enoplus communis* (syn *E. cochleatus*) by Schneider (1866), in *Chromadora poecilisoma* and *Thoracostoma figuratum* by de Man (1893), in *Porrocaecum heteroura* by Willemoes-Suhm (1869), in *Trilobus diversi-papillatus* by Daday (1905),

Trilobus gracilis by Ditlevsen (1911), W. Schneider (1922) and Linstow (1903). Hagmeier (1912), Steiner (1923) and Christie (1929) have described intersexes in mermithids. The latter author found that sex is determined by the number of parasites in a host. When one to three parasites were present in grasshoppers they were females, when four to 23, they were mixed, and when above 23 they were all males. By feeding a known number of parasite eggs to the host, Christie determined that the sex ratios were not due to selective mortality. The fact that intersexual males are unknown seems to cast doubt upon the theory that the sexes are primarily present in equal numbers and may be converted to the opposite sex by the influence of environmental factors. Nevertheless, intersexes seem to be related to crowding and when a single female is present in a host containing 10 males, it is conceivable that they might cause her to be an intersex.

General Morphology

So long ago as 1866 Bastian remarked on the difference between the reproductive systems of free-living and parasitic nemas; he felt that the relative simplicity of the system in free-living nemas was sufficient ground for the separation of that group from parasites as a separate family (Anguillulidae, equivalent in scope to our Rhabditina, Tylenchina, Chromadorida, Enoplina and Dorylaimoidea). The great complexity of the reproductive system in parasitic nemas is chiefly limited to the female sex and is correlated with increased egg production.

Division of the nemas into taxonomic groups on the basis of the reproductive system has been proposed by only one modern author, Rauther, (1918, 1930) who divided the Class Nematoda into two orders: Telogonia (includes Phasmidia, Chromadorida, Enoplina, Dorylaimoidea and Mermithoidea) and Hologonia (Trichuroidea and Dioctophymatoidea). These divisions apply to both sexes. The germinal zone in the order Hologonia (examples *Trichuris trichiura* as observed by Eberth, 1869, and *Dioctophyma renale* as observed by Leuckart, 1876, (p. 378) extends the entire length of the gonad, being composed of a series of germinal areas on one side of the gonoduct in the former and comprising the entire circumference of the gland in the latter. In neither case is a rachis present and the entire group Hologonia is characterized by the presence of a single gonad in both sexes. In the majority of nemas (Telogonia) it is a well established fact that new germ cells originate only at the end of the gonad. However, only small proportion of the Nematoda (particularly parasites) has been well studied from this standpoint. Data regarding mermithids and *Cystoopsis* should be particularly informative but are lacking to date.

Eschricht (1848) discovered that in the female ascaridids germ cells are not free in the gonoduct but are grouped around a central axis which is termed the rachis. Subsequent authors have found this structure to be a common feature of ascaridids, oxyurids, strongylins, and spirurids. Some, including Bütschli, compared the rachis to the cell of Verson or apical cell of the gonads of insects, but as pointed out by Seurat (1920) the rachis is not a constant feature of nemas even in special groups, (it is present in *Bradynema*, absent in *Ditylenchus*) and it does not supply yolk, for the nucleus and protoplasm grow in the same proportion as the oogonium passes down the tube. The function of the rachis is not understood and its apparent erratic occurrence seems to eliminate it as an ordinal character.

The genital primordium is identical in that it consists of two germ cells and two epithelial cells in the first stage larva of both sexes of all nemas studied. In forms with two gonads the primordial germ cells are thereafter separated by somatic cells, one germ cell entering each gonad, while in forms with one gonad the germ cells remain together. In two-ovaried forms the intervening cell group forms the two uteri and connects with the vagina in the female, or forms the vas deferens in the male. It would therefor, seem obvious that two gonads are the primitive nemic condition. Bütschli (1876), Steiner (1919, 1920) and Seurat (1920) have taken the view that parallel posteriorly opening gonads are primitive for both sexes in nemas. Such a condition is a contradiction to what we know of free-living nemas. First, parallel ovaries are known to occur in only two genera, not parasitizing vertebrates, said genera being the highly specialized plant parasites, *Heterodera* and *Meloidogyne* (Fig. 115 M). Second, parallel testes are known to occur in only two genera, *Anticoma* (Fig. 124 Y) in which the parallel condition appears to be a modification of the opposed condition and *Meloidogyne hapla* in which the paired testes seem to have arisen secondarily as a longitudinal splitting of a single testis Otherwise, wherever two gonads are present in free-living nemas, they are opposed. Without venturing further in the matter of primitivity at this time the following points seem notable: Admitting that from the standpoint of general comparative anatomy the paired gonads of both sexes should have opened posteriorly into the ventral side of the cloaca, the ontogeny of present day nemas indicates that in the original nema the vulva was separate and equatorial. Any other arrangement must have occurred prior to the origin of the phylum.

Since the female reproductive system is more often used in systematics and since more observations have been recorded in more diverse groups it shall be considered first.

Female Reproductive System
GROSS MORPHOLOGY

Seurat (1913-1920), in a series of papers culminating in his "Histoire Naturelle des Nématodes de la Berbérie", developed an extremely useful nomenclature for the parts of the female reproductive system and at the same time presented more useful information on this subject than all other workers combined. Seurat was fully aware that the groups formed by his classification of the female reproductive system were artificial and he himself pointed out the numerous transitions from one of his major groups to another, within families and genera. His classification was as follows:

I. Uteri opposed. Amphidelphes:

1. a: *Dictyocaulus filaria* (Metastrongylidae). Vagina short, uteri opposed, oviducts U-shaped, ovaries converging (Fig. 116 Q).

　　b: *Allodapa numidica* (Heterakidae). Similar to 1a but with longer vagina, and uteri twice reflexed in U-shape.

　　c: *Camallanus microcephalus* (Camallanidae). Similar to 1a but with longer vagina, and posterior gonad represented by a blind uterine sac, posterior ovary and oviduct absent (Fig. 117 H).

2. *Heterakis* (Heterakidae). Vagina much elongated, U-shaped; uteri opposed, reflexed; ovaries much contorted (Fig. 116 R).

3. *Haemonchus* (Trichostrongylidae). Vagina short; anterior uterus and ovary extending anteriad; posterior uterus U-shaped; posterior ovary also anterior to vulva (Fig. 116 T).

4. *Trichuris* (Trichuridae). Anterior genital tube totally absent; posterior uterus extending nearly to posterior extremity; oviduct U-shaped extending nearly to vulva where it is again reflexed; ovary much coiled, extending length of body (Fig. 117 F).

5. *Acuaria laticeps* (Acuariidae). Uteri opposed, very long, serving to contain a mass of many small eggs; oviducts and ovaries very filiform, oviducts entering uterus in esophageal and preanal regions (Fig. 116 P).

6. *Protrellus* (Thelastomatidae). Vulva shifted anteriad; vagina greatly elongated; uteri opposed (Fig. 116 K).

7. One-ovaried forms with the vulva posterior. *Acuaria invaginata, Heligmosomum laeve, Atractis dactylura* (Fig. 117 E).

II. Uteri parallel.

A. Opisthodelphes. Uteri directed posteriad; vulva usually anterior to middle of body.

8. *Ascaris lumbricoides.* Vagina short; uteri directed posteriad; oviducts and ovaries much contorted (Fig. 117 L & T).

9. Physalopterids, thelaziids, and filariids are usually opisthodelphic; the ovaries coiled in the preanal region.

10. *Maupasina.* Vulva situated in anal region; uteri directed anteriad, turn posteriad and terminate in two parallel ovaries in posterior region.

B. Prodelphes. Uteri directed anteriad.

(a) Original or primitive prodelphy.

11. *Dermatoxys.* Seurat considered the oxyurids the most primitive group and characterized them as for the most part prodelphic but with variably situated vulva. (Fig. 117 W).

(b) Secondary prodelphy.
　　aa) Vulva near anus.

12. *Chabertia.* Uteri extending anteriad. Ovaries coiled in anterior part of body. Type obviously a modification due to posterior shift of vulva of the amphidelphic type found in trichostrongyles.

13. *Tetrameres.* Uteri narrow, parallel and very long; oviducts and ovaries filiform, entwined around the uteri in the esophageal region (Fig. 117 I).
　　bb) Vulva anterior.

14. *Aprocta orbitalis.* Genital tubes parallel, U-shaped, descending posteriad then reflexed anteriad where they terminate in oviducts and ovaries entwined in the cephalic region.

Today, Seurat's terminology is accepted in general and adjectives derived therefrom are applied widely in taxonomic descriptions. Because of the difficulty in tracing uteri and ovaries certain arbitrary limitations of definition have come to be accepted. Thus, *amphidelphic* is redefined as having uteri opposed at origin regardless of location of oviducts or ovaries. Similarly, *prodelphic* is defined as having uteri parallel and anteriorly directed at origin while *opisthodelphic* is defined as having uteri parallel and posteriorly directed at origin.

Other adjectives also introduced by Seurat, Ortlepp and Schulz were as follows: *Monodelphic*, meaning: provided with one complete genital tube. (Because of variability in the degree of development of a second uterus and ovary in such forms, this word has come to be applied to the ovary. Thus, *Camallanus* and *Atractis* are now both considered monodelphic prodelphic). *Didelphic* meaning: provided with two complete genital tubes; forms may be didelphic opisthodelphic or didelphic amphidelphic. *Tetradelphic*, meaning: provided with four complete genital tubes; *Polydelphic* provided with more than four complete genital tubes.

PARTS OF THE REPRODUCTIVE SYSTEM. Seurat (1920) like Bastian (1866) noted the morphologic difference between the vagina and uteri of free-living nemas and those same structures of parasites. Primarily the genital tubes in females (Fig. 3) of free-living nemas consist of *vulva*, simple transverse *vagina*, paired opposed *uteri* without heavily muscled areas, short sometimes indistinct oviducts and short tubular *ovaries*. A definite seminal receptacle or spermatheca may be developed as an outpocketing of each uterus (Fig. 116 A) or oviduct; in other forms its function is assumed by the ovarial end of the uterus (Fig. 116 B, F & G). As noted by Filipjev (1918, 1922, 1929, 1934) the ovaries are outstretched (Fig. 116 L-N) in some large groups and reflexed (Fig. 116 C-F & O) in other large groups. Actually the ovaries themselves are seldom reflexed in free-living nemas, more commonly the flexure occurs at the junction of ovary and uterus, or of ovary and oviduct.

Vagina and Uteri. With parasitism there is generally an increase in length of the entire genital tube with coincident increased egg production, and increase in muscular development of uteri and vagina for the ejection of the eggs. The vagina (lined with cuticle) commonly becomes elongated to the form of a muscular tube. This tube may connect directly with the bifurcation of the uteri (Fig. 119 A & B), or it may be followed by a non-ectodermal tube of similar construction (Fig. 115 L). In the latter case the second section

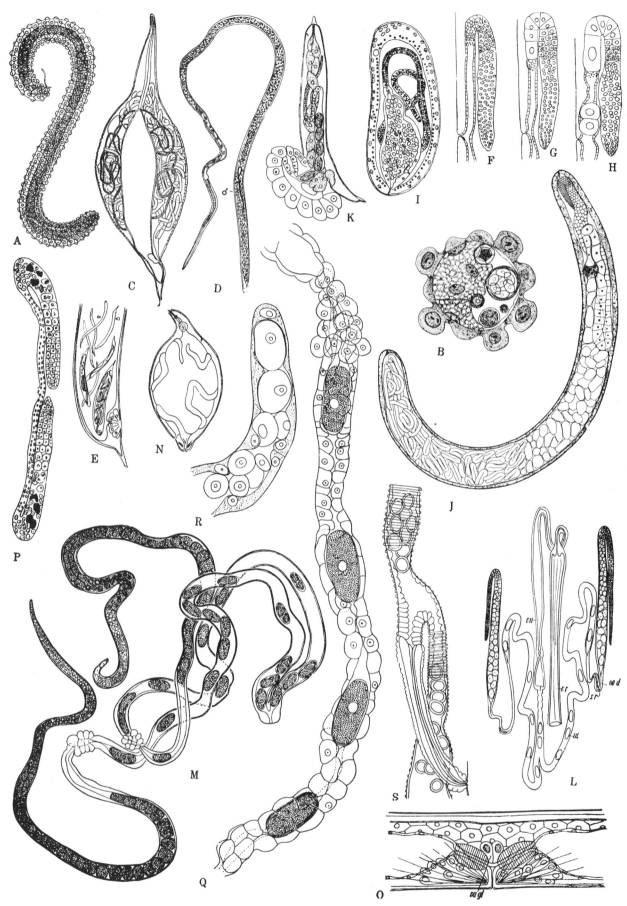

FIG. 115.

138

is termed the *vagina uterina*, the first section, *vagina vera*. The functional term *ovejector* was applied to the entire thick-walled terminal part of the reproductive system by Seurat. In some nemas (*Nematodirus*, Fig. 118 BB-DD) the ovejector bifurcates. The three parts of each ovejector were named by Maupas and Seurat (1912), vestibule, sphincter, and trunk (Trompe). Later, (1920) Seurat revised his terminology, then calling the three parts of the strongylin ovejector *vestibule, glazing gland* [Firnissdrüse of von Linstow (1878)] and *sphincter*. These parts correspond to ovejectors 1, 2 and 3 as described by Ransom (1911) and *pars haustrix* and *pars ejectrix* as described by Looss (1905). It is unfortunate that such confusion has occurred. The aplication of functional terms is always likely to lead to such. However, since the ovejector is itself a functional rather than a structural entity, we must continue at least some of the terminology. As revised by Seurat, the *trompe* is sometimes equivalent to the vagina uterina (praeuterus of Rauther, 1918) being the undivided part of the uterus in *Ascaris, Gongylonema, Spirura* (Figs. 117 L, 118 GG-KK & 118 N-O). It may also become a divided thick walled non-cuticular part of the ovejector as in *Habronema microstoma* (Fig. 118 P) and finally it may exist both in divided and undivided parts as in *Protospirura numidica*. Attempts to homologize parts on the basis of gross appearance lead to little of value. In some forms such as *Gongylonema scutatum* the vagina vera constitutes the major part of the ovejector and it is separated by a constricted region, i. e., sphincter, from the shorter vagina uterina (Fig. 118 KK). In the same genus in *G. mucronatum* the vagina vera is extremely inconspicuous, no sphincter is present, and the major part of the ovejector is vagina uterina. Undoubtedly such specializations as sphincters are of multiple origin and we see no advantage in renaming the functional sphincter of the strongyle ovejector, glazing gland (Fig. 118 CC-ovj. 2).

In parasites, as also in free-living nemas, monodelphic forms commonly have a postvulvar uterine sac (Fig. 117 H) which functions as a *seminal receptacle* or *spermatheca*. Such a uterine sac is considered the remainder of a second genital tube. In others (*Tetrameres fissispina*) a separate sac is formed as an outpocketing of the vagina vera (Fig. 118 G). More often, the distal ends of the uteri are more or less distinctly modified as seminal receptacles.

Oviduct. This part of the system is less likely to be confused than other parts. However, like other parts it is a functional rather than a structural entity, i. e., a constricted thick-walled region between uterus and ovary. In *Rhabditis strongyloides* (Fig. 3) it is hardly an entity while in *Sabatieria* (Fig. 120 B) and *Anaplectus* (Fig. 120 C) it is clearly differentiated. In the parasites it attains its greatest development in oxyurids such as *Syphacia obvelata* (Fig. 159 K-L) and occasionally, as in this form, a dilation of the oviduct serves as a seminal receptacle. Vogel (1925) was led by these structures to believe the oviduct functioned as egg former (see chapter XII).

Ovary. In telogonic forms each ovary consists of (1) a germinal zone and (2) a growth zone. The latter commonly becomes the major part of the gonad in parasitic nemas.

Abnormalities. Bivulvar specimens have been recorded by Bütschli (1874) in *Linhomoeus mirabilis*, Paramonov (1926) in *Trilobus gracilis* and Cassidy (1928) in *Dorylaimus* sp. Cassidy (1933) later recorded a specimen of *Prionchulus muscorum* with three ovaries and uteri, the third being connected with the second vulva. Chandler (1924) recorded a specimen of *Ascaris lumbricoides* with three uteri and ovaries. These must all be considered monstrosities.

Fig. 115.

Female reproductive system. A-B—*Sphaerularia bombi.* C—*Tetrameres fissispina.* D—*Trichosomoides crassicauda.* E—*Scatonema wuelkeri.* F-H—*Rhabditis sechellensis* (F, during sperm formation; G and H, later stages). I—*Allantonema mirabile.* J—*Tylenchinema oscinellae.* K—*Tripius gibbosus.* L—*Macracis monhystera.* M, Q-R—*Meloidogyne hapla* (M, entire female reproductive system; R, blind end of ovary; Q, upper part of uterus, all as seen in specimen dissected in egg albumen). N—*H. schachtii.* O—*Thoracostoma strasseni.* P—*Rhabditis aspera v. aberrans.* S—*Aplectana gigantica,* vagina showing origin of uteri and structure of ovejectors. A-B and K, after Leuckart, 1887, Abhandl. Math.-phys. Classe, Konigl. Sachs. Gesellsch. Wiss. v. 13 (8); C, after Travassos, 1919, Mem. Inst. Oswaldo Cruz, v. 11; D, after Hall, 1916, Proc. U. S. Nat. Mus. v. 50; E, after Bovien, 1932, Vidensk. Medd. Dansk. Naturh. Foren., v. 94; F-H, after Potts, 1910, Quart. J. Micr. Sc., v. 55 (3); I, after Wuelker, 1923, Ergeb. u. Fortsch. Zool., v. 5; J. after Goodey, 1930, Philosoph. Tr. Roy. Soc. Lond. s. B, v. 218; L, after Rauther, 1918, Zool. Jahrb. Abt. Morph., v. 40; N, after Strubell, 1888, Biblioth. Zool., v. 1 (2); O, after Turk, 1903, Mitt. Zool. Stat. Neapel, v. 16; P, after Kruger, 1913, Ztschr. Wiss. Zool., v. 105 (1). Remainder original.

DETAILED MORPHOLOGY

The minute anatomy of very few nemas has been adequately investigated. Those who have been major contributors to this subject are as follows: Nelson (1852), Bischoff (1855), Meissner (1855), Thompson (1857), Van Beneden (1883), Nussbaum (1884), Vogt and Jung (1888), Wasielewski (1893), Sala (1904), Domaschko (1905), Scheben (1905), Looss (1905), Kemnitz (1912), Romeis (1913), Zacharias (1913), Maupas and Seurat (1912), Seurat (1920), Maupas (1899), Pai (1927, 1928), Musso (1930), Baker (1936) and Mackin (1936). The major part of this work has been done on *Ascaris lumbricoides* and *Parascaris equorum;* less comprehensive studies have been made on other species.

(a) *Ovary.* The ovary consists of a tubular sac in which germinal cells develop. This sac consists of an epithelial layer and a germinal chord. In most nemas with the exception of hologonic forms, germinal elements do not arise from the epithelium. The epithelium consists of a single layer of greatly elongate, flat (simple squamous), spindle-shaped cells which reach 1 meter in length in *Ascaris lumbricoides* according to Musso (1930); each of these cells is multinucleate; as they near the oviduct they become shorter until they are only 2 mm. in length and have 12 to 20 nuclei (Fig. 121 N-O). Such astounding size is apparently a proportional development in keeping with the general oversize structure of *Ascaris.* The cells show the same spindle-shape but are not spectacularly long in *Spironoura* (Fig. 122 H).

In general the gonad is divisible into two regions, as follows: (1) The germinal zone: An area of rapid division of relatively small cells, often not showing clear cell boundaries. (2) The growth zone: An area of gradual increase in size of the oogonia. The first of these zones is always relatively short while the latter varies tremendously in size, sometimes amounting to the greater part of the gonad length, as in *Ascaris lumbricoides.* Although a conspicuously long growth zone is usually associated with parasitism, it is by no means limited in occurrence, being found likewise in such forms as *Metoncholaimus.*

At the blind end of the ovarial tube the epithelium becomes extremely thin so that its very existence has been denied by some authors, while others have concluded that the large *cap cell* is the epithelial cell of the ovarial terminus. This viewpoint has been questioned by Musso (1930) who regards the cap cell as an undifferentiated germinal stem cell which gives rise equally to both epithelial cells and germinal cells. In accord with the majority of observers, the writers have never seen any sign of nuclear division in a cap cell. Also, we find the protoplasm of the germinal chord rather clearly segregated from this cell in *Spironoura* (Fig. 122 G). With Maupas (1899) and Pai (1927) the writers would conclude that the cap cell is a part of the ovarial epithelium. Cell borders are often difficult to distinguish at the proximal or germinal end of the germinal chord, but sometimes they are distinct and it seems proper to regard the region as cellular rather than syntial in all instances, since cell boundaries gradually become more apparent as the cells move down the gonoduct.

The origin and significance of the rachis is as yet unsettled. Beginning at a slight distance from the extreme end of the germinal chord one finds a central strand of nonnucleated tissue which extends in *Ascaris* to the beginning of the oviduct. It would seem that there are two possible origins of the rachis, one as a continuation and product of the cap cell, the other as a residuum of enucleated plasma separated off from a germinal syncytium. Either assumption has lacked evidence. Earlier authors (Bütschli) leaned to the opinion that the rachis of nemas was comparable to the rachis found in the telotrophic insect ovary; such a view would require that it be a feeding mechanism and that it contain nuclei (nurse cells). Von Kemnitz (1912) found that the chief stored nutrient material of ascarid oogonia is glycogen and that this substance is absent from the rachis throughout the germinal region, only making its appearance in that structure *after* it has become the most conspicuous feature of the oogonium. Von Kemnitz traced the distribution of glycogen in the gonad demonstrating conclusively that it is first seen in the epithelium of the upper part of the ovary, thereafter, less concentrated in both epithelium and oogonia and finally absent from the epithelium, concentrated in the oogonia and scattered in the rachis. One concludes that it is obtained from the epithelium by absorption. Seurat (1920) pointed out that increase in nuclear and plasmatic

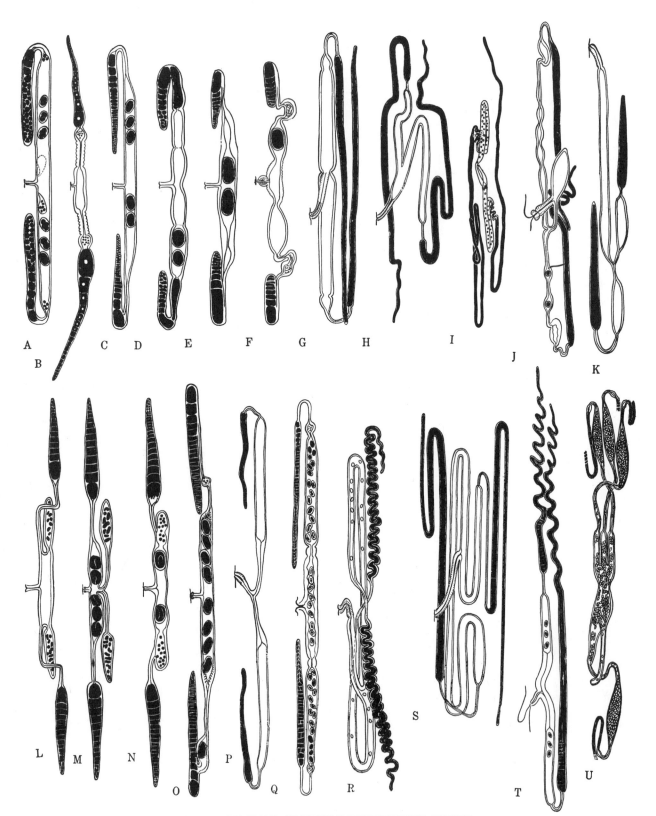

FIG. 116. DIAGRAMS OF FEMALE REPRODUCTIVE SYSTEM.

A—*Rhabditis strongyloides;* B—*Tylenchorhynchus dubius,* C—*Anaplectus granulosus.* D—*Chromadora* sp. E—*Mononchus lacustris.* F—*Trilobus pellucidus.* G—*Cephalobellus papilliger.* H—*Cucullanus micropapillatus.* I—*Aplectana gigantica.* J—*Rhigonema infectum.* K—*Protrellus kuenckeli.* L—*Dorylaimopsis metatypicus.* M—*Sabatieria hilarula.* N—*Axonolaimus spinosus.* O—*Actinolaimus* sp. P—*Acuaria laticeps.* Q—*Dictyocaulus filaria.* R—*Heterakis gallinarum.* S—*Spironoura affine.* T—*Haemonchus contortus.* U—*Tanqua tiara.*

B, after Goodey, 1932, J. Helminth., v. 10 (2-3); H, after Toernquist, 1931, Goteborgs Kungl. Vetensk . . . s. B v. 2 (3); K, after Schwenk, 1926, Sc. Medica v. 4; I, after Olsen, 1938, Tr. Am. Micr. Soc., v. 57 (2); P, after Seurat, 1920, Histoire naturelle . . . ; Q, after Augstein, 1894, Arch. Naturg. 60J., v. 1 (3); R, after Baker, 1936, Tr. Roy. Canad. Inst. v. 21 (2); U, after Monnig, 1923, 9 & 10th Rpt. Director Vet. Education & Res. Remainder original.

size in the growth zone is fairly proportionate, which would not be the case if nurse cells or a vitelline tissue were actively contributing plasma to the oogonium. The same author further noted that the rachis cannot be of great significance to nemas since it is often apparently absent in some species (*Macracis monhystera*) while it is present in closely related forms, (*Oxyuris equi*). Without in any way passing judgement on the function of the rachis, we may note that in some forms the growth zone of the ovary consists of a cylinder of oogonia surrounding the rachis (*Ascaris*), while in others (*Spironoura affine*, Fig. 119, A-B) and the majority of free-living nemas) the growth zone is a single chain of cells and the rachis disappears as the oogonia assume the single file. The rachis appears in direct continuity with the cap cell in *Spironoura* (Fig. 122 G). In countless free-living nemas (ex. *Mononchus, Trilobus* Figs. 120 D & I-P) the germinal zone is greatly reduced in length and there seems never to be a cylindrical germ chord. In such forms no distinct cap cell or rachis is present.

(b) *Oviduct.* The oviduct, when distinct, consists of a narrow tube with high columnar epithelium. According to Musso, the oviduct in *Ascaris,* like the ovary, is devoid of a muscular layer except in the zone approaching the uterus. Rauther (1918) described a muscular sphincter at the ovary-oviduct junction in *Macracis* (Fig. 115 L), and in many oxyurids the oviduct has a muscular appearance. A slight enlargement in the proximal part of the oviduct serves as a spermatheca and fertilization chamber in *Macracis monhystera* (Rauther, 1918) and *Syphacia obvelata,* (Vogel, 1925). In the latter organism (and in many other oxyurids) the ova develop their shell while in the oviduct and Vogel was of the opinion that the oviduct should be regarded as an egg former and possibly a "shell gland". In the majority of nemas (including *Macracis, Spironoura* and *Rhabditis*) shell formation is first visible in the proximal part of the uterus and the oviduct cannot be regarded as a zone of egg shell deposition.

(c) *Uterus.* The greater part of the uterus has squamous epithelium covered by a muscle layer of circular and oblique fibers, the development of which varies both in localized regions of a given form and in corresponding regions in diverse nemic groups. The distal part of the uterus commonly functions as a seminal receptacle or fertilization chamber even though it is not externally set off from the remainder. In *Ascaris* the epithelium of this region consists largely of elongated, tufted cells which were regarded by Leuckart as nurse cells to the spermatozoa. Romieu (1911), Von Kemnitz (1912) and Romeis (1913) found that these cells exercise a phagocytic activity on unused aging spermatozoa. Musso (1930) has shown that the inter-locking branched muscle fibers of the seminal receptacle rather suddenly give place to more regularly arranged circular muscle fibers of the uterus proper (Fig. 121 C). In other nemas such as *Spironoura* (Mackin, 1936) the uterine musculature is obviously spiral (119 A, 122 D) rather than circular though individual muscle cells do not reach completely around the uterus. Peristaltic contraction waves passing along the spiral serve to carry the egg along without the aid of longitudinal muscle fibers. The uterus of *Enterobius* (Fig. 122 BB-CC) has a web-like musculature. Sometimes, as in *Ascaris,* (Musso, 1930) longitudinal connective tissue fibers are apparent in the external membranes which cover the uterus, oviduct, and ovary, as well as other organs bordering on the body cavity. Concerning the histology of the uterine musculature there is little to be said. Plenk (1924) erroneously attributed striation to it as to all other musculature in *Ascaris.* Actually, the individual muscle cells seem to be either platymyarian, with the fiber layer next to the epithelium, or circomyarian.

In *Ascaris* the epithelium of the uterus proper, is composed of five to six sided, low or high cuboidal or irregular epithelial cells. Van Beneden (1883) and Martini (1916) attributed a bacillary layer to the uterine epithelium in *Ascaris* and *Oxyuris* respectively but other workers have been unable to confirm such findings in these or other nemas. In *Ascaris lumbricoides* Musso found the majority of uterine epithelium cells binucleate (Fig. 121 A-B) in all specimens, other nuclear numbers being three, four, and one in order of occurrence, four nuclei being found only in comparatively young females; nuclear division was always found to be by amitosis and cell division often unequal. In many parasitic nemas such as *Ancylostoma duodenale* (Looss, 1905), *Macracis monhystera* (Rauther, 1918) and *Cephalobellus papilliger* (Chitw. & Chitw., 1933) and in most free-living nemas the chief part of the uterus has no distinct musculature. A uni-nucleate simple squamous epithelium is the rule. In *Ancylostoma duodenale,*

Looss reported the uterine wall to be composed of only two cell rows. Possible cell limitation (oligocyty) in the uterine epithelium of other meromyarian nemas has not been investigated.

When an undivided portion of the uterus forms an egg pouch, as in *Ascaris lumbricoides* and *Abbreviata poicilometra* (Fig. 118 LL-MM), the epithelium and musculature are not particularly modified but when an ovejector is formed the epithelium usually becomes thicker and the musculature more highly developed.

(d) *Vagina.* The true vagina (vagina vera) of nemas is always distinctly recognizable histologically though it may be difficult, if not impossible to limit in gross study. Regardless of degree of development, the vagina is lined with a distinct, well-developed cuticular layer continuous with the external cuticle but differing from this cuticle in being composed of a single layer. Its epithelium is composed of relatively few large cells usually quite distinct in appearance from the epithelium of the uterus or vagina uterina. Because of its exceedingly narrow, slit-like form the vagina of free-living nemas has not been investigated histologically, but in *Trilobus* (Fig. 120 J) we judge it to have four cells situated at the junction of vagina and uterus. With elongation of the vagina, the structure becomes more obvious and Rauther (1918) was able to identify a single group of four nuclei at its proximal end corresponding to four longitudinally oriented cells forming the entire vagina vera of *Macracis monhystera* (Fig. 122 I). Chitwood and Chitwood (1933) have found that eight cells, arranged in two tandem groups of four each, compose the vagina of *Cephalobellus* and *Hystrignathus rigidus* (Fig. 122 X) and *Spironoura affine.* This arrangement is in accord with the general tetragonal symmetry of the vagina of oxyuroids, the nuclear number eight being preserved in *Atractis* though no other semblance of symmetry occurs in the vagina of that form. In ascaridids and spiruroids there is a distinct cell limitation in the vagina, but even so symmetry and cell constancy are apparently lost; Musso records the vagina of *Ascaris* as composed of 10 to 12 longitudinal rows of polyhedral squamous epithelial cells (Fig. 121 E-F), while Rauther (1918) found the vaginal epithelium of *Trichuris* to be composed of an unlimited number of cells (Fig. 122 T-U).

The musculature of the vagina is continuous with that of the uterus and is of the same general type but may form a much thicker layer; in some groups the vaginal muscle may be several cells in thickness and cause the vagina to have a distinctly laminated appearance. The cell bodies of the muscle cells are commonly pressed out of the wall assuming a bladder-like shape. These protoplasmic masses have commonly been termed vaginal glands. Structures which cannot be so interpreted but may actually be glands have been described in *Allodapa numidica* by Seurat (1914), in *Thoracostoma strasseni* by Türk (1903), seen also in *Cylicolaimus magnus* by Jägerskiöld (1901) and *Halichoanolaimus longicauda* by Ditlevsen (1919) (Fig. 123). Their function has not been ascertained.

In addition to the ordinary vaginal muscles there may be a large sphincter muscle at or near the vulvar opening; such a muscle (Fig. 122 B) is present in *Spironoura* and consists of one large muscle cell whose fibers are on its external surface. A similar cell has been observed in some free-living nemas (*Chromadora* sp. and *Sabatieria hilarula* Figs. 120 B & E). Dilator muscles of the vulva have already been mentioned. In most cases transverse or modified somatic musculature serves to dilate the vulva.

COMPARATIVE MORPHOLOGY

FREE-LIVING PHASMIDIANS. The female reproductive system exists in its most simplified condition in the Rhabditoidea. The vagina is always simple, more or less transverse to the body axis, flattened, without special musculature and the uteri are likewise unmodified. Amphidelphic reflexed genital tubes are the rule, but monodelphic forms are also common. Studies on this group include Maupas (1900) on *Rhabditis dolichura,* Krüger (1913) on *R. aspera* (*R. aberrans*), Schleip on *Rhabdias bufonis,* Goodey (1935) on *Brevibucca saprophaga,* Thorne (1937) on *Cephalobus persegnis* and Steiner (1937) on *Eucephalobus teres.* In amphidelphic forms the vulva is always more or less equatorial in position (about 40-65 percent) and oddly enough this position is retained in the characteristically monodelphic family Cephalobidae. The vulva is preanal in

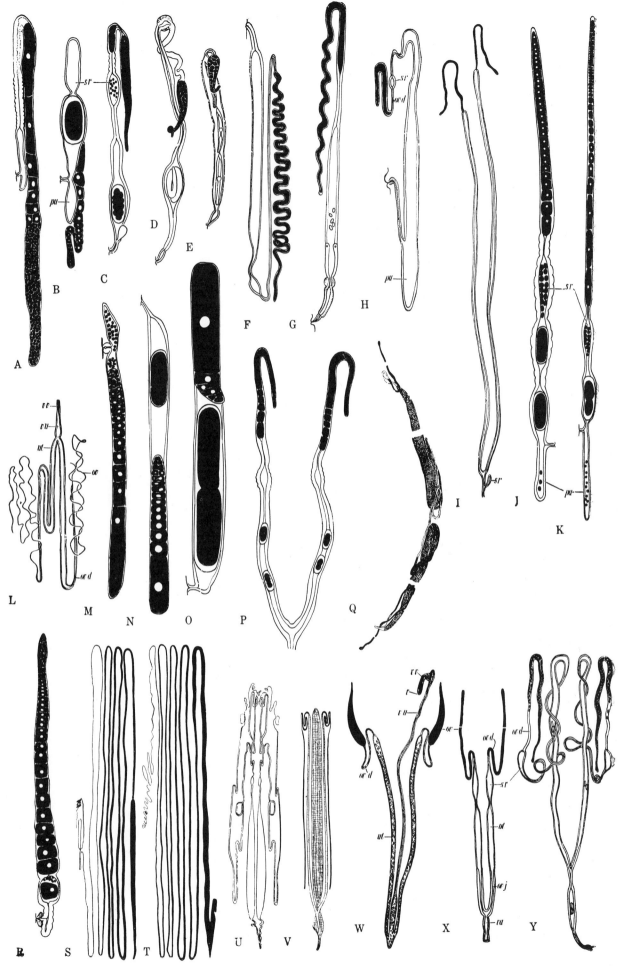

FIG. 117.

142

other monodelphic forms such as *Brevibucca saprophaga* (Fig. 117 C), *Rhabditis lambdiensis* and *Longibucca*. Apparently reduction from the amphidelphic to the monodelphic condition is a sporadic happening throughout the Nematoda, and in many other groups, such as the Rhabditidae the number of ovaries has no bearing on the degree of relationship and in that particular instance is not even a sound generic character as may be readily ascertained by comparing the males of the various monodelphic rhabditids (*R. lambdiensis*, *R. monhystera*, *R. spiculigera*, *R. ocypodis*). Prior to its use as a generic character at least one correlated, preferably non-female, character might be found. In the case of the Cephalobidae there is adequate evidence in the way of stomatal and male characters to indicate that ovarial reduction is a general character.

The uteri are usually thin-walled, composed of a low squamous epithelium in gravid females, while in quite young females the lumen may be minute, the epithelium thick. Monodelphic forms have a distinct postuterine sac (Fig. 117 B) in the Cephalobidae but not always in other families. A distinct seminal receptacle or spermatheca may or may not be present. In *Rhabditis strongyloides* (Fig. 116 A) there is a pair of seminal receptacles near the vulva; in *Brevibucca* (Fig. 117 C) Goodey described a pair of uninucleate uterine glands. The distal end of the uterus functions as a seminal receptacle and fertilization chamber in most rhabditoids (Maupas, 1900, Krüger, 1913, Goodey, 1924) even though it is not ordinarily distinctly set apart. Thorne (1937) has described a well defined spermatheca at the ovarial end of the uterus in *Cephalobus persegnis* (Fig. 117 B).

The degree of development of the oviduct is seemingly without significance in the Rhabditoidea since the existence of such an organ in some forms may be dependent on the age of the specimen. Within the superfamily diversity extends from no oviduct in *Rhabditis lambdiensis* and *R. strongyloides* to a well defined tube in *Rhabdias sphaerocephala*.

Diversity of ovarial form in the Rhabditoidea is limited chiefly to length; the ovaries never become coiled or greatly elongated organs as often occurs in the more highly evolved parasitic groups. Maupas (1900) and Krüger (1913) have demonstrated that in *Rhabditis dolichura* and *R. aspera* v. *aberrans* the ovary produces a limited number of spermatozoa prior to the appearance of oocytes. This type of reproduction is termed syngonism (Cobb, 1916) and is seemingly widespread among free-living nemas. In the hermaphroditic *Rhabdias*, Schleip (1911) found the syngone to produce spermatozoa and ova alternately.

The Tylenchoidea exhibit the same fundamental simplicities in the reproductive system as do the Rhabditoidea, but in this group monodelphic forms are much more common and are generally placed in separate genera from the amphidelphic types. The gonoducts may be either outstretched (*Ditylenchus dipsaci* (Fig. 177 K) or flexed (*Anguina tritici*) or doubly flexed (*Hexatylus intermedius*.). This group also contains the only prodelphic didelphic free-living nemas, the highly specialized root parasites of the genus *Heterodera* (Fig. 117 P) and in members of the genus *Heterodera* the genital primordium is originally equatorial as in other nemas. Some of the oddities (ex. *Sphaerularia*, *Allantonema*) of the arthropod parasites comprising the Allantonematidae have been previously mentioned. Zur Strassen (1892), Leuckart (1887), Goodey (1930) and Bovien (1932) have made careful

Fig. 117. DIAGRAMS OF FEMALE REPRODUCTIVE SYSTEMS.

A—*Panagrolaimus heterocheilus.* B—*Cephalobus persegnis.* C— *Brevibucca saprophaga.* D—*Labidurus gulosa.* E—*Atractis dactyluris.* F—*Trichuris suis.* G—*Heligmosomum laeve.* H—*Camallanus lacustris.* I—*Tetrameres fissispina.* J—*Ditylenchus dipsaci.* K—*Aphelenchoides ritzema-bosi.* L—*Ascaris lumbricoides;* M—*Oxystomina cylindraticauda.* N—*Halanonchus macramphidum.* O—*Cryptonchus nudus.* P—*Meloidogyne hapla.* Q—*Heliconema anguillula.* R—*Theristus sentiens.* S-T—*Ascaris lumbricoides* (S, male; T, female). U—*Zoniolaimus setifera.* V—*Oxyuris equi.* W—*Dermatoxys veligera.* X—*Kiluluma brevivaginata.* Y—*Hedruris armata.*

A, after Steiner, 1935, Proc. Helm. Soc. Wash. v. 2 (2); B, after Thorne, 1937, Proc. Helm. Soc. Wash. v. 4 (1); C, after Goodey, 1935, J. Helminth. v. 13 (4); D-E, after Thapar, 1925, J. Helminth. v. 3 (3-4); F. after Rauther, 1918, Zool. Jahrb. Abt. Morph., v. 40; G, after Seurat, 1915, Bull. Sc. France & Belg. 7. s., v. 48; H, after Toernquist, 1931, Goeteborgs Kungl. Vetensk. s. B, v. 2 (3); W, after Seurat, 1920, Hist. Nat. Nem. Berberie; Q, after Yamaguti, 1935, Jap. J. Zool. v. 6 (2); R, after Cobb, 1914, Tr. Am. Micr. Soc. v. 33; S-T, after Musso, 1930, Ztschr. Wiss. Zool. v. 137 (2); U, after Cobb, 1898, Dept. Agric. N. S. Wales, Misc. Publ. No. 215; V, after Martini, 1916, Ztschr. Wiss. Zool. v. 116; X, after Thapar, 1925, J. Helminth., v. 3 2; Y, after Perrier, 1871, Compt. Rend. Acad. Sc. v 72 (12). Remainder original.

studies of the reproductive systems of *Bradynema*, *Allantonema*, *Sphaerularia*, *Tripius*, *Tylenchinema* and *Scatonema* (Figs. 115 A-B, E & I-K). In all instances the reproductive system is highly developed, and often the ovary becomes coiled and the uterus enormous. Unlike most parasites these forms do not develop specialized mechanisms for the ejection of eggs but instead there is often degeneration and sometimes atrophy of the vulva and vagina.

FREE-LIVING APHASMIDIANS. Since most modifications of the reproductive system are related to increase in egg production and are associated with life habits rather than relationships, it seems best to cover the remainder of the free-living nemas before proceeding to the parasites. As previously noted, Filipjev (1918, 1934) has attached a great deal of weight to the position of the uteri and ovaries. It is true that in certain large groups the female genital tubes tend to be extended while in most nemas there is a flexure at the junction of ovary and oviduct regions in the gonoduct which caused it to be described as reflexed. Filipjev (1928, 1934) separated the Monhysterata, (equivalent to our Monhysteroidea and Axonolaimoidea and Microlaimidae) from the Chromadorata (equivalent to our Chromadoroidea, Desmodoroidea, and Plectoidea) on this basis, the female genital tubes being outstretched in the former group while they are ''reflexed'', or more properly speaking, *flexed* in the latter group. In so far as this is a tendency which is correlated with other characters, we recognize the soundness of its use as a taxonomic character. However, it must be stated that there is nothing so very fundamental in the differences, and other characteristics must also be considered. In addition, separations on the basis of the female reproductive system are not absolute since undeniable exceptions to the rule are known. Thus, the Comesomatidae, which otherwise show so many characters in common with the Axonolaimidae (Monhysterina) on the one side and with the Cyatholaimidae (Chromadorina) on the other side, present embarrassing exceptions. The genera *Dorylaimopsis* and *Laimella* with outstretched gonads must be ranged next to *Mesonchium* with flexed gonads, and *Comesoma minimum* has an anterior flexed and posterior outstretched gonad, while other members of the genus *Comesoma* supposedly have two outstretched gonads. Also, in the Microlaimidae (Chromadoroidea), *Microlaimus* and *Bolbolaimus* with outstretched gonads must be ranged next to *Achromadora*, *Ethmolaimus*, *Prodesmodora*, and *Statenia* with flexed gonads. *Halanonchus* (Fig. 117 N) is an example of a monhysterid with a flexed gonad. Similarly, the tendency of the members of the subfamilies Monhysterinae and Sphaerolaiminae to exhibit ovarial reduction is well recognized but no one would today exclude amphidelphic genera from these groups providing other characters conformed. In all nemic groups one notes the appearance of monodelphy and among free-living nemas, as also among parasites, this usually means posterior shifting of the vulva because it is the anterior ovary that persists. Exceptions to this rule are *Halanonchus macramphidum* (Monhysteridae), *Oxystomina cylindraticauda* (Enoplidae), *Dorylaimus monhystera* (Dorylaimidae) in all of which the vulva is shifted anteriad, the posterior ovary having persisted. In parasites, exceptions (Trichuroidea, Dioctophyma, Soboliphyme) to the rule seem to be as numerous as conformists but here there is probably some functional or fundamental reason as yet not understood.

If there is any single characteristic which might serve to contrast free-living aphasmidians with free-living phasmidians in the female reproductive system, it is a tendency toward the formation of larger, fewer eggs, often accompanied by marked relative shortening of the growth zone. The very large size of one or two maturing oogonia may give cause for a noticeable attenuation of the ovary, (Fig. 119 F). There are exceptional aphasmidians which produce numerous relatively small eggs (*Metoncholaimus pristiuris*, *Actinolaimus* sp. Fig. 120 G-H) and in such forms the ovaries are much more elongated, due to a more extensive growth zone.

The semi-diagrammatic illustrations speak largely for themselves. We need only call attention to the apparent absence of an oviduct in *Chromadora* (Fig. 120 E) and *Theristus* (Fig. 117 R), the individual and distinct oviduct of *Anaplectus*, *Axonolaimus*, *Sabatieria*, etc. and the modified oviducts of *Mononchus*, *Trilobus*, and *Actinolaimus*. In the latter forms there is marked flattening of the oviduct against the ovary (Fig. 120 H) and eventually the oviduct disappears as an entity, the ovarial epithelium being separated from the germ cells on one side so that the mature ovum may slip between the germ cells and the epithelium. This feature was

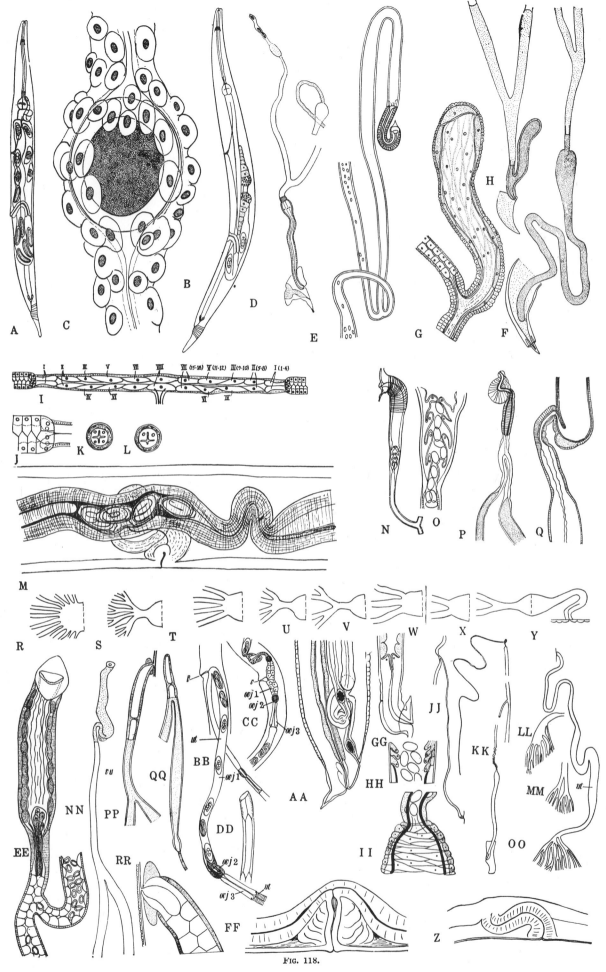

FIG. 118.

first called attention to by Jägerskiöld (1901). The varied modifications serving the function of seminal receptacle or spermatheca are interesting from the standpoint that they may usually be regarded as the ovarial end of the uterus. Uterine outpocketing or separation from the oviduct in axonolaims and comesomes (Figs. 116 L-N) is sufficiently spectacular to lend considerable weight to hypothesized comesomatid-axonolaimid relationships.

Vaginal cuticle and muscular developments in free-living aphasmidians are often quite characteristic of species and genera but are, unfortunately, not easy to describe. Laminated vaginal musculature occurs very rarely, and in the genus *Trilobus* (Fig. 120 J) it is used as a specific character, being absent in many species. Often the vaginal cuticle appears in optical section of totomounts as two paired sclerotized rods (Fig. 119 I). Mononchs and dorylaims are particularly notable in this respect.

There is a very peculiar system of organs connected with the female reproductive system of oncholaimids first described by de Man (1886) and later studied by zur Strassen (1894) and Cobb (1930) and named the demanian system by Cobb. In description we can not do better than quote Cobb.

"*Demanian Vessels:* In adult female nemas (Oncholaims) a complicated double system of efferent tubes; connecting (1), with the middle or posterior part of the intestine through an *osmosium*, and (2), with the uterus, (or uteri); these two efferents being confluent at a special glandular 'gateway', the *uvette*, and emptying thence backward and outward, through one or two ducts having more or less moniliform affluent glands. Normally, the ducts lead to exit pores in the body wall, usually lateral, one or more on each side, near the base of the tail." Cobb decided, on the basis of a systematic exploration of the theoretic possibilities, that the course of flow in the demanian system is from the intestine and uterus, through the moniliform glands to the exterior. Cobb associated this system with the formation of a "gelatin like" mass in the uterine lumen of gravid females which he believed flows out through the posterior pores and is deposited as a "sticky, non-water-soluble, nearly colorless secretion possibly utlized during agglomeration and copulation and also presumably to preserve the batches of eggs after deposition and segmentation." The writers (1938) observed orange pigmentation in the demanian exit ducts of *Oncholaimus oxyuris* which was similar to that found in the olivaceous sphaeroids of the same species.

PARASITIC NEMAS. Seurat recognized the fact that the female reproductive system increases in complexity with the degree of parasitism. He noted tendencies in this direction among rhabditids (ex. *Rhabdias*) and tyenchoids (ex. *Sphaerularia*). There is everywhere a tendency toward increased egg production, increased length and coiling of ovary, oviduct and uteri, and formation of complicated muscular ovejectors, but each parasitic group seems to have followed its own course of development.

Strongylina. Members of the Strongylina have a basically equatorial or slightly post-equatorial vulva joined to amphidelphic flexed gonoducts by means of a short transverse vagina and paired opposed ovejectors. Such a condition is approximately realized in *Dictyocaulus filaria* (Fig. 116 Q) and *Haemonchus contortus* (Fig. 116 T). Critical studies of the female organs of strongylins are due to the investigations of Augstein (1894), Looss (1905), Maupas and Seurat (1912), and Seurat (1920). In addition Ransom (1911), Seurat (1915), Veglia (1916), Thapar (1925), and Alicata (1935) have made contributions.

Varied nomenclatures applied to the ovejectors of strongylins have already been mentioned. The vagina vera in this group is always a short, more or less transverse flattened tube. As noted by Looss, the ectodermal cuticle stops suddenly at the junction of vagina and ovejector (Fig. 118 I). Seurat (1920) was apparently under the impression that the entire ovejector was of ectodermal origin because it was lined with "cuticle". However, the character of this cuticle is quite different from that of the vagina vera and external body covering. According to the observations of Alicata (1935) the ovejectors are formed from the central part of the genital primordium in *Hyostrongylus rubidus* while the vagina is an ectodermal invagination. With Looss (1905) we regard the ovejectors of strongylins as of totally uterine origin. There seems to be no evidence of glandular activity in any part of the ovejector so the term glazing (or varnishing) gland is a misnomer. Each ovejector is composed of two (Fig. 118 I) or three parts (Fig. 118 CC). Its function in *Ancylostoma duodenale* was described by Looss as follows:

"The function of the pars haustrix (infundibulum) is evident from its structure. The contraction of the spiral muscles causes it to shorten, its inner cavity at the same time widening. The decrease of pressure thus brought about cannot be compensated for from behind, this being prevented by the funnel-shaped ends of the most anterior cells (I). The diminution of pressure therefore tells on what lies in front, sucking in the egg located nearest the pars haustrix. This egg is prevented from being pushed back into the uterus by a contraction of the anterior circular musculature of the pars haustrix, which, during the enlargement, has been passively extended. So soon as the egg has passed the funnel made by cells I, its way out is unimpeded. The return of the spiral musculature of the pars haustrix to its normal condition drives it into the pars ejectrix and it is ejected by the successive contractions of the muscles of the knobs. In passing from one knob to another, the egg follows a zig-zag course, but cannot escape backward, being prevented from doing so by the prolongations of the epithelial cells directed backward. In this way the egg is forcibly propelled onward into the vagina; from this it may be ejected either by pressure of the eggs forced on behind it or through the action of the vulvar muscles."

In *Ancylostoma* and several other members of the Strongylina each ovejector consists of only two parts. The uterine end, which Looss termed the *pars haustrix* (Fig. 118 I) consists of two sets of four cells (I and II). This part apparently corresponds to the two parts of the ovejector of *Nematodirus mauritanicus* (Fig. 118 BB) described by Maupas and Seurat (1912) as *trunk* and *sphincter* (later renamed *sphincter* and *glazing gland*). Since, as Looss describes them, the four cells at the uterine end act as a suction funnel, we suggest the name *infundibulum* for them; and since the second group of cells act as a sphincter, the name is retained. The proximal part of each ovejector in *Ancylostoma* consists of four sets of two cells (III-VI) and one set of four cells (VII), making a total of 12 cells; in addition the two vaginal ends of the ovejectors are united with each other and with the vagina by four cells. For the uterine ends of the ovejector Looss' term *ejectrix* is retained. From the observations by Looss, Maupas and Seurat, and the writers examinations of *Oesophagostomum dentatum* and *Kalicephalus* sp., we judge cell constancy to be the rule for ovejectors in the Strongylina. The paired ovejectors with the union piece in *Ancylostoma* include 28 cells, the two sphincters and the two infundibula, four each or 16 cells making a total of 44. Seurat lists the ejectors as including 32, the sphincters two each, and the infundibula four each, also making a total of 44. We have found the same total in both nemas studied by us. Differences in appearance as suggested by Seurat are minor modifications; he found the same cellular arrangement in *Nematodirus mauritanicus* and *N. filicollis* but tremendous cellular hypertrophy in the ejector of the latter species. Slight modifications of cellular arrangement do exist, however, for we found tetraradiate symmetry (Fig. 122 P) throughout the ovejectors of *Oesophagostomum* while Looss found sets of two cells in the ejectrix of *Ancylostoma* and Seurat a set of two cells in the sphincters of *Nematodirus*.

In no instance have the cells of the vagina vera been identified separately from the ectodermal epithelium. The musculature of the ovejectors is the same as the type which

FIG. 118. FEMALE REPRODUCTIVE SYSTEM.

A-B—*Tachygonetria vivipara* (viviparous and oviparous females). C—*Rhigonema infectum* (Distal region of uterus). D—*Tetrameres nouveli.* E—*Acuaria anthuris* (Vagina and uteri). F—*Tetrameres inermis* (Vagina and uteri). G-H—*T. fissispina* (Vagina and uteri showing "Bursa copulatrix" or seminal receptacle). I-L—*Ancylostoma duodenale* (I—Diagram of ovejectors; J—Uterine end of ovejector; K—Cross section at level of II; L—Cross section at level of III). M—*Pseudomermis vanderlindei* (Vaginal region). N-O—*Spirura gastrophila* (Ovejector). P—*Habronema microstoma* (Ovejector). Q—*Habronema muscae* (Ovejector). R—*Physaloptera turgida.* S—*P. capensis.* T—*P. tumefaciens.* U—*P. paradoxa.* V—*Abbreviata abbreviata.* W—*Physaloptera cebi.* X—*P. retusa.* (R-X—diagrammatic representations of vagina showing uterine branching). Z—*Hydromermis leptoposthia* (Vagina). AA—*Oesophagostomum brevicaudum.* BB—*Nematodirus mauritanicus.* CC-DD—*N. filicollis* (DD, enlarged part of ovj. 1). EE—*Protospirura numidica.* FF—*Mesomermis bursata* (Vagina). GG—*Gongylonema mucronatum* (Vagina) HH-II—*G. scutatum* (HH, Junction of vagina vera and vagina uterina; II, Junction of constricted and dilated parts of vagina uterina). JJ—*G. mucronatum* (Vagina). KK—*G. scutatum* (Vagina). LL, MM & OO—*Abbreviata poicilometra* (Variations in uterine origin). NN & PP-RR—*Spirocerca lupi* (NN, adult; PP-RR, fourth stage larva). A-B, D-H, BB-EE, GG-KK, NN and PP-RR after Seurat, 1920, Hist. Nat. Nem. Berberie; M, after Steiner, 1937, Skrjabin Jubilee; R-X, after Schulz, 1927, Samml. Helminth. Arb. Prof. K. I. Skrjabin gewidmet; Z-FF, after Steiner, 1929, Zool. Jahrb., Abt. Syst. v. 57; AA, after Schwartz and Alicata, 1930, J. Agric. Res. v. 40 (6); LL-MM & OO, after Sandground, 1936, Bull. Mus. Comp. Anat. Harvard, v. 79 (6). Remainder original.

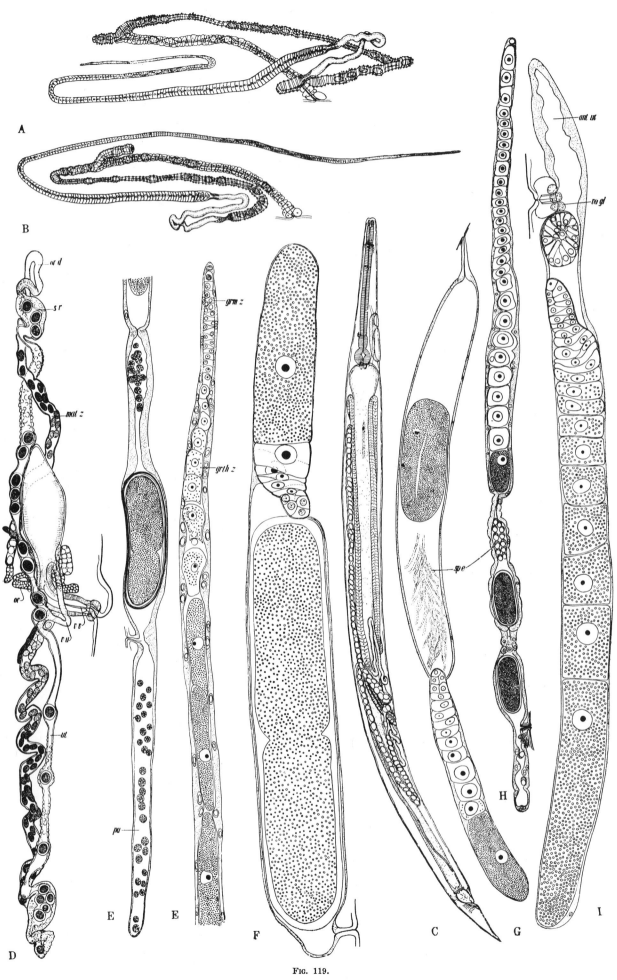

covers the uterus of nemas, that is, circular, oblique, and having spiral anastomosing fibers. When concentrated in a specific area they form a conspicuous sphincter.

The primary amphidelphic form is preserved chiefly in the families Ancylostomatidae and Trichostrongylidae (Fig. 116 T) in which the vulva tends to be only a short distance post-equatorial. In the majority of other families of the suborder Strongylina the vulva is preanal in position. In such cases the paired ovejectors may separate in an amphidelphic manner before extending anteriad and thus becoming prodelphic as in *Oesophagostomum* or they may originate in a prodelphic manner as in *Kiluluma* (Fig. 117 X) and *Zoniolaimus* (Fig. 117 U). When there is a reduction to one ovary as in heligmosomes (*Heligmosomum laeve*) there is a single ovejector having parts identical with the paired structure (Fig. 117 G).

The unspecialized sector of each uterus in *Oesophagostomum* consists of two rows of cells (Fig. 122 S) exhibiting possible glandular activity and a distinct bacillary layer. At its connection with the *infundibulum* the number of cells in a circumference is increased from two to six (Fig. 122 R). In practically all cases the ovaries are greatly elongated due to an extensive growth zone.

Ascaridina. Within the Ascaridina, vulva position varies from anterior to the base of the esophagus (*Protrellus künckeli*, Fig. 116 K) to practically anal (preanal in *Aorurus agile*, actually into the rectum in *Rondonia rondoni* 108 L). As in the Strongylina, there is usually a well-developed, muscular ovejector, particularly in the Oxyuroidea. The vagina vera is always elongated, usually tubular in the Oxyuroidea; and with the unpaired vagina uterina composes the ovejector. The internal structure of the vagina vera has been described.

For all the marked specialization in ovejector formation, the superfamily Oxyuroidea contains some representatives which closely approach the free-living nemas. In practically all parasitic nemas the ovaries are filiform with an elongated growth zone. This is true of the majority of oxyuroids and ascaroids as well, but in the families Atractidae and Oxyuridae there are several examples with short tapering ovaries, and these forms, but for the vagina, might easily be mistaken for rhabditoids. *Tachygonetria vivipara* (Pharyngodoninae) typifies the amphidelphic organization while *Heth juli* (Ransomnematinae) typifies the prodelphic, didelphic, and *Labidurus gulosa* (Labidurinae) and *Atractis dactyluris* the prodelphic monodelphic condition. Among the oxyurids vulvar position is extremely variable and the uteri are often parallel. In *Dermatoxys veligera* (Fig. 117 W), *Oxyuris equi* (Fig. 117 V) and *Syphacia obvelata* the vulva is shifted anteriad while the greatly elongated vagina uterina connects at the posterior end with the anteriorly directed uteri. Seurat (1920) interpreted this formation as modified prodelphy but Vogel (1925) has shown that on the basis of young specimens (Fig. 159 K), *Syphacia* and *Dermatoxys* must be considered primarily amphidelphic like *Tachygonetria vivipara*. Similarly *Protrellus künckeli* is an amphidelphic thelastomatid example of vulva shifting, with elongation of vagina uterina; *Cephalobellus papilliger* and *Hystrignathus rigidus* exemplify the standard condition. A well marked vagina uterina (ovejector) is characteristic of all thelastomatids and oxyurids but not of atractids. The Rhigonematidae are a group apart, in the Oxyuroidea, in female structures as in other characters. They are more like the Ascaridoidea in that only the vagina vera takes part in the formation of the heavily walled ovejector. In *Rhigonema infectum* (Fig. 119 D) the vagina vera is connected with a massive sac and through this with the amphidelphic uteri. Artigas (1930) found this chamber absent in other species and formed for them a new genus, *Dudekemia*. In *R. infectum* the ovarial end of each uterus is set apart and serves as a spermatheca and fertilization chamber; we therefore presume the larger chamber to be an egg pouch.

Within the Ascaridoidea no forms are known that approach the simplicity of *Tachygonetria vivipara*; the ovaries are always greatly elongated, commonly coiled. Within the Cosmocercidae, Kathlaniidae* and Heterakidae, most of the forms are primarily amphidelphic, with the vulva usually more or

less equatorial, while the Ascarididae contain forms which are chiefly opisthodelphic, with vulva shifted anteriad. The ovejector is seldom as prominent an organ in this superfamily. In so far as information is available the vagina of cosmocercids (Olsen, 1938) and kathlaniids (Mackin, 1936) is entirely vagina vera, i. e. the ectodermal part extends to the separation of the two uteri.

In *Heterakis gallinarum* (Heterakidae) the heavily muscled ovejector (Fig. 116 R) is vagina vera but in addition there is an elongated vagina uterina which extends posteriorly and is reflexed anteriad before connecting with the amphidelphic uteri. For such a (relatively) non-muscular modification of the vagina uterina we feel that Seurats term ''trompe'' or trunk may properly be preserved. This same term would apply in the case of *Rhigonema* (Fig. 119 D). Other heterakids including *Ascaridia lineata* (Ackert, 1931), *Allodapa numidica* (Seurat, 1915) and *Maupasina weissi†* (Seurat 1931) have a short ovejector composed of vagina vera and a long trunk composed of vagina uterina. The two former species are amphidelphic while the latter is prodelphic and like *Rhigonema*, possesses a large chamber at the end of the ovejector, Seurat named this structure a *bursa copulatrix* considering that it might function as a temporary storage place for spermatozoa. Such a structure also occurs in some spiruroids.

Ascaridids are opisthodelphic as represented by *Ascaris lumbricoides* (Fig. 117 L); the vulva is usually situated pre-equatorially, the vagina directed posteriad, followed by trunk and posteriorly directed parallel uteri. As in heterakids, the vagina uterina (trunk) is not conspicuously muscular and in many forms there is a dilation at its distal end which might serve equally well as a temporary spermatheca (*bursa copulatrix*) or as a temporary egg chamber. Polydelphy makes its appearance in the genus so named, *Polydelphis*, in which there are four parallel uteri arising at the end of the trunk while the twin genus *Hexametra* is identical except that there are six uteri.

Camallanina. In this suborder the reproductive system is chiefly amphidelphic, the vulva more or less equatorial. Within the Dracunculoidea, the vagina is never heavily muscled, nor are the uterine ovejectors developed. In *Philometra* and *Dracunculus* the vagina is functional only in young females, becoming rudimentary with gravidity; in *Micropleura* it is retained but not especially developed. The uteri of dracunculoids are great sacs dilated with embryos and larvae filling practically the entire body cavity. Camallanoids, on the contrary, retain a functional muscular ovejector formed by the vagina vera. In addition there may be a pair of short uterine ovejectors in *Cucullanus* (116 H) or an elongate tubular trunk in *Camallanus* (Fig. 117 H). The family Camallanidae is monodelphic, prodelphic but a postuterine sac is generally present. The family Cucullanidae, while developmentally amphidelphic, contains some transitions toward prodelphy.

Spirurina. In this suborder there is a well developed muscular ovejector, the very heavily muscled part of which is properly considered a vagina vera. Opisthodelphic, amphidelphic and prodelphic forms are all represented commonly within the group. Of these the opisthodelphic forms are chiefly to be found in the Filarioidea, Thelaziidae and Physalopteridae in all of which the vulva is predominantly anterior in position. However, even in these groups exceptional forms are amphidelphic (*Desmidocercella* Filarioidea), prodelphic (*Physocephalus*, Thelaziidae; *Proleptus*, Physalopteridae). The remaining families, Spiruridae, Acuariidae and Gnathostomatidae might be termed primarily amphidelphic or prodelphic but there is too much variation except in the Gnathostomatidae for any generalization. The vagina vera seldom if ever reaches the bifurcation of the uteri so that there is always a trunk. This trunk has high epithelial cells which extend to a variable degree beyond bifurcation of the uteri. The various arrangements have been studied by Seurat (1920) in *Acuaria laticeps, Tetrameres fissispina, Spirura gastrophila, Protospirura numidica* and *Abbreviata abbreviata* (Fig. 118). The Physalopteridae present more taxonomic difficulty than any other group because of their tendency toward polydelphy. Ortlepp (1922, 1937) divided the genus *Physaloptera* into four groups (Didelphys, Tridelphys, Tetradelphys and Polydelphys) on the basis of uterine and ovarial numbers. At the same time he attempted to follow out correlations according to the method of uterine origin, i. e.,

FIG. 119. FEMALE REPRODUCTIVE SYSTEM.

A-C—*Spironoura affine* (A, showing posterior uterine branch. B, anterior branch; C, entire female). D—*Rhigonema infectum*. E, *Aphelenchoides ritzema-bosi*. F—*Cryptonchus nudus*. G, *Halanonchus macramphidum*. H—*Ditylenchus dipsaci*. I, *Oxystomina cylindraticauda*.

A-C, after Mackin, 1936, Ill. Biol. Monographs v. 14 (3). Remainder original.

*The structure termed "shell gland" by Mackin (1936) is the oviduct.

†The uteri are parallel and directed anteriad, thereafter reflexed posteriad, the ovaries being situated in the caudal region. Seurat (1920) characterized this formation as opisthodelphic.

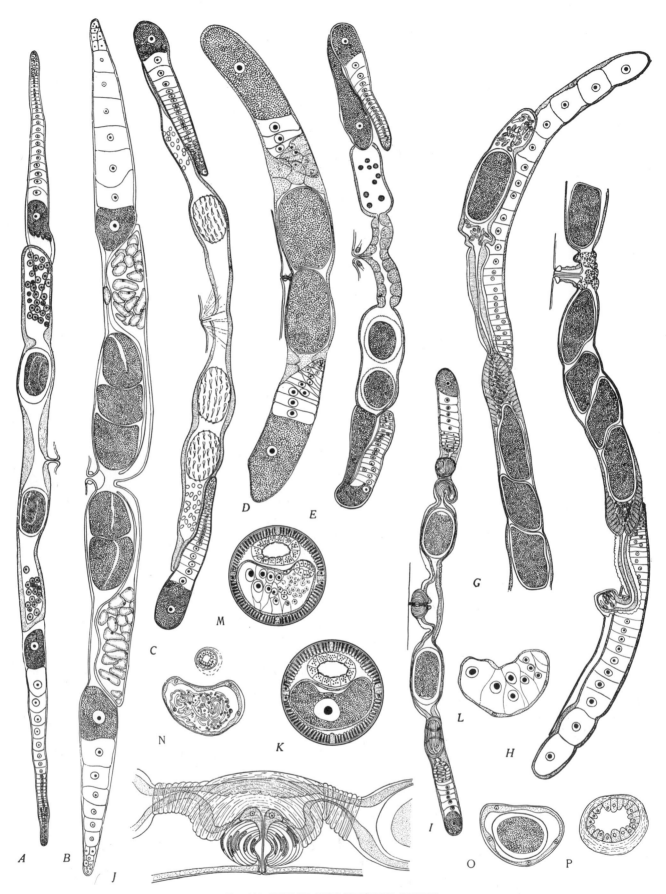

FIG. 120. FEMALE REPRODUCTIVE SYSTEM.

A—*Axonolaimus spinosus.* B—*Sabatieria hilarula.* C—*Anaplectus granulosus.* D—*Mononchus lacustris.* E—*Chromadora* sp., G-H—*Actinolaimus* sp., I-P—*Trilobus pellucidus* (I—female; J—Vaginal region; M—Germinal region of ovary; K—Distal or nearly ripe region of ovary; L—Growth zone; N—Seminal vesicle, with uterus above; O—Uterus in mid region; P—Uterus in vaginal region). Original.

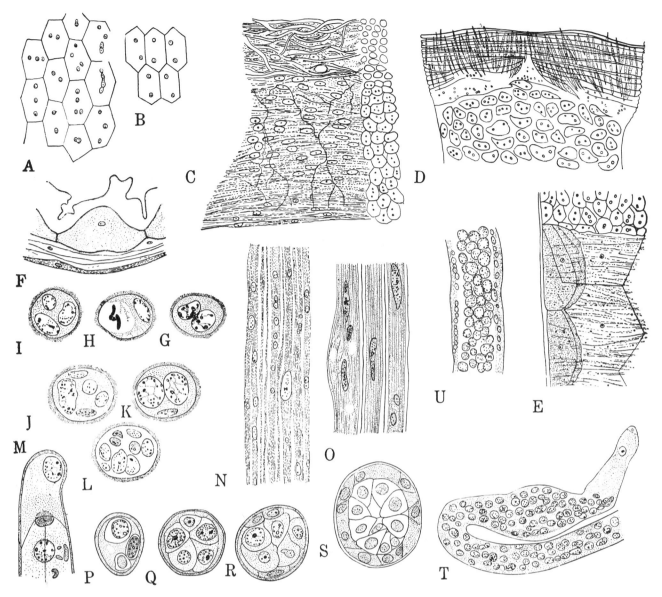

FIG. 121. *ASCARIS LUMBRICOIDES* SECTIONS.

A-B—Surface view of uterine epithelial cells. C—Junction of seminal vesicle and uterus. D—Junction of furrowed and smooth regions of oviduct. E—Junction of vagina uterina and vagina vera. F—Cross section of vagina vera. G-L—Cross sections of germinal region of testis. M—Surface view of blind end of testis. N—Surface view of epithelium in growth zone of ovary. O—Same near end of growth zone; P-S—Cross sections in germinal region of ovary. T—Totomount preparation of terminal region of ovary. U—Longitudinal section through germinal region of ovary. All after Musso, 1930, Ztschr. Wiss. Zool. v. 137 (2).

whether the uteri arose direct from the base of the egg chamber (trunk) (Fig. 118 X) or were connected to the egg chamber by a common trunk (Fig. 118 Y). In the didelphic, tetradelphic and polydelphic types the two possibilities both made their appearance. Schulz (1927) compared these data with cephalic characters and came to the conclusion that ovarial number must be considered a secondary character, and mode of origin of the uteri a tertiary character. Thus the old genus *Physaloptera* was divided into three genera on the basis of cephalic structures, each genus being subdivided on ovarial number. Apparently Sandground (1936) has substantiated Schulz in discarding mode of origin of the uteri as a character since he showed three modes of uterine origin in the species *Abbreviata poicilometra* (Fig. 118 LL-MM & OO).

Polydelphy outside the Physalopteridae is unusual, having been recorded only in *Elaeophora poeli* and *Tanqua tiara* (Fig. 116 U), Dipetalonematidae and Gnathostomatidae respectively. In the latter form one encounters polydelphy and amphidelphy associated, there being three anterior uteri and one posterior uterus all of which empty into a common egg chamber which is in turn connected with the vagina (Monnig, 1923).

Mermithoidea. The Mermithoidea are purposely taken up here, along with the vertebrate parasites because they exhibit the same characteristics of vaginal development as do the more generally considered forms. Steiner (1923, 1926, 1929) has described a remarkable series of stages in the formation of an elongate muscular vagina from the transverse slit characteristic of all dorylaims, as seen in *Mesomermis bursata* (Fig. 118 FF), enlargement as in *Bathymermis sphaerocephala* and elongation as in *Hydromermis leptoposthia* (Fig. 118 Z), to the complex type found in *Limnomermis euvaginata*. As in many other parasitic groups, Steiner (1923) found that there is elongation of the oviduct. Mermithoids are always amphidelphic, so far as known.

Trichuroidea-Dioctophymatoidea (Hologonia). The specific details of differentiation of these two groups from other vertebrate parasitic groups rests not only on reproductive system, but on hypodermis, coelomocytes and stylet formation. Just how significant is the altered type of germ cell origin, upon which Rauther (1930) based the Hologonia, we are not prepared to say. All members of the group are monodelphic and in all with the exception of the genera *Eustrongylides* and *Hystrichis* the vulva is anteriorly shifted. The muscular ovejector so far as known is of ectodermal origin.

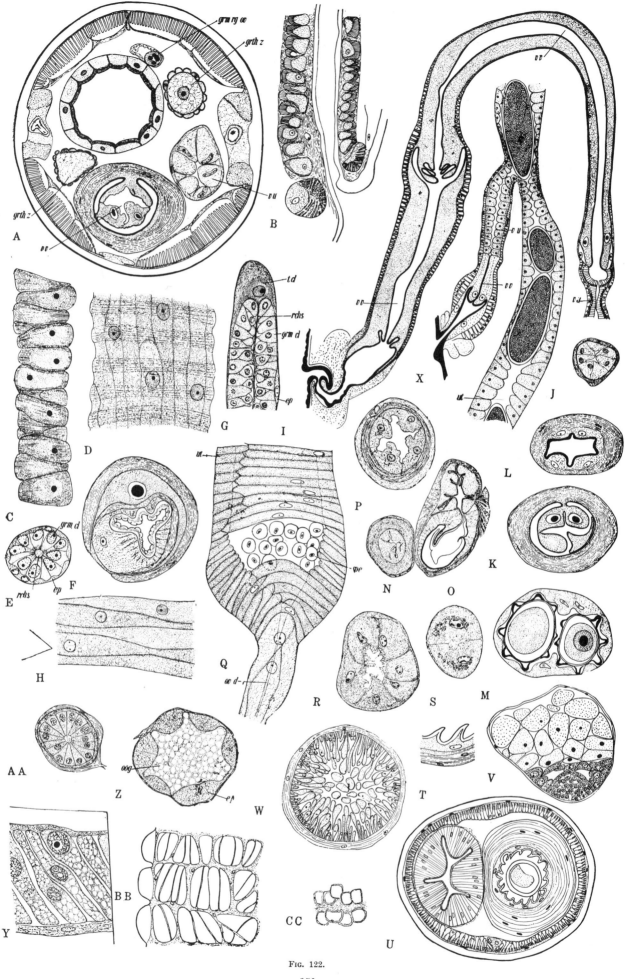

Fig. 122.

150

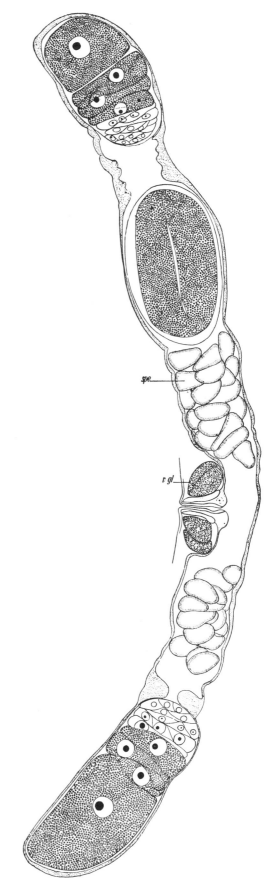

Fig. 122. FEMALE REPRODUCTIVE SYSTEM.

A-H—*Spironoura affine* (A—Cross section near vulva; B—Longitudinal section of vagina; C—Surface section of vagina showing muscles; D—Surface view of uterus showing epithelium and muscles; E—Cross section of germinal region of ovary; F—Cross section of vagina; G—Longitudinal section of blind end of ovary; H—Surface view of epithelium in growth zone of ovary). I—*Macracis monhystera* (Longitudinal reconstruction of vagina vera). J-M and X—*Hystrignathus rigidus* (J—Cross section of vagina uterina; K-L—Cross sections of vagina vera, K at valve; M—Uterus; X—Longitudinal reconstruction of vagina and uteri). N-S—*Oesophagostomum dentatum* (N-O—Adjoining parts of ovejector, N is uterine while O shows connection of vagina vera to other ovejector; P—Distal part of ovejector; R—Uterus near ovejector; S—Uterus more distal). T-U—*Trichuris ovis* (T—detail of vaginal cuticle; U—Cross section near vulva). V—*Trichuris suis* (Ovary showing germinal region on one side). W—*Dioctophyma renale* (Ovary showing germinal zone extending around surface). Y-AA—*Cephalobellus papilliger* (AA—Ovary in germinal zone; Y and Z—Ovary in growth zone). BB-CC—*Enterobius vermicularis* (Uterine musculature in living specimens, BB relaxed and CC contracted).

V, after Rauther, 1918, Zool. Jahrb. Abt. Morph. v. 40; W, after Rauther, 1930, Handb. Zool. v. 2; Y-AA, after Chitwood & Chitwood, 1933, Ztschr. Zellforsch. v. 19 (2). Remainder original.

Fig. 123. FEMALE REPRODUCTIVE SYSTEM.

Halichoanolaimus robustus. Original.

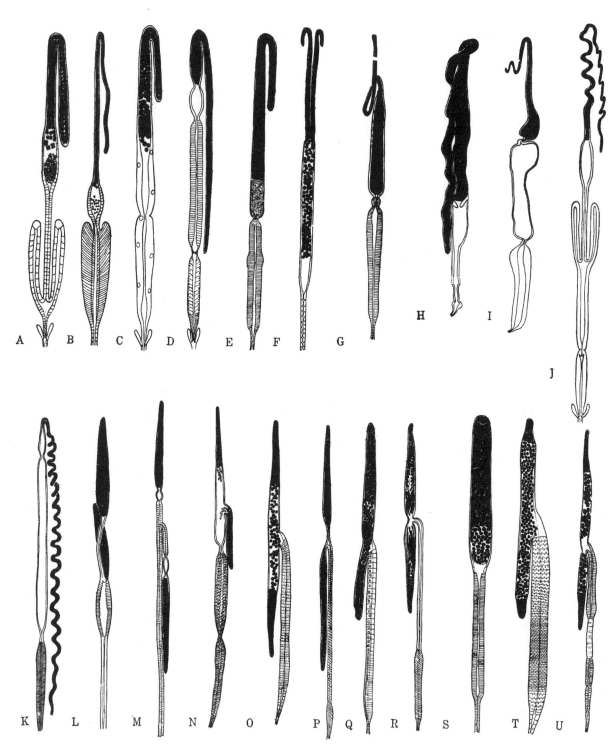

FIG. 124.

Diagrams of male reproductive system. The epithelial portions are white, the germinal portions black. When rectal (i. e. cloacal) glands are known to be attached to the ejaculatory duct (A, C, D) they are shown as lateral protuberances. A—*Rhabditis strongyloides.* B—*Oesophagostomum dentatum.* C—*Rhabditis lambdiensis.* D—*Spironoura affine.* E—*Rhigonema infectum.* F—*Meloidogyne hapla* (with two testes). G—*Heliconema anguillae.* H—*Cucullanus heterochrous.* I—*Camallanus lacustris.* J—*Heterakis gallinarum.* K—*Trichuris suis.* L—*Anticoma typica.* M—*Agamermis decaudata.* N—*Enoplus communis.*

O—*Desmolaimus zeelandicus.* P—*Metoncholaimus pristiuris.* Q—*Anaplectus granulosus.* R—*Trilobus gracilis.* S—*Chromadora quadrilinea.* T—*Halichoanolaimus robustus* (*Spilophorella paradoxa* and *Metachromadora onyxoides* are similar). U—*Sabatieria hilarula.* G. After Yamaguti, 1935, Jap. J. Zool. v. 6 (2). H & I. after Toernquist, 1931, Goeteborgs Kungl. Vetenskaps. o. Vitterhets-Samm Handl. s. 8, v. 2 (3). K, after Rauther, 1918, Zool. Jahrb Abt. Anat. v. 40. L, after Cobb, 1890, Proc. Linn. Soc. N. S. Wales, s. 2, v. 5. M. after Steiner, 1923, J. Heredity, v. 14 (4). Remainder original.

Male Reproductive System

GENERAL MORPHOLOGY

There has been no comparative morphological study on the male reproductive system comparable to that of Seurat on this system in the female. This is due, at least partially, to the structural uniformity of the male apparatus which lacks distinct adaptation coincident with parasitism; as in the female, with parasitism the male reproductive system increases in length but accessory structural differentiations are not apparent.

There may be either one or two testes. If two, they are arranged in opposite directions except in *Meloidogyne hapla* and *Anticoma typica*. In the first species there may be either one or two testes continuous posteriad with a single seminal vesicle; transitions apparently indicate that the double condition arises as a longitudinal split of a single original testis. This is quite different from the normal origin of double testes in which they grow apart as they form at the two poles of the genital primordium. The genus *Meloidogyne* contains the only examples with two testes known in the Phasmidia. In *Anticoma typica*, Cobb 1890, described a form with parallel testes in tandem. This arrangement (Fig. 124 L) may easily be considered as due to a shift of the posterior testis from a normally opposed to a tandem position.

Flexure of the testis is common in the Phasmidia (*Rhabditis*, Fig. 124 A) but just how widespread the phenomenon is we do not know. In *Meloidogyne hapla* the one or two testes may be either flexed or outstretched. In the Aphasmidia, flexure is unknown except in some of the forms with a single testis.

The same differentiation of groups according to origin of germ cells is seen in the male as in the female; hologonic forms (*Dioctophymatoidea* and *Trichuroidea*) have an extended region of germ cell formation while in the remaining nemic groups germ cell formation is confined to the end of the testis.

Many phasmidian nemas have a rachis to which the spermatogonia are attached (*Ascaris, Spironoura, Heterakis,* the Strongylina in general and rhabditids). This structure is apparently a process from the cap cell or terminal epithelial cell of the testis similar to that of the ovary.

PARTS OF THE MALE REPRODUCTIVE SYSTEM. The male reproductive system consists of three or four major divisions, the basic and more general divisions are testis, seminal vesicle and vas deferens (Fig. 3). In some of the more highly developed parasites such as *Ascaris lumbricoides* and *Trichuris suis* (syn. *T. crenatus*) a testicular duct, the *vas efferens* (Samenleiter) separates the testis from the seminal vesicle. The terminal (posterior) end of the vas deferens may be set off as an ejaculatory duct.

Testis. In telogonic nemas the testis is subdivisible into two regions, *germinal zone* and *growth zone*. In phasmidians the cells of the growth zone first take the form of a chord of six or more radiating series of cells attached centrally to the protoplasmic rachis while in aphasmidians they more often take the form of a four cell chord which rapidly gives place to a double row and finally a single chain of cells. In all cases except hologonic forms the testis is covered with an epithelium continuous with that of the seminal vesicle (or vas deferens) just as the ovary is covered with an epithelium continuous with that of the oviduct.

Vas efferens. In the few instances where such a structure has been described it is a duct separating the growth zone of the testis from the seminal vesicle. It has a rather high simple cuboidal to columnar epithelium.

Seminal vesicle. This is a dilated part of the male gonoduct which acts as a storage organ for sperms. The vas efferens may be considered as a specialized tubular part of this structure.

Vas deferens. The vas deferens is the chief part of the male gonoduct. It is generally subdivided into tubular and glandular regions and it may be covered by a muscular layer either in its terminal region (near the cloaca) or throughout its length. Thus only a part of the structure is usually the functional ejector or ejaculatory duct. The detailed anatomy of the vas deferens changes so much in the various groups that it will be considered with the description of special anatomy.

COMPARATIVE MORPHOLOGY

As previously discussed the Phasmidia characteristically have a single testis. Among the aphasmidians the Trichuroidea and Dioctophymatoidea also have a single testis. No comprehensive surveys have been made which would permit unequivocal characterizations of other groups but two testes are the typical condition, so far as known, of the remaining aphasmidian groups.

Marked development of musculature covering the ejaculatory duct does not parallel parasitism as does marked development of this layer in the female gonoduct. On the contrary, it is very markedly correlated with the taxonomic group. So far as known, a heavily and extensively muscled ejaculatory duct occurs only in the order Enoplida and holds true as a taxonomic character in all its representatives studied by the writers. This is particularly interesting since it so clearly confirms the placement of the Trichuroidea and Dioctophymatoidea in this order.

PHASMIDIA. The male reproductive system of this group has been studied by Schneider (1866) on *Rhabditis strongyloides*, by Leuckart (1876) on *Ascaris lumbricoides*, by Cobb (1888) on *Anisakis*, by Looss (1905) on *Ancylostoma*, by Jägerskiöld (1909) on *Dichelyne*, by Krüger (1913) on *Rhabditis aspera* (*R. aberrans*), by Magath (1919) on *Camallanus*, by Steiner (1923) on *Agamermis decaudata*, by Musso (1930) on *Ascaris*, by Chitwood (1930) on *Rhabditis*, by Törnquist (1931) on *Camallanus* and *Cucullanus*, by Chitwood (1931) on *Oesophagostomum*, by Chitwood & Chitwood (1933) on *Cephalobellus*, by Yamaguti (1935) on *Heliconema*, by Baker (1936) on *Heterakis* and by Mackin (1936) on *Spironoura*. Other workers have made more or less casual observations incidental to studies of spermatogenesis or postembryonic development.

The chief point of interest is the diversity in structure of the vas deferens. At its junction with the seminal vesicle there may be a well marked constriction. The duct may be incompletely differentiated into an anterior glandular region, comprising the greater part of its length and a posterior non-glandular region or ejaculatory duct, as in *Rhabditis lambdiensis, Cephalobellus papilliger* and *Heliconema anguillae*. (Figs. 124 C & G). In others the anterior part may be subdivided into two sections both of which may be glandular but the form of the epithelium and the character of the secretory masses differ in the two regions; this occurs in *Rhigonema infectum* and *Spironoura* spp. (Fig. 124 D & E). In the latter forms there is a distinct valve between anterior and posterior sections. A similar division has also been noted in *Cucullanus, Camallanus* and *Ascaris.* At the division point there is a muscular sphincter in many forms. Musso and Magath claimed to have seen at least one nucleus in the sphincter but other authors have been unable to observe such. Muscular fibers in general are always very sparse and confined to the second and third (or just the third) zone of the ejaculatory duct. As first pointed out by Voltzenlogel (1902) they appear to originate with the posterior intestinal muscles and are without nuclei of their own.

A second general type in the Phasmidia is exemplified by *Rhabditis strongyloides* (Fig. 124 A). In this form the vas deferens has two large lateral pouches which extend anteriorly on both sides. Here there has been a differentiation of purely epithelial cells (forming the anterior section of the vas deferens) and glandular cells (forming the pouches and midsection of the duct). These pouches were first described by Schneider and later studied by Chitwood (1930); they are called ejaculatory glands and are thought to form the adhesive cement deposited on the female at copulation. They seem to have arisen as a minor modification of the glandular region generally typical of phasmidian forms. Looss has described very similar though incompletely separated ejaculatory glands in *Ancylostoma* and the writers have seen such in *Oesophagostomum*. In the latter form the glandular cells are often so placed as to give a laminated appearance (Fig. 124 B) and purely epithelial cells are distinctly recognizable along the dorsal side of the vas deferens in the glandular region. Chitwood (1931) described processes from these cells which he considered possible homologues of cilia; Cobb (1888) and Rauther (1918) have previously described hairlike processes from the epithelial cells of the vas deferens of ascaridids and trichuroids respectively. None of these authors have observed vibratile movements. In *Heterakis* Baker found ejaculatory glands very similar in gross morphology to those

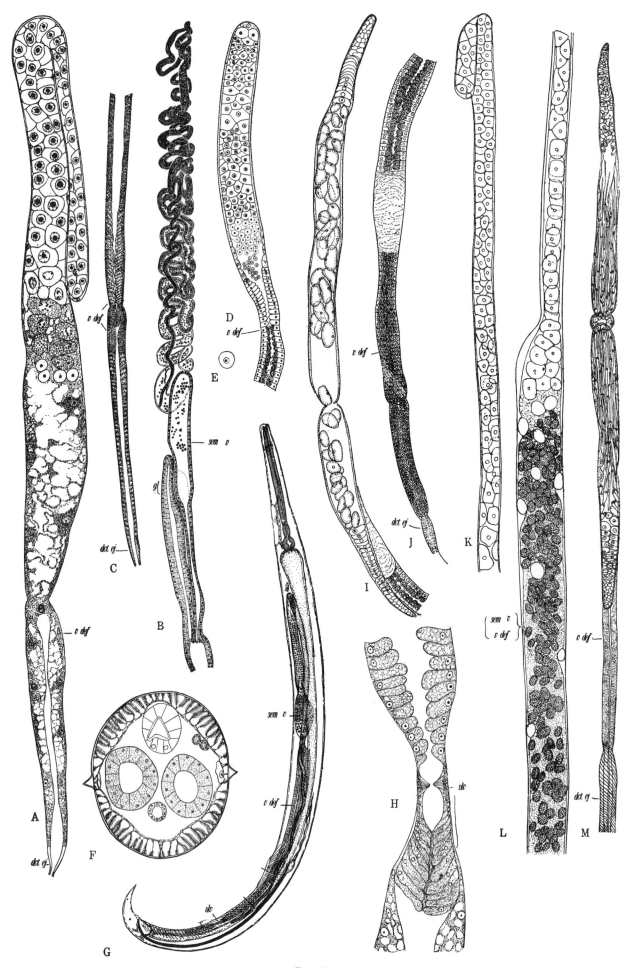

Fig. 125.

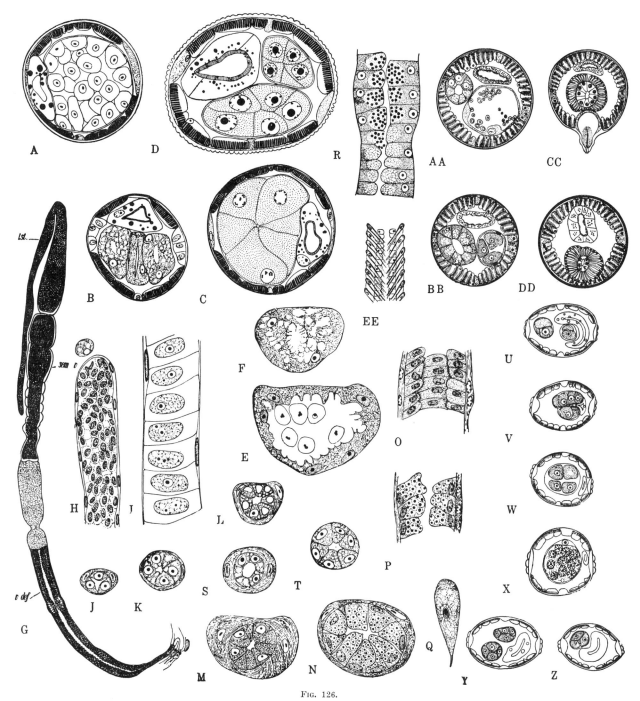

Fig. 126.

A-C—*Rhabditis terricola*. (A—Cross section at level of seminal vesicle; B—Vas deferens with ejaculatory glands on each side; C—Testis just anterior to seminal vesicle). D-F—*Rhabditis lambdiensis* (D—At level of testis; E—Seminal vesicle; F—Vas deferens). G—*Rhigonema infectum* (male gonoduct). H-K & S-T—*Metoncholaimus pristiuris* (H—Blind end of testis; I—Growth zone of testis; J—Ejaculatory duct just preanal; K—Vas deferens; S—Ejaculatory duct mid-region; T—Vas deferens opposite seminal vesicle). L-Q—*Enoplus communis* (L—Ejaculatory duct just preanal; M—Ejaculatory duct mid-region; N— Vas deferens; O—Ejaculatory duct; P—Glandular region vas deferens; Q—Sperm). R—*Chromadora quadrilinea* (Junction of glandular and non-glandular regions of vas deferens). U-Z—*Meloidogyne hapla* (U-Y—Diorchic specimen; Sections at intervals from anterior, posteriad in the order Y, U, V, W, X, [X—Seminal vesicle]; Z—Specimen monorchic). AA-EE—*Trilobus gracilis* (AA—Showing vas deferens and seminal vesicle; BB—Vas deferens and testis; CC—Mid-region of ejaculatory duct; EE—Longitudinal section of ejaculatory duct). Original.

Fig. 125.

Male reproductive system. A—*Rhabditis lambdiensis*. B-C & F—*Heterakis gallinarum*. (B—Anterior part of ejaculatory glands and vas of gonoduct; F—Cross section at level of ejaculatory glands and vas deferens). D-E—*Chromadora quadrilinea* (D—Testis and upper part of vas deferens; E—Spermatozoan). G-H—*Spironoura affine*. (G—Totomount male; H—Detail of valve indicated by *vlv* in G). I-J—*Sabatieria hilarula*. (I—Anterior part of gonoduct; J—Posterior part of gonoduct). K-L—*Meloidogyne hapla* (Monorchic form; K—Anterior part of gonoduct. L—Posterior part of gonoduct). M—*Trilobus gracilis*. G, After Mackin, 1936, Ill. Biol. Monogr., v. 14 (3). Remainder original.

of *Rhabditis strongyloides* but in this form there are four distinct sections of the vas deferens (Fig. 124 J). These sections consist of a narrow tubular anterior part continuous with the seminal vesicle, two wider glandular sections, separated from one another by a valve, and a short posterior ejaculatory part. In this case the paired ejaculatory glands open into the anterior end of the first glandular section. Whether they should be interpreted as outpocketings of that part or as differentiated lateral outpocketings of the tubular anterior section, is not known.

APHASMIDIA. Study has been made of the male reproductive system of members of this group by Eberth (1860) on *Trichuris*, by Leuckart (1876) on *Trichuris*, by de Man (1886) on *Enoplus*, *Anticoma*, *Tripyloides* and *Euchromadora*, by Cobb (1890) on *Anticoma*, by Jägerskiöld (1901) on *Cylicolaimus*, by Türk (1903) on *Thoracostoma*, by Schepotieff (1908) on *Desmoscolex*, *Greeffiella* and *Epsilonema*, by Rauther (1918) on *Trichuris*, by Steiner (1923) on *Agamermis*, by Cobb (1928) on *Spirina* and by Cobb (1929) on *Draconema*. In addition to these papers numerous authors have noted the number of testes in descriptions of species. Cobb, however, is the only author who consistently provided such information.

As previously noted above, the two aphasmidian orders, so far as existing information goes, may be separated upon

---

*Monorchic (with one testis) and *diorchic* (with two testes) are here introduced as adjectives comparable to monodelphic and didelphic which are used to denote the number of ovaries.

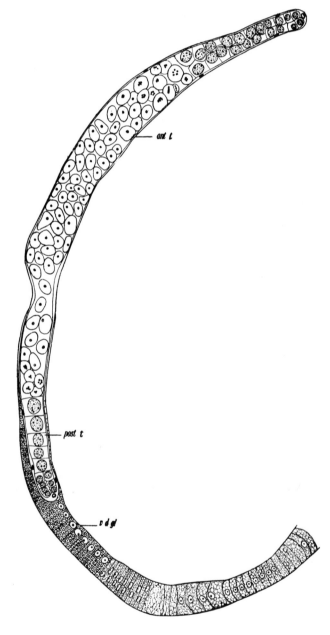

FIG. 127.

Anaplectus granulosus. Reconstruction of male reproductive system. Original.

the basis of muscular development of the ejaculatory duct.

Within the Order Chromadorida the ejaculatory duct has very little musculature, and that is discernible only under the most favorable conditions. From this standpoint, members of the group are definitely closer to phasmidians such as *Rhabditis* than are members of the other aphasmidian group, the Enoplida. Diorchic* forms are the rule but *Euchromadora*, *Chromadora*, *Monoposthia*, *Spirina*, *Epsilonema*, *Tripyloides*, *Desmoscolex* and *Greeffiella* are all monorchic. According to Cobb, some species of *Monhystera* and *Cyatholaimus* are monorchic. Examples of forms known to be diorchic include *Anaplectus granulosus*, *Aphanolaimus* spp., *Bastiania exilis*, axonolaimids, comesomatids, cyatholaimids, *Spilophorella paradoxa*, *Desmolaimus zeelandicus*, *Theristus setosus* and *Draconema cephalatum*. Flexure of the gonoduct is limited to some forms with one testis (*Anaplectus granulosus* sometimes exceptional).

The typical testicular arrangement (Fig. 124 O) consists of two testes extending in opposite directions but joined by a pair of seminal vesicles which may or may not be clearly separated from one another by a constriction. At the junction of the seminal vesicles, the vas deferens is given off as a single tubular canal directed posteriad. This structure is highly glandular throughout its length. Rarely (*Sabatieria*) the posterior part may be set off by a constriction (Fig. 124 U). A very short non-glandular terminal division acts as ejaculatory duct. In monorchic forms (*Chromadora*) the arrangement (Fig. 124 S) is so similar to that of *Rhabditis lambdiensis* that it needs no special description. De Man mentioned special copulatory glands in *Euchromadora* and Schepotieff did the same for *Desmoscolex* but in neither instance is an adequate description or illustration furnished.

Three types of spermatozoa are known in the Chromadorida, these being standard ameboid (typical of the Phasmidia), flagellate, and hollow. Of these, the flagellate type of spermatozoan is known only in *Halanonchus* and the ameboid is known in plectids, chromadorids and *Desmolaimus*. The hollow sperm is a very interesting and peculiar structure. As a rule it is large with a narrow ectoplasm radially striated by fibrils and a central vacuole. The nucleus usually lies next to the periphery and is quite small (exception *Spirina parasitifera**). This type of sperm is known to occur in axonolaimids, comesomatids, cyatholaimids, *Tripyloides*, *Microlaimus*, *Monoposthia* and *Metachromadora*.

Within the Order Enoplida the musculature of the ejaculatory duct is usually very prominent and may extend to the seminal vesicle. So far as known, all Enoplida are diorchic except the hologonic groups Trichuroidea and Dioctophymatoidea. In these the musculature of the ejaculatory duct is particularly thick, giving a laminated appearance due to the presence of several layers of muscle cells. Steiner (1923) has shown the entire vas deferens of *Agamermis decaudata* as being covered by a well developed muscular layer. The writers have found the ejaculatory duct covered by a particularly well developed layer of oblique muscles in *Metoncholaimus* (Fig. 126 S) *Trilobus* (Fig. 126 DD-EE), *Enoplus* (Fig. 126 L-N), and *Phanodermopsis*; the muscle cells in these forms extend only half way around the ejaculatory duct, extending obliquely anteriad from the medio-ventral to the medio-dorsal line. In *Metoncholaimus* the musculature extends anteriad, in a less well developed manner for over half the length of the vas deferens while in *Enoplus* the musculature extends throughout the length. Glandular activity is evident throughout the vas deferens and the anterior part of the ejaculatory duct in *Enoplus* while it is confined to the vas deferens in *Trilobus*, *Thoracostoma* and *Cylicolaimus* and is confined to the non-muscular part of the vas deferens in *Metoncholaimus*. The paired testes are continuous with paired seminal vesicles except in *Enoplus* where the junction of posterior testis and seminal vesicle appears to have been shifted anteriad. So far as known, the spermatozoa are ameboid in the parasitic groups (Mermithoidea, Trichuroidea and Dioctophymatoidea) while in the free-living forms they are definitely flagellate as in *Trilobus*, spindle to tear drop shaped as in *Mononchus*, *Tripyla*, *Phanodermopsis*, *Enoplus* and *Actinolaimus* or rounded as in *Metoncholaimus* and *Thoracostoma*.

**Cobb (1928) described the "Gametogenesis" of *Spirina parasitifera* by including in the sperm development the formation of a 128 cell "spermatophore" due to amitotic division of the spermatid. Re-examination of this species shows that what he interpreted as the spermatid is the enormous hollow spermatozoon and what he interpreted as the nuclei of the spermatophore are secretion globules in the wall of the vas deferens. The latter are arranged in transverse rows in the narrow transversely elongate epithelial cells.

Bibliography

ACKERT, J. E. 1931.—The morphology and life history of the fowl nematode, *Ascaridia lineata* (Schneider). Parasit., v. 23 (3): 360-379, figs. 1-25, pls. 13-14.

ALICATA, J. E. 1935 (1936).—Early developmental stages of nematodes occurring in swine. Tech. Bull. No. 489, U. S. Dept. Agric. 96 pp., 30 figs.

ARTIGAS, P. 1930.—Nematoides dos generôs *Rhigonema* Cobb, 1898 e *Dudekemia* n. gn., (Nematoda: Rhigonemidae n. n.). Mem. Inst. Oswaldo Cruz, v. 24 (1): 19-30, pls. 8-14, figs. 1-32.

ATKINSON, G. F. 1896.—Root galls of cotton (*Heterodera radicicola* (Greeff) Muell). (*In his* Diseases of cotton) Bull. (33), Office Exper. Stations, U. S. Dept. Agric., pp. 311-316, fig. 8.

AUGSTEIN, O. 1894.—*Strongylus filaria* R. Arch. Naturg., 60 J., v. 1 (3): 255-304, pls. 13-14.

BAKER, A. D. 1936.—Studies of *Heterakis gallinae* (Gmelin, 1790) Freeborn, 1923, a nematode parasite of domestic fowls. Tr. Roy. Canad. Inst., v. 21 (1): 51-86, pls. 1-15.

BALSS, H. H. 1906.—Mitteilungen über einen abnormen weiblichen Geschlechtsapparat von *Ascaris lumbricoides* L. Zool. Anz., v. 30: 485-487, 1 fig.

BASTIAN, H. C. 1865.—Monograph on the Anguillulidae, or free nematoids, marine, land, and fresh-water; with descriptions of 100 new species. Tr. Linn. Soc. Lond. v. 25 (2): 73-184, pls. 9-13, figs. 1-248.

　1866.—On the anatomy and physiology of the nematoids, parasitic and free; with observations on their zoological position, and affinities to the echinoderms. Phil. Tr. Roy. Soc. Lond. v. 156: 545-638, pls. 22-28.

BENEDEN, E. VAN 1883.—L'Appareil sexuel femelle de l'ascaride mégalocéphale. Arch. Biol. Gand., v. 4 (1): 95-142, pl. 3, figs. 1-17.

BENEDEN, E. VAN & JULIN, C. 1884.—La spermatogénèse chez l'ascaride mégalocéphale. Bull. Acad. Roy. Sc. Belg., 3. s., v. 7 (4): 312-342.

BISCHOFF, T. 1855.—Ueber Ei- und Samenbildung und Befruchtung bei *Ascaris mystax*. Ztschr. Wiss. Zool., v. 6 (3-4): 377-405.

BOVERI, T. 1892 (1893).—Ueber die Entstehung des Gegensatzes zwischen den Geschlechtszellen und den somatische Zellen bei *Ascaris* megalocephala. Sitz. Ber. Ges. Morph. & Phys. München, v. 8 (2-3): 114-125, figs. 1-5.

　1899.—Die Entwickelung von *Asc. megalocephala* mit besonderer Rücksicht auf die Kernverhältnisse. Festschr. Kupffer. pp. 383-430, figs. 1-6, pls. 40-45, figs. 1-45.

BOVIEN, P. 1932.—On a new nematode, *Scatonema wülkeri* gen. et. sp. n. parasitic in the body-cavity of *Scatopse fuscipes* Meig. (Diptera; Nematocera). Vidensk. Medd. Dansk. Naturh. Foren v. 94: 13-32, figs. 1-7.

BÜTSCHLI, O. 1871.—Untersuchungen über die beiden Nematoden der *Periplaneta (Blatta) orientalis* L. Ztschr. Wiss. Zool., v. 21 (2): 252-293, pls. 21-22, figs. 1-29.

　1873.—Beiträge zur Kenntniss der freilebenden Nematoden. Nova Acta K. Leop. Akad. Naturf., v. 36 (5): 124 pp., pls, 17-27, 69 figs.

　1874.—Zur Kenntnis der freilebenden Nematoden, insbesondere der des Kieler Hafens. Abhandl. Senckenb. Naturf. Gesellsch. Frankfurt, v. 9 (3): 237-292, pls. 1-9, figs. 1-39.

　1876.—Untersuchungen über freilebende Nematoden und die Gattung *Chaetonotus*. Ztschr. Wiss. Zool., v. 26 (4): 363-413, pls. 23-26.

CASSIDY, G. H. 1928.—A meristic variation in a female nematode. Nature v. 121 (3047): 476-477.

　1933.—A bivulvar specimen of the nematode *Mononchus muscorum* (Dujardin) Bastian. J. Wash. Acad. Sc., v. 23 (3): 141-144, 1 fig.

CHANDLER, A. C. 1924.—A note on *Ascaris lumbricoides* with three uteri and ovaries. J. Parasit., v. 10 (4): 208.

CHITWOOD, B. G. 1929.—Notes on the copulatory sac of *Rhabditis strongyloides*. Schneider. J. Parasit., v. 15. (4): 282, figs. 1-5.

　1930.—Studies on some physiological functions and morphological characters of *Rhabditis* (Rhabditidae, nematodes). J. Morph. & Physiol., v. 49 (1): 251-275, figs. A-H, pls. 1-3, figs. 1-24.

　1931a.—A comparative histological study of certain nematodes. Ztschr. Morph. & Oekol. Tiere, v. 23 (1-2): 237-284, figs. 1-23.

　1931b.—Flagellate spermatozoa in a nematode, *Trilobus longus*. J. Wash. Acad. Sc., v. 21 (3): 41-42, figs. 1-2.

CHITWOOD, B. G. & M. B. 1933.—The histological anatomy of *Cephalobellus papilliger* Cobb, 1920. Ztschr. Zellforsch, v. 19 (2): 309-355, figs. 1-34.

　1938.—Notes on the ''culture'' of aquatic nematodes. J. Wash. Acad. Sc. v. 28 (10): 455-460, figs. 1-2.

CHRISTIE, J. R. 1929.—Some observations on sex in the Mermithidae. J. Exper. Zool., v. 53 (1) 59-76, figs. 1-5, tables 1-3.

　1931.—Some nemic parasites (Oxyuridae) of Coleopterus larvae. J. Agric. Res., v. 42 (8): 463-482, figs. 1-14.

CIUREA, J. 1911.—Ueber *Spiroptera strongylina* Rud., Centrlbl. Bakt. I Abt., Orig. v. 61 (1/2) 128-133, fig. 1, 1 pl., figs. 1-5.

CLAPAREDE, E. 1859.—De la formation et de la fécondation des œufs che les vers nématodes. Mém. Soc. Phys. Hist. Nat. Genève, v. 15 (1): 1-101, pls. 1-8.

COBB, N. A. 1888.—Beiträge zur Anatomie und Ontogenie der Nematoden. Diss. Jena. 36 pp., 3 pls., 32 figs.

　1890a.—*Anticoma*: a genus of free-living marine nematodes. Proc. Linn. Soc. N. S. Wales s. 2, v. 5 (4): 765-774, figs. 1-2.

　1890b.—*Tylenchus* and root-gall. Agric. Gaz. N. S. Wales, v. 1 (2): 155-184.

　1898.—Extract from Ms. report on the parasites of stock. Misc. Publ. No. 215, Dept. Agric., N. S. Wales, 62 pp., 129 figs.

　1901.—Root-gall. Agric. Gaz. N. S. Wales, v. 12 (9): 1041-1052, figs. 1-8.

　1902.—Internal structure of the gall-worm. Agric. Gaz. N. S. Wales, v. 13 (10): 1031-1033, fig. 1.

　1914.—Citrus-root nematode. J. Agric. Res., v. 2 (3): 217-230, figs. 1-13.

　1916.—Notes on filter-bed nematodes. J. Parasit., v. 2 (4): 198-200, figs. 1-2.

　1928a.—*Howardula benigna;* a nema parasite of the cucumber-beetle (*Diabotrica*). Contrib. Sc. Nemat. (10): 345-352, figs. 1-8 (Also in Science, n. s., (1409) v. 54: 667-670, figs. 1-4, 1921).

　1928b.—Nemic spermatogenesis. Contrib. Sc. Nemat. (16): 375-387, figs. 1-17 (Also in J. Wash. Acad. Sc. v. 18 (2): 37-50, figs. 1-17).

　1929.—The ambulatory tubes and other features of the nema *Draconema cephalatum*. Contrib. Sc. Nemat. (22): 413-418, figs. 1-2 (Also in J. Wash. Acad. Sc. v. 19 (12): 255-260, figs. 1-2.)

　1930.—The demanian vessels in nemas of the genus *Oncholaimus;* with notes on four new oncholaims. J. Wash. Acad. Sc., v. 20 (12): 225-241, figs. 1-9.

DADAY, E. VON 1905.—Untersuchungen über die Süsswasser Microfauna Paraguays. Zoologica. v. 18: 44.

DITLEVSEN, H. 1911.—Danish freeliving nematodes. Vid. Medd. Naturh. Foren v. 63: 213-256, pls. 2-5, figs. 1-48.

　1919.—Marine freeliving nematodes from Danish waters. Vidensk. Medd. Dansk. Naturh. Foren v. 70: 147-214, pls. 1-16.

DOMASCHKO, A. 1905.—Die Wandlung der Gonade von *Ascaris megalocephala*. Arb. Zool. Inst. Wien, v. 15 (3): 257-280.

EBERTH, C. J. 1860.—Die Generationsorgane von *Trichocephalus dispar*. Ztschr. Wiss. Zool., v. 10 (3): 383-400, pl. 31, figs. 1-2.

ESCHRICHT, 1848.—Referred to by Seurat, 1920.

FILIPJEV, I. N. 1918.—(Free-living marine nematodes in the vicinity of Sevastopol. Part I.) (Russian). Trudy Osob. Zool. Lab. Sevastopol, Biol. Stantsii Ross. Akad. Nauk., s. 2 (4): 1-350, pls. 1-11, figs. 1-81.

　1922.—(Encore sur les nématodes libres de la Mer Noire.) (Russian). Acta Inst. Agron. Stauropol, v. 1 (16): 83-184, figs. a-c, pls. 1-4, figs. 1-36.

　1929.—Classification of free living Nematoda and relations to parasitic nematodes. J. Parasit. v. 15 (4): 281-282.

1934.—The classification of the free-living nematodes and their relation to the parasitic nematodes. Smithson. Misc. Coll., v. 89 (6): 1-63, pls. 1-8, figs. 1-70.

GALEB, O. 1878.—Recherches sur les entozoaires des insectes. Organisation et développement des oxyuridés. Arch. Zool. Expér. & Gén. v. 7 (2): 283-390, figs. 17-26.

GOODEY, T. 1924.—Two new species of the nematode genus *Rhabdias*. J. Helminth. v. 2 (5): 203-208, figs. 1-11.
    1930.—On a remarkable new nematode, *Tylenchinema oscinellae* gen. et. sp. n., parasitic in the frit-fly, *Oscinella frit* L., attacking oats. Phil. Tr. Roy. Soc. Lond., s. B., v. 218: 315-343, fig. 1, pls. 22-26, figs. 1-54.
    1935.—*Brevibucca saprophaga* gen. et sp. nov., a nematode from a rotting lily bulb-scale. J. Helminth., v. 13 (4): 223-228, figs. 1-7.

HAGMEIER, A. 1912.—Beiträge zur Kenntnis der Mermithiden. 1. Zool. Jahrb., Abt. Syst., v. 32 (6): 521-612, figs. a-g pls. 17-21, figs. 1-55.

HALL, M. C. 1916.—Nematode parasites of mammals of the orders Rodentia, Lagomorpha, and Hyracoidea. Proc. U. S. Nat. Mus., v. 50 (2131): 1-258, figs. 1-290, 1 pl.

JÄGERSKIÖLD, L. A. 1894.—Beiträge zur Kenntnis der Nematoden. Zool. Jahrb., Abt. Anat., v. 7 (3): 449-532, pls. 24-28.
    1901.—Weitere Beiträge zur Kenntnis der Nematoden. K. Svenska Vetensk. Akad. Handl. Stockholm., v. 35 (2): 1-80, figs. 1-8, pls. 1-6.
    1909.—Nematoden aus Aegypten und dem Sudan. Results Swedish Zool. Exped. to Egypt and White Nile, 1901. No. 25; 66 pp., 23 figs., 4 pls.

KEMNITZ, G. VON 1912.—Die Morphologie des Stoffwechsels bei *Ascaris lumbricoides*. Arch. Zellforsch., v. 7 (4): 463-603, figs. a-h, j, pls. 34-38.

KRÜGER, E. 1913.—Fortpflanzung und Keimzellenbildung von *Rhabditis aberrans*, nov. sp. Ztschr. Wiss. Zool., v. 105 (1): 87-124, pls. 3-6, figs. 1-73.

LEUCKART, R. F. 1876.—Die Menschlichen Parasiten. v. 2, 882 pp., 401, figs. Leipzig.
    1887.—Neue Beiträge zur Kenntniss des Baues und der Lebensgeschichte der Nematoden. Abhandl. Math.-Phys. Classe Königl. Sächs. Gesellsch. Wiss., v. 13 (8): 565-704, pls. 1-3.

LINSTOW, O. VON 1878.—Neue Beobachtungen an Helminthen. Arch. Naturg., 44 J., v. 1 (2): 218-245, pls. 7-9, figs. 1-35.
    1903.—Helminthologische Beobachtungen. Centrlbl. Bakt. 1 Abt., Orig., v. 34 (6): 526-531, figs. 1-7.

LOOSS, A. 1905.—The anatomy and life history of *Agchylostoma duodenale* Dub. Rec. Egypt. Govt. Sch. Med. v. 3: 1-158, pls. 1-9, figs. 1-100, pl. 10, photos 1-6.

MACKIN, J. G. 1936.—Studies on the morphology and life history of nematodes in the genus *Spironoura*. Ill. Biol. Monogr. v. 14 (3): 1-64, pls. 1-6, figs. 1-69.

MAGATH, T. B. 1919.—*Camallanus americanus* nov. spec., a monograph on a nematode species. Tr. Am. Micr. Soc., v. 38 (2): 49-170, figs. A-Q, pls. 7-16, figs. 1-134.

MAN, J. G. DE 1886.—Anatomische Untersuchungen über freilebende Nordsee-Nematoden. 82 pp., 13 pls. Leipzig.
    1893.—Cinquième note sur les nématodes libres de la mer du nord et de la manche. Mém. Soc. Zool. France, v. 6 (1-2): 81-125, pls. 5-7, figs. 1-13k.

MARION, A. F. 1870.—Nématoides non parasites marins. Ann. Sc. Nat. Paris Zool., 5. s., v. 13, art. 14, 100 pp., pls. 16-26.

MARTINI, E. 1916.—Die Anatomie der *Oxyuris curvula*. Ztschr. Wiss. Zool., v. 116: 137-534, figs. 1-121, pls. 6-20, figs. 1-269.

MAUPAS, E. 1899.—La mue et l'enkystement chez les nématodes. Arch. Zool. Expér & Gén., 3 s., v. 7: 563-628, pls. 16-18, figs. 1-29.
    1900.—Modes et formes de reproduction des nématodes. Ibid. 3. s., v. 8: 463-624, pls. 16-26.

MAUPAS, E. & SEURAT, L. G. 1912.—Sur un nématode de l'intestin grêle du dromédaire. Compt. Rend. Soc. Biol., Paris, v. 73 (36): 628-632, figs. 1-10.

MAYER, A. 1908.—Zur Kenntnis der Samenbildung bei *Ascaris megalocephala*. Zool. Jahrb., Abt. Anat., v. 25 (3): 495-546, 2 figs., pls. 15-16, 61 figs.

MEISSNER, G. 1853.—Beiträge zur Anatomie und Physiologie von *Mermis albicans*. Ztschr. Wiss. Zool., v. 5 (2-3): 207-284, pls. 11-15, figs. 1-55.

MONNIG, H. O. 1923 (1924).—South African parasitic nematodes. 9th & 10th Rpt. Dir. Vet. Educ., pp. 435-478, figs. 1-46.

MUNK, H. 1858.—Ueber Ei- und Samenbildung und Befruchtung bei den Nematoden. Ztschr. Wiss. Zool., v. 9 (3): 365-416, pls. 14-15, figs. 1-28.

MUSSO, R. 1930.—Die Genitalröhren von *Ascaris lumbricoides* und *megalocephala*. Ztschr. Wiss. Zool., v. 137 (2): 274-363, figs. 1-29, pls. 1-2, figs. 1-24.

NAGAKURA, K. 1930.—Ueber den bau und die Lebensgeschichte des *Heterodera radicicola* (Greeff) Müller. Jap. J. Zool., v. 3 (3): 95-160, figs. 1-85.

NELSON, H. 1852.—The reproduction of *Ascaris mystax*. Phil. Trans. Roy. Soc. Lond., pp. 563-594, pls. 25-30, figs. 1-92.

NUSSBAUM, M. 1884.—Ueber die Veränderungen der Geschlechtsproducte bis zur Eifurchung; ein Beitrag zur Lehre der Vererbung. Arch. Mikr. Anat., v. 23 (2): 155-213, pls. 9-11, figs. 1-75.

OLSEN, O. W. 1938.—*Aplectana gigantica* (Cosmocercidae), a new species of nematode from *Rana pretiosa*. Tr. Am. Micr. Soc., v. 57 (2): 200-203, figs. 1-9.

ORTLEPP, R. J. 1922.—The nematode genus *Physaloptera* Rud. Proc. Zool. Soc. Lond., pp. 999-1107, figs. 1-44.
    1937.—Some undescribed species of the nematode genus *Physaloptera* Rud., together with a key to the sufficiently known forms. Onderstepoort J. Vet. Sc., v. 9 (1): 71-84, figs. 1-8.

PAI, S. 1927.—Lebenszyklus der *Anguillula aceti* Ehrbg. Zool. Anz., v. 74 (11-12): 257-270, figs. 1-12.
    1928.—Die Phasen des Lebenscyklus der *Anguillula aceti* Ehrbg. und ihre experimentell-morphologische Beeinflussung. Ztschr. Wiss. Zool., v. 131 (2): 293-344, figs. 1-80.

PARAMONOV, A. 1926.—(Ueber einen Fall von ''Bivulvarität'' bei einen freilebenden Nematoden.) (Russian). Russ. Hydrobiol. Ztschr., v. 5 (10-12): 218-222.

PLENK, H. 1924.—Nachweis von Querstreifung in sämtlichen Muskelfasern von *Ascaris megalocephala*. Ztschr. Anat. & Entwicklungsgeschichte, v. 73 (3/4): 358-388, figs. 1-29.

RANSOM, B. H. 1911.—The nematodes parasitic in the alimentary tract of cattle, sheep and other ruminants. U. S. D. A., Bur. Anim. Ind. Bull. No. 127, 132 pp., figs. 1-152.

RAUTHER, M 1909.—Morphologie und Verwandtschaftsbeziehungen der Nematoden. Ergeb. & Forstschr. Zool., v. 1 (3): 491-596, figs. 1-21.
    1918.—Mitteilungen zur Nematodenkunde. Zool. Jahrb. Abt. Anat., v. 40 (4): 441-514, figs. A-P, pls. 20-24, figs. 1-40.
    1930.—Vierte Klasse des Cladus Nemathelminthes. Nematodes, Nematoidea = Fadenwürmer. Handb. Zool. (Kükenthal & Krumback) v. 2, 8 lief., 4 Teil, Bogen 23-32, pp. 249-402, figs. 267-426.

ROMEIS, B. 1912.—Beobachtungen über Degenerationserscheinungen von Chondriosomen. Nach Untersuchungen an nicht zur Befruchtung gelangten Spermiem von *Ascaris megalocephala*. Arch. Mikr. Anat., v. 80, Abt. 2 (4): 129-170, pls. 8-9, figs. 1-33.
    1913.—Ueber Plastosomen und andere Zellstrukturen in den Uterus-, Darm-, und Muskelzellen von *Ascaris megalocephala*, Anat. Anz. v. 44 (1/2): 1-14, 1 pl. figs. 1-11.

ROMIEU, M. 1911.—La Spermiogénèse chez l'*Ascaris megalocephala*. Arch. Zellforsch., v. 6 (2): 254-325, pls. 14-17, figs. 1-94.

SALA, L. 1904.—Intorna ad una particolarita di strutta delle cellule epiteliali, che tapezzano il tubo ovarico e spermatico degli Ascaridi. Reale. 1st. Lomb. Sc. e Lett, et Arti. 2. s., v. 37 (16): 874-887, 1 pl. figs. 1-7.

SANDGROUND, J. H. 1936.—Nematoda. Scientific results of an expedition to rain forest regions in Eastern Africa. v. 1. Bull Mus. Comp. Zool. v. 79 (6): 339-366, figs. 1-22.

SCHEBEN, L. 1905.—Beiträge zur Kenntnis des Spermatozoons von *Ascaris megalocephala*. Diss. Marburg 37 pp., 3 figs. Also in Ztschr. Wiss. Zool. v. 79 (3): 397-431, figs. 1-3 pls. 20-21, figs. 1-44.

SCHEPOTIEFF, A. 1908a.—*Rhabdogaster cygnoides* Metschn. Zool. Jahrb. Abt. Syst. v. 26 (3): 393-400, pl. 26, figs. 1-23.

1908b.—*Trichoderma oxycaudatum* Greeff. Ibid., v. 26 (3): 395-392, pl. 25, figs. 1-18.

1908c.—Die Desmoscoleciden. Ztschr. Wiss. Zool. v. 90: 181-204, pls. 8-10.

SCHEWIAKOFF, W. 1894.—Ein abnorm gebauter weiblicher Genitalapparat von *Ascaris lumbricoides* L. Centrlbl. Bakt. v. 15 (13-14): 473-476, figs. 1-2.

SCHLEIP, W. 1911.—Das Verhalten des Chromatins bei *Angiostoma (Rhabdonema) nigrovenosum*. Arch. Zellforsch., v. 7 (1): 87-138, pls. 4-8, figs. 1-108.

SCHNEIDER, A. 1858.—Ueber die Seitenlinien und das Gefässsystem der Nematoden. Arch. Anat. Physiol. & Wiss. Med., pp. 426-436, pl. 15, 9 figs.

1860.—Ueber eine Nematodenlarve und gewisse Verschiedenheiten in den Geschlechtsorganen der Nematoden. Ztschr. Wiss. Zool., v. 10: 176-178.

1866.—Monographie der Nematoden. 357 pp., 28 pls. Berlin.

SCHNEIDER, K. C. 1902.—Lehrbuch der vergleichenden Histologie der Tiere. 988 pp., 691 figs. Jena.

SCHNEIDER, W. 1922.—Freilebende Süsswassernematoden aus ostholsteinischen Seen. Arch. Hydrobiol., v. 13 (4): 696-753, figs. 1-2.

SCHULZ, R. ED. 1927.—(Die Familie Physalopteridae Leiper, 1908, (Nematodes) und die Prinzipien ihrer Klassifikation.) (Russian). Samml. Helminth. Arb. Prof. K. I. Skrjabin gewidmet 287-312, pl. 1.

SCHWARTZ, B. & ALICATA, J. E. 1930.—Two new species of nodular worms (*Oesophagostomum*) parasitic in the intestine of domestic swine. J. Agric. Res., v. 40 (6): 517-522, figs. 1-12.

SEURAT, L. G. 1913a.—Sur un cas de poecilogonie chez un Oxyure. Compt. Rend. Soc. Biol. Paris, v. 74 (19): 1089-1092, figs. 1-4.

1913b.—Sur l'existence d'un anneau vulvaire, consécutif a l'accouplement, chez un nématode. Ibid. v. 75 (30): 326-330, figs. 1-6.

1913c.—Observations sur le *Tropidocerca inermis* Linst. Bull. Soc. Hist. Nat. Afriq. Nord., v. 5 (9): 191-199, figs. 1-11; reprint 9 pp. 11 figs.

1914a.—Sur l'*Habronema (Spiroptera) leptoptera*. (Rud.). Compt. Rend. Soc. Biol. Paris v. 76 (1): 21-24, figs. 1-5.

1914b.—Sur la morphologie de l'ovéjector des *Tropidocerca*. Compt. Rend. Soc. Biol. Paris v. 76 (4): 173-176, figs. 1-3.

1914c.—Sur deux physaloptères tétrahystériens des reptiles. Ibid., v. 77: 433-436, figs. 1-5.

1915a.—Sur la morphologie de l'*Acuaria laticeps* (Rud.) Ibid., v. 78 (3): 41-44, figs. 1-2.

1915b.—Sur l'existence, en Algérie, du *Dermatoxys veligera* (Rud.) et sur les affinités du genre *Dermotoxys*. Ibid. v. 78 (5): 75-79, figs. 1-4.

1915c.—Sur deux nouveaux oxyures du Maroc. Bull. Soc. Hist. Nat. Afrique Nord, v. 7 (2): 24-31, figs. 1-9.

1915d.—Sur deux nouveaux parasites du renard d'Algérie. Compt. Rend. Soc. Biol. Paris, v. 78 (6): 122-126, figs. 1-4.

1915e.—Sur le cucullan de la clemmyde lépreuse et les affinités du genre *Cucullanus*. Compt. Rend. Soc. Biol. Paris, v. 78 (14): 423-426, figs. 1-4.

1915f.—Sur les conditions de la ponte du strongle lisse. Bull. Sc. France & Belg., 7. s., 48 (3): 171-177, figs. 1-4.

1920.—Histoire naturelle des nématodes de la Barbérie. 221 pp., 34 figs. Alger.

STEINER, G. 1919.—Untersuchungen über den allgemeinen Bauplan des Nematodenkörpers, 96 pp., 3 pls. Jena. Also in Zool. Jahrb. 1921 Abt. Morph. v. 43, 96 pp., 3 pls.

1921.—*Phlyctainophora lamnae* n. g., n. sp., eine neue parasitische Nematodenform aus *Lamna cornubica* (Heringshai). Centrlbl. Bakt. 1 Abt. Orig. v. 86 (7/8): 591-595, figs. 1-13.

1923a.—Intersexes in nematodes. J. Heredity, v. 14 (4): 147-158, figs. 1-11, 1 pl.

1923b.—Limicole Mermithiden aus dem Sarekgebirge und der Torne Lappmark. Naturw. Untersuch. Sarek. Schwed.-Lappland, v. 4 (8): 805-828, figs. 1-29.

1926.—A new arctic mermithid, *Limnomermis euvaginata* n. sp. From Novaya Zemlya. Norsk. Vidensk. Akad. Oslo, Rpt. Sc. Results Norweg. Exped. Novaya Zemlya 1921, No. 33, 8 pp., 9 figs.

1929.—On a collection of mermithids from the basin of the Volga River. Zool. Jahrb. Abt. Syst., v. 57 (3-4): 303-328, figs. 1-42.

1937.—Opuscula miscellanea nematologica VI. Proc. Helm. Soc. Wash., v. 4 (2): 48-52, figs. 18-19.

STEWART, F. H. 1906.—The anatomy of *Oncholaimus vulgaris* Bast., with notes on two parasitic nematodes, Quart. J. Micr. Sc. Lond., n. s., (197), v. 50 (1): 101-150, figs. 1-9, pls. 7-9, figs. 1-40.

STRASSEN, O. ZUR 1892.—*Bradynema rigidum* v. Sieb. Ztschr. Wiss. Zool., v. 54 (4): 655-747, pls. 29-33, figs. 1-98.

1894.—Ueber das röhrenförmige Organ von *Oncholaimus*. Ztschr. Wiss. Zool., v. 58 (3): 460-474, pl. 29, figs. 1-15.

STRUBELL, A. 1888.—Untersuchungen über den Bau und die Entwicklung des Rübennematoden, *Heterodera schachtii* Schmidt. Bibliotheca Zool. Orig. Gesammt. Zool., Heft. 2, 50 pp., 2 pls., figs. 1-57.

THAPAR, G. S. 1925a.—On some new members of the genus *Kiluluma* from the African rhinoceros. J. Helminth., v. 3 (2): 63-80, figs. 1-32.

1925b.—Studies on the oxyurid parasites of reptiles. J. Helminth., v. 3 (3/4): 83-150, figs. 1-132.

THOMSON, A. 1856.—Ueber die Samenkörperchen die Eier und die Befruchtung der *Ascaris mystax*. Ztschr. Wiss. Zool., v. 8 (3): 425-438.

THORNE, G. 1937.—A revision of the nematode family Cephalobidae Chitwood & Chitwood, 1934. Proc. Helm. Soc. Wash., v. 4 (1): 1-16, figs. 1-4.

TÖRNQUIST, N. 1931.—Die Nematodenfamilien Cucullanidae und Camallanidae. Göteborg's K. Vetensk.-o. Vitterhets-Samh. Handl., 5 f., s. B, v. 2 (3): 441 pp., pls. 1-17.

TRAVASSOS, L. 1914.—Sobre as especies brazileiras do genero *Tetrameres* Creplin, 1846. Mem. Inst. Oswaldo Cruz, v. 6 (3): 150-162, pls. 16-23, figs. 1-21.

TÜRK, F. 1903.—Ueber einige im Golfe von Neapel freilebenden Nematoden. Mitth. Zool. Stat. Neapel, v. 16 (3): 281-348, pls. 10-11.

VEGLIA, F. 1915.—The anatomy and life history of the *Haemonchus contortus* (Rud.). 3rd & 4th Rpt. Direct. Vet. Res. pp. 347-500, figs. 1-60.

VOGEL, R. 1925.—Zur Kenntnis der Fortpflanzung, Eireifung, Befruchtung und Furchung von *Oxyuris obvelata* Bremser. Zool. Jahrb., v. 42 (2): 243-271, figs. A-V, pl. 1, figs. 1-6.

VOGT, C. & YUNG, E. 1888.—Lehrbuch der vergleichenden Anatomie. v. 1, 906, pp., 425 figs. Braunschweig.

VOLTZENLOGEL, E. 1902.—Untersuchungen über den anatomischen und histologischen Bau des Hinterendes von *Ascaris megalocephala* und *Ascaris lumbricoides*. Zool. Jahrb. Abt. Anat. v. 16 (3): 481-510, pls. 34-36, figs. 1-20.

WASIELEWSKI, T. VON 1893.—Die Keimzone in den Genitalschläuchen von *Ascaris megalocephala*. Arch. Mikr. Anat., v. 41 (2): 324-337, pl. 19, figs. 1-16.

WILLEMOES-SUHM, R. VON 1869.—Helminthologische Notizen. Ztschr. Wiss. Zool., v. 19 (3): 469-475, pl. 35, figs. 1-5; v. 20 (1): 94-98, pl. 10.

YAMAGUTI, S. 1935.—Studies on the helminth fauna of Japan. Part 9. Nematodes of fishes I. Jap. J. Zool., v. 6 (2): 337-386, figs. 1-65.

ZACHARIAS, O. 1913.—Ueber den feineren Bau der Eiröhren von *Ascaris megalocephala*, insbesondere über zwei ausgedehnte Nervengeflechte in denselben. Anat. Anz., v. 43 (8/9): 193-211, figs. 1-2, 1 pl. figs. 1-5.

CHAPTER XI

THE NERVOUS SYSTEM

B. G. CHITWOOD and M. B. CHITWOOD

Historical

Otto (1816) is credited with the discovery of the chain of ganglia in the ventral chord which we have since come to recognize as the ventral nerve. This author examined the two large ascarids, *Ascaris lumbricoides* and *Parascaris equorum*. Not long thereafter Owen (1836-39) observed the ventral nerve and preanal ganglion in *Dioctophyma renale*. Siebold (1848) reexamined *D. renale* and was able to confirm the existence of a ventral nerve but he did not see the preanal ganglion; he also described fibers from the ventral nerve to the somatic musculature. At about the same time Blanchard (1847) studying various ascaridids and filarioids ascribed a nervous function to the dorsal and ventral lines (chords) and described two cell groups on each side of the esophagus; these latter have since been identified as the paired lateral and ventral cephalic ganglia.

Meissner (1853) working on *Hexamermis albicans* described the nervous system in great detail and, considering his pioneer position, in a comprehensive and creditable manner. Briefly, he characterized the nervous system of this species as follows:

There is a central nervous system consisting of a fibrous ring around the esophagus and connecting four submedian anterior ganglia, two lateral ganglia and two posterior ventral ganglia. (He termed this group of structures the brain) *and three posterior ganglia at the tail. The lateral and anterior submedian ganglia supply the six anterior nerves to the six cephalic papillae* (Now recognized as the amphids and various cephalic papillary groups rather than individual papillae). *Posteriorly there are four longitudinal nerves, a dorsal, two lateroventral, and one ventral. Of these the dorsal originates directly from the fibrous ring while the other three originate as two subventral nerve trunks which unite posteriorly as the ventral nerve from which the lateroventrals branch off. Three of these longitudinal nerves are connected posteriorly with ganglia. The somatic musculature is connected with the longitudinal nerves, the nerve processes and the muscle cells merging so that one cannot say where one begins and the other ends.*

So far as this description goes it is reasonably accurate but, as we shall see later, it received a great deal of criticism.

Wedl (1855) saw the nerve ring in various parasitic nemas but mistook the sarcoplasmic part of the muscle cells for ganglion cells. Walter (1856, 1862) saw the nerve ring and associated ganglia of *Cosmocerca trispinosa;* he also saw the lateral, ventral and dorsorectal ganglia at the posterior end. But he confused the nervous system with other structures. Schneider (1860) stated that the nerve ring is the central organ and that structures described by other authors either were not present or were misidentified. Bastian (1863) described two "ganglionated chords" (now recognized as the lateral chords) as the nervous system of *Dracunculus medinensis.* Schneider (1863) redescribed the nervous system of *Parascaris equorum* as having a central organ or brain in the form of a circumesophageal commissure, the nerve ring, from which six anterior nerves extend to the cephalic sensory organs and two subventral branches to the ventral chord. Ganglion cells were seen in the six anterior nerves but it was noted that the lateral nerves were not connected with the nerve ring. (This difference is of fundamental importance and is further discussed on p. 163). He further stated that no special ventral nerve exists and criticized Meissner, Wedl, and Walter severely, stating that the transverse processes from the muscles to the median chords were muscles and that the cells seen were not nerve cells. Bastian (1866) confirmed Schneider's observations and further noted that the nerve ring usually has an inclined position with the dorsal side most anteriad. He also noted that processes from the muscle cells anterior to the nerve ring enter the nerve ring directly while those posterior to the nerve ring join the dorsal and ventral nerves. Bastian furthermore described lateral and ventral ganglia posterior to the nerve ring. Thereafter Schneider (1866) admitted the existence of median nerves.

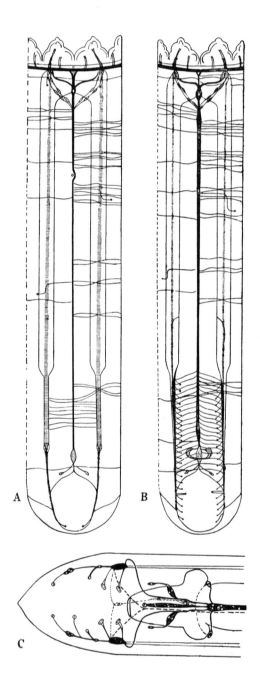

Fig. 128.

Diagrams of nervous system of *Ascaris*. A—Female dissection, ventral view. B—Male dissected, ventral view. C—Male tail, reconstruction (preanal papillae omitted). A-B after Brandes 1899, Abhandl. Naturf. Gesellsch. Halle, v. 21. C—Based on Voltzenlogel, 1902, Zool. Jahrb. Abt. Anat., v. 16 (3): 481-510.

Bütschli (1874) reinvestigated *Ascaris lumbricoides* and *Parascaris equorum* recognizing the six anterior sensory nerves with their ganglia and the lateral and ventral ganglia connected with the nerve ring. He found that some of the fibers of the lateral cephalic nerve passed the nerve ring, entered the lateral ganglia and continued through a hypodermal commissure (Cephalic lateroventral commissure I) to the nerve ring by way of the subventral nerve trunks. Bütschli also discovered first, the laterodorsal somatic nerves which originate from the dorsal part of the lateral ganglia; second, the lateroventral somatic nerves which originate from the subventral nerve trunks; and third, the ventrolateral somatic nerves which originate in the lateral ganglia and innervate the deirids or cervical papillae. In the posterior part of the body he found that the fibers from the nerve cells of the genital papillae are connected with the ventral nerve via the hypodermis. He concluded by stating that the question whether the transverse processes from the muscles to the nerves are of muscle or of nerve tissue remained unsettled but that its function seemed clear

Rohde (1883-1885) discussed the nervous system of the tail of *Ascaris* stating that the ventral nerve bifurcates anterior to the anus at which level there is a hypodermal commissure (preanal lateroventral commissure) to the lateral ganglia (lumbar ganglia) and a pair of internal commissures passing around the rectum. Joseph (1882-1884), Rohde's former teacher, attempted to claim this as his own work but the documentary evidence is sufficient to convict him of plagiarism. He obviously knew nothing about nematodes as is evidenced by his single other paper (1879) on this matter in which he identified a worm as *Anaplectus granulosus*, giving no description but remarking on its unusual size for a free-living nema (9 to 13 mm. long). This bears on our case since he later referred to the ventral nerve as being double in the same species. For various reasons, we shall later show (p. 162) that this double ventral nerve) must have been the primitive condition just as Meissner (1853) first indicated, but Joseph's publications should be ignored.

Hesse (1892) working on *Parascaris equorum* established the remaining general features of the nervous system topography as we know them today. To Bütschli's findings in the anterior part of the body Hesse added several more hypodermal commissures including the lateroventral commissure II. In the mid-region of the body of the female he discovered 30 dorsoventral hypodermal commissures on the right side, 12 on the left, while in the male he found 32 on the right side and 13 or 14 on the left. In the mid-region of the body of both sexes he saw a pair of dorsolateral papillae (postdeirids Fig. 128) connected with the lateral nerve and thence with the ventral nerve. In the caudal region he observed that the bifurcate ventral nerves unite with the lateral nerve by way of the preanal lateroventral commissure. The lateral caudal nerves formed by this union innervate the so called "caudal papillae" of the female (phasmids). In the male the preanal papillary nerve fibers join first the lateral nerve where a ganglion cell is located thence they return through the hypodermis to the ventral nerve by the genito-papillary commissures. All later contributions concerning the nematode nervous system are to be characterized as refinement work since no new major points in the topographic anatomy of the nervous system were added.

Topographic Anatomy

The most thoroughly studied species are, of course, *Ascaris lumbricoides* and *Parascaris equorum* due to the investigations of numerous early workers including Hesse (1892) and concluding with Goldschmidt (1908). We shall describe the anatomy as found by those workers in *Ascaris* in order to have a point of departure. The nervous system is divisible into four parts; the central, the peripheral, the recto-sympathetic and esophago-sympathetic systems. The latter has already been described (p. 97); it is connected with the central nervous systems by a pair of ventrolateral nerve fibers extending anteriad from the nerve ring.

Topographic Anatomy of Ascaris lumbricoides

CENTRAL NERVOUS SYSTEM. We classify as central nervous system the ganglia connected with the nerve ring, the nerve ring, and the ventral ganglion chain which is termed the ventral nerve. Attached to the anterior side of the nerve ring there are six small *cephalic papillary ganglia*—two subdorsal, two lateral and two subventral (Fig. 129 D). Fibers from the posterior side of these ganglia pass directly into the nerve ring while fibers from the anterior side form the six cephalic papillary nerves. These nerves may be compared with the cranial nerves of vertebrates since their cells are situated in the chief part of the central nervous system. Anterior to the nerve ring there is also a commissure connecting the dorsal and ventral sides of the nerve ring; this asymmetric commissure consists of a nerve fiber extending anteriad from the dorsal side of the nerve ring in the dorsal chord half way to the anterior end, then turning right, through the hypodermis to the ventral chord, thence posteriad to the anterior face of the nerve ring; this is called *dorsoventral commissure I*. Attached to the posterior side of the nerve ring there is a small *dorsal ganglion* in which the *dorsal somatic nerve* originates; there are also two small *subdorsal ganglia* and two large *ventral ganglia*. In the lateral area in the region of the nerve ring there are two large masses of nerve cells which are generally referred to as the *lateral ganglia*. Their neurones are of many different types and their fibers connect with the nerve ring in various ways. Sometimes, as in *Ascaris*, the lateral ganglia may be subdivided. Goldschmidt termed the groups of lateral cells which lie directly against the nerve ring and connect with it, the *internal lateral ganglia*. There is a pair of large hypodermal commissures, the *major lateroventral commissures* connecting the major part of the lateral ganglia with the nerve ring by way of the massive paired *subventral nerve trunks*. Several subdivisions of the lateral ganglia have processes into these commissures. The largest are the *amphidial ganglia* which connect anteriad with the amphidial nerve, the others by *posterior internolateral ganglia* and the *anterior externolateral ganglia*. The other two subdivisions of the lateral ganglia, the *median* and *posterior externolateral ganglia* are connected with the ventral nerve through a second pair of hypodermal commissures, the *minor lateroventral commissures*. One of these, the median externolateral, is also connected anteriad with the nerve ring. After giving off the major lateroventral commissures the subventral nerve trunks unite forming the ventral nerve which is in reality a chain of ganglia. The first and largest of these is called the retrovesicular ganglion; it is situated just posterior to the excretory pore at some distance posterior to the minor lateroventral commissures. Throughout the body asymmetrically placed *dorsoventral commissures* connect the dorsal and ventral nerves. A pair of symmetric dorsoventral commissures, the *anterior dorsoventral commissures*, originates from the anterior side of the nerve ring at a point posterior to where it joins with the major lateroventral commissure. There is also an asymmetric dorsoventral commissure passing by the posterior externolateral ganglion and proceeding anteriad to join the ventral nerve near the ventral ganglia; this is called the *oblique dorsoventral commissure*. Since all of these structures are rather closely associated with the nerve ring, they may be classified as central nervous system. Further groups of cells of nervous character are found throughout the length of the ventral nerve, groups being distinct in some species though not in *Ascaris*. The ventral nerve passes to the right of the excretory pore and vulva. Posteriad it gives off two internal branches, the *rectal commissures*, which extend through the body cavity and unite dorsal to the rectum; in the course of each of these commissures there is a *latero-rectal ganglion*. There is a *dorso-rectal ganglion* where the two commissures unite and from it the *median caudal nerve* extends posteriad first in the dorsal pulvillus, then in the ventral chord. Since both the commissures and ganglia are merely a branch of the ventral nerve the whole complex might be considered as a rectal sympathetic system. Posteriorly the ventral nerve ends by more or less bifid *preanal ganglia* from which the paired hypodermal *ano-lumbar commissures* extend to the lateral chords where they join the lateral nerves near the *lumbar ganglia*.

PERIPHERAL NERVOUS SYSTEM. The peripheral nervous system of nematodes consists of the *somatic nerves, the cephalic papillary nerves, amphidial nerves, genital papillae, cephalic papillae, amphids, deirids* and *postdeirids*.

SOMATIC NERVES. Those nerves which extend through the length of the body in the hypodermis are called somatic nerves. These include the *dorsal nerve* which originates at the posterior surface of the nerve ring with the dorsal ganglion and extends posteriad in the dorsal chord to the postanal region where it bifurcates, both branches extending through the hypodermis to the lumbar ganglia forming the *dorsolateral lumbar*

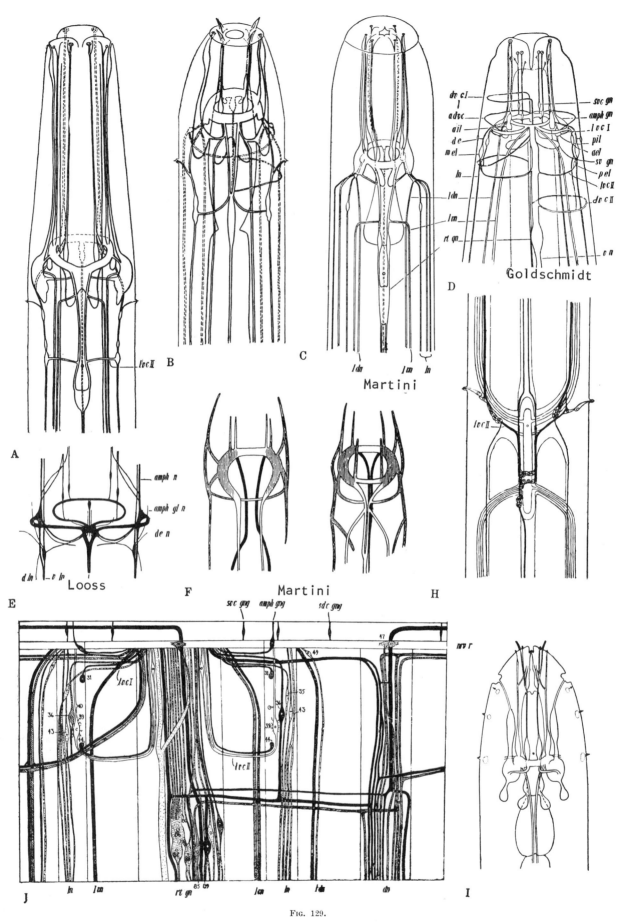

FIG. 129.

Diagrams of nervous systems. A—*Spironoura affine*. B—*Oesophagostomum dentatum*. C—*Oxyuris equi*. D—*Ascaris lumbricoides*. E—*Ancylostoma duodenale*. F—Trematoda (Central nervous system). G—Nematoda (Central nervous system). H—*Spironoura affine* (Showing postlateral ganglia and retrovesicular ganglion). I—*Siphonolaimus weismanni*. J—*Ascaris lumbricoides* (Dissection showing major nerve processes. C, based on Martini, 1916, Ztschr. Wiss. Zool., v. 116. D, after Goldschmidt, 1908, Ztschr. Wiss. Zool., v. 90. E, after Looss 1905, Rec. Egypt. Gov't. School Med., v. 3. F & G. after Martini, Deutsch. Zool. Gesellsch a. d. 23 Jahresversamml. zu Bremen. Remainder original.

commissures. Throughout its length no ganglia have been observed. It connects directly with innervation processes from the somatic muscles and is considered a motor nerve.

Laterodorsal somatic nerves originate at the dorsolateral side of the nerve ring at which point each contains one bipolar neurone. Thereafter, they extend through the hypodermis to the laterodorsal (submedian) lines where they assume a longitudinal posteriad course.

Lateroventral somatic nerves arise from the lateroventral part of the nerve ring and pass through the ventral ganglia where each has a bipolar neurone. They then leave the ventral chord via the major lateroventral commissures and reach the lateroventral (submedian) lines and extend posteriad like the laterodorsals. Neither of these paired ''submedian'' nerves has any ganglia in its course and like the dorsal nerve the submedians are considered motor nerves since they connect with innervation processes of the somatic musculature.

The *Ventral* nerve has already been described as part of the central nervous system and though it undoubtedly contains associational centers, it also acts as a motor nerve since it innervates the muscle cells in the subventral muscle sectors.

Dorsoventral commissures have previously been mentioned. These occur at intervals throughout the body and with one exception, the anterior dorsoventral commissures, they are unpaired and are supposed to coordinate the activity of the somatic muscles. Goldschmidt (1908) was of the opinion that the ''submedian'' somatic nerves were also connected with the ventral nerve by these commissures but this has not been adequately demonstrated.

Ventrolateral nerves originate from the internolateral ganglia, the anterior and the medial externolateral ganglia; the fibers from these ganglia come together posteriad and the nerves so formed extend posteriorly at the side of the excretory canal. Further ganglia have not been observed in their course until they reach the posterior extremity. In the anal region each ventrolateral nerve contains a group of nerve cells called the lumbar ganglion. At approximately the same level the ano-lumbar commissure connects it with the ventral nerve. Paired lateral nerves which continue posteriad from these ganglia are called the *lateral caudal nerves.* They innervate the phasmids in the female. In the male the ventrolateral nerve (also called the *bursal nerve*) innervates the genital papillae and paired *genito-papillary commissures* connect it with the ventral nerve. Postanal genital papillae are innervated by processes from the enlarged lumbar ganglion and branches of the lateral caudal nerves. The lateroventral somatic nerve is thus seen to be at least partly a sensory nerve.

Dorsolateral nerves have been described by Hesse (1892), Brandes (1899) and Voltzenlogel (1902). Goldschmidt (1908) apparently overlooked them; they might correspond to the short processes extending posteriad from the anterior externolateral ganglia. They unite posteriad with the ventrolateral nerve in the anal region.

Cephalic papillary nerves and papillae. There are six anterior papillary nerves arising from the anterior surface of the nerve ring and extending to the anterior extremity; unlike the somatic nerves, the papillary nerves go through the body cavity, being applied closely to the external surface of the esophagus. Each of these nerves contains a ganglion (cephalic papillary) of bipolar nerve cells at the anterior surface of the nerve ring. The sensory endings of the nerves include the cephalic papillae. Each subdorsal and subventral cephalic papillary nerve terminates anteriad in a large pair of partially fused papillae (dorsodorsal and laterodorsal or ventroventral and lateroventral) of the external circle, and a small rudimentary papilla (internodorsal or internoventral) of the internal circle. The lateral papillary nerves each terminate in one large papilla of the external circle (ventrolateral and a rudimentary papilla of the internal circle (internolateral).

Amphidial nerves and amphids. Like the cephalic papillary nerves, the amphidial nerves each innervate a cephalic sensory organ but unlike the papillary nerves their connection with the nerve ring is indirect. The nerve cells are situated in the amphidial ganglia which lie posterior to the nerve ring, and their axones pass into the major lateroventral commissures to the ventral nerve trunks before reaching the nerve ring. Anteriorly each amphidial nerve enters an *amphidial gland* and its processes break up in an elongate sac, *amphidial pouch;* the sensory elements which represent the specialized ends of individual neurones are called *terminals* and the group of elements is called a *sensilla.* From the pouch, anteriad, there is a short *amphidial duct* which opens to the outside at the *amphidial pore.*

Deirids. These are paired papillae in the cervical region commonly called *cervical papillae.* Each is innervated by a branch of the nerve trunk which connects the medial externolateral ganglia with the nerve ring; the sensory cell of each deirid lies in that ganglion.

Postdeirids. These are paired and asymmetrically placed papillae first seen by Hesse (1892), in the mid-region of the body. Each is innervated by a single nerve fiber connected with a sensory neurone in the lateral chord. The axone passes through the hypodermis to the ventral nerve.

Genital papillae and Connections. Each preanal genital papilla sends a nerve fiber through the hypodermis to the lateral chord where it enlarges to form a bipolar neurone the axone of which joins the ventrolateral somatic nerve and reaches the ventral nerve by way of a genito-papillary commissure. In addition there is a medioventral preanal papilla which, according to Voltzenlogel (1902) is innervated by four bipolar neurones connected directly with the posterior end of the ventral nerve. (Fig. 128 C). Posterior to the anus the genital papillae are innervated by processes of the lumbar ganglia. Of these, the first pair is double and has two nerve fibers.

Phasmids are known to occur in both sexes of *Ascaris* and to be innervated by processes of the lateral caudal nerves but no further details have been obtained.

Topographic Anatomy of the Nervous System of Species Other Than Ascaris

The nervous systems of only a few other nematodes have been studied, namely: *Siphonolaimus weismanni* by zur Strassen (1904), *Ancylostoma duodenale* by Looss (1905), *Hexamermis albicans* (syn. *Mermis albicans*) by Rauther (1906), *Oxyuris equi* by Martini (1916), *Rhabditis strongyloides* by Chitwood (1930), *Camallanus* spp. and *Cucullanus heterochrous* by Törnquist (1931), *Cephalobellus papilliger* by Chitwood and Chitwood (1933) and *Rhabditis terricola* by Chitwood and Wehr (1934). In preparing this volume we have made a point of studying two additional species, *Spironoura affine* and *Oesophagostomum dentatum.* Restudy of *Rhabditis strongyloides* has shown that trust was misplaced in the use of methylene blue as an intravitam stain and the publication in which it was described (Chitwood, 1930) contains so many gross errors that it should be ignored. The correct form of the nervous system of *Rhabditis* is shown in Fig. 8. The general features of the nervous system in all the forms studied is very similar even though they represent considerably diverse groups.

Central nervous system. Differences in the central nervous system lie chiefly in the degree of subdivision of the lateral ganglia, the form of the ventral ganglia and degree of fusion of the ventral nerves.

In Siphonolaimus weismanni (Fig. 129 I) the ventral nerves are paired as far as they have been traced but this tracing did not extend so far as the retrovesicular ganglion. In *Ancylostoma duodenale, Oesophagostomum dentatum, Spironoura affine, Cephalobellus papilliger,* and *Rhabditis terricola* they are fused in the retrovesicular ganglion and thereafter are apparently single but, considering the minuteness of the left fiber group, doubleness of the ventral nerve might be overlooked. In *Oxyuris equi* (Fig. 129 C) Martini found the ventral nerves paired to the posterior side of the vulva which happens to be situated distinctly pre-equatorial in this species. In the other species the ventral nerve passes to the right side of the vulva. Both Martini (1916) and Chitwood (1933) have enumerated the cells found in the ventral nerve between the retrovesicular ganglion and the posterior *preanal ganglion* but neither found sufficiently clear grouping of the cells to warrant establishment of a series of named ganglia; comparing their work with the situation in *Spironoura* (Fig. 132 D) we find that there are 25-30 neurones of which a group of 6-7 form a post-vulvar ganglion, that sometimes there is a more or less distinct prevulvar ganglion and that other neurones may or may not be grouped. Looss found a vaginal nerve, originating from the prevulvar ganglion, which forms a commissure around the vagina at its juncture with the uteri. The same author also described paired *anal nerves* originating in the paired preanal ganglia and extending posteriad to the anterior lip of the anus (Fig. 130 AA). At the preanal ganglion the ventral nerve is double in *Ancylostoma duodenale, Oxyuris equi* and *Rhabditis terricola.* The apparent doubleness in both anterior and posterior ends of the ventral nerve caused Meissner and many later authors to conclude that the entire ventral nerve was at one time double. Arguments

163

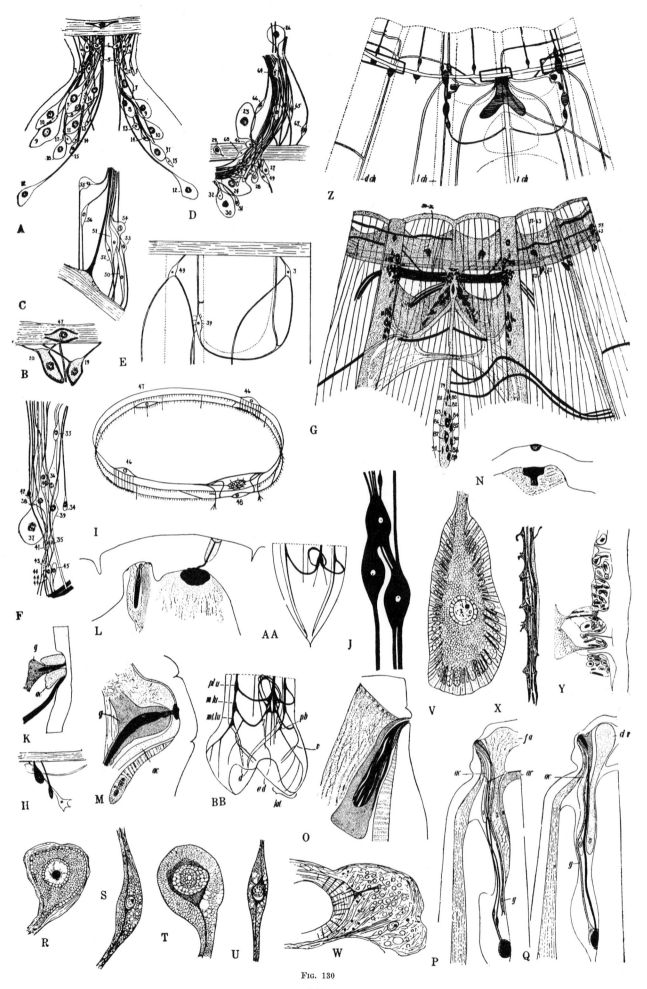

Fig. 130

164

against this view have been based on the fact that in most nemas studied the ventral nerve goes to the right of the vulva instead of dividing to go around it. *Oxyuris equi* is a conspicuous exception in this latter respect but here it might be argued that the vulva is shifted anteriad. As Looss pointed out, the asymmetric vulvar by-pass is to be expected in developmental anatomy since the ventral nerve is already in position when the vulva is formed. The writers subscribe to the primitive double ventral nerve hypothesis.

The ventral ganglia are each transversely lobed in *Siphonolaimus, Hexamermis,* and *Rhabditis* the smaller lobes being called the postventral ganglia, the term ventral ganglia being reserved for the anterior lobes. In other nemas studied, the ventral ganglia are not subdivided.

The lateral ganglia are more or less lobed in all nemas, but in none thus far studied are they subdivided into as many parts as in *Ascaris.* As a general rule there is a lobe forming the *dorsolateral ganglion* connected with the nerve ring near the origin of the laterodorsal and dorsolateral somatic nerves. The term *lateral ganglion* is reserved for the major part of the lateral ganglion while a posterior lobe, the postlateral ganglion is some distance from the major ganglion group; it connects anteriorly and posteriorly with the ventrolateral somatic nerve and ventrally with the minor lateroventral commissure.

Somatic Nerves. The dorsal somatic nerve and ventrolateral somatic nerves in the genera studied are similar to those of *Ascaris* with the exceptions that in *Camallanus* and *Cucullanus* Törnquist identified several nerve cells in the course of the dorsal nerve and in those forms without deirids (*Oxyuris, Hexamermis, Cephalobellus* and *Siphonolaimus*) the ventrolateral nerve has no branch to the surface. A dorsolateral nerve has been seen only in *Ancylostoma* and *Oxyuris* and submedian somatic nerves in *Cephalobellus, Spironoura, Rhabditis, Hexamermis* and *Oesophagostomum.* Numerous dorsoventral commissures, such as Hesse describes in *Ascaris* have not been seen in any of the other forms studied. The two anterior lateroventral commissures, the rectal and anal commissures are the only ones known to be generally existent.

Cephalic papillary nerves and *papillae.* There are always six cephalic nerves connected posteriad with six papillary ganglia and anteriad with six groups of papillae. Törnquist (1931) observed these nerves and ganglia but was unable to find the lateral cephalic papillary group and confused the, fibers and ganglia with the amphidial nerve and lateral ganglia respectively. The papillae have been previously described (pp. 56-66) in these and other forms so that there is no need of repeating the discussion here. We need only call to the readers' attention that there are primarily three papillae in each submedian group and two in each lateral group. Among parasitic phasmidians there is a general tendency for the internal circle of papillae to become reduced or rudimentary. There is also a tendency toward reduction of the external

circle either through reduction of the externomedial papillae or through the joining of the lateromedial papillae. The ventrolateral papillae may or may not become reduced or rudimentary coincident with reduction of the mediomedials.

Amphidial nerves and amphids are fundamentally like those of *Ascaris* in all forms.

Enumerative and Minute Anatomy

CEPHALIC PAPILLARY GANGLIA, NERVES, PAPILLAE AND ASSOCIATED CELLS. Goldschmidt (1903) worked out the cephalic papillary nervous system of *Ascaris* along with the nonnervous cells of the anterior end. We have already discussed many of the non-nervous cells (p. 45) but because of their rather intimate association with the papillary nerves some repetition will be necessary.

Arcade Cells. These are apparently hypodermal cells of the stomatal and labial regions. Nine such cells have been found in *Ascaris* (Goldschmidt, 1903, and Hoeppli, 1925), *Oxyuris* (Martini, 1916), *Strongylus* (Imminck, 1924), *Cephalobellus* (Chitwood & Chitwood, 1933), and *Rhabditis, Spironoura* and *Oesophagostomum* (this publication). They are not connected in any way with the nervous system but must be distinguished from it. In all except *Strongylus, Oesophagostomum* and *Spironoura* they take the form of elongate, posteriorly directed cells united anteriad in a transverse plate opposite the base of the labial region. Their cell bodies are closely applied to the esophageal surface, distributed as follows: one dorsal, four dorsolateral, two ventrolateral and two ventral (Fig. 46 A-B). In *Strongylus* (Fig. 46 E) they are distributed in the same pattern but are confined to the stomatal region and take the form of three protoplasmic bands. In *Oesophagostomum* they are situated in the body cavity but at the level of its nucleus each cell has a process inserting it into the hypodermis next to the chord with which it is associated. There are two subdorsal arcade cells beside the dorsal chord, one dorsolateral on the left side two on the right side next to the lateral chord, one on each side ventrolateral next to the lateral chords and two subventral on the right side of the ventral chord. In *Spironoura* the arrangement is more nearly typical but instead of there being four dorsolateral and two ventrolateral there are two dorsolateral and four ventrolateral.

Non-specific Connective tissue. There may be several types of connective tissue cells not associated with the nervous system but situated in the body cavity around the esophagus. In *Ascaris* Goldschmidt listed three fibril cells opposite the three esophageal radii and two "Füllzellen" (left dorsolateral and subventral) and in *Oxyuris* Martini listed three "Bindegewebe" cells, one dorsal and two ventrolateral, in addition to those already mentioned. In *Oesophagostomum* the three fibril cells have been observed but other cells could not be identified.

Submedian Papillary Ganglia, Nerves and Papillae. In *Ascaris* Goldschmidt (1903, 1908) found each of these ganglia to be composed of seven bipolar sensory neurones (Cells 50-56 times 2, subventral, and cells 57-63 times 2 subdorsal, Fig. 130 G). Posteriad a process from each cell enters the nerve ring while anteriad they come together forming the submedian papillary nerves. In addition to the nerve processes a glia process enters the corresponding nerve from each of the four giant glia cells situated on the anterior surface of the nerve ring (Fig. 130 C).

Anteriorly the processes of four of the neurones stop in the postlabial region while the other three innervate papillae. The fiber innervating each lateromedian papilla (laterodorsal or lateroventral) anteriad becomes ensheathed by the glia process from the corresponding *glia cell* on the nerve ring and this is in turn partially surrounded by an *escort cell,* the two together forming the papillary mass. At its termination the cuticle has a very deep invagination from which a fine sensory hair projects; the hair is continuous with the dendritic process of the nerve (Fig. 130 L). The medio-medial papillae are each formed by a glia and escort cell; in this case the nucleus of the glia cell (Fig. 130 P) is some distance anterior to the nerve ring. The sensory terminus differs from that of the lateromedial papilla in that it ends under the surface in a sensory plate or receptaculum (Fig. 130 L). A clavate cell accompanies each of the submedian papillary nerves but is not associated with either of the processes to the external circle of papillae; it may, perhaps, act as an escort cell of the fiber to the internomedial papilla. This papilla is greatly reduced in

FIG. 130

Details of nervous system. A-Z—*Ascaris lumbricoides* (except K-Q, *Parascaris equorum*). (A—Ventral ganglion; B—Dorsal ganglion; C—Subventral cephalic papillary ganglion; D—Amphidial and internal lobes of lateral ganglion; E—Diagrammatic representation of the tripolar neurones from which fibers of the laterodorsal (49) and the lateroventral somatic nerves originate; F—External lobe of lateral ganglion; G—General dissection of central nervous system as seen from inside, ventral chord median; H—Subdorsal ganglion (Glia cells white); I—Diagram of commissural connections of nerve ring (47 dorsal), lines posterior connect with innervation processes and somatic nerves; J—Cells 80, 86, 87 and 88 of the retrovesicular ganglion showing cells in direct continuity; K—Ventrolateral papilla showing two nerve endings, upper one forming standard papilla, lower one rudimentary; actually the lower is supposed to represent the rudimentary dorsolateral, the upper the well developed ventrolateral; L—Submedian double papilla, laterodorsal to readers' right, dorso-dorsal to readers' left (with receptaculum in white); M—Laterodorsal papillae showing sensory plate, only plate and terminus seen in L; N—Plate of laterodorsal papilla; O—Amphidial pouch and sensilla; P—Dorsodorsal papilla and associated cells; Q—Laterodorsal papilla and associated cells; R—Cell 46, showing glia capsule; S—Cell 51, showing neurofibril; T—Cell 25, showing radial network and neurofibrillar basket; U—Cell 67, showing neurofibrillar network; V—Cell 23, showing glia fibrils entering protoplasm; W—Cell 24, glia cell of nerve ring; X—Ventral nerve with innervation processes, methylene blue; Y—Ventral nerve with insertion of neurofibrils from innervation process; Z—Diagrammatic dissection of nervous system from which parts of nervous system were reconstructed (Figs. 133 & 134); AA-BB—*Ancylostoma duodenale* (Diagrams of caudal part of nervous system, AA—Female, BB—Male).

Figs. A-J, After Goldschmidt 1908, Ztschr. Wiss. Zool., v. 90. K, After Goldschmidt 1903, Zool. Jahrb. Abt. Anat., v. 18 (1). R-Y, After Goldschmidt, 1910, Festschr. Hertwig, v. 2., (X, From Deineka). Z, After Goldschmidt, 1909, Ztschr. Wiss. Zool., v. 92 (2). AA-BB, After Looss 1905, Rec. Egypt. Govt. School Med., v.3.

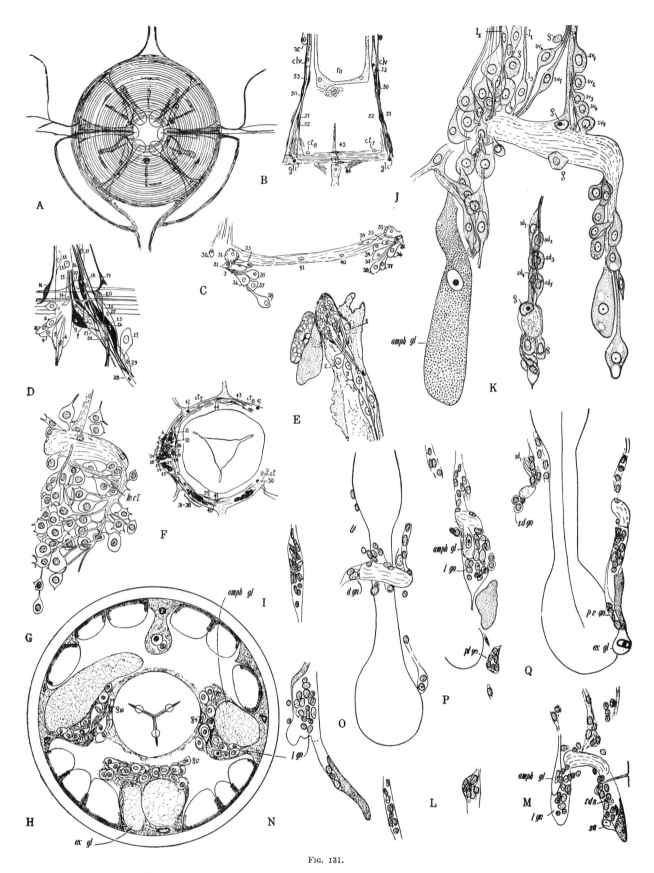

FIG. 131.

Details of nervous system. A–F—*Oxyuris equi* (A—Diagram showing cephalic part of nervous system as seen *en face;* B—Dorsal part of nerve ring and subdorsal cephalic papillary ganglia; C—Ventral part lateral lobe of lateral ganglion; F—Diagrammatic reconstruction of of nerve ring and ventral ganglion; D—Lateral ganglion; E—Dorso-lateral lobe of lateral ganglion; F—Diagrammatic reconstruction of ganglia on level of nerve ring.) G—*Metoncholaimus pristiuris* (Longi-tudinal section showing nerve ring and parts of lateral and ventral ganglia). H—*Oesophagostomum dentatum* (Cross section just posterior to nerve ring). I—*Rhabditis lambdiensis,* retrovesicular ganglion. J—*Spironoura affine* (Longitudinal section showing parts of lateral and ventral ganglia. The lateral cells (on readers' left) which terminate as circles at level of nerve ring are amphidial neurones, the circles representing cut ends of fibers to lateroventral commissure I.). K—*Spironoura affine* (Subdorsal cephalic papillary ganglion and subdorsal ganglion). L–Q—*Rhabditis terricola* (L—Postlateral ganglion; M—Longitudinal section showing parts of lateral ventral and subventral papillary ganglia; N—Section adjoining M, shaded area to right is excretory tissue, neurones next to it are postventral ganglion cells; to right are retrovesicular ganglion cells. Q, O & P follow in series.

A–F, After Martini 1916, Ztschr. Wiss. Zool., v. 116. Remainder original.

166

Ascaris; Hoeppli (1925) described a special glia cell for this structure. The clavate cell acts as its escort cell.

In *Oxyuris,* Martini (1916) found three neurones (Cells 50-52 times 2 in Fig. 131 B) in each subdorsal papillary ganglion and two glia cells (Cells 53 times 2 and ct. 7-8 in Fig. 131 B) disposed in the same manner as the glia cells of *Ascaris.* The cell whose fiber enters the nerve ring between the other two (Cells 52) becomes surrounded anteriorly by the clavate cell and then terminates as an internomedial papilla; no glia cell is associated with it. The cell whose fiber enters the nerve ring most laterally (Cell 50 retains this position throughout its course, becoming associated first with the glia cell (Cell 53), then the escort cell (Hügelzelle), all three together then forming the large laterodorsal papilla. The cell whose fiber enters the nerve ring most dorsally is probably supplied by glia tissue from the cell on the nerve ring (Ct. 7-8) for no other glia or escort cells are associated with it. It terminates under the cuticle as the rudimentary dorsodorsal papilla. The subventral papillary ganglia are similar to the subdorsal except that there are two cells connected with the ventroventral rudimentary papillae; one of these might be a glia cell.

In *Spironoura* we have found five bipolar neurones in each of the subdorsal and subventral papillary ganglia. With these, three glia and four escort cells are associated. In addition a submedian glia cell at the surface of the nerve ring also has a process into each nerve. Three of the neurones with corresponding glia and escort cells form the three well developed papillae in which each of these nerves terminate. The ultimate destination of the two other neurones and the extra glia and escort cells is unknown. In addition to the neurones just described each subventral nerve receives processes from two bipolar neurones in the region of the lateral papillary ganglia (Fig. 131 J); presumably they innervate the esophago-sympathetic system.

In *Oesophagostomum* each subdorsal papillary ganglion contains only four bipolar neurones while the subventral contains five. (Fig. 132 A). As in *Spironoura* seven cells in addition to the anterior glia cell of the nerve ring are associated with each papillary nerve, three glia cells and four escort cells. Anteriorly each nerve terminates in a reduced papilla of the internal circle and a conoid (double) papilla of the external circle (Fig. 129 B).

Lateral papillary ganglia. In *Ascaris,* Goldschmidt (1903, 1908) found four bipolar neurones (Fig. 130 D Cells 64-67), one glia cell, one escort cell and one clavate cell. Two of these neurones (Cells 66-67) innervate the ventrolateral papilla but only one is surrounded by the glia and escort cells. The most ventral cell (66) is the one surrounded by glia and forms the bulk of the sense organ; it ends in a plate or receptaculum (Fig. 130 K) like those of the medial papillae. The other fiber (67) ends under the cuticle without any special sensory terminus and must, therefore, be considered rudimentary. The most anterior cell (64)) becomes associated with the clavate cell and presumably innervates the greatly reduced internolateral papilla.

In *Oxyuris,* Martini (1916) recorded four bipolar neurones (Fig. 131 D) connected with the nerve ring, in each of the lateral papillary ganglia (Cells 11, 14, 18 and 19); with these cells two glia cells (12 and 17) and one clavate cell are associated. The most ventral of these (19) is probably supported by the corresponding glia cell (17); it terminates anteriorly under the cuticle as the rudimentary ventrolateral papilla. The most dorsal (11) is associated with the glia cell (12) and the clavate cell and terminates anteriad as the greatly reduced internolateral papilla.

In *Spironoura* there are likewise four bipolar neurones (Fig. 131 K) in the lateral papillary nerve but in addition two glia and three escort cells were observed. The most dorsal neurone innervates the internolateral papilla and is associated with the most anterior glia and escort cells while the most ventral neurone innervates the well developed ventrolateral papilla and is associated with the other glia cell and at least one of the escort cells. The lateral papillary ganglia of *Oesophagostomum* contain only three bipolar neurones which become associated with two glia and no escort cells. The nerve innervates the greatly reduced internolateral and rudimentary ventrolateral papillae.

It seems obvious from the comparative anatomy that escort cells (including clavate cells) are definitely correlated with papilla formation. In *Ascaris* and *Oxyuris* the number of such cells associated with each nerve is the same as the number of papillae reaching the body surface, for the lateral nerve two in *Ascaris* and one in *Oxyuris* and for the submedian

nerves three in *Ascaris* and two in *Oxyuris.* In *Oesophagostomum* there are none laterally and no distinct papillae. However the submedian nerves of both *Spironoura* and *Oesophagostomum* have four escort cells though there are but three papillae in *Spironoura* and one very large papilla in *Oesophagostomum.* We have previously seen (p. 58) that the large conoid papillae of *Oesophagostomum* are double; in addition we have found a small internomedial papilla. This would leave one escort cell unaccounted for in both *Spironoura* and *Oesophagostomum.* The reason for variability in number of glia cells is not apparent, neither is the variation in the number of neurones. Three neurones are required for papillary innervation and these are always present, even in the lateral nerve. If we may judge by Goldschmidt's work on the ventrolateral papilla (Fig. 130 K) we cannot help but conclude that a dorsolateral papilla (homologous to the lateromedial papillae) was present in ancestral forms. The additional cell in the subventral ganglia of *Oesophagostomum* probably connects with the esophago-sympathetic system; the cells which have this function in the other genera studied have been omitted from the descriptions. The neurones not connected with papillae in *Ascaris* and *Spironoura* must end in the lip tissues.

NERVE RING. The nerve ring is an associational structure where processes from the many ganglia of the central nervous system come into direct relationships with one another. Goldschmidt believed its primary function was to correlate motor impulses entering the various motor nerves but he did not preclude the association of sensory impulses.

In *Ascaris* there are eight cells which must be classified as cells of the nerve ring. Four of these are the anterior glia cells which send processes into the submedian papillary nerves. The other four cells, numbered 46 (paired lateral), 47 (unpaired dorsal) and 48 (unpaired ventral) are associational neurones (Fig. 130 I). These four cells have direct continuity with each other and with various motor neurones, with motor nerves and with commissural cells of the lateral ganglia. Several of the unipolar central cells (21, 27, 29) were found to divide after entering the nerve ring, having one process dorsad, the other ventrad (Fig. 134). Goldschmidt reconstructed portions of the dorsal, ventral ,and lateral areas of the nerve ring (Fig. 133, 134) showing the direct anastomoses of many of the fibers. However, he was able to identify particular fibers with cells only in the same instances (those labelled Z 21 etc.) in which the number corresponds to the cell number given in other illustrations. In other instances he merely numbered the individual fibers for descriptive purposes (small numbers opposite fibers). The fibers of the major lateroventral commissure are labelled L 1-12, those of the ventral nerve B 1-31, those of the submedian somatic nerves Subl I-IV, 1-4, and a-d. The ordinates and coordinates are also numbered but this is merely for purposes of location. Further discussion seems unnecessary since those interested in the detail may obtain it from the illutrations.

In *Oxyuris equi* Martini (1916) found eight glia cells (ct. in Fig. 130 F) and seven commissural cells corresponding to the four described by Goldschmidt for *Ascaris.* The commissural cells are located: two lateral (cell 10 r. l.), two dorsal (43, 44) and three ventral (39, 40, 41).

In *Spironoura affine* (Fig. 131 J-K) and *Oesophagostomum dentatum* (Fig. 132 A-B) we find the same arrangement as in *Ascaris* except that in both forms there is a pair of subventral glia cells on the posterior surface of the nerve ring and in *O. dentatum* there is also a dorsal posterior and a ventral anterior glia cell.

DORSAL AND SUBDORSAL GANGLIA. In *Ascaris* the dorsal ganglion consists of two central cells (Fig. 130 C) each with two processes to the nerve ring while in *Parascaris equorum* one of these cells has a single process. In the latter species Goldschmidt also saw the posterior dorsal glia cell which we have mentioned on the nerve ring of *Oesophagostomum.* The dorsal ganglion of other species also contains two nerve cells which may have a very long central cylinder as in *Spironoura;* in this case they appear to be located in the dorsal nerve. Possibly this is the reason for our having found only one such ganglion cell in *Cephalobellus* and for Martini's not finding a dorsal ganglion in Oxyuris (an extra dorsal commissural cell probably corresponds to one of the dorsal ganglion cells of *Ascaris*).

The subdorsal ganglia each consist of two central cells in *Ascaris, Spironoura, Rhabditis* and *Oesophagostomum* and with it two glia cells are associated in the first two genera but only one in the two last named. Paired posterior subdorsal glia cells in *Oxyuris* and *Cephalobellus* were attributed to the

167

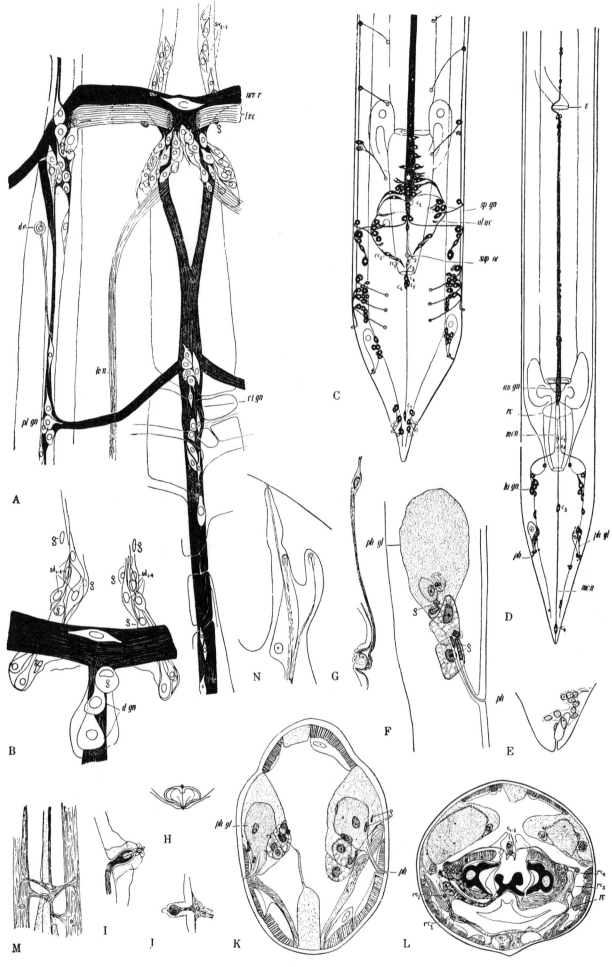

Fig. 132.

168

nerve ring, but they are undoubtedly homologous to those associated with subdorsal ganglia where such ganglia are present. Casual observation indicates that absence of subdorsal ganglia may be characteristic of the Thelastomatidae and Oxyuridae.

LATERAL GANGLIA, AMPHIDS AND ASSOCIATED STRUCTURES. In all nemas the lateral ganglia contain the largest number of cells and are so intimately associated with the amphids and deirids that it is easiest to consider them at once. In *Ascaris*, Goldschmidt found a total of 35 nerve cells while in *Oxyuris*, Martini found 21, in *Cephalobellus*, the writers found 26 cells, in *Spironoura* 42, in *Oesophagostomum* 41 and in *Rhabditis* 42.

Postlateral ganglia are absent in *Oxyuris* and *Cephalobellus* while in *Spironoura*, *Rhabditis* and *Oesophagostomum* they are the most clearly set apart. In *Spironoura* and *Oesophagostomum* each of these ganglia is composed of two groups of cells. The anterior group consists of two cells, one a sensory cell connected with the deirid, the other a glia cell. The second group consists of five cells all of which have processes into the minor lateroventral commissure. In *Rhabditis* all seven cells are in one group (Fig. 131 P).

The postlateral and mediolateral ganglia of *Ascaris* contain cells corresponding to those found in the postlateral and mediolateral ganglia of other genera studied but the grouping differs. Since Goldschmidt's grouping in the case of *Ascaris* is rather artificial, there being no distinct lobes in that species, we shall regroup the cells he described. The internolateral ganglia are in close association with the amphidial ganglia and cannot be grossly distinguished from them. Goldschmidt subdivided the internolaterals into anterior and posterior lobes, but this division is not practical. Each whole internolateral ganglion contains eleven cells, seven being unipolar with a process to the nerve ring (23-29), three unipolar with a process to the major lateroventral commissure (30-32) and one (49) tripolar with one process to the nerve ring, one to the ventrolateral nerve and one to the dorsosubmedial nerve.

In *Ascaris* the amphidial ganglia (Fig. 130 D) each contain 11 sensory neurones (68-78) connected anteriorly with the amphidial nerve and posteriorly with the major lateroventral commissure. The amphidial gland lies dorsal to the amphidial ganglion and extends anteriad eventually surrounding the amphidial nerve. At this level (slightly posterior to its terminus) the amphidial gland duct attains a cuticular lining and is slightly dilated forming the *amphidial pouch*. Within this structure there is a *sensilla* consisting in *Ascaris* of 11 rod-like sensory *terminals* (Fig. 130 O). The lumen of the pouch is in direct continuity with a short tube leading to the amphidial pore.

The externolateral lobes of the lateral ganglia were subdivided by Goldschmidt into three parts, the anterior, medial and posterior parts but since there is no real subdivision (Fig. 130 F) we will describe these structures as a single unit. Each whole ganglion contains 13 cells of which eight are unipolar, four have a process to the nerve ring (37, 40, 41, 42), two have a process to the major lateroventral commissure (33, 34) and two have a process to the minor lateroventral commissure (44, 45). The five bipolar neurones have processes as follows: two (38, 43) have processes to the nerve ring and ventrolateral nerve; one (35) has a process to the major lateroventral commissure and another to the lateral nerve; one (39) has a process to the nerve ring (Fig. 130 E) and another to the minor lateroventral commissure (a side branch of this process innervates the deirid); and one (36) has its anterior process to the nerve ring, its posterior process entering the oblique ventrodorsal commissure (right side) or the ventrodorsal commissure II (left side).

Martini (1916) found only two subdivisions of the lateral ganglia in *Oxyuris*, these being the dorsal and ventral lobes (Fig. 131 D). The dorsal lobe contains six neurones, three

unipolar to the nerve ring (1-3) and three bipolar with one process to the nerve ring and the other to the common root of the laterodorsal, dorso-and mediolateral somatic nerves. The ventral lobe consists of 15 cells, five of which are bipolar, innervating the amphids and passing through the latero-ventral commissure; the latter (9, 15, 16, 22, 26) undoubtedly correspond to Goldschmidt's amphidial ganglion (11 cells in *Ascaris*). Of the remaining 10 neurones, two are unipolar to the nerve ring (7-8), one is unipolar to the lateroventral commissure (27), three are bipolar with processes to the ventrolateral somatic nerve and the nerve ring (20, 23 ,24) and four are bipolar with processes to the lateroventral commissure and the nerve ring. The mediolateral nerve has three neurones in its course, two far back in the body region and one not far distant from the lateral ganglion. The latter might be considered part of the lateral ganglion, possibly representing a rudimentary postlateral lobe.

The writers (1933) found no marked subdivision but a distinct tendency toward separation of an amphidial and a dorsal lobe as well as a rudimentary postlateral lobe of the lateral ganglion in *Cephalobellus*. The chief lobe contained three unipolar cells to the nerve ring (4, 12, 17), two unipolar cells to the lateroventral commissure (20,23), six cells of undetermined character (14, 16, 18, 21, 22), and a glia cell (8). The dorsal lobe contained two unipolar cells to the nerve ring (6, 7) and three bipolar cells to the nerve ring and dorsosubmedian lateral nerve trunk. The amphidial lobe contained eight bipolar cells to the lateroventral commissure and amphid (13, 19, 24-28, N). The postlateral lobe contained two neurones (29, 30) attached to the lateral nerve.

In *Spironoura* there is a very distinct division of the lateral ganglia into anterior and posterior lobes, the anterior lobe containing 35, the posterior lobe seven cells. In the anterior lobe subdivision is indistinct. The courses of only part of the cells have been traced; of these six are unipolar direct central cells, seven are unipolar cells entering the major lateroventral commissure, eight dorsolateral cells are connected with the dorsosubmedian and lateral nerve trunk and eight are connected with the amphidial nerve and lateroventral commissure. The amphidial glands (Fig. 131 J) are particularly massive in this form. In the postlateral ganglion only six of the cells are neurones, the seventh being the glia cell of the deirid. At least three, possibly four neurones are bipolar with anterior processes in the lateral nerve and posterior processes through the minor lateroventral commissure (one of these has a branch to the deirid); another of the cells is bipolar with one process anterior and another posteriad entering the lateral nerve, while the sixth appears to be tripolar, with two processes entering the lateral nerve and a third the dorsolateral commissure.

In *Oesophagostomum* and *Rhabditis* practically the same subdivision of the lateral ganglia (seven cells in postlaterals) has been observed except that in *Oesophagostomum* the dorsoventral commissure originates with the dorsosubmedian nerve.

Metoncholaimus pristiuris (Fig. 131 D) is the only aphasmidian that we have casually studied and no exact information is available. However, considering Filipjev's (1912, 1934) indication that the nervous system is basically different in aphasmidians it is interesting to note the same general arrangement in the lateral ganglion with many cells entering the lateroventral commissure. The cells are not so closely packed but there is no other obvious difference.

VENTRAL GANGLION. In *Ascaris* the ventral ganglion (Fig. 130 A) is bilobed and contains 33 cells of which 15 are paired and three unpaired, cell 16 being medial (on right side in figure) and cells 17 and 18 on the right side. All except two cells are unipolar and the bipolar exceptions innervate the lateroventral somatic nerves. In *Oxyuris* the ventral ganglion is definitely paired, there being two groups of eight cells (31-38 r. l.) making a total of 16. The two extra ventral commissural cells of the nerve ring in this species probably correspond to two unipolar ventral ganglion cells in *Ascaris*; even so, only 18 ventral ganglion cells would have been accounted for and this is a marked reduction from 33. In *Cephalobellus* there are likewise eight pairs of cells but there is also a medioventral cell making 17 in all; one cell in each lobe is bipolar as in *Ascaris*. In *Spironoura* the partly bilobed ventral ganglion contains 29 neurones while in *Oesophagostomum* and *Rhabditis* 30 and 33 neurones were observed. In the latter instance the ventral ganglion is rather distinctly paired and also subdivided transversely into anterior and posterior lobes; each posterior lobe contains four cells (Fig. 131 Q).

FIG. 132.

A-B—*Oesophagostomum dentatum* (A—Dissection showing ventral and lateral ganglia with nerve trunks; B—Dissection showing dorsal and subdorsal ganglia). C-D—*Spironoura affine* (Reconstruction of posterior part of nervous system. C—Male; D—Female). E—*Rhabditis terricola* (Longitudinal section of female showing phasmidial gland). F-L—*Spironoura affine* (F—Longitudinal reconstruction of phasmids showing phasmidial glands, neurones and glia cells; G—Longitudinal section showing innervation of preanal median papilla; H—Same papilla in transverse section; I—Genital papilla; J—Deirid; K—Reconstruction of cross sections of female at level of phasmids; L—Reconstruction of cross sections of male showing spicular ganglia and rectal commissure). M—*Oxyuris equi* (Dissection showing connection of innervation processes with median nerves). N—*Oesophagostomum dentatum* (Branch of dorsal ray). M, after Martini, 1916, Ztschr. Wiss. Zool., v. 116. Remainder original.

RETROVESICULAR GANGLION. The retrovesicular ganglion in *Ascaris* contains 13 bipolar cells (79-91), several of which are in direct continuity with one another (Fig. 130 J) forming an associational center according to Goldschmidt (1908). In *Oxyuris*, Martini found the same number of cells in this ganglion as did the writers in *Cephalobellus*. In *Oesophagostomum* the retrovesicular ganglion consists of 13 cells (Fig. 132 A) followed so closely by a postretrovesicular ganglion of seven cells that sometimes they seem about to merge. This has apparently taken place in *Spironoura* where the retrovesicular ganglion is unusually far back and contains 20 cells. In *Rhabditis* the ganglia (Fig. 131 I) are also merged, there being 21 cells.

ANAL, LUMBAR and GENITAL GANGLIA and SENSORY ORGANS. In *Ascaris* these structures have not been very well studied though investigated by Hesse (1892), Voltzenlogel (1902) Goldschmidt (1903) and Deineka (1908). Each preanal genital papilla of the paired series is connected by means of a fiber (which passes through the hypodermis) to the ventrolateral nerve where a sensory neurone and a glia cell are situated. The fibers then pass posterial in the ventrolateral nerve before reaching the genito-papillary commissures through which they reach the ventral nerve. (Fig. 128 C). The anal ganglion which terminates the ventral nerve contains seven cells; the number of cells in the lumbar ganglia has not been determined. According to Voltzenlogel the medioventral preanal papilla is innervated by a process direct from the anal ganglion, this process containing four neurones. On each side of this process a branch of the ventral nerve extends posteriad and just anterior to the anus both branches bend dorsally forming the *ano-lumbar* commissures which connect the ganglia of the same names. Voltzenlogel found that both the ventral medio-caudal nerves divide at the level of the lumbar ganglia, a process passing on each side to the corresponding lumbar ganglion. From these ganglia, the laterocaudal nerves extend posteriad. In all, there are seven pairs of postanal sensory organs in the male; the first four form two pairs of double subventral papillae, one pair anterior and one posterior to the lumbar ganglion. The fifth pair, more laterally situated and externally pore-like rather than papilloid, is the phasmid; the sixth and seventh pairs are both typical simple papillae. Anterior to the last pair the laterocaudal nerves each contain three neurones. All of the postanal sensory organs are innervated by processes from the laterocaudal nerves.

In *Ancylostoma* Looss found the anal ganglion longitudinally divided in both sexes. In the female the anolumbar commissures join the paired anal and lumbar ganglia from which ventrolateral caudal nerves extend posteriad to innervate the phasmids; just anterior to this point the dorsolateral nerves join the ventrolateral. In the male (Fig. 130 BB) the ganglia are essentially similiar but the anal ganglia are each subdivided into two parts (antero-anal and postero-anal ganglia) and the lumbar ganglia each into three subdivisions (prolumbar, mesolumbar and metalumbar parts corresponding to lumbar, postlumbar and costal as named by Looss). Two ano-lumbar commissures join the subdivided anal ganglia with the two anterior subdivisions of the lumbar ganglia. In *Ancylostoma*, as in other strongyloids, all except one pair of genital papillae are situated in the bursa and terminate its rays. The prebursal papillae are the exceptions and they are innervated by processes extending directly from the second ano-lumbar commissure. The typical strongyloid (Fig. 33 I-J) bursal ray pattern consists of five bursal ray trunks, two ventral, two lateral and a single dorsal. Each ventral trunk bifurcates forming a ventroventral and a lateroventral ray; each lateral trunk trifurcates forming an externolateral (ventrolateral), mediolateral and posterolateral (dorsolateral) ray while the dorsal gives off two lateral branches, the externodorsal rays, and then bifurcates, forming the two dorsal rays. Each of the dorsal rays terminates in two or three digitations. Since the dorsal trunk contains the remnants of the lateral and ventral chords as well as the somatic muscles, it corresponds to the tail of the female. It is not really dorsal, but terminal. Looss found that all the larger bursal rays terminate in papillae. The ventral trunks are innervated by a pair of nerves from the second anolumbar commissure; a branch from each nerve extends out to the tip of each ray. A nerve extends posteriad from each metalumbar ganglion and divides into three branches, one to each of the lateral rays,

where they terminate as papillae, while another nerve extends medially from each metalumbar ganglion to the trunk of the dorsal ray, where it gives off a lateral branch into the externodorsal ray before terminating in the corresponding medial digitation of the dorsal ray. The papillae of the externolateral and externodorsal rays are situated on the dorsal side of the bursa while the remaining papillae protrude on the ventral side of the bursa. In both *Strongylus* (with tridigitate dorsal rays) and *Oesophagostomum* (with bidigitate dorsal rays) we have found three pairs of papillae on each dorsal ray. There is a papilla for each digitation in the former genus while there is one papilla on each lateral and two on each medial branch in the latter genus (Fig. 132 N). With Looss, we conclude that in the male bursa the most medial digitation of the strongyloid dorsal ray is the homologue of the phasmid or "caudal papilla" of the female. Including the prebursal papillae there are 10 pairs of sensory organs in male strongyloids; this is also true for *Rhabditis strongyloides* (Fig. 4) *R. caussanelli* (Fig. 33 A) and *R. aspera* (Fig. 33 B-C). In *R. strongyloides* all sensory organs are papilloid while in *R. caussanelli* and *R. aspera* the tenth sensory organ (phasmid) is pore-like. Some of the genital papillae in rhabditids always end on the dorsal or outer side of the bursa; the particular ones vary with the species. In two or three species, such as *Rhabditis oxyuris* the first pair of papillae is prebursal as in the strongyloids. In order to find a comparable arrangement of bursal papillae in rhabditids we must number the strongyloid papillae from the anterior end as follows: prebursal—1, ventroventral—2, lateroventral—3, externolateral—4, mediolateral—5, posteriolateral—6, externodorsal—7, laterodorsal (digitation)—8, subdorsal (digitation)—9, dorsodorsal (digitation)—10 (= phasmid). Thus, the fourth and seventh papillae terminate on the dorsal side of the bursa. The similarly numbered papillae terminate dorsally in *Rhabditis strongyloides;* in addition the ninth (or tenth, depending on relative position) is also dorsal in this species; presumably it is the phasmid. The lumbar ganglia are not subdivided in rhabditids as they are in strongyloids.

In *Spironoura* the ventral ganglion is undivided and contains nine cells in the female (Fig. 132 D) and 27 in the male (Fig. 132 C). The lumbar ganglia contain six neurones on the right side and five on the left side; of these the first cell is somewhat removed from the remainder. The anolumbar commissures are approximately adanal in position connecting with the anterior part of the lumbar ganglia. From the posterior end of the lumbar ganglia the lateral caudal nerves extend posteriad and each contains a bipolar neurone before passing through the *phasmidial ganglion* from which fibers pass to the phasmidial gland. Posterior to this ganglion the lateral caudal nerves extend nearly to the caudal extremity. In the male of this form the ano-lumbar commissure is decidedly preanal and there is a medioventral process originating with it, containing a bipolar neurone connected with the medioventral preanal papilla (Fig. 132 G). Each prolumbar ganglion contains five neurones, one slightly anterior, which is connected with the recto-sympathetic system. In the course of the ventrolateral nerves anterior to the prolumbar ganglia, there are three neurones in each nerve from which a like number of nerve processes extend to the three pairs of preanal genital papillae. No genital commissures to the ventral nerve were observed, so we presume the axones of the three neurones pass posteriad and reach the ventral nerve through the ano-lumbar commissure. If this is the case, the distantly placed sensory cells might be regarded as part of the prolumbar ganglia. Postanally there are two pairs of nerve cell groups, the mesolumbar and metalumbar ganglia in addition to the phasmidial ganglia. Processes from the mesolumbars innervate the fourth to seventh pairs of genital papillae while processes from the metalumbars innervate the eighth to tenth pairs.

Regarding the general arrangement of genital papillae in nemas one's attention is called to two standard forms in the Phasmidia. In the first there is a very definite grouping of papillae as in strongyloids and rhabditoids; it probably reflects the triple division of the lumbar ganglia. Postanal papillae number up to seven or eight pairs including the phasmids. Preanal papillae number two to three pairs. In the second standard form the postanal papillae are not so clearly grouped and there are many more serially arranged preanal papillae, as in *Ascaris;* this linear formation is due to reduplication and probably arose from a form such as *Spironoura* in which the preanal papillary neurones are separated some distance from the prolumbar ganglion. The oxyurid-thelastomatid arrangement, in which there are but four pairs of papillae besides the phasmids, may be considered a reduction of the rhabditid scheme. A medio-ventral preanal papilla, innervated directly from the ventral nerve, was described in *Ascaris* and *Spironoura;* a comparable papilla is present in a large number of other phasmidian nemas and may even be general; one is recorded in *Strongyloides, Rhabditis, Physaloptera* and we have seen it in *Oesophagostomum.* The medioventral papilla of the preanal sucker in *Heterakis* (Fig. 33 P) might also be the same structure. We regard it as a probable homologue of the preanal supplementary organs (See p. 33) in aphasmidians.

PHASMIDS and ASSOCIATED STRUCTURES. Since our (1933) proposal of the divisions Phasmidia and Aphasmidia, parasitologists have been much perturbed by these "mysterious organs." They are nothing other than the "caudal papillae" as seen in the larvae and females of most parasitic nemas. Because of their differences from ordinary sensory papillae Cobb (1923) proposed the name phasmid (= ghost thing) for them. The external manifestation is a pair of lateral or subventral pores. From each of these a short tube extends internally and in it lies the sensilla which is very similar to that of the amphid except that its elements or terminals are fewer in

of bipolar neurones ending in a sensory terminus within the phasmidial tube of *Rhabditis*, and the writers (1933) confirmed this finding in *Cephalobellus.* The illustration (Fig. 8) of the general situation in *Rhabditis terricola* is the only other information extant.

The writers have selected *Spironoura* for detailed study because it is optimum in size for histological work. Similar studies on free-living nemas are difficult; section and intravitam preparations, though beautiful, require confirmation because they are apt to be confusing. The errors of Deineka (1908) and Chitwood (1930) with methylene blue on the central nervous system make the description of an organ by section obligatory in at least one species before the use of intravitam staining methods for comparative purposes.

In the female of *Spironoura* each phasmid (Fig. 132 F & K) opens to the outside through a small postanal lateral phasmidial pore. From this pore a cuticularly lined canal extends inward and disappears in a large phasmidial gland. In the wall of the phasmidial canal there are two glia cells. The phasmidial ganglia each contains three neurones and one glia cell; processes from at least two of these enter the phasmidial gland where they form sensory terminals. The male has phasmids which are similar but the phasmidial ganglia each contains

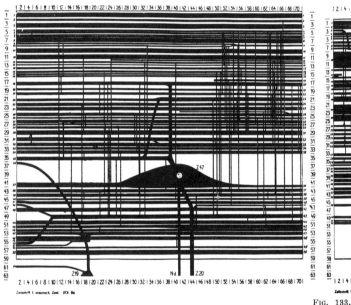 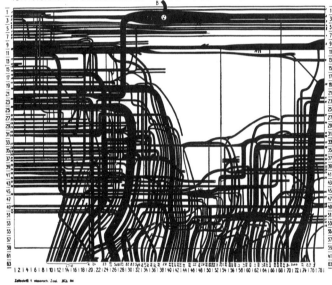

FIG. 133.

Diagrams of parts of nerve ring of *Ascaris lumbricoides.* Dorsal part of nerve ring on readers' left, ventral part on readers' right.

See also ng. 130. After Goldschmidt, 1909, Ztschr. Wiss. Zool., v. 92 (2).

number. Like the amphids, the phasmids are usually (? always) provided with a flushing gland, the phasmidial gland. In the Phasmidia, our experience indicates that the phasmids are always present in larvae and females. They are sometimes transformed into plate-like scutella (*Rotylenchus blaberus* vide Steiner, 1937) and sometimes they take the form of large pockets as in larval dracunculoids and drilonematids (*Dicelis nira*). In the male, they are often difficult to distinguish from genital papillae or they may be so faint that they are truly ghost-like. In most rhabditids, cephalobids, diplogasterids, oxyurids and thelastomatids they are easily enough recognized and were called excretory glands by Stefanski (1922). In male spiruroids and dracunculoids they are very minute. Only in the males of a few rhabditids (*R. strongyloides*) and all strongylins are the phasmids grossly indistinguishable from genital papillae and we have seen (p. 170) that in these cases they are probably represented by the most posterior papilla or dorsodorsal digit of the dorsal ray. In *Oesophagostomum* it appears that this structure terminates in a pore as in typical phasmids.

The minute structure of the organs has been sorely neglected. Looss (1905) found them to be innervated by the lateral caudal nerves in *Ancylostoma.* Chitwood (1930) described the phasmidial gland and mentioned the observation

five neurones. In both sexes axones from these neurones pass anteriad through the lateral caudal nerves to the ano-lumbar commissures.

In rhabditids (Fig. 8 & 132 E) the phasmidial glands have the same general arrangement as in *Spironoura* but there is no separation between the lumbar and phasmidial ganglia.

RECTO-SYMPATHETIC SYSTEM. This part of the nervous system received casual attention by Hesse (1892) and Voltzenlogel (1902) in *Ascaris*, by Looss (1905) in *Ancylostoma duodenale*, by Martini (1906) in *Oxyuris equi*, and by the writers (1933) in *Cephalobellus papilliger.* Diagrams have, for the most part, been inadequate. Our observations on *Spironoura affine* indicate that there is probably considerable yet to be learned. The condition in the female of *Spironoura* seems to be quite simple (Fig. 132 D), since it apparently consists merely of a pair of commissures meeting dorsally and extending posteriad in the median caudal nerve. The latter contains two neurones dorsal to the rectum and two additional neurones in its postanal course. In the male of this form, the system is much more complex. In the course of each of the two ano-rectal commissures there are two bipolar neurones ventral to the cloaca (rc1-2 and rc3-4). As in the female, the commissures meet dorsally, forming the medial caudal nerve, but in its course there are three dorsocloacal neurones and four postanal neurones (Figs. 132 C & L). Laterally each ano-rectal commissure gives off a branch which joins the laterally situated *spicular ganglion* which contains four neurones and a glia cell. Two of these neurones have processes into the spicular sheath and two have processes to the medial caudal nerve (in the ano-rectal commissure). Two processes from each of these ganglia pass posteriorly and ventrally to two bipolar *subcloacal* neurones which extend toward the anterior cloacal lip. An additional process connects the posterior part of the spicular ganglion to the ventrolateral nerve. In *Ascaris* Voltzenlogel found many errors in Hesses's description and diagrams of the male. Unfortunately no corrected diagram was supplied. We have attempted to make one on the basis of Voltzenlogel's description (Fig. 128 C) but it is necessarily liable to considerable error. According to the description each ano-rectal commissure contains two lateral bipolar neurones and the medial caudal nerve contains three dorso-cloacal neurones and one postanal neurone after which this nerve divides, sending a process to each lumbar ganglion. The two pairs of bipolar subcloacal

Finer Structure of Nervous System

Work on the finer structure of the nervous system has been confined to *Ascaris lumbricoides*. Goldschmidt (1908) classified the nerve cells of this species as follows: I. Ganglion cells. Central cells (i. e. with one or rarely two processes to nerve ring) either (a) Direct or (b) Indirect (pass through a commissure before reaching nerve ring) and II. Sensory cells.

I. CENTRAL CELLS. The direct cells (Ia) were subdivided according to size and shape; large cells (Fig. 130) termed (1) *chonoid*, (Cell 23 Fig. V); (2) *corynoid*, (Cells 7-12, Fig. A); or (3) *amphoroid*, bipolar (Cells 19-20, Fig. B). Medium sized cells were classified as (4) *lagenoid* (Cell 26, Fig. D), (2) *corynoid* (Cell 18 Fig. A) or (5) *aranoid* (Cell 40, Fig. F). Small cells were classified as (6) *pyriform* (Cells, 1, 2, 4, Fig. A) or (2") *corynoid* (Cell 13, Fig. F). The Indirect cells (Ib) consist of large chonoid and small corynoid types.

II. SENSORY CELLS. These are bipolar cells situated either peripherically or centrally. They are of four general types

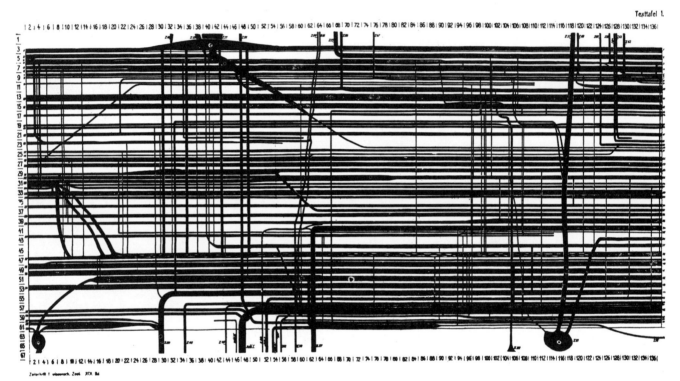

Texttafel 1.

FIG. 134.

Diagram of lateral part of nervous system of *Ascaris lumbricoides*. Dorsal part of nerve ring on readers' left, ventral part on readers' right. See also fig. 130. After Goldschmidt, 1909, Ztschr. Wiss. Zool., v. 92 (2).

neurones originate from the paired ventral nerve trunks and processes extend posteriad where a third subcloacal neurone is found. No spicular ganglia have been described. Despite the great detail of Martini's work on *Oxyuris equi* we are unable to follow his description without diagrams. Looss is more intelligible and kindly supplies the needed diagrams but is not too exact. His findings in the female of *Ancylostoma* correspond with ours on *Spironoura* except that he found four neurones of the medial caudal nerve anterior to the anus. Two processes extend posteriad from the ends of the ventral nerve to the anterior anal lip. In the male the system consists of two commissures, one of which connects with the suprarectal ganglion, dorsal to the spicules, while the other connects with the sub-rectal ganglion, ventral to the spicules; the two ganglia are connected. The commissure connected with the suprarectal ganglion joins on each side with a subcloacal ganglion from which a group of fibers passes posteriorly to the dorsal side of the gubernaculum. There is obviously a distinct similarity in findings and differences between the conditions in *Ascaris, Spironoura* and *Ancylostoma* may be superficial and not important when an adequate amount of information for critical analysis is available.

according to disposition: (1) Direct sensory cells (Bipolar neurones of cephalic papillary nerves), (2) Indirect sensory cells (Bipolar neurones of amphidial nerves and genital papillae), (3) Collateral cells (Cells essentially bipolar but having one axone direct to nerve ring and a second axone passing through a commissure such as cell 39, innervating the deirid), (4) Unclassifiable bipolar neurones of longitudinal nerves such as those occurring in submedian and ventral somatic nerves (It seems dubious that these cells should be classified as sensory without knowledge of their ultimate destinations).

SUPPORTING TISSUES. The supporting and ensheathing substance of the nervous system is called glia. Cells which form this connective tissue are of several types, some being very closely integrated into the general nervous system, the glia cells, others being less intimately associated. These latter include the *escort cells* (Geleitzellen) and the *clavate cells* (Kolbenzellen).

According to Goldschmidt (1910) the chief mass of glia which surrounds the nerve ring in *Ascaris* is two to three times as large as the nerve ring itself and contains nests of three to four small nuclei (Fig. 130 W). Some of the tissue and nuclei extend into the lateral chords. Special glia cells of the cephalic papillary nerves have already been described.

Each large ganglion cell has a specific glia hull surrounding it. According to Goldschmidt this material is often poorly fixed but in satisfactory preparations it may be seen that glia fibrils actually enter the ganglion cell (Fig. 130 V) thus causing the appearance characterized as "radially striated" ganglion cells. All nerve fibers have one or more neurofibrils and these are in direct continuity with the glia fibrils which may form a basket-work around the nucleus. The protoplasm of ganglion cells is essentially alveolar and the web conforms to the radial fibril structure. Tigroid substance (Nissl bodies, chromophil substance) is sometimes fine and generally distributed throughout the cell while at other times it is coarse and localized (Fig. 130 R-U).

MUSCLE INNERVATION. Due to the efforts of Schneider (1866) and later workers the nemic "innervation processes," by which somatic muscles are connected with longitudinal nerves, have received a great deal of publicity and it is widely accepted that nemas differ from all other animals in that "The muscle seeks the nerve" rather than the reverse. The present writers frankly do not want to become involved in this controversial subject but will present as impartial a review of present day knowledge as is possible.

Technic has played a large part in earlier discussions and the two approaches to the subject have been via impregnation (silver method chiefly) or methylene blue, either as an intravitam stain or according to the ammonium molybdate technic. Apathy (1894) entered the field chiefly due to the taunts of Rohde (1894) whose studies of the muscle cells of *Ascaris* caused him to attack the whole structure of the school of neurone cytology which Apathy was developing. It was Rohde's idea that the entire conception of the condition in other animals was fundamentally wrong and that the condition in *Ascaris* was typical. Apathy found the ascaridid muscle to contain an intimate and complex network of fibrils (Fig. 49 H) which not only converge in the "innervation" process but also continue into the hypodermis and finally reach the lateral and submedian somatic nerves. He conceived this "neurofibril" network as going from the median somatic nerve via the innervation process to the individual fibrils and thence back to the lateral nerves. Apathy explained the innervation process as an "interstitial muscle" in which a large neurofibrillar process was centrally located.

Deineka (1908), using intravitam methylene blue, observed direct union of nerve cells by contact and some intracellular neurofibrils passing from cell to cell. He described the "innervation" processes as direct branches of the somatic nerves each of which forms a wide terminal plate at its contact with the muscle (Fig. 130 X). As Rouville (1910-1911) and Goldschmidt (1910) pointed out this was a misinterpretation due to the stains flow along the process until it reached the chief part of the muscle cell. There can be no doubt that the innervation processes are directly continuous with the sarcoplasmic portion of the muscle cell.

Goldschmidt (1910) agreed with all the facts as presented by Apathy but pointed out that the central fibril of each innervation process is directly continuous (Fig. 130 Y) with a neurofibril and that the fibrillar network of the muscle cell not only passes through the hypodermis to the lateral chord but also becomes inserted into the cuticle. He did not differ with Apathy as to any facts, only interpretations. He viewed the entire fibrillar system not only within the muscle but also within the nerve cells as supportive. From his standpoint the entire theory that neurofibrils transmitted impulses was false. Invoking Koltzoff's Principle he explained the system as skeletal. According to this view nonspheroid cells develop an intracellular skeleton in order to retain their shape. Nerve cells are in direct protoplasmic continuity with one another. (This cannot be denied at least as to some cells in nemas) and nerve impulses pass through the cytoplasm.

Such discussion is pure theory of broad application and not particularly significant to nematology. No one denies that the fibrillar network exists and that neurofibrils pass *from the somatic nerves to the muscle cell* where they break up. Thus far nemas are in complete conformity with other groups of animals. The fact that the sarcoplasm of the muscle cell ensheaths the fibrils forming an "innervation process" no longer seems as important as Schneider and Rohde believed it to be.

Bibliography

APATHY, S. 1894.—Das leitende Element in den Muskelfasern von *Ascaris*. Arch. Mikr. Anat., v. 43 (4): 886-911, pl. 36.
    1897.—Das leitende Element des Nervensystems und seine Topographischen Beziehungen zu den Zellen. Erste Mitth. Zool. Station Neapel., v. 12 (4): 495-748, pls. 23-32.

BASTIAN, H. C. 1864.—On the structure and nature of the *Dracunculus*, or guineaworm. Tr. Linn. Soc. Lond., (1863) v. 24 (2): 101-134, pls. 21-22, figs. 1-60.
    1866.—On the anatomy and physiology of the nematoids, parasitic and free; with observations on their zoological position and affinities to the echinoderms. Phil. Tr. Roy. Soc. Lond., v. 156: 545-638, pls. 22-28.

BLANCHARD, E. 1847.—Recherches sur l'organisation des vers. Ann. Sc. Nat. Par., Zool., 3. s., v. 7:87-128; v. 8:119-149, 271-341, pls. 8-14.

BRANDES, G. 1899.—Das Nervensystem der als Nemathelminthen zusammengefassten Wurmtypen. Abhandl. Naturf. Gesellsch. zu Halle., v. 21 (4): 271-299, figs. 1-11.

BÜTSCHLI, O. 1874.—Beiträge zur Kenntniss des Nervensystems der Nematoden. Arch. Mikr. Anat. v. 10: 74-100, pls. 6-7, figs. 1-19.
    1885.—Zur Herleitung des Nervensystems der Nematoden. Morph. Jahrb. v. 10 (4): 486-493, pl. 23, figs. 1-3.

CHITWOOD, B. G. 1930.—Studies on some physiological functions and morphological characters of *Rhabditis* (Rhabditidae, Nematodes). J. Morph. & Physiol. v. 49 (1): 251-275, figs. A-H, pls. 1-3, figs. 1-24.
    1933.—A revised classification of the Nematoda. J. Parasit. v. 20 (2): 131.
    1937.—A revised classification of the Nematoda. Papers on Helminthology, 30 year jubileum K. J. Skrjabin. Moscau. pp. 68-79.

CHITWOOD, B. G. & CHITWOOD, M. B. 1933a.—The characters of a protonematode. J. Parasit., v. 20 (2): 130.
    1933b.—The histological anatomy of *Cephalobellus papilliger* Cobb, 1920. Ztschr. Zellforsch., v. 19 (2): 309-355, figs. 1-34.

CHITWOOD, B. G. & WEHR, E. E. 1934.—The value of cephalic structures as characters in nematode classification with special reference to the superfamily Spiruroidea. Ztschr. Parasitenk., v. 7 (3): 273-335, figs. 1-20, 1 table.

COBB, N. A. 1888.—Beiträge zur Anatomie und Ontogenie der Nematoden. Diss. 36 pp., pls. 1-3, figs. 1-32, Jena.
    1913.—[Free living nematodes.] Science, n. s., v. 37 (952): 498.
    1923.—Interesting features in the anatomy of nemas. J. Parasit., v. 9 (4): 242-243, figs. 3-4.

DEINEKA, D. 1908.—Das Nervensystem von *Ascaris*. Ztschr. Wiss. Zool., v. 89 (2): 242-307, 7 text figs., pls. 12-20, figs. 1-55.

EBERTH, C. J. 1863.—Untersuchungen über Nematoden. 77 pp., 9 pls. Leipzig.

FILIPJEV, I. N. 1912.—Zur Kenntnis des Nervensystems bei den freilebenden Nematoden. Trav. Soc. Imp. St. Petersbourg., v. 43 (4): 205-215, 220-222, figs. 1-8. Russian text with German summary).

GOLDSCHMIDT, R. 1903.—Histologische Untersuchungen an Nematoden. I. Die Sinnesorgane von *Ascaris lumbricoides* und *A. megalocephala*. Zool. Jahrb. Abt. Anat., v. 18 (1): 1-57, figs. A-D, pls. 1-5, figs. 1-40.
    1904.—Ueber die sogenannte radiärstreiften Ganglienzellen von *Ascaris*. Biol. Centrlbl., v. 24 (5): 173-182, figs. A-B.
    1907.—Einiges vom feineren Bau des Nervensystems. Verhandl. Deutsch. Zool. Gesellsch., pp. 130-131.
    1908. a.—Die Neurofibrillen im Nervensystem von *Ascaris*. Zool. Anz. v. 32 (19): 562-563.
    1908b.—Das Nervensystem von *Ascaris lumbricoides* und *megalocephala*. I. Ztschr. Wiss. Zool., v. 90: 73-136, figs. A-W, pls. 2-4, figs. 1-14.
    1909.—Idem. II. Ibid., v. 92 (2): 306-357, figs. 1-21, pls. 1-3.
    1910.—Idem. III. Festschrift Hertwigs, v. 2: 253-354, pls. 17-23, figs. 1-125.

HAMANN, O. 1895.—Die Nemathelminthen. v. 2, 120 pp., pls. 1-11. Jena.

HESSE, R. 1892.—Ueber das Nervensystem von *Ascaris megalocephala*. Diss. 23 pp., 2 pls. 20 figs. Halle. Also in Ztschr. Wiss. Zool., v. 54 (3): 548-568, pls. 23-24, figs. 1-20.

HOEPPLI, R. 1925.—Ueber das Vorderende der Ascariden. Ztschr. Zellforsch., v. 2 (1): 1-68, figs. 1-27, 1 pl., figs. 1-12.

IMMINCK, B. D. C. M. 1924.—On the microscopical anatomy of the digestive system of *Strongylus edentatus* Looss. Arch. Anat., Hist. & Embryol., v. 3 (4-6): 281-326, figs. 1-46.

JOSEPH, G. 1882.—Vorläufige Bemerkungen über Musculatur, Excretionsorgane und peripherisches Nervensystem von *Ascaris megalocephala* und *lumbricoides*. Zool. Anz., v. 5 (125): 603-609.
　　　1883a.—Erwiederung auf die Erklärungen des Herrn Dr. Rohde im Zoologischen Anzeiger No. 131. Zool. Anz., v. 6 (133): 125-127.
　　　1883b.—Zur Abwehr gegen die ferneren Angriffe des Herrn Dr. Rohde auf p. 196-199 des Zoolog. Anzeigers. Zool. Anz., v. 6 (139): 274-278.
　　　1884.—Beiträge zur Kenntnis des Nervensystems der Nematoden Zool. Anz, v. 7 (167): 264-266.

LOOSS, A. 1905.—The anatomy and life history of *Agchylostoma duodenale* Dub. Rec. Egypt. Govt. School Med. v. 3: 1-158, pls. 1-9, figs. 1-100, pl. 10, photos. 1-6.

MAGATH, T. B. 1919.—*Camallanus americanus* nov. spec., a monograph on a nematode species. Tr. Am. Micr. Soc. v. 38 (2): 49-170, figs. A-Q, pls. 7-16, figs. 1-134.

MARTINI, E. 1913.—Ueber die Stellung der Nematoden im System. Deutsch. Zool. Gesellsch. a. d. 23. Jahresversamml. zu Bremen, pp., 233-248, figs. 1-2.
　　　1916.—Die Anatomie der *Oxyuris curvula*. Ztschr. Wiss. Zool., v. 116: 137-534, figs. 1-121, pls. 6-20, figs. 1-269.

MEISSNER, G. 1853.—Beiträge zur Anatomie und Physiologie von *Mermis albicans*. Ztschr. Wiss. Zool., v. 5 (2-3): 207-284, pls. 11-15, figs. 1-55.

MIRZA, M. B. 1929.—Beiträge zur Kenntnis des Baues von *Dracunculus medinensis*, Velsch. Ztschr. Parasitenk., v. 2 (2): 129-156, figs. 1-33.

OTTO, A. 1816.—Ueber das Nervensystem der Eingeweidewürmer. Mag. Entdeck. Ges. Naturk., v. 7, 3 Quartal: 223-233, pls. 5-6, figs. 1-10.

OWEN, R. 1836-39.—Entozoa. Cycl. Anat. & Physiol., Lond., v. 2: 111-144, figs. 51-96.

RAUTHER, M. 1906.—Beiträge zur Kenntnis von *Mermis albicans* v. Sieb. mit besonderer Berücksichtigung des Haut-Nerven-Muskelsystems. Zool. Jahrb. Abt. Anat., v. 23 (1): 1-76, pls. 1-3, figs. 1-26.

ROHDE, E. 1883a.—Beiträge zur Kenntnis der Anatomie der Nematoden. Diss. 26 pp., Breslau.
　　　1883b.—Einige Erklärungen zu "Vorläufige Bemerkungen über Muskulatur (etc.)" von Dr. Gustav Joseph in No. 125 des Zoologischen Anzeigers. Zool. Anz., v. 6 (131): 71-76.
　　　1883c.—Ueber die Nematodenstudien des Herrn Dr. Joseph. Zool. Anz., v. 6 (136): 196-199.
　　　1885.—Beiträge zur Kenntniss der Anatomie der Nematoden. Zool. Beitr., Bresl., v. 1 (1): 11-32, pls. 2-6, figs. 1-35.

　　　1892.—Muskel und Nerv. 1. *Ascaris* 2. *Mermis* und *Amphioxus*. 3. *Gordius*. Zool. Beitr. Bresl., v. 3 (2): 69-106, pls. 8-13, figs. 1-38; (3): 161-192, pls. 23-25.
　　　1894.—Apathy als Reformator der Muskel- und Nerven-lehre. Zool. Anz., v. 17 (439): 38-47, figs. 1-2.

ROUVILLE, E. DE 1910-1911.—Le système nerveaux de l'*Ascaris* d'apres des travaux récents. Arch. Zool. Expér. & Gén. notes et rév., v. 45, 5. s., v. 5 (3): 81-98; figs. 1-15, v. 46, 5. s., v. 6 (1): 20-47, figs. 1-17, v. 47, 5. s., v. 7 (2): 28-49, v. 48, 5. s., v. 8 (4): 102-123, figs. 1-27.

SCHNEIDER, A. 1860.—Ueber die Muskeln und Nerven der Nematoden. Arch. Anat. Physiol. Wiss. Med., Leipzig., pp, 224-242, pl. 5, figs. 1-12.
　　　1863.—Neue Beiträge zur Anatomie und Morphologie der Nematoden. Ibid., pp. 1-25, pls. 1-2.
　　　1866.—Monographie der Nematoden. 357 pp., 122 figs., 28 pls., 343 figs. Berlin.

SEURAT, L. G. 1913.—Sur l'évolution du *Physocephalus sexalatus* (Molin). Compt. Rend. Soc. Biol. 75 (35): 517-520, figs. 1-4.
　　　1914.—Sur un nouveau gongylonème, parasite de la gerbille. Ibid., v. 77 (31): 521-524, figs. 1-4.
　　　1915.—Sur les premiers stades évolutifs des spiroptères. Ibid., v. 78 (17): 561-565, figs. 1-5.
　　　1920.—Histoire naturelle des nématodes de la Berbérie. Première partie. 221 pp., 34 figs. Alger.

SIEBOLD, C. T. E. von. 1848.—Lehrbuch der vergleichenden Anatomie der wirbellosen Thiere. 679 pp., Berlin.

STEFANSKI, W. 1922.—Excrétion chez les nématodes libres. Arch. Nauk. Biol., v. 1 (6): 1-33, figs. 1-39.

STEINER, G. 1920.—Betrachtungen zur Frage des Verwandtschafts-Verhältnisses der Rotatorien und Nematoden. Festschr. Zschokke, (31) 16 pp., 15 figs. Basel.

STRASSEN, O. zur. 1904.—*Anthraconema*, eine neue Gattung freilebender Nematoden. Zool. Jahrb., Suppl. 7: 301-346, figs. A-J, pls. 15-16, figs. 1-9.

TÖRNQUIST, N. 1931.—Die Nematodenfamilien Cucullanidae und Camallanidae etc., Göteborgs K. Vetensk.—o. Vitterhets—Samh. Handl., 5., f., s. B, v. 2 (3), 441 pp., pls. 1-17.

VOLTZENLOGEL, E. 1902.—Untersuchungen über den anatomischen und histologischen Bau des Hinterendes von *Ascaris megalocephala* und *Ascaris lumbricoides*. Diss. 32 pp., 3 pls., 20 figs. Jena. Also in Zool. Jahrb., Abt. Anat. v. 16 (3): 481-510, pls. 34-36.

WALTER, G. 1856.—Beiträge zur Anatomie und Physiologie von *Oxyuris ornata*. Ztschr. Wiss. Zool., v. 8 (2): 163-201, pls. 5-6, figs. 1-28.
　　　1858.—Fernere Beiträge zur Anatomie und Physiologie von *Oxyuris ornata*. Ibid., v. 9 (4): 485-495, pl. 19, figs. 29-34.
　　　1862.—Beiträge zur mikroskopischen Anatomie der Nematoden. Arch. Path. Anat. Berlin, 24 J, 2 F., v. 4 (1-2): 166-182, pl. 3, figs. 1-19.
　　　1863.—Mikroskopische Studien über das Central-Nervensystem wirbelloser Thiere. 56 pp., 4 pls., Bonn.

WEDL, C. 1855.—Ueber das Nervensystem der Nematoden. Sitzungsb. Königl. Akad. Wiss. Wien, Math.-Naturw. Cl., v. 16 (2): 298-312, 1 pl. figs. 1-10.

CHAPTER XII

NEMIC OVA

REED O. CHRISTENSON, formerly of Dept. Zoology-En
tomology, Alabama Experiment Station, Auburn, Alabama
with contributions by others

ACKNOWLEDGMENTS: *The writer is especially indebted to Dr. B. G. Chitwood for many helpful suggestions and criticisms. He also wishes to acknowledge his appreciation to Dean M. J. Funchess and Professor J. M. Robinson for making facilities and time available. Many of the eggs studied were supplied from the collection of the Bureau of Animal Industry through the cooperation of Dr. Benjamin Schwartz and other members. Acknowledgments are likewise due Dr. Dale A. Porter for materials and reviewing the manuscript, Dr. Harold Manter, Dr. Wilford Olsen, Dr. V. N. Moorthy, Dr. Franklin G. Wallace, Eugenia Rutland Moore, and Dr. J. E. Greene, all of whom supplied materials for study. Of considerable value were the suggestions and the aid of Alyce Mae Christenson and Mr. Ernest Rouse who assisted on certain aspects of the problem.—R. O. C.*

Layers of the Egg Envelope

HISTORICAL REVIEW

R. O. C.

The history of the development of the egg of nematodes begins with the classic work of Nelson (1852) on *Toxocara cati.* Nelson recognized three portions of the genital tube set apart from each other by sphincters, namely: (1) the ovaries, (2) the oviducts, and (3) the uteri which join to form the vagina. He noted that the egg primordia of the extremities of the ovaries enlarged as they passed down the ducts forming the germinal vesicles at the center of which the nucleus or germinal spot was located. The yolk granules were considered to be derived either in the distal portion of the ovary or from the striated part of the ovarian wall near the junction of the oviduct. The vitellus was formed when the consolidation of the yolk was completed about the germinal vesicle.

According to Nelson the production of the vitelline membrane and the chitinous shell occurred in the oviducts. Following contact with the wedge-shaped sperm the eggs began to change in form. Almost immediately the chorion (chitinous shell) began to develop, three strata being recognized. The vitelline membrane separates off from the inner stratum when the chitinous shell is completed. Nelson implies that shell production is an endogenous process since there is no increase in total diameter of the egg during its development.

Bischoff (1855) criticizes Nelson regarding the penetration of the sperm, maintaining that he saw only epithelial cells adhering to the periphery of the vitellus. The diagrammatically clear descriptions of Nelson, however, and his excellent figures, leave no doubt that what he observed were actually the sperm-cells.

Meissner (1856) divided the genital tube of the Mermithoidea into four parts: (1) the ovary (Eierstock), (2) the vitellogene Eiweissschlauch), (3) the tuba, and (4) the uterus. The vitellogene and tuba together correspond in part to the oviduct of Nelson. According to Meissner the eggs "ripened" in the ovary and during the process acquired a membrane which possessed a micropyle for the entrance of sperm. The egg was considered to be intimately related to the rachis as a sort of diverticulum which separated off as the eggs entered the vitellogene. In the vitellogene the chorion was first produced, and in the vaginal portion the protein coat was described as originating from a clear, droplet containing substance. The byssi, according to Meissner, are produced in the tuba. The uterus was considered to function only as a retention chamber for the eggs.

Küchenmeister (1857) described the formation of the chorion from a solidifying mass secreted by the walls of the uterus, laid down in constantly thickening layers about the vitelline membrane. Küchenmeister noted the difference between the appearance of the opercula and the chorion of *Trichuris.* In *Enterobius vermicularis* he described a light hood or cap at one end which he considered as an expression of the fact that the chorion had not been completed. This, in all probability, was the protein coat.

Cobbold (1864) describes the process of egg formation and membrane production in *Ascaris lumbricoides* essentially as does Nelson (*Loc. Cit.*). The ovarian portion he divided into the ovary and the vitellogen. The union of the sexual elements is immediately followed by the condensation of the yolk granules which obliterate the germinal vesicle. The ovum assumes an oval shape and the vitelline membrane and chorion form. The chorion finally assumes a regular tuberculated surface. In *Trichuris trichiura* he describes the abrupt termination of the poles at the end, and projection of a transparent inner membrane to form the opercular papillae.

Leuckart (1886) states that the eggs of *Ascaris lumbricoides* and *Trichuris trichiura* are thick-shelled, and the former further enveloped in an albuminous sheath usually colored with bile pigment. He describes the polar perforations in the latter species and notes that the opercula are albuminous plugs.

Blanchard (1889) describes the formation of two layers of the egg envelope after the egg reaches the uterus in *Ascaris,* the inner being more resistant to pressure and the outer more friable in spite of its great thickness. It is formed of concentric beds indicated by a delicate striation. The egg in the anterior (vaginal) part of the uterus comes into contact with a clear albuminous substance which is deposited over its surface. This substance is at first homogeneous but soon distributes in the form of small, hemispherical tubercles giving the characteristic appearance. The eggs are agglutinated together by their albuminous envelopes into a voluminous mass.

It thus becomes clear that some of the early workers recognized the three primary layers of the egg shell, namely: (1) the vitelline membrane, (2) the chitinous true shell, or chorion, and (3) the protein coat. The protein coat is generally accepted to originate from an exogenous process from the uterine secretion. The endogenous origin of the chitinous shell is implied by Nelson. Regarding the origin of the protein coat the general opinion is that it develops from a secretion of the uterus which adheres to the surface of the cuticular shell.

Ziegler (1895) noted that the eggs of *Diplogaster longicauda* and *Rhabditis terricola* entered the uterus and within an hour the shell is formed. In the former species he found no centrosome in unfertilized eggs and that the shell did not form. Rauther (1930) points out that there is no specially modified structure for shell formation, and that it seemed probable that the chitinous shell was formed from a secretion of the egg itself.

Fauré-Fremiet (1913) reports on the endogenous development of the vitelline membrane of *Parascaris equorum.* He states that this membrane is formed from a particular fat body which pre-exists in the oocyte under the form of refringent bodies or crystalloids. He named the pre-existent substance of the germinal region extracted from gonads ascarylic acid.

Wharton (1915) gives the best description of the eggs of *Ascaris lumbricoides.* He describes the eggs as consisting of a central mass of protoplasm and yolk with a very thin vitelline membrane, surrounded by a thick, transparent shell consisting of an inner layer of chitin and an outer layer of

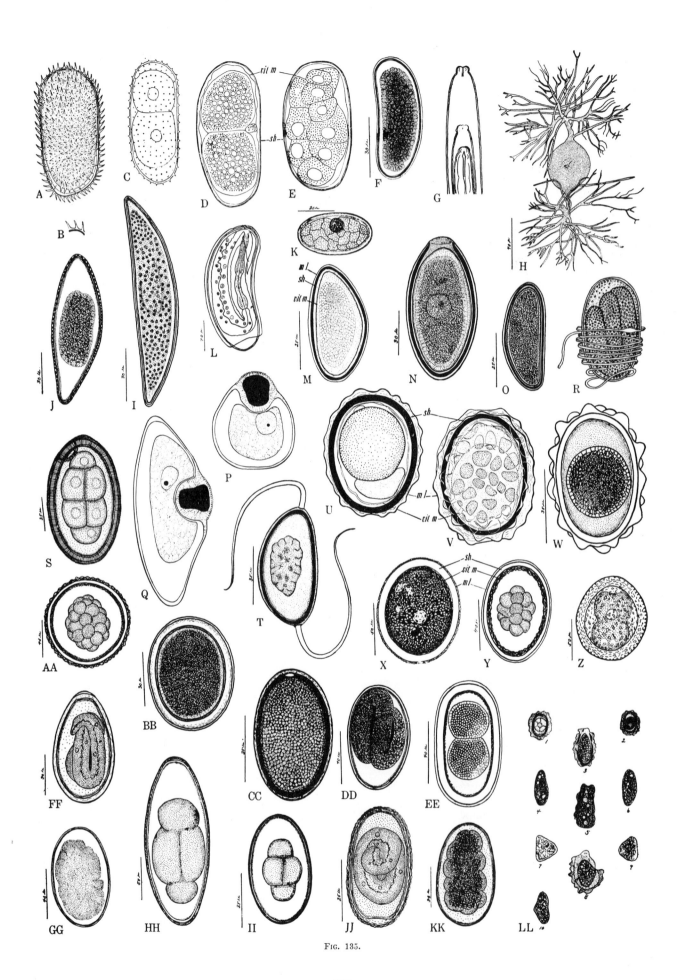

Fig. 135.

some albuminous material. Like Blanchard (1889) he considered the chitinous shell to be composed of two parts, a thin, tough very refractive layer, and a thicker, more brittle outer layer which often showed delicate striations. The egg does not completely fill the shell, but forms a round ball in the center with a clear space at each end. The polar bodies were observed in this clear space. The pigmentation of the albuminous coat is considered to be due to the absorption of bile, a view held by Blanchard, since in females kept alive in Kronecker's solution the albuminous coat was colorless at the time of oviposition.

Thomas (1924) finds three principal membranes composing the envelope of eggs of *Trichosomoides crassicauda*: (1) the outer shell which stains black with Heidenhain's hematoxylin (protein coat), (2) an almost blue-gray fertilization membrane (chitinous shell), and (3) a brownish vitelline membrane. Between the vitelline membrane and the so-called fertilization membrane a peri-vitelline space is described.

Chitwood (1930), like Nelson (Loc. Cit.), considers that the eggs receive their shells while passing through the oviduct. He does not commit himself as to whether the development is exogenous or endogenous. In unpublished manuscript Chitwood points out that shell formation varies; in some species it occurs in the oviduct, whereas in others it occurs in the uterus. The production of the cuticular shell, he says, can be observed in the uterus of *Rhabditis terricola* if a mature female is isolated and observed for a few hours. The shell first appears as a delicate line which thickens as development progresses.

Wottge (1937) offers the most critical studies on the development of the eggs of *Parascaris equorum*. He found that following the penetration of the sperm a clear, transparent shell develops very quickly. This he called the homogeneous membrane. After its appearance the egg does not increase in size. The egg-cell itself shrinks in size and there is a corresponding reduction in the diameter of the homogeneous membrane. The first polar body is thrown off when this layer is completed. A second membrane then appears between the egg-cell and the homogeneous layer which is termed the striated membrane. It also develops rapidly, and is thick, with striations running parallel to its surface. After the production of this layer a wide space, the "Saftraum" or fluid cavity, develops around the egg-cell. When this is established the second polar body is discharged. These membranes remain unchanged during the further development of the egg.

From the above survey it is clear that the vitelline membrane is a zone of condensed substance about the egg-cell (vitellus of Nelson) and that the chitinous shell forms rapidly about it. The diameter of the egg does not increase during shell production as it would if an exogenous process were involved. This was noted by both Nelson (1852) and Wottge. Both workers describe the peripheral protoplasm during shell formation as vacuolate and granular. A shrinkage of the egg protoplasm away from the vitelline membrane leaves the perivitellus space or "Saftraum" of Wottge. The production of the shell in the Ascaridoidea is in the oviduct; in some other forms it may occur in the uterus. The protein coat forms by the adherence of the uterine secretion to the surface of the true chitinous shell. The brown pigmentation of ascarid eggs, and possibly others, is due to staining by the bile pigments. The so-called fertilization membrane of Thomas is undoubtedly the chitinous shell. The following section gives the sequence of development of the egg membranes.

LAYERS OF THE EGG MEMBRANE
R. O. C.

The present conception of the egg envelope is similar to the interpretation of Blanchard. Three coverings should be recognized, namely: (1) the protein external coat which is secreted by the uterine wall, (2) the chitinous membrane or true shell which is a secretory product of the egg itself, and (3) the vitelline membrane which is formed in the oviduct and binds the elements of the initial ovum together. In some groups, such as the Rhabditoidea, Tylenchoidea and Strongyloidea, the protein coat is absent in most of the species. It is also absent in ovoviviparous species such as *Trichinella spiralis*, *Dirofilaria immitis* and *Dracunculus medinensis*. In these forms a true egg of the strongyloid type is produced and hatching occurs *in utero*. All three membranes have been demonstrated in the Mermithoidea, Ascaridoidea, Oxyuroidea, Dioctophymatoidea, Spiruroidea, the oviparous Filaroidea, and some of the Strongyloidea. They are probably also present in some representatives of the Rhabditoidea, Plectoidea and Tripyloidea.

The protein external coat is variously modified in different forms. The terminal byssi described for the Mermithoidea (Figs. 139 & 140) are products of this layer and are continuous with the polar thickenings of it (Christie, 1937). Foster (1914) considers that the terminal filaments of the eggs of *Tetrameres sp*. are not simple prolongations of the chitinous shell but are added after the shell is complete. Our studies of the eggs of *Citellina marmotae* and *Tetrameres sp*. give the same impression. The opercula of the eggs of the Trichuroidea are protein akin but not identical chemically to the protein coat (Chitwood, 1938) and like the latter, the opercula might also be products of the uterine secretion.

The true shell presents little variation in morphology although it may be stratified in appearance. Biochemically the strata of the shell are the same notwithstanding the fact that Zawadowsky (1928) *et al*, divide the shell into three layers. The shell forms the general contours of the egg. In the Trichuroidea and Dioctophymatoidea it is discontinuous at the ends and projects collar-like parallel to the long axis forming the opercular aperture. Fig. 141Z shows a longitudinal section through the egg of *Capillaria aerophila* presenting the membrane relationships to the opercula. The shell is apparently formed of chitin (Chitwood, 1938) or a closely allied substance.

The vitelline membrane varies in its morphology in the different types. In some forms (*Toxascaris leonina*, *Toxocara vulpis* and *Parascaris equorum*) it is thick and filled with reticulations in the mature ova. In developing ova of *Parascaris equorum* the reticulations are not so apparent, but the peripheral zone is filled with granules similar to the so-called "secretory granules" of gland cells. The reticulations are apparently lipoidal since they, with other elements of the vitelline membrane, are dissolved in the ordinary fat solvents. Reticulation to a marked degree can be seen in the eggs of *Toxascaris*. Similar reticulations, though to a lesser degree, are present in the eggs of some of the Trichuroidea. In some species of parasitic roundworms the vitelline membrane appears as simply a wide zone of condensed protoplasm (*Enterobius vermicularis, et al.*), which is fairly resistant to mechanical abrasives. In others (*Necator americanus, et al*) it is a very delicate membrane which is scarcely discernible microscopically.

The sequence of development is as follows: The egg cells undergo a period of multiplication in the caecal or distal portion of the ovary. As they pass down the ovarian tube along the sides of the rachis each egg accumulates yolk material forming a vitellus. The vitellus is pyramidal in shape immediately prior to its discharge into the oviduct and contains a germinal condensation and a nucleus.

Upon reaching the oviduct the egg is fertilized and the vitellus begins to assume the subglobular shape typical of the species. This change of shape may be due, in part, to the contraction of the vitellus to form a sphere or it may result from the pressure exerted following the imbibition of water (Nelson, 1852). The shell begins to form immediately after the penetration of the vitellus by the sperm, and results from an endogenous process. This is evidenced by the appearance of glandular activity on the part of the periphery of the vitellus, granules similar to the secretory granules of gland cells being present and the protoplasm being extremely vacuolate. Even in the egg production of the trematodes in which specialized glands have been described presumably functioning in shell production there is evidence (Kouri and Nauss, 1938) that the shell is derived from the granules of the vitellus. Another point of evidence of endogenous development of the chitinous shell in the nematodes is the fact that after assuming a spherical shape the egg does not increase in diameter as it would with exogenous development. Figure 136 A shows the penetration of the vitellus by the wedge-shaped sperm, the condensed granular periphery and the vacuoles. The vitellus varies slightly from the typical pyramidal shape usually seen at this time.

The chitinous shell is almost completed by the time the egg reaches the uterus. The vitelline membrane has formed within the shell but does not show the degree of reticulation seen later. The protein coat is absent but begins to form as the egg enters the uterus. At first it is weakly mammillated and very thin. By the time of the discharge of the first polar body it is a well-established membrane. Figure 136B shows a uterine egg at the time of the completion of the protein coat. The vitelline membrane is diagrammatically represented in our figure, but can be seen at this time. The vitellus shrinks leaving a fluid cavity between it and the vitelline membrane, the "Saftraum" of Wottge (1938), or the perivitellus space. The tetrads of the first polar body can be seen, and the condensed chromatin of the male nucleus surrounded by the archoplasm. Figures 136C and D show two types of mitotic figures in the production of the first polar body. After division of the polar body the dyads are left as in figure 136E. A second polar body forms in the same manner. Figure 136F shows the completely developed egg with the male and female pronuclei. The two polar bodies can be seen, one against the vitelline membrane and the other in the vitellus which at this time is surrounded by a wide perivitellus zone. The vacuoles have disappeared.

Segmentation follows the release of the eggs from the host. The first segmentation stage results from the mitotic figure formed by the coalescence of male and female pronuclear material. Cleavage and embryology occur as shown for *Ascaris lumbricoides*, figures 137 A to H.

The formation of the pattern of the protein coat of parasite eggs is difficult to hypothesize. Some workers describe molding chambers formed by constriction of the genital tube. They assume that the eggs pass along the duct single file and consequently the surface comes into contact with the epithelial walls. This may be true in some cases, but many of the most highly sculptured eggs occur in all levels of the uterus of some species in such numbers that their contact is problematical. This is certainly true in Ascaridoidea.

A more tenable theory of the development of the external sculpturing might be advanced on the basic precepts of colloidal behavior. The protein droplets in the uterus have a fairly high degree of consistency. As they come into contact with the shell they adhere at the point of contact. They congeal, possibly through loss of water absorbed by the vitellus, or possibly as a result of increase of hydrogen ion concentration in the neighborhood of the cell. Ascarid eggs have been shown by Nolf (1932) to require oxygen which results in the liberation of carbon dioxide as a waste. This might produce a pH differential between the uterine fluid and the periphery of the egg. If it is assumed that there is a specific difference in surface tension of the protein droplets, and the assumption seems reasonable, there would be a difference in size to the initial congealed particles. Subsequent addition of protein material would maintain the difference resulting in sculpturing of different degrees of prominence and different designs in the various species.

The production of specializations such as filaments and byssi are even more difficult to visualize. If a filament were present as a central cord they could be explained on the basis of adsorption phenomena but no such filament has as yet been demonstrated.

Unfertilized eggs possess no centrosome, have a granular appearing shell (Nelson) and a vitelline membrane. In some eggs the shell is reduced to a barely discernible membrane. The protoplasm is vacuolate from the time of the formation of the vitelline membrane until it degenerates. The protein coat may or may not be present (Fig. 135LL). In some cases at least, both the vitelline membrane and the true shell are ap-

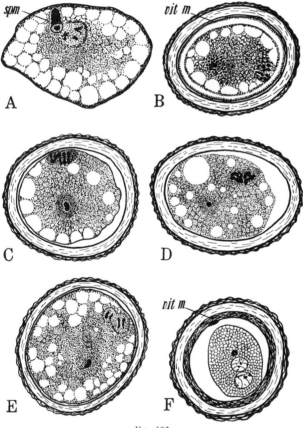

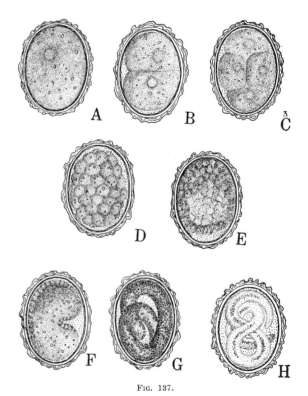

Fig. 136.

A—The penetration of the sperm into the highly vacuolate vitellus of the egg. B—The tetrad of the first polar body. The shell and the vitelline membrane have completely formed. The vitellus has condensed toward the center of the egg leaving a space, the perivitellus space, between the vitellus and the vitelline membrane which is diagrammatically shown. The chromatin of the male nucleus is condensed and surrounded by granular archoplasm. C and D—The division of tetrads to form the first polar body. E—Dyads after the first polar body is given off. These divide to form the tetrads of the second polar body. F—The male and female pronuclei and the two polar bodies, the first against the vitelline membrane and the second in the vitellus itself. Fertilization (Fig. A) occurs in the oviduct. All subsequent stages are found *in utero*. Original, Christenson.

Abbrevations: *ml*, protein layer; *sh*, shell; *vit m*, vitelline membrane; *op*, operculum.

Fig. 137.

Cleavage and early embryonic development of *Ascaris lumbricoides*. Figures A, B and C represent the 1-cell, 2-cell and 4-cell stages respectively. Figure D shows a late morula or early blastula. Figure E shows the cleavage cavity in a late blastula which is followed by gastrulation (Fig. F). Figures G and H give two stages in larval formation, the latter being the infective larva. Original, Christenson.

parently absent. In the parthenogenic species, *Rhabditis fili-formis,* no vitelline membrane is formed in the developing eggs according to Chitwood (unpublished observation). The same appears to be the case in the parasitic generation of *Strongyloides ratti,* while a vitelline membrane is present in developing eggs of the bisexual free-living generation according to Chitwood and Graham (1940).

Oviparity
R. O. C.

By far the majority of parasitic nematodes are oviparous. In some (*Trichuris trichiura, Ascaris lumbricoides, et al.*) the eggs are discharged in the unsegmented condition. After variable periods of incubation outside the host, usually of long duration, infective embryos are produced. Other species (*Necator americanus, et al.*) develop rapidly, hatching occurring in a few days. The liberated larvae undergo further development in the soil. In some species partial intra-uterine development of the larvae occurs. Typical morulae, for example, are produced in the eggs of *Contracaecum quadricuspe,* and *Habronema colaptes* before they are discharged (Walton, 1923). *In utero* development may progress until the eggs contain infective larvae in some forms. Such is the case with *Enterobius vermicularis* and many other species. Intermediate hosts are involved in the transfer of many nematodes, either the embryonated eggs or the infective larvae being taken up by annelid worms, arthropods and other essential hosts.

The condition of oviparity approximates ovoviviparity in some parasitic nematodes. Development progresses until larvae are produced and these hatch almost immediately after oviposition. This condition is seen in *Rhabdias bufonis,* a pulmonary parasite of Amphibia, or in *Rhabditella axei* (syn. *Rhabditis macrocerca*) as reported by Faust and Martinez (1933). After the discharge of the eggs in the lungs, in the case of the former species, they traverse the alimentary canal usually hatching in the rectal portion. Intra-host hatching likewise occurs in the case of *Strongyloides stercoralis* and *Spirocerca lupi* (*S. sanguinolenta*). Following gastric infections with the latter species the embryos are found free of the egg membranes in the caecum and colon of dogs. This parasite is likewise found commonly in aortic nodules from which the embryonated eggs are discharged directly into the blood stream where they hatch. The microfilaria-like, unsheathed larvae have often been confused with those of filarioid worms (Lewis, 1874, *et al*).

Ovoviviparity
R. O. C.

Ovoviviparity refers to the intra-uterine hatching of fully formed eggs. It is a wide-spread phenomenon among nematodes parasitic in tissues. Many of the so-called viviparous species are, in the strict sense of the term, actually ovoviviparous.

A good example of ovoviviparity is afforded by *Cosmocercella haberi.* In this species Steiner (1924) reports from one to three larvae, and one to five eggs in the uterus at one time. The case of *Trichinella spiralis* is complicated by a greater biotic potential but is none the less clear. Immature females possess typical eggs in the uterus, each composed of the segmenting protoplasmic mass, a vitelline membrane and a shell (Fig. 141AA). *Dirofilaria immitis* and *Dracunculus medinensis* are likewise commonly considered viviparous species which are actually ovoviviparous. In both cases typical eggs form which hatch *in utero* (Figs. 141 T & N).

Some of the other well-known, so-called viviparous forms should be included here. The eggs of *Onchocerca fasciata* are described by Badanine (1938), as covered by an excessively delicate membrane which may present either a spherical or oval outline. No attempt was made to determine the egg membranes present, but the figure given is fairly typical of the eggs of ovoviviparous species. Blacklock (1926, 1939) describes the eggs of *Onchocerca volvulus* stating that the egg membrane remained practically unstained when dried, fixed in alcohol and stained with hot haemalum. This procedure would destroy the delicate vitelline membrane leaving the shell which is undoubtedly the membrane Blacklock observed about the coiled embryos.

Augustine (1937) made some interesting observations of the early development stages of *Vagrifilaria columbigallinae.*

He observed that the developing larvae in the uterus are enclosed in a delicate membrane but that no membranes were present about the larvae in the vaginal region. He found crumpled, hyaline objects in the vaginal part of the uterus which were similar in size and shape to the egg membranes indicating intra-uterine hatching.

Some authors go so far as to include under ovoviviparity species in which the embryonated eggs hatch outside the uterus. Kreis and Faust (1933), for example, consider *Rhabditella axei* (syn. *Rhabditis macrocerca*), as being ovoviviparous but state that hatching occurs after the eggs have left the parent worm.

Whether or not viviparity in the true sense of the term actually exists in the Nematoda will have to be determined through critical research on early developmental stages of additional species.

Significance of the Embryonic Sheath
R. O. C.

Nematodes, like arthropods, grow through the process of ecdysis. Usually the cuticle is shed four times, the fifth stage thus formed being the adult. The so-called "infective stage" reached before the invasion of the definitive host in many species follows the second molt. This stage generally occurs in the soil (*Necator americanus, et al.*), or in an intermediate host. Molting usually follows shortly after the escape of the larvae from the egg envelope but it may occur in the egg itself. This was observed by Seurat (1914) in *Nematodirus mauretanicus.* Cobb, Steiner and Christie (1923) show this to be the case in the development of *Agamermis decaudata.* It has also been reported for *Ascaris lumbricoides* v. *suum* by Alicata (1935), and in *Rhabdias fuscovenosa* var. *catanensis* by Chu (1936), and in other species. Some nematodes molt twice in the egg, two cuticles being present when the larvae hatch. In some instances the embryonic sheath is retained by the larvae as a loosely-fitting coat. This is especially true of the third stage, infective larvae of certain groups of the superfamily Strongyloidea, the second cuticle being the one retained. It is apparently protective in function. Occasionally the last two cuticles may be found covering the adult stage preparatory to the last molt. Goodey (1930) reports this to be true for *Tylenchinema oscinellae,* as shown in figure 135G.

In the Filarioidea the adult parasites live in locations often not associated with the natural openings of the body. Living young are discharged into the humoral elements. These may be sheathed (*Foleyella spp., Setaria spp., Isociella spp., Wuchereria bancrofti, Thamagadia hyalina, Saurositus agamae, Loa loa, et al*) or unsheathed (*Onchocerca spp., Dirofilaria spp., Dipetalonema spp., et al*). These larvae may live in the circulatory system for long periods of time without significant morphological changes, for example Underwood and Harwood (1939) transfused larvae of *Dirofilaria immitis* into uninfected dogs and found that they would survive over two years. Even unsheathed mirofilariae can leave the blood stream and migrate through the tissues as has been demonstrated by Harwood (1932) in the case of *Litomosoides sigmodontis.*

Shortly after being taken up by the alternate host the sheath of ensheathed forms, such as *Wuchereria bancrofti,*

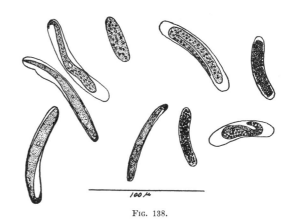

Fig. 138.

Eggs and embryos of *Habronema muscae.* After Ransom, 1913, U. S. D. A., B. A. I. Bull. No. 163.

is shed. Manson (1884) describes the process in detail as found in *Wuchereria bancrofti*, stating that it usually occurs within an hour in the mosquito vector. The shedding of the initial sheath in this species is followed by a striation of the larval integument. Yamada (1927) and Feng (1936) show that two additional molts follow the initial shedding of the sheath. In unsheathed forms, for example *Dirofilaria immitis* and *Dracunculus medinensis*, only two molts occur in the vectors before the microfilariae are ready for their transfer. The first molt involving the initial sheath of *Mf. bancrofti*, followed by two additional, has led some workers to doubt that the first "cuticle" is homologous to the other two.

Penel (1904) advanced the idea that the initial sheath of *Loa loa* was derived from the vitelline membrane. He removed developmental stages from various levels of the uterus noting that the membrane enlarged with the growth of the larvae. In the vaginal portion of the uterus he recovered the sheathed microfilariae. His conclusion was that the larval investment was derived from the egg membrane. Penel's idea has been carried over to account for the origin of the sheath of *Mf. bancrofti* (See Fülleborn, 1928) and other species.

The modification of the egg membrane accompanying the development of the embryo has been seen in other species. Ransom (1904) shows that as the eggs of *Habronema muscae* develop the egg membrane follows the larval contours producing a sheath-like covering (Fig. 138). Faust (1928) notes a similar modification in *Thelazia callipaeda*. In the case of this species, however, there is a peculiar ballooning at the end, a large vesicle being formed. Faust states that the covering is retained for some time and that it is protective in function. Neither of these authors refer to a possible relationship between the modified egg membrane and the initial larval sheath. They, similarly, do not make a statement as to which of the egg membranes is involved.

Penel's idea that it is the vitelline membrane which forms the microfilarial sheath is not tenable. He, like most workers of his period, did not attempt a critical study of the membranes. Our studies show that the developing larvae of such so-called viviparous species as *Dirofilaria immitis* (Fig. 141T) and *Dracunculus medinensis* (Fig. 141N) are covered by a true shell within which there is a delicate vitelline membrane. This membrane is discernible only by careful study with oil immersion lenses in formalin-preserved materials. It is visible only at points of separation from the chitinous shell. Furthermore, Blacklock (1939) noted the presence of the egg membrane of *Onchocerca volvulus* following fixation in alcohol and staining with hot hemalum. It is safe to assume that the external membrane in this species, also, is the chitinous shell since the procedure used would destroy the delicate vitelline membrane. Augustine (1937) observed the shedding of the egg membranes in the uterus of *Vagrifilaria columbigallinae*, remnants of the shell being found in the vaginal portion.

It is thus clear that the chitinous shell is present in the eggs of ovoviviparous Filarioidea and Dracunculoidea which have been studied critically. The presence of a chitinous shell implies the presence of a vitelline membrane since the two appear almost simultaneously in the developing egg, and the early vitelline membrane must be considered to be simply the zone of shell formation (p. 178). That the chitinous shell has some elasticity and can adjust to the contours of the developing larva has been demonstrated by Penel, Ransom and Faust. This does not necessarily mean that the chitinous shell forms the initial microfilarial sheath. Biochemically the chitinous shell is not easy to distinguish from the embryonic cuticle (Chitwood, 1938) since both are highly insoluble. The critical test, solubility in hot alkalis, is often uncertain with such small materials.

Some workers express the view that the sheath of *Mf. bancrofti* is derived from the shed cuticle as is true of subsequent stages. As early as 1874, Lewis noted the presence of the sheath in *Mf. bancrofti* and its absence in *Mf. immitis*. He advanced two possibilities of formation, either it was derived as the shed cuticle, or it was derived from the egg envelope. More recently Augustine (1937) points out that in the fresh blood the sheath of *Mf. bancrofti* is extremely difficult to see, but that as heparinized blood dried the "formation" of the sheath could be followed. The efforts of the microfilariae to push on and back out caused a stretching of the once close-fitting, inconspicuous outer covering. He further states that it is quite possible that some stretching may occur in the circulation when the microfilariae are temporarily trapped in the capillaries. Augustine concludes that the sheath of *Mf. bancrofti* is comparable to that of infective hookworm larvae—namely, the result of an incomplete ecdysis.

He advances the possibility that the absence of the sheath in some species might be explained on the basis of delayed molting in which the formation of the membrane takes place in the vector.

The development of *Trichinella spiralis* should be considered briefly from the standpoint of sheath formation. Schwartz (1918) has demonstrated that trichinae removed from the alimentary canal of rats will undergo molting outside of the host. He does not state the number of molts undergone but implies that there are at least two before the adult stage is reached. So far as we are aware no one has attempted to determine the possibility of molting occurring before encapsulation of the larvae. Raffensperger (1918) found that trichina larvae were not infective 15, 17 and 18 days after infection, but were after 21 days. Our experiments showed that the point of infectivity was between 18 and 20 days. Nolf and Edney (1935) found the larvae infective after 17 days. Undoubtedly at this time some biological change occurs in the parasite, probably associated with molting, which results in their infectivity. The explanation of some workers that infectivity is the result of encapsulation will not hold since it is inconceivable that the connective tissue capsule, which is dissolved in the stomach, could offer any protection for the larvae against the stomach barrier. It is more probable that infectivity follows the second molt as is the general rule with other nematodes. The presence of the shed cuticles in the blood stream might be a contributory factor in producing the eosinophilia associated with trichina infection.

Special Morphology
R. O. C.

Eggs of parasitic nematodes are variously modified to adapt them to ofttimes complicated life cycles. The more obvious of these specializations are the presence of byssi, terminal or polar filaments, equatorial filaments, opercula, and the mammillation of the shell. Byssi (Fig. 139) are branched, polar cords forming tassel-like structures on the eggs of the Mermithoidea. Apparently their function is to hold the eggs to the pubescent surfaces of plants until they are eaten by the essential hosts. Terminal filaments are found in certain genera of the superfamilies Spiruroidea and Oxyuroidea. They are unbranched structures which may occur singly, in pairs, or as tufts. These filaments may be unipolar or bipolar in distribution. Foster (1914) advances the idea that in *Tetrameres* (Fig. 141E) they function in holding the eggs together to insure massive infection of the hosts. Subpolar filaments are sometimes distributed over the surface of the egg accompanying the polar tufts. Equatorial filaments (Fig. 135R) occur in the genus *Pseudonymus* of the superfamily Oxyuroidea. They may function in the manner suggested by Foster for the filaments of *Tetrameres*. Chitwood (by correspondence) points out that species with filaments on the eggs are basically associated with an aquatic habitat, and that the filaments may function by entangling the egg in the vegetation thus preventing it from settling into the débris of the substratum which would reduce its chances of survival.

Among free living nemas hooks (*Anaplectus granulosus*) and minor excrescences (*Rhabditis filiformis, Mononchus punctatus, Trilobus pellucidus*) of the protein layer occur in some aquatic species. It is notable that *Rhabditis filiformis* is one of the very few species of aquatic *Rhabditis* and it is the only one known with a protein layer.

Opercula are zones of escape by which the embryos leave the egg membranes. They may have bipolar distribution and appear plug-like as seen in the Trichuroidea, and, to a lesser extent, in some of the other major groups. Some nemas possess a single operculum, others an opercular spot marked by the thinning of the membrane in a certain region, or the shell in some species may have lines of fracture indicating the area from which the embryo leaves the egg.

Mammillation, in the strict sense of the term, refers to the rounded excrescences over the surface of the protein coat as seen in eggs of *Ascaris lumbricoides* (Fig. 135U-W) and *Dioctophyma renale* (Fig. 141BB-CC). Since there is a gradation between this type of external modification into pitting, ridging, and simple rugosity it seems proper to include all of these under a single general term. Mammillation, to some extent, occurs in most of the superfamilies of parasitic nematodes.

The foregoing specializations of the eggs will be considered in detail under the superfamily in which they are found.

Eggs of Free-living and Plant Parasitic Nemas. M. B. Chitwood. Very little information is available regarding the eggs of free-living and plant parasitic nemas since they are seldom used as taxonomic characters. In the Rhabditoidea the eggs are usually of about the shape of the egg of *Rhabditis strongyloides* Fig. 135E), both shell and vitelline membrane being simple and no protein layer being present. In a few species, however (*Diploscapter coronata, Rhabditis filiformis* and some cephalobids) the shell bears small protuberances (Fig. 135C) apparently formed by an exceedingly fine protein layer. In some parthenogenic species (*Strongyloides ratti* and *Rhabditis filiformis*) Chitwood and Graham (1940), found no vitelline membrane. In the Tylenchoidea, particularly the family Tylenchidae, there is quite a bit of variation in the relative length of the eggs, some species having short, thick eggs (*Aphelenchoides parietinus*), others rather narrow elongate eggs (*Meloidogyne hapla*, Fig. 115 M & Q, and *Aphelenchoides fragariae* Fig. 119E). In the subfamily Heteroderinae a mucoid is commonly produced as a protective mass for the entire egg brood. This material is hygroscopic and is undoubtedly important in the life of the organism.

Eggs with protuberances or spines are also known to occur in a few species of free-living aphasmidians. These include *Anaplectus granulosus* (Fig. 120C), *Mononchus punctatus* according to Cobb (1917), and *Trilobus pellucidus* according to Steiner (1916). Some chromadorids have shells giving a punctate appearance. These species with modified egg shells are the only members of the Aphasmidia known to produce a protein layer. The eggs of few other nemas have been observed outside the parent so that, except for shape and relative size, little is known. In some forms such as *Chromadora sp.* (Fig. 120E) and *Microlaimus fluviatilis* the eggs are nearly spheroid while in others they are long and narrow as in *Achromadora minima,* described by Cobb, (1918), in which the mature egg is about one sixth the body length. All sizes and shapes between these extremes are known.

Strongylina. (R. O. C.). Eggs usually with a thin, smooth shell, composed of two layers, the chitinous shell externally and the delicate vitelline membrane internally. In some forms a mammillated protein coat is present, for example *Metastrongylus salmi* (Fig. 135JJ). Usually there is a wide fluid cavity between the vitelline membrane and the cell mass. The shape is that of a regular ellipse in most species, but truncation has been reported for some eggs. While most of the ova measure below 100 microns, in the family Trichostrongylidae measurements approximating 300 microns have been reported. Operculation is a rare condition but has been reported. In most forms the ova are in early segmentation stages when discharged; embryonated eggs are rare. A few viviparous forms have been reported. A typical egg is that of *Necator americanus* (Fig. 135II).

Certain modifications of the egg envelopes occur in this group which deserve mention. Eggs of *Cyathostoma* possess an operculum at one pole as shown for *C. americanum* (Fig. 135FF). The chitinous shell of this species is relatively thick and truncated at one pole. Beyond it is a delicate extension enclosing a lenticular space. No line of fracture is evident microscopically but it is at this point the embryo leaves the egg envelope. The eggs are unembryonated *in utero*. Operculation has similarly been reported in other representatives of the family Syngamidae.

As in the case of *Cyathostoma americanum* the shape of strongyloid eggs often varies from the regular ellipse characteristic of most species. In *Nematodirus aspinosus* the egg is attenuated at one pole and truncate at the other; similar truncation exists in one pole of the egg of *Stephanurus dentatus* (Fig. 135KK). Eggs of *Nematodirus roscidus* possess a rugose or alveolate shell. A slight degree of striation appears in the shell of *Nematodirus fillicolis*. Some eggs are distinctly mammillated (*Metastrongylus salmi, et al*).

Eggs of the strongyloid worms are fairly uniformly below 100 microns in length. In three genera of the family Trichostrongylidae mammoth forms occur ranging upwards of 300 microns. The largest egg of which we have record is that of *Nematodirus orientianus* which ranges from 255 to 272 microns in length. The genera *Marshallagia, Nematodirus,* and *Nematodirella* all contain species with large eggs.

There is some variation in the degree of uterine development of the embryos in the superfamily *Strongyloidea*. The usual condition is that seen in eggs of *Necator americanus* (Fig. 135II), *Nematodirus fillicolis* (Fig. 135HH), or *Stephanurus dentatus* (Fig. 135KK), in which early segmentation

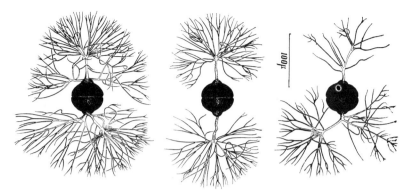

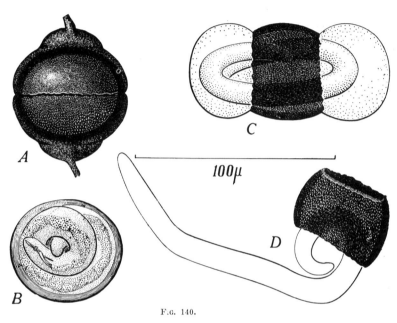

F.G. 140.

Egg of *Mermis subnigrescens.* A—showing outer and inner layers of shell; B—with outer layer of shell removed, showing larva within; C—in process of hatching; D—with larva emerging. After Christie, 1937, J. Agri. Res., v. 55 (5): 353-364.

stages are seen. In eggs of such forms as *Dictyocaulus filaria* and *Metastrongylus salmi* (Fig. 135JJ) intra-uterine development of the embryo occurs. Viviparity has been reported for the genus *Crenosoma* and a few other genera.

Figures 135DD and 135FF show the eggs of *Oesophagostomum radiatum* and *Cyathostoma americanum* respectively after incubation outside the host.

Oxyuroidea. (R. O. C.). The eggs of the superfamily Oxyuroidea are usually described as possessing a double-contoured shell flattened on one side. The double-contoured appearance is due to the presence of the protein coat and the chitinous shell forming the external investment. Internally is the vitelline membrane of considerable thickness and durability in some species. In general the eggs are small but some species run above 100 microns in length, for example *Syphacea obvelata.* (Fig. 135 I). The eggs of some species possess polar filaments, equatorial filaments, opercula and mammillation of the shell.

Polar filaments are found in the eggs of *Citellina marmotae* and other species of the genus. Mantor (1930) describes the eggs of this species as thin-shelled, elongate oval in outline, and somewhat flattened on one side. From each pole there arises a long filament, broader at its base where it is attached to the outer shell and tapering to a fine thread. The filaments are equal in length and normally possessed by all eggs (Fig. 135T).

Peculiar equatorial filaments occur on the eggs of *Pseudonymous sp.*, a parasite of certain aquatic beetles. The normal number is two, but they may occur singly. They originate from a cone-like papilla on the protein coat and coil spring-like about the egg *in utero*. At oviposition the filaments unwind, rotating the egg on its long axis. The eggs are embryonated at the time of discharge (Fig. 135R).

Operculation is quite commonly seen in the Oxyuridae. Bipolar operculation is possesed by eggs of *Pharyngodon spinicauda*. The asymetrical eggs of this species have prominent polar caps. In *Oxyuris equi* (Fig. 135N) a single operculum is present, the egg being somewhat similar to those of typical trematodes. The chitinous shell in this species is incomplete at one end, as is seen in the opercular apertures of the Trichuroidea, the end being closed by a thin, protein deposition. The egg of *Dermatoxys veligera* has a subpolar operculum marked by a prominent line of fracture (Fig. 135S). In the case of eggs of *Citellina marmotae* the operculum is likewise subterminal. The protein coat is evenly striated over its entire surface except at a point near one pole. Here the striation is lost and the membranes become thin. This marks the opercular spot described by Mantor (1930), appearing as shown in figure 135T.

Mature eggs of *Dermatoxys veligera* have excrescences over the surface giving the shell a striated appearance (Fig. 135 S). Freitas and Almeida (1936) describe two longitudinal ridges, which are strongly striated transversely, for the eggs of *Heteroxynema wernicki*. Chitwood (1932) describes a lateral crest on the flattened eggs of *Protrellus aureus* (Fig. 135P-Q).

Some workers lay much stress on the shape of the egg as a diagnostic character. The case which comes to mind is Kofoid and White's "*Oxyuris incognita*" which was described on the basis of asymmetrical eggs passed by soldiers in Texas. The eggs seen (Kofoid and White, 1919) were of the flattened type usually possessing large, bluish-green globules at the end of the protoplasmic mass. Sandground (1923) pointed out that these eggs belonged to *Meloidogyne*, a group of plant inhabiting nematodes whose eggs will pass through the alimentary canal unaltered. Figure 135F shows the egg of *Meloidogyne sp.* which is similar to those found by Kofoid and White. These were extracted from worms occurring in nodules of potatoes. We have encountered similar eggs following gastric expression of a patient suffering chronic stomach trouble. A lettuce sandwich had been eaten shortly prior to the examination which was the cause of the so-called "nemic infection." Though often asymmetrical, the eggs of the Oxyuroidea may be oval as in *Oxyuronema atelophora* (Fig. 135K), spheroid as in some of the Thelastomatidae as reported by Chitwood (1932), or truncate as in *Oxyuris equi*.

The degree of development at the time of deposition shows considerable variation; *Oxyuronema atelophora* eggs are discharged in early segmentation stages, *Enterobius vermicularis*, and others, are embryonated at the time of discharge.

Zawadowsky and Schalimov (1929) figure the egg of *Enterobius vermicularis* as possessing a wide protein coat extending far beyond the chitinous shell. The appearance of the protoplasmic mass indicates that the egg figured was derived from the ovarian portion of the uterus. In all probability the "albuminous" coat had not condensed as it would later. Reardon (1938) shows the protein coat in this species to be much thinner at the time of oviposition, which is supported by our studies (Fig. 135M).

ASCARIDOIDEA. (R. O. C.). The eggs of the ascaroid worms have been studied by biologists since the middle of the past century. The classic work of Nelson (1852) demonstrating the penetration of the egg by the sperm and its subsequent development, and the cytoligical studies of Boveri (1888) have added much to our knowledge of the phenomena associated with sex cells. During more recent years studies on longevity, environmental factors influencing development, and on the permeability of the layers have been directed upon the eggs of this group. Ascarid eggs have become to cytology and cellular physiology what the frog is to general physiology and morphology, an indispensible laboratory subject.

In general the eggs of the Ascaridoidea are thick-shelled and possessed of three membranes in the egg envelope, namly: (1) the external protein coat which may be mammillated and deeply pigmented, giving a brown color; (2) the chitinous shell which is thick and quite transparent, and (3) the vitelline membrane which may contain coarse reticulation of the periphery. The typical shape is that of a regular ellipse but spherical or subglobular forms are well-known (*Toxocara canis*, Fig. 135AA). In some species the eggs are almost oblong in outline (*Ascaridia lineata*, Fig. 135EE). Usually development does not progress beyond early segmentation stages *in utero*, the eggs requiring relatively long periods of incubation in the soil before they become embryonated.

The protein coat of the ascaroid eggs is a thick layer in some species and coarsely mammillated, for example, *Ascaris lumbricoides* (Fig. 135W). In other forms it is thinner and marked by less prominent irregularities as seen in *Toxocara canis* (Fig. 135AA). Baylis (1936) describes the condition in *Toxocara pteropodis* as pitting (Fig. 135Z). *Parascaris equorum* has eggs in which the mammillations are even less pronounced. Upon surface view the exterior of the latter three species may give a honey-comb appearance. Other eggs present a perfectly smooth outline, the protein coat being barely discernible about the margin of the thick shell.

Variations occur in the appearance of the protein coat in normal eggs, but they are especially noticeable in unfertilized eggs. Unfertilized eggs of *Ascaris lumbricoides* have received much study because of their clinical importance (Miura and Nishiuchi, 1902; Foster, 1914; Otto, 1932; Keller, 1933; Matuda, 1939, et al). In some instances the protein coat and the chitinous shell are apparently lacking, the vitelline membrane alone covering the highly vacuolated protoplasmic mass. In others the protein coat may be lacking, with the shell and the vitelline membrane appearing to be perfectly normal. In still other cases the protein coat may be thickened and distorted to produce grotesque shapes. Blanchard (1888) reports a stringy appearance in the mammillation occurring at times in normal fertilized eggs of *Ascaris lumbricoides*. This is thought to be due to the cohesion of eggs entering the vaginal portion of the uterus. Fig. 135 LL shows some of the variations found in unfertilized eggs of *Ascaris lumbricoides* as observed by Otto, 1932. Matuda (1939) points out that an anomalous condition is sometimes found in this species in which several egg-cells are bound together by a single chitinous shell. Stratification of the chitinous shell has been noted by some authors. Nelson (1852) observed this in *Toxocara cati*, and it has been reported more recently by Zawadowsky (1928) in several species of ascaroid worms.

The vitelline membrane is relatively thick and, in some species, filled with coarse reticulations. This is seen most clearly in the eggs of *Toxascaris leonina* (Fig. 135Y), but also occurs in *Parascaris equorum* (Fig. 135X). The reticulations disappear following treatment with ordinary fat solvents and are considered to be lipoidal in nature. The vitelline membranes in *Toxocara canis* and *Ascaridea lineata* have a dense, granulated appearance which may run toward reticulation.

Opercula have been reported for some of the ascaroid eggs. Dorman (1928) records two opercular plugs for the eggs of *Heterakis papillosa*. These, he states, are seen most prominently in the second membrane, but usually extend through all three. The eggs of *Heterakis gallinarum* possess no plug-like structures but a lenticular clear space is present at one pole (Fig. 135CC). The shell may be somewhat thickened in this area.

Ackert (1931) presents a classic study on the morphology and development of the eggs of *Ascaridia galli*. He states that the shells begin to form in the distal portion of the uterus, and that they are composed of three membranes: (1) an inner, highly permeable vitelline membrane, (2) the thick, resistant shell, and (3) a thin, "albuminous" covering. In one end a structure resembling a micropyle was seen which on micromanipulation was found to be a solid, conical appendage of the vitelline membrane. Baylis (1929) considered this structure to be an internal thickening of the shell.

Aside from the structures already mentioned the shell of Ascaridoidea eggs presents some additional modifications of interest. Olsen (1938) reports a thickening of one side of the shell in *Aplectana gigantica*. This egg is spherical, or subglobular in outline, and the thickening is confined to the chitinous shell and does not involve the other membranes (Fig. 135BB).

In general the eggs of the Ascarididae and Heterakidae require incubation outside the host before they are embryonated. There are exceptions to this, however, as in the case of *Ascaris phacochoeris* from the wart hog. Ortlepp (1939) finds intra-uterine development of the eggs to the embryonic stage in this species. *Cosmocercella haberi* is an example of an ovoviviparous form.

DRACUNCULOIDEA. (R. O. C.) But few observations have been made upon the intra-uterine stages of the Dracunculoidea. *Dracunculus medinensis*, a commonly reported viviparous species, is actually ovoviviparous since typical eggs are formed *in utero* covered by a vitelline membrane, a chitinous shell, and possibly a very thin protein coat presenting a slight degree of rugosity (Fig. 141N). Intra-uterine hatching occurs since the vaginal portion of the uterus is filled with the characteristic, long-tailed larvae.

True viviparity may occur in this group in the case of *Micropleura vivipara*. Baylis and Daubney (1922) state that the development of the embryos appears to be very rapid, the uterus being entirely filled, from end to end, with young apparently fully formed and not inclosed in membranes. Thomas (1929) similarly does not mention the presence of eggs in *Philometra nodulosa*. Van Cleave and Mueller (1934) likewise made no mention of the eggs in the latter species.

SPIRUROIDEA. (R. O. C.) Little can be said regarding the general charateristics of the eggs of the superfamily Spiruroidea. Usually they possess smooth, thick shells and are embryonated at the time of discharge. The egg envelope is composed of the three typical membranes; the protein coat, the chitinous true shell, and the vitelline membrane. The general shape is that of a regular ellipse, but various specialized shapes exist. Terminal filaments, opercula, and mammillations have been reported in this group.

Terminal filaments have been observed in the genera *Tetrameres*, *Cystidicola*, *Metabronema*, *Ascarophis*, *Rhabdochona*, and *Spinitectus*. They are not uniformly present, however, in all species of the genera in which they occur. The number of filaments, and their length, is variable in some species while it is constant in others.

The filaments of *Tetrameres* were described by Seurat (1914) and Foster (1914). Seurat described an egg of the flattened oxyurid type possessing a tuft of filaments at each pole (Fig. 141E). He did not attempt an analysis of the membranes of the egg envelope. Foster noted similar filaments in a species of *Tetrameres* (*Tropidocerca*) from the American woodcock. He found seventeen to twenty-three filaments forming the polar tufts. Most of them were not over half the length of the egg, but one or two at each pole were twice as long (Fig. 141D). Foster points out that these filaments are not prolongations of the chitinous shell but are added after the shell is complete. He considers them analogous to the mammillations seen in the shell of *Ascaris lumbricoides*, a view which is supported by our studies. Sandground (1928) observed polar filaments on the eggs of *Tetrameres paucispina*.

Skinker (1931) gives a good discussion of the polar filaments that are found on the eggs of *Cystidicola stigmatura*. In no case were fewer than four present, and the majority of the eggs possessed from eight to twenty. The variable number and differing length of these structures, is apparent from Skinker's figures. Not all members of the genus *Cystidicola* possess filaments, however, since Hunter and Bangham (1933) report their absence in *Cystidicola lepisostei*. Figure 141F shows the egg of *Cystidicola stigmatura* drawn from paratype material. The filaments, as in the case of *Tetrameres*, are derived from the protein coat. Van den Berghe (1935) observed both terminal and lateral filaments on *Cystidicola farionis*. They are similarly present over the surface of the egg of *Rhabdochona ovifilamenta* as well as occurring in polar tufts (Weller, 1938; Fig. 141M).

Van Beneden (1871) was the first to observe polar filaments in species of *Ascarophis*. He states that the eggs are distinguished from those of other nematodes by the presence of two filaments which garnish one of the poles. Nicoll (1907) likewise observed these structures. Cobb (1928) notes their absence in the uterine eggs of *Ascarophis helix*. Baylis (1933) includes them as a generic character in his recharacterization of the genus *Ascarophis* in spite of their apparent absence in the eggs of some species.

Polar filaments have likewise been observed in the genera *Spinitectus* and *Metabronema*. The condition in *Metabronema magnum* is worthy of mention since in this species two filaments arise from button-like opercula at each pole (Fig. 141H).

Operculation is a fairly common phenomenon among the eggs of the Spiruroidea. The button-like opercula of *Metabronema magnum* have already been mentioned. The eggs of *Hedruris siredonis* have opercula suggestive of the Trichuridae (Chandler, 1919). Baylis (1931) finds a similar condition in *Hedruris spinigera*. Some spiruroid eggs are truncate at both poles and the opercula demarcated by sub-terminal lines of fracture. Ransom (1904) reported this to be the case for *Oxyspirura mansoni* (Fig. 141L). It can be seen even more clearly in the eggs of *Ascarops strongylina* (Fig. 141K). In *Gongylonema pulchrum* the opercula are indicated by a thinning of the egg envelope with no lines of fracture being visible (Fig. 141O). Foster (1912) reports bipolar lines of fracture in the eggs of *Physocephalus sexalatus* (Fig. 141J) but they were not visible in our studies. The egg of this species is truncate at one end and possesses a lenticular operculum. The other end is somewhat attenuated and has a zone

of separation between the chitinous shell and the protein coat. The area is not, however, suggestive of an operculum.

In the majority of the Spiruroidea the egg envelope is smooth and the contours regular, but mammillation and rugosity of the shell occur in the eggs of some species. Chandler (1919) reports the presence of two prominent, longitudinal ridges of "mammillae" running down the sides of the eggs of *Hedruris siredonis* (Fig. 141G). Mammillation in *Physaloptera ortleppi ortleppi* is expressed in the form of spinulation of the outer membrane. (Fig. 141P). The eggs of *Protospirura numidica* (Fig. 141C) are weakly mammillated, suggestive of the condition seen in certain well-known ascarid eggs, for example *Toxocara cati* or *Parascaris equorum*. A lesser degree of mammillation is seen in the rugosity of the shell of eggs of *Haploncma hamulatum*.

Brumpt (1931) shows a peculiar condition in the egg of *Protospirura bonnei*. These eggs are described as enclosed in a gelatinous sheath which sharply separates the egg from the surrounding débris. Apparently the sheath described is the slightly condensed uterine secretion forming a gelatinous protein coat similar to the condition found in many nematodes in the ovarian portion of the uterus. For example in the eggs of the filarioid worm *Hastospiculum sp.* from *Erpetodryas fuscus*, the eggs from the ovarian region of the uterus contain a slightly condensed, irregular protein coat (Fig. 141R-S) while those from the vaginal portion have the protein coat condensed as in the eggs of other species. It is conceivable that varying degrees of consolidation of the protein coat at oviposition are to be found among the different species of roundworms.

FILARIOIDEA. (R. O. C.). Wide bionomic differences occur in the development of the Filarioidea. Many so-called viviparous species are known, some of which have been discussed in the section on ovoviviparity. The extent to which true viviparity occurs is a question which can be determined only on the basis of further research.

The eggs of ovoviviparous species are thin-shelled and contain usually but two membranes, the chitinous shell and the very delicate vitelline membrane. They usually have a smooth surface and are spheroid or ellipsoid in outline, as in *Dirofilaria immitis* (Fig. 141T), *Onchocerca fasciata* (Vide Badanine, 1938) or *Onchocerca volvulus* (Vide Blacklock, 1939). Intra-uterine shedding of the egg membranes has been observed by Augustine (1937) in the case of *Vagrifilaria columbigallinae*, and probably occurs in other species.

Oviparous species of Filarioidea are about equally as numerous as the ovoviviparous forms. The eggs are similar to those of the Spiruroidea in that they possess a protein coat, contain a coiled embryo at the time of discharge and are usually thick-shelled. As in the Spiruroidea considerable variation of morphology exists.

Baylis and Daubney (1922) describe the peculiar egg of *Hastospiculum macrophallos*. The eggs have a characteristic barrel shape, are thick-shelled and embryonated. Curious annular thickenings are present at each pole giving a superficial resemblance to the trichuroid type.

Chitwood (1932) describes the eggs of *Hastospiculum setiferum* as nearly spherical and embryonated *in utero*. *Hastospiculum onchocercum* has eggs of the same type with a "simple" shell. Chandler (1929) describes the eggs of *Hastospiculum spinigerum* as having a thick shell, further thickened into a collar near each end, the ends being covered by thin opercula; they contain developed embryos while still in the uteri. Figures 141R and S show the eggs of *Hastospiculum sp.* from *Erpetodryas fuscus*. Two different stages of intra-uterine development are seen. In Figure 141R the eggs were taken from the ovarian portion showing the partially condensed protein coat. The coat of the formed egg is seen in Figure 141 S.

Some genera apparently contain both oviparous and "viviparous" species. Walton (1929) describes two new species of the genus *Foleyella*. *Foleyella ranae* is reported as being viviparous, well developed embryos being present in the uterus. *Foleyella americanus* is implied to be oviparous since embryonated eggs are described as occurring in the lower uterine region. Similarly Walton (1935) noted eggs occurring in the uterus of *Isociella neglecta* whereas most of the members of the group have been considered to be viviparous. It is our opinion that in the case of both *Foleyella americanum* and *Isociella neglecta* that Walton's specimens were not entirely mature and both species are actually ovoviviparous.

Ransom (1904) describes the eggs of *Aprocta cylindrica* as elliptical and those of *Aprocta orbitalis* as thick-shelled, the mature eggs being embryonated. Caballero (1938) notes a double, thin shell for the oval eggs of *Aprocta travassosi*.

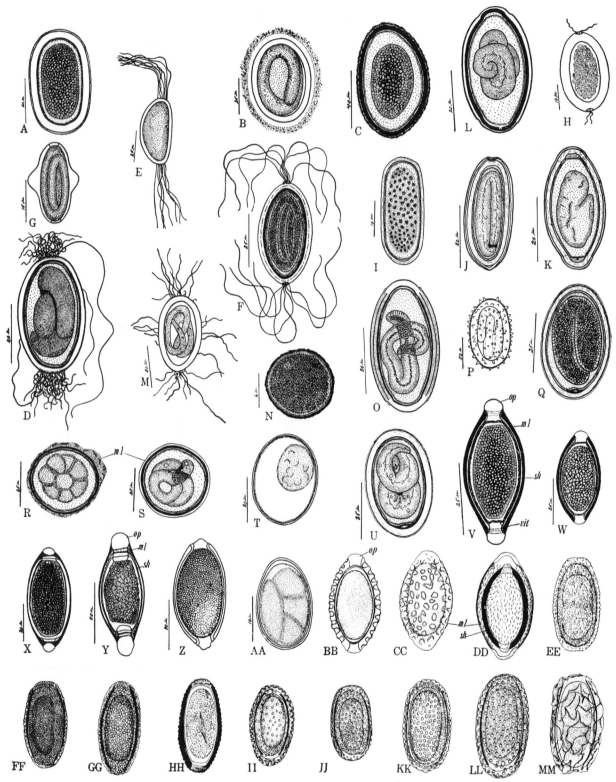

Fig. 141.

Nemic ova continued. A—*Mastophorus muris.* B—*Protospirura bonnei.* C—*Protospirura numidica.* D—*Tetrameres* sp. from American woodcock. E—*Tetrameres nouveli.* F—*Cystidicola stigmatura.* G—*Hedruris siredonis.* H—*Metabronema magnum.* I—*Spirocerca lupi.* J—*Physocephalus sexalatus.* K—*Ascarops strongylina.* L—*Oxyspirura mansoni.* M—*Rhabdochona ovifilamenta.* N—*Dracunculus medinensis.* O—*Gongylonema pulchrum.* P—*Physaloptera ortleppi.* Q—*Hamatospiculum cylindricum.* R-S—*Hastospiculum* sp. from *Erpetodryas fuscus.* T—*Dirofilaria immitis.* U—*Diplotriaena tricuspis.* V—*Trichuris leporis.* W—*Trichuris trichiura.* X—*Trichuris vulpis.* Y—*Trichuris ovis.* Z—*Capillaria aerophila.* AA—*Trichinella spiralis.* BB-CC—*Dioctophyma renale* (BB-optical section, CC-surface). DD—*Soboliphyme baturini.* EE—*Eustrongylides tubifex.* FF—*Eustrongylides elegans.* GG-HH—*Eustrongylides africanus* (GG-surface, HH-optical section). II—*Eustrongylides ignotus.* JJ—*Eustrongylides perpapillatus.* KK—*Hystrichis tricolor.* LL—*Hystrichis tricolor.* LL—*Hystrichis neglectus.* MM—*Hystrichis acanthocephalicus.* B, after Brumpt, 1931. E. after Seurat, 1914. G, after Chandler, 1919. H, after Yorke and Maplestone, 1926. M, after Weller, 1938. P, after Baylis, 1937. DD, after Petrov, 1930. EE-MM, after Jaegerskioeld, 1909. BB, CC original, Wallace. Remainder original, Christenson.

184

The eggs of *Filaria martis* have been described as having remarkably thick shells with external shagreening.

TRICHUROIDEA. (R. O. C.). The eggs of the Trichuroidea possess three membranes; an outer protein coat which may be deeply pigmented presenting a brownish color; an intermediate true shell which is usually transparent, and an internal vitelline membrane which may be granular or possess reticulations (Fig. 141V). The most characteristic structures are the plug-like opercula at either pole. These penetrate the protein coat and the true shell, but not the vitelline membrane. In some species the opercula are very prominent, projecting well beyond the protein coat externally, and well into the egg cavity internally (*Trichuris ovis*, Fig. 141Y). In others they conform in length with the polar thickness of the egg envelope presenting even contours both externally and internally (*Trichuris vulpis*, Fig. 141X). The cuticular shell projects along the sides of the opercula forming collar-shaped sockets into which the opercula fit. The internal limits of the opercula widen beyond the diameter of the opercular apertures making them difficult to dislodge by mechanical means. Under pressure the egg-capsule itself will often break before the opercula are dislodged. The eggs are usually unsegmented at the time of discharge.

The appearance of the protein coat varies considerably. Thomas (1924) describes the eggs of *Trichosomoides crassicauda* as rugose. When the eggs of this species are discharged they are held together in stringy masses by a sticky secretion. This condition has also been reported by Walton (1923) for *Capillaria longistriata* and it occurs in the spiruroid genus *Gongylonema*. Pologentsev (1935) notes a striated appearance in the protein coat of ova of *Trichuris busulka*. A similar striation exists in *Capillaria aerophila* as shown by our studies (Christenson, 1935).

The eggs of *Capillaria magalhaesi* are described by Lent and Freitas (1937) as marked by circular and oblique striae. Baylis (1934) notes a punctate appearance in the shell of *Capillaria lophortygis*, and in *Capillaria bervicollis* and *Capillaria inequalis* it is mammillated (Walton, 1935). Faust and Martinez (1935) describe a striated or channeled sculptoring externally fo rthe eggs of *Capillaria hepatica*.

Size has little diagnostic value in the ova of this group. Species of *Trichuris* give measurements which overlap and are therefore not significant (Chandler, 1930). The same is true for the genus *Capillaria*. It is of some value in separating species of Trichuroidea which may be present in a single host (Christenson, 1935).

Although usually considered viviparous our studies show that *Trichinella spiralis* is actually ovoviviparous. Immature females removed from the intestine of rats possess typical, thin-shelled eggs of the type shown in Figure 141 AA. The thin chitinous shell presents a slightly yellowish tinge and in spots the vitelline membrane was observed to have pulled away from it. Hatching occurs *in utero*, the minute larvae being discharged with no noticeable embryonic investments.

Figures 141V to AA show the eggs of some of the common trichuroid species occurring in man and domesticated animals.

MERMITHOIDEA. (R. O. C.). The bizarre eggs of the mermithoid worms have received much attention from biologists since they were first noted. Dujardin (1842) describes the eggs of *Mermis nigrescens* observing the peculiar, branched filaments termed byssi. (Fig. 139-140). Meissner (1856) further described the eggs of a form he considered to be the same species, noting their lenticular shape, the brownish color, and the two membranes composing the egg envelope enclosing a developed larva. He observed the transverse line of juncture of the outer shell, the polar thickenings, and the byssi arising from them as cords ending in tassel-like branches a short distance from the egg. He described the outer shell as essentially colorless, the brown color of the egg envelope being due to the pigmentation of the cuticular shell, or chorion.

Cobb (1926) expressed the view that the species studied by Meissner was not *Mermis nigrescens* of Dujardin but a different species which should be termed *Mermis meissneri*. He described a new species of *Mermis* under the name *M. subnigrescens* and presented an excellent figure of the egg (Fig. 135 H). He described the byssi as flexible, branched, entangling filaments which arise from polar elevations. He, also, noted the equatorial line of juncture between the two halves of the outer shell. Within the shell he shows the outline of the coiled larva with its three-pronged spear.

Christie (1937), like Meissner, observed the concentration of the pigment in the inner shell. This envelope is described as being spherical in outline and slightly compressed at the poles. No mention is made of the presence of a vitelline membrane but such a structure is shown in one of his figures (Fig. 139-140).

From the foregoing review it is apparent that the outer "shell" described by different authors is comparable to the protein coat, and that the byssi associated with it are not alien in origin to the terminal filaments of other forms. This view is supported by Meissner's description of the intrauterine formation of the outer shell. The inner shell, or chorion, is the same as the chitinous shell of other groups, and that a vitelline membrane is present is indicated by Christie. Christie shows that hatching is accompanied by the fracture of the eggs at the polar thickenings (Fig. 140) similar to the condition reported by Ransom (1904) in the hatching of *Oxyspirura mansoni*, leaving the barrel-shaped shell remnants.

Some workers have been impressed with the similarity between the eggs of the Mermithoidea and those of the Trichuroidea. They point out that the thickened portion of the outer shell containing the byssi might be compared to the opercula in the latter group, and that in other details the eggs are similar. The analogy is even more apparent in non-byssate forms. Steiner (1938) states that the eggs of *Pseudomermis vanderlindei* are oval in shape, with heavy shells having both ends truncate, and containing fully developed embryos. He points out that they distantly resemble the eggs of the Trichuridae.

The lenticular shape assigned to eggs of some of the Mermithoidea is well seen in those of *Tetradonema plicans* as pointed out by Hungerford (1919). The eggs of this form are thick-shelled, somewhat testaceous in color, and disc-shaped in outline. When on edge an oblong contour is presented. These peculiar eggs are retained under the cuticula of the female after oviposition. All stages of embryonic development are seen *in utero*.

In some nematodes it has been observed that the first embryonic molt occurs within the egg envelope. This has been noted to be the case in the egg of *Agamermis decaudata* by Cobb, Steiner and Christie (1923).

DIOCTOPHYMATOIDEA. (F. G. WALLACE). The ova of the superfamily Dioctophymatoidea are unique among nematode ova in the form of the outer coat of the egg shell which is deeply pitted with funnel-shaped depressions. As representative of the group we may take the egg of *Dioctophyma renale*, which is symmetrically oval, 64 to 80 microns in length, and 36 to 48 microns in width (author's measurements). The shells of specimens dissected from the uterus of preserved forms are colorless, while those taken from urine-contaminated feces are brown. The surface of the shell is pitted with irregular-shaped depressions 4 to 7 microns each in greatest diameter. (Fig. 141BB-CC). The wall of each pit, when seen either in surface view or in optical section, appears to have a double contour. The surface of the shell between the pits is smooth. At either end of the egg, the shell bulges slightly, is free from pits, and is colorless. These clear ends are spoken of as terminal plugs.

The shell appears to consist of three layers; the external pitted coat or cortical layer, the inner shell, and the vitelline membrane. The terminal plugs though somewhat different in chemical composition, belong structurally to the outer coat. The entire shell measures 6.8 to 8 microns in thickness along the equator of the egg and reaches 12 microns in the terminal plugs. The greater part of this thickness (4.5 microns at the equator and 8 to 9 microns at the terminal plugs) is occupied by the cortical layer.

The studies of Chitwood (1938) and Lukasiak (1930) on the chemical composition of the various layers are in part inconclusive as formalin-preserved material was used, but it appears that the inner shell is probably chitinous. The terminal plugs differ from the rest of the cortical layer in being more soluble in KOH, sulphuric acid, sodium hypochlorite, and nitric acid. According to Balbiani (1870), whose figures of the egg have been copied directly or indirectly by most subsequent authors, the terminal plugs are the weakest points in the shell, as when embryonated eggs were subjected to coverglass pressure the embryos escaped at the ends.

Table 2 gives the egg sizes taken from the literature and from the author's measurements of 50 eggs each from the

Author	Length			Width		
	Min.	Max.	Av.	Min.	Max.	Av.
Present (dog)	72	80	76.7	40	48	44.9
Present (mink)	64	76	70.8	36	44	41.4
Balbiani (1870)			68			42
Blanchard (1889)	64	68		42	44	
Leuckart (1876)			64			44

Table 2. Ova of the Dioctophymatoidea. Egg measurements of *Dioctophyma renale* in microns.

uterus of a formalin-preserved specimen from a dog and from a formalin-preserved fecal sample from a mink.

The ova of the other genera of Dioctophymatoidea, namely *Eustrongylides*, *Hystrichis*, and *Soboliphyme*, all show considerable resemblance to the egg of *D. renale* in general size and in the pitted cortical layer. The drawings given by Jägerskiöld show clear terminal plugs for only a few species of [*E. africanus*, *E. ignotus*, *E. perpapillatus*, and *H. neglectus* (Fig. 141 EE-LL)]. The egg of *H. acanthocephalicus* differs from others of the group in having a network of ridges over the surface instead of the usual small pits (Fig. 141MM). The figures of the egg of *Soboliphyme baturini* Petrov show the pitted cortical and the inner layers of the shell (Fig. 141DD) but show in addition a structure not seen in the egg of *D. renale*, a plug at either end of the inner shell of such form that the egg, without the cortical layer, would resemble that of *Trichuris*.

Species	Author	Length Min.	Max.	Width Min.	Max.
Eustrongylides					
africanus	Jägerskiöld 1909	70	76	36	42
E. elegans	Jägerskiöld 1909	70	76	33	38
E. ignotus	Jägerskiöld 1909	58	66	35	44
E. perpapillatus	Jägerskiöld 1909	53	61	31	33
E. tubifex	Jägerskiöld 1909	65	75	37	44
Hystrichis					
acanthocephalicus	Jägerskiöld 1909	75	79	40	44
H. neglectus	Jägerskiöld 1909	71	74	41	44
H. tricolor	Jägerskiöld 1909	71	74	41	44
Soboliphyme					
baturini	Petrov 1930	80.6	89.9	43.4	46.5

Table 3. Ova of the Dioctophymatoidea. Egg measurements of the various species.

THE CHEMISTRY OF THE EGG MEMBRANES
LEON JACOBS
Revised by B. G. Chitwood

There has been much confusion concerning the number and kinds of membranes surrounding the developing nematode embryo in the egg. Zawadowsky and his collaborators (1914, 1928, 1929a, 1929a, 1929c) have written a number of papers describing the egg membranes of various species of *Ascarididae*, *Trichostrongylidae*, and of *Enterobius vermicularis* (*Oxyuridae*) and *Nematodirus spathiger* (*Trichostrongylidae*). In dealing with eggs of *Ascaris* and related species these authors speak of five membranes, an inner lipoidal layer, three middle layers which are designated *membrana lucida*, and an outer albuminous membrane which coats the egg; in dealing with the eggs of the other species they studied, they describe four membranes and homologize them with the membranes of ascarid eggs. Dinnik (1930), a worker in Zawadowsky's laboratory, figures four membranes, excluding the plugs, on the egg of *Trichuris trichiura*. On the other hand, Wottge (1937) and Chitwood (1938) could demonstrate chemically only three layers on the egg of *Ascaris lumbricoides*, and Jacobs and Jones (1939) found only three membranes on the egg of *Enterobius vermicularis* by means of similar chemical tests. Chitwood (1938) also described three membranes, excluding the plugs, on the eggs of *Dioctophyma renale*. It is difficult to interpret the homologies drawn by Zawadowsky and his coworkers relating the five membranes of the ascarid eggs to the four membranes they describe on the ova of the other species. Nevertheless, the fact that they designated the three middle layers of the *Ascaris* egg as *membrana lucida* and based their descriptions on the results produced with alcohol treatment and on the optical effects observed during the penetration of cedar oil into the egg, indicates the solution of the problem. The *membrana lucida* undoubtedly corresponds to the refractive part of the shell, which has been called the *homogeneous membrane* by Wottge, and the *shell proper* by Chitwood, and Jacobs and Jones. The latter have pointed out that the chitinous shell of *Enterobius* eggs is composed of two layers and that this layering is probably the reason Zawadowsky and his collaborators saw four different interfaces during the penetration of cedar oil into the pinworm egg. Biedermann (1912) has noted that the chitin found on various animal forms may be layered, and Schmidt (1936) has described different effects produced on plane polarized light by two layers of the chitin of *Parascaris equorum* eggs. The difference in the number of membranes described by various authors can therefore be attributed to the variety of techniques used in studying the eggs. Chemical tests show the presence of three

chemically different membranes; physical tests detect the lamellation of these membranes.

The picture becomes clearer if the above interpretation is applied. The membrane immediately enclosing the embryo is the *vitelline membrane*, *fibrous membrane*, or *lipoidal layer(s)*. It corresponds to membranes "D" of Zawadowsky. The hard refractive membrane surrounding this is the *shell proper* or *homogeneous membrane*, corresponding to the *membrana lucida* (the middle layers) described by Zawadowsky. The outer covering is generally designated the *albuminous membrane*; it has been called the *proteinaceous membrane* by Jacobs and Jones. The latter is a more satisfactory term in view of its chemical nature, but for the sake of brevity it will hereafter be referred to as the *protein membrane*.

Ditylenchus dipsaci has been described by Chitwood as having only two membranes, the outer membrane being lacking.

Only a few authors have concerned themselves with the question of the origin of the various membranes. Fauré-Fremiet (1912a, 1912b, 1913) and Wottge (1937) have described the formation of the shell proper and the vitelline membrane following fertilization. Two consecutive series of vacuoles appear in the egg cell and move towards the surface. The first series transports glycogen to the surface where it is converted into glucosamine and finally laid down as chitin in the shell proper. The second series of vacuoles carries saponified lipoidal constituents of the ovum to the surface where they are transformed and deposited as the vitelline membrane. There is disagreement between the two workers (Wottge and Chitwood) who have studied the origin of the outermost protein membrane. Wottge, on the basis of the presence of this covering on unfertilized eggs, expresses the opinion that it originally exists as a sort of "ectoplasm" and is raised away from the egg by the secondary secretion of the other two membranes. The presence of rugose markings (*Ascaris*, etc.) and spirals (*Pseudonymus*) in the protein membrane could easily be explained by the assumption that regional secretion of the protein takes place on the surface of the egg cell. Chitwood, on the other hand, has found *Ascaris* eggs without this layer in the upper part of the uterus and with advancing stages in the deposition of this covering over the shell proper on eggs nearer the vagina. He suggests that the outermost membrane is a uterine secretion, added after the two inner membranes have been formed. Support for this view is found in the report of Wharton (1915) who obtained eggs without the protein membrane from female ascarids which oviposited for several days *in vitro* in Kronecker's solution. He attributed the absence of the membrane to "some physiological condition which prevents formation and deposition of the required substance by the uterine glands."

Protein Coatings: The eggs of some nematodes are covered (*Ascaris*, *Enterobius*) or deposited in a protein (*Meloidogyne* spp.). This material has been studied by Yoshida and Takano (1923), Kosmin (1928), Wottge (1938), and Chitwood (1938) in *Ascaris*, by Lukasiak (1930) and Chitwood (1938) in *Dioctophyma renale*, by Chitwood (1938) in *Meloidogyne hapla* and by Jacobs and Jones (1939) in *Enterobius vermicularis*. Such materials as the byssi of *Mermis* (Fig. 139) and the covering an dfilaments of *Binema*, *Pseudonymus*, *Citellina* and *Cystidicola* (Fig. 141) are probably of a similar nature. In all cases these materials give standard protein reactions (xanthoproteic, nitrite, ninhydrin). The solubility and digestibility of the materials differs widely according to organism and condition of materials.

Recent (unpublished) investigations by Rita Buckner confirm and extend previous observations on the external layer of the egg shell of *Ascaris lumbricoides* v. *suis*. The material as observed on eggs dissected from fresh uteri, is insoluble in neutral water, dissolved by artificial gastric juice, Fairchild's Trypsin, 0.2 percent of conc. HCl, 1 percent acetic acid, 1 percent KOH, picric acid and picric acid-alcohol at room temperature. As obtained in 0.2 percent HCl it is reprecipitated upon neutralization. This precipitate gives the standard tests indicating a mucoid. The digestibility in trypsin is peculiar since we know that the eggs bear an external covering in the feces. However, the external layer also has some coloration in the feces and we may well suspect a tanning or oxidation phenomenon.

Jacobs and Jones (1938) found the external covering of *Enterobius vermicularis* digested in gastric and pancreatic juices, became swollen but did not dissolve in dilute acids and alkalis, but became orange when exposed to fuming nitric acid and ammonium hydroxide successively. These results also indicate a protein.

The so-called "gelatinous egg mass" of *Meloidogyne* spp. belongs in the same group of compounds though it is minutely fibrous and not associated with the individual eggs. This material is clear and colorless when first deposited but turns yellow to orange on oxidation, indicating possible tanning. Unlike the mucoids previously discussed it is incompletely digested by artificial gastric juice and Fairchild's trypsin though many eggs are released. The coloration of the oxidized material is removed by treatment with sulfurous acid and many more eggs are removed. Alternate exposure to gastric juice and sulfurous acid is moderately effective in dissolving the jelly. Like the material from *Ascaris*, it gives positive xanthoproteic and ninhydrin reactions but it is much less soluble in dilute acids and alkalis. It is definitely hygroscopic and when the exterior surface becomes orange the contained eggs dry much less readily. It also acts as a barrier to penetration by fat solvents since they must first penetrate the contained water.

In general, the exterior coatings appear to be composed of a series of mucoids which probably undergo something similar to quinone tanning. The latter would certainly reduce water loss from the eggs. The hygroscopic nature is a great advantage to the organisms also.

Shell proper: This is the first structure of the egg to have received attention. Apparently Krakow (1892) was the first to note the interesting fact that while the cuticle of nematodes is not chitin, their egg shells are composed of this substance. Early tests for chitin depended mainly on the insolubility of the substance in hot concentrated KOH. Fauré-Fremiet's descriptions of the origin of the shell proper indicate that he had chemical proof of its composition. Schulze (1924) apparently applied more specific tests in identifying chitin in the eggs of *Ascaris sp.* Later workers have made use of the van Wisselingh tests for chitin which are discussed by Kunike (1925) Kühnelt (1928) and Campbell (1929). The procedure is superheating the substance with concentrated KOH under pressure, any chitin being converted into chitosan by this process. Chitosan so produced turns brown on treatment with iodine-potassium iodide solution and then reddish-violet upon the addition of dilute sulphuric acid. It is soluble in dilute (3 percent) acetic acid and can by recrystallized from the acetic acid solution as minute sphaerocrystals by 1 percent sulphuric acid. It is soluble in 75 percent sulphuric acid from which it can be recrystallized by dilution. The sphaerocrystals thus produced are stained red by 0.1 percent Rose Bengal. These tests distinguish the substance as chitin; cellulose, the only other organic skeletal substance which will withstand the KOH treatment, does not stain with iodine-dilute acid, does not dissolve in dilute acetic acid, and does not yield the sparerocrystals. Chitin is probably a general constituent of the egg shell throughout the phylum Nematoda.

Vitelline membrane: This structure was first studied by Fauré-Fremiet (1913) in *Parascaris equorum*, later by Zawadowsky, Wottge, Chitwood (1938), Jacobs and Jones (1939) and Timm (1950) in various nematodes. As first noted by Zawadowsky, it is the chief barrier to penetration of many chemicals used for disinfection. Fauré-Fremiet found the layer to be a fatty material with a melting point of 72°C and called it acid ascarilique. Timm extracted it from eggs of *Ascaris lumbricoides* and mixed it with refined beeswax obtaining no change in melting point. On this basis and other tests it was identified as myricyl palmitate. In recent unpublished investigations Marie D. Chitwood has crystallized the vitelline membrane from *Meloidogyne javanica* (melting point 70.2°C) and compared the crystals with those formed by precipitation of refined beeswax in cold acetone. Apparently the materials so obtained are identical in nature. Myricyl palmitate is very slightly soluble in acetone depending on the temperature. When crystallized it takes the form of fine birefringent needles which from burrs if present in considerable concentration.

Economic importance: The economic importance of the various egg membranes has previously been stated several times but constructive work has lacked support. Soil fumigants are usually volatile hydrocarbons. The molecules must be both hydrophilic and hydrophobic in order to penetrate mucoids and waxes. A model can be made by blowing a thin bubble of myricyl palmitate under water. Such a bubble, attached to a glass tube immersed in a thin sheet of water simulates the protective mechanism of the nematode. When placed in a gas chamber the penetration of fumigants can be studied.

Bibliography
(Morphology)

ACKERT, J. E., 1931.—The morphology and life history of the fowl nematode, *Ascaridia lineata* (Schneider). Parasitol., v. 23 (3): 360-379, figs. 1-25, pls. 13-14.

ALICATA, J. E. 1936.—Early developmental stages of nematodes occurring in swine. U. S. D. A. Bur. An. Ind. Bull. No. 489, 96 pp., figs. 1-30.

AUGUSTINE, D. L. 1937.—Observations on living "sheathed" microfilariae in the capillary circulation. Trans. Roy. Soc. Trop. Med. and Hyg., v. 31 (1): 55-60.

BADANINE, N. V. 1938.—Sur la question d'helminthofaune du chameau en Turkmenie. Livro Jubilare Prof. Lauro Travassos, pp. 61-74, 1 pl.

BAYLIS, H. A. 1929.—A manual of helminthology, medical and veterinary. 303 pp., 200 figs. London.
    1931.—A species of the nematode genus *Hedruris* occurring in the trout in New Zealand. Ann. and Mag. Nat. Hist., s. 10, v. 7: 105-114, figs. 1-4.
    1933.—The nematode genus *Ascarophis* van Beneden. Ann. and Mag. Nat. Hist., s. 10, v. 11: 111-117, fig. 1.
    1934.—Some parasitic worms from Australia. Parasitol., v. 26 (1): 129-132, figs. 1-3.
    1936.—A new ascarid from a bat. Ann. and Mag. Nat. Hist., s. 10, v. 17: 360-365, figs. 1-4.

BAYLIS, H. A. and DAUBNEY, R. 1922.—Report on the parasitic nematodes in the collection of the Zoological Survey of India. Mem. Ind. Mus., Calcutta, v. 7 (14): 263-347, figs. 1-75.

BENEDEN, P. J. VAN, 1871.—Les poissons des cotes de Belgique, leurs parasites et leurs commensaux. Mém. Acad. Roy. Sc. Belgique, v. 38 (4): xx + 100 pp., pls. 1-6.

BERGHE, L. VAN DEN 1935.—Sur l'existence de filaments latéraux sur la coque l'œuf de *Cystidicola farionis* Fischer, 1798. Ann. Parasitol., v. 13 (5): 435-438, figs. 1-2.

BISCHOFF, T. L. W. 1855.—Ueber Ei- und Samenbildung und Befruchtung bei *Ascaris mystax*. Ztschr. Wiss. Zool. v. 6 (3-4): 377-405.

BLACKLOCK, D. B. 1939.—The development of *Onchocerca volvulus* in *Simulium damnosum*. Vol. Jub. pro Sadao Yoshida, v. 2: 295-306, figs. 1-8.

BLANCHARD, R. 1889.—Traité de zoologie médicale. v. 1, viii + 808 pp., 387 figs. Paris.

BOVERI, T. 1887.—Zellen-Studien. Heft. 1: Die Bildung der Richtungskörper bie *Ascaris megalocephala* und *Ascaris lumbricoides*. 93 pp., pls. 1-4. Jena.
    1888.—Zellen-Studien. Heft 2: Die Befruchtung und Teilung des Eies von *Ascaris megalocephala*. 198 pp., pls. 1-5, figs. 1-94, Jena.

BRUMPT, E. 1931.—Némathelminthes parasites des rats sauvages (*Epimys norvegicus*) de Caracus. I. *Protospirura bonnei*. Infections expérimentales et spontanées. Formes adultes et larvaires. Ann. Parasitol., v. 9 (4): 344-358, figs. 1-9, pl. 8, figs. 1-4.

CABALLERO, E. 1938.—Contribucion al conocomiente de los nematodes de las aves de México. Liv. Jub. Prof. Lauro Travassos, pp. 91-98, pls. 2.

CHANDLER, A. C. 1919.—On a species of *Hedruris* occurring commonly in the western newt, *Notophthalmus torosus*. J. Parasitol., v. 5 (3): 116-122, pl. 9, figs. 1-9.
    1929.—Some new genera and species of nematode worms, Filarioidea from animals dying in the Calcutta Zoological Garden. Proc. U. S. Nat. Mus., (2777) v. 75 (6): 1-10, pls. 1-3., figs. 1-15.

CHITWOOD, B. G. 1930.—A recharacterization of the nematode genus *Blatticola* Schwenk, 1926. Trans. Am. Micr. Soc., v. 49 (2): 178-183, pl. 21, figs. 1-4.
    1932.—A review of the nematodes of the genus *Hastospiculum*, with descriptions of two new species. Proc. U. S. Nat. Mus., (2919), v. 80 (19): 1-9, pls. 1-3.
    1932.—A synopsis of the nematodes parasitic in insects of the family Blattidae. Ztschr. Parasitenk., v. 5 (1): 14-50, figs. 1-59.
    1938.—Further studies on nemic skeletoids and their significance in the chemical control of nemic pests. Proc. Helm. Soc. Wash., v. 5 (2): 68-75.

CHRISTENSON, R. O. 1935.—Remarques sur les différences qui existant entre les œufs de *Capillaria aerophila* et de *Trichuris vulpis*, parasites du renard. Ann. Parasitol., v. 13 (4): 318-321, figs. 1-3.

CHRISTIE, J. R. 1937.—*Mermis subnigrescens*, a nematode parasite of grasshoppers. J. Agric. Res., v. 55 (5): 353-364, figs. 1-6.

CHU, TSO-CHIH, 1936.—Studies on the life history of *Rhabdias fuscovenosa* var. *catanensis* (Rizzo, 1902). J. Parasitol., v. 22 (2): 140-160, 9 figs.

COBB, N. A. 1926.—The species of *Mermis*—a group of very remarkable nemas infesting insects. J. Parasitol., v. 13 (1): 66-72, pl. 2, fig. 1.
    1928.—The screw nemas, *Ascarophis* van Beneden, 1871; parasites of codfish, haddock, and other fishes. J. Wash. Acad. Sci., v. 18 (4): 96-102, figs. 1-8.

COBB, N. A., STEINER, G. and CHRISTIE, J. R. 1923.—*Agamermis decaudata* Cobb, Steiner and Christie; a nema parasite of grasshoppers and other insects. J. Agric. Res., v. 28 (11): 921-926.

COBBOLD, T. S. 1864.—Entozoa; an introduction to the study of helminthology, xxvii + 480 pp., 82 figs., 21 pls. London.

DORMAN, H. P. 1928.—Studies on the life cycle of *Heterakis papillosa* (Bloch). Trans. Am. Micr. Soc., v. 47 (4): 379-413, figs. 1-10, pl. 54, figs. 11-22.

DUJARDIN, F. 1842.—Mémoire sur la structure anatomique des *Gordius* et d'un autre helminthe, le *Mermis*, qu'on a confondu avec eux. Ann. Sc. Nat., s. 2, v. 18: 129-151, pl. 6, figs. 1-16.

FAURE-FREMIET, E. 1913.—La formation de la membrane interne de l'œuf d'*Ascaris megalocephala*. Compt. Rend. Soc. Biol. Paris, v. 4 (20): 1183-1184.

FAUST, E. C. 1928.—Studies on *Thelazia callipaeda* Railliet and Henry, 1910. J. Parasitol., v. 15 (2): 75-86, figs. 1-4.

FAUST, E. C. and WM. H. MARTINEZ, 1935.—Notes on helminths from Panama. II. Rare human nematode eggs in the feces of individuals from the Chagres River, Panama. J. Parasitol., v. 21 (5): 332-336, figs. 1-3.

FENG, L. C. 1936.—The development of *Microfilaria malayi* in *A. hyrcanus* var. *sinensis*. Chinese Med. J., Suppl. 1, pp. 345-367, figs. 1, pls. 4.

FOSTER, W. D. 1912.—The roundworms of domestic swine with special reference to two species parasitic in the stomach. U. S. D. A. Bur. An. Ind. Bull. 158, pp. 1-47, figs. 1-28, pl. 1.
    1914a.—Observations on the eggs of *Ascaris lumbricoides*. J. Parasitol., v. 1 (1): 31-36, figs. 1-4.
    1914b.—A peculiar morphologic development of an egg of the genus *Tropidocerca* and its probable significance. J. Parasitol., v. 1 (1): 45-47, fig. 1.

FREITAS, J. F. and LENT, H. 1935.—Capillariinae de animaes de sangue frio (Nematoda: Trichuroidea). Mem. Inst. Oswaldo Cruz, v. 30 (2): 241-284, pls. 1-11, figs. 1-102.
    1936.—Estudo sobre os Capillariinae parasitos de mammiferos (Nematoda: Trichuroidea). Mem. Inst. Oswaldo Cruz, v. 31 (1): 85-160, pls. 1-16, figs. 1-130.

FREITAS, J. F. and LINS DE ALMEIDA, J. 1935.—Sobre os Nematoda Capillariinae parasitas de esophago e papo de aves. Mem. Inst. Oswaldo Cruz, v. 30 (2): 123-156, pls. 1-6, figs. 1-47.
    1936.—Segunda contribuição ao conhecimento da fauna helminthologica da Argentina: *Heteroxynema wernecki* n. sp. Mem. Inst. Oswaldo Cruz, v. 31 (2): 185-188, pls. 1-3.

GOODEY, T. 1930.—On a remarkable new nematode, *Tylenchinema oscinellae* gen. et. sp. n., parasitic in the frit-fly, *Oscinella frit* L., attacking wheat. Phil. Trans. Roy. Soc. London, s. B, v. 218: 315-343, fig. 1, pls. 22-26, figs. 1-54.

HAGMEIER, A. 1912.—Beiträge zur Kenntnis der Mermithiden. Zool. Jahrb., Abt. Syst. v. 32 (6): 521-612, figs. a-g, pls. 17-21, figs. 1-55.

HARWOOD, P. D. 1932.—A note on the tissue-penetrating abilities of sheathed microfilariae. Trans. Am. Micr. Soc., v. 51 (2): 153-154, figs. 1-2.

HUNGERFORD, H. B. 1919.—Biological notes on *Tetradonema plicans* Cobb, a nematode parasite of *Sciara coprophila* Lintner. J. Parasitol., v. 5 (4): 186-192, 2 figs. pl. 19, figs. 1-2.

HUNTER, G. W. III and BANGHAM, R. V. 1933.—Studies on the fish parasites of Lake Erie. II. New Cestoda and Nematoda. J. Parasitol., v. 19 (4): 304-311, pl. 5, figs. 1-12.

KELLER, A. E. 1933.—A study of the occurrence of unfertilized *Ascaris* eggs. J. Lab. and Clin. Med., v. 18 (4): 371-374, figs. 1-11.

KOFOID, C. A. and WHITE, W. A. 1919.—A new nematode infection of man. J. Am. Med. Assoc., v. 72 (8): 567-569.

KOURI, P. and NAUSS, R. W. 1938.—Formation of the egg shell in *Fasciola hepatica as* demonstrated by histological methods. J. Parasitol., v. 24 (4): 291-310, pls. 1-8.

KREIS, H. A. 1932.—A new pathogenic nematode of the family Oxyuroidea, *Oxyuronema atelophora* n. g., n. sp. in the red-spider monkey, *Ateles geoffroyi*. J. Parasitol., v. 18 (4): 295-302, pls. 26-27, figs. 1-19.

KREIS, H. A. and FAUST, E. C. 1933.—Two new species of *Rhabditis* (*Rhabditis macrocerca* and *R. clavopapillata*) associated with dogs and monkeys in experimental *Strongyloides* studies. Trans. Am. Micr. Soc., v. 52 (2): 162-172, pls. 27-28, figs. A-K.

KÜCHENMEISTER, F. 1857.—On animal and vegetable parasites of the human body. xix + 452 pp., 8 pls. London.

LENT, H. and FREITAS, J. F. T. DE 1937.—Alguns helminthos da collecçao de Pedro Severiano de Magalhaes. Mem. Inst. Oswaldo Cruz, v. 32 (2): 305-309, pl. 1.

LEUCKART, R. 1886.—The parasites of man, and the diseases which proceed from them. xxvi + 771 pp., figs. 1-404. Edinburgh. (Transl. by W. E. Hoyle).

LEWIS, T. R. 1874.—A report on the pathological significance of nematode Haematozoa. 10th Ann. Rep. San. Com. India, Calcutta (1873), App. B: 111-133, figs. 1-2, pls. 1-3.

MANSON, P. 1884.—The metamorphosis of *Filaria sanguinis hominis* in the mosquito. Trans. Linn. Soc. London, 2. Ser. Zoology, v. 2 (10) 367-388, pl. 9, figs. 1-46.

MANTER, H. W. 1930.—Two new nematodes from the woodchuck, *Marmota monax canadensis*. Trans. Am. Micr. Soc., v. 49 (1): 26-33, pl. 4, figs. 1-13.

MATUDA, S. 1939.—Some abnormal eggs of *Ascaris lumbricoides* Linnaeus. Vol. Jub. pro Prof. Sadao Yoshida, v. 2: 311-314, pls. 1-2.

MEISSNER, G. 1855.—Beiträge zur Anatomie und Physiologie der Gordiaceen. Ztschr. Wiss. Zool., v. 7 (1-2): 1-140, pls. 1-7.

MIURA, K. and NISHIUCHI, N. 1902.—Ueber befruchtete und unbefruchtete Ascarideneier im menschlichen Kote. Centralbl. Bakt. Parasit. v. 32 (8-9): 637-641, figs. a-b.

NELSON, H. 1852.—On the reproduction of *Ascaris mystax*. Philos. Trans. London, 563-594 pp., pls. 25-30, figs. 1-92.

NICOLL, W. 1907.—A contribution toward a knowledge of the Entozoa of British marine fishes. Pt. 1. Ann. and Mag. Nat. Hist., s. 7 (109) v. 19: 66-94, pls. 1-4, figs. 1-16.

NOLF, L. O., 1932.—Experimental studies on certain factors influencing the development and viability of the ova of the human *Trichuris* as compared with those of the human *Ascaris*. Am. J. Hyg., v. 16 (1): 288-322.

NOLF, L. O. and EDNEY, J. M. 1935.—Minimum time required by *Trichinella spiralis* to produce infective larvae. J. Parasitol., v. 21 (4): 313-314.

OLSEN, O. W. 1938.—*Aplectana gigantica* (Cosmocercidae), a new species of nematode from *Rana pretiosa*. Trans. Am. Micr. Soc., v. 57 (2): 200-203, pl. 1.

ORTLEPP, R. J. 1939.—Observation on *Ascaris phacochoeri* Gedoelst, 1916, a little known nematode parasite from the wart hog. Vol. Jub. pro Prof. Sadao Yoshida, v. 2: 307-309., figs. 1-3.

OTTO, G. F. 1932.—The appearance and significance of the unfertilized eggs. of *Ascaris lumbricoides* (Linn.) J. Parasitol., v. 18 (4): 269-273, pl. 24, figs. 1-10.

PENEL, R. 1904.—Les filaires du sang de l'homme. Thèse, X + 156 pp., figs. 1-20. Paris.

POLOGENTSEV, P. A. 1935.—On the nematode fauna of the shrew-mouse, *Sorex araneus* L. J. Parasitol., v. 21 (2): 95-98, figs. 1-5.

PRENDEL, A. R. 1928.—(Zur Kenntniss der Darmhelminthen einiger-Nagetiere). (Russian) Rev. Microbiol., Epidémiol. et Parasitol., v. 7 (4): 410-416, figs. 1-4. (German summary, pp. 460-461).

RAFFENSPERGER, H. B. 1918.—Experiments in the transmission of trichinae. J. Amer. Vet. Med. Assoc., v. 53, n. s., v. 6 (3): 363-367.

RANSOM, B. H. 1904.—Manson's eye worm of chickens. (*Oxyspirura mansoni*), with a general review of nematodes parasitic in the eyes of birds. U. S. D. A. Bur. Anim. Ind. Bull. 60, 1-54 pp., figs. 1-40.

RANSOM, B. H. 1913.—The life history of *Habronema muscae* (Carter), a parasite of the horse transmitted by the house-fly. U. S. D. A. Bur. Anim. Ind. Bull. 163, 1-36 pp., figs. 1-41.

RAUTHER, M. 1930.—Vierte Klasse des Cladus Nemathelminthes. Nematodes, Nematoidea - Fadenwürmer. Handb. Zool. (Kükenthal and Krumback) v. 2, 8 Lief., 4 Teil, Bogen 23-32, pp. 249-402, figs. 267-426.

REARDON, L. 1938.—Studies on oxyuriasis. X. Artifacts in "cellophane" simulating pinworm ova. Am. J. Trop. Med., v. 18 (4): 427-430, pls. 1-2.

SANDGROUND, J. H. 1923.—"*Oxyuris incognita*" or *Heterodera radicicola?* J. Parasitol., v. 10 (2): 92-94, figs. 1-4.
———— 1928.—A new nematode parasite, *Tetrameres paucispina*, from a South American bird, *Amblyramphus holocericeus*. J. Parasitol., v. 14 (4): 265-268, pls. 12-13, figs. 1-6.

SCHWARTZ, B. 1918.—Observations and experiments on intestinal trichinae. J. Agric. Res., v. 15 (8): 467-482, figs. 1-3.

SCHWARTZ, B. and ALICATA, J. E. 1931.—Concerning the life history of lungworms of swine. J. Parasitol., v. 18 (1): 21-27, pl. 1, figs. 1-8.

SEURAT, L. G. 1914.—Sur un *Tropidocerca* parasite d'un echassier. Compt. Rend. Soc. Biol. Paris, v. 76 (16): 778-781, figs. 1-8.

SKINKER, M. S. 1931.—A redescription of *Cystidicola stigmatura* (Leidy), a nematode parasitic in the swim bladder of salmoid fishes, and a description of a new nematode genus. Trans. Am. Micr. Soc., v. 50 (4): 373-379, pls. 41-42, figs. 1-8.

STEINER, G. 1924.—Some nemas from the alimentary tract of the Carolina tree frog (*Hyla carolinensis* Pennant). J. Parasitol., v. 11 (1): 1-32, fig. A., pls. 1-11, figs. 1-65.
———— 1937.—Intersexuality in two new parasitic nematodes, *Pseudomermis vanderlindei* n. sp. (Mermithidae) and *Tetanonema strongylurus* n. g., n. sp. (Filariidae). Papers on Helminthology, 30 Year Jubilum, K. J. Skrjabin, pp., 681-688, figs. 1-4.

THOMAS, L. J. 1924.—Studies on the life history of *Trichosomoides crassicauda* (Bellingham). J. Parasitol., v. 10 (3): 105-136, fig. A, pls. 14-18, figs. 1-34.

UNDERWOOD, P. C. and HARWOOD, P. D. 1939.—Survival and location of the microfilariae of *Dirofilaria immitis* in the dog. J. Parasitol., v. 25 (1): 23-33.

WALTON, A. C. 1923.—Some new and little known nematodes. J. Parasitol., v. 10 (2): 59-70, pls. 6-7, figs. 1-25.
———— 1929.—Studies on some nematodes of North American frogs. J. Parasitol., v. 15 (4): 227-240, pls. 16-20, figs. 1-43.
———— 1935.—The Nematoda as parasites of Amphibia. II. J. Parasitol., v. 21 (1): 27-50, figs. 1-4.

WEHR, E. E. 1937.—Observations on the development of the poultry gapeworm, *Syngamus trachea*. Trans. Am. Micr. Soc., v. 56 (1): 72-78, 1 pl.

WELLER, T. H. 1938.—Description of *Rhabdochona ovifilamenta* n. sp. (Nematoda: Thelaziidae) with a note on the life history. J. Parasitol., v. 24 (5): 403-408, figs. 1-7.

WHARTON, L. D. 1915.—The eggs of *Ascaris lumbricoides*, Philippine J. Sci., v. 10 (2): 111-115.

WOTTGE, K. 1937.—Die stofflichen veränderungen in der Eizelle von *Ascaris megalocephala* nach der Befruchtung. Protoplasma, v. 29 (1): 31-59, figs. 1-16.

YAMADA, S. I. 1927 (1928).—An experimental study on twenty four species of Japanese mosquitoes regarding their suitability as intermediate hosts for *Filaria bancrofti* Cobbold. Sci. Rept. Gov't. Inst. Inf. Dis., Tokyo, (1927), v. 6: 559-622, pls. 20-22.

ZAWADOWSKY, M. M. 1928.—[The nature of the egg-shell of various species of *Ascaris* eggs (*Toxascaris limbata* Railliet & Henry, *Belascaris mystax* Zeder, *Belascaris marginata* Rud., *Ascaris suilla* Duj.)] (Russian text) Trans. Lab. Exp. Biol., Zoopark, Moscow, v. 4: 201-206, figs. 1-5, pl. 1, figs. 1-4.

ZAWADOWSKY, M. M. and SHALIMOV, L. G. 1929.—Die Eier von *Oxyuris vermicularis* und ihre Entwicklungsbedihungen, sowie, über die Bedingungen, unter denen eine Autoinfektion bei Oxyuriasis unmöglich ist. Ztschr. Wiss. Biol. Abt. F., Ztschr. Parasitenk., v. 2 (1): 12-43, figs. 1-17.

ZIEGLER, H. E. 1895.—Untersuchungen über die ersten Entwicklungsvorgänge der Nematoden. Zugleich ein Beitrag zur Zellenlehre. Ztschr. Wiss. Zool., v. 60 (3): 351-410, pls. 17-19.

(CHEMISTRY)

BIEDERMANN, W. 1912.—Physiologie der Stütz- und Skelettsubstanzen. Winterstein's Handbuch der vergleichenden Pnysiologie. v. 3: 803-893.

CAMPBELL, P. L. 1929.—The detection and estimation of insect chitin; and the irrelation of 'chitinization' to hardness and pigmentation of the cuticula of the American cockroach, *Periplaneta americana*. Ann. Ent. Soc. Amer., v. 22: 401-426.

CHITWOOD, B. G. 1938.—Further studies on nemic skeletoids and their significance in the chemical control of nemic pests. Proc. Helm. Soc. Wash., v. 5 (2): 68-75.
1950.—Nematocidal action of halogenated hydrocarbons. Symposia, Petroleum Division, Americ. Chem. Soc. In press.

CHITWOOD, MARIE, D. 1950.—Notes on the physiology of *Meloidogyne javanica*. In press.

DINNIK, J. A. & DINNIK, N. N. 1937.—The structure of the shell and resistance of the eggs of *Trichocephalus trichiurus*. Papers on Helminthology, 30 year Jubileum, K. J. Skrjabin pp. 132-138 (Russian).

DYRDOWSKA, M. 1931.—Recherches sur le comportement der glycogène et de graisses dans l'œuf d'*Ascaris megalocephala* a l'état normale et dans une atmosphere d'azote. Compt. Rend. Soc. Biol., (Paris) v. 108: 593-595.

FAURE-FREMIET, E. 1912a.—Sur la maturation et la fécondation chez l'*Ascaris megalocephala* (Note préliminaire). Bull. Soc. Zool. France, v. 37 (2): 83-84.

FAURE-FREMIET, E. 1912b.—Graisse et glycogene dans le développement de l'*Ascaris megalocephala*. Bull. Soc. Zool. France, v. 37 (6): 233-234.

FAURE-FREMIET, E. 1913.—La formation de la membrane interne de l'œuf d'*Ascaris megalocephala*. Compt. Rend. Soc. Biol. (Paris), v. 74 (20): 1183-1184.

FAURE-FREMIET, E. 1925.—La cinétique du développement. Paris. (Not seen).

HUFF, G. C. 1936.—Experimental studies of factors influencing the development of the eggs of the pig ascarid, (*Ascaris suum* Goeze). Jour. Parasit., v. 22 (5): 455-463.

JACOBS, L. & JONES, M. F. 1939.—Studies on oxyuriasis. XXI. The chemistry of the membranes of the pinworm egg. Proc. Helm. Soc. Wash., v. 6 (2): 57-60.

JAMMES, L. & MARTIN, A. 1907a.—Sur les propriétés de la coque de l'*Ascaris vitulorum*. Compt. Rend. Soc. Biol. (Paris, v. 62 (1): 15-17.

JAMMES, L. & MARTIN, A. 1907b.—Sur le déterminisme de l'infestation par l'*Ascaris vitulorum*. Compt. Rend. Soc. Biol. (Paris), v. 62 (3): 137-139.

JAMMES, L. & MARTIN, A. 1910.—Rôle de la chitine dans le développment des nématodes parasites. Compt. Rend. Acad. d. Sci. Paris, v. 151 (3): 250-251.

JONES, M. F. & JACOBS, L. 1939.—Studies on the survival of eggs of *Enterobius vermicularis* under known conditions of humidity and temperature. Program and Abstracts of the 15th Annual Meeting, Amer. Soc. Parasitol., Columbus, Ohio. Jour. Parasit., Suppl. v. 25 (6): 32.

KOSMIN, N. 1928.—Zur Frage ueber den Stickstoffwechsel der Eier von *Ascaris megalocephala*. Trans. Lab. Expt. Biol. Zoopark Moscow, v. 4: 207-218.

KRAKOW, N. P. 1892.—Ueber verschiedenartige Chitine. Ztschr. Biol., v. 29: 177-198, pl. 3.

KÜHNELT, W. 1928.—Studien ueber den mikrochemischen Nachweis des Chitins. Biol. Zentralbl., v. 48 (6): 374-382.

KUNIKE, G. 1925.—Nachweis und Verbreitung organischer Skeletsubstanzen bei Tieren. Ztschr. Vergleichenden Physiol., v. 2: 233-253.

LUKASIAK, J. 1930.—Anatomische und Entwicklungsgeschichtliche Untersuchungen an *Dioctophyme renale* (Goeze, 1782) (*Eustrongylus gigas* Rud.) Arch. Biol. Soc. Sci. et Lett. Varsovie v. 3 (3): 1-99, pls. 1-6.

SCHMIDT, W. I. 1936.—Doppelbrechung und Feinbau der Eischale von *Ascaris megalocephala*. Ztschr. Zellforsch., v. 25 (2): 181-203, figs. 1-3.

SCHULZE, P. 1924.—Der Nachweis und die Verbreitung des Chitins mit einem Anhang ueber das komplizierte Verdauungsystem der Ophryoscoleciden. Ztschr. Morph. u. Oekol. Tiere, v. 2: 643-666, figs. 1-3.

SWEJKOWSKA, G. 1928.—Recherches sur la physiologie de la maturation de l'œuf d'*Ascaris*. Bull. internat. Acad. polon. d. sc. et d. lett., Cracovie, v. 2: 489-519, 2 figs. (not seen).

TIMM, R. W. 1950.—A note on the chemical composition of the vitelline membrane of *Ascaris lumbricoides* v. *suis*. Science. In press.

WHARTON, L. D. 1915.—The development of the eggs of *Ascaris lumbricoides*. Phil. Jour. Sci. sec. B., v. 10: 19-23.

WOTTGE, K. 1937.—Die stoffliche Veränderungen in der Eizelle von *Ascaris megalocephala* nach der Befruchtung. Protoplasma, v. 29 (1): 31-59.

ZAWADOWSKY, M. M. 1914.—Ueber die lipoide semipermeable Membran der Eier von *Ascaris megalocephala*. Mitt Univ. Schanjawsky (not seen).

ZAWADOWSKY, M. M. 1928.—The nature of the egg shell of various species of *Ascaris* eggs. Trans. Lab. Expt. Biol. Zoopark Moscow v. 4: 201-206, figs. 1-5 (Russian with English summary).

ZAWADOWSKY, M. M. & SCHALIMOV, L. G. 1929a.—Die Eier von *Oxyuris vermicularis* und ihre Entwicklungsbedingungen, sowie ueber die Bedingungen unter denen eine Autoinfektion bei Oxyuriasis unmöglich ist. Ztschr. Parasitenk., v. 2 (1): 12-43, figs. 17.

ZAWADOWSKY, M. M. VOROBIEVA, E. I. & PETROVA, M. I. 1929b.—The eggs of *Nematodirus spatiger* and the properties of their shell. Trans. Lab. Expt. Biol. Zoopark Moscow v. 5: 251-254, fig. 1 (Russian with English summary).

ZAWADOWSKY, M. M. IVANOVA, S. A. VOROBIEVA, E. I. and STRELKOVA, O. I. 1929.—The biology of the Trichostrongylidae, parasitizing in hoofed animals. Trans. Lab. Expt. Biol. Zoopark Moscow v. 5: 43-83. (Russian with English summary).

YOSHIDA, S. & TAKANO, R. 1923.—Some notes on the albuminous covering of *Ascaris* (*sic*) *eggs*. Mitt. Med Gesellsch. Osaka v. 22 (2): 1-3.

CHAPTER XIII

GAMETOGENESIS

A. C. WALTON, KNOX COLLEGE, GALESBURG, ILL.

The history of the formation of the germ cells among nematodes is so closely bound up with the processes of meiosis and fertilization that consideration of any one of these phenomena involves a discussion of all three.

The process of meiosis, or reduction division, was first announced by Van Beneden in 1883 in his report of studies on the egg and spermatozoon of *Parascaris equorum* (*Ascaris megalocephala*) and the fact that the gametes contained only one-half the number of chromosomes found in the body cells, equally divided as to origin from each parent, is one of the most fundamental concepts of the fields of Evolution and Heredity. The realization that *Parascaris* germ cells were large, easily obtained, and very simple in their nuclear organization, led to their use as study material in the rapid advances of Cytology during the last decade of the nineteenth century.

The germinal cells of nematodes are differentiated during the very early cleavage divisions of the zygote and furnish a very clear history of germ-cell isolation, especially in those forms which show the "diminution" phenomenon. Ignoring for the time this peculiar process, the mitotic activity of the somatic and of the germinal cells has afforded a fruitful source for cytological investigations. It was from the study of *Parascaris* (*Ascaris megalocephala*) that Van Beneden (1883), Boveri (1887, 1888, 1890), Herla (1893), and Zoja (1896) laid the foundation work that established the doctrine of the genetic continuity of chromosomes, not only as to material, but also as to individual size and shape. The same material allowed Boveri (1909), Bonnevie (1908, 1912), and Vejdovsky (1912) to work out the structure of the individual chromosomes; a result that later workers on other materials have largely substantiated as to the main interpretations. (For a review of the literature up to 1923 see Walton, 1924).

As a result of these and other studies on nematode materials, the process of somatic mitosis seems to fall in line with the general system as follows: The reticulum of the nucleus becomes organized into a number of fine chromidial threads (Brauer, 1893) during the early prophase; these undergo an accurate longitudinal splitting; shorten and thicken, and take their places as individual chromosomes in the equatorial plate at the end of the prophase. The metaphase proper is practically absent, as splitting occurs early in the prophase. During the anaphase the chromosomes separate along the line of longitudinal splitting and pass to the two poles of the achromatic spindle. During the telophase each group of chromosomes becomes transformed into a new nuclear reticulum in which the individual chromosomes may lose their visible outlines, but not their actual identity, through vacuolization (Van Beneden, 1883, 1887), branching (Rabl, 1889; Boveri, 1887), or chromonema formation (Vejdovsky, 1912). The somatic number of chromosomes remains constant although they are divided equationally at each division and, since they are all descendants of the chromosomes of the zygote nucleus, the chromatic material of every germ cell and of every body cell is directly derived from that which was brought into the zygote nucleus by the egg and sperm nuclei of the preceding generation; a fact of enormous importance in the study of heredity and development.

The achromatic as well as the chromatic elements of the cell have been studied carefully in nematode material. Van Beneden (1887) and Boveri (1887) established the thesis that the centrosome is a permanent and genetically individual cell structure. Although usually regarded as extra-nuclear in position, it is reported as of intra-nuclear origin in *P. equorum* var. *univalens* (Brauer, 1893) and in *P. e.* var. *bivalens* (Sturdivant, 1931). In spite of much criticism, modern workers in the same field have substantiated this conclusion, at least as to cells of *Parascaris equorum* (Fogg, 1931; Sturdivant, 1934), although the exact nature of the structure is still unknown. The centriole divides (Boveri, 1900; Sturdivant, 1934) before any other visible evidence of mitosis appears, and migrates to opposite sides of the nucleus to form the poles of the next spindle figure. The spindle proper (first seen in nematode materials by Auerbach, 1874), the mitome ring, and the astral rays appear to be composed of granules and fibers which probably are the result of chemical fixation of what, in the living cells, are delimited currents of nuclear material in reaction with certain cytoplasmic elements which center at the centrosomal points, and are not fibers of actual material identity as stated by Boveri (1888). The fibers appear before the nuclear membrane disappears, and their extra- or intra-nuclear origin may depend upon the differential permeability of the membrane; streaming may first begin either in the cytoplasm or the karyoplasm, depending upon the physiological condition of the two substances.

Cytokinesis, as opposed to karyokinesis, is usually accomplished by a process of constrictive furrowing caused by differential surface tension and surface streaming phenomena (Spek, 1918 and 1920 in *Rhabditis pellio* and *R. dolichura*) which seem to depend upon the changes in the permeability of the cell membrane. These phenomena seem to be correlated with karyokinesis through the medium of the achromatic spindle.

Meiosis, as a phenomenon, accomplishes the "reduction of chromosomes" in that it affords an opportunity for the numerical reduction of the constant somatic complement of chromosomes (the diploid number) to the gametic (haploid number), and also separates the members of each pair of homologous chromosomes present in the somatic complex. In such a process, two forms of chromosome division occur; (a) separation equationally of split chromosomes, and (b) disjunction of homologous (paired) structures. As in most animals, meiosis occurs in connection with gametogenesis among the nematodes. In the male those cells (spermatogonia) destined to give rise to the spermatozoa undergo a series of ordinary equational divisions until a certain definite number is reached. The last generation of these cells undergoes a growth period during which the homologous male- and female- derived chromosomes are paired. The resultant cells (the primary spermatocytes) have the *haploid* number of chromosome *pairs*. Two successive meiotic divisions, one disjunctive and the other equational, follow without complete nuclear reorganization during the interphase. The first division gives rise to two secondary spermatocytes and the second divides the two secondary spermatocytes into four spermatids. Normally each spermatid metamorphoses into a spermatozoon, giving four spermatozoa (male gametes) as the end result of the two meiotic divisions. In certain of the free living nematodes Cobb (1925, 1928) reports the intercallation of a number of equational divisions of the spermatid before the ultimate differentiation of the spermatozoa (*spermules*). In *Spirina parasitifera** each spermatid eventually gives rise to one hundred and twenty-eight spermatozoa, the final differentiation occurring only after the spermatogenous tissue reaches the oviducts of the female. Certain arachnids (Warren, 1930), which produce two to four spermatozoa from each spermatid, show an approach to this condition. A somewhat similar phenomenon is also reported from certain snails.

In the female a similar program is followed to a great extent, except that the meiotic divisions frequently occur only after the spermatozoon has entered the primary oocyte (Boveri 1887; Sala, 1895; and modern observers). The spindle of the first meiotic division is always eccentric in position and the resultant cells are extremely unequal as to cytoplasmic content. The first polar body is separated from the large secondary oocyte by this division. The second division similarly forms a single large functional ovum and a second small polar cell. The first polar body occasionally divides into two equal cells. None of the polar cells are functional as far as is known among nematodes, although cases of entrance of the sperm into such cells have been noted.

As stated above, the significance of meiosis lies in the reduction to n/2 of the chromosomes which have undergone synapsis during the formation of *bivalent* chromosomes (homologous pairs). This synapsis is now regarded as being always "side by side" (para-synapsis) among the nematodes. If such members of homologous pairs are compound chromosomes, their synapsis may give rise to four-parted *bivalents* during the prophase of the first meiotic division. This occurs during the growth period following the last gonial division of each germ cell of either sex. These bivalents are separated during one of the two following divisions and hence that division is reductional since it separates (disjunction process) homologous structures. The other division, being equational, means that the four resulting nuclei each have a haploid set made up of one chromosome of each kind. In many nematodes the prophase chromosomes of the first division show both the plane of synapsis and that of a future longitudinal splitting, making "tetrad" chromosomes (if the chromosomes are compound they may thus form "di-tetrads", or, as in *Spirina parasitifera*, they may form 56-parted bodies). When the first division separates homologous pairs, it is termed "prereductional"; when it is the second division which causes disjunction, the process is termed "postreductional". Most nematodes show "prereduction". If the original bivalent chromosomes were "tetrads" the resultant chromosomes are "monads"; if "di-tetrads", they become "dyads" in the mature germ cells and in the polar cells.

During the formation of the prophase chromosomes the stages known as leptotène, zygotène, pachytène, diplotène and strepsitene are poorly differentiated except possibly in the races of *Parascaris equorum* (Van Beneden, 1887; Boveri, 1888; Griggs, 1906; Bonnevie, 1908, 1912); in most cases the chromosomes behave as quite solid units derived in a very early stage from a segmented spireme thread (Vejdovsky, 1912; Walton, 1918, 1924; Sturdivant, 1934).

During the process of spermatogenesis extra-nuclear bodies such as the centrioles (?), chondriosomes (Meves, 1911; Held, 1912, 1916; Hirschler, 1913; Romeis, 1912 Sturdivant, 1931, 1934), "yolk granules" (Sturdivant, 1934; Wildman, 1912; Walton, 1916a), and Golgi bodies (Sturdivant, 1934) are more or less evenly distributed so that each spermatid receives its complement of each of these elements in addition to the haploid number of chromosomes. The "yolk granules" are thought to be largely composed of glycogen and to be low in protein and lipoids (Kemnitz, 1913), although Bowen (1925) believes that further analyses are needed. These "yolk granules", or refringent globules, are apparently derived through the activity of the Golgi bodies, and therefore are pro-acrosomal in nature. The contained bodies in the center of each globule disappear during the spermatid metamorphosis and are perhaps to be regarded as temporary indications of precocious acrosomal granules, structures quite characteristic of insect spermatozoa (Sturdivant, 1934). They are not mitochondrial in nature

as earlier reported. The refringent globules eventually fuse to form the "refringent body" of the mature spermatozoon, which thus contains a structure homologous with the acrosome of other types of spermatozoa (Bowen, 1925). The Golgi remnants are cast off during the cytoplasmic reduction and cytophore formation of the maturing spermatid (Sturdivant, 1934).

The spermatozoa of nematodes are described as non-flagellated, frequently amoeboid cells, containing a considerable amount of stored material in the "refringent body", or acrosome. This type of spermatozoon is usually regarded as a simple modification of the fundamental structural plan of a flagellate sperm and has arisen secondarily during the evolution of this phylum. The fact that the acrosome is not always at the morphologically anterior end of the spermatozoon is not of particular significance. Certain of the acrosomal bodies are hollow (*Nemotospira turgida*), and this may very doubtfully represent the position of an axial tail filament in this pseudo-flagellate form. No other evidence concerning any axial filament is apparently available for the nematodes. *Passalurus ambiguus* (*Oxyuris ambigua*) has spermatozoa that may almost be considered as flagellate (Meves, 1911; Bowen, 1925). Recently Chitwood (1931) has described the spermatozoa from *Trilobus longus* which seem to be of truly flagellate form. This is to be expected since *Trilobus* is very close to the hypothetical ancestral nematode form which is believed to have possessed a typically flagellate type of sperm. *Passalurus* (*Oxyuris*) and *Trilobus* have the acrosomal body at the morphologic-

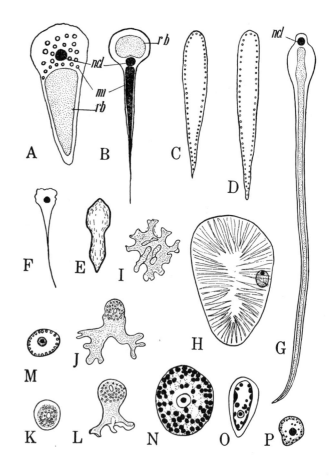

Fig. 147.

Nemic spermatozoa. A.—*Parascaris equorum;* B.—*Passalurus ambiguus;* C.—*Anticoma pellucida;* D.—*A. eberthi;* E.—*Trichosomoides crassicauda;* F.—*Tetradonema plicans;* G.—*Trilobus longus;* H.—*Dorylaimopsis metatypicus;* I-L.—*Rhabditis strongyloides* (I, J, L, ameboid stage, various views; K, resting stage); M.—*Paracanthonchus viviparus;* N.—*Halichoanolaimus robustus;* O.—*Tripyla papillata;* P.—*Axonolaimus spinosus.* B, after Meves, 1920, Arch. Mikr. Anat. v., 94; C, after de Man, 1886, Nordsee-Nematoden, Leipzig; D, after de Man, 1889, Mem. Soc. Zool. France, v. 2; F, after Cobb, 1919, J. Parasit., v. 5; G-P, original, Chitwood; A, E, original, Walton.

---

*Chitwood has re-examined this form and reports that the above observation was based on a misinterpretation of the structures present. (see page 125 for his explanation).

192

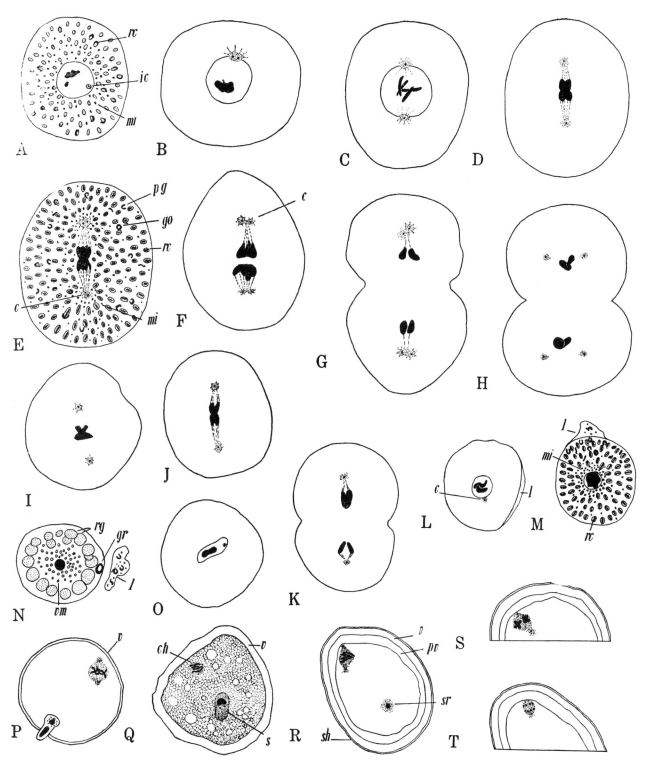

Fig. 148.

Gametogenesis. A.—*Parascaris equorum;* Spermatogonium (showing intranuclear centrosome, mitochondria, refringent corpuscles with golgi bodies, and nuclear contents). B.—*Parascaris equorum;* Early prophase of 1st. spermatocyte (extrusion of intranuclear centrosome). C.—*Parascaris equorum;* Late prophase of 1st spermatocyte. D.—*Parascaris equorum;* Metaphase of 1st. spermatocyte. E.—*Parascaris equorum;* Late metaphase of 1st. spermatocyte (centrosomes dividing). F.—*Parascaris equorum;* Anaphase of 1st. spermatocyte. G.—*Parascaris equorum;* Early telophase of 1st. spermatocyte. H.—*Parascaris equorum;* Telophase of 1st. spermatocyte (centrosomes divided). I.—*Parascaris equorum;* Late prophase of IInd. spermatocyte. J.—*Parascaris equorum;* Metaphase of IInd. spermatocyte. K.—*Parascaris equorum;* Telophase of IInd. spermatocyte. L.—*Parascaris equorum;* Early spermatid (cytoplasmic lobe forming). M.—*Parascaris equorum;* Later spermatid (cytoplasmic structures indicated). N.—*Parascaris equorum;* Spermatid (cytoplasmic reduction completed). O.—*Parascaris equorum;* Oogonium of last generation (intranuclear centrosome). P.—*Parascaris equorum;* Late prophase of Ist. oocyte (penetration of spermatozoon). Q.—*Parascaris equorum;* Prophase of Ist oocyte (sperm with divided centrosome). R.—*Parascaris equorum;* Metaphase of Ist. oocyte ("tetrad" formation). T.—*Parascaris equorum;* Telophase of 1st. oocyte. All drawings original.

193

ally anterior end of the spermatozoon. It is perhaps to be expected that in *Parascaris*, and in related genera where chromosomal behavior as well as other criteria point to a high degree of specialization, the acrosome likewise would tend to vary from the normal, as perhaps is shown by its unusual position behind the nucleus. Some nematodes show distinct polymorphism in sperm size, a condition believed to be correlated with the difference in chromosomal numbers between the "male-producing" and the "female-producing" male gametes (Goodrich, 1916; Meves, 1903; Mulsow, 1911). This chromosome variation is most clearly demonstrable in species in which there is a complex type of "X" chromosome, often involving a large number of chromatin elements (Walton, 1924).

The completion of the germ cycle involves the process of syngamy by which the union of gamete nuclei and the restoration of the diploid number of chromosomes is accomplished. Syngamy in nematodes is complicated by the fact that the maturation of the egg and fertilization proceed simultaneously, the spermatozoon frequently entering the egg during the prophase of the first meiotic division.

The entire spermatozoon, at least among those that are amoeboid in form, enters the egg and immediately a thick fertilization membrane forms, appearing first near the point of entrance and finally enclosing the entire egg. The reticulated male pronucleus gradually forms from the condensed spermatozoon nucleus, the mitochondrial elements slowly fade into the egg cytoplasm as the male cell wall disappears, and the remnant of the mass of acrosomal material eventually loses its separate identity.

In the case of many nematodes (*Rhabdias bufonis*, *R. ranae*, *Rhabditis terricola*, *Syphacia obvelata* (*Oxyuris obvelata*), *Turbatrix aceti*) the shell membrane is reported as being applied to the egg before the entrance of the spermatozoon†. In such cases a micropylar opening has been described, usually at the end of the egg which was originally attached to the rhachis, and opposite to the pole at which the polar cells are normally extruded. No such structure is necessary in the forms in which sperm penetration precedes egg-shell formation. A structure resembling a micropyle has been described in forms which have the egg shell formed after sperm entrance has occurred. In *Ascaridia galli* (*A. lineata*) Ackert (1931) has shown that this is not a true micropyle, and perhaps similar micro-dissection studies might necessitate the revision of the descriptions of the presence of a micropyle in several forms. If a true micropyle is present, it seems obvious that the sperm entrance is fixed at what may be regarded as the vegetative pole of the egg, since many observers have determined that the first polar cell is eliminated at a point opposite the entrance path of the spermatozoon. Probably the same statement holds for those forms in which sperm entrance precedes shell formation, inasmuch as sperm entrance and first polar cell positions are directly opposite in most nematode eggs, and the point of sperm entrance in *Parascaris equorum* has been shown by Schleip (1924) to be at the originally attached end. This problem is tied up with that of the polarity of the egg which is discussed elsewhere.

The two pronuclei come to lie side by side, the first cleavage spindle is established, and division follows. During this process the male and female chromosomes occupy opposite sides of the spindle and it is not until the second cleavage division that the two sets of chromosomes are indistinguishably mixed, although in some cases complete intermingling may be delayed until later in the cleavage phenomenon.

The development of the egg without fertilization (true parthenogenesis) is rare among nematodes although two species of *Rhabditis* (Belar, 1923) have been described as showing only a single maturation division and no reduction in the chromosome number. Krüger (1913) reports that the hermaphroditic *Rhabditis aberrans* (probably a variety of *R. aspera*) produces eggs that are apparently parthenogenetic of the diploid type (one polar cell and no chromosome reduction of the somatic number of 18) although frequently the sperm actually enters the egg but degenerates and fails to enter the cleavage nucleus. In a normally dioecious *Rhabditis pellio* culture, P. Hertwig (1920) found a mutant which produced only one polar cell without reduction, and thus retained the diploid number (14). None of these eggs would develop unless entered by a sperm, but again in no case did the sperm contribute to the cleavage nucleus. These two cases bridge the gap between normal fertilization and normal parthenogenesis.

Many nematode species show a "diminution" phenomenon (Walton, 1918, 1924) in the non "stem-cells" of early cleavage, examples occurring from the second to the sixth division, and then ceasing, as by the sixty-four-cell stage the primordial germ cells are entirely differentiated. The process of "diminution" which involves the elimination of a portion of each of the chromosomes in the nucleus is best known in the embryonic cells of *Parascaris equorum*. In this form the process may begin in the second cleavage of the soma cells although it usually first appears in the third cleavage, and then is found in the division of each new soma cell separated from the "stem" cell until the "germ line" cells are definitely isolated. In *P. equorum* this process is completed during the fifth cleavage. All germ cells retain the undiminished amount of chromatin, while all soma cells have the reduced amount as the result of "diminution". During the prophase of the "diminution division" the chromosomes of the soma cell break up, the center forming a definite number of small chromosomes and the ends several blobs of material. The small chromosomes divide equationally while the larger masses are left behind. The daughter nuclei reorganize without the extruded remnants, which then ultimately degenerate and disappear. The process is quite similiar in other species of nematodes (Meyer, 1895; Bonnevie, 1901; Walton, 1918, 1924) except that there is frequently no increase in number of chromosomes during the process inasmuch as the gametic chromosomes in many species are not as complex as they are in *Parascaris* spp.

† The formation of a shell before fertilization is dubious. See Sect. 1, Part 3, Chapter 12. B. G. C.

Fig. 149.

Gametogenesis. A.—*Parascaris equorum;* Prophase of IInd. oocyte (1st. polar body and "dyad" formation). B.—*Parascaris equorum;* Metaphase of IInd. oocyte. C.—*Parascaris equorum;* Anaphase of IInd. oocyte (1st. polar body). D.—*Parascaris equorum;* Telophase of IInd. oocyte (1st. polar body and "monad" formation). E.—*Parascaris equorum;* Formation of pronuclei (1st. and 2nd. polar bodies). F.—*Parascaris equorum;* Ovum (two pronuclei, centrosome dividing, egg membranes omitted). G.—*Parascaris equorum;* Ovum (pronuclei approaching, centrosomes at poles, egg membranes omitted). H.—*Parascaris equorum;* Ovum (pronuclei fusing, discrete chromosomes, egg membranes omitted). I — *Parascaris equorum;* Prophase of 1st. cleavage spindle. J.—*Parascaris equorum;* Polar view of metaphase of 1st. cleavage spindle. K.—*Parascaris equorum;* Polar view of metaphase of 1st. cleavage spindle (detached heterochromosomes). L.—*Parascaris equorum;* Metaphase of 1st. cleavage, side view. M.—*Parascaris equorum;* Metaphase of 2nd. cleavage (2-celled embryo). N.—*Parascaris equorum;* "T" embryo (4-celled) with regular chromosome structure in cells P2 and EM, and "diminution" divisions in cells A and B. O.—*Parascaris equorum;* Metaphase of "diminution" division in cell S1. P.—*Parascaris equorum;* Anaphase of "diminution" division in cell EM. Q.—*Parascaris equorum;* "Lozenge-Shaped" embryo (4-celled) with only cell P2 not showing "diminution". R.—*Parascaris equorum;* 3rd. cleavage, cell EM with "diminution" spindle. S.—*Parascaris equorum;* P2 undivided, the other cells showing chromatin elimination following "diminution" divisions. T.—*Parascaris equorum;* Blastula in section (only "germ cells" not showing evidence of "diminution" division). U.—*Rhabdias bufonis;* Prophase of 1st. spermatocyte. V.—*Rhabdias bufonis;* Metaphase plate of 1st. spermatocyte. W.—*Rhabdias bufonis;* Anaphase of 1st. spermatocyte (heterochromosomes lagging). X.—*Rhabdias bufonis;* IInd. spermatocytes (sister cells). Y.—*Rhabdias bufonis;* Metaphase of IInd. spermatocyte. Z.—*Rhabdias bufonis;* Anaphase of IInd. spermatocyte (heterochromosomes lagging). AA.—*Rhabdias bufonis;* Telophase of IInd. spermatocyte (heterochromosomes lagging). BB.—*Rhabdias bufonis;* Spermatids showing cytoplasmic reduction (heterochromosome lost with the lobe in half of the cells). CC.—*Rhabdias bufonis;* Dimorphic spermatozoon (large one retains the heterochromosome; n = 6). DD.—*Rhabdias bufonis;* Dimorphic spermatozoon (small one loses the heterochromosome; n = 5). EE.—*Rhabdias bufonis;* Metaphase plate of last generation oogonium (12 chromosomes). FF.—*Rhabdias bufonis;* Prophase of 1st. oocyte. GG.—*Rhabdias bufonis;* Metaphase plate of 1st. oocyte (6 chromosomes). II.—*Rhabdias bufonis;* Anaphase nucleus of 1st. oocyte. JJ.—*Rhabdias bufonis;* Prophase of IInd. oocyte (1st. polar body). KK.—*Rhabdias bufonis;* Metaphase of IInd. oocyte. LL.—*Rhabdias bufonis;* Ootid (female pronucleus, and polar bodies 1 and 2). MM.—*Rhabdias bufonis;* Ovum (male pronucleus with 5, and female pronucleus with 6 chromosomes). NN.—*Rhabdias bufonis;* Ovum (both pronuclei with 6 chromosomes). OO.—*Rhabdias bufonis;* 1st. cleavage spindle (5 chromosomes of male, and 6 chromosomes of female origin). PP. — *Rhabdias bufonis;* Embryonic "germ cell", nucleus with 11 chromosomes (male pronucleus). QQ.—*Rhabdias bufonis;* Embryonic "germ cell" nucleus with 12 chromosomes (female). U-PP, modified after Schleip, 1911, Arch. Zellf., v. 7; others original.

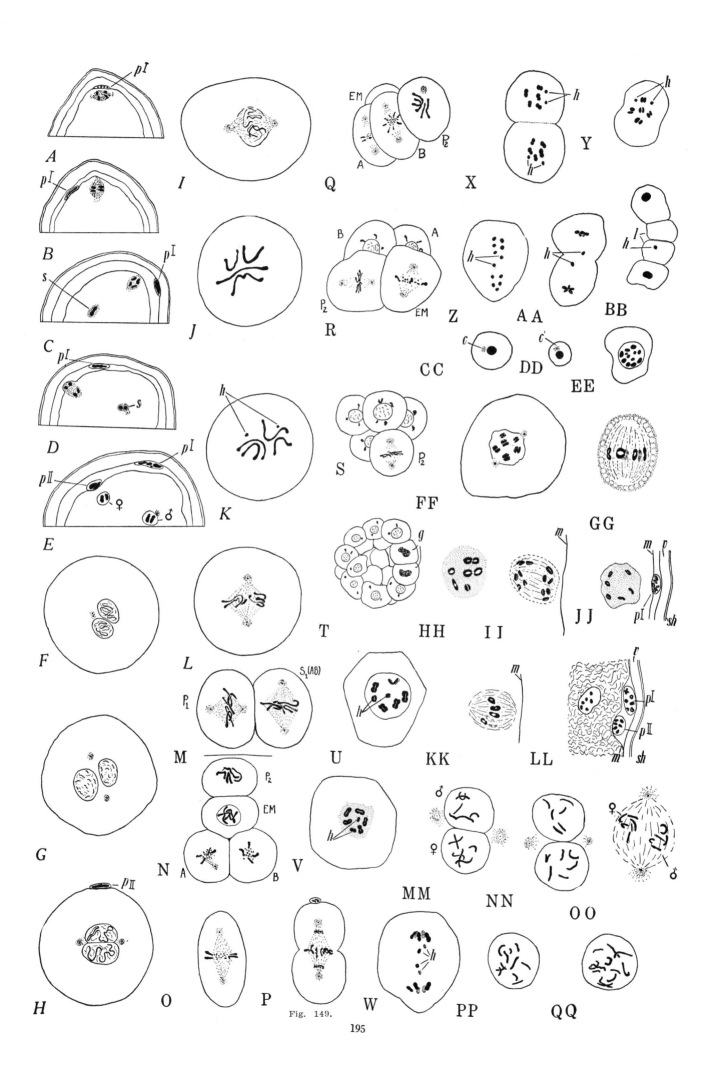

Fig. 149.

The process of diminution is not confined to the nematodes. Members of the Diptera (*Miastor*), Coleoptera (*Dytiscus* and *Colymbetes*), and Lepidoptera (*Lymantria, Orgyra, Phragmatobia, Ephestia, Philosamia,* etc.) also show a similar phenomenon. In the nematodes the process always accompanies the localization of the germinal "stem cell" and is confined to those cells which are derived from the "stem cell", but whose descendants become "soma cells". Only the cells which contain the cytoplasmic area destined to become germ cells fail to undergo diminution. Diminution is thus early in somatic history. The same is perhaps true in the case of *Miastor*, where diminution is confined to the last oogonial divisions. It might seem to separate chromatin useful in germ cells, though not in soma cells. In the Coleoptera the process comes late in the germ-line history and does not separate somatic from germinal chromatin. Among the Lepidoptera, diminution occurs after the chromosomes are set free from the nucleus (during metaphase time) and frequently is found only during the maturation divisions of the egg. This variation in the time of occurrence prevents any interpretation of the phenomenon as one of separation of somatic and germinal types of chromatin. The only generally accepted fact common to all cases of diminution is that it is oxychromatin which is lost and basichromatin which is retained. According to Fogg (1930) "The only safe conclusion that now seems admissible is that diminution plays no primary or essential part in differentiating the germ-line from the somatic. It is rather a by-product of conditions existing in the cytoplasm which may vary widely in different species in respect to the time of its occurrence, its modus operandi, and its physiological significance".

Several workers have reported the loss of portions of chromatic material by means other than "diminution." Chief of these methods is through the "cytophore" form-ation which accompanies the metamorphosis of the spermatid in many animal groups. This phenomenon has been reported for *Cystidicola farionis* (*Ancyracanthus cystidicola*) (Mulsow, 1912), *Toxocara vulpis* (*Belascaris triquetra*) (Marcus, 1906a, Walton, 1918), *Parascaris equorum* (Hertwig, 1890; Mayer, 1908; Sturdivant, 1934), *Ascaris lumbricoides* (Hirschler, 1913), *Rhabdias bufonis* (Boveri, 1911; Schleip, 1911), and *Spirina parasitifera* (Cobb, 1925, 1928).

Among the nematodes the two sexes are normally separate, although a number of hermaphroditic forms are known, particularly among those species which are free-living (Maupas, 1900 Potts, 1910; Cobb, etc.) or those which alternate between free-living and parasitic generations (Boveri, 1911; Schleip, 1911). In such cases the parasitic generation is the one showing hermaphroditism. In many of the bisexual forms the males show an "XO" type of sex chromosome (occasionally an "X" complex) and the females an "XX" condition. A similar condition is known in the hermaphroditic generation of *Rhabdias bufonis* and in a single unusual specimen of *P. equorum* var. *bivalens* (Goulliart, 1932). In both of these cases the spermatozoa are "XO" and the eggs "XX" in type. The formation of the hermaphroditic generation of *R. bufonis* is probably due to the non-viability of the non "X"-bearing spermatozoon when produced by the free-living males, but in the hermaphroditic generation both types of sperm ("X" and "O") are viable and hence union with the "X"-bearing eggs produces the free-living generation males, "XO", and the females, "XX". The "XY" and "XX" condition is doubtfully reported from several species. The only clear-cut case is one of a multiple "X" and simple "Y", and multiple "XX", from a single species (*Contracaecum incurvum = Ascaris incurva*) by Goodrich (1916). In the great majority of nematodes, the hetero-chromosome has not been recognized, possibly because, as

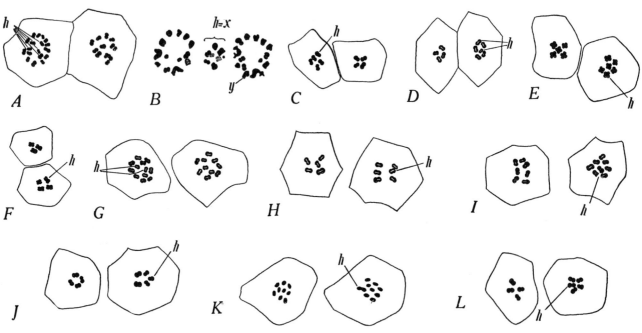

Fig. 150.

A.—*Toxocara canis;* IInd. spermatocytes (12 & 18 "dyad" chromosomes; X = 6). B.—*Contracaecum incurvum;* Anaphase of 1st. spermatocyte (13 + lagging X-group, & 13 + Y; X = 8, Y = 1). C.—*Heterakis papillosa;* IInd. spermatocytes (4 & 5 "dyad' chromosomes; X = 1). D.—*Heterakis spumosa;* IInd. spermatocytes (4 & 6 "tetrad" chromosomes; X = 2). E.—*Nematospira turgida;* IInd. spermatocytes (5 & 6 "tetrad" chromosomes; X = 1). F.—*Trichosomoides crassicauda;* IInd. spermatocytes (3 & 4 "tetrad" chromosomes; X = 1). G.—*Toxocara vulpis;* IInd. spermatocytes (10 & 12 "tetrad" chromosomes; X = 2). H.—*Cruzia tentaculata;* IInd. spermatocytes (5 & 6 "tetrad" chromosomes; X = 1). I—*Contracaecum spiculigerum;* IInd. spermatocytes (7 & 8 "tetrad' chromosomes; X = 1). J.—*Mastophorus muris;* IInd. spermatocytes (4 & 5 "dyad" chromosomes; X = 1). K.—*Toxocara cati;* IInd. spermatocytes (9 & 9 "monad" chromosomes; heterochromosome, X = 1, attached to one autosome). L.—*Physaloptera turgida;* IInd. spermatocytes (4 & 5 "dyad" chromosomes; X = 1). C, after Goodrich, 1916, J. Exper. Zool., v. 21; others original.

is so frequently the case in *P. equorum*, it is attached to the end of an autosome and only occasionally is distinct enough for positive identification. In most cases the heterochromosome undergoes "pre-reduction" as do the autosomes, but in some cases it shows "postreduction" although the autosomes seem to show the *differential division* as being the first. This may point to the primitive condition being actually one of "postreduction" (Edwards, 1910; Wilson, 1925, p. 757). In many instances the heterochromosome (or heterochromosome complex) either precedes the others or lags behind during one or both of the meiotic divisions and in some cases forms a separate chromatin nucleolus during the interphase stage. Where both "X" and "Y" are present, they are separated most frequently at the first division, each undergoing equational splitting at the second. In the early Spermatocyte I growth period nuclei, the "XY" group is differentiated from a single chromatin nucleolus and the "XY" pair assumes a "tetrad" form, usually asymmetrical because of the small bulk of the "Y" element. Even when the "X" is multiple it differentiates from a single nuclear body (Walton, 1916, 1924), just as it does when "X" = 1.

The nematodes therefore afford a wide variation of heterochromosome types, varying from a single "X" and no "Y" in *Heterakis dispar* and *Cystidicola farionis* (*Ancyracanthus cystidicola*) to forms like *Toxocara vulpis* (*Belascaris triquetra*) and *Heterakis spumosa* (*Gangulet-erakis spumosa*) with an "X"-complex of two, *Ascaris lumbricoides* with one of five, *Toxocara canis* (*Toxascaris canis*) with one of six, and *Parascaris equorum* with one of eight to nine, and no "Y", and thence to forms such as *Contracaecum incurvum* with an "X"-complex of eight and a single "Y". Peculiarly, no established case of an "XY" pair has been definitely recognized. The "X" and "Y"-chromatin may form a single body or a single unit during meiosis, just as frequently the autosomes may conceal their complexity temporarily in single bodies under the same circumstances (*P. equorum*).*

The following chart gives the majority of the examples of the species which have furnished material for the study of nematode gametogenesis. In each case the haploid and diploid chromosome numbers are indicated, the somatic number is given, and the form of the chromosomes at each stage (di-tetrad, tetrad, dyad, monad) is noted. Wherever the germinal chromosomes are plurivalent, their unit value in terms of the somatic chromosomes is pointed out. The nature and number of the heterochromosomes is indicated for each species as far as it is known. The presence or absence of the "diminution" process is also stated for such species as have been examined for that phenomenon.

---

*Jeffrey and Haertl (1938) have recently questioned the whole subject of "sex chromosomes" in nematodes in their study on *Ascaris lumbricoides, Toxocara cati, T. canis,* and *"Ascaris" sp?* from a seal. They fail to find any evidence of a consistant differential distribution of what might be called "X-chromatin", and show that, while certain chromosomes lag behind in each meiotic division, these are not necessarily distributed as "X" and "O", or "X" and "Y" materials must be. Analogous behavior of chromosomes is known to occur in various hybrid forms of plants and animals, and is evidence of their mixed ancestry. The authors argue that these nematodes, and probably all similar forms, are likewise hybrids because of the evidence presented by the behavior of their chromosomes, particularly during the process of meiosis. If the nematodes are hybrids, then the uneven distribution of the chromosomes during the maturation divisions is to be expected, and is a clue to their hybrid ancestry, not a proof of the presence of "sex chromosomes" as has been the usually accepted interpretation.

List of Abbreviations

A, Cell formed by the division of the 1st generation Soma cell.
B, Cell formed by the division of the 1st generation Soma cell.
c, Centrosome.
ch, Chromosomes.
EM, Cell formed by the division of the 1st generation Stem cell.
g, Germ cells.
go, Golgi body.
cr, Golgi ring.
h, Heterochromosome.
ic, Intranuclear centrosome.
l, Cytoplasmic lobe.
m, Cell membrane.
mi, Mitochondria.
ncl, Nucleus.
P I, 1st. polar body.
P II, 2nd. polar body.
P1, 1st. generation Stem cell.
P2, 2nd. generation Stem cell.
pg, Pre-acrosomal granule.
Pv, Perivitelline space.
rb, Refringent body, or Acrosome.
rc, Refringent corpuscle.
rg, Refringent globule.
S1, 1st. generation Soma cell.
s, Spermatozoon.
sh, Egg shell.
sr, Sperm remnant.
v, Fertilization membrane.
vm, Vacuolated mitochondria.
X, X-chromosome.
Y, Y-chromosome.
Male sign, Male pronucleus, or of male origin.
Female sign, Female pronucleus, or of female origin.

Table 7. Chromosome numbers of nematodes.

Name	Haploid number of chromosomes	Diploid number of chromosomes	Diminuation	Somatic number of chromosomes	Heterochromosome number	Autosome Unit Value
1. *Ascaridia galli* (*Heterakis inflexa*)	5 (tetrads)	9-10 (monads)	undetermined	9-10	X = 1	1
2. Ascarid (from dog)	4 (di-tetrads)	8 (dyads)	undetermined	16	?	2
3. *Ascaris anguillae* (*Ascaris labiata*)	?	?	present	?	?	?
4. *Ascaris lumbricoides*	24 (tetrads)	43-48 (monads)	present	43-48	X = 5	1
5. *Camallanus lacustris* (*Cucullanus elegans*)	6 (tetrads)	12 (monads)	undetermined	12	?	1
6. *Contracaecum clavatum* (*Ascaris clavata*)	12 (di-tetrads)	24 (dyads)	undetermined	48	?	2
7. *Contracaecum incurvum* (*Ascaris incurva*)	21 (tetrads)	35-42 (monads)	present	35-42	X = 8 Y = 1	1
8. *Contracaecum spiculigerum* (*Ascaris spiculigera*)	5 (di-tetrads)	9-10 (dyads)	absent	18-20	X = 1	2
9. *Cruzia tentaculata*	6 (di-tetrads)	11-12 (dyads)	undetermined	22-24	X = 1	2
10. *Cyclostomum tetracanthum* (*Strongylus tetracanthus*)	6 (tetrads)	12 (monads)	undetermined	12	?	1
11. *Cystidicola farionis* (*Ancyracanthus cystidicola*)	6 (tetrads)	11-12 (monads)	undetermined	11-12	X = 1	1
12. *Dictyocaulus filaria* (*Strongylus filaria*)	6 (tetrads)	11-12 (monads)	undetermined	11-12	X = 1*	1
13. *Dictyocaulus viviparus* (*Strongylus micruris*)	6 (tetrads)	11-12 (monads)	undetermined	11-12	X = 1	1
14. *Dispharynx spiralis* (*Acuaria spiralis*)	6 (di-tetrads)	11-12 (dyads)	undetermined	22-24	X = 1	2
15. *Filaroides mustelarum*	8 (tetrads)	16 (monads)	undetermined	16	?	1
16. *Heterakis dispar*	5 (tetrads)	9-10 (monads)	undetermined	9-10	X = 1	1
17. *Heterakis gallinae* (*Heterakis vesicularis*)	5 (tetrads)	9-10 (monads)	undetermined	9-10	X = 1	1
18. *Heterakis papillosa*	5 (di-tetrads)	9-10 (dyads)	undetermined	18-20	X = 1	2
19. *Heterakis spumosa* (*Ganguleterakis spumosa*)	6 (di-tetrads)	10-12 (dyads)	absent	20-24	X = 2	2
20. Heterakid (from pheasant)	5 (tetrads)	9-10 (monads)	undetermined	9-10	X = 1	1
21. *Mastophorus muris* (*Protospirura muris*)	5 (di-tetrads)	9-10 (dyads)	absent	9-10	X = 1	1
22. *Metastrongylus elongatus* (*Strongylus paradoxus*)	6 (di-tetrads)	11-12 (monads)	absent	11-12	X = 1	1
23. *Nematospira turgida*	6 (di-tetrads)	11-12 (dyads)	absent	22-24	X = 1	2
24. *Ophidascaris filaria* (*Ascaris rubicunda*)	?	?	present	?	?	?
25. *Ophiostoma mucronatum*	6 (di-tetrads)	12 (dyads)	undetermined	24	?	2
26. *Parascaris equorum univalens* (*Ascaris· megalocephala*)	1 (rod-shaped)	2 (rod-shaped)	present	51-60	X = 1	26 (X = 9)

Table 7. (Continued)

27. *Parascaris equorum bivalens*	2 (rod-shaped)	4 (rod-shaped)	present	96-104	X = 1	22 (X = 8)
28. *Parascaris equorum trivalens***	3 (rod-shaped)	6 (rod-shaped)	present	?	X̄ = 1	22-26 (hybrid)
29. *Passalurus ambiguus* (*Oxyuris ambigua*)	4 (tetrads)	7- 8 (monads)	undetermined	7-8	X = 1	1
30. *Physaloptera turgida*	5 (di-tetrads)	9-10 (dyads)	absent	18-20	X = 1	2
31. *Proleptus robustus* (*Cornilla robusta*)	8 (tetrads)	16 (monads)	undetermined	16	?	1
32 *Rhabdias bufonis* (*Rhabditis nigrovenosa*)	6 (tetrads) (di-tetrads?)	11-12 (monads) (dyads?)	absent	22-24	X = 1*	2
33. *Rhabdias fulleborni*	6 (tetrads) (di-tetrads?)	11-12 (monads) (dyads?)	absent (?)	22-24	X = 1	2
34. *Rhabditis aberrans* (female - parthenogenetic) (male - non-funtional) (var. of *R. aspera?*)	18 (dyads) 9 (tetrads)	18 (monads) 17-18 (monads)	undetermined	18 18	X = 1 X = 1*	1 1
35. *Rhabditis aspera*	7 (tetrads)	13-14 (monads)	undetermined	13-14	X = 1	1
36. *Rhabditis pellio* Butschli (*R. maupasi*)	7 (tetrads)	13-14 (monads)	undetermined	13-14	X = 1	1
37. *Rhabditis pellio* Schneider	7 (tetrads)	13-14 (monads)	undetermined	13-14	X = 1	1
38. *Rhabditis pellio* Schneider (mutant parthenogenetic) (female)	14 (dyads)	14 (monads)	undetermined	14	X = 1	1
39. *Setaria equina* (*Filaria papillosa*)	6 (tetrads)	11-12 (monads)	undetermined	11-12	X = 1	1
40. *Spirina parasitifera*	7 (compound)	14 (compound)	absent	14 (comp.)	No. "X"	Undetermined, but many
41. *Spirura talpae* (*Spiroptera strumosa*)	8 (tetrads)	16 (monads)	undetermined	16	?	1
42. *Strongylus edentatus* (*Sclerostomum edentatum*)	6 (tetrads)	11-12 (monads)	absent	11-12	X = 1	1
43. *Strongylus equinus* (*Sclerostomum equinum*)	6 (tetrads)	11-12 (monads)	absent	11-12	X = 1	1
44. *Strongylus vulgaris* (*Sclerostomum vulgare*)	6 (tetrads)	11-12 (monads)	absent	11-12	X = 1	1
45. *Syphacia obvelata* (*Oxyurus obvelata*)	8 (tetrads)	15-16 (monads)	absent	15-16	X = 1	1
46. *Toxocara canis* (*Toxascaris canis*)	18 (di-tetrads)	30-36 (dyads)	present	60-72	X = 6	2
47. *Toxocara cati* (*Belascaris mystax*)	9 (tetrads)	18 (monads)	present	18	X = 1*	1
48. *Toxocara vulpis* (*Belascaris triquetra*)	12 (di-tetrads)	22-24 (dyads)	present	44-48	X = 2*	2
49. *Trichosomoides crassicauda*	4 (di-tetrads)	7- 8 (dyads)	absent	7- 8 (dyads)	X = 1	2
50. *Trichostrongylus tenuis* (*Strongylus tenuis*)	6 (tetrads)	11-12 (monads)	undetermined	11-12	X = 1	1

* A "Y" chromosome has been reported from these species, but the accuracy of the interpretations is questionable.

**Li (1934, 1937) reports a six-chromosome and a nine-chromosome variety of *P. equorum* as occurring in Chinese Mongolian horses. He suggests that these either are examples of polyploidy or represent the primitive racial strains of the species that are still found in the most primitive of living horses. Cytological study of the six-chromosome form shows a behavior normal for *P. equorum* except that the autosomal number present after "diminution" is less than in those found in the better known varieties.

Bibliography

ACKERT, J. E. 1931.—The morphology and life history of the fowl nematode, *Ascaridia lineata* (Schneider). Parasitol., v. 23 (3): 360-379, 25 figs., pls. 13-14.

AUERBACH, L. 1874.—Organologische Studien. Heft. 1-2. Zur Characteristik und Lebensgeschichte der Zellkerne. 262 pp., pls. 1-4. Breslau.

BELAR, K. 1923.—Ueber den Chromosomenzyklus von parthenogenetischen Erdnematoden. Biol. Zentralbl., v. 43 (5): 513-518, 3 figs.

BENEDEN, E. VAN, 1883.—Recherches sur la maturation de l'œuf, la fécondation et la division cellulaire. 424 pp., pls. 3, 10-19 bis. Gand and Leipzig, Paris.

BENEDEN, E. VAN, and NEYT, A. 1887.—Nouvelles recherches sur la fécondation et la division mitosique chez l'ascaride mégalocéphale. Bull. Acad. Roy. Sc. Belgique, 3 Ser., v. 14 (8): 215-295, 6 pls.

BONNIE, K. 1901. — Ueber Chromatindiminution bei Nematoden. Jena. Zeitschr., 36 n. F., v. 29 (1-2): 275-285, 2 pls. figs. 1-21.
1908.—Chromosomenstudien. Arch. Zellforsch., v. 1 (2-3): 450-514, 2 figs., pls. 11-15, 99 figs.
1913.—Ueber die Struktur und Genese der Ascarischromosomen. Arch. Zellforsch., v. 9 (3): 433-457, 7 figs.

BOVERI, Th. 1887.—Zellen-Studien. Heft 1. Die Bildung der Richtungskörper bei *Ascaris megalocephala* und *Ascaris lumbricoides*. 93 pp., 4 pls. Jena.
1888.—Zellen-Studien. Heft 2. Die Befruchtung und Teilung des Eies von *Ascaris megalocephala*. 198 pp., 5 pls., figs. 1-94. Jena.
1890.—Zellen-Studien. Heft 3. Ueber das Verhalten der chromatischen Kernsubstanz bei der Bildung der Richtungskörper und bei der Befruchtung. 88 pp., 3 pls. Jena.
1900.—Zellen-Studien. Heft 4. Ueber die Natur der Centrosomen. 220 pp., figs. A-C, 8 pls. figs. 1-111. Jena.
1909.— Die Blastomerenkerne von *Ascaris megalocephala* und die Theorie der Chromosomenindividualität. Arch. Zellforsch., v. 3 (1-2): 181-268, figs. 1-7, pls. 7-11, figs. 1-51.
1911.—Ueber das Verhalten des Geschlechtschromosomen bei Hermaphroditismus. Beobachtungen an *Rhabditis nigrovenosa*. Verhandl. Phys.-Med. Gesellsch., Würzburg, n. F., v. 41 (5): 83-97, 19 figs.

BOWEN, R. H. 1925.—Further notes on the acrosome of animal sperm. The homologies of non-flagellate sperms. Anat. Rec., v. 31: 201-231, 5 figs.

BRAUER, A. 1893.—Zur Kenntniss der Spermatogenese von *Ascaris megalocephala*.. Arch. Mikr. Anat., v. 42 (1): 153-213, 3 pls., figs. 1-228.

CHITWOOD, B. G. 1931.—Flagellate spermatozoa in a nematode (*Trilobus longus*). J. Wash. Acad. Sci., v. 21 (3): 41-42, 2 figs.

COBB, N. A. 1925.—Nemic spermatogenesis. J. Heredity, v. 16 (10): 357-359, 1 fig.
1928.—Nemic spermatogenesis: with a suggested discussion of simple organisms, -litobionts. J. Wash. Acad. Sci., v. 18 (2): 37-50, 17 figs.

DREYFUS, A. 1937.—Contribuiçao para o estudo do cyclo chromosomico e da determinacao do sexo de *Rhabdias fülleborni* Trav., 1926. Bol. Face. Philos., Sci., Let., Univ. Sao Paulo. III. Biol. Geral No. 1., 145 pp., 92 figs.

EDWARDS, C. L. 1910.—The idiochromosomes in *Ascaris megalocephala* and *Ascaris lumbricoides*. Arch. Zellforsch., v. 5 (3): 422-429, pls. 21-22, figs. 1-39.

FOGG, L. C. 1930.—A study in chromatin diminution in *Ascaris* and *Ephestia*. J. Morph., v. 50 (2): 413-451, 4 pls., figs. 1-49.
1931.—A review of the history of the centriole in the second cleavage of *Ascaris megalocephala bivalens*. Anat. Rec., v. 49 (3): 251-264, 2 pls., figs. 1-13.

FÜRST, E. 1898 —Ueber Centrosomen bei *Ascaris megalocephala*. Arch. Mikr. Anat., v. 52 (1): 97-133, pls. 8-9, figs. 1-35.

GOODRICH, H. B. 1914.—The maturation divisions in *Ascaris incurva*. Biol. Bull., v. 27 (3): 147-150, 1 pl., figs. 1-13.

1916.—The germ cells in *Ascaris incurva*. J. Exp. Zool., v. 21 (1): 61-99, figs, a-k, 3 pls., figs. 1-49.

GOULLIART, M. 1932.—Le comportement de l'hétérochromosome dans la spermatogenèse et dans l'ovogenèse chez un *Ascaris megalocephala* hermaphrodite. Compt. Rend. Soc. Biol., Paris. v. 110 (28): 1176-1179, 9 figs.

GRIGGS, R. F. 1906.—A reducing division in *Ascaris*. Ohio Naturalist, v. 6 (7): 519-527, pl. 33, figs. 1-12.

HELD, H. 1912.—Ueber den Vorgang der Befruchtung bei *Ascaris megalocephala*. (In Verhandl. Anat. Gesellecsh. 26. Versamml.) Anat. Anz., v. 41: 242-248.
1917.—Untersuchungen über den Vorgang der Befruchtung. I. Der Anteil des Protoplasmas an der Befruchtung von *Ascaris megalocephala*. Arch. Mikr. Anat., v. 89 (2): 59-224, 6 pls.

HERLA, V. 1894.—Etude des variations de la mitose chez l'ascaride mégalocéphale. Arch. Biol., Gand. v. 13 (3): 423-520, pls. 15-19, figs. 1-103.

HERTWIG, O. 1890.—Vergleich der Ei- und Samenbildung bei Nematoden. Arch. Mikr. Anat., v. 36 (1): 1-38, 4 pls., figs. 1-18.

HERTWIG, P. 1920.—Abweichende Form der Parthogenese bei einer Mutation von *Rhabditis pellio*. Arch. Mikr. Anat., v. 94: 303-337, pl. 21, figs. 1-14.

HIRSCHLER, J. 1913.—Ueber die Plasmastrucktturen (Mitochondrien, Golgischer Apparat, u. a.) in den Geschlechtszellen der Ascariden. Arch. Zellforsch., v. 9 (3): 351-398, pls. 20-21, figs. 1-41.

JEFFREY, E. C., and HAERTL, E. J. 1938.—The nature of certain so-called sex chromosomes in *Ascaris*. La Cellule, v. 47 (2): 237-244, 2 text-figs., 1 pl.

KEMNITZ, G. A. 1913.—Eibildung, Eireifung, Samenreifung und Befruchtung von *Brachycoelium salamandrae* (*Brachycoelium crassicolle* [Rud.]. Arch. Zellforsch., v. 10 (14): 470-506, pl. 39, 37 figs.

KRÜGER E. 1913.—Fortpflanzung und Keimzellenbildung von *Rhabditis aberrans*, nov. sp. Zeit. Wiss. Zool., v. 105 (1): 87-124, pls. 3-6, figs. 1-73.

LI, J. C. 1934.—A six-chromosome *Ascaris* found in Chinese horses. Peking Nat. Hist. Bull. v. 9 (2): 131-132, 5 figs.
1937.—Studies of the chromosomes of *Ascaris megalocephala trivalens*. I. The occurrence and possible origin of nine-chromosome forms. Peking Nat. Hist. Bull., v. 11 (4): 373-379, 1 pl.

MAN. J. G. DE 1886.—Anatomische Untersuchungen über freilebende Nordsee-Nematoden. 82 pp., 13 pls. Leipzig.
1889a.—Espèces et genres nouveaux de nématodes libres de la mer du nord et de la manche. Mém. Soc. Zool. France, v. 2 (1): 1-10.
1889b.—Troisième note sur les nématodes libres de la mer du nord et de la manche. Mém. Soc. Zool. France, v. 2 (3): 182-216, pls. 5-8, figs. 1-12.

MARCUS, H. 1906a.—Ueber die Beweglichkeit der *Ascaris*-Spermien. Biol. Centralbl., v. 26 (13-15): 427-430, figs. 1-5.
1906b.—Ei- und Samenreife bei *Ascaris canis* Werner (*Ascaris mystax*). Arch. Mikr. Anat., v. 68 (3): 441-490, 10 figs., pls. 29-30, figs. 1-57.

MAUPAS, E. 1900.—Modes et formes de réproduction des nématodes. Arch. Zool. Exp. et Gen., Ser. 3, v. 8: 463-624, pls. 16-26.

MAYER, A. 1908.—Zur Kenntnis der Samenbildung bei *Ascaris megalocephala*. Zool. Jahrb., Abt. Anat., v. 25 (3): 495-546, 2 figs., pls. 15-16, 61 figs.

MEVES, F. 1903.—Ueber oligopyrene und apyrene Spermien und ihre Entstehung nach Beobachtungen an *Paludina* und *Pygaera*. Arch. Mikr. Anat., v. 61: 1-85, 3 pls.
1911.—Ueber die Beteiligung der Plastochondrien an der Befruchtung des Eies von *Ascaris megalocephala*. Arch. Mikr. Anat., v. 76 (4): 683-713, pls. 27-29, figs. 1-18.
1915.—Ueber Mitwirkung der Plastosomen bei der Befruchtung des Eies von *Filaria papillosa*. Arch. Mikr. Anat., v. 87 (2): 12-46, 4 pls., figs. 1-77.
1920.—Ueber Samenbildung und Befruchtung bei *Oxyuris ambigua*. Arch. Mikr. Anat., v. 94: 135-184, pls. 9-13, 72 figs.

MEYER, O. 1895.—Celluläre Untersuchungen an Nematoden Eiern. Jena. Zeitschr., v. 29 (N. F., v· 22): 391-410, 2 pls.

MULSOW, K. 1911.—Chromosomenverhaltnisse bei *Ancyracanthus cystidicola*. Zool. Anz., v. 38 (22-23): 484-486, 6 figs.

    1912.—Der Chromosomencyclus bei *Ancyracanthus cystidicola* (Rud.). Arch. Zellforsch., v. 9 (1): 63-72, 5 figs., pls. 5-6, figs. 1-45.

POTTS, F. A. 1910.—Notes on the free-living nematodes. Quart. J. Micr. Sc., (n. s.) v. 55 (219): 433-484, 11 figs.

RABL, C. 1889.—Ueber Zellteilung. Anat. Anz., v. 4 (1): 21-30, 2 figs.

ROMEIS, B. 1912.—Beobachtungen über Degenerationserscheinungen von Chondriosomen. Nach untersuchungen an nicht zur Befruchtung gelängten Spermien von *Ascaris megalocephala*. Arch. Mikr. Anat., v. 80 (Abt. 2) (4): 129-170, 2 pls.

ROMIEU, M. 1911. — Le Spermiogénèse chez l'*Ascaris megalocephala*. Arch. Zellforsch., v. 6 (2): 254-325, pls. 14-17, figs. 1-94.

SALA, L. 1895.—Experimentelle Untersuchungen über die Reifung und Befruchtung der Eier bei *Ascaris megalocephala*. Arch. Mikr. Anat., v. 44 (3): 422-498, pls. 25-29, figs. 1-89.

SCHEBEN, L. 1905.—Beiträge zur Kenntnis des Spermatozoons von *Ascaris megalocephala*. Ztschr. Wiss. Zool., v. 79 (3): 397-431, 3 figs., pls. 20-21, figs. 1-44.

SCHLEIP, W. 1911.—Das Verhalten des Chromatins bei *Angiostomum (Rhabdonema) nigrovenosum*. Ein Beitrag zur Kenntnis der Beziehungen zwischen Chromatin und Geschlechtsbestimmung. Arch. Zellforsch., v. 7 (1): 87-138, pls. 4-8, figs. 1-108.

    1924.—Der Herkunft der Polarität des Eies von *Ascaris megalocephala*. Arch. Mikr. Anat. Entw.-Mech., v. 100 (3): 573-598, 17 figs.

SPEK, J. 1918a. — Oberflächenspannungsdifferenzen als Ursache der Zellteilung. Arch. Entw.-Mech., v. 44 (1): 1-113, 25 figs.

    1918b.—Die amöboiden Bewegungen und Strömungen in den Eizellen einiger Nematoden während der vereinigung der Vorkerne. Arch. Entw.-Mech., v. 44 (2): 217-255, 15 figs.

    1920.—Experimentalle Beiträge zur Kolloidchemie der Zellteilung. Kolloidchem. Beih., v. 12 (1-3): 1-91.

STRUCKMANN, C. 1905.—Eibildung, Samenbildung und Befruchtung von *Strongylus filaria*. Zool. Jahrb., Abt. Anat., v. 22 (3): 577-628, pls. 29-31, figs. 1-105.

STURDIVANT, H. P. 1931.—Central bodies in the sperm-forming divisions of *Ascaris*. Science, n. s. v. 73 (1894): 417-418.

    1934.—Studies on the spermatocyte divisions in *Ascaris megalocephala*; with special reference to the central bodies, Golgi complex and Mitochondria. J. Morph., v. 55 (3): 435-475, pls. 1-5, 81 figs.

TRETIAKOV, D. 1904.—Die Spermatogenese bei *Ascaris megalocephala*. Arch. Mikr. Anat., v. 65 (2): 383-438, 1 fig., pls. 22-24, figs. 1-130.

VEJDOVSKY, F. 1911-1912.—Zum Problem der Vererbungsträger. 184 pp., 16 figs., 12 pls. Prag.

WALTON, A. C. 1916a.—*Ascaris canis* (Werner) and *Ascaris felis* (Goeze). A taxonomic and a cytological comparison. Biol. Bull., v. 31 (5): 364-372, figs. A-F, 1 pl., figs. 1-12.

    1916b.—The "refractive body" and the "mitochondria" of *Ascaris canis* Werner. Proc. Amer. Acad. Arts and Sci., v. 52 (5): 253-266, 2 pls., figs. 1-16.

    1918.—The oogenesis and early embryology of *Ascaris canis* Werner. J. Morph., v. 30 (2): 527-603, 1 fig., 9 pls., figs. 1-81.

    1924.—Studies on Nematode Gametogenesis. Ztschr. Zell.-u. Geweb., v. 1 (2): 167-239, figs. A-B, pls. 8-11, figs. 1-118.

WARREN, E. 1930.—Multiple spermatozoa and the chromosome hypothesis of heredity. Nature (Lond.), v. 125 (3165): 973-974, figs. 1-9.

WILDMAN, E. E. 1913.—The spermatogenesis of *Ascaris megalocephala*, with special reference to the two cytoplasmic inclusions, the refractive body and the "mitochondria", their origin, nature and rôle in fertilization. J. Morph., v. 24 (3): 421-457, 3 pls., 48 figs.

WILSON, E. B. 1925.—The cell in development and heredity. 3d. ed. xxxvii & 1232 pp., 529 figs. New York.

ZOJA, R. 1896.—Untersuchungen über die Entwicklung der *Ascaris megalocephala*. Arch. Mikr. Anat., v. 47 (2): 218-260, pls. 13-14.

CHAPTER XIV

NEMIC EMBRYOLOGY

B. G. CHITWOOD

Nemic embryology is a subject which has stimulated much research, especially because of the fact that the cells designed to form particular organs are laid down in the very early cleavages. This type of development is termed determinate cleavage and in substance means that each blastomere may be identified in the egg as the stem cell of a particular organ or part of an organ. In other words, the fate of each cell is foreordained from the first division.

The regularity with which division takes place in nematodes was observed by the earliest workers on the subject. No attempt will be made to give an historical account of the development of our knowledge other than to point out a few of the steps. Bütschli (1875), Galeb (1878), Goette (1882), and Hallez (1885) were among the pioneers in the field and to them the later workers are indebted for breaking the ice, but in the light of present day knowledge their observations appear rather casual. The publication of Boveri in 1892 on the embryology of *Parascaris equorum* was the foundation of modern nemic embryology. His investigations were followed by those of zur Strassen (1892, 1896), Spemann (1895), Zoja (1896), Neuhaus (1903), Müeller (1903), Martini (1903, 1909), and Pai (1927, 1928) as well as many less comprehensive studies by other authors. It should be stated that Boveri's work directly initiated the precise study of the subject by later workers; all these investigations have given us information equalled in few other groups of animals.

Nemic embryology consists of the study of individual cells; in the early cleavages each cell is differentiated to such an extent that it is capable of giving rise only to certain parts of the organism; sister cells as a rule differ to some degree in their potentialities. While there is some difference in opinion as to what some particular cells may give rise to in the mature organism, these differences appear to be based more upon conceptions of authors than actual conditions in given species. It is not surprising that misinterpretations should arise in the study of cell lineage where one must follow the course of literally hundreds of cells.

In nemas the development of germ layers as observed in other animals is highly modified. In fact one can hardly speak of germ layers in reference to nemas. In the course of the first cleavages a number of so called "primordial" or "stem" cells are formed (Fig. 151). These are highly differentiated as to their potentialties. Each of them will form a certain organ or organ system, e. g., the anterior cell is destined to form the greater part of the ectodermal epithelium and is designated *S1*, which means first "somatic" stem cell; this cell therefore is the ancestor of the primary ectoderm. The other cell, posterior in position, however, is a less differentiated cell in its potentiality. It forms the remainder of the embryo; for this reason it is designated *P1*, or first parental germinal cell, the fertilized ovum being designated *P0*. This first cleavage is a transverse one. These "primordial" or "stem" cells are therefore unequal in their prospective potencies.

Beginning with the second cleavage the cells of each given family have their own cleavage "rhythm", that is, cells descendent from each primordial or stem cell divide at the same rate but they often differ in the rate from those descended from another primordial or stem cell. Ordinarily one would expect this to be due to a difference in the amount of yolk or the size of the cells but in nematodes this is not the case; instead the cell has an inherent rate of cleavage without regard to size or yolk and it cannot be explained as caused by mechanical forces.

The second cleavage is transverse in the *S1* cell forming an anterior dorsal cell *A* and a posterior dorsal cell *B*. This is followed by transverse division of the blastomere P1 forming an anterior ventral cell *S2* and a posterior cell *P2*. The second somatic stem (*S2*) cell is destined to form the somatic musculature, part of the esophagus, and the entire intestine or mesenteron.

At the third cleavage the *S1* cell group, *A* and *B*, divides longitudinally forming four cells—two on the right and two on the left side of the embryo; the cells on the right are designated by Roman letters a and b, while those on the left are designated by Greek letters Alpha and Beta, for which the small capitals A and B are substituted in the text. Following this the second somatic stem cell, *S2*, divides transversely, the anterior cell being destined to form the greater part of the mesoderm of the body wall and esophagus, termed *MST*, and the posterior cell the entire entoderm, designated *E*. The undifferentiated stem cell *P2* divides transversely, the dorsal posterior daughter being termed *S3* or the third somatic stem cell, and the posterior ventral cell *P3*. The third somatic stem cell is destined to form the ectodermal epithelium of the posterior part of the body and is therefore secondary ectoderm. It also forms a part of the mesoderm in some nemas.

The next or fourth cleavage is commonly said to complete the formation of stem cells going into the formation of the soma or body. The cleavage of the S1 group is transverse forming *a I, a II, A I, A II, b I, b II, B I,* and *B II*; of the *S2* group *M* divides obliquely, transversely or longitudinally forming either two cells one behind the other *St* and *M* or two cells side by side *mst* and *MST*; *E* divides transversely forming *E I* and *E II*; the *S3* cell group divides longitudinally forming *c* and *C*; and finally the *P3* cell divides transversely forming *P4*, ventral, and *S4*, posterior. The descendants of *S4* are destined to form the proctodeum or rectum and sometimes the mesoderm or ectoderm of the posterior ventral part and therefore either tertiary ectoderm, or possibly secondary mesoderm (see description of embryology of *Parascaris equorum*).

In further cleavages the descendants of given cells are designated by Roman and Greek letters (small capitals in the text) if they go to opposite sides of the embryo, i. e., *a* and *A*; by Roman numerals if serial, i. e., *I* and *II*; and by Arabic numerals if neither, i. e., *1* and *2*. Thus a longitudinal division of *S4* (*D*) forms cells *d* and *D*; transverse division of them forms *d I, d II, D I* and *D II*. Additional labels such as *a', a''* etc., are sometimes necessary. *P4* is generally termed the primordial germ cell but recent investigations indicate this may not be the case. The epithelium of the reproductive system is not germinal tissue but somatic tissue and is probably laid down by a later cleavage. At the fifth cleavage two cells are formed which are variously termed *G I* and *G II* and *S5* and *P5*; these cells appear alike. The earliest differentiated genital primordium contains four cells,— two epithelial and two germinal (*G I* and *G II*). The former may be of mesenchymatous origin or they may have been derived from the *P* line at either the fifth or sixth cleavage.

The various cleavages as outlined above do not take place in all cells simultaneously so we do not have a regular doubling of cells. Though the cleavage is total or holoblastic, it is usually unequal; since it is neither radial nor spiral there is no typical morula stage. There may or may not be a segmentation cavity or blastocoele and when such is present it is usually of small size. As Martini (1908) showed, the absence of a segmentation cavity is a negative point being in these forms entirely dependent upon the depth of the cleavage furrows and the depth of furrows does not appear to be correlated with anything else in the cleavage of nemic ova and embryos so that it must be regarded as without significance. Generally speaking we may say that the blastular stage begins in the 12 to 16 cell stages of the embryo since it is at this time that a blastocoele appears in certain species, *Parascaris equorum, Rhabdias bufonis,* and *Nematoxys ornatus,* while embryos composed of homologous cells in the species *Camallanus lacustris* and

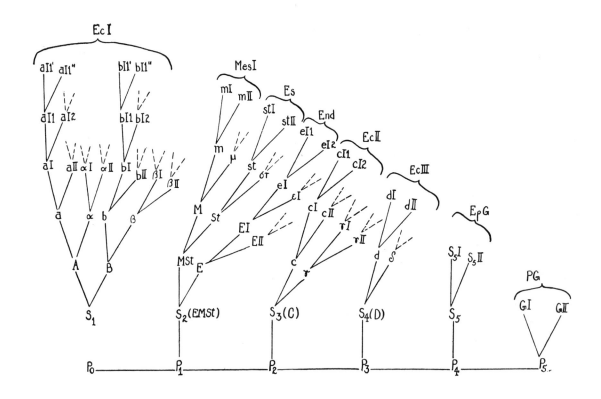

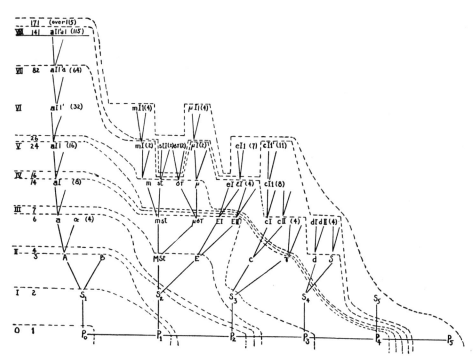

Fig. 151. DIAGRAMS OF CELL LINEAGE.

Upper figure is a general diagram showing the standard numbering system of nemic blastomeres.

Lower figure is a diagram of cleavage in *Rhabdias bufonis*. Roman numerals in the first vertical column indicate number of cleavage. Arabic numerals in first vertical column give total number of cells in embryo. Numbers in parenthesis indicate number of cells in a given germ line. Broken curves indicate corresponding levels of cleavage in the various germinal lines. Original.

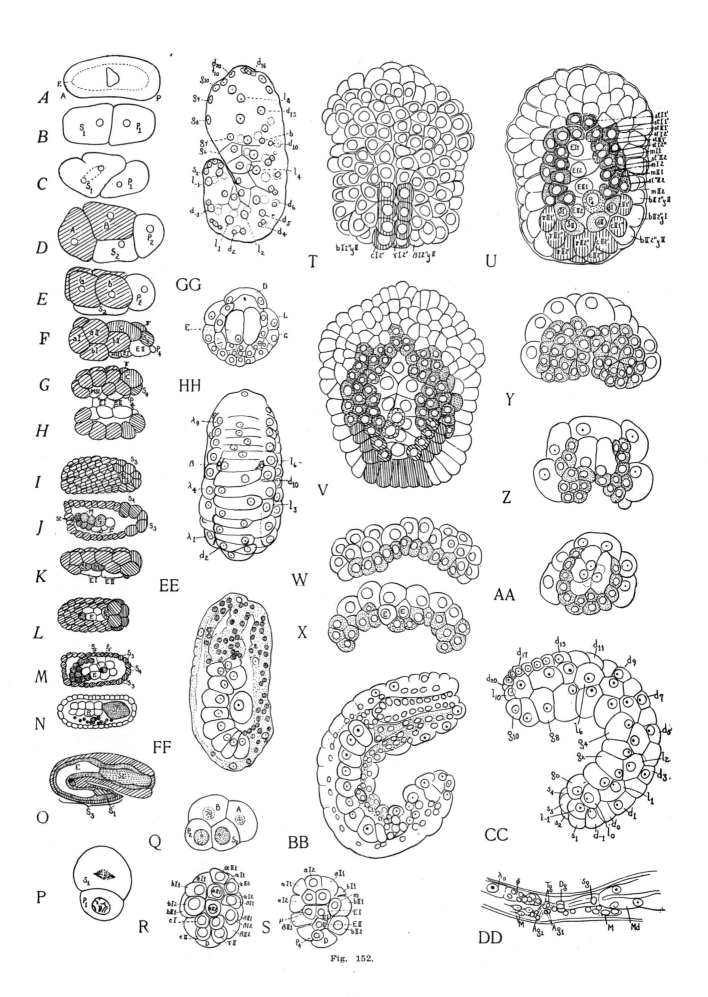

Fig. 152.

Pseudalius minor do not have a blastocoele. Two layered cell plates such as occur in the latter instance were at one time considered a type of gastrula, sterrogastrula, but since this stage is followed by epiboly characteristic of gastrulation it must be considered a type of blastula for which the term *placula* has been used.

Gastrulation (Figs. 153 L-Q), on the other hand, is the entrance of the entoderm and mesoderm into a hull surrounded by ectoderm, this being completed at the closure of the blastopore. Dependent upon the presence and size or the absence of a blastocoele there are two possible ways in which the ectoderm may come to surround entoderm (Martini, 1908); (1) the cells may retain their relative positions (synectic) or they may not retain their relative positions (apolytic). In the absence of a blastocoele the ectoderm may grow over the entoderm either with or without change in cell positions so we may have epibolic synectic or epibolic apolytic gastrulation. Invagination, that is embolic gastrulation is possible only if there is a blastocoele and in this case it may be either apolytic or synectic· Epibolic apolytic gastrulation is not known among nematodes but the other possibilities are represented. *Parascaris equorum* undergoes embolic apolytic gastrulation, *Rhabdias* and *Nematoxys* embolic synectic, and *Camallanus* and *Pseudalius* epibolic synectic (Fig. 153 P).

Regarding the development of the mesoderm in nematodes there are certain points of interest. The mesodermal stem cells at the time of gastrulation are arranged in rows on either side of the entoderm as well as anterior to it. They follow the entoderm in sinking into the primary body cavity and later form two subdorsal and two subventral strings on either side of the entoderm. Since the individual cells maintain their identity and no cavity is formed between them it would not be proper to say that nematodes had a true coelome. The individual cells would properly be termed a mesenchyme. However, certain mesodermal cells do cover the organs in the body cavity and for that reason we may say the body cavity is analogous but not homologous to a coelome, i. e., a pseudocoelome. The mesoderm may be said to have been derived from the entoderm since it comes chiefly from a cell (*S2*) which also forms the entoderm; but in *Parascaris equorum* as well as some other nematodes part of the derivatives of the mesodermal stem cell (*MSt*) enter into the formation of the ectoderm according to some authors. Strictly speaking it would seem preferable to consider six germ layers, the ectoderm, mesoderm (somatic musculature and isolation tissue), entoderm, esophagus (*St-S1*), somatic part of gonad (*S5*) and germinal layer (*P5*).

The *S1* cell group in nematodes forms the greater part of the epithelium which is usually so arranged that the nuclei and the cell bodies of the cells are situated in the dorsal, ventral, and lateral chords. It also contributes to the formation of the esophagus, the nervous system, and the excretory system.

The *S2* cell group forms the greater part of the mesoderm, that is the longitudinal muscles, transverse muscles, and probably the isolation tissue, as well as the musculature of the esophagus; it also forms the mesenteron or intestine.

The *S3* cell group forms the ectodermal epithelium of the posterior part of the body and contributes to the formation of the nervous system and the musculature of that part of the body.

The *S4* cell group is known to form the rectum and rectal glands and may also form some muscular tissue.

The *S5* cell group according to Pai (1928) forms the outer covering of the gonads and the epithelium of the

gonoducts. If nematodes can be said to have the homologue of a true coelome it would be the lumen of the gonoducts since it is formed as a cavity between cells, but the positional relationship of the S5 group with the entoderm makes its consideration as **mesoderm rather** questionable.

At the time of hatching from the egg, nematodes are fully formed, the tissues differentiated and functional with the exception of the reproductive system. In some nematodes no further division of cells takes place except in the gonads and structures either directly or indirectly connected with reproduction. In all instances known the chords are cellular rather than syncytial, there being five rows of cells in the anterior part of the body of most nematodes studied (a dorsal, two lateral, and two ventral) while in the remainder of the body there are two dorsolateral, two lateral, two ventrolateral, and two ventral, the dorsal being absent but there is a thickening of the hypodermis in the dorsal region. Changes from this condition take place during later development or not at all.

The somatic musculature is platymyarian and meromyarian in the newly hatched larva but it may become coelomyarian and polymyarian later. In such an instance we have to recognize the division of functional muscle cells, certainly highly differentiated.

The esophagus of the larva is similar to that of the adult in some forms; whether or not multiplication of cells may later take place is unknown.

The intestine of the larva is composed usually of 2 rows of cells, the lumen being formed by a separation of parallel rows of cells rather than from the archenteron. This formation of the lumen seems to be of no general significance since it is the only means by which a lumen could develop in forms with so little blastocoele.

The nervous system, at least in some forms, appears to be of the same number of cells in the larva as in the adult, with the exception of cells innervating genital papillae.

The excretory system of the larva is the one system of which our knowledge is entirely inadequate. It is usually stated to be formed by a single cell of the ectoderm (*S1*) but its development has not been satisfactorily traced.

Regarding the embryology of particular nematodes, we find that thus far no member of the Aphasmidia has been studied though many members of the Phasmidia have. These belong to several diverse groups, *Rhabdias bufonis*, *Rhabditis terricola* (partially), *Diplogaster longicauda* (partially), and *Turbatrix aceti* among the Rhabditina; *Metastrongylus elongatus*, *Pseudalius inflexus*, and *P. minor* among the Strongylina; *Ascaris lumbricoides* (partially), *Toxocara canis* (partially), *Parascaris equorum*, *Nematoxys ornatus*, and *Syphacia obvelata* among the Ascaridina; and *Camallanus lacustris* of the Camallanina. Of these forms it appears best to limit our descriptions to *Turbatrix aceti*, *Rhabdias bufonis*, *Parascaris equorum*, and *Camallanus lacustris*, comparing other forms with them whenever it appears advisable.

Turbatrix aceti (Fig. 152). Pai (1927) worked out the development of this species in a very complete manner. He found that the end of the ovum at which the sperm entered is destined to form the anterior end of the embryo. From the first cleavage which is transverse, two very slightly unequal cells result, a larger anterior one designated as *S1*, and a smaller posterior one, *P1*. By the second cleavage *S1*, dividing somewhat horizontally and obliquely, gives rise to a dorsal cell *B* and an anterior cell *A*. Division of *P1* follows shortly, a ventral cell *S2* and a posterior cell, *P2* resulting. The second somatic stem cell, *S2*, is said to form the esophagus, the intestine or mesenteron, and the mesoderm, and for that reason may be designated *EMSt*, i. e., entoderm, mesoderm, stomodeal stem cell. At the third cleavage the *S1* cells, *A* and *B*, divide longitudinally giving the embryo a distinct bilateral symmetry which it retains throughout the remainder of its development. This is followed by a transverse division of the ventral cell, *S2*, forming an anterior ventral cell *MSt* and a posterior ventral cell *E*. Thus in the seven cell stage there are four dorsal ectodermal cells, two ventral ento-mesodermal cells derived from the second somatic stem cell, and a single large posterior undifferentiated cell, *P2*. In later divisions the blastomeres procede at an even more unequal rate, there being a distinct tendency for the *S2*

Fig. 152.

A-O—*Turbatrix aceti* (A—Fertilized ovum; B—2 cell stage; C—Beginning second cleavage; D—4 cell stage; E—6 cell; F—16 cell; G-H—24 cell; I-J—141 cell; K—26 cell; L—82 cell; M—171 cell; N—Early definitive embryo; O—Tadpole stage) P-DD—*Camallanus lacustris* (P—2 cell stage; Q—4 cell; R—28 cell, dorsal view; S—28 cell, ventral view; T—177 cell, dorsal view; U—177 cell, ventral view; V—354 cell, ventral view; W-X—Cross sections of embryo slightly older than V, W—anterior region, X—Posterior region; Y-AA—Cross sections of anterior, mid and posterior regions of still older embryo; BB-CC—Sagittal and surface views of early definitive embryo; DD—Anal region of mature larva); EE-HH—*Rhabdias bufonis* (EE-FF—Surface and sagittal views of early definitive embryo; GG—Surface view of tadpole stage; HH—Cross section of stage shown· in EE and FF).

A-O, After Pai, 1928, Ztschr. Wiss. Zool., v. 131 (2); P-AA, After Martini, 1903, Idem., v. 74 (4); BB-DD, After Martini, 1906, Idem., v. 81 (4); EE-HH, After Martini, 1907, v. 86 (1).

and *P* series to lag behind the other cells. The *S1* cell group again divides, A and B on the left side, *a* and *b* on the right side, each giving rise through an oblique division to two cells, making four on each side of the body. At about the same time *P2* undergoes an unequal division, the larger daughter cell (the third somatic stem cell, *S3*) being dorsal, and the small germ cell (P3) being ventral; the ventral cells *MSt* and *E* both divide, *MSt* longitudinally giving rise to MST (left) and *mst* (right), and *E* dividing transversely gives rise to *E I* (anterior) and *E II* (posterior). The posterior dorsal cell, *S3* like *S1* is an ectodermal stem cell. It divides longitudinally forming C (left) and *c* (right); these cells produce the hypodermis of the posterior part of the embryo. The germ cell *P3* again undergoes unequal transverse division, forming a larger dorsal cell, *S4*, and a smaller ventral cell, *P4*.

These cleavages, four in number, bring the embryo to the 16 cell stage at which time all of the somatic stem cells have been formed (Fig. 152 F). The embryo is an elongate blastula with an inconspicuous blastocoele. Cells destined to form the ectoderm cover the anterior and dorsal surfaces of the embryo, cells destined to form the mesoderm and entoderm cover the posterior ventral surfaces. In later cleavages it is somewhat easier to follow the fate of the cells of each stem line separately.

At the 24 cell stage *S1* consists of 16 cells (Fig. 152 G-H): six on the left side of the embryo, four in the center, and six on the right side. The origin of the mediodorsal row of four cells has not been determined. Between this stage and the 141 cell stage (Fig. 152 I-J) gastrulation is completed; descendants of *S1* come to cover the anterior two-thirds of the embryo, the cells small and numbering 115. The remainder of the embryo is covered by descendants of *S3*. In further development *S1* comes to make up four-fifths of the hypodermis, gives rise to the nervous system and in postembryonic development to the vulva (vagina).

Returning to the *S2* line of somatic cells, the four cells present in the 16 cell stage, MST, *mst*, *E I* and *E II*, proceed at unequal rates. At the 26 cell stage there are four cells derived from *MSt*, namely, ST, *st*, M, and *m*, formed by transverse division of the bilaterally symmetrical cells MST and *mst*. By the 82 cell stage the six *S2* cells form a total of ten, in paired bilateral rows of 5 cells as follows: ST, *st*, M I, *m I*, M II, *m II*, E I, *e I*, E II, *e II*. At this time the entire *S2* cell group is somewhat sunken inward, making a ventral groove which is completely covered by ectoderm at the 141 cell stage, and gastrulation is completed. By the 171 cell stage there are four cells derived from ST and *st*, eight formed from M, and seven from E. Shortly thereafter there are eight entodermal cells. The *M* line (mesoderm) lies in the body cavity. By the time the embryo takes a definite vermiform shape (Fig. 152 O) there are 12 entodermal cells (*E*).

At the termination of gastrulation or the 141 cell stage, the third somatic stem cell, *S3*, consists of 11 cells. The first stage larva, at hatching, contains 15 cells of *S3* origin which form approximately one-fifth of the hypodermis, for they cover dorsal-posterior, and postanal parts of the body.

The fourth line of somatic stem cells, *S4*, which originated at the fourth cleavage or 16 cell stage consists of four cells at the termination of gastrulation. It forms the tertiary ectoderm (*Ec III*) which gives rise to the proctodeum or rectum.

The germ cell line represented by *P4* passes a quiescent stage during gastrulation. A cleavage takes place shortly before the 171 cell stage forming two cells, one of which (P5) produces the reproductive cells, while the other (S5) produces the somatic part of the reproductive system. Shortly before hatching both P5 and S5 divide, forming *S5 I* anterior, *S5 II* posterior, G I and G II. The *S5* cells surround the *P5* cells and are generally termed "terminal" or "cap" cells. In the male the entire vas deferens and seminal vesicle are formed later on by *S5 I*, while *S5 II* forms the epithelium of the testis. In the female *S5 I* forms the somatic part of the ovary while *S5 II* forms the oviduct, uterus, and seminal receptacle.

The entire development from fertilization to formation of the larva within the egg requires two days. The larva is "born" three days later. At this time the larva

possesses the same number of cells as the adult in all systems except the hypodermis and reproductive system. The female is mature on the 6th to 7th, the male on the 9th day after birth; specimens of both sexes may live 49 and 48 days respectively.

The gastrulation being somewhat atypical, there is difference of opinion as to the names of germ layers to be applied to the various somatic stem cells. Pai regards the *St* cell group which later forms the esophagus as secondary entoderm but since it is comparable to the *M* cell group it is better termed mesoderm. The *S5* group, forming the somatic part of the reproductive system, he also terms entoderm, though mesoderm would appear preferable. Study beyond the vermiform or "tadpole" stage (Fig. 152 O) is difficult and thus far has been carried out only by means of totomount preparations which lends uncertainty as to the results. The following is a catalogue of the cells of the adult.

Table 8. Derivation of cells in *Turbatrix aceti*.

Stem cell	Structure	Number of cells
S1	4/5 of ectoderm	?
	Nervous system.	
	Dorsal anterior to nerve ring	23
	" posterior to nerve ring	9
	" above bulb	5
	" subdorsal cephalic ganglia 2 @ 6	12
	Lateral ganglia 2 @ 28	56
	Ventral subventral ganglia posterior to nerve ring	38
	" retrovesicular ganglion	17
	" anterior to nerve ring	13
	" subventral cephalic ganglia 2 @ 7	14
	" nerve	64
	Excretory cell	1
S2	(EMSt)	
	Esophagus	
	Corpus	35
	Isthmus	0
	Bulb	24
	Esophago-intestinal valve	5
	Intestine	18
	Musculature	64
	Connective tissue	16
S3	(Secondary ectoderm)	
	Hypodermis of dorsal and postanal regions about	5
S4	(Tertiary ectoderm)	
	Rectum	20
S5	(Only partially determined)	
	Seminal vesicle	24
	Ejaculatory duct	8
	Other structures	?

Rhabdias bufonis. The embryology of this species was described by Metschnikoff (1865), Goette (1882), Neuhaus (1903), Ziegler (1895), and Martini (1907). Of these studies those of Metschnikoff were rather casual. Goette committed an unfortunate error in incorrectly orienting the early stages of the embryo, the anterior end being considered the posterior and vice-versa.

The first cleavage of this species differs from that of *Turbatrix* because it is more nearly transverse, due to the difference in shape of the egg. Ziegler traced the embryology partially through the eighth cleavage. In most respects his results correspond to those obtained by Pai. However, there are some differences. Martini has given more exact data regarding the late embryology than are known in the case of *Turbatrix aceti*.

During the early stages, the embryology of *Rhabdias* is nearly identical with that of *Turbatrix*. The following differences have been noted: *P2* divides (third cleavage) before the fourth cleavage begins in the *S1* group (Fig. 151); *S3* divides longitudinally instead of transversely at the fourth cleavage forming C I and C II. At the fifth cleavage the embryo consists of a 30 cell blastula (16 *S1* cells, 2 *St* cells, 2 *M* cells, 4 *E* cells, 4 *S3*, *S4* and *P4* cells. The fourth somatic stem cell does not divide until after the sixth cleavage has taken place in the *S1* group forming 32 primary ectodermal cells. At this time *S4* divides longitudinally forming D and *d*. At the seventh cleavage, the *S1* line totals 64 cells, the *M* line 8, the *St* line 4, the *E* line 4, *C* line 8, *D* line 4 and *P* line (*S5* plus *P5* or *P4 I* and *P4 II*) 2 cells, giving a total of 94 cells. Apparently the eighth cleavage is limited to the *S1* group at this time. Subsequently a ninth cleavage and at least a partial tenth cleavage takes place.

Before becoming elongated the *S1* group probably is composed of 248 cells. At this time there is a rest from cleavage in the *S1* group during which the other groups (*C*, *D* and *E*) evidently pass through at least some of the cleavages which they have missed, for the mature larva ready to hatch is composed of between 400 and 500 cells (Martini, 1907).

The posterior extremity of the embryo (Fig. 152 EE-HH) begins to bend ventrally and anteriorly at which time the esophagus is well formed, the intestine composed of two rows of seven cells; there are lateral mesodermal chords and a pair of ventral mesodermal chords; the genital primordium is ventral to the intestine and the two cells lie beside one another. The ectodermal cells forming the dorsal and anterior parts of the embryo are larger than the others.

During this period of elongation (Fig. 153 D-I & RR) several changes take place. The first two intestinal cells divide longitudinally forming a lumen surrounded by four cells. At this time the genital primordium lies under the sixth and seventh entodermal cells. At hatching the intestine is composed of 20 cells, four at the base of the esophagus and two rows of eight cells behind them. The lumen is zigzag but it later becomes wavy. The cells are in two more or less dorsal and ventral rows. At this time according to Martini (1907) the gonad is situated between the twelfth and thirteenth intestinal cells and is composed of about 10 cells, (Fig. 153 D).

Further differentiation of the epithelium takes place during the same period. Whereas before elongation the embryo is surrounded by two subdorsal, two dorsolateral, and two lateral to ventrolateral rows of large cells and numerous small ventral cells, a distinct rearrangement now takes place. The subdorsal rows which are at first opposite come to be alternate (Fig. 152 EE). (It should be noted that subsequent lettering of cells has no correlation with the lettering used to refer to cells during the first seven cleavages). There is a gradual pushing of the subventral rows squeezing some of the ventral small cells into the body cavity. Slightly later the embryo takes a vermiform appearance commonly called the "tadpole stage" (Fig. 152 GG). By this time the dorsal row of cells is split; it extends from what is now the swollen region, corresponding roughly to the position of the nerve ring, to slightly anterior to the level of the anus. The embryo is left without nuclei in the dorsal line in this whole region. The lateral mesodermal chords (Fig. 153 G) become dorsosubmedian and form the submedian muscle fields. Some of the cells previously ventral form the subventral muscle fields. These mesodermal tissues pressing against the epithelial cells in the submedian areas cause the six rows of previously mentioned large ectodermal cells to be pressed laterally forming the lateral chords. In the stage shown in Fig. 152 GG there is an anterior mediodorsal row of seven cells (derived from *S3* group) which do not separate (*d 14-20*) but remain as the nuclei of the dorsal chord; the next posterior-most cell, *d 13*, goes to the right, *d 12* to the left and so on. They form the dorsolateral parts of the lateral chords. Two small epithelial cells, *b* and B are covered by *d 10*. Posterior to the twenty-second dorsal cell, *d-1* and *g o*, G *o*, *l-1*, and L-1, there are four unpaired cells. The cells destined to form the lateral cell rows of the lateral chords (*l 1, 2* etc. of Fig. 152 GG) number eleven on each side, *l 7-10* being in the cephalic region, *l 1-6* being in the mid region, and *l-1*, postanal. The cells destined to form the ventrolateral part of the lateral chords (*g 0-10*) also number eleven, *g 8-10* forming subventral rows in the cephalic region while the other cells are already in final position. The anus is posterior to *g 1* and G *1*. The greater part of the nerve cells and the cells forming the ventral chord come from the small cells on the ventral surface of the embryo.

At hatching the dorsal chord has a single row of nuclei confined to the cephalic region, in which region the lateral chords also have a single row of nuclei, the ventral, two rows. Posterior to the cephalic region the lateral chords have three rows of nuclei each, the ventral a questionable number.

Camallanus lacustris. (Fig. 152 & 153). The embryology of this form described under the name *Cucullanus elegans* has been studied by Bütschli (1875) and Martini (1903, 1906). There are several minor variations in the form of early cleavage from that seen in the previously studied forms and the development has been followed somewhat further.

The first cleavage forming *S1* and *P1* is very unequal (Fig. 152 P). The four-cell stage is rhomboid at its formation as also in *Rhabdias*. All of the following cleavages are characterized by smaller furrows than in previous forms and no blastocoele is developed. Subsequent cleavages (Fig. 152 Q-S) are similar to those in *Turbatrix aceti* and will be omitted up to the initiation of the ninth cleavage.

The ninth cleavage of the *S1* group forms 256 cells, ninth of the *C* group 32 cells, ninth of the *St* group 32, eighth *M* 16 cells, seventh of *D* 8 cells which, together with the 8 cells of the *E* group and 2 of the *P4* group, forms an embryo of 354 cells. This represents a gastrula (Fig. 152 V) the rim of which is formed by several rows made up of the *St* and *M* groups anteriad, and by the *D* and *C* groups laterad and posteriad.

Following the 354-cell stage the *St* cell group divides (10th cleavage) forming 64 cells, the *M* groups divide forming 32 cells (9th cleavage). The *E* group divides forming 16 cells (7th cleavage), the *C* cell groups divide in part forming about 48 cells (9th cleavage), the *D* cell groups divide to form 16 cells (8th cleavage), and the primordial germ cell *P4* divides. It is said that a part of the *S1* cell group may also divide but this is uncertain. The resulting embryo consists of approximately 486 cells.

Gastrulation occurs between the 354 and 486 cell stages. The dorsal surface is convex, the ventral surface concave; in the anterior part of the embryo the curve is most pronounced in the median line (Fig. 152 W) while in the posterior part it is most marked toward the edges (Fig. 152 X). This becomes more outspoken with age and is correlated with swelling of the dorsal ectodermal cell rows. At this time the dorsal surface of the embryo is covered by 6 longitudinal rows of large cells, derivatives of *S1* and *C* and the sides, anterior and posterior ends are covered by smaller cells derived from the same cell groups. (Fig. 152 Y-AA). Gradually these 6 dorsal cell rows come to cover the smaller ventral cell rows which themselves cover the *M*, *St*, and *E* cell groups. The gastrulation is thus through epiboly. The *St* and *M* cell groups are pushed in the groove becoming closed at the anterior end, and there are some cells of the *S 1* group which enter the inside of the embryo at the anterior end. At the stage represented in figure 152 Z, the posterior part of the ventral groove is open. Finally closure of the posterior part of the ventral groove takes place, the large dorsal cell rows coming to surround the small ventral cells of the *D* cell group and some of the *C* cell group. (Fig. 152 CC). At the same time the two most dorsal cell rows fuse so that the embryo is covered by 5 cell rows.

Organogenesis. It has not been possible to follow the history of individual cells during their rearrangement at the completion of gastrulation. Because of this, a new nomenclature is adopted to mark the shapes of further development. The embryo apparently does not increase in number of cells but the cells become differentiated into organs. The embryo which is elongate or sausage shaped is covered by a mediodorsal cell row (*d 1-20* and *s 1-4*), 2 lateral rows (*l* and L *-1-10*), and 2 subventral rows (*g* and G *0-10*). All of these large cell rows are probably derived from *C*. The "0" designates the position or level of the future proctodeum (Fig. 152 CC). The ventral and anterior small cells of the *S1* cell group, the ventral small cells of the *S3* cell group, the *E*, *St*, and *M* cell groups as well as the descendents of *P4* are all enclosed by the epithelium formed by the above mentioned cell rows. The anterior end of the embryo is covered by five cells, *d 20*, *l 10*, L 10, *g 10*, and G 10. The primordia of all organs are definitely recognizable.

Formation of the dorsal, ventral, and lateral chords takes place in the following manner. The mesodermal strands push against the outer cell layers in the four

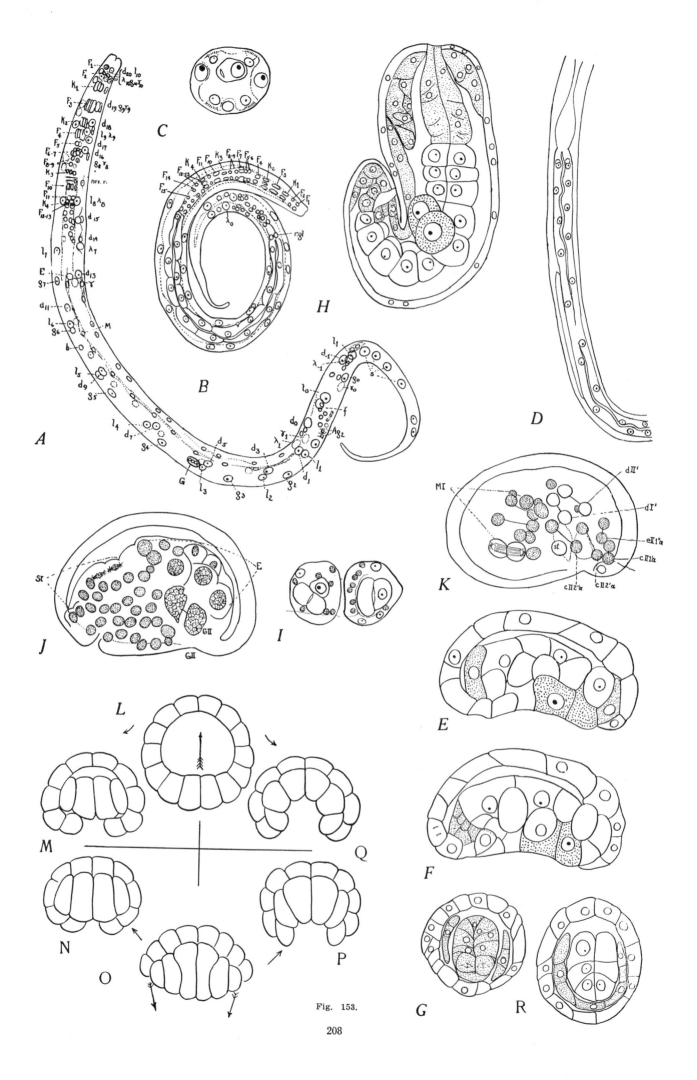

Fig. 153.

submedian regions. In the anterior part of the body this causes the *l* and L cell groups to be incompletely separated from the *d* cell row dorsally and the *g* and G cell rows ventrally. The result is that there are four chords, the dorsal and two lateral chords consisting of one cell row each, and the ventral chord of two cell rows. This takes place in the section of the embryo covered by *d 13-20*, *l* and L *8-10*, *g* and G *8-10*. In the remainder of the embryo the primordia of the muscles press and separate the alternate cells of the dorsal cell row and the cells of the two ventral cell rows. This causes the formation of two lateral rows of three cells each, the dorsolateral cell rows being formed by *d* and D cells, the ventrolateral by *g* and G cells, the result being two lateral chords of three cell rows. There remains a thickening of the mediodorsal part of the epidermis which is free of nuclei, the dorsal chord, and a ventral thickening which contains the small *S1* and *S3* cells forming the ventral chord with its ganglia.

The mesoderm giving rise to the subdorsal and subventral muscle bands is derived chiefly from cells of the *M* group but posterior cells of the *St* group also contribute. As they push out between the covering epithelial cells they become completely differentiated, and form overlapping double rows of platymyarian muscles in each sector.

Immediately after the closure of the ventral groove we find a nearly solid mass of cells anterior to the intestine. This is the primordium of the esophagus (Fig. 152 BB). It has apparently arisen from two cell groups, the *St* and small cells of the *S1* group. The lumen of the esophagus has in its origin no connection whatever with the ventral groove. The cells enter the body cavity as a mass, becoming arranged in a triradiate pattern. The lumen is formed by separation of the cells. Already in the stage represented above, the various cells may be recognized which are later present in the adult. The nuclei are very closely placed behind one another, there being a total of 66.

As the embryo becomes more elongated the nuclei are separated and at hatching come to form an esophagus consisting of an anterior part containing two groups of three marginal nuclei, two groups of six radial nuclei (Fig. 153 A); and a posterior part containing two groups of·three marginal nuclei, two groups of six radial nuclei, two groups of two subventral radial nuclei, six groups of three radial nuclei (Fig. 153 A). The same number of nuclei was observed in the adult stage. Part of the radial nuclei are probably nuclei of the esophageal glands and part nuclei of the esophago-sympathetic nervous system. Martini considers that the gland cells, marginal cells, and nerve cells of the esophagus are derived from small cells of the *S1* cell group while the radial muscle cells are derived from the *St* cell group.

The esophago-intestinal valve is formed from the same general tissues as the esophagus. In the early postgastrular stage (Fig. 152 BB) five nuclei may be seen between the esophagus and the intestine; at hatching (Fig. 153 A) these five nuclei comprise a large dorsal nucleus, two subdorsal, one left lateral, and one ventral.

The intestine in the early postgastrular stage (Fig. 152 BB) consists of 2 lateral rows of 8 large cells derived from *E*. With elongation there is a slight torsion of these cells and the two rows separate in the middle forming an irregular zigzag lumen surrounded by a dorsal and a ventral cell row.

The rectum, in so far as known, is derived from the S4 cell group, this group being entirely enclosed at gastrulation. The proctodeum is formed (Fig. 152 DD) through the separation of cells in this region. A group of 11 small cells lies between the posterior end of the intestine and the ventral side of the body. As in the case of the esophagus the nuclei later separate through elongation of the organism. Four cells surround the proctodeum at its junction with the body wall (*AG1* and *AG2*) two being dorsal and two ventral; two lateral cells are anterior to these (*Tg*); a group of three large cells one dorsal and two ventral (*Dg*) lies anterior to these; and there are two additional cells, one dorsal and one ventral, connecting the intestine and rectum. No increase in number of cells takes place in later development.

Soon after the completion of gastrulation the genital primordium is recognizable as four cells, two of which (the terminal cells) cover the other two (the primordial germ cells). Martini considers the terminal cells as probably originating from the M cell group. It seems more probable, in the light of Pai's observations on *Turbatrix aceti* (See p. 220), that the anterior cell resulting from the fifth cleavage in the *P* cell line (so called *P4 I* or *S5*) formed this layer. In case Pai is correct, the two primordial germ cells present at hatching resulted from the sixth cleavage of the *P* stem cell.

Regarding the development of the nervous system little is known except that it may form the small cells of the S1 and S3 cell groups. Nothing whatsoever is known regarding the origin of the excretory system.

Parascaris equorum. (Fig. 154). The embryology of the horse ascarid usually called *Ascaris megalocephala*, has been worked on by Boveri (1892, 1899, 1909, 1910 a, b), zur Strassen (1896, 1899 a, b), Müller (1903), Zoja (1896), Bonfig, (1925), Girgoloff (1911), Hogue (1911), Schleip (1924) and Stevens (1925) and in most of the results there is entire agreement. The lineage has been followed up to the 802 cell stage by Müller at which stage the embryo is completely developed and somewhat elongate, but has not reached the first larval stage. The large number of cells and the difficulty of following postembryonic stages, due to the life history of the species, makes it impractical to follow the differentiation of particular tissues.

In the general embryology *Parascaris equorum* is nearly identical with *Turbatrix aceti* but Boveri's beautifully illustrated work (1899) shows that chromatin material is lost from the nuclei during the division of somatic stem cells, a fact which indicates a very definite cytological basis for the unequal potentiality of the embryonic blastomeres. Chromatin diminution is not known in other groups of nemas though the same differentiations in potentialities are present.

The cleavage pattern of *Parascaris equorum* (Fig. 154) is identical with that of the previously described species. At the 56 cell stage (sixth cleavage) the cells are as follows: 32 of the *S1* group, 4 *St*, 4 *M*, 4 *E*, 8 *C*, 2 *D* and 2 *P*. Thereafter all of the cells except the C and P lines divide (seventh cleavage) forming a 102 cell stage at which time there is a well formed gastrula (Fig. 154 AA-EE) the anterior lip of which is bordered by 8 stomodeal cells (*St*) while the posterior lip is bordered by 4 proctodeal cells (*D*). During this division some of the *S1* cells divide unequally and to those which have been more carefully traced in subsequent divisions Mueller (1903) gave a simplified terminology, *g* and G corresponding to pairs of cells as the fifth cleavage (such as *A Il'*) **further divisions forming *ga*, *gb*, then *gar*, *gal*, *gbr*, and *gbl*; others were similarly renamed *oyr*, *uyr*, *oyl*, *uyl*, *xrl*, *kbr*, etc. These cells contribute a large part of the

---

Fig. 153.

A-CC—*Camallanus lacustris* (A—Mature larva, showing various nuclei; B—Slightly younger larva showing digestive tract; F, flat nuclei, i. e., between esophageal radii; K, corner nuclei. i. e., opposite esophageal radii; E, Intestinal nuclei; d, subdorsal lambda & l, lateral hypodermal nuclei, g & gamma subventral hypodermal nuclei; Ag, last cell pair of rectum. C—Cross section of larva at stage shown in A); D-I and R—*Rhabdias bufonis* (D—Intestinal region of larva; E-F—Tangential and sagittal sections of embryo with wide open blastopore; G-H and R—Tadpole stage cross sections of anterior and posterior regions, H, sagittal; I—Cross section of nearly mature larva showing intestine and primordial germ cell); J-K—*Parascaris equorum* (Sagittal and tangential views of 102-202 cell stage); L-Q—Methods of gastrulation (L—Coeloblastula; M—Epibolicsynectic gastrulation as in *Ascaris* and *Parascaris*; N—Epibolicapolytic gastrulation, unknown in Nematoda; O—Placula or sterroblastula; P—Epibolic-synectic gastrulation as in *Camallanus* and *Pseudalius*; Q—Embolic-synectic gastrulation as in *Rhabdias* and *Nematoxys*; Those above the horizontal line are embolic, those below epibolic while those on the readers' left are apolytic and on the right synectic).

A-C, After Martini, 1906, Ztschr. Wiss. Zool., v. 81 (4); D & I, After Martini, 1907, Idem. v. 86 (1); L-Q, After Martini, 1908, Idem. v. 91 (2); E-G & R, After Neuhaus, 1903, Jena Ztschr. Naturw., v. 37, n. F. v. 30 (4); J-K, After H. Mueller, 1903, Zoologica (41).

**The student who desires to trace individual cell lines will find a very complicated terminology, especially since zur Strassen used the system *I* for the first stem cell, *A* for the first divison thereof, *r* for right, *l* for left etc., so that *IarlBby* corresponds to *gar* in the new nomenclature. In his later work, Martini used a simplification which involved some of the same letters as those used by Mueller but for different cells. See *Camallanus lacustris*.

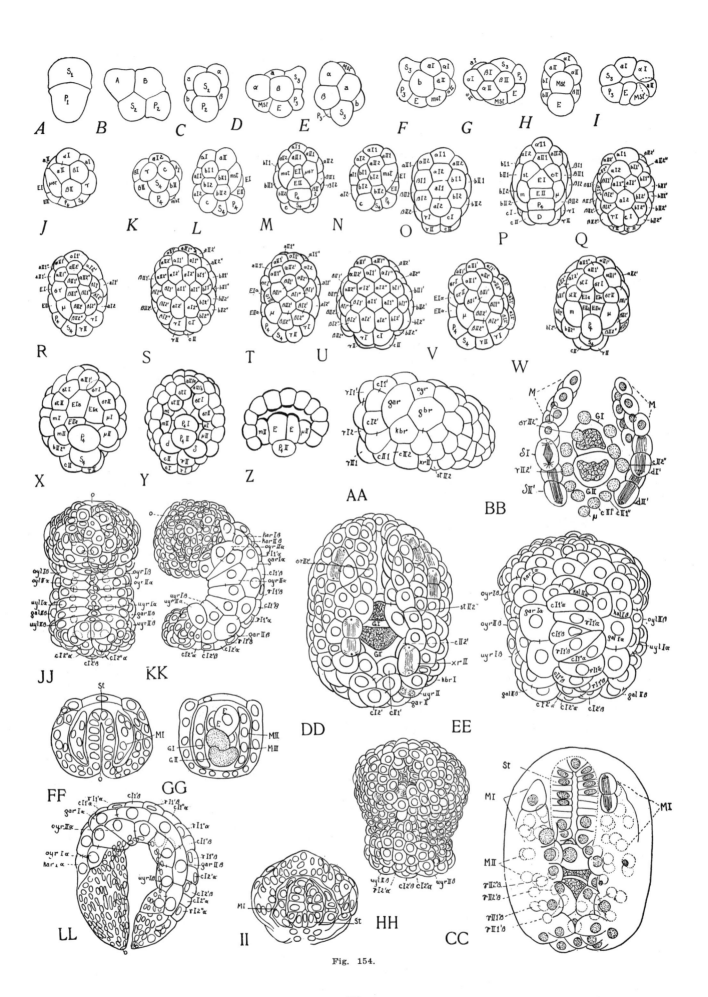

Fig. 154.

final body surface (Fig. 154 JJ-KK). By the end of the seventh cleavage the eight large E cells nearly completely fill the blastocoele.

At the eighth cleavage all cell groups with the exception of P4 divide so that a 202 cell embryo is formed. Its anterior surface is covered by small cells of the $S1$ group while much of the posterior part of the embryo is covered by the larger k, oy, uy, g and x cell groups (Fig. 154 AA & EE). Cells of the C line ($c~II'$ c II' etc.) form paired posterior subdorsal rows of cells while those of the D line ($d~II'$ D II') enter the ventral groove. The St cell group now consists of 16 cells, some of which extend as far posterior as the genital primordium (Fig. 154 BB) while the remainder form an anterior groove in continuation with the primary ectoderm ($S1$). The M cell group consists of two irregular lateral groups of eight cells each, entirely enclosed as is the 16 cell E group by $S1$ and $S3$ cell groups. The gastrular cavity is sharply V-shaped anteriad, lined by cells of the St group while it is U-shaped posteriad, the large genital primordium cells ($P4~I$ and II') forming the ventral surface (Fig. 154 CC). Parts of the $S3$ cell group ($c~II$ and c II) are definitely mesodermal.

At the ninth cleavage the embryo is 402-celled, being composed as follows: $S1$ cell group 256, St group 32, M group 32, E group 32, $c~I$ and c I (Secondary ectoderm) group together 16, $c~II$ and c II (Tertiary mesoderm) group together 16, $S4$ (D) group 16, and $P4$ (G) group 2. The two subdorsal surface cell rows (Fig. 154 FF-GG) are formed from $c~I$ and c I; two large lateral mesodermal bands are formed from M, $c~II$ and c II and part of St.

The anterior part of the ventral groove has completely closed, the esophageal primordium forming a solid cell mass in contact with the ventral and anterior small cells of the primary ectoderm. A terminal cavity, the stomodeum is then formed in the esophageal primordium. The most posterior St cells (Fig. 154 BB) do not become a part of the esophageal primordium; though at this stage the maximum number of St cells should be 32 and though not all of them enter into the formation of the esophagus, there are about twice that number in the primordium. The cells not accounted for were probably small $S1$ cells which entered the primordium during formation of the stomodeum.

During the latter part of the ninth cleavage the dorsal cells of the C group come to form a single row of 10 very large cells covering the dorsal and posterior surfaces of the embryo, this being accomplished through median movement of alternate cells (Fig. 154 JJ-KK). Anteriorly this row is continued by the cells designated $kar~II~B$, $kal~II~A$, their sister cells being lateral to them. At the sides of this dorsal cell row there are 2 large subdorsal cell rows formed from the gar, gal, kal, kar, cell groups, and at the sides of these, 2 lateral large cell rows formed from the oy and uy cell groups. These large cells are of particular significance for they swell in size and then cover most of the posterior and ventral cells of the $S1$ group, thus forming the epithelium of much of the body.

At this time the embryo begins to elongate definitely, and becomes ventrally curved, this probably being due to swelling of the 5 large cell rows. The lateral cell rows nearly come together, ventrally forcing some of the small superficial cells anteriad. This is considered the completion of gastrulation. The anterior end of the embryo and the ventral surface are covered by small cells of the $S1$ group. We now find the mediodorsal and posterior medioventral parts of the embryo covered by cells derived from $S3$ ($c~I$ and c I); the sides by cells of the $S1$ cell group (oy, uy, kar, kal, gar, gal, uy, and oy); and the

anterior and ventral part of the body by $S1$. A further division at least of the large surface cells takes place after elongation of the embryo into definite vermiform shape.

Other species.—The embryology of *Rhabditis terricola*, *Diplogaster longicauda*, and *Nematoxys ornatus* is, so far as known, similar to that of *Rhabdias bufonis*. In *Pseudalius minor* no blastocoele is developed, the embryology being very similar to that of *Camallanus lacustris*. In the case of *Syphacia obvelata* (*Oxyuris obvelata*) early cleavage is somewhat modified through the elongate "banana" form of the ovum. The first cleavage is extremely unequal, $S1$ being nearly twice as long as $P1$. This type of ovum is very common in oxyurids and thelastomatids. The first cleavage of *Metastrongylus elongatus* appears equal but must be unequal since $P1$ contains a large amount of yolk material while $S1$ does not. As in *Camallanus*, no blastocoele develops.

Abnormal development. Development is strongly determinate as would be indicated from the previous discussion. Sometimes variations occur in the early cleavages, particularly in *Parascaris*. Normal formation of rhomboid embryos in the four-cell stage is assured in most nematodes but in this form, due to the planes of the second cleavage, arrangement of the cells is observed which becomes rhomboid by passing through an]-shaped stage. Sometimes however by passing through an [-shaped stage the positions of the blastomeres are reversed, B being anterior to A. In such cases the entire development of the embryo is reversed; both A and B develop normally like B and A; the third somatic stem cell is formed at the opposite end of P; S2 divides normally and development proceeds to the blastula stage; development of Mst is probably influenced since gastrulation does not occur. Injury of P2 in the 4-cell stage does not stop further development of the $S1$ and $S2$ cells up to the blastula stage, which is abnormal; injury through loss of cytoplasm in the $S1$ cell at the two-cell stage does not stop further development of the $P1$ cell in a normal manner. The position of the spindle of the first cleavage may be changed through centrifuging or by multispermy; in either case the first cleavage may give rise to equipotential blastomeres which result in the formation of $P1$ and $S1$ cells, showing that the potentialities are dependent upon cytoplasmic material and that probably the occurrence of chromatin diminution in a blastomere is also dependent upon the cytoplasm. Separation of $P1$ and $S1$ in *Turbatrix aceti* results in the degeneration of $S1$ while $P1$ continues development to the 4-cell, to 16-cell or gastrular stage.

These observations appear to indicate that nemic embryos are essentially of mosaic structure, and that the unequal potentialities of the blastomeres are due to some differences in the cytoplasm but probably also to other factors such as influence from surrounding cells and differences in chromatin.

Bibliography

AUERBACH, L. 1874.—Organologische Studien. Zur Charakteristik und Lebensgeschichte der Zellkerne. 262 pp., pls. 1-4. Breslau.

BONFIG, R. 1925.—Die Determination der Hauptrichtungen des Embryos von *Ascaris megalocephala*. Ztschr. Wiss. Zool., v. 124 (3-4): 407-456, figs. 1-25.

BOVERI, T. 1893.—Ueber die Entstehung des Gegensatzes zwischen den Geschlechtszellen und den somatischen Zellen bei *Ascaris megalocephala*. Sitz. Gesellsch. Morph. & Physiol., v. 8 (2-3): 114-125, figs. 1-5.

1899.—Die Entwickelung von *Ascaris megalocephala* mit besonderer Rücksicht auf die Kernverhältnisse. Festschr. Kupffer, Jena: 383-430, figs. 1-6, pls. 40-45, figs. 1-45.

1909.—Die Blastomerenkerne von *Ascaris megalocephala* und die Theorie der Chromosomenindividualität. Arch. Zellforsch., v. 3 (1/2): 181-268, figs. 1-7, pls. 7-11, figs. 1-51.

1910a.—Ueber die Teilung centrifugierter Eier von *Ascaris megalocephala*. Festschr. W. Roux. Arch. Entwicklungsmech., v. 30 (2): 101-125, figs. 1-32.

1910b.—Die Potenzen der *Ascaris*-Blastomeren bei abgeänderter Furchung. Festschr. R. Hertwigs. v. 3: 131-214, figs. A-Y, pls. 11-16, figs. 1-39.

Fig. 154.

Parascaris equorum. A—2 cell stage; B—4 cell; C—6 cell; D—8 cell; E—8 cell; F—10 cell; G—12 cell, lateral view; H—12 cell, ventral view; I—12 cell, sagittal section; J—16 cell; K—16 cell; L—22 cell; M—24 cell; ventral view; N—24 cell, lateral view; O—26 cell, dorsal view; P—28 cell; Q—41 cell, dorsal view; R—41 cell, lateral view; S—44 cell dorsal view; T—48 cell, dorsal view; V—48 cell lateral view; W—54 cell, ventral view; X — 56 cell, ventral view; Y — 92 cell ventral view; Z — 92 cell, cross section; AA — 202 cell, lateral view (8th cleavage); BB 102-202 cell, horizontal section (in 8th cleavage); CC-EE—102-202 cell, ventral, ventral, and dorsal views FF-GG—202-402 cell, cross sections (in 9th cleavage); HH—402-802 cell, ventral view; II—Esophageal region of late embryo; JJ-KK—Ventral and lateral views of same; LL—Prelarval stage, surface view.

A-W, After zur Strassen, 1896, Arch. Entwickelungsmechanik., v. 3 (1-2); X-Z, After Boveri, 1892, Sitz. Gesellsch. Morph. & Physiol., v. 8; AA-KK, after H. Mueller, 1903, Zoologica (41).

BÜTSCHLI, O. 1875a.—Vorläufige Mittheilung über Untersuchungen betreffend die ersten Entwicklungsvorgänge im befruchteten Ei von Nematoden und Schnecken. Ztschr. Wiss. Zool., v. 25 (2): 201-213.

1875b. — Zur Entwicklungsgeschichte des *Cucullanus elegans* Zed. Ztschr. Wiss. Zool., v. 26 (7): 103-110, pl. 5, figs. 1-8.

CONTE, A. 1902. — Contributions a l'embryologie des nématodes. Ann. Univ. Lyon, n. s., I. Sc. Med. (8): 133 pp., figs. 1-137.

GALEB, O. 1878.—Recherches sur les entozoaires des insectes. Organisation et développement des oxyuridés. Arch. Zool. Expér. & Gén., v. 7 (2): 283-390, pls. 17-26.

GANIN, M. S. 1877a.—[Mittheilungen über die embryonale Entwickelung von *Pelodera teres*] (in Hoyer, H. Protocolle der Sitzungen der Section für Zoologie und vergleichende Anatomie der V. Versammlung russischer Naturforscher und Aerzte in Warschau im September 1876.) Ztschr. Wiss. Zool., v. 28 (3): 412-413.

1877b.—[Ueber die Untersuchungen von Natanson, betreffend die embryonale Entwickelung von drei Arten von Oxyuris]. Ztschr. Wiss. Zool., v. 28 (3): 413-415.

GIRGOLOFF, S. S. 1911. — Kompressionsversuche am befruchteten Ei von *Ascaris megalocephala*. Arch. Mikrosk. Anat., v. 76.

GOETTE, A. 1882.—Abhandlungen zur Entwickelungsgeschichte der Tiere. Erstes Heft. Untersuchungen zur Entwickelungsgeschichte der Würmer. Beschreibender Teil. 104. pp., 4 figs., pls. 1-6. Hamburg u. Leipzig.

HALLEZ, P. 1885.—Recherches zur l'embryologénie et sur les conditions du développement de quelques nématodes. 71 pp., 4 pls. Paris.

HOGUE, M. J. 1910.—Ueber die Wirchung der Centrifugalkraft auf die Eier von *Ascaris megalocephala*. Arch. Entwicklungsmech., v. 29 (1): 109-145, figs. 1-42.

JAMMES, L. 1894.—Recherches sur l'organisation et le développement des nématodes. These. 205 pp. Paris.

LIST, T. 1893.—Zur Entwicklungsgeschichte von *Pseudalius inflexus* Duj. Biol. Centrlbl. v. 13 (9/10): 312-313, 1 fig.

1894.—Beiträge zur Entwicklungsgeschichte der Nematoden. Diss. 32 pp. Jena.

MARTINI, E. 1903.—Ueber Furchung und Gastrulation bei *Cucullanus elegans* Zed. Ztschr. Wiss. Zool., v. 74 (4): 501-556, figs. 1-8, pls. 26-28, figs. 1-35.

1906.—Ueber Subcuticula and Seitenfelder einiger Nematoden. I. Ztschr. Wiss. Zool., v. 81 (4): 699-766, pls. 31-33, figs. 1-34.

1907.—Idem. II. Ibid. v. 86 (1): 1-54, pls. 1-3, figs. 1-82.

1908a.—Die Konstanz histologischer Elemente bei Nematoden nach Abschluss der Entwickelungsperiode. Anat. Anz., v. 32, Ergänz.-Heft: Verhandl. Anat. Gesellsch. 22 Vers: 132-134.

1908b.—Ueber Subcuticula etc. III. (Mit Bemerkungen über determinierte Entwicklung). Ztschr. Wiss. Zool., v. 91 (2): 191-235, 13 figs.

1909.—Ibid. Vergleichend histologische Teil IV. Tatsächliches Teil V. Zusammende und theoretische Betrachtungen. Ztschr. Wiss. Zool., v. 93 (4): 535-624, figs. z-uu, pls. 25-26, figs. 82-106.

1923.—Die Zellkonstanz und ihre Beziehungen zu andern zoologischen Vorwürfen. Ztschr. Anat. & Entwick. 1 Abt. v. 70 (1/3): 179-259.

MÜLLER, H. 1903.—Beitrag zur Embryonalentwickelung der *Ascaris megalocephala*. Zoologica Stuttg. Heft 41, v. 17: 1-30, figs. 1-12, pls. 1-5, figs. 1-24.

NATANSON, 1876.—Zur Entwickelungsgeschichte der Nematoden. Arb. der 5. Versamml. russ. Naturf. u. Aerzte. Warschau. Ztschr. Wiss. Zool., v. 28. see Ganin, 1877b.

NEUHAUS, C. 1903.—Die postembryonale Entwickelung der *Rhabditis nigrovenosa*. Jena. Ztschr. Naturw. v. 37, n. F. v. 30 (4): 653-690, 1 fig., pls. 30-32, figs. 1-40.

PAI, S. 1927.—Lebenzyklus der *Anguillula aceti*. Ehrbg. Zool. Anz. v. 74 (11-12): 257-270, figs. 1-12.

1928.—Die Phasen der Lebenscyclus der *Anguillula aceti*. Ehrbg. und ihre experimentell-mophologische Beeinflussung. Ztschr. Wiss. Zool., v. 131 (2): 293-344, figs. 1-80, tables 1-3.

SCHLEIP, W. 1924.—Die Herkunft der Polarität des Eies von *Ascaris megalocephala*. Arch. Mikrosk. Anat. & Entwicklungsmech. v. 100 (3/4): 573-598, figs. 1-17.

SPEMANN, H. 1895.—Zur Entwicklung des *Strongylus paradoxus*. Zool. Jahrb. Abt. Anat., v. 8 (3): 301-317, pls. 19-21, figs. 1-20.

STEVENS, N. M. 1909.—The effect of ultra-violet light upon the developing eggs of *Ascaris megalocephala*. Arch. Entwicklungsmech., v. 27 (4): 622-639, pls. 19-21, figs. 1-67.

STRASSEN, O. zur. 1892.—*Bradynema rigidum* v. Sieb. Ztschr. Wiss. Zool., v. 54 (4): 655-747, pls. 29-33, figs. 1-98.

1896. — Embryonalentwickelung der *Ascaris megalocephala*. Arch. Entwicklungsmech., v. 3 (1): 27-105, figs. 1-24, pls. 5-9, 49 figs. (2): 133-190, figs. 25-26.

1903.—Geschichte der T-Reisen von *Ascaris megalocephala*. Teil I. Zoologica Stuttg., v. 17, Heft 40: 1-37, figs. A-M pls. 1-5, figs. 1-67.

1906.—Die Geschichte der T-Reisen von *Ascaris megalocephala* als Grundlage zu einer Entwickelungsmechanik dieser Spezies [Continuation of 1903]. Zoologica, Stuttg. v. 17 (40): 39-342, figs. N-YYYY.

VOGEL, R. 1925.—Zur Kenntnis der Fortspflanzung, Eireifung Befruchtung und Furchung von *Oxyuris obvelata* Bremser. Zool. Jahrb. Abt. Allg. Zool. v. 42 (2): 243-271, figs. A-V, pl. 1, figs. 1-6.

WANDOLLECK, B. 1892.—Zur Embryonalentwicklung der *Strongylus paradoxus*. Arch. Naturg. 58 J., v. 1 (2): 123-148, pl. 9, figs 1-30.

ZIEGLER, H. E. 1895.—Untersuchungen über die ersten Entwicklungsvorgänge der Nematoden. Zugleich ein Beitrag zur Zellenlehre. Ztschr. Wiss. Zool., v. 60 (3): 351-410, pls 17-19.

ZOJA, R. 1896.—Untersuchungen über die Entwicklung der *Ascaris megalocephala*. Arch. Mikrosk. Anat., v. 47 (2): 218-260, pls. 13-14.

CHAPTER XV

POSTEMBRYONIC DEVELOPMENT

M. B. CHITWOOD

Except for size, reproductive organs and related structures the majority of nemas are fully developed at the time of hatching. The primitive nemas, having no increase in cell number in most organs, undergo no gross morphological changes. However some of the more highly specialized groups undergo changes in the character of the labial region, stoma and esophagus, as well as changes in internal structure.

There is no true metamorphosis in nemic development comparable to that occurring in insects since tissues are not destroyed and rebuilt. Changes in gross body form are, for the most part, of proportion rather than structure. *Heterodera* is the most striking example of modified body form. In this genus the first stage larvae are typical "eelworms" while the preadults are thickened and sac-like. The female continues enlargement, assuming a pear-shape in the adult stage (Fig. 115 N) while the male returns to the previous thread-like appearance (Fig. 163 N).

Most of the developmental changes are *palingenetic*, that is, features are derived from long evolution and concerned with adult existence. However, many *cenogenetic* features occur which are purely larval adaptive features, interpolated into development to aid the larva in coping with its own separate existence. Developmental changes may best be considered system by system.

CUTICLE. Ordinary postembryonic developmental changes of the cuticle are limited to the thickening of the cuticle and the development of such structures as caudal alae in the male. Many cenogenetic features appear after hatching, disappearing before or with the last moult, while the majority of palingenetic modifications appear with the last moult.

The caudal alae and papillae of the male develop in the fourth stage larva when the tissues of the body draw away from the old cuticle and the adult structures are formed beneath it (Figs. 156 S-T, AA & DD). The spines of adult *Hystrignathus* and cuticular plaques (Fig. 23, SSS p. 22) of adult *Gongylonema* (according to illustrations by Alicata, 1935) appear in the fourth stage larvae while the collarettes of *Spinitectus* (Fig. 23, p. 22), spines of *Hystrichis* and the large trifid lateral alae of *Physocephalus sexalatus*, according to Seurat, 1913, first appear at the last moult. Cuticular inflation around the head (Fig. 23 RRR) of trichostrongyloids (*Longistriata hassalli*) first appears in the pre-adult stage. Roberts (1934) reports that in the third stage larvae of *Ascaris lumbricoides* well developed lateral "membranes" are present, becoming very broad and fin-like in the fourth stage, diminishing in the fifth stage. As has been previously mentioned (p. 25), lateral alae are often much larger and wider in the larvae than in the adult, this being particularly true in the members of the Oxyuridae and Thelastomatidae. It is interesting to note that such wide alae are not present when the larva is pressed from the egg shell. Lateral alae probably function as "wings" or "immovable fins" to assist the larvae in locomotion and are no longer necessary once the nema has settled in a suitable place.

The remaining cuticular modifications appear to be purely cenogenetic. Wetzel (1931) described four large hooks placed in dorsolateral and ventrolateral positions in the fourth stage larvae of *Dermatoxys veligera*. These hooks were lost at the subsequent moult.

Change in size and shape of the tail is, perhaps, the most common post-embryonic phenomenon. In *Hystrignathus* the tail is distally bifurcate during the first three larval stages (Fig. 155) while in *Strongyloides* it is digitate only in the infective larvae. Buds appear on the tail of the third stage larvae of *Strongyloides* before its emergence from the cuticle of the second stage. Among the Strongyloidea the occurrence of a long thin filiform tail in the second stage larva causes the infective larva to

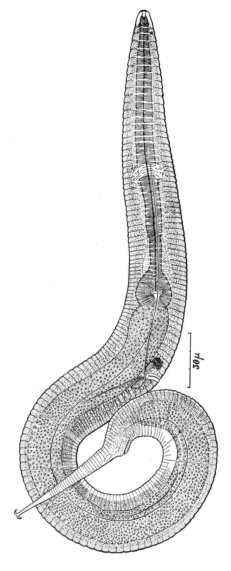

Fig. 155.

Hystrignathus? rigidus, larval female showing forked tail. **After** Christie, 1934, Proc. Helm. Soc. Wash., v. 1 (2).

have an even longer whip-like tail which is practically diagnostic for some groups (Fig. 99); the tail of the third stage larva itself usually is quite short even conoid. The tail of the second stage larva of trichostrongyles is shorter, conoid, attenuated with the tail of the third stage even more conoid inside it (*Trichostrongylus axei*, Fig. 158 D) or it may bear papilla-like digitations, (*Ornithostrongylus quadriradiatus* Fig. 158 U) which are subsequently lost while the pronged tail persists to the adult in *Ollulanus* and *Tricholeiperia*. In the Metastrongyloidea (Figs. 156 U-V & 158 T) peculiar and characteristic notching of the tail under the cuticle is evident in the first or second stage becoming very well marked in the third stage and disappearing with the third moult.

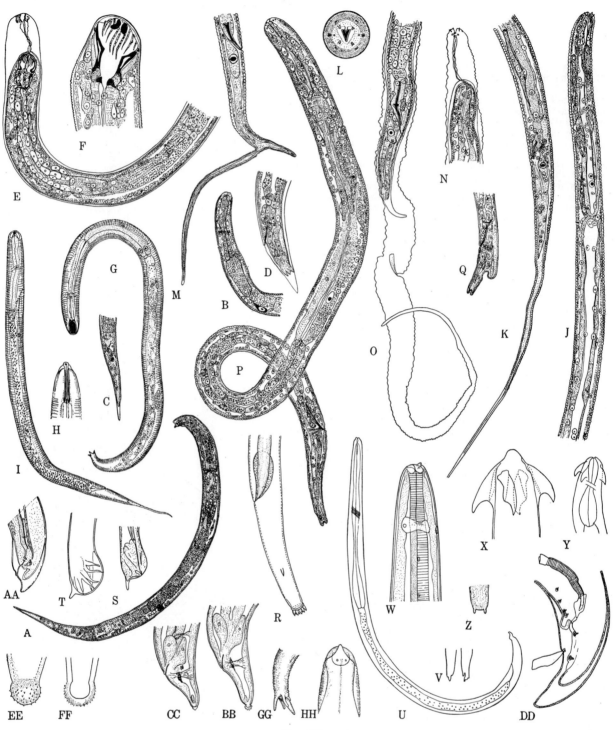

Fig. 156.

Postembryonic development. A-F—*Camallanus sweeti* (A—First stage larva; B—Anterior end, early second stage; C—Early second stage, posterior end; D-E—Larva undergoing second moult, posterior and anterior end, respectively; F—Third stage larva, anterior end). G-I—*Procamallanus fulvidraconis* (G—14 days old larva; H—Anterior end of larva, before first moult; I—Larva 6 days old). J-Q—*Dracunculus medinensis* (J—Anterior end of first stage larva; K—Posterior end of first stage larva; L—en face view of first stage larva; M—Posterior end of first stage larva; N—Cephalic region; O—Posterior end of normal larva undergoing first moult; P—Normal third stage larva; Q—Tail of abnormal third stage larva). R—*Tetrameres crami*, third stage larva. S-T—*Longistriata musculi* (S—Tail of preadult male; T—Tail of preadult male in final moult). U-V—*Aelurostrongylus abstrusus*, ? second stage larva. W—*Acanthocheilus rotundatus*, head ? third or fourth stage larva. X-Y—*Dermatoxys veligera*, head, fourth stage. Z—*Gongylonema pulchrum*, tail third stage larva. AA-BB—*Physocephalus sexalatus* (AA-Tail fourth stage male; BB-Tail of third stage larva). CC—*Ascarops strongylina*, tail third stage. DD—*Spirura gastrophila*, Tail male during fourth moult. EE-FF—*Seurocyrnea colini*, tail third stage. GG-HH—*Cheilospirura hamulosa*, third stage, tail and head. A-F, after Moorthy, 1938, J. Parasit., v. 24 (4). G-I, after Li, 1935, J. Parasit. v. 21 (2). J-Q, after Moorthy, 1938, Amer. J. Hyg. v. 27 (2). R, after Swales, 1936, Canad. J. Res. Sec. D., v. 14. S-T, after Schwartz & Alicata, 1935, J. Wash. Acad. Sc., v. 25 (3). U-V, after Cameron, 1927, J. Helminth., v. 5 (2). W, after Wuelker, in Wuelker and Stekhoven. 1933, Die Tierw. Nord-u. Ostsee, v. 5a. X, after Dikmans, 1931, Tr. Amer. Micr. Soc., v. 50 (4). Y, after Wetzel, 1931, J. Parasit. v. 18. Z-BB-CC, after Alicata, 1935, U. S. D. A. Tech. Bull. 489. AA, after Seurat, 1913, Compt. Rend. Soc. Biol., Paris, v. 75. DD, after Seurat, 1914, Ninth Congres Internat. Zool. EE-HH, after Cram, 1931, U. S. D. A. Tech. Bull. 227.

Among many representatives of the Spiruroidea and Filarioidea the first stage larva bears hooks and spines in the cephalic region. These same forms sometimes have an attenuated tail (*Gongylonema pulchrum*, *Ascarops strongylina*, *Physocephalus sexalatus* and *Dicheilonema rhea*). Both cephalic and caudal modifications are lost at the first moult. Since this is the stage of entry into the intermediate host these structures are probably used for boring through the tissues of the host. The occurrence of caudal specializations is characteristic of certain species of spiruroids. In the Thelaziidae, the tail is terminated by two or four small digitations in *Gongylonema pulchrum* (Fig. 156 Z) while in *Ascarops*, (156 CC) *Spirocerca* and *Physocephalus* (156 BB) it takes the form of a round knob, said knob being unarmed in *Ascarops strongylina* bearing a few spines in *Spirocerca lupi* and many rounded protuberances in *Physocephalus sexalatus*. According to Swales (1936) the tail of the third stage larva of *Tetrameres crami* (Fig. 156 R) is abruptly truncate bearing a circle of nine digitations of equal size and one subsequal median digitation; the tail of *Habronema muscae* has a rounded tip with many small spines while *Seurocyrnea colini* (Fig. 156 EE-FF) has a similar tail but the knob is larger and the spines relatively smaller. *Cheilospirura hamulosa* of the Acuariidae has a multi-pronged tail (Fig. 156 GG) in the third stage while in the same stage *Physaloptera turgida* has a bluntly conoid tail.

In the Camallanoidea and Dracunculoidea (Fig. 156) the first stage larva has a dorsal denticle (except *Micropleura*) on the head, *and an attenuated tail with large pocket-like phasmids. The dorsal denticle which is lost at the first moult, might be considered homologous to the hook present in the first stage *Gongylonema*. (Fig. 157 B-C). The tail of the third stage larva of *Dracunculus* has four large mucrones while that of *Camallanus* has three prongs, the tail of the adults being conoid and conically rounded respectively. However in *Procamallanus fulvidraconis* the three prongs persist to the adult.

The postembryonic changes in the tail of *Agamermis decaudata* Christie, (1936) are not limited to the cuticle. The division between the anterior and posterior portions of the body is marked by a node which is evident early in the preparasitic larva. The posterior part, about 2/5 of the entire length, becomes detached at the time of the entrance of the nema into the host.

Hypodermis. The discussion of postembryonic development in the hypodermis may be found on pp. 34-35 & 37. Briefly, it may be noted that in the more generalized nematodes little or no increase in number of cells or nuclei occurs after hatching while in more highly evolved forms many cell divisions or syncytial development may occur.

Teunissen (in Stekhoven, 1939) has found that the number of hypodermal glands in young individuals of *Anaplectus granulosus* is from 46 to 70 in a quadrant or a total of 184 to 280. The number in juvenile males varies from 204 to 312 while in young females they number 244 to 356. Before sex can be determined, young specimens can be divided into two groups, those with 50 to 60 and those with 60 to 70 glands per quadrant. Cells increase in number as the nema grows, the number being constant only in the adult. The number of glands in females of 600 to 2000 microns length varies from 60 to 85 and the number in the male of the same size varies from 50 to 75. One can be nearly certain that young specimens with less than 55 hypodermal glands in a row will develop into males and that those having more than 65 cells will develop into females. Cell constancy in the individual is not reached until it is 1200 to 1400 microns in length (or adult) since the number of hypodermal glands is definitely larger in sexually mature specimens than in larvae.

Musculature. see p. 219.

Nervous System. No changes known.

Labial region and stoma. The labial region of most nemas is not as distinctly set off from the remainder of the body at the time of hatching as it is in the adult stage, though there are some notable exceptions to this

rule. Usually the cephalic papillae in parasitic nematodes are relatively larger and better developed at the time of hatching than in the adult. In rhabditids the labial region is distinctly set off and the amphids are oval, somewhat further posterior, and relatively larger than in the adult stage (Fig. 158 W-Z). In strongylins three or six indistinct lips are usually present in the first stage larvae though lips may be totally absent in the adult (*Ancylostoma caninum*). Furthermore, the cephalic papillae show a much more generalized pattern at this time, the internal circle and dorsodorsal and ventroventral papillae of the external circle being better developed than in the adult stage in which these papillae are greatly reduced. In ascaridids several changes take place in the labial structures. Young ascaridid larvae, broken out of the eggs have (*Ascaris lumbricoides* vide Alicata, 1935) three small lips bearing the full component of separate and well developed papillae (Fig. 158 J) while the adult has large circumscribed lips with greatly reduced and partially fused papillae (Fig. 57 Y, p. 60). What

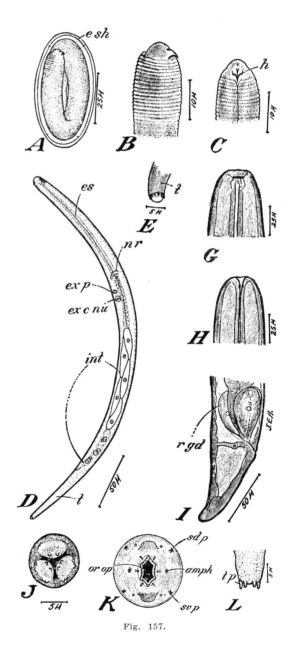

Fig. 157.

A-H—Developmental stages of *Gongylonema pulchrum*. (A—Fully developed larva in egg; B—First stage larvae, anterior end, lateral view; C—Same, ventral view; D—Larva, four days after experimental infection; E—Same, tail, lateral view; G—Second stage larva, anterior end, lateral view; H—Same, dorsal view; I—Tail, lateral view). J—*Ascaris lumbricoides* larva from egg, en face K—*Physocephalus sexalatus*, third stage larva, en face. L—*Gongylonema pulchrum*, posterior end, ventral view. After Alicata, 1935, U. S. D. A. Tech. Bull. 489.

---

*Moorthy, 1938, described a dorsal appendage on some specimens of *Dracunculus medinensis*. In the first stage larva the appendage is long and filiform; while it persists throughout both second and third stages it is greatly reduced in size disappearing entirely at the third moult.

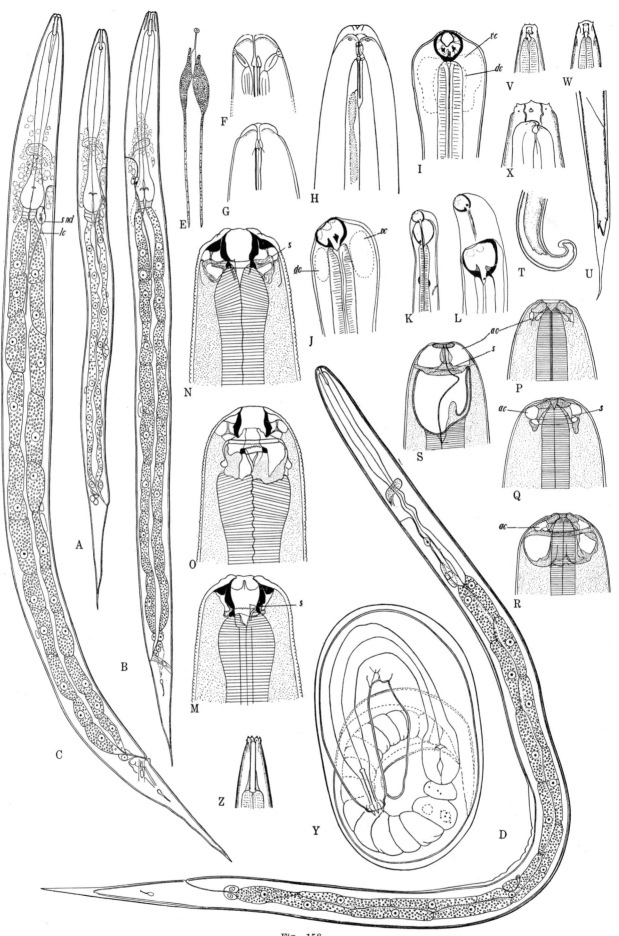

Fig. 158.

changes may take place in the intervening period we do not know, but in other members of the group there is some information. Steiner (1924) described a larval ascarid (*Agamascaris odontocephala*) in which lips are absent, the cephalic papillae well developed, and in addition there is a mediodorsal tooth, or agamodontium (Fig. 156 W). A similar tooth has been commonly observed in larval ascaridids of fish (*Anacanthocheilus rotundatus* (Wülker, 1930). Presumably this tooth aids in the migration of such larvae through tissues and is shed at the last moult.

Spiruroids may pass through stages possibly indicating phylogenetic development but certain cenogenetic features sometimes occur. Thus in the first stage larvae of *Gongylonema*, *Ascarops* and *Physocephalus*, Alicata has shown that there is a peculiar group of ventral cephalic hooks (Fig. 157 C) and in the third stage larvae there are six indistinct rudiments of lips and all except the ventrolateral cephalic papillae are well developed. Between the two dorsodorsal and the two ventroventral lips there is a pair of median cuticular elevations (Fig. 157 K). Corresponding structures are present in the adult stage of *Simondsia* (Fig. 58 R). However, all spirurids do not have the hooks and cuticular elevations. In *Physaloptera* (Alicata 1937) the third stage larva has been found to have papillae and lips approximately as in the adult stage (Fig. 58, p. 63); unfortunately the first stage larva has not been studied.* The larva of *Habronema* (Hsü and Chow, 1938) has six lips instead of the adult two. The first stage larva of *Camallanus* has an hexagonal oral opening and the labial region continues to be of larval form until the last moult at which two "lateral jaws" appear.

Little is known of the development of labial structures in the Aphasmidia. Crossman (1932) found that the larvae of *Tylocephalus* have a head resembling *Plectus* in that there are four large setose papillae and during development membranes or "cushions" form a web between these structures.

Changes occurring during development of the stoma are often very marked in specialized nemas though little change takes place in the generalized forms. In rhabditids the only noticeable change is in the diameter of the stoma which becomes wider with age; the absolute length may not change in postembryonic development. Only the cheilorhabdions and prorhabdions are cast at moulting. Related nemas such as diplogasterids may show some changes during development of the stoma. As a rule, nematodes having short and proportionately wide stoma in the adult stage have a much more narrow stoma tending to be cylindrical or prismoidal in the larval stage (Fig. 158 V-X). Thus in the development of strongyloids and trichostrongyloids the stoma in the first stage larvae is rhabditiform while in the second stage larvae it collapses and the cheilorhabdions and prorhabdions may simulate a stylet in the third stage (Fig. 158 H). In the Strongyloidea there is a rather extensive reformation of stomata between the fourth and adult stages.

The stoma of the fourth stage larva is usually rather short and wide and is termed a provisional buccal capsule. In the late fourth stage a cavity is formed around this

structure. Looss (1897) observed two such cavities in *Ancylostoma* one dorsal and one ventral, which gradually fused. In *Strongylus* (Fig. 158 P-S) and *Cylicostomum* (Fig. 158 M-O) Ihle and van Oordt (1923, 1924) observed a single anterior cavity (*a-c*) completely surrounding the provisional buccal capsule. At the base of this a septum (*s*) is formed separating the anterior cavity and provisional buccal capsule from the remainder of the body. Behind the septum a new cavity is formed outside the anterior end of the esophagus. Around it the adult buccal capsule forms (Fig. 158 N & R). The esophagus then is withdrawn and becomes attached to the base of the adult buccal capsule (Fig. 158 O & S). The old lining of the esophagus is attached to the provisional buccal capsule. In other strongylins the stoma may remain cylindrical to the adult stage (*Cylindropharynx*, Fig. 56 C). Metastrongyloids differ from the foregoing in that the stoma is never rhabditiform so far as is known; mesorhabdions and telorhabdions are degenerate in the first stage. In the later development of such forms the stoma may disappear (*Metastrongylus elongatus*). Young thelastomatids have a longer, more cylindrical stoma than adults and the same may be said of ransomnematids while in oxyurids remarkable changes have been described. Ihle and van Oordt (1921) found that the larvae of *Oxyuris equi* have a massive pseudostom (Fig. 97) formed by a dilation of the corpus the dilation being entirely absent in the adult. One would judge this to be a purely cenogenetic feature related to feeding habits. Larval subulurids have approximately the same type of stoma as do the adults while the larval ascarids, like the adults, have none. In the Spiruroidea Chitwood and Wehr (1934) found that the stoma in some forms appears to go through stages which are known to occur in the adult of other forms. Thus in the case of *Physocephalus* six cuticular projections of the prostom in the third stage larvae appear to form the lips of the adult stage while they retain their original larval position through development to the adult stage in *Ascarops*. It has also been found that the stoma is more cylindrical in the third stage larva than in the adult. Ransom (1913) describes the mouth cavity of the first stage larvae of *Habronema* as shallow becoming longer and cylindrical by the third stage. Passing now to the Filarioidea we find that in some forms (*Dirofilaria immitis*) the third stage larva has a well developed cylindrical stoma while the adult has no distinct stoma. However, in the related genus *Litomosoides* the stoma in the adult stage is practically the same form as it is in the larvae of *Dirofilaria*. Regarding stomatal changes in the Aphasmidia a little is known only in the cases of mermithids, trichuroids and dioctophymatoids. Christie (1936) has found that the stoma of the embryo of *Agamermis* is represented by two small plates posterior to which there is a long narrow cuticular tube surrounded by esophageal tissue. Within the esophageal tissue a large onchiostyl develops and gradually comes to surround and replace a part of the stoma or esophageal lumen at moulting. In trichuroids, Fuelleborn (1923) described similar onchiostyls as developing in late embryonic stages and Lukasiak (1930) described a stylet in larval dioctophymatoids which had been removed from the egg shell.

ESOPHAGUS. Postembryonic changes in the esophagus of nemas are limited to parasitic forms, some changes occurring in nearly all the large parasitic groups. Presumably the changes are correlated with the development of new feeding habits. In general the earlier stages of the esophagus are more similar to that structure in *Rhabditis* than is the esophagus of the adult, but very little or no change takes place in the number of cells during development except in those nemas with reduplicate esophageal glands (see p. 233). At the time of hatching thelastomatids usually have an esophagus consisting of a cylindrical corpus, isthmus and valvulated bulb but in a few genera of the Thelastomatidae (*Hammerschmidtiella*, *Leidynema*, *Aorurus*) the metacorpus becomes enlarged in the adult female. Peculiarly, no such change in form takes place in the development of the males. Because of the late appearance of the swelling it is not considered a homologue of the swelling present in the rhabditoid esophagus. Oxyurids usually have an esophagus like that of the adult during all stages of development (exception *Oxyuris equi* see p. 78). No particular developmental changes have been noted in the esophagus of heterakids but ascaridids present many

---

Fig. 158.

Postembryonic development of members of the Rhabditina and Strongylina. A-D—*Trichostrongylus axei* larvae [Trichostrongylidae] (A—First stage; B—Early second stage; C—Late second stage; D—Third stage). E-H—*Ancyclostoma caninum* [Ancylostomatidae] (E — Third stage larva, excretory apparatus; F-G-H — Head, F — dorsal view and G-H — lateral view). I-L — *Gaigeria pachyscelis* [Ancylostomatidae] Head. (I — Ventral view, fourth stage; J — Lateral view, fourth stage; K — Late fourth stage; L—Moulting specimen). M-O—*Cylicostomum* sp. [Strongylidae] head of larva, lateral view (M—Fourth stage; N— Late fourth stage; O—During fourth moult). P-S—*Strongylus vulgaris* [Strongylidae], fourth stage larval female anterior end. (P-R—Stages in formation of buccal capsule; S—Moulting specimen). T—*Metastrongylus elongatus* [Metastrongylidae] (Posterior end of larva in second moult). U—*Ornithostrongylus quadriradiatus* [Trichostrongylidae] third stage larva, (Lateral view of tail). V-X—*Pristionchus sp.*, [Diplogasteridae], stomatal region (V-? first stage larva, lateral view; W—Same specimen, dorsal view; X—Adult). Y-Z—*Rhabditis strongyloides* [Rhabditidae] (Y—Embryo in egg shell; Z—Stomatal region, first stage larva). E-H, after Stekhoven, 1927, Proc. Roy. Acad. Amsterdam., v. 30. I-L, after Ortlepp, 1937, Onderstepoort J. Vet. Sc., v. 8 (1). M-O, after Ihle and Oordt, 1923, Ann. Trop. Med. & Parasit. v. 17 (1). P-S, after Ihle and Oordt, 1924, Koninklijke Akad. Wetensch. Amsterdam. v. 27 (3-4). T, after Schwartz and Alicata, 1931, J. Parasit. v. 28. U, after Cuvillier, 1937, U. S. D. A. Tech. Bull. 569. Remainder original.

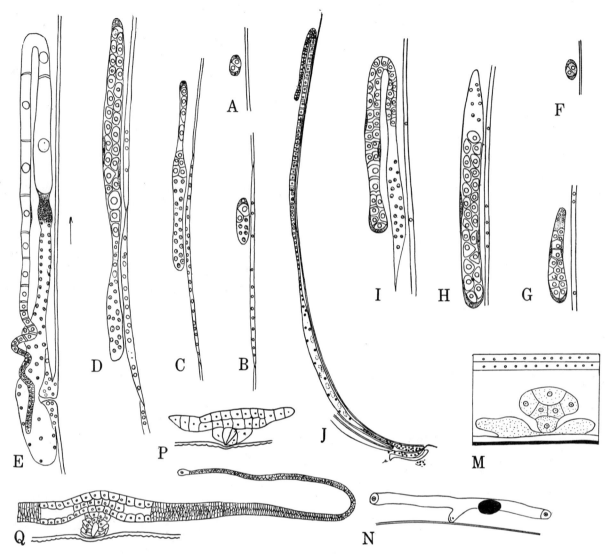

Fig. 159.

Postembryonic development of the reproductive system. A-J—*Turbatrix aceti* [Diplogasteridae] (A—Genital primordium of newly hatched female; B—24 hours; C—Three days; D—Five days; E—Ovary of nine day old female; F—Genital primordium of newly hatched male; G—Second day; H—Four days; I—Five days; J—Testis). K-L—*Syphacia obvelata*, female reproductive system, immature and adult. M-O—*Gongylonema scutatum*, genital primordium (M—Third stage larval female; N—Late third stage female; O—Fourth stage). P-Q—*Gaigeria pachyscelis* genital primordium (P—Fourth stage larval female; Q—Late fourth stage female reproductive system, 3 weeks old). A-J, after Pai, 1928, Ztschr. Wiss. Zool., v. 131 (2). K-L, after Vogel, 1925, Zool. Jahrb. Abt. Zool. & Phys., v. 42. M-O, after Seurat, 1920, Hist. Nat. Nem. P-Q, after Ortlepp, 1937, Onderstepoort J. Vet. Sci., v. 8 (1).

interesting variations. The first stage larvae of *Ascaris*, *Toxocara* and *Toxascaris* all have an esophagus which s'ightly resembles that of rhabditids or more precisely *Angiostoma plethodontis*. It consists of a somewhat clavate corpus, an indistinct isthmus, and a short pyriform bulbar region. Information on the later development is lacking but the adults have a cylindrical esophagus which in the case cf *Ascaris* and *Toxascaris* is not grossly subdivisible into separate regions. In the case of *Toxocara* the posterior end is set off as a muscular *ventriculus*, which apparently corresponds to the reduced bulb of the first stage larva. When an esophagael diverticulum is formed (*Contracaecum*), it develops as an evagination of the ventral side of the bulbar region or ventriculus. The esophagi of strongylins also pass through a very interesting series of changes. In two superfamilies, the Strongyloidea and Trichostrongyloidea the esophagus of the first stage larva is usually identical with that of rhabditids (Fig. 158 A); during the second larval stage the valves degenerate (Fig. 158 C) and in the third stage the esophagus becomes long and narrow resembling the esophagus of *Diplogaster* except that the swelling at the base of the

corpus is very indistinct (Fig. 158 D); in later development it becomes more or less clavate, obliterating nearly all signs of former division. Similar changes take place in the Metastrongyloidea except that the phylogenetic reminiscence of rhabditoid affinities is not so marked since even the esophagus of the first stage larva does not have a valvulated bulb but resembles more closely that of third stage larvae of strongyloids and metastrongyloids. Two families of the Rhabditoidea, the Rhabdiasidae and the Strongyloididae, undergo change in the form of the esophagus during development of the parasitic generation. In the first family the esophagus of the free-living generation and of the first stage of the parasitic generation is rhabditiform while the later stages of development of the parasitic generation show changes entirely comparable to the strongyloids. In the second family the esophagus of the free-living generation and of the first stage larvae of the parasitic generation is rhabditiphaniform while in the later development of the parasitic generation changes comparable to those of rhabdiasids occur except that the esophagus of the adult remains much

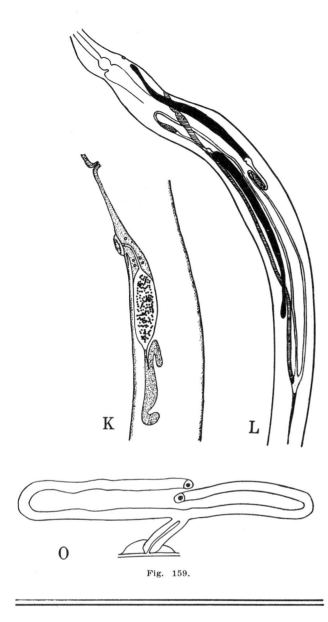

K L

O

Fig. 159.

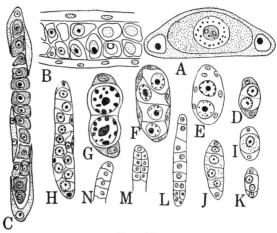

F:g. 160.

Genital primordia. A—*Bradynema rigidum*; B-C—*Rhabdias bufonis.* D-H—*Rhabditis aberrans.* I-J—*Allantonema mirabile.* K-L-N—*Bradynema strasseni.* M—*Allantonema mirabile.* All from Musso, 1930, Ztschr. Wiss. Zool., v. 137 (2). A, After Zur Strassen, 1892. B-C, after Neuhaus, 1903. D-H, after Krueger, 1913. M, after Wuelker, 1923.

as in the third stage strongyloid larvae, hence strongyliform.

Little is known regarding the character of the esophagus of the first stage larvae of spirurins. In the first stage larvae of the genera *Ascarops* and *Gongylonema* Alicata (1935) found faint indications of a division into corpus, isthmus and bulbar region (Fig. 157 D) which entirely disappear during later stages and are replaced by a division into a short narrow anterior part and a long, wide posterior part, both parts being cylindrical. The first stage larvae of mermithids. (*Agamermis decaudata*) may have an esophagus consisting of five regions; (1) and (2) equivalent to corpus, (3) a narrow part (? isthmus), (4) a swelling and (5) a long narrow posterior part. Two small subventral and a large dorsal esophageal gland are situated posterior to (4). In addition, two subventral rows of eight smaller cells, the stichocytes are situated along side the posterior narrow region (Fig. 93). During later development the esophagus narrows, the anterior swelling (2) disappears and the posterior part (5) becomes more or less surrounded by the large paresophageal body, the stichosome, which retains the two-cell-row form throughout later development. Each stichocyte is a large unicellular gland with a separate orifice. The three original esophageal glands atrophy after the nematode enters the host (See p. 92).

Very little is known about the esophagus of first stage *Trichuris* larvae but Fuelleborn (1923)

illustrated it as being composed of an anterior part terminated by a glandular swelling and a posterior part extending between two rows of stichocytes. According to Wehr (1939) the esophagus of the first stage larva of *Capillaria columbae* consists of a long narrow anterior part, a slight swelling and a cell body, or stichosome region, consisting of a double row of seven stichocytes (Fig. 163 O). In the late first stage the stichosome has greatly increased in size and number of cells (Fig. 163 P). By the third stage the two rows have fused forming a single row of cells (Fig. 163 Q). The approximate ratios of the length of the esophagus to the length of the intestine in each stage were given as follows: First stage 3. 5 : 1; Second stage 2 : 1; third stage 1. 8 : 1; fourth stage, 1. 1 : 1 and adult, 1 : 1.4.

Considering the esophagi of both trichuroids and mermithoids it seems reasonable to conclude the double row stichosome of first stage trichuroid larvae is palingenetic. The intraesophageal character of the primary esophageal glands of trichuroids is undoubtedly primitive while their extraesophageal position in mermithids is recent and their hypertrophy at the period of penetration must be considered cenogenetic.

INTESTINE. Information on the postembryonic development of the intestine is strangely lacking. A few bold writers have admitted the presence of cells in the intestine but as a rule the intestine merely forms a connecting link between esophagus and rectum. However, changes both interesting and extensive do occur in some forms. In the more primitive nematode groups the multiplication of cells must be very limited and sometimes is probably confined to certain regions of the intestine. This appears to be true of most rhabditids, oxyuroids and similar forms. Numerous cell divisions must take place in the more highly evolved aphasmidian forms and in ascaridoids, spirurins, trichuroids and dioctophymatoids. The intestine of first stage *Ascaris lumbricoides* consists of innumerable cells. According to Moorthy, 1938, the intestine of the late first stage larva of *Camallanus sweeti* has about 35 cells, the number increasing until in the early adult stage there are about 200. The same author records 12-15 cells in first stage *Dracunculus medinensis* while in the late third stage there are 35-40 cells. Of strongylins, known to possess few intestinal cells in the adult stage, Lucker (1935, 1936, 1938) offers considerable information. *Cylicodontophorus bicoronatus*, *Cylicocercus pateratus* and *Cylicocyclus insigne*, each have only eight intestinal cells in the infective larvae while the genera *Gyalocephalus* and *Strongylus* have 12 in the first genus and 16 to 32 in the latter. Lucker also definitely established the existence of a lumen in those species with an eight cell intestine; according to his observations the intestine of second stage larvae of these forms have a 22 cell intestine, the number

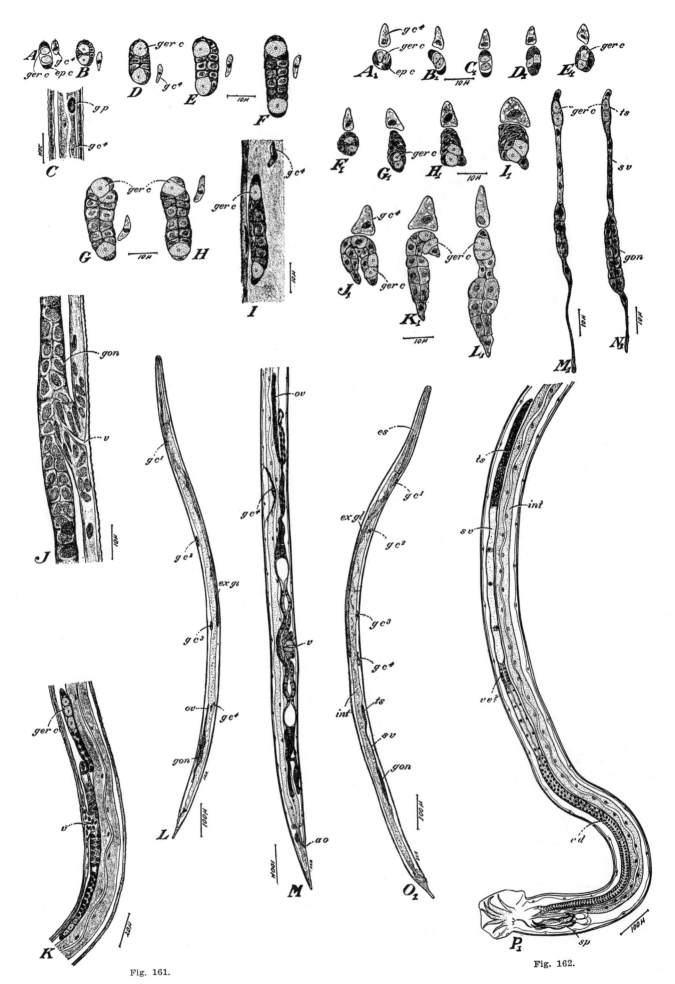

Fig. 161.

Fig. 162.

being reduced thereafter. Alicata (1935) records eight dorsal and eight ventral intestinal cells in both first and third stage larvae of *Hyostrongylus rubidus*. The present writer found seven dorsal and seven ventral cells in the intestine of the first stage larvae of *Trichostrongylus axei*, 22 cells in the second stage and only 16 in the intestine of the infective third stage larvae. The writer makes no attempt to explain the reduction in number of cells but the data were verified with numerous specimens. Nuclear division without cell wall formation must occur later in the development of strongylins (Fig. 102, p. 102). In the case of mermithoids our information is more definite. In many of these species cell division is followed by nuclear division without cell wall formation and finally fusion of syncytia and obliteration of the lumen may occur.

Outpocketings or cecae have been previously described (p. 100) in diverse groups of nematodes. In ascaridins such cecae have been found to arise as evaginations of the intestinal epithelium during late larval development. Similarly, Christie (1936) has found that the trophosome (the fat body or intestine) of mermithids grows anteriorly during larval development of *Agamermis decaudata*. On this basis we may consider the trophosome in the esophageal region of mermithids as a caecum without a lumen.

RECTUM, CLOACA and PERTAINING STRUCTURES. In so far as is known, cell division does not take place in the postembryonic development of the posterior gut of the female. It does, however, in the male for a small ventral growth of cells forms which is later joined by the vas deferens when it comes to open in the cloaca. Similarly, there is a mass (or two masses) of cells from which the spicules develop. The gubernaculum, on the other hand, is a cuticular thickening of the dorsal lining of the cloaca. One may interpret the spicular sheaths as first an evagination of the dorsal wall of the cloaca, then an invagination of this structure (Fig. 118 U). Both spicules and gubernaculum generally develop during the fourth stage.

EXCRETORY SYSTEM. Conclusive evidence is lacking with regard to the postembryonic development of the excretory system despite the numerous observations which have been made. The primary cause of this failure is that all workers have proceeded on the assumption that a single ventral gland cell is the entire system. Cobb (1890) described the excretory system of larval *Enterobius vermicularis* as a single invaginated cell from which the lateral canals and excretory vesicle developed. In 1925 the same author described the first stage larva of *Rhabditis icosiensis* as similar to that of the adult except that the ventral gland was unpaired and the lateral canals free in the body cavity; the unpaired ventral cell was then supposed to divide forming the double glands of the adult. The sinus and terminal duct nuclei were not accounted for. Stekhoven (1927), Lucker (1935) and others have described a sinus (no nucleus seen) two subventral gland cells and no lateral canals in third stage larvae of the strongyloids. The writer found the excretory pore, terminal duct, sinus, subventral gland cells and lateral canals all very plain in the first stage larva of *Trichostrongylus axei* (Fig. 158 A-B). Before theorizing too much on the development of the excretory system, it would seem necessary that more critical data be obtained on the actual conditions existent in first stage larvae. It seems possible that the so-called ventral gland or excretory cell usually described in larvae of parasitic nemas is actually the sinus cell and the terminal duct cell may be present but overlooked. If this is the case, the system may originate from two germ lines in the Phasmidia. The primary sinus nucleus might easily give rise to a secondary sinus nucleus and the paired subventral glands of the Strongylina and some rhabditids (*R. icosiensis*, *R. terricola*, *R. strongyloides*). This would still not account for the lateral canals. The theory that they develop from the sinus cell may be correct but it has not been demonstrated.

FEMALE REPRODUCTIVE SYSTEM. Development of the female reproductive system may be of two types, dependent upon the number of ovaries present in the adult. In either instance the genital primordium of the first stage larvae consists of the same number of cells, four, arranged in the same manner as in the males. *Turbatrix aceti* is the only one ovaried form that has been studied. Pai (1928) found that after 24 hours the posterior somatic cell group (S5 II) has multiplied considerably, forming a mass of cells while the other cell groups (S5 I and P5) remained constant (Fig. 159° B). Later all cell groups multiply (Fig. 159 C-D) and the anterior end of the gonad bends posteriad while the posterior end (S5 II) grows posteriad also (Fig. 159 E). The anterior somatic cell group forms the epithelium of the ovary while the posterior somatic cell group forms the oviduct, uterus, and seminal receptacle. At this time an invagination of the hypodermis of the ventral chord meets the uterus forming the vagina and vulva.

The gonad of *Tylenchinema oscinellae* was found by Goodey (1930) to develop in the same manner except that a twist occurs in the oviduct. The vagina and uterus of *Sphaerularia* and *Atractonema* were found by Leuckart (1887) to become everted or prolapsed growing until the uterus is hundreds of times larger than the body in *Sphaerularia* (Fig. 115A).

The development of the female reproductive system in nematodes with two ovaries (*Falcaustra lambdiensis*, *Gaigeria pachyscelis* and *Hyostrongylus rubidus*) differs in that both of the somatic cells form ovarian epithelium and both contribute to the formation of the uterus. Division of the terminal cells results in an epithelial tissue covering the germinal cells with a terminal cell at the end of each ovary and a mass of somatic cells separating the two groups of cells resulting from the divisions of P5 I and P5 II (Fig. 161). This mass of cells through further division and enlargement pushes the germinal cell groups apart and finally forms the uterus and oviducts. At this time the middle part of the S5 group is joined to an invagination of the hypodermis forming the vulva and vagina, (Fig. 161 J). Ortlepp (1937) found that the ovejectors of *Gaigeria pachyscelis* (Fig. 159 P-Q) originate from the genital primordium and not the vulvar invagination. Free-living nematodes with outstretched ovaries undergo no further development unless parts of the uteri are set off as seminal receptacles. Free-living nematodes with opposed reflexed ovaries differ only in that the ends of the ovaries grow towards each other. Seurat (1920) found that in the case of parallel ovaries or uteri in parasitic nematodes that the ovaries and uteri are at first outstretched; coiling and twisting of ovaries, uteri or both occur in very late larval or early adult development. Another peculiarity in parasitic nematodes is that the uteri may become fused for part of their length forming either a continuation of the vagina (vagina uterina) or a common uterus. The transformation from opposed to parallel oviducts and coincident development of a long uterine vagina was particularly well illustrated by Vogel (1925) in the development of *Syphacia obvelata* (Fig. 115 K-L).

MALE REPRODUCTIVE SYSTEM. The genital primordium of the male nematode consists of four cells at the time of hatching in all known cases. Two of these cells, the "terminal cells" cover the other two, the germinal cells. Unfortunately the development has been traced only in nemas having a single testis in the adult. Seurat (1918) discovered that the anterior end of the gonad of *Falcaustra lambdiensis* first grows anteriad, thereafter turning posteriad and extending to the cloaca, thus forming the vas deferens with the result that the gonad is flexed.

Pai (1928) found that in *Turbatrix aceti* after 48 hours the genital primordium consisted of three groups of cells the primordial germ cells (P5), the anterior somatic cells (S5 I) forming a solid mass derived from the anterior terminal cell, and the posterior somatic cells (S5 II)

Fig. 161.

Postembryonic development of female reproductive system of *Hyostrongylus rubidus*, with position of coelomocyte adjoining genital primordium. A—First stage; B—third stage; C—Preparasitic third stage larva. D—Third stage larva recovered 2 days after experimental infection. E-H—Third stage 4 days after experimental infection. I—5 days after experimental infection, larva on verge of third moult. J—Vulvar region showing differentiation of ovary and gonoduct at 9 days. M—Young adult female, posterior end. All after Alicata, 1935, U. S. D. A. Tech. Bull. 489.

Fig. 162.

Postembryonic development of male reproductive system of *Hyostrongylus rubidus*. A1—First stage larva. B1—Later first stage larva. C1—Late first stage larva. D1-E1—Second stage larva. F1—Preparasitic larva. G1—2 days after experimental infection. H1-J1—4 days after experimental infection. K1—5 days after infection. L1—6 days after. M1-N1—Fourth stage larva 9 and 11 days after. O1—11 days. P1—Young adult male, posterior end. All after Alicata, 1935, U. S. D. A. Tech. Bull. 489.

forming a covering for the germinal cells and a terminal cell (Fig. 159 G). After 96 hours, no change had taken place except multiplication of cells (Fig. 159 H) but after 120 hours the anterior somatic cells had grown posteriorly drawing the anterior part of the gonad with them (Fig. 159 I). Thus the flexure of the testis takes place. Finally the anterior somatic cells grow posteriorly and join the rectum forming the vas deferens and cloaca (Fig. 159 J). Alicata (1935) found the development of the male reproductive system of *Hyostrongylus rubidus* to be essentially similar except that multiplication of germinal cells (P5) is delayed. As in *Turbatrix*, the anterior group of somatic cells (S5 I) bend posteriorly forming the vas deferens and seminal vesicle but the germinal zone is shifted around so that it is anterior and the gonad consequently is not flexed (Fig. 162). Goodey (1930) on the contrary found that in *Tylenchinema oscinellae* the posterior terminal cell group extends posteriorly forming the vas deferens and joining with the rectum. In this species the testis is not flexed.

MICROFILARIA AND FILARIAL DEVELOPMENT. Work on microfilaria has been developed as a separate science with little or no relationship to general nematology. Since the pioneers were chiefly interested in identification of forms found in the blood of various species they developed a separate nomenclature for parts of the body. Recent workers have made rapid strides in the identification of the parts of microfilaria with other nematodes.

Some filarioids give birth to well formed first stage larvae or deposit well formed eggs containing such larvae. These were placed in the family Filariidae by Wehr (1935). The larvae of such forms often have the cephalic hook and transverse rows of spines as seen in *Gongylonema pulchrum* (Fig. 157 B); some of these have attenuated tails, others rounded and spinate tails as in *Gongylonema*. They are moderately well differentiated first stage larvae and in many cases, at least, should not be called microfilaria. This term should be reserved for the rather unformed or embryonic young produced by the genera placed by Wehr in the Filariidae. In these, stoma, esophagus, intestine and other organs are not completely differentiated.

Microfilaria may be classified as to presence or absence of a sheath (Fig. 163 J-K). The sheath is a very delicate membrane surrounding the larva, which some authors have considered a cuticle, indicative of the first moult, others a modified egg shell. The fact that the sheath resists chemicals in which a vitelline membrane would be dissolved, eliminates that possibility. Evidence that the sheath develops from an egg membrane was presented by Penel (1904) and Seurat (1917) in the cases of *Loa loa* and *Thamugadia hyalina* (Fig. 163 B-G). Its insolubility in alcohol and oils would signify that if it is an egg membrane, it is the shell. There seems to be no morphologic difference correlated with presence and absence of a sheath.

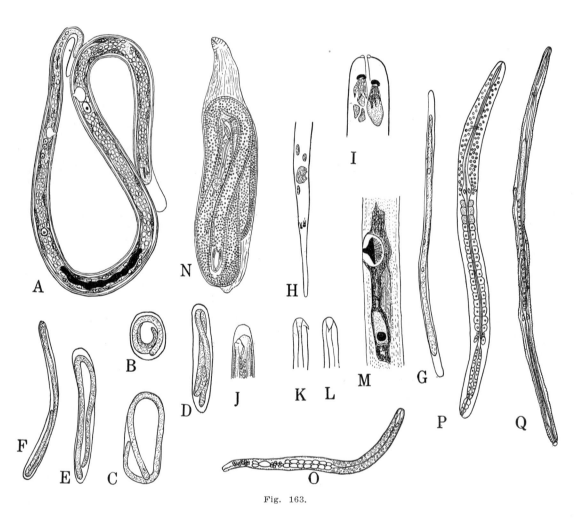

Fig. 163.

Postembryonic development continued. A, H-I, and M—*Microfilaria loa* (A—Entire larva; H—Tail showing phasmids; I—Head; M—Excretory pore and cell). B-G—*Thamugadia hyalina*, successive stages. J.—*Microfilaria bancrofti*, head. K-L—*Mf. recondita*, head, lateral and ventral views. N—*Heterodera schachtii*, fourth stage male in preparation for moult. O-Q—*Capillaria columbae*. A, H-M, after Fuelleborn, 1929, Handbuchen Path. Micro. v. 6 (28). B-G, after Seurat, 1920, Hist. Nat. Nem. N, after Strubell, 1888, Biblio. Zool. Orgi. Abh. Gesammt. Zool., v. 2. O-Q, after Wehr, 1939, U. S. D. A. Tech. Bull. 679.

General Anatomy of Microfilaria. For a comprehensive survey of the comparative anatomy of microfilarial species the reader is referred to Fuelleborn (1929). Regardless of the presence or absence of a sheath the embryo is covered by a delicate cuticle with distinct but minute striae (Rodenwaldt, 1908). The anterior end may bear a delicate hook (Fig. 163 K) much like that of the first stage larvae of spiruroids like *Gongylonema* and filariids like *Dicheilonema*. There is apparently no oral opening and the stoma, at most, is indicated by a primordium or vacuole. Paired "Mundgebilde" and "Schwanzgebilde" have been demonstrated by intravitam stains (Fig. 163 H-I); these seem to correspond to the amphids and phasmids.

The remainder of the internal anatomy appears to consist of a rather disorganized mass of cells or nuclei some of which were sufficiently outstanding to have received specific names. Manson (1903) named a clear spot with a massive adjoining cell, situated in the cervical region, the V spot. This was identified by Looss (1914) as the excretory pore and excretory cell (Fig. 163 A & M). Rodenwaldt (1908) named four other large densely staining cells G1-4 considering them as genital cells and the vesicle with which they are associated, he termed GP or the genital pore. Looss considered only the anterior-most of these (G1) as a genital cell and the remainder he named Z1-3 associating them with the rectum and correctly identifying the vesicle as proctodeum. More recently Yamada (1927) and Feng (1936) have found that all four G cells take part in the development of the rectum (Fig. 164 5-9). Beneath the cuticle there are four rows of spindle-shaped cells with elongate nuclei which Rodenwaldt named the matrix cells of the subcuticle. According to more recent observations, these are somatic muscle cells and total 14 to 62 in various species. The remaining nuclei of the body constitute the *nuclear column*. These take part in the formation of the esophagus, intestine, nervous system and chords. In the cervical region they are particularly numerous, interrupted only by the clear area indicative of the nerve ring. In some species of microfilaria there is a clear area beginning some distance posterior to the excretory cell and ending anterior to G1; this structure, termed the "Innenkörper" appears to be a yolk rest and corresponds to the lumen of the future intestine (Fig. 163 A). Near the tip of the tail, there is sometimes a pair of tail nuclei which are lost at the first moult according to Feng.

Later development. For the later development of microfilariae, we will use as an example *M. malayi* which has been very nicely worked out by Feng (1936) as far as the third stage (Fig. 164). In this species the sheath is cast in the stomach of the mosquito on the first day. Thereafter the organism is termed a first stage larva

until the fourth day at which the first moult occurs in the body of the mosquito. The second stage is terminated by a moult on the sixth day and the third stage continues until after emergence from the intermediate host and entry to the final host. During this period the nema grows from a length of 210 microns and diameter of 4.8 microns to a length of 1.3 mm. and a diameter of 20 microns. Growth is not uniform; after entry into the mosquito, there is first a shortening and widening, chiefly in the mid-region, the minimum length occurring after one and a half days (Fig. 164 7). On the second day the mid-region starts to grow and after five and a half days, the esophagus begins suddenly to increase in length. The width increases up to five and a half days, then diminishes. The tail of the first stage larva is attenuated, that of the second stage short and conical and that of the third stage truncate, with two ventral and dorsal "papillae". (The ventral "papillae" are probably phasmids, while the dorsal is probably the tip of the tail). In *Dirofilaria immitis* two massive protuberant phasmids have been observed and no dorsal "papilla" exists. The stoma begins to form on the second day. (Fig. 164 7) and by the fourth day (second stage) is indicated by refractile rods (protorhabdions), (Fig. 164 9). In the third stage the stoma is conoid and the head bears the eight cephalic papillae and amphids characteristic of adult filariids.

The esophagus first becomes distinct (Fig. 164 8) on the third day and takes the typical two part glandular appearance of the adult by the fifth day. The intestinal primordium is distinctly visible after 24 hours (Fig. 164 6) and is well formed on the third day (Fig. 164 8); by the fifth day there is a distinct lumen and the intestine is several cells in circumference. The rectal primordium, G1-4, begins with a division of G1 after 24 hours (Fig. 164 6) and a division of G2-4 accompanied by a second division of G1 after 34 hours; the rectal cuticle completes the formation of the 10 cell rectum on the third day (Fig. 164 8). The genital primordium is first apparent on the fifth day (Fig. 164 9) and is composed of seven cells (Fig. 164 12); apparently it originates from cells in the intestinal region but its earlier existence has not been traced. In both the first and second ecdyses the lining of the esophagus and rectum are moulted.

Yamada (1927) had previously demonstrated two moults after exsheathment for *Wuchereria bancrofti* while the same number have been demonstrated in the intermediate host of non-ensheathed microfilaria such as *Dirofilaria immitis*. It seems clear that the ensheathed microfilaria is an embryo rather than a larva. Even after exsheathment the first stage larva is still rather embryonic in character and hardly deserves the term larva before the organs are demonstrable, this being on the third day for *Microfilaria malayi* (Fig. 164 8).

Bibliography

ALICATA, J. E. 1935.—The tail structure of the infective *Strongyloides* larvae. J. Parasit., v. 21 (6): 450-451, 1 fig.

 1935 (1936). — Early developmental stages of nematodes occurring in swine. U. S. Dept. Agric. Tech. Bull. 489, 96 pp., 30 figs.

 1937 (1938).—Larval development of the spirurid nematode, *Physaloptera turgida*, in the cockroach, *Blatella germanica*. Papers on Helminthology, 30 Year Jubileum, K. J. Skrjabin; pp. 11-14, figs. 1-13.

BOVIEN, P. 1932.—On a new nematode, *Scatonema wülkeri*, gen. et sp. n. parasitic in the body cavity of *Scatopse fuscipes* Meig. (*Diptera nematocera*). Vidensk. Medd. Dansk., Naturh. Foren., v. 94: 13-32, figs. 1-7.

 1937.—Some types of association between nematodes and insects. Vidensk. Medd. Dansk. Naturh. Foren., v. 101: 1-114, figs. 1-31.

CAMERON, T. W. M. 1927.—Observations on the life history of *Aelurostrongylus abstrusus* (Railliet), the lungworm of the cat. J. Helminth., v. 5 (2): 55-66, figs. 1-2.

CHITWOOD, B. G. and WEHR, E. E. 1934.—The value of cephalic structures as characters in nematode classification, with special reference to the superfamily Spiruroidea. Ztschr. Parasit., v. 7 (3): 273-335, figs. 1-20, 1 pl.

CHRISTIE, J. R. 1934.—The nematode genera *Hystrignathus* Leidy, *Lepidonema* Cobb and *Artigasia*, n. g. (Thelastomatidae). Proc. Helm. Soc. Wash., v. 1 (2): 43-48, figs. 15-17.

 1936.—Life history of *Agamermis decaudata*, a nematode parasite of grasshoppers and other insects. J. Agric. Res., v. 52 (3): 161-198; figs. 1-20.

 1937.—*Mermis subnigrescens*, a nematode parasite of grasshoppers. J. Agric. Res. v: 55 (5): 353-364, figs. 1-6.

COBB, N. A. 1890.—*Oxyuris*-larvae hatched in the human stomach under normal conditions. Proc. Linn. Soc. N. S. Wales, 2. s., v. 5: 168-185, 1 pl.

 1925.—*Rhabditis icosiensis*. J. Parasit., v. 11 (4): 219-220, figs. A-B.

CRAM, E. B. 1931.—Developmental stages of some nematodes of the Spiruroidea parasitic in poultry and game birds. U. S. Dept. Agric. Tech. Bull. No. 227, 27 pp., figs. 1-25, 1 pl.

CROSSMAN, L. 1933.—Preliminary observations on the life history and morphology of *Tylocephalus bacillivorus* n. g., n. sp., a nematode related to the genus *Wilsonema*. J. Parasit. v. 23: 106-107.

CUVILLIER, E. 1937. — The nematode, *Ornithostrongylus quadriradiatus*, a parasite of the domesticated pigeon. U. S. Dept. Agric. Tech. Bull. No. 569, 36 pp., figs. 1-6.

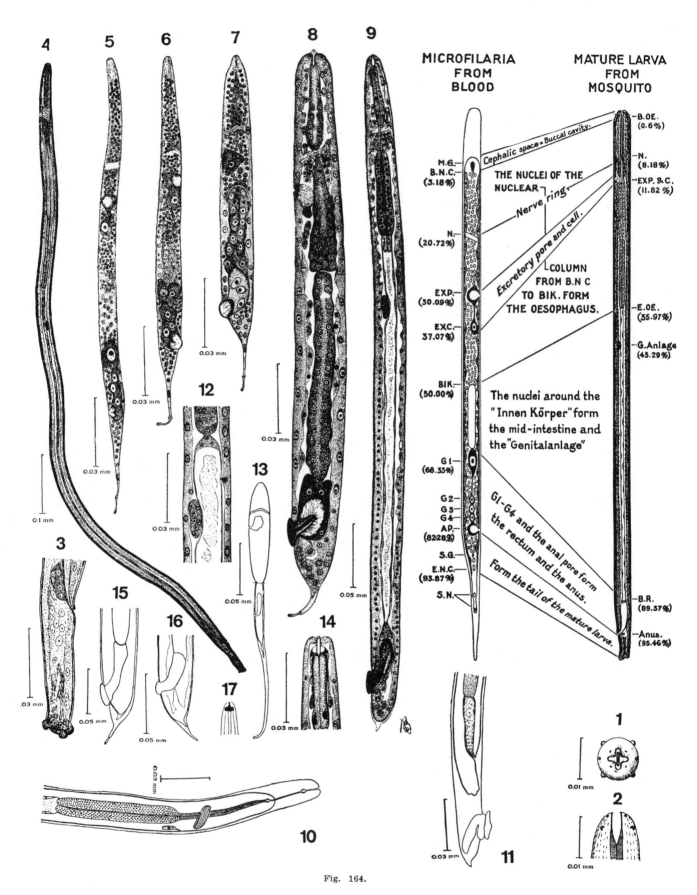

Fig. 164.

Development of *Microfilaria malayi*. 1-15, 17—*Mf. malayi* (1-4.—mature third stage larva, 1—en face; 2—lateral view of head; 3—tail, lateral view; 4—lateral view; 5—larva 8 hours after ingestion by mosquito; 6—24 hours in mosquito; 7—1½ days; 8—2½ days; 9—4½ days; 10-11—5½ days, anterior and posterior parts; 12—4½ days, showing genital primordium; 13—4½ days, abnormal development; 14—3½ days, showing buccal cavity and gland like structures; 15—4½ days, posterior end, lateral view). 16—*Mf. bancrofti*, 4½ days posterior end, lateral view. 17—*Mf. malayi*, mature larva, dorsal view. After Feng. 1936, Chinese Med. 1, Suppl. 1.

DIKMANS, G. 1931.—An interesting larval stage of *Dermatoxys veligera*. Tr. Amer. Micr. Soc., v. 50 (4): 364-365, pl. 29, figs. 1-5.

DIKMANS, G. and ANDREWS, J. S. 1933.—A comparative morphological study of the infective larvae of the common nematodes parasitic in the alimentary tract of sheep. Tr. Amer. Micr. Soc., v. 52 (1): 1-25, pls. 1-6.

DOBROVOLNY, C. G. and ACKERT, J. E. 1934.—The life history of *Leidynema appendiculata* (Leidy), a nematode of cockroaches. Parasit., v. 26 (4): 468-480, figs. 1-10, pl. 23, figs. 1-3.

ENICK, K. 1938.—Ein Beitrag zur Physiologie und zum Wirt-Parasitverhältnis von *Graphidium strigosum* (Trichostrongylidae, Nematoda). Ztschr. Parasit. 10 (3): 386-414, 1 fig.

FENG, L. C. 1933.—A comparative study of the anatomy of *Microfilaria malayi* Brug, 1927 and *Mf. bancrofti* Cobbold, 1877. Chinese Med. J. v. 47: 1214-1246, figs. 1-6, pls. 1-3.
    1936.—The development of *Microfilaria malayi* in *A. hyrcanus* var. *sinensis* Wied. Chinese Med. J., Suppl. 1: 345-367, pls. 1-4.
    1937 (1938).—Studies on the development of *Microfilariae*. Papers on Helminthology, 30 Year Jubileum, K. J. Skrjabin. Moscow, pp. 310-318, 1 text fig., pl. 1, figs. 1-17.

FÜLLEBORN, F. 1923.—Ueber den "Mundstächel" der Trichotracheliden- Larven und Bemerkungen über die jüngsten Stadien von *Trichocephalus trichiurus*. Arch. Schiffs. & Tropenhyg., v. 27: 421-425, pl. 11, figs. 1-18.
    1924.—Technic der Filarienuntersuchung. Handb. Mikr. Technik, pp. 2273-2304, figs. 691-698, pls. 13-14.
    1929.—Filariosen des Menschen. Handbuch der pathogenen Mikroorganismen. v. 6 (28): 1043-1224, figs. 1-77, pls. 1-3.

GOODEY, T. 1930.—On a remarkable new nematode, *Tylenchinema oscinellae* gen. et sp. n., parasitic in the Frit-fly, *Oscinella frit* L., attacking oats. Trans. R. Soc. Lond. B. v. 218: 315-343, pls. 22-26, 1 fig.

HSÜ, H. F. and CHOW, C. Y. 1938.—On the intermediate host and larva of *Habronema mansoni* Seurat, 1914 (Nematoda) Chinese Med. J. Suppl. 2: 419-422.

IHLE, J. E. W. and OORDT, G. J. VAN 1921.—On the larval development of *Oxyuris equi* (Schrank). Proc. Sc. K. Akad. Wetensch. Amsterdam. v. 23: 603-612 figs. 1-6.
    1923.—On some strongylid larvae in the horse, especially those of *Cylicostomum*. Ann. Trop. Med. & Parasit., v. 17 (1): 31-45, figs. 1-9.
    1924.—Over de ontwikkeling van de larvae van het vierde stadium van *Strongylus edentatus* (Looss). Koninklijke Akad. Wetensch. Amsterdam, v. 33 (9): 865-872, figs. 1-5.
    1924.—On the development of the larva of the fourth stage of *Strongylus vulgaris* (Looss). Koninklijke Akad. Wetensch. Amsterdam. v. 27 (3-4): 194-200, figs. 1-5.

LEUCKART, R. 1887.—Neue Beiträge zur Kenntniss des Baues und der Lebensgeschichte der Nematoden. Abhandl. Math.-Phys. Cl. K. Sächs. Gesellsch. Wiss., v. 13 (8): 565-704, pls. 1-3.

LI, H. C. 1935.—The taxonomy and early development of *Procamallanus fulvidraconis* n. sp. J. Parasit., v. 21 (2): 103-113, pls. 1-2, figs. 1-10.

LOOSS, A. 1897.—Notizen zur Helminthologie Egyptens. 2. Centrlbl. Bakt. 1 Abt., v. 21 (24-25): 913-926, figs. 1-10.
    1905. — Von Würmern und Arthropoden hervorgerufene Erkrankungen. Handb. Tropenkrankh. v. 1: 162. Vide Feng, 1936.
    1911.—The anatomy and life history of *Anchylostoma duodenale* Dub. part 2. Rec. Egypt. Gov't. School Med., v. 4: 159-613, pls. 11-19, figs. 101-208, photographs 1-41.
    1914.—Würmer und die von ihnen hervorgerufen Erkrankungen. Handb. Tropenkrank., 2 ed.: 433. vide Feng 1936.

LUCKER, J. T. 1934.—The morphology and development of the preparasitic larvae of *Poteriostomum ratzii*. J. Wash. Acad. Sci., v. 24 (7): 302-310, figs. 1-12.
    1934.—Development of the swine nematode *Strongyloides ransomi* and the behavior of its infective larvae. U. S. Dept. Agric. Tech. Bull. No. 437, 30 pp., figs. 1-5.
    1935.—The morphology and development of the infective larvae of *Cylicodontophorus ultrajectinus* (Ihle). J. Parasit., v. 21 (5): 381-385, figs. 1-3.
    1936.—Comparative morphology and development of infective larvae of some horse strongyles. Proc. Helm. Soc. Wash., v. 3 (1): 22-25, fig. 9.
    1938.—Description and differentiation of infective larvae of three species of horse strongyles. Proc. Helm. Soc., Wash., v. 5 (1): 1-5, figs. 1-2.

LUKASIAK, J. 1930.—Anatomische und Entwicklungsgeschichtliche Untersuchungen an *Dioctophyme renale* (Goeze, 1782) [*Eustrongylus gigas* Rud.]. Arch. Biol. Soc. Sc. & Let., Varsovie, v. 3 (3): 1-100, pls. 1-6, figs. 1-30.

MANSON, P. 1903.—Tropical Diseases; a manual of the diseases of warm climates. 756 pp., illus. London.

MARTINI, E. 1920.—Bemerkungen zur Anatomie der Microfilarien. Arch. Schiffs. & Tropen. -Hyg., v. 24: 364-370.

MOORTHY, V. N. 1938a.—Observations on the life history of *Camallanus sweeti*. J. Parasit., v. 24 (4): 323-342, pls. 1-4, figs. 1-20.
    1938b.—Observations on the development of *Dracunculus medinensis* larvae in Cyclops. Amer. J. Hyg., v. 27 (2): 437-460, pls. 1-5, figs. 1-19.

MOORTHY, V. N. and SWEET, W. C. 1936.—A peculiar type of guinea-worm embryo. Ind. J. Med. Res., v. 24 (2): 531-534, figs. 1-3.

MUSSO, R. 1930.—Die Genitalröhren von *Ascaris lumbricoides* und *megalocephala*. Ztschr. Wiss. Zool., v. 137 (2): 247-363, figs. 1-29, pls. 1-2, figs. 1-24.

OLDHAM, J. N. 1933.—On *Howardula phyllotretae* n. sp., a nematode parasite of flea beetles (Chrysomelidae; Coleoptera), with some observations on its incidence. J. Helminth., v. 11 (3): 119-136, figs. 1-3.

ORTLEPP, R. J. 1923.—The life history of *Syngamus trachealis* (Montagu) von Siebold, the gapeworm of chickens. J. Helminth., v. 1: 119-140.
    1925.—Observations on the life history of *Triodontophorus tenuicollis* a nematode parasite of the horse. J. Helminth., v. 3: 1-14, figs. 1-9.
    1937.—Observations on the morphology and life history of *Gaigeria pachyscelis* Raill. and Henry, 1910: a hookworm parasite of sheep and goats. Onderstepoort J. Vet. Sc., v. 8 (1): 183-212, figs. 1-18.

PAI, S. 1928.—Die Phasen des Lebenscyclus der *Anguillula aceti* Ehrbg. und ihre experimentell-morphologische Beeinflussung. Ztschr. Wiss. Zool., v. 131 (2): 293-344, figs. 1-80.

PENEL, 1905.—Les filaires du sang de l'homme. Paris.

RANSOM, B. H. 1913.—The life history of *Habronema muscae* (Carter), a parasite of the horse transmitted by the house fly. U. S. Dept. Agric. B. A. I. Bull. No. 163, 36 pp., figs. 1-41.

RANSOM, B. H. and FOSTER, W. D. 1920.—Observations on the life history of *Ascaris lumbricoides*. U. S. Dept. Agric. Bull. No. 817; 47 pp., figs. 1-6.

ROBERTS, F. H. S. 1934.—The large roundworm of pigs, *Ascaris lumbricoides* L., 1758; its life history in Queensland, economic importance and control. Bull. No. I Animal Health Station. Yeerongpilly, pp. 1-81; 11 figs., 2 pls.

RODENWALDT, E. 1908.—Die Verteilung der Mikrofilarien im Körper und die Ursachen des Turnus bei *Mf. nocturna* und *diurna*. Studien zur Morphologie der Mikrofilarien. Arch. Schiffs. & Tropenhyg. v. 12 (10): 18-30.
    1933.—Zur Morphologie von *Microfilaria malayi*. Meded. Dienst. Volksgezondh. Ned. -Ind. Batavia, v. 22: 54-60, figs. 1-2.

SAISAWA, 1913.—Untersuchungen über Hundefilarien. Centralbl. Bakt. Orig., v. 67: 68-75, 2 pls. 1 text fig.

SCHWARTZ, B. and ALICATA, J. E. 1931.—Concerning the life history of lung-worms of swine. J. Parasit., v. 28: 21-27, pl. 1, figs. 1-8.
    1935.—Life history of *Longistriati musculi*, a nematode parasitic in mice. J. Wash. Acad. Sc., v. 25 (3): 128-146, figs. 1-14.

SEURAT, L. G. 1913.—Sur l'évolution du *Physocephalus sexalatus*. Compt. Rend. Soc. Biol. Paris, v. 75: 517-520, figs. 1-4.
    1914.—Sur l'évolution des nématodes parasites. 9th Congrès. Internat. Zool., Monaco. pp. 623-643, illus.
    1915.—Sur les premiers stades évolutifs des spiroptères. Compt. Rend. Soc. Biol., Paris, v. 78: 561-564, figs. 1-5.
    1916.—Contributions a l'étude des formes larvaires des nématodes parasites hetéroxénes. Bull. Sci. France & Belg., s. 7, v. 49: 297-377, illus.
    1917.—Filaires des reptiles et des Batraciens. Bull. Soc. Hist. Nat. Afr. Nord. v. 8: 236.
    1919.—Contributions nouvelles a l'études des formes larvaires des nématodes parasites hetéroxénes. Bull. Biol. France et Belg. (1918) 52: 344-378, illus.
    1920.—Histoire naturelle des nématodes de la Berbérie. Première partie. Morphologie, développement éthologie et affinitiés des nématodes. 221 pp., 34 figs. Alger.

STEINER, G. 1924.—Some nemas from the alimentary tract of the Carolina tree frog (*Hyla carolinensis* Pennant). J. Parasit. v. 11: 1-32, pls. 1-11, figs. 1-65.

STEKHOVEN, J. H. S. 1927.—The nemas *Anchylostoma* and *Necator*. Proc. Roy. Acad. Amsterdam. v. 30: 113-124, figs. A-B, pls. 1-3.

SWALES, W. E. 1936.—*Tetrameres crami* Swales, 1933, a nematode parasite of ducks in Canada. Morphological and biological studies. Canad. J. Res. Sec. D, v. 14: 151-164, figs. 1-10, 1 pl.

TEUNISSEN, 1939.—In Stekhoven (1939) Nematodes. Bronn's Klassen und Ordnungen des Tierreichs, v. 4, Abt. 2, Buch 3, Lieferung 6, pp. 499-511, figs. 94-98.

VEVERS, G. M. 1921.—On some developmental stages of *Ancylostoma ceylanicum* Looss, 1911. Proc. Roy. Soc. Med., v. 14: 25-27.

VOGEL, R. 1925.—Zur Kenntnis der Fortpflanzung, Eireifung, Befruchtung und Furchung von *Oxyuris obvelata* Bremser. Zool. Jahrb., Abt. Zool. & Phys., v. 42: 243-270, Figs. A-T, 1 pl.

WEHR, E. E. 1935.—A revised classification of the nematode superfamily Filarioidae. Proc. Helm. Soc. Wash., v. 2 (2): 84-88.
    1939.—Studies on the development of the pigeon capillarid, *Capillaria columbae*. U. S. Dept. Agric. Tech. Bull. 679, 19 pp., 3 figs.

WETZEL, R. 1931.—On the biology of the fourth stage larva of *Dermatoxys veligera* (Rudolphi 1819) Schneider 1866, an oxyurid parasitic in the hare. J. Parasit., v. 18: 40-43, figs. A-B.

WÜLKER, G. 1929.—Ueber Nematoden aus Nordseetieren, I, II; in Zool. Anz., v. 87 & 88.

WÜLKER, G. and STEKHOVEN, J. H. S. 1933.—Nematoda. Allgemeiner Teil. Die Tierwelt der Nord- und Ostsee. Teil 5a. 64 pp., figs. 1-65.

YAMADA, S. 1927.—An experimental study on twenty-four species of Japanese mosquitoes regarding their suitability as intermediate hosts for *Filaria bancrofti* Cobbold. Sci. Reports Govt. Inst. Inf. Disease. v. 6: 559-622, 3 pls.

CHAPTER XVI

NEMIC RELATIONSHIPS

B. G. CHITWOOD

Even to enter a discussion of nemic relationships invites criticism and focuses a spotlight on one's ignorance of zoology. It is difficult for a scientist to obtain sufficient knowledge to speak with accuracy concerning his specialty and whether or not he can do so outside his specialty remains to be seen. Undoubtedly there are as many separate theories as there will be readers of this chapter. Past and present theories, with their evidence will be presented. We hope and expect future workers will feel as much superior to us as we do to Bastian when we read ''Having now pretty fully explained the anatomy of the Nematoids, we shall be able, with the aid of the many new facts revealed concerning their structure, to consider the question of their affinities and homologies with more chance of success than formerly, so that we may hope to throw some light upon this difficult subject.''

Past Theories

TAXONOMIC ERA. The original relationships attributed to zoologic groups were based on casual knowledge of gross anatomy and were typically taxonomic in character. Thus Linnaeus (1758) divided the Animal Kingdom into six classes; Mammalia, Aves, Amphibia, Pisces, Insects, and Vermes. In the last mentioned group were included all the then known helminths as well as the Protozoa, Mollusca, and Annelida.

In 1800 Zeder established the groups of parasitic worms essentially as they are recognized today, the groups being Roundworms, Hooked-worms (spiny heads), Suckered worms, Flatworms, and Bladderworms; the same groups were named by Rudolphi (1808-09) Nematoidea, Acanthocephala, Trematoda, Cestoidea, and Cystica, all except the last of which have survived until the present time.

With the discovery in these animal groups of excretory and nervous systems, and with superficial knowledge of the body cavity and ontogeny, the classifications were expanded by Huxley and Haekel who suggested approximately all of the possible relationships. Huxley's first classification (1856) divided the subkingdom into two phyla, Articulata and Annuloidea, the later group including the classes Annelida, Echinodermata and Scolecida. The entire Entozoa, as well as the Rotifera and Turbellaria, were included in the Scolecida, thus the Scolecida would correspond to the term unsegmented worms in the broadest sense. The phylum Annuloidea was then characterized as having a water-vascular system. Shortly thereafter (1864) Huxley redivided and renamed his groups, characterizing the Annulosa as having a double chain of ventral ganglia and including therein the Arthropoda (= Articulata) and Annelida, leaving those without such a nervous system, Echinodermata and Scolecida, in the Annuloidea; he listed the classes Rotifera, Turbellaria, Trematoda, Taeniada (Cestoda), Nematoidea, Acanthocephala, and Gordiacea as comprising the Scolecida. On the basis of presence or absence of external ciliation he put forward, in 1878, the subdivisions Trichoscolecida for the first four classes and Nematoscolecida for the remaining classes plus the Gastrotricha *despite* ciliation in some members of the latter.

The most conspicuous feature in all discourses on nemic relationships from the time of Huxley to the present, is that the Nemathelminthes concept (including classes Nematoda, Acanthocephala, and Gordiacea or Nematomorpha) is not and never has been accepted by comparative anatomists though it is seemingly entrenched in zoologic literature.

Without attempting to go into detail, we shall outline the general theories with the points brought forward by the various authors, then give a comprehensive description of the primitive nematode, followed by an evaluation of the interrelationships of the various groups.

1. ECHINODERMATA-SCOLECIDA THEORY. This grouping, originating with Huxley (1856, -64, -78), received its only support from Bastian (1866) who saw the Nematoda as bridging the gap between the echinoderms on one side and through Acanthocephala to other scolecids on the other side. This theory was based entirely on the homology of the ambulacral and water-vascular systems. Though the ambulacral system opens on the dorsal side in many echinoderms (Holothuroidea and Crinoidea exceptional) and the water-vascular system on the same side in rotifers, the corresponding system opens on the ventral side of nematodes. Bastian states ''for purposes of transcendental anatomy'' it makes no difference which side is dorsal as long as the other organs (anus, reproductive openings) retain their relative positions.

2. ANNELID-CHAETOGNATH-NEMATHELMINTH THEORY. The earth-worm-ascarid comparison is by far the most natural since here two of the oldest known worms were originally considered congeneric and were subsequently separated. In 1866 Schneider classified all Vermes, as he knew them, on the basis of musculature. These groups were:

I. Nemathelminthes—skin and muscle tissue of body wall in two layers.
  1. With one layer of longitudinal muscles, or two layers, the outer being circular and the inner longitudinal; lateral chords absent in latter.
     (a) Unsegmented. Longitudinal muscles only.
        Nematoidea
        Chaetognatha
     (b) Segmented
        (1) Longitudinal fibres only, median chords only —Gymnotoma (for *Ramphogordius* ''a segmented *Gordius*'').
        (2) Longitudinal and transverse fibres—Chaetopoda.
  2. Inner longitudinal and outer circular muscles; lateral chords absent.
     Acanthocephala
     Gephyrea

II. Plathyhelminthes—Muscle fibres embedded in skin tissue, longitudinal, circular, and sagittal all one layer.
  1. Oblique cross fibers present.
     Trematoda, Dendrocoela, Hirudinea, Onycophora.
  2. Oblique cross fibers absent.
     Cestoidea, Rhabdocoela.

As Schneider later (1873) stated, he viewed *Lumbricus* and *Nereis* as annulated ascarids just as he considered *Polygordius* an annulated *Gordius*. The idea of chaetognath-nemathelminth relationships has only persisted in a few of the archaic textbooks and we understand the chaetognaths are now regarded by many workers as having lower chordate relationships. Bütschli (1875) pointed out that nematodes are embryologically unsegmented while *Sagitta* (Chaetognatha) is composed of three segments.

Vejdovsky (1886) proposed the group named Nematomorpha for the Gordiacea and the odd marine genus *Nectonema* and related the whole class to the annelids rather than to the nematodes. J. Thiel (1902) was particularly impressed by Nematomorpha-''annelid'' relationships pointing out that the female solenogaster, *Neomenia*, has two posteriorly opening gonocoeles into which numerous germ sacs empty. Such relationship was apparently approved by Rauther (1909) who stated that the only similarity between nematodes and gordiids is that both are long thin worms. For the acanthocephalans Rauther suggested gephyrean relations dwelling to considerable extent upon *Sipunculus*. As will be seen later, Rauther viewed nematodes as primarily related to arthropods.

Greeff (1869) was so impressed by the secondary and superficial segmentation of *Desmoscolex* that he felt this genus to be, indeed, the connecting link between annelids and nematodes and for many years the Desmoscolecidae were excluded from the Nematoda or considered an aberrant appendage. Not until Schepotieff (1908) studied the group and found its members to be internally of typical nemic structure, were the Desmoscolecidae accepted as part of the Nematoda.

3. ARTHROPOD THEORIES. In reality there seem to be two schools of thought regarding nemic-arthropod relationships; the first school provides for the nematodes through degeneracy, the other through ascendancy, but at times the two theories seem to merge!

A. *Degeneracy.* Perrier (1897) seems to have originated the idea that free-living nematodes should be regarded as having originated secondarily from parasitic forms. The viewpoint was at that time more or less natural since the majority of zoologists were ignorant of free-living nematodes. Perrier's series Chitinophores, based on a "chitinous covering" and molts, included the Nemathelminthes with classes Nematoda, Acanthocephala and Gordiacea. This group was set entirely apart from the series Néphridés which included Rotatoria, Gastrotricha, Turbellaria, Trematoda and Cestoda.

Hubrecht (1904) followed with a theory in which it was plainly stated that free-living nematodes are much too "simple" to be archaic stem forms and must, instead, be secondarily degenerate products of the Arthropoda.

Rauther (1909) thereafter frankly invoked neoteny* to explain all things in anatomy and relationship of worms. A direct quotation relating to possible gordiid-echinoderid relationship will serve best to illustrate this course of thought (p. 502). "der *Gordius*-Larve schreibe ich ebsowenig die bedeutung einer Rekapitulation einer primitiven Gordien-Stammform zu, wie ich in den äquivalenten Larven der höhern Würmer und Mollusken Abbilder von deren Vorfahren sehe. Die Auffassung der Echinoderiden als neotenischer Articulaten-Abkömmlinge wird ferner gerade durch ihre unzweifelhaften Aehnlichkeiten mit Nematoden gestützt (s.u.) welche letztern, wie sich im Verlauf dieser Ausführungen ergeben dürfte, selbst in analogem Verhältnis zu Articulaten-Vorfahren stehen." Further on, in the same publication he determined that the rotatorian is not a primitive form but the neotenic branch of a higher group reduced to the trochophore stage; that the nematode developed from a land-dwelling arthropod and that the search for ancestral groups is useless since they are phantoms.

Rauther's comparison of the triradiate pharynx of *Anopheles* larvae with the triradiate nematode esophagus and tardigrade pharynx is ingenious. He placed considerable weight on the stylet of tylenchids as compared to the paired stylets of tardigrades and dipterans; molting, rachis development of gonad, and rectal glands (as compared with malpighian tubules) were also cited as evidence of relationship. The Pentastomida (Linguatulida) were placed by Rauther as an intermediate group constituting a connecting link between nematodes and dipteran larvae.

Martini (1908) after previously (1903, 1907) establishing that meromyarian nematodes are ontogenetically primitive, switched to the view that the rhabditoid larva is an ontogenetic stage and *Rhabditis* is neotenic rather than primitive.

In the face of growing knowledge of free-living nematodes Keilin (1926) reiterated the "theory of degeneracy" without adding any new facts or thoughts.

This theory has cast its shadow over all nemic classification since it simplifies grouping so greatly. If, as Rauther suggests, ontogenetic stages are of no significance, then one may sidestep all difficult points. Concerning the invocation of neotony to explain evolution within the Nematoda, Chitwood (1937) has stated, "Ontogeny supports the view that few-celled forms are more primitive than many celled forms. In the writer's opinion, the converse assumption removes the study of nemic phylogeny from the realm of logical thought."

B. *Ascendancy.* Those who have subscribed to the ascendancy idea have not necessarily believed that nematodes gave rise to arthropods but only that nematodes may be a branch coming off from the stock which eventually gave rise to insects or that free-living nemas developed from an insect-like ancestor and later gave rise to parasitic nemas. It should be noted that in no instance has an attempt been made by such observers to explain aquatic arthropods. Instead, it is apparently assumed that the Arthropoda must be a polyphyletic group.

Greeff (1869) being impressed with the secondary and superficial segmentation of marine nemas such as *Desmoscolex*, *Greeffiella*, and *Draconema* believed that these forms together with the Echinodera provided a connecting link between the Nematoda and the Arthropoda.

Bütschli (1876) supplied a much more substantial argument for common parentage of the two groups. As like characters he cited the absence of ciliation, occurrence of molting, presence of a nerve ring and ventral median nerve; he, also

homologized the trachea of insects with the lateral canals of nematodes and the nephridia of annelids. Bütschli further suggested that the caudal furcae of echinoderids may correspond to the arthropod foot and incidentally placed the Tardigrada as low arthropods. Differences in the mode of jointing in arthropods and annelids, differences in cleavage and in the vascular systems caused him to separate them into two stem lines.

Ganin (1877) studied the nervous system of *Rhabditis* and because of the circum-esophageal commissure, double subventral nerve trunks and ventral chain of ganglia, he placed the Nematoda in the general arthropod series.

Maupas (1899) emphasized that nematodes molt in their life cycle as do arthropods and he even went so far as to compare the ontogeny of nemas with that of heteroceran lepidopterans in which four molts occur during larval development and two in adulthood. The encysted stage (ensheathed larva, resistant stage, dauer-larva) of nematodes and its ability to withstand adverse environmental conditions was compared with similar stages in tardigrades and rotatorians. However, Maupas could not accept rotatorian-nemic relationship because he was unable to find rotatorians molting.

Seurat (1920) presented by far the most comprehensive and well-founded arguments for common ancestry of nematodes and arthropods. First he defined the primitive nematode on the basis of habitat and comparative anatomy as follows: saprozoic, humid media, bilateral symmetry, mouth subterminal and ventral, three lips, tail thick and conical with three caudal glands, cuticle smooth, sensory papillae sparse, epidermis composed of four bands, four muscle fields, meromyarian, unicellular lateral glands. Mouth tubular, esophagus, triradiate, and terminated by a valved bulb, intestine composed of few large cells, occasionally a caecum, rectum short with three glands. Excretory system with paired lateral canals opening laterally or without lateral canals but with a unicellular gland opening ventrally; sometimes with a secondary system opening posteriorly. Sexes separate, males with numerous genital papillae, two testes extending parallel anteriorly and opening posteriorly in a vas deferens and ejaculatory duct; a little anterior to anus large paired cement glands opening into ejaculatory duct; two spicules and a gubernaculum. Female with two parallel ovaries opening posteriorly through simple vagina, oocytes produced in small numbers. Segmentation total, unequal; four molts to adulthood.

Having stated his concept of the primitive nematode he recognized that such a form would combine characters of oxyurids and rhabditids. A common ancestral stem line with rotatorians and turbellarians would, according to Seurat's view, be eliminated due to the molts, chitinous cuticle, type of musculature, lack of cilia, form of gut, type of reproductive organs, and separation of sexes. All of these characters clearly point toward arthropods. The cuticular lining of the anterior arthropod gut and its differentiation posteriorly into a proventriculus, the existence of a caecum in the mid-gut (=chyliferous diverticulae), insignificance of hind gut, and the existence of malpighian tubules (as compared with anal glands) must (vide Seurat) be explained otherwise than by convergence with nematodes. The male reproductive system and spicules of nemas are compared with the testes and penis of insects. In the female each entire nemic ovary is compared to a single unit of the insect ovary; reduction in number of ovarian tubules is cited as evidence of relationship, such reduction having been reported by Cholodowsky (1908) in the dipteran, *Theria muscaria* Meig. Finally Seurat differed with Bütschli (1876) in that he homologized the lateral canals with the sericigenous glands of microgasters rather than with trachea. He pointed out the fallacy of Rauther's comparison of the triradiate pharynx of Anopheles larvae, Seurat's argument being that the three muscle bands or sectors extend from the *body wall* to the *external surface* of the pharynx and are hardly comparable to the radial muscles of nemic esophagi; rarity of the occurrence of a stylet in free-living nemas and its obviously secondary adaptive function in feeding were pointed out as evidence of the lack of phylogenetic significance of stylets. Seurat then asked, if in the presence of the evidence, one must not think of the primitive nematode as being a larva of the Holometabola adapted to detriticolous life, having lost all segmentation, become adult and sexual after having fulfilled its normal molts but preserving infantile characters. As a parallel instance of neoteny, the coleopterous malacoderm, *Phengodes*, according to Haase (1888), has a larval female possessing normally constituted genitalia and producing fertilized eggs. As contrary evidence Seurat cited the following differences between nematodes and insects:

---

* De Coninck (1938) has recently resurrected the term "Eutely" given by Martini (1923) to the phenomenon of cell constancy. He also made clear the distinction between this phenomenon and neoteny, attributing eutely to "very rapid segregation of all potentialities of the egg" while neoteny "is the result of hormonal deficiency". As to the distinction, the writers are in complete agreement.

Absence in nematodes of any trace of segmentation or articulate appendages and anything corresponding to the trachea, as well as the divergence in cleavage of the egg. (The latter he felt could be explained through the presence of an abundance of yolk in the insect egg).

Baylis (1924) reviewed the theories extant and concluded that nemic-insect (arthropod) relationship is probable on the basis of the common cuticular esophageal (pharyngeal) lining, malpighian tubule-rectal gland homology, tubular form of gonads, homology of penis and spicules, metameric arrangement of setae in nematodes, common absence of cilia, paedogenesis of insect larvae, and molting. More recently (1938) Baylis expressed the view that the origin of nematodes is uncertain, perhaps in a very remote period nematodes and arthropods had a common ancestor but it would be unwise to press the suggestion since at the present time we do not know whether the conditions exhibited by dipterous larvae are primitive or secondarily adapted.

It will be seen from the above résumé that of all of the proponents of common nemic-arthropod relationships, Bütschli alone proposed a theory of descent presupposing direct rather than regressive evolution and he placed the Tardigrada as primitive arthropods. The majority of Seurat's points would be as acceptable to the concept of progressive as to the concept of regressive evolution. Paedogenesis might be even considered an atavistic tendency of insects. It would still be necessary to account for the origin of aquatic arthropods in order to accept progressive nemic-arthropod relationships but such an explanation is entirely unnecessary to the regressivists.

4. SCOLECIDAN (PROTONEPHRIDIAL) THEORY. As previously noted, this theory is traceable directly to Huxley (1856) but it has undergone many modifications both by the original author (1864, 1878) and by other workers, the chief of whom was Bütschli (1876). This theory in substance, provides for the union of all "unsegmented worms" in one superphyletic group just above the coelenterates and ctenophores. All higher forms of life are supposed to have arisen from lower ancestral (primitive, extinct, rhabdocoele or rotatorian-like) scolecidans. Such a view presumes phylogenetic significance of the trochophore larva and is very close to the consensus of present day zoologic opinion. Disagreements relate to the subdivision of the "Scolecida" into its major series, phyla and classes.

Haeckel (1872, 1896) revised the Animal Kingdom on the basis of his "Gastraea Theory" placing the forms with neither body cavity nor anus in the Acoelomati; he accepted the common Platyhelminthes (renamed Platodes) grouping (Turbellaria, Trematoda, Cestoda) considering the platyhelminths as coelenterates and for them hypothesizing a simple gastrula-like ancestor with protonephridia. The Acoela were placed as the most primitive living worms and the Rhabdocoela as ancestors of all higher animals. Such an ancestral form is described as having two testes and two ovaries, a muscular stoma, no anus, a parenchymatous body cavity, an epithelial brain and an incompletely differentiated mesoderm. The formation of a body cavity, according to Haeckel, should be considered as a regression from a previously parenchymatous state. In his earlier revision (1872) he listed Rotatoria and Nematoda in the Coelomati. Later he revised this group into the "True Vermes" composed of the following phyla: 1. Rotatoria (with Gastrotricha as oldest and in turn descendent from Rhabdocoela). 2. Strongylaria (a) Echinocephala [(=Echinodera), ancestral group, descendant in turn from Gastrotricha]; (b) Nematoda (with gordiids as most primitive forms because of parenchyma): (c) Acanthocephala: (d) Chaetognatha. 3. Prosopygia (=Molluscoidea), and 4. Frontonia (=Nemertea). Rotatorian-like Trochozoa ancestors were assumed for the Molluscoidea, Nemertea, and Echinodermata, while annelids were derived from nemerteans. Arthropods were derived biphyletically from the Chaetopoda in two lines, —one the Crustacea, the other the Tracheata. Chordates were derived from trochophore-like ancestors in common with those which gave rise to the Nemertea. Pentastomes and tardigrades were both included in the Annelida.

Bütschli formed the group Nematorhyncha to include the Gastrotricha and Echinodera (Atricha); he related both of these groups to rotatorians on the basis that the somatic musculature in all three groups does not form a tube but consists of isolated cells extending through the body cavity as in a Pilidium. He also considered these forms as close to the ancestors of arthropods. Nematodes and nematorhynchs were closely associated with each other because of the superficial similarity of echinoderids to gordiid larvae, the similarity of the water vascular (excretory) systems of nematorhynchs and nematodes, and the similarity of the musculature

of meromyarian nematodes to the musculature of gastrotrichs. Complete absence of circular muscles was pointed out as the chief factor separating rotatorian-nematode-nematorhynch series from annelid-gephyrean-platyhelminth series. The uniting of reproductive and digestive systems in male nematodes, in both sexes in gordiids, tardigrades and low arthropods he cited as evidence of their common ancestry. The excretory systems of platyhelminths, rotatorians, gastrotrichs and nematodes were considered undoubtedly homologous, while the tracheal system of insects and the metamerically segmented organs of annelids were considered divergent offshoots of the same system.

Stimulated by Gaffron's diagram of the nervous system of an ectoparasitic trematode, Bütschli (1885) compared it with that of a nematode and judged therefrom that a common ancestor must have existed. The dorsal brain and lateral nerves of the trematode need only to have bent ventrally forming a commissure in order to form a plan like that of a nematode. The lateral (amphidial) by-pass (lateroventral commissure) of nematodes pre-exists in trematodes.

Zelinka (1896) supported the opinion that gastrotrichs and rotatorians must have been derived from a trochophore ancestor and that echinoderids and nematodes probably arose from gastrotrichs.

Zacharias (1885) felt that he had established beyond doubt the common ancestry of nematodes and rotatorians on the basis of similar development (? bilateral cleavage).

Grobben (1910) crystallized the formation of a rotatorian-nematode group naming it the Aschelminthes and differentiating it from the Platyhelminthes on the basis of body cavity vs. parenchyma. In this group he included Rotatoria, Gastrotricha, Echinodera, Nematoda, Nematomorpha and Acanthocephala.

Martini (1913) considered the possible relationship of nematodes to both platyhelminths and arthropods and concluded that nematodes by possessing a hind gut are higher than platyhelminths, that the rectal glands of nematodes are homologues of tardigrade and insect malpighian tubules and that the excretory system of nemas might have had a separate origin and might not be homologous with that of platyhelminths.

Steiner (1919, 1920) subscribed to the general concept of Bütschli (1876) but was more explicit in the comparison of organs in nematodes and rotatorians. In general, it was his conception that nematodes developed from organisms similar to the philodinid rotatorians. He described the primitive nematode as a partially sedentary form with an H-shaped excretory system, valved esophageal bulb, caudal glands, and having vulva, anus, and excretory pore opening together (a separate orifice of the excretory pore somewhat posterior to the position in present day nematodes was considered as a possibility). He considered the mixture of radial and bilateral symmetry in nemas as due to their change from a mobile to a semisessile life. Bilaterality is associated with the mode of locomotion (dorso-ventral oscillation) of nematodes and radial symmetry with a sessile mode of life. Bilaterality occurs in the excretory system, musculature, and chords while radial symmetry occurs in the lips, mouth, and esophagus. As he conceived it, the appearance of radial symmetry coincided with the loss of cilia in the anterior part (corpus) of the esophagus and occurred when the original nematode was formed. This was a semi-sessile form. A plurality of present day free-living nematodes are partially sessile and when they move it is on the longitudinal axis; neither side is flattened for they do not normally rest in a prone position; in a dish of water they lie on their side abnormally. Apparently, he hypothesized, there was a primary motility, a secondary sessility (at origin of nematodes) and a tertiary remotility (within the Nematoda). Only this hypothesis could explain the mixtures of symmetries known to occur. Comparing the primitive nematode with the rotatorian he found it to conform in cuticle, hypodermis, presence of primary body cavity, divisions of gut (anterior, middle, and posterior), presence of caudal glands, original round form of body, and that possibly the musculature traces back to common ancestry. He considered the possibility of an homology of the dorsal side of the rotatorian with the ventral side of the nemic body, but rejected this hypothesis because of the embryology. The ventral side corresponds to the open side of the gastrula, identical in both groups. Measured from the posterior, the relative positions of the anus, reproductive and excretory system orifices are the same in the two groups; hence the orifice of the excretory system could first have separated from the cloaca and thereafter the vulva of the female separated from the gut orifice.

The variability in position of the vulva was cited as evidence of recentness of its separation. Paired parallel gonads were considered primitive for both sexes although no such example is known in free-living nemas. Ciliation of the anterior gut was considered primary; the mastax homologized with the esophageal bulb of *Rhabditis*, radial symmetry developing from bilateral. The amphids and accompanying glands were homologized with the retrocerebral organ and subcerebral glands of *Callidina*. On the whole, his comparison seems apt but the mastax is actually triradiate in symmetry, secondarily bilateral, and the ciliated anterior gut of rotatorians is a recent acquisition (secondary invagination), the primary stomodeum forming the mastax and esophagus (=esophago-intestinal valve of nemas).

More recently Remane (1928, 1929) has expressed the view that gastrotrichs are near the ancestor of both nematodes and echinoderids and that they in turn developed from archiannelid-like (trochophore) progenitors. A trochophore origin was also suggested for rotatorians but the unsuitability of the Trochhelminthes (Zelinka included rotatorians, gastrotrichs and echinoderids) as a systematic group was pointed out. Rotatorians differ from gastrotrichs and echinoderids, as well as from nematodes, by having circular muscles, a ciliated foregut, and bilateral mastax all of which are secondary developments. Remane subscribed to the Aschelminthes concept.

The Primitive Nema

Having brought forward the various theories to account for nemic origin we need examine them in the light of present knowledge. In order to do this a picture should be formed of the primitive nema. Considering the various types of evidence customarily available we find some forms of knowledge conspicuous by their absence, others by their richness.

GEOLOGY. Taylor (1935) reviewed the knowledge of fossil nemas and listed a total of six species of which two were mermithids while the remainder were free-living nemas. The mermithids [*Heydonius antiquus* (v. Heydon, 1862) and *H. matutinus* (Menge, 1866)] were collected from Rhine lignite (Eocene) and Baltic amber and the free-living nemas [*Oligo-*

plectus succini (v. Duisburg, 1862), *Vetus duisburgi* Taylor 1935, *V. pristinus* (Menge, 1866) and *V. capillaceus* (Menge, 1866)] were all found in Baltic amber (lower Oligocene). As no morphologic details can be distinguished the records are of no advantage from that standpoint; neither are they of value in determining the origin of nemas since the group obviously must predate the Tertiary. These records, however, help us to understand why fossil nemas are so rare. The nema has no hard parts such as insects have for preservation and a nema must be included in a deposit such as rosin or something of exceedingly fine grain. Repeated examinations of fossil shales and ambers by the writer have been uniformly negative.

DISTRIBUTION: Among the free-living nematodes there seems to be little correlation between geographic distribution and species. Both saprophagous and predatory forms seem to be more specialized as to medium than to continent. Thorne (1929) found certain aquatic European species common on western mountain peaks in this country. Under such circumstances one could hardly assume introduction of the organisms by man. However, great advances in taxonomy have been made during the past 20 years and today such a finding would have to be verified before we could consider the natural occurrence of species on two continents as a fact. Likewise, identical marine species have been collected from European and American Atlantic coasts but this happening is relatively rare.

We can accept the wide general distribution of some saprophagous and plant parasitic species quite readily when we know that they move in or on seeds, nursery stock, tools, equipment, and in soil. The same thing may be said for the majority of nematode parasites of domestic animals. Ready world-wide dispersal is only limited by such factors as climate, topography, or intermediate hosts. It is rather difficult to establish the nativity of these organisms unless one studies the fauna of virgin lands. Very little work of this nature has been done. We do know many nematode parasites of animals are practically coincidental with their hosts. This was formerly believed to be the case with most of the plant parasitic nematodes but we have recently learned that organisms thought to be the same species, with world-wide distribution, are actually quite separate and distinct. The entire matter must be reconsidered.

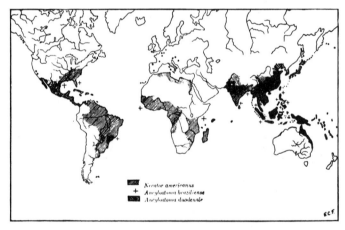

FIG. 143.

Geographic distribution of *Necator americanus* and *Ancylostoma duodenale*. From Craig & Faust, Clinical Parasitology, Lea & Febiger.

The parasites of man have received considerable attention from the standpoint of distribution. The "American" hookworm, *Necator americanus*, is distributed over a large part of the earth and it could easily have been imported to the Americas with negroes from Africa (Fig. 143). Two filariids of man, *Wuchereria bancrofti* and *Dipetalonema perstans* (syn. *Acanthocheilonema perstans*) seem to have originated in Europe or Asia and later to have been brought to the Western World (Fig. 144).

Evidence of the origin of "physiological races" of a species in two or more hosts (ex. *Ascaris lumbricoides* in pig and man) may indicate at least one line of evolution. Even this parasite should be reinvestigated by critical morphologic study and experimental work.

Fossil nemas. A—*Heydonius antiquus*; B-C—*Heydonius matutinus*; D—*Vetus pristinus*; E—*Vetus capillaceus*; F—*Oligoplectus succini*; G-H—*Vetus duisburgi*. After Taylor, 1935, Proc. Helm. Soc. Wash. v. 2 (1).

F.G. 142.

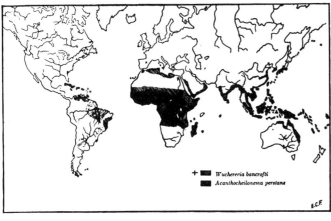

FIG. 144.

Geographic distribution of *Wuchereria bancrofti* and *Dipetalonema perstans* (*Acanthocheilonema* p.). From Craig & Faust, Clinical Parasitology, Lea & Febiger.

"Physiologic races" have often been postulated to explain differences in host range, symptoms, or climatic behavior of nematode parasites of plants. We have passed through this stage of knowledge in nearly every genus of plant parasites, i.e., *Heterodera, Meloidogyne, Aphelenchoides, Ditylenchus*. Such postulations have, in every case, been followed by morphologic studies which resulted in the recognition of species. Some of these species are exceedingly variable, others quite stable. In general, the more stable the morphologic characters, the narrower the host range. Quite recently we have found morphologic variations in the progeny of individual female nematodes. It seems possible that they may represent multiple genetic factors. Host selection may easily be the key to formation of "specialized races." In the genus *Meloidogyne* we encountered such strange happenings as the reproduction in but one female out of 10,000 specimens on the wrong host. Left on such a host they have died out. What would happen if they were then propagated on a more favorable host and reintroduced to the unfavorable host? If the factors are cumulative we should be able to establish a higher rate of reproduction on the abnormal host. Alternation of hosts is the usual occurrence in the field. It is also the natural step in host adaption.

Aside from the natural selection through the accumulation of multiple factors, we are also probably dealing with some mutants or sports. Thus we have found individuals which differed in superficial appearance on the two sides of the body which may well be compared to a vegetative hybrid or chimera. If such somatic mutants are possible, we must accept the fact that genetic mutants are also possible. Evidence of this was presented by Graham in the case of *Strongyloides ratti*, by Dougherty (1949) in the case of *Rhabditis briggsae*. Thomas (1949) has artificially produced mutants in the cases of *Ascaridia galli* and *Rhabditis strongyloides*. While rare productive cases of interspecific hybridization must occur and contribute to the recombination of heredity, these have not as yet been proven in the Nematoda. We may well suspect some mutants to be a result of chromosomal aberrations such as translocation of the gene, gain or loss of parts of or of while chromosomes, or failure of a maturation division. Evidence of polyploidy is at hand in the case of *Parascaris equorum* where three varieties, namely *univalens, bivalens* and *trivalens* have been described.

In order to better understand the mechanism of evolution we must think in terms of multiple factors, mutants, ploidy, interspecific hybridization, and natural selection as combined and interacting factors. Organisms vary in all possible directions within the scope of their heredity. Environment in the form of natural selection provides for the survival only of those organisms best adapted to a given set of conditions. Thus, the bulk of a population may be destroyed by an alteration of the environment. The rare survivors set up a new center of variation. Like their predecessors, they also vary in all directions within their capacity but their capacity has been changed. Successions of climactic changes would, of necessity,

cause evolutionary series which would give the appearance of purposive evolution.

We have such strange cases as that of *Dracunculus dahomensis* originally described from pythons from the old world, later collected in American Zoos from South American constrictors. Whether or not these organisms represent Zoo host transfers of the parasite, we cannot say. According to herpetologists, these two groups of constrictor snakes had separate evolutionary origins. The nematodes would seem to be of a contrary opinion. The distribution of *Dracunculus medinensis* represents a similar disturbing problem. This species is known in the old world as a parasite of man. Domestic and wild mammals are not supposed to be involved in its normal epidemiology despite the fact that Moorthy artificially infested dogs with the species. The *Dracunculus* of mammals in this country has been called *D. insigne* by Chandler primarily because it is not known as a parasite of man. Careful investigation provides no valid means of distinguishing the females or larvae and the males differ from those described by Moorthy only in the presence of one additional pair of male genital papillae and a somewhat longer gubernaculum. Pending comprehensive quantitative study a subspecific rank for the American organism would seem to be preferable.

Steiner (1917) discussed the habitat distribution of free-living nemas, pointing out that there are two chief faunas, one terrestrial and the other marine. The fresh water fauna is more closely related to the terrestrial than to the marine and interchange is not particularly common, there being no typical brackish water fauna. Following the suggestion of Simroth (1891) he concluded that terrestrial nemas are more primitive than marine. This view also presumes that other early forms of life (i. e., Bacteria, Algae, Rotatorians, etc.) were originally terrestrial or fresh water and subsequently marine.

HOST GROUPS. Among the parasites of animals we recognize several evolutionary series. The Strongylina make their first appearance in amphibians and reptiles and are found in birds and mammals but not in fish. All strongylins of marine mammals belong to the Metastrongyloidea. Furthermore the Strongyloidea and Trichostrongyloidea (both without intermediate hosts) are parasites of all groups above fish while the Metastrongyloidea are confined to birds and mammals. On that basis one would say that without doubt the Strongylina originated with the amphibians or reptiles and that the adaptation to intermediate hosts was secondary. We would presume the Strongylina to have arisen with the Amphibia and probably from a rhabditoid with two subventral excretory glands similar to those of *Rhabditis strongyloides*.

In the Ascaridina, the Oxyuroidea include parasites of insects, millepeds, amphibians, reptiles and mammals with a few aberrant species (*Rondonia rondoni, Monhysterides piscicola*) inhabiting fresh water fish. These latter infections seem to have occurred secondarily. The Oxyuroidea seem to have originated with the millepeds, scarabeids and cockroaches. It is particularly interesting to note apparently identical species of nemic parasites in millepeds from Australia and from the United States. In the Ascaridoidea, which occur in land molluscs, fish, amphibia, reptiles, birds, and mammals, one comes again to the problem of intermediate hosts. All nemic parasites of fish and marine mammals have intermediate hosts but intermediate hosts occur in the development of ascaridoids parasitic in all other vertebrate groups. The intermediate host is known to occur only in one subfamily of the Ascaridoidea, the Anisakinae. Tiner (1949) has shown that *Ascaris columnaris* may utilize various rodents as facultative intermediate hosts. It would appear that here we have the first step in the evolution of an indirect life cycle in the Ascaridoidea. The family Cosmocercidae which is morphologically the simplest (i. e., most rhabditoid) includes species that occur in amphibians and reptiles, with occasional aberrant forms in snails. The Kathlaniidae are more typically parasitic in reptiles, while the Heterakidae are reptile, bird and mammal parasites. Excluding morphologic considerations, one might argue that the Anisakinae (Ascarididae) are the most primitive since they occur in fish. However, secondary entrance of Anisakinae into fish which came up rivers to feed would be ample explanation of their parasitism. Absence of any aquatic (marine) nema with lateral excretory canals makes fresh water or more typically land origin mandatory.

231

We presume the Ascaridoidea arose from the Oxyuroidea, hence from insect parasites. These, in turn, probably originated from a rhabditoid with an H-shaped excretory system similar to that of *Rhabditis dolichura*.

The origin of the Spirurida is indeed dubious and possibly lost in antiquity. No adult forms are found in insects, millepeds or other invertebrates. The Camallanoidea are characteristically parasites of fish and reptiles while the Dracunculoidea and Spiruroidea are parasites of fish, amphibians, reptiles, birds and mammals, and the Filarioidea are parasites of all vertebrate groups except fish. The latter (fish) are the lowest normal final host of the Spirurida. Since all members of the order require an intermediate host, it might be logical to assume an aquatic origin for the group. However, there is no reason for assuming the intermediate host antedates the final host, but quite the reverse, closely related species may have extremely diverse intermediate hosts according to the opportunities and habits of the final host. If we were to suggest any group as being similar to possible ancestors of the Spirurida, it would be the Cylindrocorporidae (Rhabditoidea).

Dioctophymatoids all have intermediate hosts; examples are known that parasitize only birds and mammals in the adult stage while fish, amphibians, and reptiles may act as secondary intermediate hosts. Evidently the Dioctophymatoidea are primarily aquatic forms which arose after the origin of birds and mammals—hence an extremely recent group. All groups of vertebrates harbor trichuroids. Those that occur in aquatic animals have intermediate hosts but species of *Capillaria* from terrestrial mammals differ from one another in this respect. One might suspect the Trichuroidea as being as old as the Vertebrata but in that case one would need to assume that the direct-life-cycle Trichurinae (parasites of mammals) developed from the partially indirect-life-cycle Capillariinae. As in the Dioctophymatoidea, one would trace them back to an aquatic nematode. There is no clear evidence as to whether this nemic group was marine or fresh water but from the fact that speciation continues in land animals while marine species are relatively less numerous, one could easily conceive a fresh-water origin. *Cystoopsis*, a parasite of fish, is probably the most primitive genus. The Mermithoidea are parasitic mainly in fresh water and terrestrial arthropods (Tracheata) but do occur in Crustacea. Steiner (1917) derived the Mermithoidea from the Dorylaimoidea while Fülleborn (1923) gave the same origin for the Trichuroidea. Evidence from comparative anatomy indicates the Trichuroidea developed from forms very similar to present day mermithoids.

Baylis (1938) supplies a further discussion of the host distribution of nemas.

EMBRYOLOGY. In all of the nematodes thus far investigated the embryology may be considered identical for comparative purposes. Bilateral determinate cleavage must certainly have been a characteristic of the nemic ancestor; the resultant adult must have been composed of few large cells, probably about 600, exclusive of the reproductive system. Only one to two eggs were produced at a time. Furthermore, a very close phylogenetic relationship should be supported by similar tissue origins from the primary somatic stem cells arising at the first five cleavages (see Section II Chapter II).

Bilateral determinate cleavage with little or no blastocoele is the rule in all species of nematodes studied embryologically. However, certain nematodes have a spherical blastocoele and the cleavage appears to be radial. We may expect many changes in concept when critical studies are made. Particular emphasis should be given to aphasmidians.

ONTOGENY. The primitivity of meromyarian nemas has been concluded (p. 48) on the basis of embryonic development as well as comparative anatomy. Oligocyty in the intestinal epithelium (p. 104) and cylindricity of the stoma (p. 67) seem also to have been characters of the primitive nema.

COMPARATIVE ANATOMY. The primitive nema will be described, in so far as present evidence permits. Universal attributes of nemas must be included. While they do not limit the ancestor, they define potentialities which must have existed at the origin of the group. Among the universal attributes are a layered, protein cuticle formed as a deposit (differentiated) by the hypodermis, longitudinal somatic muscles, a pseudocoelome, a triradiate esophagus, and a tubular intestine composed of one cell layer. In addition, it seems clear that the hypodermis (p. 42) must have been composed of a few large cells, the cell bodies grouped in four longitudinal chords, modifications therefrom (*Trichuris*) being secondary. Sublateral hypodermal glands and caudal glands must be considered as possible attributes of the nemic ancestor. Paired spicules, probably with a gubernaculum, are also to be listed in our hypothetical organism. The somatic musculature is presumed to have been meromyarian (p. 48). Transverse somatic muscles and specialized muscles (p. 50) might conceivably be remnants of a more widespread system of muscles; hence a double system of musculature is not out of the question. Whether or not the coelomocytes should be interpreted as a system on the wane (reduced parenchyma) or more ancient but homologous to the parenchyma system of the Platyhelmintha is not clear. We tend to accept the latter view. The esophagus and esophago-intestinal valve may be considered primary invagination (stomodeum) while the stoma is secondary invagination (hence more comparable to the ciliated "Schlundrohr" of rotatorians). Cuticular lining of esophageal and rectal structures must be presupposed in the primitive nema. As to the intestine, a few large cells lined by a bacillary layer (possibly reduced cilia) seems obvious. Rectal glands are by no means universal but their existence in the majority of phasmidian groups and at least in some aphasmidians (*Anaplectus granulosus*) indicates a primary existence.

As to the nervous system, much can be said, but little proved. Apparently all nemas have paired amphids connected indirectly (through a commissure) with the nerve ring; these amphids consist of a terminal pocket into which sensory terminals and a gland empty. Postembryonic development indicates that they were postlabial in position. Circumoral tactile papillae are also universal and they connect directly with the nerve ring. Serial sublateral tactile organs are, for the most pare part, confined to one class, the Aphasmidia, but might well be primitive. Specialized lateral tactile organs (deirids) are confined to the other class (Phasmidia). Posterior lateral organs, phasmids, similar in structure to the amphids are also confined to the Phasmidia but might well be primitive. Male tactile organs are practically universal but whether the paired system of the Phasmidia or the unpaired system of the Aphasmidia is primitive we cannot say. The fundamental architecture of the central nervous system is so set that it leaves little room for question. An anterior commissure, the nerve ring, connects the dorsal, subdorsal, lateral and subventral cephalic ganglia with each other and with the chief nerve which is a more or less double system of incompletely separated ganglia. The nerve ring is not as completely closed ventrally as dorsally indicating that closure may have occurred recently. From the nerve ring six anterior cephalic nerves and a lateroventral commissure connect with anterior sensory organs, while eight direct innervation processes connect with anterior muscles and six posterior motor nerves (the mediodorsal and subventral as well as the laterodorsal and lateroventral somatic nerves function as motor nerves). One or two lateral sensory nerves are also connected with the nerve ring. The ventral or subventral nerves are the chief nerves of the body and connect through commissures with the reproductive system, rectum, copulatory organs and other nerves.

The excretory system (p. 126) has been considered as of double ontogenetic origin. One could interpret the one-cell aphasmidian system as most primitive but it would not account for the tubular phasmidian system. The terminal duct cell of the H-shaped system is apparently the homologue of the aphasmidian system and, if so, the H-shaped system should be considered primitive.

The reproductive system is a very real stumbling block. Seurat's concept of simplicity (p. 192) is easily accepted but ontogeny and comparative anatomy of free-living nemas entirely oppose both Steiner's and Seurat's ideas concerning parallel paired gonads opening posteriorly in both sexes. Even in the male, we must presume opposed gonads as primitive, as least in the time nemas originated.

Sublateral hypodermal glands and caudal glands must be considered possible attributes of the nemic ancestor. In conclusion, we will reemphasize the point that from each organ system studied, we concluded that the primitive nema must have been a composite of plectoid and rhabditoid characters and only that with such an organism could the later progeny be explained. If this is true, we might expect a saprophagous life in moist soil or swampy habitat.

Relationships with Other Groups

The above requirements with respect to cuticle (in some representatives at least), esophagus and intestine are met with in the Rotatoria, Gastrotricha, Echinodera and Tardigrada. In all of these groups, except the Echinodera, the pharynx (i. e., muscular foregut) is triradiate, a term which would be applicable to the nemic esophagus; the esophagus in these forms corresponds to the esophago-intestinal valve in nemas. The musculature and body cavity conform in the first three groups. There is as yet no parallel to, or explanation of, the innervation process of the nemic muscle cell which seeks the nerve rather than vice versa. This apparent contradiction may not be as basic as it seems superficially (see p. 173). The nervous system presents some contrasting points. If we assume a "gravitational hypothesis" the nemic nervous system might be rationalized. The ventral nerve is partly paired and the amphidial nerve connection would be direct were the subventral nerves separate and rather lateral in position. If the bulk of the cephalic ganglia were similarly shifted dorsad, the nemic nervous system would correspond remarkably not only with that of the above mentioned groups but also with that of the Platyhelmintha. Even in rotatorians, which are usually characterized as having a "dorsal" nervous system, the two subventral nerves are the chief nerves of the body and the anterior ganglia (brain) are often lobed so that they are lateral as well as dorsal. A cylindroid body form and vibratile to serpentine locomotion could have caused the ventral gravitation of the system. The acanthocephalan and nematomorphan systems could be rationalized with the Rotatoria-Platyhelmintha system in the same manner. There is no especial similarity of either with that of the nematode. We would rather assume convergence. This gravitational view is the one adopted by annelid and arthropod morphologists (see Snodgrass, 1935) for the formation of paired ventral nerves in those groups. It also helps to explain the unpaired ventral nerve of echinoderids (Fig. 145) which are otherwise so much like gastrotrichs.

The excretory systems of rotatorians, gastrotrichs, echinoderids, acanthocephalans and platyhelminths are all true protonephridial types including terminal flame cells. In all these groups the system opens posteriorly through either a single orifice, paired separate orifices, or in a cloaca. The nemic tubular system presents a very striking resemblance to the protonephridial system but despite conscientious efforts terminal flame cells have not been demonstrated. The canalicular H-shaped system could be homologized (barring flame cells) with the rotatorian system if we conceive of a new terminal duct connecting with the anterior junction (Huxley's anastomosis) of the lateral canals of Monogononta rotatorians (see Remane, 1929). The auxiliary excretory system of certain female rhabditids has a striking superficial resemblance to the paired system of gastrotrichs.

The amphids of nemas seem to check well with the retrocerebral organs of rotatorians and the lateral organs of gastrotrichs, rhabdocoele turbellarians and trematode larvae. Paired lateral or sublateral glands or glandular setae have their counterparts in the adhesive gland tubes of gastrotrichs; the deirids (cervical papillae) have been compared with the lateral antennae of rotatorians. The unpaired, three cell caudal glands of the Aphasmidia may be readily homologized with the paired adhesive toes or caudal appendages of rotatorians, gastrotrichs and echinoders. It may be more than coincidence that phasmids and caudal glands are not known to occur on the same organism. The origin of phasmids has been one of the most perplexing problems in nematology. They might be compared with either a single pair of hypodermal glands or with the unpaired caudal glands. If the latter is true, then we no longer have a problem since phasmids would also be homologous to the paired caudal structures of rotatorians, gastrotrichs and echinoders.

According to Bresslau, the genus *Gnosonesima* (Alloeocoelida, Tubellaria) has two pairs of opposed ovaries each producing both eggs and yolk cells. The more common condition in turbellarians is for two of the gonads, either the anterior or the posterior pair, to specialize in production of yolk cells and the other two in the production of egg cells. Here, we believe, is the explanation of the opposed gonad condition which was such a puzzling problem when nematodes were merely compared with gastrotrichs and rotatorians. The rhabdocoele turbellaria also appear to hold the key to spicule formation for in the genus *Gyratrix* we find the only structures comparable to the paired spicules and gubernaculum of nematodes.

Throughout this publication we have tended to question total absence of cilia in nemas so-called arthropod character*), there being so many suspicious cases where a ciliate structure seems clearly present but vibration has not been observed. Contrast with arthropod and annelid characteristics (lack of true metamerism and of a true coelome, musculature, lack of arthropod chitin in the adult exoskeleton, different symmetry etc.) seems so obvious we feel no need for special discussion. One point, the rectal glands of nemas and their comparison with malpighian tubules, warrants thought, especially when one considers multiplication of these glands (p. 118) and the resultant similarity to those of tardigrades. If there were anything to this we would have to hypothesize arthropods as derived from primitive nemas via the Tardigrada, which view would be objectionable to all good entomologists. It is probably a matter of convergence. Oddly enough, the only other counterpart for the rectal glands is in the anal glands of gastrotrichs.

The Tardigrada certainly furnish much evidence to link the Nematoda with the Onycophora, Pentostomida, Arachnida and particularly the Insecta. The pharynx, intestine, rectum, cuticle molts and somatic musculature and absence of the coelome all bespeak nemic origin. The paired ventral series of ganglia connected with 4 pairs of distinct appendages definitely allocate them to the Articulata in most texts. Origin of the mesoderm as four gut pockets (see Cuenot) is definitely contrary to all classifications based on embryology. So far as we can see, Hyman would have to place them in the Deuterostoma series. Still the assemblage of evidence appears to warrant a placement directly between the Nematoda and the Insecta. If this is sound the Arthropoda might be considered a polyphyletic group without sound foundation. The trochophore theory is supposed to account for union of the phyla Annelida, Mollusca, Sipunculoidea, Gephyrea, and Echiurida.

Tabular Comparison of Groups

Purely for supplementary consideration a table is given in which there are brief characterizations of the various organs of some of the invertebrates (Table 4). The writer has used this table in teaching and found it of some value. A conscientious attempt was made to attain accuracy but doubtless there are many errors.

Based on the anatomical characters given in table 4, a system was devised whereby the degree of similarity of each group with every other group could be expressed numerically. Each anatomical category was assigned a value of 6. Thus in table 5 the Trematodea are compared with the other groups in respect to each of the 16 anatomical characters given in table 4. Recorded in column 1 of table 5 are scores based on a comparison of the exoskeleton. When the exoskeleton of the Trematodea was compared with that of the Nematoda a score of 0 was given, indicating no similarity; when compared to the Rotatoria a score of 4 was given, indicating a certain degree of similarity, etc. In the next vertical column (2) scores are recorded based on the ectodermal epithelium, in the third vertical column (3) scores based on the somatic musculature, etc. The scores are totaled in the right hand vertical column. In this table, it will be noted, the Trematoda are compared *with the Trematoda*, resulting in a total score of 96 which is, of course, the highest possible score. In the following discussion the score of a given group when compared with itself will be referred to as the *base score*.

A similar table was prepared for each of the 12 other invertebrate groups under consideration. These tables are not reproduced herein but the totals, i. e., the scores in the right hand vertical column of each are assembled in table 6. Thus in vertical column "N" of table 6 are recorded the scores resulting from a comparison of the Nematoda with the other groups. In the following discussion these scores will be referred to as the *Nematoda score series*, those in the next vertical column as the *Rotatorian score series*, etc. When information regarding one or more anatomical features was lacking the base score is less than 96, as for example, the base score for the Echinodera, Rotatoria and Acanthocephala is, in each case, 90, and for the Gastrotricha, 84.

*Dr. H. J. Van Cleave states Foster reported cilia from the gonoducts of crabs.
†In the following discussion (Tabular comparison of groups) it will be seen that an evaluation of morphological similarities removes the Acanthocephala to the Platyhelmintha.

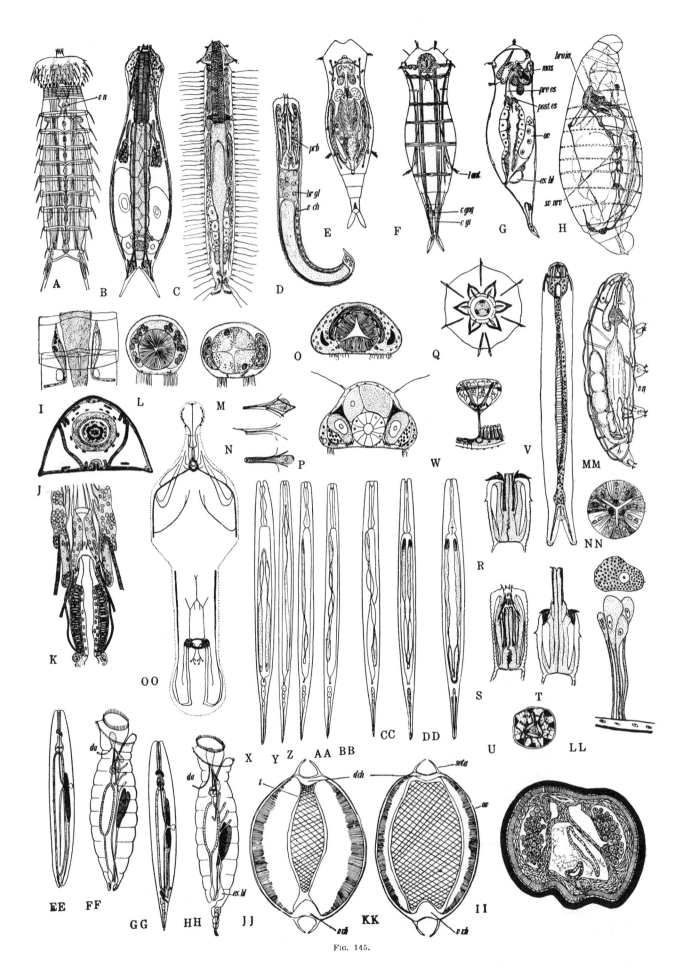

Fig. 145.

Whether or not this system is valid, it gives an interesting numerical comparison of apparent similarities. By using all organs the stress of possible convergent characters is lowered. We regard the relative as well as the actual scores as having significance. In each score series, the groups are arranged in order of score, and breaks in the score series are indicated by transverse lines. It is very similar to the modified grade curve often used in small college classes. Groups necessarily show reciprocal scores but the position in the series is regarded as indicating *degree of similarity*.

The remarkable homogeneity of the Platyhelmintha is emphasized by the minimum score of 80 (with 96 as base score) scoring two groups on the basis of the third group (Turbellara, Trematoda, Cestoda). In all three instances the position of these groups in the score series is 1, 2, 3. The Nemerta, included as a class in the Platyhelmintha by the Pearse Classification (1936) have fourth high score in two series Turbellarea and Trematodea) with a score of 74, and fifth high score in the third series (Cestoda) with a score of 54. The Nemerta also shows reciprocally close similarity to these groups since in their own score series the Turbellarea, Trematoda, and Cestoda are placed 2, 3, 4. Only in the annelid score series are the Nemerta disassociated from the other three and this may be due to either real or superficial similarity because of circulatory system and body cavity. The Acanthocephala are associated with the Cestoda and Nemerta in five score series. They are placed fourth in the Cestoda series and fifth in the Turbellara, Trematoda, and Nemerta series. In their own series they place the Cestoda second and Nemerta third. Considering the breaks in the various series, the Series Parenchymata would appear to contain three phyla, namely, Platyhelmintha, Nemerta, and Acanthocephala; the first of these would contain three classes Turbellara, Trematoda, and Cestoda. Of the latter classes, the Turbellarea show the greatest similarity to the Nemerta while the Cestoda show the greatest similarity to the Acanthocephala.

The Gastrotricha, Echinodera, Rotatoria and Nematoda also seem to show reciprocal relationships but not as close as those of the classes of the Platyhelmintha. Thus the Rotatoria (90 base score) provide a basis for scoring Gastrotricha, 64; Echinodera, 63; and Nematoda, 57. The Echinodera (90 base score) provides scores of Gastrotricha, 63; Rotatoria, 63; and Nematoda, 56. The Gastrotricha (84 base score) provides scores of Rotatoria, 64; Echinodera, 62; Nematoda, 45; and Turbellara, 45. And the Nematoda (96 base score with Rotatoria and Echinodera at 6 point disadvantage and Gastrotricha at 12 point disadvantage††) provides the scores Rotatoria, 57; Echinodera, 56; Tardigrada, 52; and Gastrotricha, 45. The Tardigrada appear in the table as fifth group in the Rotatoria score series, third in the Annelida series, sixth in the Echinodera series and second group in the Arthropoda series (the latter position is reciprocal). It would appear, therefore, that the Tardigrada more correctly belong near the Arthropoda.

The Arthropoda (base score 96) might be said to lay claim to only two close relatives, the Tardigrada and Annelida; the Annelida (base score 90) by the same token, claim distant kinship with the Nemerta with 44 (see discussion above); the Arthropoda, 48; Tardigrada, 44; and Nematomorpha and Acanthocephala, 34. The Tardigrada (base score 96) show similarity to Arthropoda, 61; Nematoda 52; Annelida, 44; Rotatoria, 43; Echinodera, 42; Gastrotricha, 39; Nematomorpha, 34. The Nematomorpha (No) score series (base score 96) provides us with the odd series Nematoda and

Rotatoria, 42; Acanthocephala and Echinodera, 41; Gastrotricha, 39; Turbellara 36; Annelida and Tardigrada, 34; Nemerta and Trematoda, 32; Cestoda, 29; and Arthropoda, 16. This is practically a uniform series with no obvious break.

According to Schepotieff's (1908) illustrations (Fig. 145 U) the proboscis of the larva of *Gordius* is triradiate with one ray directed dorsad instead of ventrad as in nemas. If this is true, the non-radiate symmetry of the adult esophagus may be considered secondary degeneracy due to parasitism such as happens in the development of mermithids (p. 95). The inverted triradiate symmetry also occurs in one group of gastrotrichs. The echinoderid non-radiate esophagus may be interpreted as a modification of the triradiate, which it resembles more closely than that of the turbellarean. May (1919) found that the larval stylets of gordiids and adjoining cuticle are lost at the time the parasite leaves its host; whether or not this should be interpreted as a molt seems questionable. The earlier view that the "praesoma" did not take part in later development was disproved by May. These points all support the inclusion of the Nematomorpha in the Aschelmintha.*

If the Nematomorpha were credited unqualifiedly as having molts (No!) then five points more would be scored in comparison with some groups and five deducted from others so that the score series would be Nematoda, 47; Echinodera, 43; Rotatoria, 42; Tardigrada, 39; Gastrotricha, 39; Acanthocephala, 36; Annelida, 34; Turbellarea, 31; Trematodea, 27; Nemerta, 27; Cestodea, 24; and Arthropoda, 21. In the reciprocal scoring series the Nematomorpha would be included as showing similarities to the Nematoda and Echinodera and just under the break of the Rotatoria, Gastrotricha and Annelida series. These indications of similarity support the two diverse schools of thought, the one which is the more popular (and here the more distinct) placing the Nematomorpha in the Aschelmintha or Nemathelminthes, and the other less popular one which relates the Nematomorpha to the Annelida.

On the basis of the information obtained from tables 4 to 6, it appears that there are three major groups among the organisms considered. These groups are as follows: Parenchymata (containing the phyla Nemerta, Acanthocephala, and Platyhelmintha), Aschelmintha, and Annelida-Tardigrada-Arthropoda.

The Nematoda (synonym Nemata Cobb, 1919) appears to be a sound phylum, properly associated with the Rotatoria, Gastrotricha and Echinodera. Pending more exact information relative to the anatomy of the larval stages the Nematomorpha is recognized as a separate phylum, perhaps distantly related to the Echinodera.

The position of the Tardigrada within the Annelid-Arthropod series is very clearly substantiated. It is unfortunate that students of the larger segmented forms do not direct some of their attention to the Tardigrada.

Upon the basis of sundry authorities, it would appear that there are three bilaterally symmetric, triplobastic subkingdoms or series of Metazoa. These have been named the Scolecida (Vermes Amera or Unsegmented Worms), the Annulosa (Vermes Polymera, Arthropoda, Tardigrada and Onycophora) and the Eucoelomata. The subkingdom or series Scolecida may be defined as follows:

---

†† By disadvantage we mean that due to lack of information there is no statement in one or more brackets in table 4 and the forms are given zero as though the information indicated total dissimilarity, which it does not.

---

* Woodhead (1950) has described a very astounding life cycle for *Dioctophyma renale*. His description involves stages which do not differ substantially from those of larval gordiids. If all of his stages are correctly identified as *D. renale*, we will have no alternative but to place the Gordiacea as a suborder of the Enoplida and adjacent to the Dioctophymatina. Natural infection of his intermediate host with gordiids weakens his arguments as do morphologic errors in his descriptions. The case certainly requires verification. His description of the ovic larva showing stylet and dorylaimoid esophagus are unquestionable and a distinct contribution, tying the Dioctophymatina to the Dorylaimina, as previously indicated by many other characters.

FIG. 145. RELATIVES OF NEMATODES

A—Cyclorhagae (Echinodera), ventral view. B—*Chaetonotus* (Gastrotricha), ventral view. C—*Turbanella cornuta* (Gastrotricha), ventral view. D—*Gordius aquaticus* (Nematomorpha), lateral view of larva. E-G—Monogononta (Rotatoria) (E—Dorsal view; F—Ventral view; G—Lateral view). H—*Zelinkiella synaptae* (Rotatoria), nervous system. I-K—*Pycnophes communis* (Echinodera) (I—Excretory system; J—Cross section in pharyngeal region; K—Horizontal section of pharyngeal region). L-M—*Chaetonotus maximus* (Cross sections, L—At level of pharynx; M—At level of intestine). M—*Macrodasys* (Gastrotricha) (Adhesive tube setae of various types). O-P—*Turbanella cornuta* (Cross section, O—At level of pharynx; P—At level of intestine). Q-T—*Gordius aquaticus* larva (Q—En face view; R-T—Proboscis region in various stages of contraction). U—*Gordius* larva (Cross section near base of proboscis). V-W—*Gordius* adult (V—Cross section of ventral chord showing ganglion cell groups; W—Diagram of nervous system). X-DD—Diagrams of female reproductive system of nemas (DD—Hypothetical form). EE-HH—Comparison of nemas and rotifers (EE-FF—without caudal glands; GG-HH—with caudal glands). II—*Gordius tolosanus* (Cross section of female). JJ-LL—*Nectonema agile* (Nematomorpha) (JJ—Cross section of male; KK—Cross section of female; LL—Muscles and oocyte). MM-NN—*Macrobiotus hufelandi* (Tardigrada) (MM—Female; NN—Cross section of pharynx). OO—Diagram of nervous system in Acanthocephala. A & I-K, after Remane, 1928, Die Tierwelt der Nord-u. Ostsee, Part 7d 2; B & L-P after Remane 1929, Handb. Zool. v. 2 (6); D & Q-T after Dorier, 1930, Recherches biologiques et systematiques sur les Gordiaces, Thesis Grenoble; E-G, after Remane, 1929, Die Tierwelt der Nord-u. Ostsee part 7 e; U, after Schepotieff, 1908, Ztschr. Wiss. Zool. v. 89; V-W & OO, after Brandes, 1899, Abhandl. Naturf. Gesellsch. Halle, v. 21; X-DD, After Steiner, 1919, Untersuchungen ueber den allgemeinen Bauplan des Nematodenkoerpers, Jena; EE-HH after Steiner 1919, Festschrift f. Zschokke (31); II, after Rauther 1905, Jena Ztschr. v. 40, n. f. v. 33 (1); JJ-LL after Feyel, 1936, Arch. Anat. Microsc. v. 32; MM-NN after Cuenot, 1932. Faune de France (24).

Table 4. Comparison of nemas with other animals. (Cont. p. 201).

Group	Exoskeleton	Ectodermal Epithelium	Somatic Musculature	Metamerism	Body cavity
Nematoda	Cuticle (protein); striated or rarely annulated	Chords-hypodermis	Longitudinal layer attached; transverse few attached ends.	Pseudometamerism in transverse musculature; sometimes in mesenterial cells; sometimes in external annulation and seta distribution also in ganglia.	Pseudocoelome containing fluid and fixed coelomocytes; mesenteries present.
Rotatoria	Cuticle, cilia at head, some external annulation	Hypodermis not differentiated	Longitudinal and transverse (circular) attached only at ends. No complete layer	Pseudometamerism of transverse muscles and ganglia in some	Pseudocoelome containing fluid and amebocytes, filaments present.
Gastrotricha	Cuticle sometimes entire surface ciliated	Hypodermis not differentiated	Longitudinal attached at ends; No complete layer.	None except paired setae	Pseudocoelome containing fluid and free amebocytes
Echinodera	Cuticle annulated	Hypodermis not differentiated	Longitudinal and transverse both attached at ends; No complete layer	Pseudometamerism of exoskeleton; transverse musculature and ventral ganglia	Blastocoele containing fluid and mesenchymatous conn. tissue
Acanthocephala	Cuticle striated	Hypodermis pseudostratified columnar	Longitudinal internal and circular external in 2 layers attached throughout	Pseudometamerism in some	Pseudocoelome containing free cells
Nematomorpha	Cuticle	Pseudostratified columnar ventral chord	Longitudinal attached throughout layer	Serial pouches of gonads; Pseudometamerism of larva	Pseudocoelome usually containing fixed parenchyma cells in fluid
Tardigrada	Cuticle some annulation	Squamous.	Attached at ends 1 muscle 1 cell longitudinal and transverse	Probably true; 4 pairs ventral ganglia and 4 pairs legs	Coelome formed by 4 pairs of evaginations of intestine
Arthropoda	Chitin, annulated	Squamous to columnar	Attached at ends; striated	True paired jointed appendages; metameric ganglia	Coelome filled with fluid
Annelida	Cuticle annulated	Chords and columnar	External circular; internal longitudinal; 2 layers	True internal metameric ganglia	Coelome filled with fluid
Turbellara	Epidermis often ciliated	Pseudostratified columnar	Circular and longitudinal layers	None	Parenchymatous
Trematoda	Cuticle	Do.	Do.	Do.	Do.
Cestoda	Cuticle	Do.	Do.	Reduplicate in Cestoda, not in Cestodaria	Do.
Nemerta	Epidermis ciliated	Do.	Ext. circular; int. longitudinal; int. circular; int. longitudinal layers	Pseudometamerism of gonads and gut pockets	Rhynchocoelome and parenchyma

Table 4. (Continued).

	Stomodeum	Mesenteron	Proctodeum	Excretory system	Nervous system
Nematoda	"Esophagus" triradiate, stoma 2nd invagination	Cuboidal 2 cell-rows to columnar epithelium bacillary layer simple	Cuticularly lined rectal glands often, ventral anus	Lateral canals joined anteriorly; ventral secondary orifice; no flame cells	Partially fused vent. nrv. indistinct series of ganglia; nrv. ring with large lateral & ventral cephalic ganglia
Rotatoria	Triradiate "mastax" grossly bilateral	Columnar or cuboidal epithelium, cilia. Simple.	Cuticle lined dorsal anus	Lateral canals joined anteriorly, open together posteriorly into cloaca; flame cells	Ventrolateral paired nerves; with serial ganglia dorsal and dorsolateral cephalic ganglia
Gastrotricha	Triradiate "pharynx"	Cuboidal epithelium 4-8 cell-rows up; no lining. Simple.	Cuticularly lined; anal glands sometimes; anus dorsal or ventral	Lateral canals not joined, open separately in midregion on ventral side; flame cells	Dorsal and lateral cephalic ganglia and lateral nerves
Echinodera	Non-radiate	Columnar epithelium, no lining. Simple.	Cuticularly lined terminal or ventral	Short lateral canals; not joined; open separately, laterally; flame cells	Dorsal and lateral cephalic ganglia, lateral and unpaired ventral with series of ganglia
Acanthocephala	None	None	Cuticularly lined cloaca;	Lateral canals open together posteriorly; flame cells	Nerve ring† with associated ganglia; lateral somatic nerves. †Questioned by Van Cleave
Nematomorpha	Non-radiate. ? Proboscis of larva might be triradiate	Columnar epithelium. Simple.	Cuticularly lined terminal	None	Nerve ring with circum esophageal ganglion mass and ventral ganglion mass with serial nerve branching.
Tardigrada	Triradiate "bulb"	Cuboidal; with bacillary layer. Simple	Cuticularly lined with "Malpighian" tubules. Ventral anus	? Malpighian tubule	Nerve ring with subdorsal and subventral ganglia and paired ventral nerve with 4 pairs of ganglia
Arthropoda	Non-radiate or superficially triradiate pharynx	columnar with bacillary layer; often highly differentiated and with 2 muscle layers	Chitin lining partial, malpighian tubules	Malpighian tubules or head glands	Circumesophageal commissure and paired ventral nerves with metameric ganglia
Annelida	Non-radiate pharynx	columnar with cilia or bacillary layer	Cuticularly lined	Open metameric nephridia	Circumpharyngeal commissure with ganglia, and paired with metameric ganglia
Turbellara	Non-radiate pharynx	Columnar with cilia or bacillary layer	None or ? atavistic	Lateral canals open separately posterior; flame cells	No ring; subdorsal ganglia; lateral and lateral cephalic subdorsal and subventral somatic nerves
Trematoda	Do.	Do.	Do.	Do.	Do.
Cestoda	None	None	None	Do.	Do.
Nemerta	Non-radiate	Cecum; lateral sacs	Present	Open or closed protonephridia; longitudinal vessels with sublateral orifices	Circum-pharyngeal commissure with dorsal and ventral ganglia and lateral nerves

237

Table 4. (Continued).

	Reproductive system	Cleavage	Molts	Respiratory	Circulatory system	Appendages
Nematoda	Tubular paired gonads, sexes separate, female ventral separate orifice male ventral cloacal orifice	Bilateral determinate; no regenerate; no true metamorphosis	Present	None	None	Tube glands; posterior spinerette single
Rotatoria	Tubular paired gonads, sexes separate; orifice into cloaca in both sexes	Bilateral determinate; no regeneration; no true metamorphosis	Unknown	None	None	None; posterior paired adhesive toes
Gastrotricha	Tubular paired gonads; hermaphrodites or Orifices separate, ventral.	Unknown	Unknown	None	None	Lateral tube glands; posterior paired adhesive toes
Echinodera	Tubular paired gonads, sexes separate. Orifice paired, lateral or ventrolateral posterior	Unknown	Present	None	None	None, paired caudal adhesive setae
Acanthocephala	Somewhat tubular or ovid, paired; sexes separate; orifice into cloaca in both sexes	Superficially bilateral, actually radial determinate; metamorphosis	None	None	Lymphatic superficially	None
Nematomorpha	Sac-like paired gonads with linear pockets; sexes separate; open into cloaca in both sexes	? Radial determinate; Metamorphosis	Absent in usual sense. Does shed larval stylets	None	None	None
Tardigrada	Tubular paired gonads; hermaphroditic; orifices of both sexes into cloaca	Radial. No regeneration. No metamorphosis	Present	None	None	4 pairs of legs
Arthropoda	Paired multitubular gonads, sexes separate, female orifice cloacal; male separate or cloacal	Spiral determinate; regenerate; usually metamorphosis	Present	Varied types	Highly developed	Varied jointed appendages
Annelida	Metameric with separate orifices; hermaphroditic	Spiral determinate; regenerate; metamorphosis		None	Highly developed	Metameric gills and setae
Turbellara	Paired tubular gonads; hermaphroditic; separate orifices ventral	Spiral determinate; regenerate; metamorphosis	None	None	None	None
Trematoda	Do. except gonads saccate	Cleavage? eggs include yolk cells; metamorphosis	None	None	? None	None
Cestoda	Do.	Do.	None	None	None	None
Nemerta	Saccate, opening directly to outside; bisexual or hermaphroditic	Spiral determinate; regeneration; metamorphosis	None	None	Highly developed	None

	Exoskeleton	Ectodermal epithelium	Somatic musculature	Metamerism	Body cavity	Stomodeum	Mesenteron	Proctodeum	Excretory system	Nervous system	Reproductive system	Cleavage	Molts	Respiratory system	Circulatory system	Appendages	Total
	1	2	3	4	5	6	7	8	9	10	11	12	13	14	15	16	Total
N	0	0	3	2	1	0	5	0	2	2	0	0	0	6	0	0	21
R	4	3	0	3	2	0	5	1	5	5	2	0	X	6	0	0	36
G	3	3	0	5	2	0	5	0	5	5	6	X	X	6	0	0	40
E	0	3	0	2	2	4	5	0	5	3	5	X	0	6	2	0	37
Ac	0	6	6	0	3	0	0	1	5	4	4	1	X	6	6	6	48
No	0	4	3	1	3	4	5	0	0	1	2	1	0	6	2	0	32
Ta	0	3	0	0	0	0	5	0	0	0	2	1	0	6	2	0	19
Ar	0	3	0	0	0	2	2	0	0	0	0	2	0	0	2	0	11
An	0	3	6	0	0	3	2	1	2	0	3	2	X	6	1	0	29
Tu	6	6	6	6	6	6	6	6	6	6	5	4	6	6	2	6	89
Tr	6	6	6	6	6	6	6	6	6	6	6	6	6	6	6	6	96
Ce	6	6	6	5	6	0	0	6	6	6	6	6	6	6	6	6	83
Ne	6	6	4	5	6	6	5	0	5	6	5	2	6	6	0	6	74

TABLE 5.

Trematodea Score Series

Abbreviations: N, Nematoda, R, Rotatoria; G, Gastrotricha; E, Echinodera; Ac, Acanthocephala; No, Nematomorpha; Ta, Tardigrada; Ar, Arthropoda; An, Annelida; Tu, Turbellara; Tr, Trematoda; Ce, Cestoda; Ne, Nemerta.

N	R	G	E	No	No!	Tu	Tr	Ce	Ac	Ne	An	Ta	Ar
N 96	R 90	G 84	E 90	No 96	No 96	Tu 96	Tr 96	Ce 96	Ac 90	Ne 96	An 90	Ta 96	Ar 96
R 57	G 64	R 64	G 63	N 42	N 47	Tr 89	Tu 89	Tr 83	Ce 63	Tu 74	Ar 48	Ar 61	Ta 61
E 56	E 63	E 63	R 63	R 42	E 43	Ce 80	Ce 83	Tu 80	Ne 54	Tr 74	Ta 44	N 52	An 48
Ta 52	N 57	N 45	N 56	Ac 41	R 42	Ne 74	Ne 74	Ac 63	Tr 48	Ce 54	Ne 44	An 44	N 25
G 45	Ta 43	Tu 45	Tu 42	E 41	Ta 39	Ac 45	Ac 48	Ne 54	Tu 45	Ac 52	No 34	R 43	E 23
No 42	Tu 43	Tr 40	Ta 42	G 38	G 39	G 45	G 40	G 39	No 41	An 44	Ac 34	E 42	Ne 23
An 34	No 42	Ce 39	No 41	Tu 36	Ac 36	R 43	E 37	E 30	R 41	No 32	N 34	G 39	No 16
Ac 32	Ac 41	Ta 39	Tr 37	An 34	An 34	E 42	R 36	No 29	An 36	G 32	Tu 32	No 34	Tu 16
Ar 25	Tr 36	No 39	Ac 33	Ta 34	Tu 31	No 36	No 32	R 29	G 33	E 31	Tr 29	Ac 29	Ac 15
Tu 23	Ce 29	Ac 33	Ne 31	Ne 32	Tr 27	An 32	An 29	An 22	E 33	Ar 23	E 25	Tu 21	G 13
Tr 21	An 21	Ne 32	Ce 30	Tr 32	Ne 27	N 23	N 21	N 21	N 32	Ta 19	Ce 22	Tr 19	R 12
Ce 21	Ne 16	An 19	An 25	Ce 29	Ce 24	Ta 21	Ta 19	Ta 19	Ta 29	R 16	R 21	Ce 19	Tr 11
Ne 14	Ar 12	Ar 13	Ar 23	Ar 16	Ar 21	Ar 16	Ar 11	Ar 7	Ar 19	N 14	G 19	Ne 19	Ce 7

TABLE 6.

Nemic Relationships.   Table of Scores.

SUBKINGDOM SCOLECIDA

Triploblastic, bilaterally symmetric metazoa without a true coelome (mesodermal lining of body cavity arising as migratory cells or mesenchyme and either remaining sparse or tending to grow together); mesoderm never arising as gut pockets; muscle not striated; nervous system not metameric; intestine sometimes absent (Acoela, Cestoda, Acanthocephala); circulatory system absent or very unorganized, no "hearts"; protonephridia present or absent (sometimes modified, see Nemerta); cleavage determinate.

Within the subkingdom Scolecida we recognize the phyla Platyhelmintha, Nemerta, Acanthocephala, Gastrotricha, Nematoda, Echinodera, Nematomorpha, Rotatoria and Entoprocta (Bryozoa).

The rhabdocoele or alloeocoele turbellarians seem to account for the origin of trematodes, cestodes (a chain of zooids), acanthocephalans, nemerteans and a primitive generalized pseudocoelomate. This latter organism must have given rise to two main branches, one passing through the Rotatoria to the Entoprocta, the other branch subdividing with one limb passing through the Gastrotricha to the Nematoda, the other passing through the Echinodera to the Nematomorpha. The Entoprocta may account for the origin of two main lines, one the Annulosa giving rise to the Ectoprocta, Tardigrada, Onycophora, Annelida, Sipunculida, Echiurida, Arthropoda and Mollusca, the other giving rise to the Echinodermata, Chaetognatha, Hemichordata and Chordata.

Bibliography

BASTIAN, H. C. 1866.—On the anatomy and physiology of the nematoids, parasitic and free; with observations on their zoological position and affinities to the echinoderms. Phil. Tr. Roy. Soc. Lond., v. 156: 545-638, pls. 22-28.

BAYLIS, H. A. 1924.—The systematic position of the Nematoda. Ann. & Mag. Nat. Hist., s. 9, v. 13: 165-173.
   1938.—Helminths and Evolution. From "Evolution." Essays on aspects of Evolutionary Biology presented to Prof. E. S. Goodrich (edited by G. R. de Beer), pp. 249-270. Oxford (Clarendon Press) 1938.

BERGH, R. S. 1885.—Die Excretionsorgane der Würmer. Eine Uebersicht. Kosmos J. 9, v. 17, v. 2: 97-122, pl. 2, figs. 1-10. Stuttg.

BOISSEZON, P. DE 1930.—Contribution a l'étude de la biologie et de l'histophysiologie de Culex pipiens. L. Arch. Zool. Exper. & Gen., v. 70 (4): 281-431, figs. 1-17, pls. 14-17, figs. 1-36.

BRANDES, 1899.—Das Nervensystem der als Nemathelminthen zusammengefassten Wurmtypen. Abhandl. Naturforsch Gesellsch. Halle, v. 21 (4): 271-299, 11 figs.

BRESSLAU, E. 1928-1933.—Erste Klasse des Cladus Platyhelminthes. Turbellaria. Handb. d.Zool. Kükenthal u.Krumbach. v. 2 (1).

BÜRGER, O. 1891.—Zur Kenntnis von Nectonema agile Verr. Zool. Jahrb. Abt Anat., v. 4 (4): 631-652, 1 fig., pl. 38, figs. 1-34.
   1897.—Nemertini. (Schnurwürmer) Bronn's Klassen u. Ordnung Tierreichs. v. 4, Suppl. (1-4): 1-64, Figs. 1-3, pls. 1-4.

BÜTSCHLI, O. 1872.—Freilebende und parasitische Nematoden in ihren gegenseitigen Beziehungen. Berlin. Senck. Naturf. Gesellsch., pp., 56-73.
   1875.—Zur Entwicklungsgeschichte des Cucullanus elegans Zed. Ztschr. Wiss. Zool., v. 26 (7): 103-110, pl. 5, figs. 1-8.
   1876.—Untersuchungen über freilebenden Nematoden und die Gattung Chaetonotus. Ztschr. Wiss. Zool., v. 26 (4): 363-413, pls. 23-26.
   1885.—Zur Herleitung des Nervensystems der Nematoden. Morph. Jahrb., v. 10 (4): 486-493, pl. 23, figs. 1-3.

CHITWOOD, B. G. 1937. [1938].—A revised classification of the Nematoda. Papers on Helminthology, 30 year Jubileum K. J. Skrjabin pp. 69-80.

CHITWOOD, B. G. and CHITWOOD, M. B. 1933.—The characters of a protonematode. J. Parasit., v. 20 (2): 130.

CHOLODOWSKY, N. 1908.—Ueber den Weiblichen Geschlechtsapparat einiger viviparen Fliegen. Zool. Anz., v. 33 (11): 367-376, figs. 1-13.

COBB, N. A. 1917.—Segmentation in nematodes. Science, v. 45 n. s. (1171): 593, figs. 1-2.

CONINCK, L. DE 1938.—Eutelie. Naturwet. Tijd., v. 20: 226-234.

CUENOT, L. 1932.—Faune de France No. 24 Tardigrades. 95 pp. 98 figs.

DORIER, A. 1930.—Recherches biologiques et systématiques sur les gordiacés. Thesis Grenoble, 183 pp., 139 figs. pls. 4.

DOUGHERTY, E. C. 1949.—A new species of the free-living nematode genus Rhabditis of interest in comparative physiology and genetics. J. Parasit. v. 35 (6-2): 11.

DUISBURG, H. v. 1862.—Schr. Koniglichen Phys.—Okonom. Gesellsch. v. 3: 31-36, pl. 1.

FEYEL, TH. 1936.—Recherches histologiques sur Nectonema agile Verr. Arch. Anat. Micr., v. 32 (2): 197-234, figs. 1-25.

FILIPJEV, I. N. 1934.—The classification of the free-living nematodes and their relation to the parasitic nematodes. Smithsonian Misc. Coll., v. 89 (6): 1-63, 8 pls. 70 figs.

FÜLLEBORN, F. 1923.—Ueber den "Mundstächel" der Trichotracheliden Larven und Bemerkungen über die jüngsten Stadien von Trichocephalus trichiurus. Arch. Schiffs. & Tropenhyg., v. 27 (11): 421-425, pl. 11, figs. 1-18.

GANIN, 1899.—The development of Pelodera teres Confer. Zool. Rec., v. 14. [Not seen].

GOLDSCHMIDT, R. 1906.—Mitteilungen zur Histologie von Ascaris. Zool. Anz., v. 29 (24): 719-737, figs. 1-13.

GREEF, R. 1869.—Untersuchungen über einige merkwürdige Formen des Arthropoden und Wurm-Typus. Arch. Naturg. 35 J., v. 1 (1): 71-121, pls. 4-7.

GROBBEN, K. 1909.—Die systematische Einteilung des Tierreiches. Verhandl. K. K. Zool. Bot. Gesellsch. Wien (1908) v. 58 (10): 491-511, 3 figs.

HAASE, 1888.—See Seurat, 1920, p. 206.

HAECKEL, E. 1866.—Generelle Morphologie der Organismen. Berlin.
   1872.—Gastraea-Theorie. (See Haekel, 1896).
   1896.—Systematische Phylogenie der wirbellosen Thiere (Invertebrata). Berlin. 720 pp.

HATSCHECK, B. 1888.—Lehrbuch der Zoologie. Jena.

HEIDER, K. 1920.—Ueber die Stellung der Gordiiden. Sitzungsb. Preuss. Akad. Wiss., v. 26 (20): 464-477, figs. 1-5.

HEINLEIN, E. & WACHOWSKI, H. E. 1944.—Studies on the flatworm Catenula virginia. Am. Midland Naturalist v. 3 (1): 150-158.

HEYDON, C. VON. 1862.—Gliederthiere aus der Braunkohle des Niederrhein's, der Wetterau und der Rohn. Palaeontographica. v. 10 (2): 62-82, pl. 10, figs. 1-39.

HUBRECHT, A. A. W. 1904.—On the relationship of various invertebrate phyla. Proc. Acad. Sci., Amsterdam, v. 6: 839-846.
   1904.—Die Abstammung der Anneliden und Chordaten und die Stellung der Ctenophoren und Plathelminthen im System. Jena. Ztschr. Naturw., v. 39: 141-176.

HUXLEY, T. H. 1852.—Lacinularia socialis. A contribution to the anatomy and physiology of the Rotatoria. Trans. Micr. Soc. Lond., n. s., v. 1: 1-19 (See Bütschli, 1876).
   1856.—Lectures on General Natural History. Med. Times. London, v. 33, n. s., v. 13: 27-30, figs. 1-3.
   1864.—Lectures on the elements of Comparative Anatomy. 303 pp. (p. 75 cited) London.
   1878.—Grundzüge den Anatomie der wirbellosen Tiere (Ref. by Rauther 1909).

KEILIN, D. 1926.—The problem of the origin of nematodes. Parasit. v. 18 (4): 370-374.

KORSCHELT, E. 1933.—Sipunculidea. Handwörterbuch d. Naturwiss. v. 2: 85-100.

LEUCKART, R. 1879.—Die parasiten des Menschen und die von Ihnen herührenden Krankheiten. Ein Hand. -und Lehrbuch für Naturforscher und Aerzte. 2 ed., v. 1. (1): 336 pp., 130 figs.

LINNAEUS, C. 1758.—Systema naturae. 10 ed. v. 1, 823 pp. Holmiae.

MARCUS, E. 1929.—Tardigrada. Bronn's Klassen u. Ordnung des Tierreichs, v. 5, 4 Abt., Buch 3: 1-608, 398 figs., 1 pl.

MARTINI, E. 1903.—Ueber Furchung und Gastrulation bei *Cucullanus elegans* Zed. Ztschr. Wiss. Zool., v. 74 (4): 501-556, figs. 1-8, pls. 26-28, figs. 1-35.

1907.—Ueber Subcuticula und Seitenfelder einiger Nematoden. II. Ztschr. Wiss. Zool., v. 86 (1): 1-54, pls. 1-3, figs. 1-82.

1908.—Die Konstanz histologischer Elemente bei Nematoden nach Abschluss der Entwickelungsperiode. Verhandl. Anat. Gesellsch., v. 22: 132-134.

1913.—Ueber die Stellung der Nematoden im System. Verhandl. Deutsch. Zool. Gesellsch. v. 23: 233-248.

1923.—Die Zellkonstanz und ihre Beziehungen zu anderen zoologischen Vorwürfen. Ztschr. Anat. & Entwicklungeschichte. 1 Abt., v. 70 (1/3): 179-259.

MAUPAS, E. 1899.—La mue et l'enkystement chez les nématodes. Arch. Zool. Expér. & Gén. ser. 3, v. 7: 563-628, pls. 16-18, figs. 1-29.

MAY, H. G. 1920.—Contributions to the life histories of *Gordius robustus* Leidy and *Paragordius varius* (Leidy). Ill. Biol. Monogr. v. 5 (2): 123-238, pls. 1-21, figs. 1-174.

MENGE, A. 1866.—Ueber ein Rhipidopteren und einige Helminthen im Bernstein. Schrift Naturf. Gesellsch. Danzig. v. 1, (3-4): 8 pp., figs. 1-23.

MEYER, A. 1931.—Das urogenitale Organ von *Oligacanthorhynchus taenioides* (Dies.), ein neuer Nephridialtypus bei den Acanthocephalen. Ztschr. Wiss. Zool., v. 138 (1): 88-98, figs. 1-6.

1932-1933.—Acanthocephala. Bronn's Klassen u. Ordnung des Tierreichs. v. 4, Abt. 2, Buch 2, Lief. 1-2: 1-582, figs. 1-383.

MICHEL, A. 1888.—De l'existence d'un véritable épiderme cellulaire chez les nématodes, en spécialement des gordiens. Compt. Rend. Acad. Sc., Par., v. 107 (27): 1175-1177, 1 fig.

MONTGOMERY, T. H. 1903.—The adult organisation of *Paragordius varius* (Leidy). Zool. Jahrb., Abt. Anat., v. 18 (3): 387-474, pls. 37-43, figs. 1-91.

NIERSTRASZ, H. F. 1907.—Die Nematomorpha der Siboga-Expedition. Siboga - Exped. Uitkom. Zool., Nederl. Ost-Indie (1899-1900), Livr. 34, Monogr. 20, 21 pp., 3 pls., 51 figs.

PEARSE, A. S. 1936.—Zoological names. A list of phyla, classes, and orders. 24 pp. Durham, N. C.

PERRIER, E. 1897.—Traité de Zoologie. Fasc. IV. pp. 1345-2136, figs. 980-1547. Paris.

RAUTHER, M. 1905.—Beiträge zur Kenntnis der Morphologie und der phylogenetischen Beziehungen der Gordiiden. Jena. Ztschr. Natur. Wiss., v. 40, n. f., v. 33 (1): 1-94, pls. 1-4, figs. 1-40.

1909.—Morphologie und Verwandtschaftsbeziehungen der Nematoden und einiger ihnem nahe gestellter Vermalien. Ergeb. & Fortschr. Zool., v. 1 (3): 491-596, figs. 1-21.

1914.—Zur Kenntnis und Beurteilung von *Nectonema*. Zool. Anz., v. 43 (12): 561-576, figs. 1-8.

RÉMANE, A. 1928.—Kinorhyncha. Die Tierwelt der Nord und Ostsee. Teil VII d. 2: 77-84, figs. 1-9.

1929.—Zweite Klasse des Cladus Plathelminthes, Gastrotrichs. Handb. Zool., Kükenthal & Krumbach, v. 2 (5): Teil 4: 121-186, figs. 132-210.

1929.—Dritte Klasse des Cladus Nemathelminthes, Kinorhincha=Echinodera. Handb. Zool. (Kükenthal & Krumbach) v. 2 (5): Teil 4: 187-248, figs. 211-266.

1929.—Rotatoria. Die Tierwelt der Nord und Ostsee Teil VII; 1-156, figs. 1-198.

1929-1933.—Rotatoria. Bronn's Klassen u. Ordnung des Tierreichs. Abt. 2, Buch 1, Teil. 1, Lief. 1-4: 1-576, figs. 1-332.

1935-1936.—Gastrotricha und Kinorhyncha. Bronn's Klassen u. Ordnung. v. 4, Abt. 2, Buch 1, Teil 2, Lief. 1-2: 1-385, figs. 1-297.

RUDOLPHI, C. A.1808.—Entozoorum sive vermium intestinalium historia naturalis. v. 1: 527, 6 pls. Amstelaedami.

1809.—Idem. v. 2 (1), 457 pp., pls. 7-12. Amstelaedami.

SCHEPOTIEFF, A. 1907.—Zur Systematik der Nematoideen. Zool. Anz., v. 31 (5/6): 132-161, figs. 1-24.

1907.—Die Echinoderiden. Ztschr. Wiss. Zool., v. 88.

1908.—Ueber den feineren Bau der *Gordius* larven. Ztschr. Wiss. Zool., v. 89 (2): 230-241, pl. 11, figs. 1-32.

SCHNEIDER, A. 1866.—Monographie der Nematoden. 357 pp., 122 figs., 28 pls., 343 figs. Berlin.

1868.—Ueber Bau und Entwickelung von *Polygordius*. Arch. Anat. Physiol. & Wiss. (1): 51-60, pls. 2-3, figs. 1-14.

1873.—Untersuchungen über Plathelminthen. 14 Ber. Oberhess. Gesellsch. Nat. -u. Heilk., Giessen, pp. 69-140, pls. 3-7.

SEURAT, L. G. 1920.—Histoire naturelle des nématodes de la Berbérie. Première partie. Morphologie, developpement, éthologie et affinitiés des nématodes. 221 pp., 34 figs. Alger.

SIMROTH, H. 1891.—Die Entstehung der Landtiere. Ein biologischer Versuch. Leipzig. 492 pp., 254 figs.

SNODGRASS, R. E. 1935.—Principles of insect morphology. 667 pp., 319 figs. New York.

STEINER, G. 1917.—Ueber die Verwandtschaftsverhältnisse und die systematische Stellung der Mermithiden. Zool. Anz., v. 48 (9): 263-267.

1917.—Ueber das Verhältnis der marinen freilebenden Nematoden zu denen des Süsswassers und das Landes. Biol. Zentrlbl. v. 37 (4): 196-210.

1919.—Zur Kenntnis der Kinorhyncha, nebst Bemerkungen über ihr Verwandtschaftsverhältnis zu den Nematoden. Zool. Anz., v. 50 (8): 177-187, figs. 1-4.

1920.—Betrachtungen zur Frage des Verwandtschaftsverhältnisses der Rotatorien und Nematoden. Festschr. Zschokke, 16 pp., 15 figs. Basel.

1921.—Untersuchungen über den allgemeinen Bauplan des Nematodenkorpers. Zool. Jahrb. Abt. Anat. v. 43, 96 pp., figs. A-E, 3 pls., figs. 1-16.

STEKHOVEN, J. H. SCHUURMANS, 1937 [1938].—Interrelation between free-living and parasitic nematodes. Papers on Helminthology, 30 year Jubileum K. J. Skrjabin pp. 637-639.

STOSSBURG, K. 1932.—Zur Morphologie der Rädertiergattung *Euchlanis*, *Brachionus*, und *Rhinoglena*. Ztschr. Wiss. Zool. Abt., A. v. 142 (3): 313-424, figs. 1-70.

TAYLOR, A. L. 1935.—A review of the fossil nematodes. Proc. Helm. Soc. Wash., v. 2 (1): 47-49, figs. 6.

THIELE, J. 1902.—Zur Cölomfrage. Zool. Anz., v. 25 (661): 82-84.

THOMAS, L. J. & QUASTLER, H. 1949.—The effect of X-ray on *Rhabditis* species. J. Parasit. v. 35 (6-2): 20.

THORNE, G. 1929.—Nematodes from the summit of Long's Peak, Colorado. Trans. Am. Micr. Soc., v. 49 (2): 181-195, figs. 1-18.

TINER, J. D. 1949.—Preliminary observations on the life history of *Ascaris columnaris*. J. Parasit. v. 35 (6-2): 13.

VAN CLEAVE, H. J. 1941.—Relationships of the Acanthocephala. Am. Naturalist v. 75: 31-47.

1948.—Expanding horizons in the recognition of a phylum. J. Parasit. v. 34 (1): 1-20.

VEJDOVSKY, 1886.—Zur Morphologie der Gordiiden. Ztschr. Wiss. Zool., v. 43 (3): 369-433, pls. 15-16.

WARD, H. B. 1892.—On *Nectonema agile* Verrill. Bull. Mus. Comp. Zool. Cambridge, v. 23 (3): 135-188, pls. 1-8, figs. 1-102.

WILLEMOES-SUHM, R. VON, 1871.—Ueber einige Trematoden und Nemathelminthen. Ztschr. Wiss. Zool., v. 21 (2): 175-203, pls. 11-13.

WOODHEAD, A. E. 1950.—Life history of the giant kidney worm, *Dioctophyma* renale (Nematoda), of man and many other mammals. Tr. Am. Micr. Soc. v. 69 (1): 21-46.

WRIGHT, R. R. and MACCALLUM, A. B. 1887.—*Sphyranura osleri*; a contribution to American helminthology. J. Morph., v. 1 (1): 1-48, 1 pl., figs. 1-18.

ZACHARIAS, O. 1885a.—Ueber die Bedeutung des Palmform-Stadiums in der Entwicklung von Rotatorien und Nematoden. Biol. Centralbl., v. 5 (8): 229-233.

1885b.—Ueber die Verwandtschaftsbeziehungen zwischen Rotiferen und Nematoden. J. Roy. Micr. Soc., s. 2, v. 5: 1006 (Abstract of 1885 a above).

ZEDER, J. G. H. 1860.—Erster Nachtrag zur Naturgeschichte der Eingeweidewürmer. 320 pp., 6 pls. Leipzig.

ZELINKA, C. 1896.—Echinoderes - Monographie. Verh. Deutsch. Zool. Gesellsch.

1908.—Zur Anatomie der Echinoderen Zool. Anz., v. 33 (19-20): 629-647, figs. 1-11.

CHAPTER XVII

LIFE HISTORY. GENERAL DISCUSSION

B. G. CHITWOOD

The development of nematodes in its simplest form is direct, or not marked by a metamorphosis such as occurs in the insects. In general the newly hatched nematode resembles the adult in all gross morphologic characters with the exception of the reproductive system and secondary sexual characters. The various growth stages, except the adult stage, are terminated by molts (or ecdyses), the number of molts being four, the number of stages five. Internal changes do not occur to any marked extent in the simplest form of life history. We should, therefore, speak of the stages previous to the adult as nymphs, if a terminology were used similar to that employed in the Arthropoda, but usage has made larva, as applied to such stages, the accepted term.

The number of molts occurring in the course of development is common for nearly all nematodes, and it appears to be the generalized or primitive number for the class. Development may be outlined as follows:

> Egg
> First stage (larva)
> (molt)
> Second stage (larva)
> (molt)
> Third stage (larva)
> (molt)
> Fourth stage (larva)
> (molt)
> Fifth stage (adult)

Correlated with mode of life, various adaptations or modifications have taken place in the life history, these adaptations having arisen through the need for food and a means of dissemination. With free-living nematodes, living either upon decaying matter or preying upon other microscopic organisms, these factors seem to have played a smaller part than with those living as parasites.

Probably need for dissemination was the earlier influence; at any rate, it has caused the simplest modifications of life history. The action of this factor on some free-living nematodes is evidenced by the occurrence of a persistent stage, the cuticle of one of the larval molts being retained as a protective sheath or "cyst." It is not uncommon for such species to have two types of larva, environmental conditions determining whether or not larvae will be of the persistent type. The significance of these persistent larvae is indicated by their negatively geotropic tendencies, ror they crawl to the highest surface available and "standing" on their tails swing about, catching upon any moving object. The climax of this type of development is found in species where an encysted stage on some arthropod (*Rhabditis coarctata*, see cover page, sec. 1, part 1), or annelid is obligatory before the adult stage can be reached.

The need for obtaining food plays a much more striking role, being evidenced by all conceivable degrees of parasitism both on and in plants and animals. In the group of "herbivors" life cycles may be of numerous types, depending on whether the nematodes are "grazers," passing from host to host, or sedentary forms, entering the host and there undergoing all or most of the stages of development. Life histories may be further modified by the factor of dissemination and the growth habits of the host. Among the parasites of animals the factors of dissemination and nourishment also play their roles. We have forms that are parasitic only during a particular larval stage, the third, which, incidentally, is usually sheathed or "encysted." Certain parasites of annelids (i.e., *Rhabditis pellio*) pass the third stage in the nephridia of their host and can only develop to adults in the decomposing tissues of their host. Other species (mermithids) enter their hosts either as eggs or larvae and develop to preadults (fourth stage) within the body cavity, finally leaving their host before maturing. In such instances the nourishment necessary for the entire life cycle is obtained from the host and stored during the parasitic stages.

The type of life history in parasitic nematodes being entirely correlated with the degree of parasitism, we find, with more advanced parasitism, more complicated life cycles and more morphologic changes taking place during the course of development. Seurat (1916; 1920), recognizing this, proposed a terminology for the different types of life cycles based on the mode of development.

Some forms have an alternation of generations, one generation being free living, the other parasitic. This type of life cycle is termed *heterogenous*. In such forms we find free-living adults giving rise to larvae which enter the host and develop to parasitic adults. These larvae may or may not be ensheathed (third stage), i.e., the cuticle of one or more larval molts retained though separated from the body. The stage ready to enter the host is termed the *infective* stage. Nematodes with no alteration of generations are termed *monogenous*. These are by far the most common.

Parasites of animals may also be classified according to the number of hosts necessary for completion of the life cycle. Species in which there is a single host are termed *monoxenous*, those in which there are two or more hosts, *heteroxenous*.

Some nematodes have both free-living and parasitic stages, the free-living stages being larvae, the parasitic stages late larvae and adults. In these we find young larval stages (the first and second) feeding upon bacteria and similar organic matter, the third stage usually ensheathed or persistent, this commonly being the infective stage. Upon entering the host these species develop through the fourth stage to the sexually mature adult.

A further development of parasitism is indicated by the absence of the free-living stages. Eggs of the parasite pass out of the host and, outside, undergo only embryonic development within the egg shell. In such instances the egg shell is often covered by a protein layer (p. 178), and the embryo often contains more yolk than forms in which the eggs hatch before entering the host. With such a completely parasitic mode of existence, the factor of dissemination again becomes manifest and we find still other modifications in the life cycle.

Some nematode parasites of vertebrates pass through larval stages in invertebrates, this course of development being either obligatory or facultative; still others undergo larval development in other vertebrates, such development usually being obligatory, rarely facultative. The host in which such a parasite develops to infectivity is termed the *intermediate* or *secondary host* while that in which it develops to sexual maturity is called the *terminal, definitive* or *primary host*. Sometimes the intermediate host is eaten by another animal (secondary intermediate host) in which the parasite can continue its existence but cannot reach maturity. When this second animal is, in turn, eaten by the primary host the life cycle is completed. If the parasite neither feeds nor undergoes growth within an animal, that host is termed a *transport host*. This type of intermediate host serves chiefly as a means of dissemination and is facultative rather than obligatory.

We have attempted to extend Seurat's outline to include all nematodes. With the recent and extensive increase in information on the life histories of vertebrate parasites it has become very difficult to adjust Seurat's classification to the many variations in life cycles. For example it is hardly proper to speak of a form as being heteroxenous when the use of an intermediate host (*Dictyocaulus filaria*) yet other nematodes in the same taxonomic group may be truly heteroxenous requiring an intermediate host (*Metastrongylus elongatus*). In general one can say that the Spiruroidea, Filarioidea, Camallanoidea, Dracunculoidea and Dioctophymatoidea are heteroxenous. The Strongyloidea, Trichostrongyloidea and Oxyuroidea are monoxenous while the Metastrongyloidea and Ascaridoidea contain forms with both monoxenous and heteroxenous life cycles. Some exceptional forms do not fit into any part of the classification. *Neoaplectana glaseri* (Rhabditoidea) and *Probstmayria vivipara* (Oxyuroidea) reproduce through several consecutive generations within the host. Some strains of a *Strongyloides* species may reproduce without an alternation of generations while other strains of the same species may be predominantly heterogenetic. The difficulty in fitting life histories into a well defined classification appears to be due to the adaptation of each species to its host which entails a means of dissemination suitable to the host's environment and habits.

A large assembly of nematodes have been found in more or less close association with vertebrates or invertebrates. Some of these merely use the "host" as a means of transportation (*Rhabditis coarctata* which may pass an encysted stage on the surface of dung beetles). Such nematodes are not considered parasitic unless they actually penetrate the host. Some well known free-living nematodes have been reported also existing under parasitic conditions. Thus *Rhabditis strongyloides* has been repeatedly taken, in the larval stage, from diseased skin of dogs and *Diploscapter coronata* from the ahydrochloric acid stomaches of human beings. Yet these forms are free-living nematodes and it would not be proper to classify them otherwise. If it happens that they are adaptable to unusual environments it is but an evidence of the nature of the group to which they belong.

Because of the numerous difficulties and inconsistencies apparent in any classification of nematode life histories, each of the authors has followed the system which seemed most logical to himself. Thus, the nematode parasites of invertebrates are grouped according to the manner and site of parasitism, beginning with the semiparasitic forms that mature at the death of their host and feed upon the carcass, then taking up the intestinal parasites and finally the parasites of the body cavity. Most of the invertebrate parasites belong to the Rhabditoidea and Tylenchoidea in which groups parasitism has arisen so many times and adaptations are so numerous that life cycles have little in common with systematics. The vertebrate parasites are taken up according to their systematic position since the large groups show some consistency within themselves and distinct trends are apparent.

For those who desire an outline after the manner of Seurat, we have revised his system to include groups with which he did not deal. The classification is entirely artificial. Nematodes are divided into the Vagantia or wanderers and the Parasitica. The Vagantia includes members of the Rhabditoidea, Tylenchoidea, Monhysterina, Chromadorina, Enoplina and Dorylaimoidea. Some representatives of most, if not all, of these groups have been found in more or less close association as semiparasites or parasites of plants or animals but the groups are basically free living. The only known modification in the life history of such free living forms is the existence of a persistent stage. Thus far, this stage is known only in terrestrial and semiterrestrial forms.

The Parasitica is subdivided into Phytoparasitica and Zooparasitica. All the known nematode parasites of plants belong to the Tylenchoidea though certain members of the Rhabditoidea and Dorylaimoidea are commonly found in close association with plants. In the Zooparasitica the heteroxenous group consists exclusively of parasites of vertebrates including all members of the order Spirurida, the suborder Dioctophymatina, and representatives of the Trichuroidea, Ascaridoidea and Metastrongyloidea. Those monoxenous nematodes in which the adult is wholly or partially free living belong to the Rhabditoidea, Tylenchoidea and Mermithoidea and are all parasites of invertebrates. The monoxenous nematodes in which the adult is wholly parasitic include the Strongyloidea, Trichostrongyloidea, Oxyuroidea and representatives of the Rhabditoidea, Metastrongyloidea, Ascaridoidea and Monhysteroidea. One commonly thinks of the groups with this type of life cycle as vertebrate parasites yet *Neoaplectana*, and *Cephalobium microbivorum* are rhabditoid parasites of invertebrates, the Thelastomatidae (*Leidynema*, *Pseudonymous*), Rhigonematidae and Ransomnematinae are oxyuroid parasites of invertebrates while *Longibucca*, *Rhabdias*, and *Strongyloides* are rhabditoid parasites of vertebrates and *Odontobius* is the lone monhysterid parasite of vertebrates.

CLASSIFICATION OF NEMATODES ACCORDING TO LIFE HISTORY*

I. Vagantia (Free-living nematodes).
1. Without persistent stage.
   Enoplidae
   (1) *Enoplus communis* (Marine)
2. With persistent stage.
   Rhabditidae
   (1) *Rhabditis strongyloides* (Soil, sometimes causing dermatitis in dogs).
   (2) *Rhabditis coarctata* (Dung, encysting on dung beetles).
II. Parasitica (Nematodes deriving nourishment from their host).
1. Phytoparasitica (Nematode parasites of plants).
   A. Vagrant parasites. More or less migratory, often feed externally, do not permanently localize in part of plant.
   Tylenchidae
   (1) *Criconemoides mutabile*—*Tagetes erecta* (External, roots).
   (2) *Pratylenchus pratensis*—Cowpea (Internal, roots).
   (3) *Aphelenchoides ritzema-bosi* — Chrysanthemums (Leaf and bud).
   (4) *Ditylenchus dipsaci*—Narcissus, onions, clover (Stem, leaf, and bulb).
   B. Semivagrant parasites. (Localize during definite period of life history.)
   Tylenchidae
   (1) *Anguina tritici*—wheat (Stem and seed).

*In this outline no attempt is made to supply all hosts or to include all nematode life histories. Only examples are given.

C. Sedentary parasites. (Female does not migrate after maturity.)
Tylenchidae
(1) *Heterodera marioni*—Tomatoes, potatoes, tobacco (Roots and tubers).
(2) *Heterodera schachtii* — Sugar beets, potatoes (Roots and tubers).
(3) *Tylenchulus semipenetrans*—Citrus plants (Roots).
(4) *Rotylenchulus reniformis*—Cowpea (Roots).
2. *Zooparasitica* (Nematode parasites of animals).
A. Monoxenous (Only 1 animal host in life cycle).
AA. Adult stage wholly or partially free-living.
a. Only larval stages parasitic or semiparasitic.
   aa. Feed in adult stage usually on carcass of host.
   Rhabditidae
   (1) *Rhabditis pellio*—Earthworms (Nephridia).
   Diplogasteridae
   (2) *Pristionchus aerivora*—Termites (Head).
   (3) *Alloionema appendiculatum*—*Limax ater* (Foot, alternation of generations reported).
   Steinernematidae
   (4) *Neoaplectana bibionis*—flies (Intestine).
   bb. Do not feed in adult stage.
   Mermithidae
   (1) *Agamermis decaudata*—Grasshoppers (Body cavity).
   (2) *Mermis subnigrescens*—Grasshoppers (Body cavity).
   (3) *Allomermis myrmecophila* — *Lasius* spp. (Body cavity).
   Allantonematidae
   (4) *Chondronema passali*—*Popilius interruptus* (Body cavity).
   Tetradonematidae
   (5) *Tetradonema plicans*—*Sciara coprophila* (Body cavity).
b. Adult stage partially parasitic, partially free-living.
   aa. Monogenetic (Without alternation of generations).
   Allantonematidae
   (1) *Allantonema mirabile*—*Hylobius abietus* (Body cavity).
   (2) *Tylenchinema oscinellae* — Frit-fly (Body cavity).
   (3) *Howardula benigna*—Cucumber beetle (Body cavity).
   (4) *Scatonema wülkeri*—*Scatopse fuscipes* (Body cavity; sometimes reproduces several generations in host).
   (5) *Aphelenchulus diplogaster*—*Ips typographus* (Body cavity).
   (6) *Parasitylenchus dispar* — *Ips typographus* (Body cavity).
   (7) *Sphaerularia bombi* — *Bombus terrestris* (Body cavity).
   (8) *Tripius gibbosus*—*Cecidomyia pini* (Body cavity).
   bb. Heterogenetic (With alternation of generations).
   Allantonematidae
   (1) *Fergusobia curriei*—One generation in plant, *Eucalyptus macrorrhynchia* (Leaf and flower) other in fly *Ferusonina nicholsonia* (Body cavity).
   (2) *Heterotylenchus abberans*—One generation bisexual, other parthenogenetic, both in body cavity *Hylemyia antiqua*.
BB. Adult stage wholly parasitic.
a. Heterogenetic (Free-living generation sometimes suppressed).
   Strongyloididae
   (1) *Strongyloides stercoralis*—Man (Small intestine).
   Rhabdiasidae
   (2) *Rhabdias bufonis*—*Bufo americanus* (Lung).
b. Monogenetic.
   aa. Reproduce in the host.
   Atractidae
   (1) *Probstmayria vivipara*—Equines (Intestine).
   Steinernematidae
   (2) *Neoaplectana glaseri*—Japanese beetle (Body cavity).
   Cylindrogasteridae
   (3) *Longibucca lasiura* — *Lasiurus borealis* (Small intestine).
   Diplogasteridae
   (4) *Cephalobium microbivorum*—*Gryllus assimi-*

lis (Intestine).

Monhysteridae.

(5) *Odontobius ceti*—Whale (Baleen).

(6) *Monhystera cambari*—Crawfish (Gills).

(7) *Tripylium carcinicolum*—*Gecarcinus lateralis* (Gills).

Myenchidac

(8) *Myenchus botelhoi*—Leeches (Muscle & connective tissue).

bb. Do not reproduce in host.

aaa. First three larval stages free-living.

Ancylostomatidae

(1) *Ancylostoma duodenale*—Man (Small intestine).

Trichostrongylidae

(2) *Haemonchus contortus*—Sheep (Abomasum).

(3) *Oswaldocruzia filiformis*—Amphibians (Intestine).

Syngamidae

(4) *Syngamus trachea*—Poultry (Bronchi or trachea) [Invertebrate, annelid, mollusc or insect transport host facultative].

(5) *Ollulanus tricuspis*—Cats (Stomach). [1st moult in parent worm.]

Metastrongylidae

(6) *Dictyocaulus filaria*—Sheep (Bronchi). [Annelid transport host facultative; 1st 2 larval stages do not feed.]

Cosmocercidae

(7) *Cosmocercoides dukae* — Amphibians and snails (Intestine).

bbb. Eggs infective to host.

Thelastomatidae

(1) *Leidynema appendiculatum* — *Periplaneta americana* (Intestine).

(2) *Pseudonymous spirotheca*—*Hydrophilus piceus* (Intestine).

Oxyuridae

(3) *Enterobius vermicularis* — Man (Appendix, caecum).

(4) *Oxyuris equi*—Equines (Colon).

Heterakidae

(5) *Heterakis gallinae*—Poultry (Intestine).

(6) *Ascaridia galli*—Poultry (Intestine).

Ascarididae

(7) *Ascaris lumbricoides*—Man (Intestine).

Trichuridae

(8) *Capillaria columbae*—Pigeons (Small intestine).

(9) *Trichuris trichiura*—Man (Caecum).

B. Heteroxenous (Two or more animal hosts in life cycle).

a. Eggs infective to intermediate host.

Metastrongylidae

(1) *Metastrongylus elongatus* — Earthworms — Swine (Lung).

Heterakidae

(2) *Subulura brumpti*—Various insects—Poultry (Cecum).

Ascarididae

(3) *Raphidascaris canadensis*—*Erogon* nymphs —Minnows—*Esox lucius* (Intestine). 2 intermediate hosts, mandatory.

Thelaziidae

(4) *Gongylonema pulchrum*—Beetles, roaches— Pig, sheep, deer (Esophagus and mouth).

(5) *Spirocerca lupi*—Dung beetles—Dog (Esophagus).

(6) *Ascarops strongylina*—Dung beetles—Swine (Stomach).

(7) *Physocephalus sexalatus* — Dung beetles— Swine (Stomach).

Spiruridae

(8) *Tetrameres crami*—Amphipods—Duck (Proventriculus).

Acuariidae

(9) *Cheilospirura hamulosa* — Grasshoppers — Poultry (Gizzard).

(10) *Echinuria uncinata*—Cladocera (*Daphnia*)— Duck (Fore and mid-gut).

(11) *Dispharynx spiralis* — Isopods — Poultry (Esophagus and crop).

Gnathostomatidae

(12) *Hartertia gallinarum* — Termites — Poultry (Small intestine).

Trichuridae

(13) *Capillaria annulata*—Annelid transport host obligatory—Chickens (Crop).

Cystoopsidae

(14) *Cystoopsis acipenseri* — Amphipods—Sturgeons (Skin).

Eustrongylididae

(15) *Eustrongylides ignotus—?* Crustacean—*Fundulus diaphanus*—*Ardea herodias* (Gizzard).

Dioctophymatidae

(16) *Dioctophyma renale*—? Crustacean—? fish —Man, dogs, mink (Kidney).

b. Larvae infective to intermediate host.

aa. Enter final host per os.

Dracunculidae

(1) *Dracunculus medinensis* — Cyclops — Man (Under skin).

Philometridae

(2) *Philometra nodulosa*—Cyclops—*Catostomus commersonii* (Lip).

(3) *Philometra fujimotoi*—Cyclops—*Ophicephalus argus* (Fin).

Camallanidae

(4) *Camallanus sweeti*—Cyclops—*Ophicephalus gachua* (Intestine). Second intermediate host, small fish? obligatory.

Pseudaliidae

(5) *Muellerius capillaris*—Molluscs—Sheep and goats (Lung).

Spiruridae

(6) *Habronema muscae* — *Musca domestica* — Equines (Stomach).

(7) *Habronema microstoma* — *Stomoxys* spp. —Equines (Stomach).

(8) *Draschia megastoma*—*Musca domestica*— Equines (Stomach).

Gnathostomatidae

(9) *Spiroxys contorta* — Cyclops — Minnows — Turtles (Stomach). Second intermediate host not mandatory.

(10) *Gnathostoma spinigerum*—Cyclops—Fish or snakes—Felidae (Stomach). Second intermediate host mandatory.

Ascarididae

(11) *Contracaecum spiculigerum*—Minnows—Carnivorous fish—Cormorant (Proventriculus). Second intermediate host mandatory.

(12) *Raphidascaris canadensis*—*Erogon* nymphs— Minnows—*Esox lucius* (Intestine). Second intermediate host? mandatory.

Trichinellidae

(13) *Trichinella spiralis*—Rat, pig, man (Intestine). Hosts serve both as intermediate and final host.

bb. Enter final host through skin.

Dipetalonematidae

(1) *Wuchereria bancrofti* — Mosquitoes — Man (Lymphatic system).

(2) *Onchocerca volvulus*—*Simulium damnosum*— Man (Subcutaneous).

(3) *Onchocerca cervicalis*—*Culicoides nebeculosis*—Equines (Cervical ligament).

(4) *Dirofilaria immitis* — Mosquitoes — Dogs (Heart).

CHAPTER XVIII
LIFE HISTORY (ZOOPARASITICA)
Parasites of Invertebrates
J. R. CHRISTIE, U. S. Horticultural Station, Beltsville, Md.

Introduction

There are many different types of association between nematodes and other invertebrates and it is difficult to draw a line between what should and what should not be regarded as parasitism. Most of the nematodes that live within the bodies of invertebrates are customarily referred to as parasites though there is little evidence that some of them interfere materially with the well-being of their "hosts." We know very little, however, about the effects of these nematodes on the animals that harbor them unless the manifestations are pronounced and obvious. The only feasible procedure is to regard as eligible for inclusion in this chapter all nematodes that regularly spend part of the life cycle within the bodies of invertebrates regardless of the precise character of the association. Species for which vertebrates serve as definitive hosts and invertebrates only as intermediate hosts are dealt with in the following chapter.

In general the parasites of invertebrates and those of vertebrates are not found in the same phylogenetic groups and in those cases where both belong to the same group the vertebrates involved are almost always amphibians and reptiles. However, the Thelastomatidae and the Oxyuridae have very close affinities.

Arthropods, annelids and mollusks are the invertebrates most commonly parasitized by nematodes though scattered cases have been reported where other invertebrates, even nematodes themselves, serve as hosts. There are surprisingly few records of marine invertebrates harboring nematodes and most of these apparently deal with cases where the association is erratic or accidental or where some vertebrate serves as definitive host.

Included among the nematodes harbored by invertebrates are species where a parasitic mode of life is only now being acquired and others where it is of great antiquity. There is great diversity in the types of life cycles and to simplify discussion and facilitate comparison the nematodes are divided into three groups.

The first of these groups is made up of nematodes that are more or less closely related to free-living species and in the life cycles we often find a combination of saprophagous and "parasitic" habits. In one line of evolutionary development the nematodes live and reproduce in the carcass of the "host," to the death of which they may or may not have contributed. Life cycles are simple, perhaps the most outstanding feature being the frequent occurrence of dauer larvae,* a characteristic that has been carried over from a free-living to a parasitic mode of life. Another line of evolutionary development seems to have culminated in a life cycle where the nematode may pass through one or more free-living generations, then gain entrance to the host and pass through one or more parasitic generations.

The second group comprises those nematodes, not included in the first group, that inhabit the alimentary tract. Life cycles, so far as known, are simple. With perhaps an occasional exception (i. g., *Cephalobium microbivorum*), only the egg stage occurs outside the host, a characteristic shared by very few species in the other two groups.

The third group includes the body-cavity and tissue parasites. In contrast to the first group, these nematodes are highly specialized, obligate parasites and, in contrast to the second group, they pass, at the most, only a transitory period in the alimentary tract of the host. Five families are included in this group. The Drilonematidae and Myenchidae have received little attention and our knowledge regarding life cycles is very meager. The Tetradonematidae, Mermithidae and Allantonematidae have been somewhat more adequately studied. The nematodes belonging to these three families have been parasites for a very long time and many of them have complicated life cycles that are highly adapted to individual requirements. Of the various factors that have influenced these life cycles, two stand out as being of great importance.

One of these factors is the necessity for the infective stage to reach and gain entrance to the host. This, of course, is a requisite in the life cycle of every parasite but for the allantonematids and mermithids there are certain restricting conditions with which many of the others do not have to contend, at least not to an equal extent. Some of the hosts are insects that develop in seasonal cycles and where the total life span of the individual may be only a few months. It is frequently necessary that the parasite enter when the host is in a particular stage and this stage may be available only at restricted times of the year. As a result the life cycles of many of these parasites have become closely correlated with the life cycles of their respective hosts.

The other factor is the ability of the nematode to take food only during restricted periods. The fact that for many of these parasites the free-living stage may be of considerable duration and that during this period the nematodes take no food, but, nevertheless, pass through important phases of the life cycle, has had a profound effect on development. In many cases the larval mermithid, during a comparatively short period of parasitic life, must make a phenomenal growth and store sufficient nutrient materials to carry the adult through its relatively long, free-living period of sexual activity and reproduction. The larval allantonematid that develops to maturity outside the host after only a very brief period of parasitic life, must exercise the strictest economy in the utilization of its limited supply of stored nutrients. Since, as a rule, only the female again becomes parasitic, the male must produce and mature its spermatozoa though the production and maturation of the eggs by the female is postponed. There can be little or no increase in body size during this free-living period, hence the adult, impregnated female, after entering a new host, undergoes a period of rapid growth. In the Sphaerulariinae a prolapsus of the uterus has resulted through the inability of the small, underdeveloped body of the young female to keep pace with the rapidly growing reproductive organs.

Novitious Parasites and Semiparasites

Among these nematodes two lines of evolutionary development seem to stand out more or less distinctly though it is obviously improbable that they account for the origin of all the different types of parasitism or semiparasitism encountered in this heterogeneous group.

One line of evolutionary development appears to have been initiated when certain saprophagous nematodes utilized other invertebrates, frequently saprophagous insects, as vehicles for transportation. These "hitch-hikers," first seeking protection from desiccation in crevices on the external surface, eventually entered the bodies of their "hosts." In the life histories of species representing an intermediate step in this line of development, larval nematodes, after gaining entrance to the body of the "host" and becoming established therein, do not at once grow to maturity and reproduce but remain in a more or less quiescent condition. These larvae do not appear to interfere materially with the life processes of the animal that harbors them but when the animal dies from other causes the nematodes immediately resume development and reproduce in the carcass.

In some cases, however, this type of relationship has evolved to a point where it is no longer passive but where the nematodes are an important factor in bringing about the death of the animal whose body they enter. Even though present in small numbers, some species of *Neoaplectana* are said to kill their insect hosts in a very short time.

The parasitic or semiparasitic relationship between these nematodes and their respective "hosts" is not always obligatory. Johnson (1913) concluded that entrance into the body of an earthworm is not necessary in the life cycle of *Rhabditis maupasi* but if larvae, during their sojourn in the soil, find suitable decaying organic matter they will develop and reproduce therein. Neither is *Pristionchus aerivora* dependent on entrance into a termite or some other insect to complete its development as it has been found reproducing in a number of different habitats including decaying plant tissues. *Neoaplectana glaseri*, on the other hand, appears to be an obligate parasite that, in nature, develops only after entering the living body of its insect host.

---

*The term *dauer larva* is used in this text to designate a larva, in a particular stage of development, that is especially adapted to withstand adverse conditions and, when a dauer stage is not obligatory, that differs from a larva of the same stage that develops when conditions are favorable and food is abundant. The term is not new, having been used by Fuchs and others with approximately this same meaning and, while not of classic origin, it is short, expressive, appropriate and useful. Dauer larvae are of common occurrence in the Rhabditidae and Diplogasteridae and are more characteristic of free-living than of parasitic species, hence the term is not synonymous with "infective larva."

Most of these nematodes are bisexual and females produce fertile eggs only after copulation. Males are usually somewhat less numerous than females, reach maturity a little quicker, and do not live quite so long. According to Johnson, females of *Rhabditis maupasi* usually, though not always, reproduce without males. In many species of this group a female may be oviparous when young but toward the end of life some of the last eggs produced may be retained and hatch in the uterus. The resulting larvae may not escape through the vulva but undergo part of their development within the mother nematode, consuming her internal organs and converting her into a brood sac. Incidentally, this same mode of reproduction is characteristic of many free-living species of *Diplogaster*, *Rhabditis* and related genera.

The second line of evolutionary development referred to above may have been initiated when, during periods of adversity, certain saprophagous nematodes, seeking refuge and succor, entered and temporarily dwelt within the bodies of other invertebrates. In the case of nematodes in this category parasitism apparently does not ordinarily result in the death of the host nor are the parasites able to live in a decaying carcass. Usually these nematodes either inhabit the alimentary tract of the host (e. g., *Angiostoma limacis*) or are associated with its reproductive organs (e. g., "*Angiostoma*" *helicis*). For at least one species (i. e., *Alloionema appendiculata*) an alternation of one or more parasitic generations with one or more free-living generations has become a more or less regular procedure.

RHABDITIS MAUPASI Caullery and Seurat, 1919 (Syn. *R. pellio* Bütschli, 1873; not Schneider, 1866). Larvae of *Rhabditis maupasi* are found in the nephridia and coelom of living earthworms. For *Lumbricus terrestris* L. the incidence of infection is frequently very high and at least several and perhaps many other species harbor these nematodes more or less frequently.

Larvae are found near the nephridiopore in the dilated, muscular termination or "bladder" of the nephridial tube. Often nearly every tube is inhabited, the number of worms in each varying from 2 or 3 to 12 or more. Also larvae may occasionally be found in the seminal vesicles. When in these above mentioned locations larvae are in an active condition and not ensheathed. Johnson concluded that these inhabitants of the nephridia are not necessarily confined to this location throughout the life of the earthworm but may move out into the soil and later go back through the nephridiopores into the same or a different earthworm.

Larvae occur also in the coelom and these are usually ensheathed and inactive (Fig. 165C). Occasionally a larva may be embedded in the muscles of the body wall or encysted on a septum. Frequently several larvae are embedded in a brown, oval body composed of cysts of the sporozoan, *Monocystus*, and various earthworm tissues. Such bodies are most common at the posterior end of the coelom.

There is no evidence that the presence of these larval nematodes is detrimental to the annelid. So long as the earthworm is alive the nematodes remain in a larval stage but when the earthworm dies they quickly grow to adults (Fig. 165 A & B) and reproduce in the carcass. Otter (1933) concluded that a female lives from 7 to 10 days after reaching maturity and lays from 150 to 300 eggs. Males, in his opinion, live about a third as long as females. No doubt several generations occur before the food supply is exhausted though Johnson was uncertain on this point. After the body of the earthworm is consumed large numbers of larvae move out into the soil where they live awaiting the opportunity to enter another earthworm. Larvae from the soil are said to be in the same stage as those from the nephridia, but what this stage is has not been stated.

With regard to the method of entering the earthworm, Johnson writes: "Those that enter by the nephridiopores take up their position in the terminal, bladder-like part of the nephridia. Those that use the spermiducal apertures travel up the vasa deferentia and occupy the seminal vesicles. Lastly, those that pass in by the dorsal pores and the oviducal apertures find themselves in the coelom, where, being attacked by the amoebocytes, they encyst. These encysted larvae coated with amoebocytes are worked backward by the movement of the worm till they come to rest in the tail end of the worm, where, together with other foreign bodies, such as cysts of *Monocystis* and discarded setae, and with masses of dead brown-colored amoebocytes, they are compressed and cemented into the brown bodies which are found there."

According to Keilin (1925) the accumulation of foreign bodies in the posterior segment of an earthworm may induce the development of a stricture that will sever this distended terminal portion from the rest of the body. The detached portion then decomposes and in this manner *R. maupasi* and other coelomic parasites of the earthworm may be liberated.

Males of *R. maupasi* are much fewer in number than females. Although Johnson did not observe copulation, his rearing ex-

periments lead him to conclude that most females are hermaphroditic but that occasionally females occur that are able to reproduce only after being fertilized by males. Otter, who observed copulation and agrees, in the main, with Johnson, writes that *R. maupasi* "may thus be considered to be one of those species of *Rhabditis* in which hermaphroditism is in a very early stage, and in which functional males, females, and hermaphrodite females, exist side by side in fluctuating proportions."

PRISTIONCHUS AERIVORA (Cobb, 1916), was first found by Merrill and Ford in the heads of termites, *Leucotermes lucifugus* Rossi,* collected near Manhattan, Kansas. Under natural conditions the nematodes varied from 0 to about 75 per insect. After experimental termites had been kept for 4 days in soil heavily infested with *P. aerivora*, the average number of nematodes per insect was 46.6 while termites used as controls averaged about 3 nematodes per insect. How the nematodes enter or why, in living termites, they are found only in the head are points that have not been determined. The parasites do not reach maturity in living hosts but when the termites are heavily infected they become sluggish and die, whereupon the nematodes reproduce in the carcass. Hence, in this instance, the relationship is not purely passive.

Merrill and Ford were able to rear this nematode in water cultures with various substances supplied for food, preferably the macerated bodies of insects. Eggs hatched in about 18 hours and the adult stage (Fig. 165 J) was reached in about 2 days. The complete life cycle from egg to egg required about 4 to 5 days but after beginning to lay eggs an adult female usually lived for 12 to 13 days. During a period of 13 days one female, while under observation, copulated with 7 males and deposited 317 fertile eggs and 14 infertile eggs. Males were somewhat less numerous than females. They lived for about 19 days and one male, while under observation, copulated with 10 different females.

Toward the end of life a female becomes sluggish and eggs are not extruded but hatch in the uterus. While the resulting larvae may sometimes escape through the vulva they usually remain in the mother nematode, feeding on her internal organs.

Since Merrill and Ford's investigations nematodes identified as *P. aerivora* have been reported from various other habitats. They have been found in other termites, usually located in the head while the insect is alive. They have been found in dead pupae of the corn ear worm, *Heliothis armigera* (Hübn.), and in dead pupae of the rose leaf beetle, *Nodonota puncticollis* (Say). They have been found in grasshopper egg masses where they were reported to have been destroying the eggs. On several occasions they have been found in decaying plant tissues. However, the populations from these different habitats may represent different strains or, perhaps, even different, though closely related, species.

The peculiar habit of swallowing air, to which this nematode owes its specific name, is shared by several species of *Diplogaster* and *Rhabditis*. When mounted in water on a microscope slide, one of these nematodes may place its head against the surface of an entrapped air bubble and air can be seen as it passes down the esophagus to the anterior end of the intestine where it is quickly absorbed. According to Cobb (1915) some of these nematodes can ingest their own volume of air in the course of an hour or two. The swallowing of air is accomplished by the usual rhythmic muscular movements of the esophagus. During the first muscular movement a small bubble of air passes quickly from the mouth to the median pseudobulb where it stops. At the next muscular movement the bubble passes on into the intestine while another simultaneously passes from the mouth to the median pseudobulb. This may continue uninterrupted for a considerable period of time.

NEOAPLECTANA BIBIONIS Bovien, 1937, was studied by Bovien (1937) who found it in Denmark associated with the dipterous insects *Bibio ferruginatus* (L.), *B. hortulanus* (L.) and *Dilophus vulgaris* Meig.

An interesting and significant point in the life cycle of this nematode is the occurrence of dauer larvae (Fig. 165 G). These, according to Bovien, are in the third stage. A dauer stage is not obligatory but occurs only when environmental conditions are unfavorable to enable the nematode to persist through periods of adversity. Dauer larvae are relatively sluggish and are usually enclosed in a partly separated cuticle though this may be lost before the end of the dauer stage. These larvae are easily distinguished from third-stage larvae that develop under favorable conditions being slenderer and differing in other morphological details. Bovien found dauer larvae clinging to the surface of adult flies and being transported by them. The various host insects become infected by swallowing these dauer larvae which, on reaching the alimentary tract, remain

*Regarded by Snyder, according to Van Zwaluwenburg (1928, p. 9), as either *Reticulitermes tibialis* Banks or *R. claripennis* Banks.

247

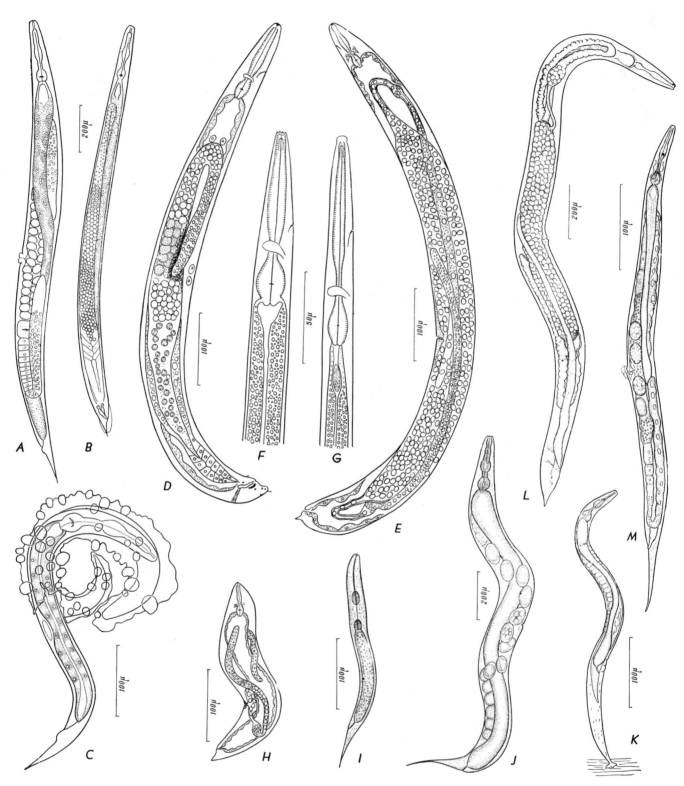

Fig. 165. NOVITIOUS PARASITES AND SEMIPARASITES

A-C—*Rhabditis maupasi* (A—Adult female; B—Adult male; C—Larva escaping from "cyst"). D-H—*Neoaplectana bibionis* (D—Adult male; E—Adult female; F—Larva that developed under favorable conditions; G—Dauer larva of same stage as F; H—Pigmy female). I-J—*Pristionchus aerivora* (1—Newly hatched larva; J—Adult female). K—*Diplogaster labiata,* dauer larva. L & M—*Alloionema appendiculatum* (L—Adult female of parasitic generation; M—Adult female of free-living generation). A-C, after Johnson, 1913; D-H, after Bovien, 1937; I-K, after Merrill and Ford, 1916; L & M, after Claus, 1868.

unchanged, apparently having no adverse effect on the insect. When eventually the insect dies, presumably from other causes, the larvae move into its tissues and proceed in development, passing through several generations and quickly building up a large population. When the carcass has been consumed young nematodes move out into the soil and develop into dauer larvae.

Although flies become infected while in the larval stage, if the nematodes, on reaching the intestine, are wholly innocuous, many of the insects must carry their infection on into the adult stage. Bovien is not very lucid on this point but he mentions finding nematodes in living fly pupae, on one occasion in the body cavity.

Bovien concluded that development from egg to egg-laying female requires about 4 days. A young female is oviparous but some of the last eggs laid by an old female hatch within the uterus. Each female usually produces somewhat in excess of 200 eggs.

It is not strictly necessary that *N. bibionis* enter living insects as Bovien was able to rear several generations on dead insects of different species if fresh cadavers were periodically provided. Several generations could sometimes be reared on egg albumen.

Gravid females (Fig. 165 E) usually attain a length of up to 5 mm. but Bovien reports finding mature, reproducing females that failed to reach a length of 1 mm. (Fig. 165 H), perhaps due to some nutritional deficiency. Between these dwarfs and females of maximum stature numerous intermediate sizes were found.

NEOAPLECTANA GLASERI Steiner, 1929, was first found in dead larvae of the Japanese beetle, *Popillia japonica* Newm., collected in New Jersey and is best known as a parasite of this insect. However it has been demonstrated that this nematode will infect larvae of the European corn borer, *Pyrausta nubilalis* (Hübn.) and coleopterous larvae belonging to at least nine genera including the white-fringed beetle, *Pantomorus leucoloma* (Boh.).

The life history of *N. glaseri* has been investigated by Glaser (1932) and by Glaser, McCoy and Girth (1940). The following account is based on their results that were secured, in part by using Japanese beetle larvae as experimental hosts and in part by rearing on culture media. It is believed that the behavior of this nematode is not materially different whether growing on culture media or in the various susceptible insect hosts.

Japanese beetle grubs acquire their infection by ingesting third-stage, infective larvae of the parasite. On reaching the alimentary tract the larvae immediately develop to maturity and copulate. A female will not produce offspring unless fertilized by a male. The female is ovoviviparous, eggs hatching within the uterus. Larvae may remain within the uterus and move about for a considerable period but eventually pass out through the vulva one at a time. If a female dies before all larvae are born those remaining may undergo partial development within the dead body. Each normal-sized female produces a total of about 15 offspring and a generation under optimum conditions requires about 5 to 7 days. By the time the first generation of offspring have matured the insect is usually dead whereupon its entire body is invaded. The nematodes usually pass through two more generations consuming the carcass and leaving only a sac formed by the skin and head capsule and filled with a thin fluid swarming with larval parasites. In a few cases Glaser was able to infect newly killed beetle grubs but he concluded that the nematodes do not enter and multiply as readily in cadavers as in living insects.

With regard to the virulence of this parasite, Glaser, McCoy and Girth (1940) write that "occasionally an insect host becomes parasitized very lightly, so that only one nematode becomes successfully established. This individual may be of either sex, and while it frequently (if not always) causes the death of the host, there is no reproduction."

McCoy, Girth and Glaser (1938) report that exceptionally large females of *N. glaseri* are occasionally found in beetle larvae though never on cultures. Such individuals may develop an enormous number of eggs, one giant female producing 1,420 larvae. When offsprings of a giant are reared to maturity on cultures only normal-sized females are obtained. McCoy, Girth and Glaser concluded that fecundation at a late period in development and abundant food are factors contributing to the production of these giant females.

So long as conditions are favorable and abundant food is available, the life cycle of *N. glaseri*, according to Glaser, McCoy and Girth (1940), is completed in three molts the third stage being omitted. When conditions are unfavorable, as when the carcass of the beetle larva has been consumed and food is exhausted, the young parasites develop into third-stage, dauer larvae. At the end of the second stage growth ceases, the alimentary tract is emptied, and, as a result of certain morphological changes, the body becomes more slender. The second molted

cuticle is retained hence the dauer larva is ensheathed though the sheath is not very tenacious and may soon be lost. These dauer larvae escape into the soil where they are able to persist, in a more or less active condition, for at least 2½ years.

Glaser and his coworkers have reared this nematode successfully on Petri dish plates of veal infusion agar flooded with living yeast, on potato culture medium, and on veal pulp medium. These investigators found that "distinct cultural characteristics occur in nematodes from different insect cadavers, . . . There is a slow decline in fecundity of the cultured nematodes, some 'strains' dying out after 5 or 6 transfers, while others continue to yield good cultures after 20 or more transfers." If beetle larvae are infected with nematodes from cultures that are dying out and several generations are passed in the natural host, the nematodes can again be reared successfully on cultures, the length of time before the cultures again die out depending, to some extent, on the number of generations passed in beetle larvae.

ALLOIONEMA APPENDICULATUM Schneider, 1859, has on several occasions been found within the bodies of slugs. Schneider found it originally in *Arion ater* (L.) and Claus (1896), who investigated its life history, secured his material from the same host. The life cycle of this nematode appears to represent a somewhat different line of evolutionary development than the life cycles already discussed. According to Claus (1896), one or more free-living generations alternate with one or more parasitic generations, both males and females (Fig. 165 L & M) developing in each instance. Individuals of the parasitic generations leave the host just before reaching maturity by boring their way out through the foot. On reaching the exterior they mature, copulate and produce progeny that usually develop as free-living individuals. Maupas (1899) found that larvae of the free-living generation undergo the usual four molts and reach maturity in about 3½ days.

A regular alternation of a free-living with a parasitic generation does not necessarily follow, however, as there may be several consecutive free-living or several consecutive parasitic generations. There are usually consecutive free-living generations as long as conditions are favorable but when conditions become unfavorable the nematodes "encyst" and these "encysted" larvae will continue development only when taken into the body of a slug. According to Maupas, "encysted" larvae that fail to gain entrance to a slug become exhausted and die in about 4 months. Precisely how the nematodes enter the slugs and whether or not, in event of consecutive parasitic generations, females mature without leaving the host, are points that seem to need further elucidation.

Claus found certain morphological differences between corresponding stages of the two generations. Adults of the parasitic generations are much larger than adults of the free-living generations and parasitic larvae, in the later stages of development, are said to possess two long, ribbon-like, caudal appendages not present on free-living larvae of the corresponding stage.

OTHER SPECIES. *Diplogaster labiata* Cobb (in Merrill and Ford, 1916) was found in the elm borer, *Saperda tridentata* Oliv., collected near Manhattan, Kansas. This nematode reproduces in the intestine of the living, adult borer and may accumulate in sufficient numbers to rupture the gut and kill the insect. Infected female beetles are usually sterile. When reared on cultures, Merrill and Ford (1916) found that eggs hatched in from 30 to 32 hours and the nematodes matured in 7 to 10 days. Oviposition began from 2 to 4 hours after copulation and lasted for about 2 days with an average output of seven eggs per female. Only a few individuals were seen copulating a second time. Apparently dauer larvae (Fig. 165 K) develop when conditions are unfavorable.

Neoaplectana affinis, Bovien, 1937, was found in Denmark where it infects larvae of the same insects that harbor *Neoaplectana bibionis*, i.e., *Bibio ferruginatus*, *B. hortulanus* and *Dilophus vulgaris*. These two nematodes were differentiated morphologically by Bovien (1937) only on the basis of males and dauer larvae, the life cycles and behavior of the two being almost identical. Bovien made one observation, however, that deserves mention. When in the intestine of any of its three natural hosts mentioned above, *N. affinis* remained in the dauer stage and was apparently innocuous so long as the insect remained alive. When two larvae of a beetle, *Telephorus* sp., were experimentally infected, they became moribund in a few days and dissection revealed several adult and preadult nematodes in the body cavity of each beetle. This observation suggests that whether or not *N. affinis* remains passively in the intestines depends on the insect involved.

A mode of life on the border line between saprophagous and parasitic is characteristic of other nematodes, probably of a considerable number. Other species of *Neoaplectana* are known to exist but life cycles have not been investigated. *Steinernema kraussei* (Steiner, 1923), found in the intestine of the wasp, *Cephalcia abietis* (L.), is so closely related to the genus *Neoaplectana* that a similar mode of life is suggested but verifying information is lacking.

Among the rather numerous and diverse nematodes that have been reported from snails and slugs are representatives of the Angiostomatidae and Cosmocercidae, two families that include also parasites of Amphibia. The four species mentioned below will serve as examples but very little information is available about life cycles. *Angiostoma limacis* belongs to the Angiostomatidae while the other three, according to Chitwood and Chitwood (1937), probably belong to the Cosmocercidae.

Angiostoma limacis Dujardin, 1845, has, on at least two occasions, been found in the intestine of *Arion ater* (L.) (Syn. *Limax rufa*) where, apparently, it reaches maturity. Chitwood and Chitwood (1937) report finding a very closely related species in the intestine of a salamander, *Plethodon cinereus*.

Ascaroides limacis Barthélemy, 1858, was found in eggs of *Deroceras agrestis* var. *cineracea* Moq. Tand. (Syn. *Limax griseas*), each infected egg containing one to four larval parasites. Barthélemy (1856) determined that the nematodes were already present when the eggs were deposited. Apparently the adult of this parasite has not yet been studied.

"*Angiostoma*" *helicis* Conte and Bonnet, 1903, was secured by its discoverers from the slug, *Helix aspersa* (Müll.), where it occurred in the genital organs, especially the oviducts and seminal vesicle, but not elsewhere in the body. Conte and Bonnet (1903) concluded that the parasite is passed from host to host during copulation.

Trionchonema rusticum Kreis, 1932, was secured from the land snail, *Polygyra espicola* Bland. Presumably this parasite is an inhabitant of the alimentary tract though the location within the host was not specified. Kreis (1932) refers to the development of a "filariform" larva and suggests the possibility "that there is still another stage of development, perhaps a rhabditiform larva, which could not be found and which may perhaps be free-living."

Parasites of the Alimentary Tract

All nematodes belonging to the families Thelastomatidae and Rhigonematidae and to the subfamily Ransomnematinae are parasites of the alimentary tract and one finds an occasional species of the family Diplogasteridae that has acquired this mode of life.

The thelastomatids are parasites of insects and myriapods and scattered through the literature are descriptions of between 60 and 70 species but usually not much other information. However, studies by Galeb (1878), Dobrovolny and Ackert (1934), and others indicate that most of these species probably have about the same type of life cycle and that it is comparatively simple. Eggs pass out of the host with the feces. Eggs do not hatch in the intestine to reinfect the same host but must first undergo some development on the outside to reach an infective stage. The various arthropod hosts acquire their parasites by swallowing these infective eggs.

In the genus *Pseudonymous*, the species of which are parasites of aquatic beetles, the egg is provided with two entangling appendages, the so-called spiral filament (Fig. 135 R, p. 176) which, presumably, enables the egg to hang on aquatic vegetation thus increasing its chance of being ingested. From two to four eggs of *Binema binema* and *B. ornata* (Fig. 166G) are enclosed in an outer capsule or case of loose texture formed, apparently, by the entangling and anastomosing of polar filaments. The purpose of this adaptation is obscure.

The Rhigonematidae and Ransomnematinae are small groups with only a few species each. It seems probable that life cycles of these nematodes are not materially different from the type of life cycle characteristic of many thelastomatids though, admittedly, such a statement is wholly conjectural.

CEPHALOBIUM MICROBIVORUM Cobb, 1920, a member of the Diplogasteridae, inhabits the intestine of the black field cricket, *Gryllus assimilis* (Fab.), where it may occur in numbers up to 30 or more. Infected crickets have been collected in Virginia and Kansas. In the region of Manhattan, Kansas, according to Ackert and Wadley (1921), there are two races of this insect each having one brood a year. One race matures during April and May and overwinters in the nymph stage while the other race matures during August and September and overwinters in the egg stage. These investigators found that in autumn over 85 per cent of the adults of the latter generation were infected, the incidence being somewhat higher in female (about 90 percent) than in male crickets (about 70 percent).

Eggs of *C. microbivorum* are usually deposited in a four-cell stage and pass out of the host with the feces. Ackert and Wadley concluded that probably eggs hatch after being voided and that a cricket becomes infected by ingesting larval nematodes

perhaps after these have undergone a brief period of free-living development. The two races of crickets provide the parasite with suitable hosts throughout most of the year and, no doubt, some of the nematodes pass the cold season in overwintering nymphs. The presence of this nematode has no obvious effect on the well-being of the cricket.

LEIDYNEMA APPENDICULATUM (Leidy, 1850) Chitwood, 1932. —The life history of *Leidynema appendiculatum*, which was investigated by Dobrovolny and Ackert (1934), is probably more or less typical of many thelastomatids and will serve as an example of the family. This nematode is a parasite of the cockroaches, *Blatta orientalis* (L.) and *Periplaneta americana* (L.). Out of 259 individuals of *P. americana* collected by Dobrovolny and Ackert at Manhattan, Kansas, 90 harbored this parasite in numbers of from 1 to 36 per host.

The egg, deposited in a one to a four-cell stage, passes out of the insect with the feces. After extrusion it undergoes a short period of development and a tadpole-like larva (Fig. 166 A) is formed. The larva is at first motile, wiggling and squirming about, but becomes inactive as the infective stage (Fig. 166 B) is reached. Dobrovolny and Ackert found that at 37° C. eggs reach this infective stage in 3 to 7 days and

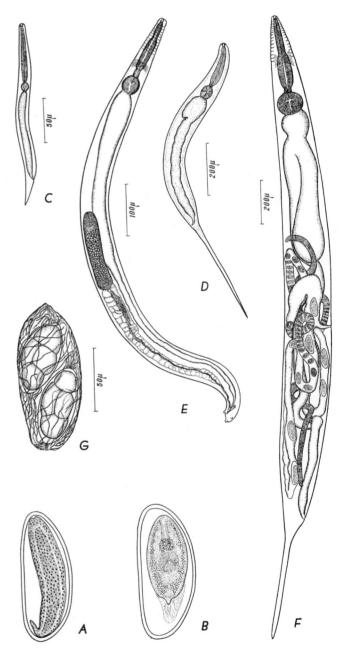

Fig. 166. PARASITES OF THE INTESTINE

A-F—*Leidynema appendiculatum* (A—Egg with active embryo; B—Egg with larva in resting stage; C—An early stage larva, presumably second stage; D—Larval female showing intestinal diverticulum beginning to form; E—Adult male; F—Adult female). G—*Binema ornata*, egg capsule. A-F, after Dobrovolny and Ackert, 1934.

that eggs containing motile larvae are not infective. Alicata (1934) found that the larva of *Blatticola blattae*, a closely related thelastomatid, molts in the egg before reaching the infective stage and while Dobrovolny and Ackert do not mention the matter their figures indicate that *L. appendiculatum* undergoes a similar molt. If kept moist at room temperature and in subdued light infective eggs remain viable for a considerable time but are killed by prolonged exposure to direct sunlight. Infective eggs are ingested by the insects and hatch in the posterior part of the midgut.

Dobrovolny and Ackert kept heavily infected cockroaches in captivity for more than a year and saw no evidence that the insects were markedly affected by the parasites.

Body Cavity and Tissue Parasites

This group includes what are, perhaps, our oldest parasitic nematodes in the sense that their progenitors were the first to assume a parasitic mode of life and through the ages they have become very highly adapted to this way of living. Some of the allantonematids have become almost incredulously specialized in morphology, behavior and host-parasite relationships and among them are to be found some of the most unusual nematodes known.

MYENCHIDAE

This is a small and comparatively little known group of nematodes that are parasites of amphibians and leeches. The systematic position of the family is somewhat questionable but investigators who have studied the group regard it as probably related to the Tylenchidae. Both sexes are characterized by a medium-sized stylet without basal swellings and by a peculiar, sucker-like organ situated on the mid-ventral surface about one-fifth of the distance from head to tail, this latter presumably marking the position of the excretory pore. Two species have been reported from leeches.

MYENCHUS BOTHRYOPHORUŠ Schuberg, 1904, was found in Germany parasitizing the leech, *Eropobdella octoculata* (L.) (Syn. *Nephelis vulgaris* (Müller) Moq. Tand.). Different stages of the nematode, including sexually mature individuals (Fig. 167 A & B), occurred in the connective tissues and larvae were found within the muscle cells (Fig. 167 C). Adults were also found in the cocoons of the leech. All the details of the life cycle are not known with certainty but Schuberg and Schröder (1904) concluded that larvae undergo the first part of their development within the muscle cells, then leave this location and enter the connective tissues where they continue development to sexual maturity. From this point on the life cycle is apparently continued outside the host, presumably in the cocoons. Schuberg and Schröder suggest that the nematodes reach the cocoons either by penetrating into the gonads and passing out with the reproduction products ·or by penetrating directly through the body wall and entering the cocoon while this structure still encompasses the body of the leech. The fact that the parasites are frequently found in the connective tissues immediately underlying the epidermis of the leech seems to make the latter alternative all the more probable. Schuberg and Schröder concluded that the females lay their eggs within the cocoons and that the resulting larvae infect the young leeches. How the parasite enters the host has not been determined.

MYENCHUS BOTELHOI Pereira, 1931, is a parasite of the leech, *Limnobdella brasiliensis* Pinto, and was found and studied in Brasil. According to Pereira (1931), infected leeches harbored the nematode in all stages of development. The epididymus was a favored location but the parasite was found in other connective tissues though rarely in the muscles and never in the alimentary tract. Apparently the worms occurred between but not within the cells. The outstanding point of interest regarding this nematode is the fact that Pereira found it regularly within the spermatophores of the leech. It would appear, therefore, that the parasite enters the spermatophores at some time during their formation or passage out of the leech and uses them as a vehicle for transmission from host to host.

DRILONEMATIDAE

This is a small family of about a dozen genera that are either monotypic or contain only a few species each. These nematodes are parasites of earthworms and occur in the coelomic cavities, in or associated with the reproductive organs or embedded in the muscles. Many of the species are characterized by large, sometimes almost sucker-like, phasmids and some of the species by large cephalic hooks. Very little is known about life cycles.

DICELIS FILARIA Dujardin, 1845.—Of the specimens of *Lumbricus rubellus* Hoff., collected by Wülker (1926) in Germany near Frankfort a. M., about 25 percent harbored this parasite (Fig. 167 E & F) but other species of earthworms collected in the same region were not infected. The usual number of nematodes per host was 6 to 8 with a maximum of 22, females generally outnumbering males. The parasites occurred in the body cavity of the host in the region of the reproductive organs but not in the nephridia.

The covering of the egg (Fig. 167 D) is thick with a rough outer surface indicating that the shell proper is probably covered by an external coat and suggesting that the egg is equipped to resist adverse conditions and persist in the soil for a considerable period. Eggs are laid in the body cavity of the host but do not continue development in this location. Wülker did not find larval stages either in earthworms or in surrounding soil and was unable to follow the life cycle. It is not known how eggs are expelled from the host, or in what stage, the parasites enter. Wülker demonstrated that if the earthworm dies these nematodes are unable to reproduce in the carcass but perish with the host.

TETRADONEMATIDAE AND MERMITHIDAE

To the family Tetradonematidae there have, as yet, been assigned only two species, *Tetradonema plicans* and *Aproctonema entomophagum*. These, essentially, are primitive mermithids and must be included in any general consideration of life cycles in this group.

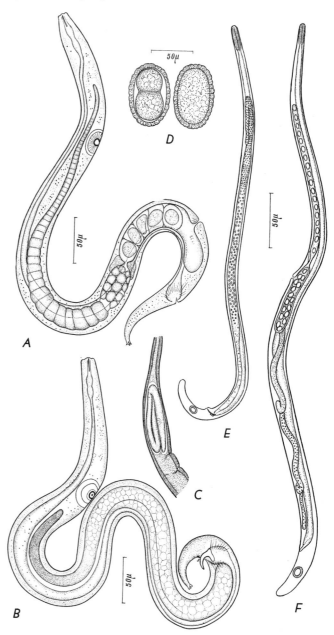

Fig. 167. MYENCHIDAE AND DRILONEMATIDAE

A-C—*Myenchus bothryophorus* (A—Adult female; B—Adult male; C—Larva in muscle cell). D-F—*Dicelis filaria* (D—Eggs; E—Adult male; F—Adult female). A-C, after Schuberg and Schröder, 1904; D-F, after Wülker, 1926.

The mermithids are preeminently insect parasites although crustaceans, spiders, snails, and some other invertebrates are included among their hosts. Most of our knowledge regarding life cycles and habits has been derived from a study of species that infect insects and the following discussion has, of necessity, been written with these hosts in mind.

Eggs may hatch outside the host and larvae reach the body cavity of the young insect by penetrating its body wall or eggs may be ingested and larvae reach the body cavity by penetrating the wall of the gut. In the former type of life cycle there is a tendency for larvae to enter while their hosts are young and for each host to harbor a small number of parasites. In the latter type of life cycle the chances of the host becoming infected are likely to increase with its age and food consumption and the number of parasites per host is likely to be greater.

Tetradonema plicans, after reaching the body cavity of its dipterous host, develops to maturity, copulates and lays its eggs as an internal parasite. This is a simpler and probably a more primitive life cycle than that known for any mermithid. Most mermithids, after completing growth, force their way out of the host and are free living during the adult stage. For *Agamermis decaudata*, *Mermis subnigrescens*, and probably some other species, the free living period is of two years' duration and during it the worms undergo their last molt, copulate, and females lay their eggs. *Aproctonema entomophagum* develops to maturity and copulates within its host but females emerge to lay eggs, while *Paramermis contorta* undergoes its final molt within the host but emerges before copulation. For both these species the free-living stage is of very short duration and these life cycles seem to represent intermediate steps between the life cycle of *Tetradonema plicans* and that of such species as *Agamermis decaudata*.

If a species enters its host by penetrating the body wall the posterior portion of the larva is often modified to serve as a propelling organ. In some cases, as for example species of *Agamermis*, this modified posterior portion, which may constitute as much as four-fifths of the total body length, is detached during the act of penetration and remains on the outside. In other species this modified portion is relatively shorter and persists to form a horn-like appendage at the posterior terminus of the fully grown larva.

In most mermithids, especially those having an adult, free-living stage of considerable duration, the intestine grows rapidly during parasitic development until it fills nearly all the space in the body not occupied by other organs. This modified intestine, filled with reserve nutrient materials and frequently referred to as the "fat body," is largely responsible for the opaqueness of the fully grown larva. The adult becomes increasingly transparent as these stored nutrients are consumed and life ends when they are exhausted.

Most mermithids are represented by both sexes but the sex ratio is subject to a good deal of variation, not only as between different species but in the same species. Males of *Amphimermis zuimushi* and of *Agamermis decaudata* considerably outnumber females while males of *Mermis nigrescens* and *M. subnigrescens* are rarely found. The sex ratio of some species is influenced by environmental conditions during parasitic development. One or a few parasites per host results in a preponderance of the larvae developing into females while a large number of parasites per host results in all, or nearly all, developing into males. Convincing data demonstrating this environmental influence on sex ratios have been presented by Caullery and Comas (1928) for *Paramermis contorta*, by Christie (1929) for *Mermis subnigrescens*, and by Kaburaki and Iyatomi (1933) for *Amphimermis zuimushi*. There is evidence suggesting that some other species behave in a similar manner.

Functional females that possess such male characters as caudal papillae, male copulatory muscles, and even rudimentary spicules have been reported from numerous species. It seems probable that there is some correlation, as yet not understood, between the influence of environment on sex and the occurrence of these so-called "intersexes."

Females of *Hexamermis* sp. (parasite of the ant, *Pheidole pallidula*) and of *Agamermis decaudata* lay eggs only after copulation. Females of *Allomermis myrmecophilia* and of *Mermis subnigrescens* produce viable eggs in the absence of males though individuals of the latter species have been observed *in copula*.

The presence of mermithid parasites affects insects in various ways; development of the gonads, especially the ovaries, is usually suppressed resulting in sterility; wing muscles are sometimes weakly developed reducing ability to fly; internal fat deposits are largely consumed; development of the body as a whole may be retarded and metamorphosis delayed; and infected individuals may be sluggish or, in the case of ants, have a voracious appetite. As a rule external morphological characters are not appreciably modified but there are exceptions, that of

ants being the most outstanding. The emergence of the parasite usually results in the death of the host.

Numerous species of ants are rather commonly infected with mermithids. Males, females, workers and soldiers have been reported as harboring these parasites and there is wide variation in the effects of the mermithids on the external anatomy of the hosts. In some instances infected ants show little recognizable difference from normal individuals of the same sex or caste, except, perhaps, a somewhat more distended gaster and slight variations in color. This seems frequently to be the case with infected males but sometimes, according to Gösswald (1930) and Vandel (1934), infected females, workers or soldiers are not materially modified. In some instances, on the other hand, the external anatomy is greatly modified (Fig. 169 C-G) and infected ants are not identical to any normal caste but show female, worker and soldier characters in varying degrees. Such individuals are called *intercastes*.

In the genus *Lasius* infected females resemble normal females but are easily recognized, at least in many instances, by a smaller head, shorter wings, and a somewhat more distended gaster. Intercastes of this type have been designated *mermithogynes*.

In the genus *Pheidole*, Wheeler (1928) found a variety of different intercastes with mixtures of soldier, worker and female characters. He recognized five more or less distinct types based on the degree of resemblance to one or another of these three normal castes. In all these types the resemblance was more especially to workers and soldiers and for these intercastes Wheeler proposed the term *mermithergates*.

To Vandel (1930), working with *Pheidole pallidula*, the situation was somewhat simpler as he was able to recognize only two types of intercastes. One type showed no very pronounced difference from normal workers except a somewhat more distended gaster. The other type he believed to be modified soldiers and for these he proposed the term *mermithostratiotes* reserving the term mermithergates for those intercastes where resemblance to workers predominates.

Gösswald (1930) found young mermithid larvae in ants at various times of the year and concluded that there may be considerable variation in the time when these insects acquire their parasites. Although mermithids have been found in larval ants, only a few such cases have been reported, and Vandel concluded that the infection is usually acquired during or just prior to the pupal stage. Based on the size and development of larval mermithids from young ants, Gösswald concluded that the parasites may be acquired when the immature insects are in different stages of development and that the stage when the parasites are acquired determines, in a large measure, the degree to which the adult host will be modified.

How ants acquire these parasites is a question that has aroused considerable interest but stimulated little actual investigation. Gösswald (1930) conducted infection experiments with *Lasius alienus* and used eggs of what was, presumably, *Allomermis myrmecophilia*. His results indicate that the ant acquires this parasite by ingesting the eggs. As ant-infecting mermithids belong to several genera (*Agamermis*, *Hexamermis*, *Allomermis*, etc.) life cycles and behavior undoubtedly differ and all may not necessarily enter the host at the same time or in the same manner. It would be surprising if an ant became infected with a species of *Agamermis* by ingesting its eggs.

TETRADONEMA PLICANS Cobb, 1919, is a parasite of the dipterous insect, *Sciara coprophila* Lint. It has been found in only one collection of these insects made by Hungerford (1919) at Manhattan, Kansas, in which every individual was infected. It occurred in larval, pupal, and adult flies each insect harboring from 2 to 20 parasites with an average of about 10, the number of males slightly exceeding the number of females. *T. plicans* passes its adult stage and lays eggs within its host, differing in this respect from any mermithid of which the life history is known.

How the insects acquire their infection has not been determined. Eggs (Fig. 168 G) secured by Hungerford from around females dissected out of fly maggots hatched in a few hours when placed in water and the larvae that emerged seemed to be identical with the youngest larvae found within the insects. These larvae were of two types, a slender type about 125μ long with a curved caudal end and a plumper type about 90μ long. This difference, presumably, is sexual dimorphism. Hungerford found eggs of the parasite in the digestive tract of small *Sciara* larvae and concluded that eggs are probably swallowed and nematode larvae, after hatching, penetrate through the wall of the gut into the body cavity. He noted, however, that "the older maggots are much less susceptible to infestation than the younger ones" and he figures the tail of the adult parasite with a horn-like projection which suggests that the larva has a caudal propelling organ, two characteristics that one is inclined

252

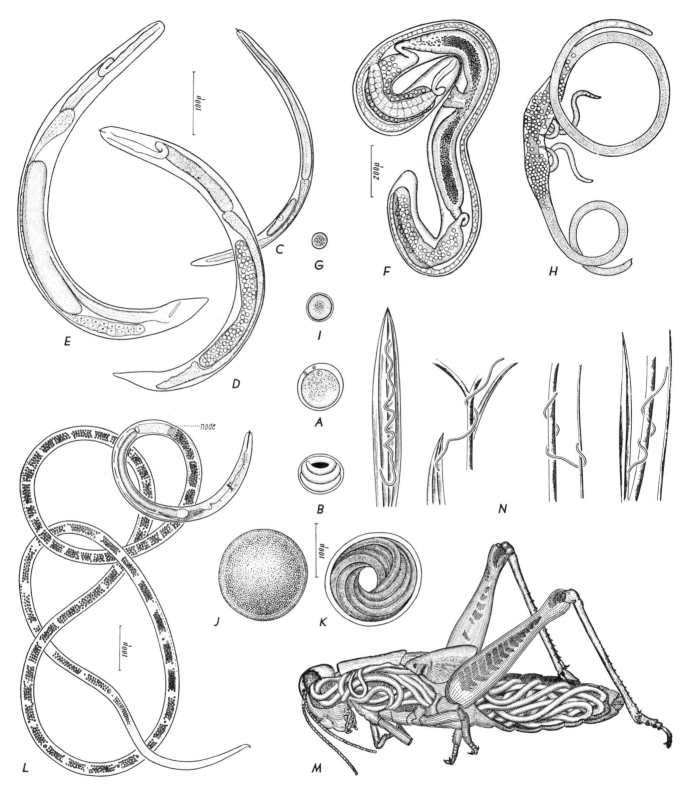

Fig. 168. TETRADONEMATIDAE AND MERMITHIDAE

A-F—*Aproctonema entomophagum* (A—Fertilized egg; B—Egg containing ovic larva; C—Very young larval female; D—Older larval female; E—Larval male; F—Spermatised female). G & H—*Tetradonema plicans* (G—Egg; H—Egg-laying female with males attached). I—*Allomermis mermicophyla*, egg. J-M—*Agamermis decaudata* (J—Newly deposited egg; K—Egg containing ovic larva; L—Infective, preparasitic larva; M—Grasshopper nymph containing one fully grown parasitic female). N—*Mermis subnigrescens*, females depositing eggs on vegetation. (All eggs, A, B, G, I, J, & K, drawn to same scale). A-F, after Keilin and Robinson, 1933 (C-F, drawn from compressed specimens); G & H, after Hungerford, 1919; I, after Crawley and Baylis, 1921; J, K & M, after Christie, 1936; L, after Christie, 1929; N, after Christie, 1937.

to associate with a species that enters its host by penetrating the body wall.

After arriving in the body cavity of the host the larval nematodes grow rapidly and development is usually so timed that females copulate and deposit eggs before the fly pupates. Adult flies were found that contained egg-laying female nematodes and also small larvae that approximated the size usually reached after a few days of parasitic life. Hungerford believed these small individuals had been arrested in development by the growth and maturity of the other worms.

Eggs pass through the vulva and are retained within the separated but unshed cuticle of the final molt which, near the middle region of the body, becomes distended to form a more or less spindle-shaped egg capsule (Fig. 168 H). Eggs are not normally discharged from this capsule prior to the death of the female. This final molt of the female is the only one mentioned by Hungerford. How the males circumvent this encompassing cuticle and effect coition is not explained. The host insect is eventually killed and the body disintegrates to set free a residual mass of nematode eggs.

Hungerford found that the internal fat deposits of infected fly larvae were largely consumed leaving the body much more transparent than that of a normal individual. Most infected fly larvae died before pupating but where the infection was acquired late or the parasites were few in number the fly might endeavor to pupate. Many such pupae died being little more than nematode-filled shells but some succeeded in casting off the larval skins. The emerging, infected adults were able to fly and differed very little in appearance from normal individuals but they lacked functional reproductive organs.

APROCTONEMA ENTOMOPHAGUM Keilin, 1917, was found in England where it is a parasite of the dipterous insect, *Sciara pullula* Winn., the larval stages of which inhabit decaying wood. The morphology and life history of the nematode are discussed in a paper by Keilin and Robinson (1933) upon which the following account is based. It will be noted that the host of this parasite belongs to the same genus as the host of *Tetradonema plicans* and the two nematodes have many points in common.

Each infected larval fly usually harbors several females of *A. entomophagum* (Fig. 168 F) and a varying number of smaller males (Fig. 168 E). Mention is made of one larval fly that contained 2 females and 10 males. The parasites reach maturity in the body cavity of the host and copulate whereupon the males die and the females emerge forcing their way out of the host in the manner of most mermithids. Egg laying begins almost immediately after emergence. Each female deposits somewhat over 200 eggs and when egg laying is completed the female dies. Hence only the female has a free-living, postparasitic stage and it is of very short duration.

The egg (Fig. 168 A) is laid before cleavage but develops rapidly and in a few days contains a coiled larva (Fig. 168B) that molts before hatching. There seems little reason to doubt that the larval mermithids enter the young fly larvae by penetrating the body wall though actual penetration was not observed.

If infection occurs late in the development of the fly larva the parasites may be carried through the pupal stage and infected adult female flies were found though not infected adult males. The parasites delay the metamorphosis of the insects and infected adult female flies lack functional reproductive organs.

PARAMERMIS CONTORTA (Linstow, 1889) Kohn, 1913, is one of the aquatic mermithids of which there are a considerable number. It is a parasite of *Chironomus* larvae and was discovered and studied in Europe. Each host usually harbors one parasite but sometimes two to three or more. The sex ratio, as reported by different investigators, varies a great deal but in most cases females have considerably outnumbered males.

According to Kohn (1905), *P. contorta* molts before leaving its host. This, undoubtedly, is the last molt and the uteri are already filled with eggs. The parasite may issue through the anus or force its way directly through the body wall, the majority emerging just before their insect hosts would normally pupate. The worms settle into the mud at the bottom of the pool and copulation soon takes place to be followed immediately by egg laying. According to Comas (1927), the uteri are emptied and egg laying completed in 4 or 5 days whereupon the female dies.

Eggs are laid before cleavage but develop immediately and hatch in the course of a few weeks. The mermithid larvae swim in the water and seek young *Chironomus* larvae which they enter by penetrating the body wall. Comas states that these mermithid larvae do not appear capable of living long in water and, if unable to find and enter a host, will die in a few hours. Comas recounts that if a mermithid larva attempts to penetrate between the more posterior abdominal segments of its prospective host, the *Chironomus* larva may reach back and

with its mandibles pull the nematode away or bite it in two. If penetration is attempted nearer the middle of the body the insect will be unable to reach the nematode and penetration is more likely to take place.

ALLOMERMIS MYRMECOPHILIA (Crawley and Baylis, 1921) Steiner, 1924, was named and described by Baylis and its life history was studied by Crawley (Crawley and Baylis, 1921). The specimens were from two species of ants collected in England, *Lasius alienus* (Först), and *L. flavus* (F.) and a third ant, *L. niger* (L.), was reported as a host. Observations on a mermithid identified as this species and secured from the same ants were made in Germany by Gösswald (1929; 1930).

After completing its parasitic development this mermithid, according to Crawley, emerges from the ant, sometimes through the anus and sometimes between two of the ventral plates of the gaster, whereupon it enters the soil. As with many other mermithids, emergence apparently occurs over a considerable period during summer and autumn; Baylis mentions specimens that emerged during July. Crawley first saw eggs in the uteri of experimentally reared females on December 5. Egg laying begins before completion of the final molt and many eggs are retained within the separated but uncast cuticle after the manner of *Tetradonema plicans*. As mention is made of four experimental females that had molted by November 20, one might infer that two molts take place after emergence but Crawley and Baylis are not explicit on this point. Some of Gösswald's (1930) ant-infecting mermithids molted twice after emergence but presumably these were not *A. myrmecophilia*. By actual count Crawley found that one cast cuticle contained 6,560 eggs and another 5,900 eggs. Oviposition continues after the cuticle is cast off and probably at least as many more eggs are laid making a total egg output of 12,000 or more. Eggs (Fig. 168 I) are embedded in a ''gelatinous'' matrix that causes them to collect in masses around the vulva or sometimes to be extruded in the form of a ribbon. Crawley and Baylis failed to find males of this mermithid and Gösswald demonstrated that females develop and lay viable eggs without copulation.

Crawley believed that ants become infected while in the larval stage and Gösswald's infection experiments seem to indicate that eggs of the parasite are ingested. Crawley and Baylis reported finding only mermithogynes which, when present in a colony, rarely exceeded the normal females in number and usually were much fewer. One series of colonies showed an average proportion of about 1 to 12. Gösswald found infected males and workers of *Lasius alienus* and *L. flavus* and one infected male of *L. niger*. Each infected ant usually harbors one mermithid though sometimes as many as three.

Infected males and workers, according to Gösswald, show, at the most, only very slight external differences from normal ants. The ovaries and wing muscles of mermithogynes fail to develop normally, according to Crawley and Baylis, but, except for a marked reduction in the size of the wings and a more distended gaster, the external characters show no pronounced difference from those of normal females (Fig. 169 C & D).

HEXAMERMIS SP. This unidentified species of the genus *Hexamermis* is a parasite of the ant, *Pheidole pallidula* (Nyl.), and its life history was studied in France by Vandel (1934). Most individuals complete parasitic development by late summer or autumn and emerge from the ant by forcing their way out through the anus. They do not remain in the ant galleries but penetrate a short distance into surrounding soil where they occupy small cavities. The final molt occurs about a month after emergence and is followed, within the next month, by copulation and egg laying. One of Vandel's experimental females had begun to lay eggs by December 23 and was still laying eggs on March 15. Females exhaust their reserve nutrient materials, stop laying eggs, and die by the end of March or soon thereafter. Hence there is one generation each season with no postparasitic individuals in the soil during late spring and early summer.

The infected individuals of this ant are mermithergates and mermithostratiotes and, with at most very few exceptions, each harbors one parasite. It is not known how the parasite enters the host. Vandel concluded that the infection is acquired either immediately prior to, or during the pupal stage. The location where eggs are laid, the small number of parasites per host, and the vestigeal caudal appendage of the adult is circumstantial evidence suggesting that the larva penetrates the body wall of the young ant.

Copulation is necessary in the reproduction of this mermithid. Experimental females reared in the absence of males by Vandel failed to lay eggs. These females lost their opaque appearance very slowly and some lived for from 22 to 33 months after emergence whereas females that were allowed to copulate and that layed eggs lost their opaque appearance much more quickly and lived for only about 5 months after emergence.

AGAMERMIS DECAUDATA Cobb, Steiner and Christie, 1923, occurs in the north central and northeastern United States where it is a common parasite of grasshoppers including both Acrididae and Tettigoniidae. It sometimes infects crickets (Gryllidae) and has been found, occasionally, in leaf hoppers and beetles. The life history of this mermithid was studied by Christie (1936) upon whose work the following account is based and which applies to the soil and climatic conditions of northeastern Virginia.

The free-living stages of this nematode occupy small cavities in the soil usually from 5 to 15 cm. below the surface (probably deeper in sandy or loose soil). When inhabited by adults each cavity, almost without exception, contains one female and several males, generally two or three, sometimes as many as eight, coiled and intertwined to form a "knot." Copulation is necessary and females reared in the absence of males fail to lay eggs. Egg laying begins about the first of July, continues until interrupted by the advent of cold weather, and eggs (Fig. 168 J & K) accumulate over the surface of the soil cavities and over the parent nematodes. For the most part eggs laid during a given summer do not hatch until the following spring. Cleavage and embryonic development take place after deposition and the first molt occurs within the egg shell.

At the time of hatching the second-stage larva is immediately infective. The body, which shows a high degree of organization and development, is divided into two parts by the *node* (Fig. 168 L). In the anterior part, which constitutes about one-fifth of the total length, one finds most of the organs common to nematodes including esophagus and esophageal glands, intestine, nerve ring, and excretory pore. The posterior part of the body serves as a propelling and food storage organ and contains a row of cylindrical cells, probably modified intestinal cells. An anus is apparently lacking.

During late fall and winter a female is surrounded by her total egg output of the season. Egg counts on six females made during the winter showed the total number of eggs present to vary from 2,625 to 6,530. As will be noted later, a female lays eggs during two summers hence these figures represent roughly about half the total egg output.

Although some larvae may begin to emerge from the eggs fairly early in spring, a greater part of them hatch during a short period at about the middle to the latter part of June. The species of grasshoppers that most commonly serve as hosts (*Melanoplus femur-rubrum* and *Conocephalus brevipennis* (Scudder) in northeastern Virginia) also hatch at about this time. The larval nematodes migrate to the surface of the soil and climb grass and other low vegetation when it is wet with dew or rain. They seek newly hatched grasshopper nymphs and enter their body cavity by penetrating the body wall. Penetration takes place under the edges of the pronotum, between the abdominal segments, or at other places where the chitinous covering is thin. Penetration is effected by the use of the stylet probably aided by the dissolving action of a chitin solvent secreted by one or more of the most anterior esophageal glands.

After the anterior end is inserted into the host the body of the larva breaks at the node and the postnodal portion is left on the outside. If the body fails to break, as occasionally happens, the postnodal part undergoes no development in the host but remains as a vestigeal appendage that eventually sloughs off. The *nodal scar* (Fig. 93, p. 89) persists throughout the parasitic stage as convincing evidence that no molt takes place during this period. The number of parasites per host is usually one (Fig. 168 M), sometimes two, rarely three or more.

Once inside the body cavity of the host the parasite undergoes a period of phenomenal growth accompanied by pronounced morphological changes. The stychocytes (see p. 92) are a conspicuous anatomical feature of larvae that have been in the host from 4 to 10 days (Fig. 93, p. 89). As the body increases rapidly in length it becomes filled by the intestine, in fact intestinal tissue eventually fills all available space not occupied by other organs even growing past the base of the esophagus and extending into the neck region. Apparently this modified intestine performs no digestive function but serves as a reservoir for nutrient materials. Males remain in the host for from 1 to 1½ months and females from 2 to 3 months. The mermithids emerge head foremost forcing their way through the body wall between the segments, fall to the surface of the ground, and enter the soil.

During the first winter in the soil males and females remain isolated each individual forming a separate "knot." The final molt takes place the following spring about the latter part of June and at this time males seek the females. It will be noted that only two molts have been observed. Egg laying begins soon after the final molt, usually about the first of July, and continues until interrupted by cold weather. The following spring a year-old female begins laying eggs slightly earlier than one that has just molted. By the end of the second sum-

mer of egg laying the reserve food has become exhausted and the transparency of the body is in sharp contrast to its opaqueness at the time of emergence from the host. Most females probably fail to survive a third winter in the soil. Information regarding the longevity of males is not very satisfactory but it seems probable that they live for about the same length of time as females.

A. decaudata causes no noticeable change in the external anatomy of grasshoppers. Infected individuals sometimes have distended abdomens and are likely to appear sluggish, adults being incapable of sustained flight. The most pronounced effect of this parasite is on the gonads of the host (Fig. 169 A & B). It is doubtful if infected female grasshoppers are capable of laying eggs as the ovaries are always greatly reduced in size. The effect on the testes is less pronounced and infected male grasshoppers have been observed *in copula*. The emergence of the parasite invariably results in the death of the host.

MERMIS SUBNIGRESCENS Cobb, 1936, appears to be strictly a grasshopper parasite. It occurs in the United States over about the same range as *Agamermis decaudata* where it has been found infecting nine different species of grasshoppers including both Acrididae and Tettigoniidae. Several other species have

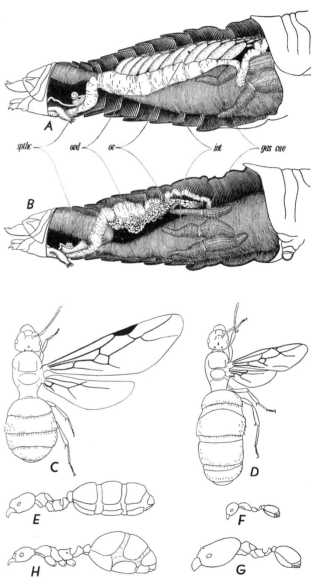

Fig. 169. EFFECTS OF MERMITHIDS ON THEIR HOSTS

A & B—Dissections of adult female grasshoppers, *Melanoplus femur-rubrum*, showing reproductive organs (A—Normal grasshopper; B—Grasshopper parasitized by *Agamermis decaudata*). *gas cae*, gastric caeca; *int*, intestine; *ov*, ovary; *ovd*, oviduct; *spthc*, spermatheca. C & D—Females of the ant, *Lasius alienus* (C—Normal female; D—Female parasitized by *Allomermis mermicophyla*, i.e., a mermithogyne). E-G—The ant, *Pheidole absurda* (E—Individual parasitized by a mermithid, i.e., a mermithergate; F—Normal worker; G—Normal soldier). H—The ant, *Pheidole gauldi*, a mermithergate. A & B, after Christie, 1936; C & D, after Crawley and Baylis, 1921; E-G, from Wheeler, 1928, after Emery; H, after Wheeler, 1928.

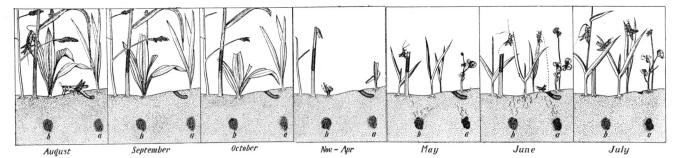

| August | September | October | Nov – Apr | May | June | July |

Fig. 170a. LIFE CYCLE OF *AGAMERMIS DECAUDATA*

Diagram illustrating 12-month period, August to July, inclusive. *a*, "knot" composed of one female and several males that emerged from hosts 2 years previous to beginning of period represented. *b*, "knot" composed of one female and several males that emerged from hosts 1 year previous to period represented. Mermithids that would emerge during September and October of period represented are omitted for simplicity. Female in "knot" *a* has, by October, completed its second summer of egg laying and dies during ensuing winter but the accumulated eggs, deposited during previous summer, hatch May to June. Female in "knot" *b* has, by October, completed its first summer of egg laying and the accumulated eggs hatch May to June while the second summer of egg laying is begun during May.

been experimentally infected. Attempts to infect other insects including crickets (Gryllidae), mole crickets (Gryllotalpinae), and larvae of several species of Lepidoptera have been unsuccessful. The following account of the life history is based on investigations by Christie (1937) conducted, for the most part, in Massachusetts.

Grasshoppers become infected with *M. subnigrescens* by swallowing the eggs. In order to bring this about the egg-laying habits of this nematode are radically different from those of *Agamermis decaudata*, otherwise the two life cycles are somewhat similar. Eggs of *M. subnigrescens* are never laid in the soil. Gravid females climb low vegetation on which they lay their eggs (Fig. 168 N) and to which the eggs cling by means of the entangling appendages or bissi.

The egg, when deposited, contains a fully developed infective larva (Fig. 140 B, p. 181). The shell proper is protected by an outer covering that is divided into two cup-like halves by a groove at the equator (Fig. 140 A, p. 181). At each pole there is a raised or thickened area formed by the attachment of the entangling appendages (Fig. 139, p. 181). The outer covering breaks apart along the groove at the equator and the two cup-like halves are easily removed. In the shell proper there are two opposite areas at the equator where the color is lighter than elsewhere and these areas are partly dissolved by the digestive action of the host thus facilitating the escape of the larva. Both the outer covering and the shell proper contain brown pigment, presumably to protect the larva from the action of sunlight. Eggs deposited on foliage remain viable throughout the summer. When eggs were kept experimentally in a moist chamber some remained viable for a year.

When an egg reaches the alimentary tract of its host the outer covering has usually been rubbed off. The two opposite areas of the shell at the equator gradually become clearer and begin to protrude until they appear as colorless hemispherical projections (Fig. 140 C, p. 181) that finally rupture and provide openings for the escape of the larva. The larva itself does not appear to aid in its own liberation. When first freed it is rather sluggish but soon becomes active, penetrates the wall of the gut and enters the body cavity. Penetration through the intestinal wall is aided by the stylet which is rhythmically protruded.

From 1 to 5 parasites per host is the number most frequently encountered but there is great variation and grasshoppers harboring 100 or more parasites of widely different ages are not uncommon in some localities. As a nymph grows older and its food consumption increases, its chance of becoming infected is correspondingly greater. The sex ratio of *M. subnigrescens* is influenced by the number of parasites per host. When a grasshopper harbors a large number, all develop into males but when a grasshopper harbors only 1 or 2 these usually develop into females (Christie, 1929).

The parasitic development of *M. subnigrescens* is essentially the same as that of *Agamermis decaudata*. There is the same rapid increase in size and the same extensive proliferation of intestinal tissue. Males remain in the host from 4 to 6 weeks and females from 8 to 10 weeks. At the end of this time the parasites force their way through the body wall of the host and enter the soil. When a grasshopper harbors parasites of different ages, all that are too immature to escape and survive in the soil perish with the host when the older ones emerge.

Postparasitic individuals of *M. subnigrescens* are found in the soil down to about 60 cm., the majority occurring from 15 to 45 cm. below the surface. They usually remain isolated and one rarely finds a "knot" composed of a female and one or more males as is characteristic of *Agamermis decaudata*. Most individuals emerge from the host during summer and autumn and molt the following April. This is the final molt and the only one that has been observed. Copulation may take place and has been seen on several occasions but copulation is not necessary as females reared in the absence of males produce viable eggs. By July females begin to exhibit a brownish color due to accumulating eggs and by September they are nearly black except for a short region at each extremity of the body. At this time most of the eggs are viable but they are not laid until the following spring. Before ovipositing, a gravid female 85 mm. long contains about 14,000 eggs.

Egg laying usually begins in May and may continue throughout July or even into August, depending on weather conditions. Eggs are laid during rain and should the early summer months be dry egg laying will be delayed. Gravid females climb grass and other low vegetation over which they constantly move while eggs are being laid. If rain continues egg deposition goes on throughout the day but if the rain stops and the foliage becomes dry females coil up, fall to the surface of the ground and enter the soil, presumably to resume egg laying during the next rain.

It is not known how long females live after the uteri are emptied of eggs but by this time their stored food is nearly exhausted and it seems highly improbable that they are able to survive a third winter or to develop more eggs. However, if prevented from coming to the surface to deposit eggs they are able to survive a third winter and to lay eggs the following spring. Females that normally would have deposited eggs in 1932 were buried in containers and prevented from coming to the surface (Christie, 1937). When examined during May, 1933, many of these females were alive, in good condition, and filled with eggs. There was no evidence that eggs had been deposited, although these females promptly began laying eggs when brought to the surface and placed in the light.

Apparently eggs are not laid at night. Egg laying is controlled, at least in part, by light stimuli. When an ovipositing female is placed in the dark, egg laying promptly stops, but is resumed just as promptly when the female is again placed in the light. The head of the adult female is colored with areas of reddish-brown pigment which, presumably, is an organ for light perception. The male, which never comes to the surface, lacks this pigment.

Mermis subnigrescens has about the same effects on its host as does *Agamermis decaudata*. These effects are suppression of the gonads, especially the ovaries, and death of the host when the parasite emerges. With *M. subnigrescens* the effect on the gonads of the host is much more variable than with *A. decaudata* due to variations in the number of parasites per host and the time the parasites are acquired.

ALLANTONEMATIDAE

The Allantonematidae is a group of insect parasites that are closely related to the preeminently plant-infecting Tylenchidae. The species that have been studied and named probably constitute but a small part of the number that exist but in nearly every instance where the life cycle is known it follows the same general plan and differs from that found in any other group of nematodes.

Adult gravid females occupy the body cavity (haemocoel) of the insect, frequently in small numbers, often one per host. Here larvae accumulate and develop to a certain stage, molting at least once (probably twice in most species); then they escape from the host either by entering the alimentary tract and passing out through the anus or by entering the female reproductive system and passing out through the genital aperture. Most species infect both males and females of their host insect. In some cases the only known way by which larvae are able to

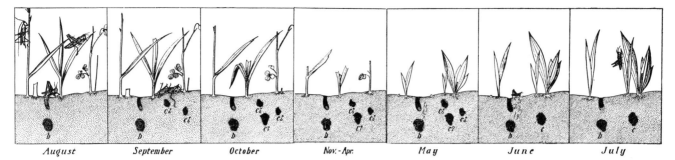

| August | September | October | Nov.-Apr. | May | June | July |

Fig. 170b. LIFE CYCLE OF *AGAMERMIS DECAUDATA*—Continued

Diagram illustrating ensuing 12-month period, August to July, inclusive. *c*, mermithids emerging from hosts September to October. Female and males remain in separate cavities until latter part of May when males seek female, copulation takes place, and, during June, the female begins its first summer of egg laying. Female in "knot" *b* has, by October, completed its second summer of egg laying and dies during the ensuing winter but the accumulated eggs hatch May to June. (While the above diagram is essentially correct for a majority of individuals where grasshoppers serve as hosts, the various life-cycle changes are actually spread out over somewhat greater periods of time than the diagram indicates. A few eggs hatch before May and June. A few preparasitic larvae enter hosts as early as April and as late as August and the time of emergence is correspondingly affected. A few individuals, especially males, emerge from hosts at least as early as July.)

leave the host is via the female reproductive system and the fate of larvae that inhabit the body cavity of male insects is not yet fully understood.

A free living stage is passed wherever the host insect undergoes its early development and during this period the nematodes molt at least once (as a rule probably twice) and become adults. In the adult male the stylet is usually either absent or weakly developed and the esophageal glands are inconspicuous and apparently lacking. In the preparasitic adult female the stylet is usually well developed and at least one of the esophageal glands is large and conspicuous. Exceptions to these morphological differences between the sexes are usually correlated with deviations from the more typical life cycle. The apparent absence of esophageal glands and the somewhat more rapid development of the genital primordium in the male usually make it possible to distinguish sex at an early stage sometimes while a larva is still within the egg.

The ovary of the adult preparasitic female is small and composed of only a few cells the extent of its development differing somewhat with different species. When copulation takes place the uterus is packed with small, more or less spherical spermatozoa. After copulation males usually die and impregnated females enter their respective hosts, usually by penetrating the body wall while the insect is still in the larval stage. The fact that in most species only the female possesses an effective stylet and at least one well-developed esophageal gland has been regarded as evidence that these structures function in connection with penetration into the host. There is little reason to doubt that the stylet is employed for this purpose. It has been suggested that penetration is further facilitated by a secretion of the esophageal glands which may serve as a chitin solvent. The validity of this suggestion does not rest entirely on morphological evidence for Bovien (1932) demonstrated that *Scatonema wülkeri* does, in fact, exude a rather copious secretion through the stylet at the time of penetration.

The free-living stage is usually of short duration. There is no evidence that the nematodes feed during this period (with the exception of *Fergusobia curriei*) and larvae, at the time they leave the host, are at least nearly as large as young adults. However, after entering a new host, the female undergoes a very great increase in size. The fully grown gravid female of most species is curved ventrad and assumes a form usually referred to as "sausage-shaped." There are exceptions, however, and, for example, *Allantonema mirable* is oval while in many species of *Aphelenchulus* the body is bent dorsad with the vulva on the outside of the curve.

Some species deposit eggs in the body cavity of the host but in many species eggs hatch before deposition and the uterus becomes distended with developing eggs and larvae that gradually fill the greater part of the body and push the ovary into the anterior end. As a rule larvae eventually pass through the vulva into the body cavity of the host. There is a tendency for the other internal organs of the female to degenerate, the extent of this degeneration differing in different species.

In most species the rapid increase in the size of the female after becoming parasitic provides space for the reproductive organs. In one group, the Sphaerulariinae, adequate space for the developing reproductive organs is not provided by a corre-

sponding increase in body size. The uterus of *Sphaerularia bombi* is everted through the vulva and the entire reproductive system develops outside the body proper. This prolapsed uterus increases enormously in size and the body proper remains attached to one end as a vestigial and apparently functionless structure. *Tripius gibbosus* (Syn. *Atractonema gibbosum*) represents an intermediate stage in the evolutionary development of this peculiar adaptation and the size of the body and of the prolapsed uterus is less disproportionate. In both these species the life cycle, so far as known, is essentially the same as that of most allantonematids.

There are, nevertheless, several deviations from this typical life cycle. Young adult males, as well as young adult females, of *Parasitylenchus dispar typographi* enter the body cavity of their host insect where they are found in large numbers, while neither adult males nor adult females of *Chondronema passali* become parasitic, only larval stages being found in the host insect. *Chondronema passali* enters its host, not as young adults, but as young larvae, probably by being ingested.

Two species of this family have heterogeneous life cycles. There is interpolated into the life cycle of *Heterotylenchus aberrans* a parasitic, parthenogenetic generation and into the life cycle of *Fergusobia curriei* several, consecutive, "free-living," parthenogenetic generations. The parthenogenetic females of *Fergusobia curriei* occur, associated with their "host" insect, in plant galls where they feed on plant cells and are, in fact, plant parasites. In each of these heterogeneous species, however, the gamogenetic generation still follows the typical allantonematid plan of development.

TYLENCHINEMA OSCINELLAE Goodey, 1930, is a body-cavity parasite of the frit-fly, *Oscinella frit*. (L.). The life history of this nematode was studied in England by Goodey (1930, 1931).

The frit-fly has three generations a year. Eggs are laid on small oat plants generally during May and fly larvae penetrate the shoots, destroying the central tissues. This is the first or stem generation. Adult flies appear by mid-July and deposit eggs on the panicles of oats where the larvae attack the tissues of the inflorescence. This is the second or panicle generation. Adult flies again appear during August or early September and lay eggs on various species of wild grasses. This is the third or grass generation; also it is the overwintering generation and winter is passed in the larval stage. The life cycle of the nematode is, of necessity, closely correlated with that of the frit-fly and like it, undergoes three generations a year (Fig. 175).

Infected flies harbor usually one, sometimes two or three, more rarely four to eight, adult female nematodes that give birth to living young. Eggs pass into the uterus of the mother nematode where they undergo development. As more and more eggs are produced the uterus becomes distended, pushing the ovary into the anterior end and finally occupying most of the space within the body. Larvae (Fig. 171 A & C), escaping from the egg membranes, pack the posterior end of the uterus and finally pass through the vulva into the body cavity of the host. Here they accumulate and continue development. Goodey observed one molt that takes place when a larva is about 460μ long and which he believed to be the second suspecting that the first molt takes place while the larva is still within the uterus of the mother. The gonads undergo considerable development and show differences that make it possible to distinguish sex. The wall of the intestine becomes well stocked with reserve food globules.

After attaining a size nearly as large as free-living adults, the larvae escape from the host. To accomplish this they penetrate the food reservoir of the fly's digestive system from which they migrate through the intestine to the rectum and are ejected through the anus. With regard to this escape of the larvae, Goodey writes as follows: "Parasitized flies of both sexes, having failed to develop their sex cells, fly about and instead of taking part in the normal process of reproduction are able only to deposit larvae of the nematode parasite. Normal females go to oat panicles and there lay eggs; similarly, the parasitized flies responding to the same urge of the life-cycle rhythm also fly to oat panicles, but, instead of eggs, deposit larvae of the nematode parasite. These find their way, possibly in response to some chemotactic stimulus, into the plant tissues surrounding the fly larvae.'' In this environment the nematode larvae continue development and two final molts take place, the last cuticle separating while the larva is still within the cuticle of the preceding molt.

Larvae that Goodey removed from the gut of infected flies and kept in tap water completed their final molt in about 41 hours. Males remained alive for about 14 days and females for about 29 days but copulation did not take place while the worms remained in water. In nature copulation follows the final molt and the uterus of the female is distended with spermatozoa. The preparasitic female (Fig. 171 F & G) has a well developed stylet with basal swellings and a large dorsal esophageal gland. These structures are inconspicuous or lacking in the male. (Fig. 171 E).

The male does not again become parasitic but the impregnated, precocious female enters the body cavity of a frit-fly larva. Goodey did not actually observe the entrance but assumed, no doubt correctly, that it is accomplished by penetrating the body wall. The incidence of infection is about the same for male and female flies except possibly in the grass or overwintering generation, where Goodey found that about two-thirds of the infected flies were females.

After becoming parasitic the female nematode increases very greatly in size and is about fully grown when the host emerges from its pupal case. The body has assumed the characteristic ''sausage shape'' and the ovary has completed its development (Fig. 115 J, p. 136). The stylet is retained and Goodey believes that probably the parasite continues to take food via the alimentary canal.

Tylenchinema oscinellae produces no noticeable effect on the external characters of its host but it prevents the normal development of the gonads and both male and female flies are sterilized. Occasionally, however, parasitized flies of both sexes develop normal sex organs and when this happens the parasite fails to undergo normal development. In regard to this Goodey (1931) writes: ''In the great majority of cases the worm manages to get the upper hand and grows to sexual maturity within the host, but occasionally the fly, during its final metamorphosis, is able, by some means, to build up its gonads in the normal manner. When this happens the worm fails to grow, remains non-functional and becomes degenerate. . . . These relationships may possibly be explained on the supposition that the worm secretes or excretes something, perhaps from the intesti-

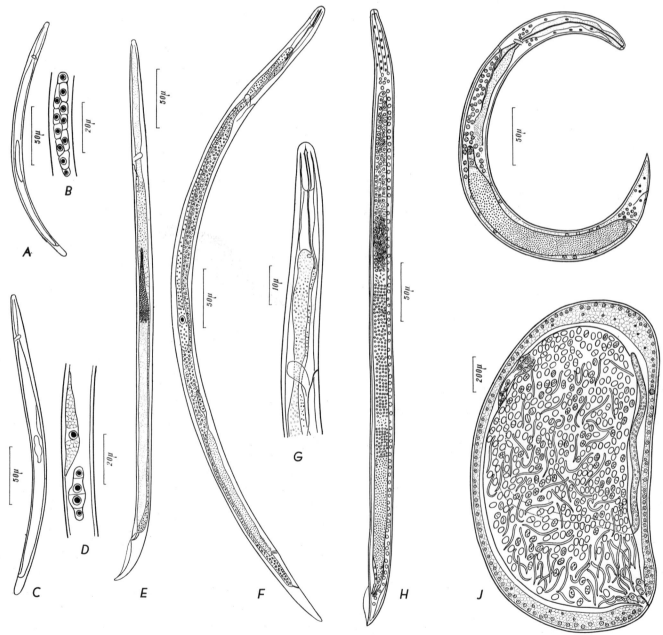

Fig. 171. ALLANTONEMATIDAE

A-G—*Tylenchinema oscinellae* (A—Very young male larva and, B, genital primordium of same; C—Very young female and, D, genital primordium and esophageal gland of same; E—Adult male; F—Adult, preparasitic female and, G, anterior end of same. For fully grown, adult, parasitic female, see Fig. 115 J, p. 136). H-J—*Allantonema mirabile* (H—Adult male; I—Adult, preparasitic female after copulation; J—Fully grown, adult, parasitic female. For stage intermediate between I and J, see Fig. 115 I, p. 136). A-G, after Goodey, 1930; H-J, after Wülker, 1923.

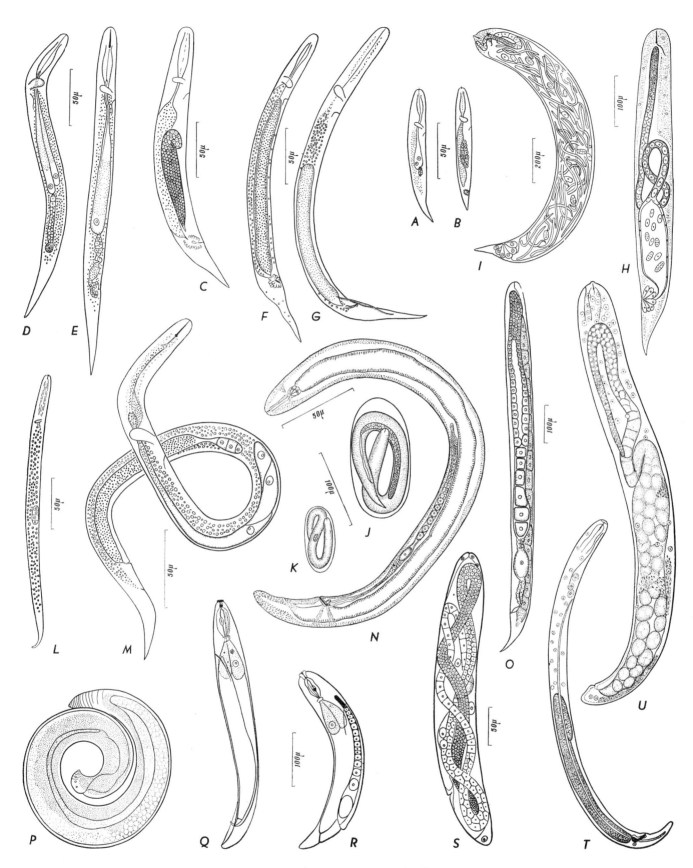

Fig. 172. ALLANTONEMATIDAE

A-I—*Scatonema wülkeri.* (A—Newly hatched female larva; B—Newly hatched male larva; C—Partly grown male larva; D—Female just
before final molt; E—Adult preparasitic female after copulation; F—
Male just before final molt; G—Adult male; H—Adult parasitic female
a few days after entering host; I—Fully grown gravid female). J-O—
Heterotylenchus aberrans (J—Egg laid by female of gamogenetic generation; K—Egg laid by female of parthenogenetic generation; L—Newly hatched larva of gamogenetic generation; M—Adult preparasitic

female of gamogenetic generation; O—Fully grown female of parthenogenetic generation). P—*Aphelenchulus diplogaster,* adult parasitic female. Q-S—*Fergusobia curriei* (Q & R—Adult male and adult
female, respectively, of "free-living" generation, i.e., from *Eucalyptus*
galls; S—Gravid female of "parasitic" generation, i.e., from body cavity
of gall fly). T & U—*Parasitylenchus dispar typographi,* adult parasitic
male and adult parasitic female, respectively. A-I, after Bovien, 1932;
J-O, after Bovien, 1937; P, T & U, after Fuchs, 1915; Q-S, after Currie, 1937.

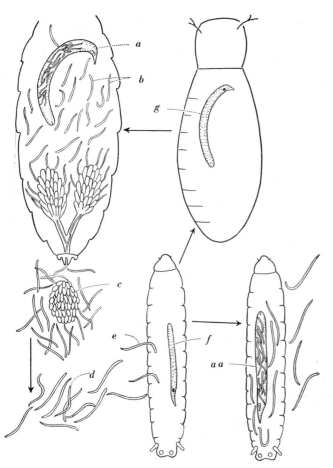

Fig. 173

Diagram illustrating life cycle of *Scatonema wülkeri*. The adult parasitic female (*a*) produces offspring (*b*) that eventually enter the female fly's reproductive organs and are extruded with the eggs (*c*). Outside the host these larvae develop into adults (*d*) and copulate whereupon males die and impregnated females (*e*) enter fly larvae. These females then undergo a period of growth (*f*) and may begin producing offspring (*aa*) while the fly is still in the larval stage or females may be only partly grown (though adult) (*g*) when the fly pupates and begin laying eggs (*a*) when the fly becomes adult. After Bovien, 1937.

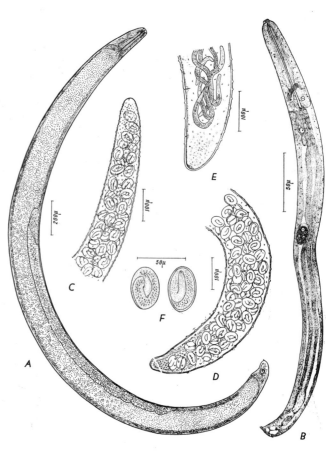

Fig. 174. *CHONDRONEMA PASSALI*

A—Oldest larva found within body cavity of host; B—Youngest larva found within body cavity of host; C-E—Portions of body of adult female filled with eggs or larvae and serving as brood sac; F—Eggs. All figures after Christie and Chitwood, 1931.

nal [esophageal] gland, which prevents the normal growth of the host's sex-cells. At the same time it is quite likely that the same may be true of the host; if once its reproductive organs become sufficiently developed, then it is able to pour out some substance which definitely inhibits the growth of the worm.''

ALLANTONEMA MIRABLE Leuckart, 1884, is a body cavity parasite of the pine weevil, *Hylobius abiëtis* (L.) and occurs in Europe but has not been found elsewhere. This nematode differs from *Tylenchinema oscinellae*, not so much in its life cycle, which is essentially the same, as in the form and degeneration of the gravid female. Unlike most allantonematids, the fully grown female (Fig. 171 J) is oval, some 1.5 to 2 mm. in length and about half as wide as long. Its body is virtually a sac largely filled by the uterus as it becomes distended with eggs and larvae. The other internal organs degenerate to such an extent that if vestiges persist their identity has not been recognized.

Eggs hatch in the uterus where they begin to accumulate during late summer and where they remain during the winter undergoing little development. In the spring larvae begin to pass through the vulva into the body cavity of the weevil where they undergo two molts. Larvae finally leave the host by penetrating its alimentary tract and passing out through the anus.

The adult female of *Hylobius abiëtis* eats small holes in the bark on the trunk and roots of fir and certain other coniferous trees. In this cavity eggs are laid and hatch, the young weevils tunneling into adjacent tissues. In order to pass their free-living stages in the immediate vicinity of newly hatched weevils, the larval nematodes must escape when and where female weevils are laying eggs, albeit not through the genital aperture of the insect. Wülker (1923) observed only one molt during free-living development which took place after 8 to 10 days. Bovien (1937) found that larvae, taken from the rectum of adult weevils and placed in hanging drops of water, became

adult in 5 to 6 days. The final molt is followed very soon by copulation after which males die and impregnated females enter the body cavity of weevil larvae that, in the meantime, have hatched.

The adult preparasitic female of *A. mirable* (Fig. 171 I) has a well developed stylet and Bovien (1937) figures two esophageal glands, one opening into the esophagus on the dorsal side near the base of the stylet and the other on the ventral side farther back. In the adult male (Fig. 171 H) a stylet is present though somewhat more weakly developed than in the female but the esophageal glands are inconspicuous or lacking.

By about July, when the weevils are pupating, the parasitic female nematode is producing ova. Fuchs (1915) states that *Hylobius abiëtis* lives for at least 31 months and, finding females of *A. mirable* in 2-year-old weevils, he concludes that the nematode lives for at least 2 years.

SCATONEMA WÜLKERI Bovien, 1932, is a body cavity parasite of the dipterous insect, *Scatopse fuscipes* Meig., the immature stages of which develop in manure and other putrescent material. Eggs of this nematode hatch in the uterus where larvae (Fig. 172 A & B) undergo early development, the extent of this development varying considerably. In some cases, which Bovien (1932) regards as exceptional, an individual, while still within the uterus, may reach maturity and, in turn, develop larvae within its uterus, thus creating three generations, one within another. Most of the progeny, however, pass through the vulva into the body cavity of the host as partly grown larvae. These larvae enter the reproductive system of the insect and pass out with the eggs. When an infected fly dies not all the harbored parasites necessarily perish but some larvae may complete development, molt, and copulate after which impregnated females escape from the dead body. As male flies die soon after copulation Bovien concludes that, in moist surroundings, part of the nematodes may be able to

260

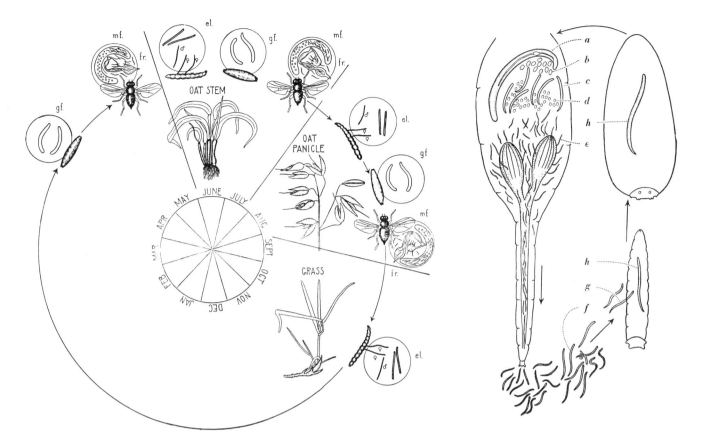

Fig. 175. *TYLENCHINEMA OSCINELLAE*

Schematic drawing illustrating life history of frit-fly in its three seasonal generations, and the approximate time occupied by each, linked with that of its parasite, *Tylenchinema oscinellae*. The various stages of the fly and worm are greatly enlarged whilst the oats and grass are smaller than natural size. Although the female fly only is shown it is to be understood that the male also carries the parasite. The dotted circles contain stages of the parasite related to the corresponding stage of the host. The circles cut into the pupa and imago but not into the fly larva in each case, thus showing that the parasite is within the pupa and fly but not within the larvae. *e l*, ensheathed larvae; *f r*, food reservoir of fly with larvae passing in; *g f*, growing female worms; *m f*, mature female worms. After Goodey, 1931, explanation quoted verbatim.

Fig. 176. HETEROTYLENCHUS ABERRANS

Diagram illustrating life cycle of *Heterotylenchus aberrans*. The adult parasitic female of the gamogenetic generation (*a*) lays eggs (*b*) that develop into females of the parthenogenetic generation (*c*). These females lay eggs (*d*) and the resulting larvae (*e*) enter the reproductive organs of the female fly and pass out through the genital aperture. Outside the host these larvae develop into adults of the gamogenetic generation (*f*) and copulate whereupon males die and impregnated females (*g*) enter fly larvae. While the fly matures and pupates the female grows (*h*) to reach full stature (*a*) and lay eggs (*b*). After Bovien, 1937.

escape from the decaying insects. Otherwise the fate of larvae harbored by male flies is unknown.

Bovien observed only one molt, the last, which may occur before, but usually not until after, emergence from the host. As this molt takes place not later than 24 hours after emergence and is followed immediately by copulation, the free-living stage is of short duration. In the adult, preparasitic female (Fig. 172 F) the stylet and esophageal glands (Bovien figures two) are well developed while in the adult male (Fig. 172 G) these structures are inconspicuous or absent. By the less well-developed genital primordium and the presence of esophageal glands one can distinguish female larvae while they are still within the egg.

Regarding penetration of preparasitic females into the body cavity of a fly larva Bovien (1932) writes as follows: "In many cases I found the nematodes in the act of entering the body of the larva. In a few cases I saw dead nematodes, which had not succeeded in penetrating the body-wall, held fast by it. The penetration may take place through all parts of the surface of the larva and no preference seems to be given to any particular region. The very beginning of this act, however, was not observed. I placed female worms in hanging drops together with *Scatopse*-larvae, the presence of which had an unmistakably attractive influence on the nemas. The nematodes slung themselves around the body of the larva, pressing their mouths against the skin without being able to puncture it. I suppose this failure may be ascribed to the lack of supporting surfaces. On the third day the worms were dead. An oblong, somewhat spiral-wound, coagulated mass of secretion had been ejected from the aperture of the buccal stylet, and the salivary [esophageal] glands appeared to be empty." The presence of the parasite does not result in sterility of the host.

CHONDRONEMA PASSALI (Leidy, 1852) Christie and Chitwood, 1931, is a body cavity parasite of the beetle *Popilius interrup-*

tus (L.) (Syn. *Passalus cornutus* Fab.). This beetle occupies galleries in decaying stumps and logs where eggs are laid and where larvae develop and pupate. Leidy (1852) found 90 percent of the adult beetles infected and Christie and Chitwood (1931) estimated that each beetle usually harbors from 500 to 1,000 parasites. In the body cavity of the insect one finds larvae in all stages of development from young, newly hatched individuals (Fig. 174 B) to those that are fully grown (Fig. 174 A) but never adults.

Larval nematodes of both sexes taken from the body cavity of a beetle have a minute stylet, a moderately large esophageal gland (presumably the dorsal), and exceptionally large and conspicuous phasmids. Sex can be distinguished at a rather early stage partly through differences in the genital primordia but more especially through differences in the general appearance of the body, females being more opaque than males. Movement is sluggish.

The mode of exit from the host has not been determined. Once the nematodes have escaped neither males nor females again become parasitic but remain in the beetle galleries throughout the remainder of their lives. The mouth, anus, and vulva of the female become vestigeal. If the vulva functions it is only during copulation. Eggs (Fig. 174 F) are retained within the body where they accumulate and hatch pushing aside the internal organs and converting the female into a brood sac (Fig. 174 C-E).

C. passali enters its host as a very young larva but it is not known how this is accomplished. Larvae of all sizes may be found in old beetles at any time of the year when the insects can be collected. The incidence of infection seems to be very much lower in larval beetles and pupae than in adults. These circumstances, together with the exceedingly large number of parasites usually harbored by a beetle, caused Christie and Chitwood (1931) to suggest that the larval nematodes enter *per os*, possibly the gravid female and her entire progeny being swallowed.

261

HETEROTYLENCHUS ABERRANS Bovien, 1937, is a body cavity parasite of the onion fly, *Hylemya antiqua* (Meig.), and its life history was studied by Bovien (1937) at Lyngby, Denmark. The onion fly hibernates in the pupal stage and emerges in May to lay eggs on onion plants or in nearby cracks in the soil. The young fly larvae move down the plant usually inside the sheath and finally burrow into the bulb. Pupation takes place in the soil or occasionally in the bulb. There are two or perhaps, occasionally, three broods a year with considerable overlapping.

In the body cavity of infected flies one finds from one to four large, adult females of *H. aberrans* (Fig. 172 N) and a greater number of smaller, adult females (Fig. 172 O). The larger individuals are females of the gamogenetic generation and the smaller ones are females of the parthenogenetic generation. The reproductive organs of a gamogenetic female, as compared with these structures in most allantonematids, are

exceptionally small. Much of the space within the body is occupied by the intestine which, according to Bovien, is without a lumen. A small stylet is present and the three esophageal glands, empty and reduced in size, are grouped around the base of the esophagus. Eggs (Fig. 172 J) are deposited in the body cavity of the host where they hatch and where the larvae develop into parthenogenetic females.

The outstretched reproductive organs of a parthenogenetic female are relatively much larger than those of a gamogenetic female. The esophagus and esophageal glands have almost completely degenerated but a small stylet is present and, according to Bovien, the intestine is represented by a single row of large, binuclear cells. Eggs (Fig. 172 K), which are smaller than those of the preceding generation, are deposited in the body cavity of the host and from them develop larvae of both sexes. These larvae remain in the host until they are ready to undergo their final molt when they penetrate the fly's ovaries,

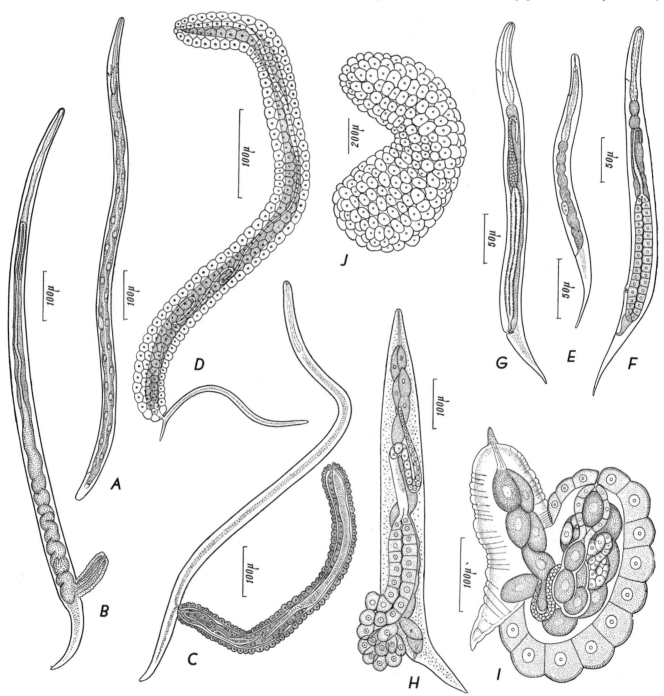

Fig. 177. SPHAERULARIINAE

A-D—*Sphaerularia bombi* (A—Fully grown larva; B-D—Adult parasitic females showing progressive stages in prolapsus of uterus. For fully grown adult parasitic female, see Fig. 115 A, p. 136). E-I—*Tripius gibbosus* (E—Newly born larva; F—Young but sexually mature female; G—Sexually mature male; H—Adult parasitic female showing early stage in prolapsus of uterus; I—Adult parasitic female showing late stage in prolapsus of uterus. For stage intermediate between H and I, see Fig. 115 K, p. 136). J—"*Tylenchus sulphureus piceae*," adult parasitic female. A-I, after Leuckart, 1887; J, after Fuchs, 1929

migrate to and assemble in the oviducts and escape through the genital aperture.

Both male and female flies are parasitized. Bovien found no evidence that male flies are rendered sterile but the ovaries of infected females fail to develop and because of this the nematode larvae can scarcely be transferred to onion plants with the eggs of their host. However, Bovien noted that infected female flies "stretched out the ovipositor, as if they wanted to lay eggs." As in the life cycle of *Tylenchinema oscinellae*, infected female flies probably accompany normal females to the place where eggs are laid but deposit there, not eggs, but larval nematodes. Bovien found no evidence that the genitalia of male flies are invaded and he concluded that larvae have no way of escaping from male flies.

The free-living stages of the nematode are passed in close association with the larval insects, either in the onion plant or in nearby soil. The nematodes reach maturity, copulate and the males die. Impregnated females (Fig. 172 M) enter larval flies presumably by penetrating the body wall though the act of penetration was not observed. The adult male has a stylet with basal swellings comparable to that of the female but somewhat more slender. All three esophageal glands are well developed in the preparasitic female but inconspicuous or lacking in the male. After entering the host a female develops into the large, parasitic individual of the gamogenetic generation.

The life cycle of *H. aberrans* (Fig. 176), therefore, consists of a parthenogenetic generation, passed entirely within the host, alternating with a gamogenetic generation that has both parasitic and free-living stages. If the parthenogenetic generation was omitted the life cycle would be essentially the same as that of most other members of this family.

FERGUSOBIA CURRIEI (Currie, 1937) n. comb. [Synonyms: *Anguillulina (Fergusobia) tumifaciens* Currie, 1937; *Anguillulina (Fergusobia) curriei* (Currie, 1937) Johnston, 1938; not *Anguillulina tumefaciens* (Cobb, 1932)].

Fergusobia curriei occurs in Australia where, in association with flies of the genus *Fergusonina*, it produces galls on *Eucalyptus* trees. This association was discovered by Morgan (1933) and later investigated by Currie (1937), the following account being based on the latter's observations. There are many species of *Fergusonina* that attack *Eucalyptus* trees in Australia and, according to Currie, all are probably associated with a nematode. Several species of *Eucalyptus* are attacked and galls may be formed on leaf buds, axil buds, stem tips, and flower buds, depending on the species involved. The following account of the life history of *Fergusobia curriei* is based on a study of flower galls on *Eucalyptus macrorrhynchia* and of the associated fly, *Fergusonina nicholsonia*. Currie regards the stages of the nematode found in the galls as true plant parasites but is inclined to regard the relationship with the gall flies as symbiosis rather than parasitism.

Each female fly harbors two gravid female nematodes (Fig. 172 S). These nematodes deposit eggs and the resulting larvae, on reaching the proper stage, leave the body cavity and enter the reproductive system of the "host." Adult flies emerge during summer and females, after mating, lay eggs in young flower buds, depositing with each egg from 1 to 50 nematode larvae. The same fly or different flies may lay numerous eggs in a single bud as many as 74 eggs and 227 nematode larvae having been found. The larval nematodes immediately start to feed on the anther primordial cells that form a ring around the inner wall of the bud cavity and under this stimulus the tissue proliferates rapidly forming masses of large, thin-walled parenchymatous cells full of mucilagious cell sap. The fly eggs hatch in about 6 weeks and by this time masses of gall tissue are already present in the bud. On hatching a fly larva moves in between two of these cell masses and tears out a small crypt in which to lie. The larval nematodes migrate into this crypt and quickly develop into adults all of which are parthenogenetic females (Fig. 172 R). Apparently the nematode passes through several parthenogenetic generations feeding on surrounding plant cells and in no way injuring the insect. During its first two instars the fly larva feeds on the viscous cell sap which oozes from surrounding cells that have been punctured by nematode stylets. During its last larval instar the fly larva tears down the walls of the cavity and feeds on the ruptured cells.

In autumn both male (Fig. 172 Q) and female nematodes appear that become the adults of the gamogenetic generation. This "preparasitic" female does not differ materially from the female of the preceding parthenogenetic generations (Fig. 172 R). Both the adult male and the adult "preparasitic" female have a stylet and three well developed esophageal glands. As in most other allantonematids, the male does not become parasitic. Just before pupating, if the fly larva is a female two adult, fertilized female nematodes enter its body cavity, presumably by penetrating the body wall. Male flies are never infected, female flies invariably so. Once in the body cavity of the "host" the

female nematodes proceed in their development and, by the time the fly has emerged as an adult, they are depositing eggs.

The life cycle of *Fergusobia curriei*, therefore, consists of several parthenogenetic generations passed entirely outside the "host" alternating with a gamogenetic generation that has both a "free-living" and a "parasitic" stage. If the parthenogenetic generations were omitted the life cycle would be essentially the same as that of most allantonematids.

In the case of *Fergusobia curriei* associated with *Fergusonina nicholsonia* in galls of *Eucalyptus macrorrhynchia*, only two gravid "parasites" are normally found in each female fly but in some other species of *Fergusonina*, usually those of larger size, a female fly may harbor a greater number. As Currie suggests, further work may demonstrate that the nematodes associated with different species of flies are themselves specifically distinct.

TRIPIUS GIBBOSUS (Leuckart, 1886) Chitwood, 1935 [Synonum, *Atractonema gibbosum* (Leuckart, 1886)] is a parasite of the dipterous insect *Cecidomyia pini* (Degeêr). Since the investigations by Leuckart (1887), following its original discovery in Germany, this nematode has not been reported elsewhere or received further study. Each infected larva of *C. pini* usually harbors a dozen or more, sometimes as many as 50, adult female parasites showing different degrees of development.

Eggs are laid in the body cavity of the host where they hatch and where larvae (Fig. 177 E) accumulate in great numbers. Leuckart could never find larvae in the alimentary tract or secure other evidence that they pass out through the anus and he was inclined to believe that they are liberated by the death and decomposition of the insect. The extrusion of larvae along with the eggs of the host when adult flies are ovipositing seems to be an uninvestigated possibility.

The free-living period, passed in the soil, is of short duration and in a few days after leaving the host the larval nematodes have developed to adult males and females (Fig. 177 F & G). Leuckart mentions one molt, apparently the last, but noted that sometimes the cuticle shed by the male is double. After copulation males die and females enter new hosts. Leuckart did not determine how the young females reach the body cavity of larval flies but suggests entrance through the mouth or anus as a possibility. In the light of our present knowledge of this group, penetration directly through the body wall seems more probable. Fly larvae are susceptible to infection from the time they hatch until they go into the pupal stage.

During parasitic development of the female the uterus is gradually everted through the vulva (Fig. 177 H) and develops on the outside eventually forming an oval structure, somewhat exceeding in size, but always firmly attached to, the body proper that, in the meantime, has become greatly foreshortened. The remainder of the reproductive system and part of the modified intestine occupy this prolapsed uterus (Fig. 177 I).

The effect on the host is not pronounced and when the nematodes are present in moderate numbers fly larvae are able to pupate and become adults. However, Leuckart concluded that this parasite is not harmless and that heavily infected flies frequently die in the pupal stage.

SPHAERULARIA BOMBI Dufour, 1837. This remarkable nematode is a parasite of queen bumble bees. It has been reported from several species of *Bombus*, each host usually harboring one or, at the most, only a few adult female parasites though Leuckart (1887) found 32 in one bee. *Vespa rufa* and *V. vulgaris* have also been reported as hosts. This parasite has been found in several localities in Europe and North America and is apparently widespread.

S. bombi, in so far as information is available, has the typical allantonematid life cycle. Eggs are laid and hatch in the body cavity of the host and larvae, after a period of parasitic development, pass out by way of the anus and enter the soil. Here the nematodes reach maturity and copulate whereupon the males die and the impregnated females enter their new hosts. The free-living period, according to Leuckart, is of several months' duration.

Queen bees hibernate in the soil and Leuckart found that under coniferous trees where the soil is moist and covered with humus and moss is a favored place. Leuckart concluded that the bees become infected in autumn when they are penetrating the soil preparatory to hibernation and that this explains why only queens are parasitized. Infected queens, due to retarded development of the ovaries, are either unable to produce eggs or produce only a few and both Schneider (1885) and Leuckart were convinced that such queens never found colonies.

The interesting and unusual feature about this nematode is not its life cycle but the morphological development of the parasitic female. After entrance into the new host the body of the young female undergoes little or no increase in size. Instead the uterus is everted through the vulva, carries within it the other reproductive organs as well as the modified intestine or "fat body," and develops outside the body proper (Fig. 177 B-D). This prolapsed uterus increases enormously in size while the body proper remains a relatively minute, functionless structure that may sometimes become detached (Fig. 115 A, p. 136).

OTHER SPECIES. In addition to the species already discussed, the number of allantonematids that have as yet been named and described is not great and for most of these information about life cycles, especially regarding free-living stages, is meager or lacking. A few exceptions, however, may be noted briefly.

Bradynema rigidum (v. Siebold, 1836) zur Strassen, 1892, is a parasite of the dung beetle, *Aphodius fimetarius* (L.) and *Bradynema strasseni* Wülker, 1921, is a parasite of the wood-boring beetle, *Spondylis buprestoides* (L.). These two nematodes have been rather extensively investigated in Europe (zur Strassen, 1892; Wülker, 1923) though the free-living stages of *B. rigidum* are still imperfectly known. Both have the typical allantonematid life cycle, larvae passing out of the hosts by way of the anus.

Howardula benigna Cobb, 1921, is a parasite of the cucumber beetle, *Diabrotica vittata* (Fab.) and, less commonly, of the related beetles, *D. trivittata* (Mann) and *D. duodecimpunctata* (Fab.). This nematode has the typical allantonematid life cycle, larvae passing out with the eggs of the host (Fig. 178). Beetles of both sexes are infected and the fate of larvae that find themselves in male beetles is not known. Cobb (1928) was of the opinion that these larvae may be transferred to female beetles during copulation. He found considerable numbers of larvae in the proximal end of the male genitalia but was unable to demonstrate experimentally that such larvae are actually transferred to female beetles.

In the genus *Aphelenchulus* the adult, parasitic female is usually characterized by being curved dorsad with the vulva on the outside of the curve. *A. diplogaster* (Linstow, 1890), Filipjev, 1934 (Fig. 172 P) is a parasite of the bark beetle, *Ips typographus* (L.) and *A. tomici* Bovien, 1937, is a parasite of the bark beetle, *Pityogenes bidentatus* (Hbst.) (Syn. *Tomicus bidens* (F.)). These two nematodes are very closely related and both have the typical allantonematid life cycle, larvae passing out of the host by way of the anus (Fuchs, 1915; Bovien, 1937) to undergo free-living development in the frass of the beetle galleries.

"*Tylenchus*" *aptini* Sharga, 1932, was found in Scotland, where it is a parasite of the thrips, *Aptinothrips rufus* (Gmelin). Eggs of this parasite are deposited in the body cavity of the host and larvae leave by way of the anus. Males remain in the host until bursa, spicules and gubernaculum are formed and Sharga (1932), finding no evidence that males enter the gut or pass out through the anus, suggests that copulation takes place before the parasites leave the host. Furthermore, Sharga states that "after several ecdyses the mature female stage is reached" and his discussion and drawings seem to indicate that one or more of these ecdyses take place after the female has passed through the free-living stage and entered a new host. If copulation takes place before this parasite leaves the first host and the female molts after entering the second host, the life cycle is, indeed, a departure from that known for any other allantonematid.

Parasitylenchus dispar (Fuchs, 1915) Micoletzky, 1922, sub-species, *typographi* (Fuchs, 1915) is a parasite of the bark beetle, *Ips typographus* (L.). In general this nematode has the typical allantonematid life cycle. The adult, parasitic female gives birth to larvae, large numbers of which accumulate in the body cavity of the host to eventually enter the gut and pass out through the anus. In one respect, however, the life cycle differs from that of most allantonematids. After completing free-living development young adults of both sexes enter the new host. One finds in the body cavity of infected beetles 200 to 300 adult parasitic females (Fig. 172 U) and an even greater number of adult parasitic males (Fig. 172 T).

Fuchs (1915) did not observe copulation or determine whether it takes place before or after entering the new host. However, he was able to rear to maturity larvae taken from the rectum of a beetle, the adult stage being reached in 7 to 10 days. The experimentally reared females did not lay eggs and it seems probable that eggs are not laid until after entrance into a new host. If this is true we have, not a free-living generation, as Fuchs called it, but a free-living stage.

Ostensibly, *Tripius gibbosus* and *Sphaerularia bombi* are the only members of the Sphaerulariinae that have as yet been reported. It may be noted, however, that Fuchs (1929) has described two very unusual nematodes from bark beetles, viz., "*Tylenchus sulphureus piceae*" and "*Tylenchus sulphureus pini*." Fuchs maintained that in the case of these two nematodes the gravid female is not a prolapsed uterus basing his

contention, in part, on a failure to find any vestige of the body proper or transitional stages showing the uterus in the process of prolapsus. But in the case of *Sphaerularia bombi*, as Leuckart points out, the body proper is sometimes detached and one wonders if Fuchs' material included a sufficiently complete series of developmental stages. If the gravid female of "*Tylenchus sulphureus piceae*" (Fig. 177 J) is not a prolapsed uterus its resemblance to the gravid female of *Sphaerularia bombi* is, to say the least, very remarkable.

Fig. 178.   *HOWARDULA BENIGNA*

Showing relative size of beetle, *Diabrotica vittata*, and of its parasites (line XY indicates actual length of beetle); also egg of beetle and larval nematodes deposited with it.   After Cobb, 1921.

Bibliography

ACKERT, J. E. and WADLEY, F. M. 1921.—Observations on the distribution and life history of *Cephalobium microbivorum* Cobb and of its host, *Gryllus assimilis* Fabricius. Tr. Am. Micr. Soc., v. 40(3):97-115, 15 figs.

ALICATA, J. E. 1934.—Observations on the period required for *Ascaris* eggs to reach infectivity. Proc. Helm. Soc., Wash., v. 1(1):12.

BARTHÉLEMY, A. 1858.—Études sur le développement et les migrations d'un nématoide parasite de l'oeuf de la limace grise. Ann. Sc. Nat., Zool., 4. s., v. 10(1):41-48, pl. 5, figs. 8-15.

BOVIEN, PROSPER. 1932.—On a new nematode, *Scatonema wülkeri* gen. et sp. n. parasitic in the body-cavity of *Scatopse fuscipes* Meig. (Diptera nematocera). Vidensk. Medd. Dansk. Naturh. Foren, Kjøbenhavn, v. 94:13-32, figs. 1-7.
      1937.—Some types of association between nematodes and insects. Ibid., v. 101:1-114, figs. 1-31.

CLAUS, C. 1868.—Beobachtungen über die Organisation und Fortpflanzung von *Leptodera appendiculata*. Schrift. Gesellsch. Beförd Ges. Naturw. zu Marburg, Suppl. Heft 3, 24 pp., 3 pls., 31 figs.

CAULLERY, M. and COMAS, M. 1928.—Le déterminise du sexe chez un nématode (*Paramermis contorta*), parasite des larves de chironomes. Compt. Rend. Acad. Sc., Paris, v. 186:646-648.

CHITWOOD, B. G. and CHITWOOD, M. B. 1937.—Snails as hosts and carriers of nematodes and Nematomorpha. Nautilus, Quart. J. Devoted to Interests of Conchologists, v. 50(4):130-135.

CHRISTIE, J. R. 1929.—Some observations on sex in the mermithidae. J. Exper. Zool., v. 53(1):59-76, figs. 1-5.
      1936.—Life history of *Agamermis decaudata*, a nematode parasite of grasshoppers and other insects. J. Agric. Res., v. 52(3):161-198, figs. 1-20.
      1937.—*Mermis subnigrescens*, a nematode parasite of grasshoppers. Ibid., v. 55(5):353-364, figs. 1-6.

CHRISTIE, J. R. and CHITWOOD, B. G. 1931.—*Chondronema passali* (Leidy, 1852) n. g. (Nematoda), with notes on its life history. J. Wash. Acad. Sc., v. 21(15):356-364, figs. 1-17.

COBB, N. A. 1915.—[Note dealing with a new species of free-living nematode.] J. Parasit., v. 1(3):154.

1921.—*Howardula benigna;* a nema parasite of the cucumber-beetle. Science, n. s., v. 54(1409):667-670, figs. 1-4.

1921.—Idem [Reprinted with only minor changes], Contrib. Sc. Nemat. (10):1-4, figs. 1-4.

1928.—Idem [Reprinted with additions] Ibid. (10): 345-352, figs. 1-8.

COMAS, M. 1927.—Sur le mode de pénétration de *Paramermis contorta* v. Linst. dans la larve de *Chironomus rhummi* Kief. Compt. Rend. Soc. Biol., Paris, v. 96(10):673-675.

CONTE, A. and BONNETT, A. 1903.—Sur un nématode nouveau (*Angiostoma helicis* n. sp.) parasite de l'appareil génital d'*Helix aspersa* (Muell.). Compt. Rend. Soc. Biol., Paris, v. 55(5):198-199.

CRAWLEY, W. C., and BAYLIS, H. A. 1921.—*Mermis* parasitic on ants of the genus *Lasius*. J. Roy. Mier. Soc., 1921:353-372, figs. 1-12.

CURRIE, G. A. 1937.—Galls on *Eucalyptus* trees. A new type of association between flies and nematodes. Proc. Linn. Soc. N. S. Wales, v. 62(3-4):147-174, figs. 1-31, pls. 6-7.

DOBROVOLNY, C. G. and ACKERT, J. E. 1934.—The life history of *Leidynema appendiculata* (Leidy), a nematode of cockroaches. Parasit., v. 26(4):468-480, figs. 1-10, pl. 23, figs. 1-3.

DUJARDIN, FÉLIX. 1845.—Histoire naturelle des helminthes ou vers intestinaux. xvi + 654 + 15 pp. 12 pls. Paris.

FUCHS, GILBERT. 1915.—Die Naturgeschichte der Nematoden und einiger anderer Parasiten. 1. des *Ips typographus* L. 2. des *Hylobius abietis* L. Zool. Jahrb., Abt. System., v. 38(3-4):109-222, figs. a-b, pls. 17 21, figs. 1-82.

1929.—Die Parasiten einiger Rüssel- und Borkenkäfer. Ztschr. Parasitenk., Abt. F, Ztschr. Wiss. Biol., v. 2(2): 248-285, figs. 1-36.

GALEB, OSMAN. 1878.—Recherches sur les entozoaires des insectes. Organisation et développement des oxyuridés. Arch. Zool. Expér. & Gén., v. 7(2):283-390, pls. 17-26.

GLASER, R. W. 1932.—Studies on *Neoaplectana glaseri*, a nematode parasite of the Japanese beetle (*Popillia japonica*). N. J. Dept. Agric., Circ. No. 211, 34 pp., 3 pls., 17 figs.

GLASER, R. W., McCOY, E. E. and GIRTH, H. B. 1940.—The biology and economic importance of a nematode parasitic in insects. J. Parasit., v. 26(6):479-495, figs. 1-8.

GOODEY, T. 1930.—On a remarkable new nematode, *Tylenchinema oscinellae* gen. et sp. n., parasitic in the frit-fly, *Oscinella frit* L., attacking oats. Phil. Tr. Roy. Soc. London, s. B, v. 218:315-343, fig. 1, pls. 22-26, figs. 1-54.

1931.—Further observations on *Tylenchinema oscinellae* Goodey, 1930, a nematode parasite of the frit-fly. J. Helm., v. 9(3):157-174, figs. 1-2.

GÖSSWALD, KARL. 1929.—Mermithogynen von *Lasius alienus*, gefunden in der Umgebung von Würzburg. Zool. Anz., v. 84(7-8):202-204.

1930.—Weitere Beiträge zur Verbreitung der Mermithiden bei Ameisen. Ibid., v. 90(1-2):13-27, fig. 1.

HAGMEIER, ARTHUR. 1912.—Beiträge zur Kenntnis der Mermithiden. I. Biologische Notizen und systematische Beschreibung einiger alter und neuer Arten. Zool. Jahrb., Abt. System., v. 32(6):521-612, figs. a-g, pls. 17-21, figs. 1-55.

HUNGERFORD, H. B. 1919.—Biological notes on *Tetradonema plicans* Cobb, a nematode parasite of *Sciara coprophila* Lintner. J. Parasit., v. 5:186-192, 2 text figs., 1 pl., figs. 1-6.

JOHNSON, G. E. 1913.—On the nematodes of the common earthworm. Quart. J. Mier. Sc., v. 58(4):605-652, figs. 1-2, pl. 35, figs. 1-10.

KABURAKI, TOKIO and IMAMURA, SHIGEMOTO. 1932.—Mermithid-worm parasitic in leaf-hoppers, with notes on its life history and habits. Proc. Imp. Acad., v. 8(4):139-141, figs. 1-6.

KABURAKI, TOKIO and IYATOMI, KISABU. 1933.—Notes on sex in *Amphimermis zuimushi* Kab. et Im. Proc. Imp. Acad., v. 9(7):333-336.

KEILIN, D. 1925.—Parasitic autotomy of the host as a mode of liberation of coelomic parasites from the body of the earthworm. Parasit., v. 17(2):170-172.

KEILIN, D. and ROBINSON, V. C. 1933.—On the morphology and life history of *Aproctonema entomophagum* Keilin, a nematode parasite in the larvae of *Sciara pullula* Winn. (Diptera-Nematocera). Parasit., v. 25(3):285-294, figs. 1-2, pls. 19-20, figs. 1-19.

KOHN, F. G. 1905.—Einiges über *Paramermis contorta* (v. Linstow) (*Mermis contorta* v. Linstow). Arb. Zool. Inst. Univ. Wien, v. 15(3):213-256, pl. 1, figs. 1-21.

KREIS, H. A. 1932.—*Trionchonema rusticum* n. g. n. sp., a parasitic nematode from the land snail, *Polygyra espicola* Bland (Helicidae). Tr. Am. Mier. Soc., v. 51(1):48-56, pls. 8-9, figs. 1-13.

LEIDY, JOSEPH. 1852.—Some observations on *Nematoidea imperfecta*, and descriptions of three parasitic infusoriae. Tr. Am. Phil. Soc., Phila., n. s., v. 10(2):241-244, pl. 11, figs. 42-51.

LEUCKART, R. 1887.—Neue Beiträge zur Kenntnis des Baues und der Lebensgeschichte der Nematoden. Abhandl. Math.-Phys. Cl. K. Sächs. Gessellsch. Wiss., v. 13(8):565-704, pls. 1-3.

LINSTOW, O. v. 1890.—Ueber *Allantonema* und *Diplogaster*. Centralbl. Bakt. u. Parasitenk., v. 8(16):487-497, figs. a-f.

MAUPAS, E. 1899.—La mue et l'enkystement chez les nématodes. Arch. Zool. Expér. & Gén., 3. s., v. 7:563-628, pls. 16-18, figs. 1-29.

McCOY, E. E., GIRTH, H. B. and GLASER, R. W. 1938.—Notes on a giant form of the nematode *Neoaplectana glaseri*. J. Parasit., v. 24(5):471-472.

MERRILL, J. H. and FORD, A. L. 1916.—Life history and habits of two new nematodes parasitic on insects. J. Agric. Res., v. 6(3):115-127, figs. 1-3.

MORGAN, W. L. 1933.—Flies and nematodes associated in flower bud galls of spotted gum. Agric. Gaz. N. S. Wales, v. 44(2):125-127.

OTTER, G. W. 1933.—On the biology and life history of *Rhabditis pellio* (Nematoda). Parasit., v. 25(3):296-307.

PEREIRA, CLEMENTE. 1931.—*Myenchus botelhoi* n. sp., curioso nematoide parasito de *Limnobdella brasiliensis* Pinto (Hirudinea). Tese, Fac. Med., 29 pp., 5 figs., 2 pls., Sao Paulo, Brasil.

SCHNEIDER, ANTON. 1859.—Über eine Nematodenlarve und gewisse Verschiedenheiten in den Geschlechtsorganen der Nematoden. Ztschr. Wiss. Zool., v. 10(1):176-178.

1885a.—Über die Entwickelung der *Sphärularia bombi*. Zool. Beitr., Bresl., v. 1(1):1-10, pl. 1, figs. 1-5.

1885b.—Fortgesetzte Untersuchungen über *Sphärularia bombi*. Ibid., v. 1(3):247-251, 1 fig.

SCHUBERG, A. and SCHRÖDER, O. 1904.—*Myenchus bothryophorus*, ein in den Muskelzellen von *Nephelis* schmarotzender neuer Nematode. Ztschr. Wiss. Zool., v. 76(4):509-521, pl. 30, figs. 1-11.

SEURAT, L. G. 1920.—Histoire naturelle des nématodes de la Berbérie. 1st part. Morphologie, développement éthologie et affinités des nématodes. vi + 220 pp., 34 figs., Algiers.

SHARGA, U. S. 1932.—A new nematode, *Tylenchus aptini* n. sp., parasite of Thysanoptera (Insecta: *Aptinothrips rufus* Gamelin). Parasit., v. 24:268-279, figs. 1-26.

ZUR STRASSEN, O. 1892.—*Bradynema rigidum* v. Sieb. Ztschr. Wiss. Zool., v. 54(4):655-747, pls. 29-33, figs. 1-98.

STRICKLAND, E. H. 1911.—Some parasites of *Simulium* larvae and their effects on the development of the host. Biol. Bull. Mar. Biol. Lab., Woods Hole, v. 21(5):302-338, pls. 1-5.

VANDEL, A. 1930.—La production d'intercastes chez la fourmi *Pheidole pallidula* sous l'action de parasites du genre *Mermis*. Bull. Biol. France & Belg., v. 64(4):458-494, figs. 1-14, pl. 17.
    1934.—Le cycle évolutif d'*Hexamermis* sp., parasite de la fourmi (*Pheidole pallidula*). Ann. Sc. Nat., Zool., 10. s., v. 17:47-58, 1 fig.

VAN ZWALUWENBURG, R. H. 1928.—The interrelationships of insects and roundworms. Hawaiian Sugar Planters' Assoc. Exper. Sta. Bull. (Ent. Ser.) No. 20, 68 pp.

WHEELER, W. M. 1928.—*Mermis* parasitism and intercastes among ants. J. Exper. Zool., v. 50(2):165-237, figs. 1-17.

WÜLKER, G. 1923.—Über Fortpflanzung und Entwicklung von *Allantonema* und verwandten Nematoden. Ergeb. u. Fortschr. Zool., v. 5(4):389-507, figs. 1-53.
    1926.—Über geschlechtsreife Nematoden im Regenwurm. Arch. Schiffs- u. Tropen-Hyg., v. 30:610-623, figs. 1-6.

CHAPTER XIX

LIFE HISTORY (ZOOPARASITICA)

II PARASITES OF VERTEBRATES

ASA C. CHANDLER, Rice Institute, Houston, Texas; J. E. ALICATA, University of Hawaii,
Honolulu, T. H.; and M. B. CHITWOOD, Babylon, N. Y.

Introduction

The life cycles of the nematodes parasitic in vertebrates differ in no essential from those of free-living nematodes, but are subject to a number of modifications which enable the parasites to gain access to new hosts with more facility and greater certainty. With a few exceptions these nematodes have five stages of development separated by four moults as do most free-living nematodes, but in a few forms (e.g., *Contracaecum* and *Trichinella*) the number of molts is said to be increased, and in some forms one or more of them is suppressed to the extent of being passed through rapidly in the egg, or in hatched larvae with no intervening period of growth.

The outstanding feature in the life cycle of parasitic nematodes is a cessation of development of the young worms after reaching an infective stage, while they await an opportunity to gain access to a new definitive host. In most cases the organisms pass through this period of waiting outside the body of the original host, either (1) as embryos inside the egg shells (oxyurids, ascaridids, trichurids); (2) as free-living but nonfeeding third-stage larvae, often enclosed in the shed cuticle of stage two (*Strongyloides*, many strongylins); or (3) as third-stage larvae, usually encysted, in the body of an intermediate host, which in some cases is obligatory (e.g., spiruroids, camallanins, some metastrongylids) but in other cases is optional (e.g., *Capillaria annulata*, *Syngamus trachea*). In many cases such larvae are capable of re-encystment, sometimes over and over again, in other hosts—transport hosts—in which development to maturity does not occur. In a few cases such secondary intermediate hosts have become necessary parts of the life cycle (e.g., *Gnathostoma spinigerum*). In the filariae and a few other nematodes (e.g., *Habronema*) the infective larvae do not become encysted, and habitually emerge through a break in the labium of the vector as a result of their own activities. A striking exception to the usual waiting period outside the body of the host occurs in the case of *Trichinella spiralis*, which passes its waiting period encysted in the flesh of the parental host.

In considering the life cycles of parasitic nematodes from an evolutionary standpoint it is necessary to consider possible ways in which the nematodes may have developed into parasites of vertebrates. One method was presumably the result of ingestion by the host, followed by adaptation to the environment encountered inside the alimentary canal. It seems probable that the Oxyuridae, for instance, became parasitic in this manner. Such nematodes might be expected to have the simplest possible type of life cycle, reproducing generation after generation in the lumen of some part of the alimentary canal, with enough eggs or larvae escaping with the feces to allow for spread to other hosts through the medium of contaminated food or water. It seems remarkable that only a single instance (*Probstmayria vivipara*) is known of a parasite which has unequivocally adapted itself to this type of life. The nearest approach, with the exception of *Probstmayria*, is the facultative parasitism of a number of species whose congeners are saprozoic free-living forms, e.g., a species of *Longibucca* in the stomach of a snake (Chitwood, 1933), and another species of *Longibucca* in the stomach and intestines of a bat (McIntosh and Chitwood, 1934); *Diploscapter coronata* in ahydrochloric human stomachs (Chandler, 1938); and *Cephalobus parasiticus* in the stomachs and intestines of monkeys (Sandground, 1939). In addition to these cases, it is claimed by a number of writers (Koeh, 1925; Penso, 1932) that *Enterobius vermicularis*, *Passalurus ambiguus* and other oxyurids are capable of reproducing, generation after generation, in the lumen *and walls* of the intestine. This is denied by others (Zowadowsky and Schalimov, 1929; Lentze, 1935) because of the demonstrated need of oxygen by the embryos before they can complete their development. Even if the larvae can occasionally develop to maturity in the gut walls, such an occurrence can certainly be considered the exception rather than the rule. One other instance of repeated generations in a single host has been claimed for *Strongyloides stercoralis* (Nishigori, 1928; Faust, 1931) but this is a case of short-circuited rather than continuous development, and occurs only under exceptional conditions. The offspring of worms in the intestine do not grow to maturity directly in the intestinal lumen, but migrate

through the body as they would if they had infected from outside.

With the few exceptions mentioned above, the simplest type of life cycle in the case of obligatory parasites is that exhibited by most species of oxyurids, in which the eggs fail to develop beyond a certain point (morula stage in some, "tadpole" in others) until exposure to oxygen outside the body of the host, followed by reentrance of the embryonated eggs or hatched larvae into the same or another host with food or water contaminated by them. In many cases, possibly in all, this simple cycle is modified further by a stage in with the larvae attach themselves to the mucous membrane, bury their heads in it, or actually burrow into the walls of the gut before they take up their residence in the lumen as adults.

Few parasites other than the Oxyuroidea have as simple a life cycle as that described in the last paragraph. Most of them have an instinct for burrowing at some time during the course of their development and exercise it either (1) by burrowing through the skin and going on a tour of the body via the circulatory system, lungs and throat before reaching the intestine; (2) by burrowing into the mucous membranes of the alimentary canal, either being content to live buried in the gut wall for a few days, or entering the circulatory system and going on a tour of the body similar to that of the skin penetrators or, in some cases, burrowing directly through into the body cavity or through mesenteries, parenteral tissues, etc.; or (3) by burrowing into the body cavity or tissues of an intermediate host, either through the surface or through the walls of the gut after being ingested.

Two possible origins of this burrowing habit suggest themselves. One possible origin is as a useful instinct on the part of gut parasites to serve either one or both of two purposes, (1) to protect the young worms from being swept out of the intestine with the feces, and (2) to provide a better type of nourishment for the period of rapid growth and development. There can be little doubt but that the burying of the head of fourth-stage larvae of *Dermatoxys veligera* (Wetzel, 1931) and the use of the "corpus" of the esophagus of *Oxyuris equi* as a mouth capsule (Wetzel, 1930) are steps in this direction. One could then visualize as further developments complete burrowing into the gut wall, penetration into the circulatory system, and the circuit through the body that would necessarily be entailed.

The alternative explanation is that the worms which migrate through the body originally became vertebrate parasites by burrowing through the skin. An initial step in this direction can be observed today in the occasional invasion of the skin of dogs and sometimes of other animals by *Rhabditis strongyloides*, the adults of which live in soiled straw bedding. Successful development of adult parasitism by this method would necessitate an ultimate location in the body whence the eggs or embryos could escape in order to reach new hosts. This condition would be fulfilled in the case of those parasites which, after penetration of the skin, reach the circulatory system and eventually arrive in the lungs, where, still imbued with an instinct for burrowing, they would escape into the air spaces. Here they could successfully reach maturity and reproduce (e.g., Metastrongylidae) or they could be carried passively, via trachea and throat, to the alimentary canal. Successful parasitism would, of course, be dependent upon *loss* of the burrowing instinct after the third moult, which actually occurs.

The temporary burrowing into the intestinal mucosa of the larvae of such worms as *Ascaridia*, *Haemonchus* and *Oesophagostomum* might, then, be construed either as a step in the direction of a more extended migration, as practiced by related worms, or as a step in the direction of the simple oxyurid type of life cycle, with abandonment of a primitive but no longer necessary migration from skin to intestine. In those species which usually perform the entire migration, the failure of some individuals to do so (e.g., *Ascaris*, hookworms) could be either atavistic or progressive. The fact that the curtailed migration is more likely to occur in the normal than in abnormal hosts is of little help in the matter, for it could be argued either that the reason for the failure of migration in normal hosts is due to less restlessness in such hosts and a consequent slipping back to ancestral ways, or that it is due to more perfect adaptation and therefore more advanced evolution.

It is, as a matter of fact, probable that the migration is primitive for some worms and secondarily acquired for others. Worms which may be assumed normally to develop directly in the intestine are at least occasionally able to reach the intestine even if injected under the skin. This was demonstrated by Harwood (1930) in the case *Cosmocercoides dukae,* for although he found the larvae of this worm to be incapable of skin penetration, he succeeded in recovering a few worms from lungs and intestine after subcutaneous injection. There is some reason to believe that the Strongyloididae and Rhabdiasidae, the latter of which never establish themselves in the intestine at all, may be primitively skin penetrators, whereas it is very unlikely that the ascarids are. Whether parenteral migration is primitive or secondary among the Strongylina is not so easy to guess.

An interesting derivative of the migratory type of life cycle is the course of development of *Trichinella.* The unique life cycle of this worm has apparently resulted from a precocious development and hatching of the eggs in the uterus of the mother, accompanied by early acquisition of the burrowing instinct, the result being the invasion of the parental host instead of a new host. The *Strongyloides* life cycle is another derivative in which the parasitic worms have become parthenogenetic and a free-living sexually-reproducing generation may be interpolated in the course of a cycle of development which is otherwise similar to that of hookworms.

In the case of nematodes whose larvae hatch outside the body and have an instinct for burrowing it is easy to conceive of the accidental or, in time, routine invasion of intermediate or transport hosts. This might come about by invasion from the outside (e.g., Protostrongylinae), or by penetration through the gut wall after being swallowed (e.g., Anisakinae). Such penetration of hosts other than the definitive one, and subsequent encapsulation in parenteral tissues, is an extremely common phenomenon, and occurs in all the major groups of parasitic nematodes. In some instances it is a more or less exceptional phenomenon, e.g., the encystment of *Toxocara* larvae in mice (Fülleborn, 1921); in others it constitutes an important but not absolutely essential factor in the epidemiology, e.g., *Syngamus trachea;* and in still others it has become obligatory, the invaded hosts then becoming true intermediate hosts rather than transport hosts, e.g., spiruroids.

The frequent encapsulation of some nematodes in transport hosts and its non-occurrence in others is probably dependent upon the behavior of the larvae in the hosts concerned. Larvae that keep on the move do not become encapsulated. It is for this reason that most spiruroids are encapsulated, whereas *Habronema* and filariae are not. No encapsulation of hookworm or *Ascaris lumbricoides* larvae occurs when these enter rodents since the larvae complete the migration to the intestine, and are then evacuated because the environment is unsuitable for growth to maturity. On the other hand, since *Toxocara* is encapsulated in mice, it must be assumed that this worm loses its burrowing instinct before it has regained the alimentary canal, and then becomes quiet enough to be encapsulated by the host.

In the Metastrongylidae alone all gradations can be found from more or less accidental and unnecessary penetration of an intermediate host (e.g., *Dictyocaulus filaria*) to obligatory development in specific invertebrates (e.g., *Metastrongylus, Protostrongylus* and *Muellerius*). Similar obligatory dependence upon specific intermediate hosts has become the lot of the entire group of spiruroids, the Camallanina, the Dioctophymatina, and apparently at least one ascaridoid, *Subulura brumpti* (Alicata, 1939).

A clue to the origin of the filarioid type of life cycle, in which the microfilariae are withdrawn from blood or skin by blood sucking arthropods, and are eventually given an opportunity for reinvasion of the skin by these same arthropods, after development within them, is afforded by the habronemas (see p. 286). In these the larvae show a definite step towards the filarial type in that they fail to become encapsulated in the intermediate host, there to await passive transfer to the definite host, but instead remain free and active, and leave the intermediate host, under suitable conditions, of their own volition. The further steps to a filarial life cycle are merely (1) substitution of a parenteral for a gastro-intestinal habitat for the adult worms, and consequent liberation of the embryos into the blood or tissues whence blood-sucking insects can withdraw them; and (2) successful penetration of the skin by the infective larvae to reach their definitive location.

It will be seen that in no case is there reason to believe that intermediate hosts of nematodes are ancestral hosts, as is the case with intermediate hosts of flukes.

The same modifications in life cycle have a tendency to reappear over and over again in the various groups of parasitic nematodes, and sometimes several of the principal types may occur within a group of closely related genera. In the genus *Habronema,* for instance, the species parasitic in the stomachs of horses are deposited by the intermediate hosts on the lips or skin and they reach their destination by way of the mouth, either by direct migration into it or by being licked from the skin. In the habronemas parasitic in insectivorous birds, on the other hand, there can be little doubt but that they reach their destination in the orthodox spiruroid fashion, by the intermediate hosts harboring them being swallowed. In the species found in raptorial birds, however, a secondary transport host is usually if not always involved. Because of this lack of uniformity within even nearly related forms, and because of the endless number of minor variations by which one type of life cycle grades into another, we believe that a clearer picture of the life cycles of parasitic nematodes can be given by discussing the outstanding types and principal variations in each natural group, than by discussing types of life cycles irrespective of the natural groups in which they occur. By way of summary, however, we suggest the following classification of the principal life cycle types:

A. Monoxenous or Direct (no intermediate host required).

  1. Continuous reproduction within host, generation after generation; various stages of worms occasionally carried out of body and infect other hosts through contaminated food. Ex., *Probstmayria;* facultative rhabditoid parasites.

  2. Discontinuous, eggs or embryos escaping habitat of adults, and usually leaving parental host.

    (1) Without free-living phase.

      a. Simple. Eggs leave body of host, usually becoming embryonated outside, reënter via the mouth usually before hatching, and grow to maturity in the alimentary canal. Ex., *Enterobius, Trichuris.*

      b. With temporary burrowing into mucosa. Ex., *Ascaridia.*

      c. With parenteral migration via blood system to heart and lungs, returning to intestine via throat. Ex., *Ascaris lumbricoides.*

      d. With parenteral migration via blood system to definitive locations elsewhere in body. Ex., *Capillaria hepatica.*

    (2) With free-living phase. Eggs usually hatch outside body of host into first-stage larvae which grow to third (infective) stage while free, but in some forms may develop to third stage before hatching.

      a. With skin penetration and migration to intestinal tract via heart and lungs.

        (a) Free-living forms larvae only. Ex., *Necator.*

        (b) With possible development of an alternative generation of free-living adult males and females. Ex., *Strongyloides.*

      b. Without skin penetration; infection by mouth.

        (a) With temporary burrowing of larvae into mucosa. Ex., *Haemonchus.*

        (b) With more extended burrowing and formation of nodules in intestinal wall. Ex., *Oesophagostomum.*

        (c) Migration through intestinal wall and formation of nodules in parenteral locations. Ex., *Strongylus.*

      c. With optional use of transport host. Infective larvae when ingested by various invertebrates become encysted and reach final host when transport host is eaten. Ex., *Syngamus; Dictyocaulus.*

B. Heteroxenous or Indirect (development occurs only in an intermediate host)

  1. Passive Indirect. Embryonated eggs or larvae enter an intermediate host and become infective upon reaching third stage. Final host reached when intermediate host is eaten. Migration in definitive host, if any, via tissues or natural passages, not via blood system.

    (1) Eggs or larvae leave host with feces.

      a. Embryonated eggs or larvae are swallowed by intermediate host. Ex., *Metastrongylus,* spiruroids.

      b. Larvae superficially penetrate foot of molluscs. Ex., Protostrongylinae.

    (2) Larvae leave host through skin or by other parenteral routes. Develop after being swallowed by intermediate host. Ex., Dracunculoidea.

  2. Active Indirect. Larvae actively leave intermediate host to reach skin of definitive host.

    (1) Larvae reach intermediate host by eggs being eaten. Ex., *Habronema.*

    (2) Larvae reach intermediate host by being sucked from blood or skin. Ex., Filariae.

  3. Double Indirect. Larvae utilize two or more successive intermediate hosts.

(1) Second and subsequent intermediate hosts optional. Possible variation in many subdivisions above.

(2) Two successive intermediate hosts obligatory; definitive host reached by eating of second intermediate host. Ex., *Gnathostoma*.

RHABDITOIDEA

As noted on p. 267, there are a considerable number of nematodes belonging to this group which are facultative parasites of vertebrates, but only the Rhabdiasidae and Strongyloididae have become *obligatory* parasites of vertebrates. In both of these families there is a tendency for an alternation between free-living and parasitic generations, and in both, except in one species, *Parastrongyloides winchesi*, there is a suppression of males in the parasitic generation. The larvae of most species penetrate the skin or mucous membranes and migrate to the lungs before growing to maturity. The Rhabdiasidae mature in the lungs, and this is probably the more primitive condition; the Strongyloididae only exceptionally mature in the lungs, ordinarily returning to the intestine before maturing.

STRONGYLOIDIDAE

The life cycle of *Strongyloides stercoralis* of man, the main features of which were first elucidated by Grassi (1878) and Leuckart (1882), has been studied by a large number of prominent parasitologists, yet even today there is no unanimity of opinion about some phases of it. Grassi observed direct development into filariform* larvae; Leuckart discovered that an alternation of generations might occur; van Durme (1902) first demonstrated that infection resulted from skin penetration; and Looss (1905) showed that the migration of the larvae after penetration paralleled that of *Ancylostoma*. Important additional details or interpretations with respect to this or related species have been added by Leichtenstern (1899, 1905), Fülleborn (1914), Sandground (1926), Nishigori (1928), Kreis (1932), Faust (1933), Lucker (1934), Graham (1936-1939), Beach, T. D., (1935-1936) and Chitwood and Graham (1940).

The parasitic females live more or less deeply imbedded in the mucous membrane of the small intestine where they produce embryonated eggs which in this species hatch promptly within the host. The embryos are rhabditiform, and resemble those of hookworms except for the very short stoma. They normally pass out of the body with the feces of the host, and then begin to feed and grow. From this point on they may follow either one of two courses of development, known respectively as the direct or homogonic type, and the indirect or heterogonic type.

In the homogonic type of development the rhabditiform larvae grow and transform into filariform larvae, sometimes in 24 hours or less. Looss reported two molts in the course of the development of rhabditiform to filariform larvae in the human and other species, and Lucker (1934) observed two molts in the larvae of *S. ransomi* of pigs, but other observers have not mentioned more than one molt. The second-stage larva, according to Lucker, does not at first differ in any morphological characters from the first-stage larva, but transition to the filariform type of larva begins soon afterwards. The first molt occurs 12 to 18 hours after hatching, the second within 48 hours, at room temperatures. The filariform larvae are unsheathed and constitute the infective stage. They creep up on points of vantage in the soil or culture, often clustering together in brush-like groups to await an opportunity to burrow through the skin of a host.

In the heterogonic type of development the rhabditiform larvae, instead of developing into filariform larvae, change into adult free-living males and females. With the exception of Lucker (1934), no observer has mentioned or suggested more than a single molt in the course of this development, but Lucker has been able to trace the usual four molts and five stages. The only morphological changes are in size and in growth of the genital primordium until the fourth stage is reached, when the structure of the head simulates that of the adult, and the male and female characters are gradually assumed. At the time of the final molt the ovaries of the females and spicules of the males are fully formed. Adults may be found after 36 to 48 hours at room temperature.

The impregnated females produce eggs soon after they reach maturity, and these hatch soon after being deposited. These first-stage rhabditiform larvae are morphologically similar to those hatching from eggs deposited by the parasitic females, although a few small differences have been mentioned

---

*The term "filariform" is used here to denote the third stage larva of *Strongyloides* to distinguish it from the third stage larva of strongylins which is called "strongyliform" larva. Its use in no way signifies a similarity to any stage of filariids. Actually, the esophagus is very similar to that of an infective strongyl larva.—B. G. C.

by Kreis (1932). Although Kreis, like all others before him, fails to mention more than a single molt, Lucker was able to trace the orthodox two molts before the infective filariform larva was produced, just as in the case of filariform larvae of direct development. No difference between the two types of infective larvae has been noted.

Beach (1935-1936) showed that under particularly favorable conditions there may be several generations of free-living bisexual forms. Kourí, Basnuevo and Arenas (1936) reported that *S. stercoralis*, after numerous free-living generations, becomes entirely free-living; the females become parthenogenetic and there are no males, but the fecundity of the females gradually decreases until the cultures become sterile.

Nishigori (1928) first demonstrated the opposite extreme in the life cycle of *S. stercoralis*—internal auto-infection (called hyperinfection by Faust), with complete elimination of a free-living stage. Nishigori also suggested circumstances under which this might occur. Faust (1931) and Faust and Kagy (1933) confirmed Nishigori's observations, but the evidence was inconclusive for many until Faust and deGroat (1940) made observations at autopsy of a case which left no room for doubt but that under exceptional conditions in human beings auto-infection by filariform larvae of *S. stercoralis* through the walls of the colon can occur. There is no certainty, however, that it occurs in other species or hosts.

The filariform larvae of *S. stercoralis* and most other species normally penetrate the skin. If swallowed they burrow through the mucous membranes of mouth, esophagus or stomach. Mönnig (1930), however, states that sheep are usually infected with *S. papillosus* by mouth, this species being a poor skin-penetrator, and that larvae administered by mouth do not migrate to the lungs. Lucker's (1934) experiments with *S. ransomi* suggest that in this species also migration to the lungs may not take place after oral infection.

The larvae of *S. stercoralis*, after penetrating skin or mucous membranes, enter the circulatory system and are carried to the lungs. Faust (1933) states that they reach the lungs unchanged; they are sometimes recovered as early as the third day, and sometimes as late as the thirtieth day. Although no molts are mentioned, Faust distinguishes post-filariform, preadolescent, adolescent, and mature female *and male* forms. The post-filariform type of larvae is found most commonly in the lungs about the fifth day; if carried to the digestive tract they seem unable to establish themselves. These larvae are slenderer than infective larvae, with longer esophagus, and are more plastic. The preadolescent forms also occur principally in the lung tissue and bronchioles, and are believed to be too immature to establish themselves in the intestine. At this stage, according to Faust, sexual differences are observable for the first time, the female being still more slender than the post-filariform type and with a longer esophagus, whereas the male shows decided resemblances to the rhabditiform larva. The adolescent forms are migratory, and are commonly found not only in the lungs but also in the upper parts of the respiratory tree, esophagus and intestines. Both mature females and males were reported from lung tissue and bronchioles, but only mature females from the intestine, where they burrow into the walls. Lucker (1934), studying *S. ransomi*, observed only a single molt after entering the body of an animal, this occurring in the intestine about 6 days after infection; Looss (1911) also reported only a single molt, but Fülleborn (1914) apparently considers that two molts occur. By analogy with other nematodes, and with the development of the free-living adults of *Strongyloides* itself, it would seem more probable that two molts do occur in the course of the development in the host.

There has been much difference of opinion on several points in connection with the life cycle of *Strongyloides*, particularly (1) the factors determining whether the development is homogonic or heterogonic: (2) the reproduction status of the parasitic females, and (3), since the work of Kreis (1932) and Faust (1933), the occurrence and function of parasitic males.

Sandground (1926) gave a brief but valuable summary of views up to the time of his writing on the factors determining direct or indirect development. Environmental factors were first thought to be the cause, but Braun (1899) and others showed that such was not the case; Sandground felt that there remained no substantial reason for questioning the generally accepted idea that the direction of development was fixed before the larvae entered their period of free life. Leichtenstern (1905) advanced the view that there were two genetically different varieties of the human species, differing in their life cycles, the indirectly developing variety being confined to the tropics, the directly developing one being especially characteristic of the temperate zone. Leichtenstern considered the heterogonic type to be the more primitive and gave a very plausible explanation for the evolution of the homogonic type. Darling (1911) suggested as a cause environmental effects on the rhabditiform larvae prior to leaving the host, and Brumpt

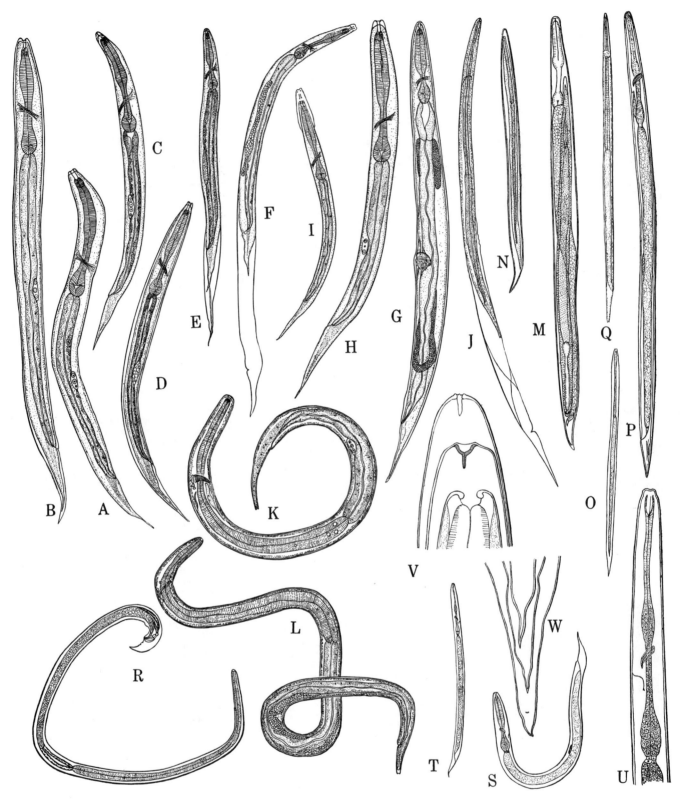

Fig. 179. DEVELOPMENT OF *STRONGYLOIDES AND RHABDIAS*

A-L—*Strongyloides ransomi* (A-B—Direct cycle, A—First stage larva, newly hatched; B—Larva in first molt; C-J—Indirect cycle; C—First stage larva, newly hatched; D—Larva in first molt; E—Larva in second molt; F—Larva in third molt; G—Larval female in fourth molt, H—First stage larva from free-living female, newly hatched; I—Second stage larva, immediately after first molt; J—"Filariform" larva undergoing second molt; K-L—Parasitic generation, K—Larva from small intestine of pig about 4 days after percutaneous infection; L—Larva from pig, showing early final molt). M-O—*Rhabdias fulleborni* (M—Free-living female with larva, the genital organs already destroyed; N—Cuticular hull of female with only one filariform larva, 115 hr. old culture; O—Filariform larva (Same as N) from which cuticle of fe-

male has been carefully removed). P-Q—*Rhabdias fuscovenosa* (P—Infective rhabditiform larva 72 hours after hatching from egg of parasitic generation; Q—Filariform larva (infective larva) from free-living generation). R—*Parastrongyloides winchesi* male. S-W—*Rhabdias fuscovenosa*, direct development (S—Rhabditiform larva; T—Ensheathed infective larva; U—Anterior end; V—Anterior end during third and fourth molts; W—Posterior end of same). A-L, after Lucker, J. T., 1934, U. S. D. A. Bull. no. 437. M-O, after Travassos, L., 1926. Arch. Schiffs. u. Tropen-Hyg. v. 30. P-Q, after Chu, T. C., 1936, J. Parasit. v. 22 (2). R, after Morgan, D. O., 1928, J. Helm. (6). S-W, after Goodey, T., 1924, J. Helm. 2.

(1921) pushed the determination back still further, to the developing eggs. Sandground (1926), who reported the finding of sperms in female worms and concluded that the worms were hermaphroditic, suggested that the direction of development is determined by the chromosomal constitution of the eggs subsequent to fertilization. Faust (1933), having found what he interpreted as parasitic males, suggested that fertilized eggs give rise to heterogonic and unfertilized eggs to homogonic progeny. Subsequently Beach (1935-1936), working in Faust's laboratory, showed conclusively that the course of development can be influenced by nutritional conditions; as these become less favorable more and more of the rhabditiform larvae undergo direct development to filariform larvae instead of becoming males and females. The evidence indicated that the potential females are influenced in this way more readily than the potential males.

Meanwhile Graham (1936-1939) started two pure lines of *S. ratti* of rats from original single-larva infections of the homogonic and heterogonic type, respectively, and found marked inherent differences between them. In each line over 85 percent of the total progeny were of its own type, with an extreme difference in the number of males produced. Graham also observed that there was a falling off in heterogonic larvae in winter as compared with summer, brought about by climatic effects on the host, not on the developing larvae. The conclusion seems warranted, therefore, that the course of development is dependent upon nutrition or other environmental conditions and not upon genetic constitution, but that there are genetic differences in the extent to which different strains are influenced towards homogony by a given degree of unfavorableness in the environment.

The reproductive status of the parasitic females was brought into question by Sandground (1926); prior to that time it had been generally accepted that they were parthenogenetic, although Leuckart apparently suspected that they were hermaphroditic, by analogy with the condition in the parasitic generation of *Rhabdias*. Sandground believed them to be protandrous hermaphrodites; he described what he interpreted as sperms and observed what seemed to be fertilization in specimens of *S. ratti*. Faust (1933), having found male worms in the lungs, concluded that the sperms observed by Sandground probably were the result of copulation. He considered the worms to be bisexual early in life, later becoming parthenogenetic. Chitwood and Graham (1940) concluded that *S. ratti* was parthenogenetic since they were unable to find sperms and also unable to find fertilization membranes. The weight of evidence is therefore in favor of parthenogenesis.

The occurrence of parasitic males described by Kreis (1932) and Faust (1933) has not been confirmed by others. In an unpublished observation, one of us (J. E. A.) has noted adult rhabditiform males in the fresh feces from a case of human strongyloidiasis but it was unknown whether these were parasitic males or males developing from eggs of parasitic females. The fact that the supposed parasitic males of Kreis and of Faust were rhabditiform and practically identical with free-living males is sufficient cause for doubt that they are really males of the parasitic generation, for in the one other member of the Strongyloididae in which males have been found—*Parastrongyloides winchesi*, Morgan 1928—the parasitic males are filariform like the females. We suggest that, since Faust not only observed eggs and rhabditiform larvae, but also filariform larvae which he interpreted as the progeny of the parasitic worms, in the lungs and bronchioles of infected hosts, the males observed were free-living males produced precociously in the lungs. The observations of Beach (l. c.) that males will develop more readily than females under suboptimal conditions would account for the failure to find free-living females. Graham's work with single-larva infections has shown clearly that males are at least unnecessary in *S. ratti*, though no conclusions can be drawn from this, for it is, of course, possible that there might be differences between species in this respect. For the present the occurrence of rhabditiform parasitic males in members of the genus Strongyloides must certainly be considered *sub judice*.

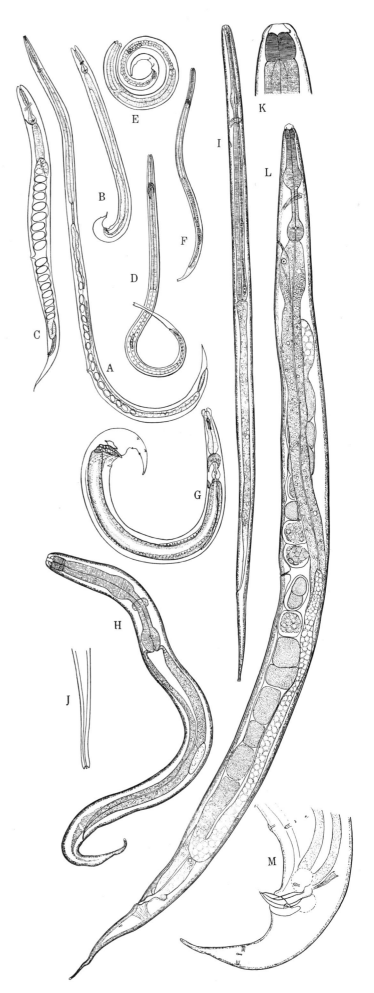

Fig. 180. DEVELOPMENT OF *STRONGYLOIDES*

A-G—*Strongyloides stercoralis* (A—Parasitic female; B—Free living male; C—Free living female; D—Filariform larva (human strain A), from lung tissue of experimental dog three days after skin inoculation; E—Post-filariform larva from same; F—Preadolescent female from lung of experimental dog 11 days after skin inoculation, note developing genital primordium; G—Adult male from lung tissue of experimental dog 57 days after skin inoculation). H-M—*Strongyloides* sp. from dog, free living generation (H—First stage larva; I—Infective larva; J—Tail of infective larva; K—Head, adult female; L—Adult female; M—Tail, male). A-C, after Faust, E. C., Human Helminthology, 1939. D-G, after Faust, E. C., 1933, Am. J. Hyg. v. 18 (1). Remainder original drawings by M. B. C.

RHABDIASIDAE

The members of the genera *Rhabdias* and *Entomelas*, now separated into a separate family from *Strongyloides*, resemble Strongyloididae in having an alternation of generations, at least in some species. This double life cycle was first demonstrated by Mecznikov (1865) in the case of *R. bufonis*. Unlike *Strongyloides*, the parasitic generation, at least of some species of *Rhabdias*, consists of hermaphroditic females, possessing a well-developed seminal receptacle. Seurat (1920a), however, thinks that the parasitic forms of *Entomelas dujardini* and *E. entomelas* from *Anguis fragilis* are parthenogenetic rather than protandrous hermaphrodites, since he was unable to find seminal receptacles or to detect sperms.

As in *Strongyloides*, both homogonic and heterogonic types of development may occur in the free-living phase of the life cycle of *Rhabdias*. In most of the species one type or the other strongly predominates or may even occur exclusively, though in some of the forms in which one type of development was long thought to occur exclusively, the alternative type has more recently been observed. Travassos (1926) called attention to the fact that the species found in Amphibia and Lacertilia have indirect development, while those found in snakes have direct development. Chu (1936), however, reported some unpublished observations of Chitwood's, and also some of his own, in which both types of development were found in several amphibian and reptilian species, (*ranae, eustreptos, fülleborni,* and *fuscovenosa* var. *catanensis*). In the last-named species Chu observed only homogonic development except when an especially favorable culture medium was used, whereupon a small percentage of free-living adults, predominantly males, were usually found. The offspring of these adults failed, however, to infect snakes. It seems evident from this data that the course of development of *Rhabdias* is determined by factors similar to those operating in the case of *Strongyloides*.

Whereas in *Strongyloides* both direct and indirect infective larvae are filariform, in the Rhabdiasidae the direct larvae are rhabditiform while the indirect ones are filariform (cf. Figs. 179, P-Q). The free-living adults of different species vary considerably in their mode of reproduction. Travassos describes the free-living female of *R. fülleborni* of frogs as producing only one or two larvae, which may become fully developed within the mother, destroying her tissues, whereas Chu (1936) describes *R. fuscovenosa* var. *catanenis* as having a few eggs in each horn of the uterus, which are usually laid when little or no development has occurred.

According to Goodey (1924) the homogonic larvae of *R. fuscovenosa* undergo two ecdyses outside the body of the host, the second shed cuticle being retained as a tight-fitting sheath for the infective larvae. The sheath is shed upon gaining entry to the host. The larvae molt twice more during development in the host's parenteral tissues, but both shed cuticles are retained as sheaths. Although the infective larvae of *R. bufonis* were reported by Fülleborn (1920) to penetrate the skin, and by the same writer (1928) to migrate to the lungs via the circulatory system, Goodey (1924) failed to get the infective larvae of *R. fuscovenosa* to penetrate skin, although their behavior outside the body was like that of skin-penetrating larvae, and he also thought it probable, from their distribution in the body, that they migrated to the lungs, after penetrating the gut wall, by direct migration through the mesentery and not via the blood stream. Fülleborn (1928) called attention to the fact that larvae of *R. bufonis* would also penetrate snails and possibly other invertebrates, where they remain unchanged for weeks, capable of infecting a frog when the snail is eaten. Similarly the larvae may sometimes became encapsulated parenterally in frogs which may then act as "transport hosts" for infection of larger frogs which eat them. Fülleborn suggests that since the skin of snakes is hard to penetrate transport hosts may constitute the principal method of infection for these hosts.

STRONGYLINA

I. STRONGYLOIDEA AND TRICHOSTRONGYLOIDEA

Three general types of life cycles, which more or less merge into each other, occur in the superfamilies Strongyloidea and Trichostrongyloidea of the suborder Strongylina. One of these, characteristic of the Ancylostomatidae and a few other forms in the Strongyloidea and Trichostrongyloidea, is essentially the same as the homogonic cycle of *Rhabdias*, except that the parasitic worms are bisexual. It involves development to the third (infective) stage outside the body of the host, skin penetration, and parenteral migration via the circulatory system after infection. The second, characteristic of most of the

Trichostrongyloidea and many of the Strongyloidea, differs in that there is no skin penetration, and no migration in the host beyond the walls of the alimentary canal. The third, characteristic of the Syngamidae in the Strongyloidea, involves development and molting within the egg, without feeding or growth, with at least optional establishment in an invertebrate transport host, and a parenteral migration which leads to the respiratory system but not beyond. We think it probable that both types 2 and 3 were derived from type 1, although it is also possible that type 2 is the most primitive, and that types 1 and 3 were both derived from this.

1. ANCYLOSTOMA SPP.

The genus *Ancylostoma* will serve as a typical example of the first type, involving skin penetration and parenteral migration. The eggs of these worms are deposited by the adult females in the lumen of the small intestine, whence they make their exit with the feces; at the time of leaving the body of the host they are nearly always in the four-celled stage of development, normally being unable to progress beyond this point without free oxygen. Under optimum conditions of oxygen, moisture and warmth (75° to 85° F.) the eggs proceed with

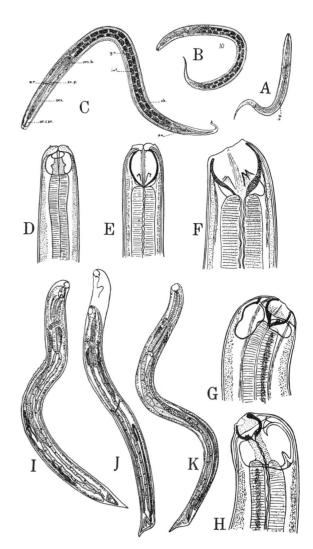

Fig. 181. DEVELOPMENT OF HOOKWORMS

Ancylostoma duodenale. A—First stage larva; B—Second stage; C—Third stage. D-H—Development of primitive and definitive capsules; (D—Bladder-like structures forming around the larval oral cavity, with beginning of formation of the primitive larval teeth; E—Nearly completed primitive capsule, with triangular teeth at base, and old larval oral cavity still running through center of primitive capsule; F—Fully developed primitive capsule with beginning of formation of bladders at its base; G—Later stage in development of dorsal and ventral bladders which will eventually form the definitive capsule; H—Later stage in development of definitive capsule, with primitive capsule still connected with esophagus by a strand of tissue; I—Female larva with definitive capsule formed but primitive capsule still attached; J—Male after final moult but last cuticle still enclosing it; K—Male larva with primitive capsule. After Looss, Chandler, A. C., 1929, Hookworm Disease.

272

their development rapidly, and a rhabditiform larva may hatch within 24 hours. This larva is about 250 μ long, with an elongated buccal cavity and a typical rhabditiform esophagus possessing esophageal valves. These larvae were shown by Mc-Coy (1929) to develop normally with only pure cultures of certain species of living bacteria as food. Under favorable conditions the larvae undergo the first ecdysis or molt within 48 hours after hatching, but second stage larvae show very slight morphological differences from first-stage larvae, although they are about 400 to 430 μ long. After a minimum of about 2 more days the larvae cease feeding, undergo a second ecdysis, and enter the third or infective stage. The cuticle shed at this molt is normally retained as a protective sheath, though it may occasionally be lost. The most important morphological changes in the infective larva are noted in the shape of the tail and the structure of the esophagus. The tail is shorter and more stumpy than that of the preceding stages. The esophagus is "filariform," or preferably "strongyliform," i.e., it is more uniform in width with tapering anterior portion, and the esophageal valves are lacking. The anterior portion of the lumen of the stoma is closed and the remaining posterior portion remains open in a characteristic shape. According to Alicata (1935) the various third-stage larvae of strongylid nematodes parasitic in swine can be differentiated, among other ways, by the form of the stoma; there is a possibility that this characteristic may hold true for other members of the Strongylina.

The infective larvae climb up on objects as high as a film of moisture extends, and show positive thermotropism and thigmotropism. They retire from excessive warmth in direct sunlight. They migrate vertically if buried in soil, but migrate laterally to a very slight extent (Chandler, 1925).

Although Leuckart (1866) showed that A. caninum of dogs could be transmitted per os, and Leichtenstern (1886) proved the same thing for A. duodenale of man, the usual mode of infection is by penetration of the skin; this method of infection was first demonstrated by Looss (1898). In subsequent work Looss (1905) established the course which the larvae follow in the body to reach the intestine. Skin penetration is accomplished in a few minutes when the larvae are able to obtain leverage, as in mud, to help them in their burrowing, but they are unable to penetrate when submerged in water. Within 35 to 40 minutes the larvae, having left their sheaths behind them, have reached the dermis and within a few hours are in the subcutaneous tissue. Many find their way into superficial lymphatic capillaries, and a few directly enter blood vessels. Some larvae are slow in entering the circulation, and may be encapsulated in the skin, especially in hosts sensitized by previous exposure. Certain "foreign" species of hookworms, e.g. Ancylostoma braziliense and Uncinaria stenocephala in man, commonly fail to enter the circulation at all but wander aimlessly in the skin, causing "creeping eruption." Although the larvae may remain in the skin for considerable periods no development takes place there (Fülleborn, 1927).

When larvae enter the lymphatics they are carried first to the regional lymph glands, and then to the main lymph channels leading to the thoracic duct, through which they enter the circulation. Such larvae, as well as those which entered the blood system directly, eventually reach the right heart, whence they are pumped out to the lungs. Here the majority burrow into the air spaces (Fülleborn, 1925), and are then mechanically carried in mucus, helped by epithelial cilia, to the trachea and throat. If swallowed they now pass to the alimentary canal, and grow to maturity in the intestine.

Although skin penetration is undoubtedly the usual mode of infection, infection by mouth can also occur. There has been considerable controversy as to whether swallowed larvae had of necessity to penetrate the mucosa and migrate to the lungs before growing to maturity, or whether they could develop to maturity without such migration. Yokogawa (1926) investigated the matter and found that when A. caninum larvae are fed to puppies a few penetrate the walls of the alimentary canal and enter the circulation, but the great majority of those which develop at all do so directly, without migration. In abnormal hosts, however, such as rodents, most of them perform the usual migration via the circulatory system, and a few migrate through the tissues to the body cavity whence they enter the liver, or go through the diaphragm to the pleural cavity, whence they enter the lungs. This work was confirmed by Scott (1928). Fülleborn (1926-1927) showed that the larvae of Uncinaria stenocephala of dogs also develop directly after oral infection, few migrating even in abnormal hosts. Several Japanese workers, however (Myiagawa, 1916; Myiagawa and Okada, 1930, 1931; Okada 1931) have persisted in the belief that lung migration is a biological necessity for hookworms. Foster and Cross (1934) carried through some further experiments which conclusively confirm the earlier work, showing that the lung journey is not a biological necessity for these worms (though it apparently is for Strongyloides stercoralis.) Swallowed larvae rarely migrate in susceptible normal hosts, but commonly do so in abnormal hosts and in resistant normal ones. Looss (1911) and Yokogawa (1926) observed that swallowed hookworm larvae remain in the stomach at least 2 days, and Fülleborn (1927) found they could remain there at least 5 days, partly in the lumen, partly deep in the mucous glands. He demonstrated that the larvae have an initial tendency to burrow into the glands, later to return to the lumen, as is the case with Ascaridia. He thinks that something in the secretion of the mucous glands causes the larvae to lose their mobility; possibly the same mechanism is responsible for the loss of the burrowing instinct in the larvae reaching the intestine from the lungs after skin penetration (see below).

The minimum time required for the larvae to reach the trachea after skin penetration is usually about 3 days, but the majority require 4 or 5 days, and some still longer. By the time the larvae appear in the bronchioles and trachea they have grown slightly in length, have developed a provisional mouth capsule, and are ready for the third molt, although there is no evidence that they ever complete it before reaching the digestive tract. The formation of the provisional, and subsequently of the definitive, mouth capsules is accomplished by the development of dorsal and ventral bladder-like structures posterior to the already existing mouth. These spread around the sides and finally unite (Looss, 1905) (Figs. 181).

Up to the time of the third molt the larvae grow very little in length, but increase from about 20 μ to 30 μ in diameter. The molt usually occurs very soon after the larvae reach the intestine, and the larvae at this time lose their tendency to burrow, so remain in the intestine. There is no evidence that they temporarily burrow into the glands of the stomach as do larvae that are directly swallowed. The young worms now grow very rapidly. They may reach a length of 2.5 to 3 mm within a few days. Sexual differentiation now begins, and in from 4 to 6 days after the third molt the definitive mouth capsule is developed. By the time the worms have reached a length of from 3 to 5 mm. the fourth molt takes place. Thereafter the worms grow to maturity, copulate, and begin egg production. In the case of Ancylostoma duodenale in man the eggs first appear in the feces 5 to 6 weeks after infection, whereas in A caninum of dogs, eggs may appear as early as 15 days (Herrick, 1928).

2. HAEMONCHUS CONTORTUS

The life cycle of this worm as worked out by Ransom (1906), Veglia (1916) and others is essentially the same as that of the ancylostomas in its free-living phase. The infective, ensheathed third-stage larvae, however, are not skin-penetrators, but have a tendency to climb up on vegetation or other objects where they are in a favorable position to be ingested by their herbivorous definitive hosts. Here they curl up, and are remarkably resistant to cold and to moderate desiccation. Upon being ingested by the final host the larvae bury themselves in the mucous glands and crypts of the abomasum, where they undergo the third and fourth molts; the adult stage is reached after about the 9th to 11th days, and the worms emerge to live in the lumen of the organ, beginning egg production about 3 weeks after infection. Although there is no evidence that the worms perform a parenteral migration in sheep, Ransom (1920) showed that they do migrate to the lungs in guinea pigs.

3. SYNGAMUS TRACHEA

The life cycle of this worm was first experimentally worked out by Ortlepp (1923). The eggs of the worm are laid in the bronchi or trachea of the host in an advanced stage of segmentation. Under favorable conditions the first-stage larva is developed in about 3 days, but the egg does not become infective until after 1 to 2 weeks, whereupon they may or may not hatch. Ortlepp observed only a single molt during the course of development and interpreted the infective larva as a second-stage larva but Wehr (1937) demonstrated that the developing larva undergoes two molts within the egg. Buckley (1934), studying S. ierei of cats, also observed the usual two molts. Yokogawa (1922) also missed the first molt in the case of Nippostrongylus muris, and in spite of the large amount of experimental work done with that worm the missed molt was not discovered until 1936, when Lucker demonstrated it. The first cuticle in both these worms is extremely thin, and the second ecdysis may be in progress before it is completely shed.

Infective larvae, whether hatched or still in the eggs, are infective when directly swallowed by susceptible hosts, but very often they are swallowed by various invertebrates; when this happens they penetrate the gut wall and become encapsulated in the body cavity. Walker (1886) and Waite (1920) both called attention, on epidemiological grounds, to the importance of earthworms in the dissemination of this parasite, but Clapham (1934) first experimentally worked out the rôle played by these annelids. Subsequently Taylor (1935) showed that

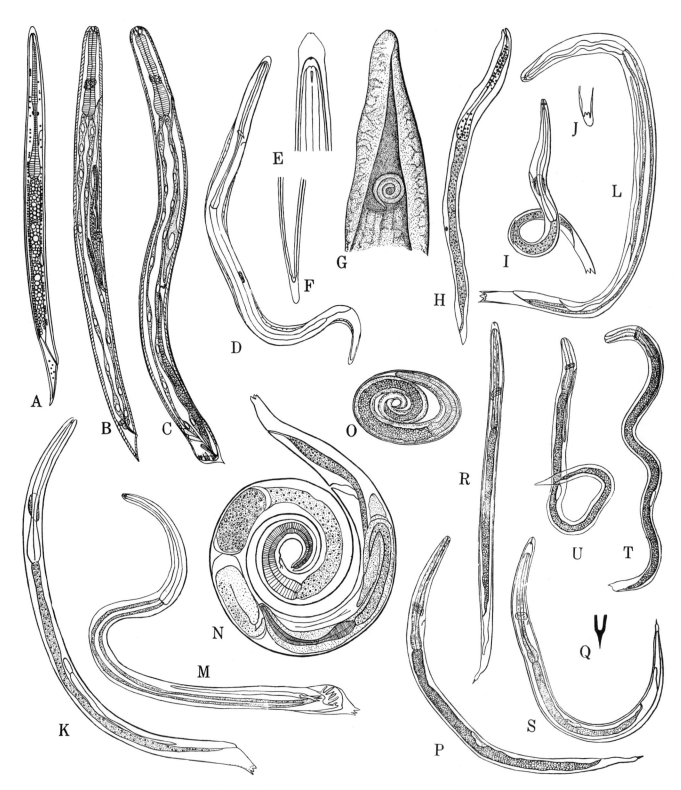

Fig. 182. DEVELOPMENT OF THE STRONGYLINA

A-C—*Syngamus trachealis.* (A—Ensheathed second stage larva; B— Third stage larval female; C—Ensheathed young fourth stage larval male). D-F—*Syngamus ierei* (D—Third stage larva; E—Anterior end of third stage larva; F—Tail of third stage larva). G—*Haemonchus contortus* on blade of grass. H-N—*Ollulanus tricuspis* (H—First stage larva; I—Second stage larva; J—Tail between first and second stage; K—Third stage (infective) larva; L—Fourth stage female; M—Fourth stage male; N—Gravid female). O-U—*Dictyocaulus arnfieldi* (O—Egg from the feces; P-Q—First stage; R—Second stage; S—Third stage; T —Fourth stage male; U—Fourth stage female). A-C, after Ortlepp, 1923, J. Helm. v. 11. D-F, after Buckley, 1934, J. Helm. v. 72. G, after Ransom, 1906, U. S. Bur. An. Ind. Circ. 93. H-N, after Cameron, 1927, J. Helm. v. 5. O-U, after Wetzel and Enigk, 1938, Arch. Wiss. u. prakt. Tierheilk, 73(2).

snails and slugs would also serve as transport hosts, and later found that the encysted larvae would remain viable in these molluscs for several years. More recently Clapham (1939a, 1939b) showed that maggots, crane fly larvae, spring tails and centipedes would serve in a similar capacity, and that the worms were able to survive metamorphosis in the tissues of flies.

When ingested by these hosts the infective larvae hatch from the eggs if they have not already done so, penetrate the gut wall, and enter the body cavity, where they are eventually encapsulated by the host tissues. Clapham has shown that the life cycle is completed somewhat more readily with the aid of a vector than without, and was able to infect chickens readily with a starling strain when an earthworm vector was used, whereas Taylor (1928) had had difficulty in doing so by direct infection. Clapham calls attention to the fact that *Syngamus trachea* is evidently undergoing evolution in its life cycle; at present it can still develop without an intermediate host, and has not as yet adapted its requirements to any *particular* intermediate host, but can use almost any that happens to swallow it. She makes the reasonable suggestion, however, that in time different strains may adapt themselves to different intermediate hosts, as determined by the food habits of the final hosts, and thus perhaps give rise to new species. At present, however, the effect of living in a transport host seems to be to *eliminate* physiological differences; for example, in the case of starling strains developing in chickens. It is possible that some species of *Syngamus* may already have reached the stage of *requiring* an intermediate host, since Buckley (1934) was unable to infect cats with eggs containing third-stage larvae of *S. ierei*.

After infection by swallowing eggs, free larvae, or larvae contained in invertebrate transport hosts, *S. trachea* apparently reaches the lungs via the circulatory system. Ortlepp (1923) found the larvae in the lung tissues within 24 hours and Wehr as early as 17 hours after infection. Wehr found fourth-stage larvae after 3 days and immature adults after 7 days; some of the latter were already *in copula* even before entering the trachea.

VARIATIONS IN THE LIFE CYCLE IN OTHER STRONGYLOIDEA AND TRICHOSTRONGYLOIDEA

The preparasitic stages of nearly all the members of the Strongyloidea and Trichostrongyloidea, except the Syngamidae, are remarkably similar, involving two free-living rhabditiform stages separated by a molt, and a strongyliform third stage, in which the shed cuticle is usually retained as a sheath. The time intervals between the molts and the total time required to reach maturity vary considerably; in some species, e.g. *Ornithostrongylus quadriradiatus*, the infective third stage may be reached within 3 days. The infective larvae are distinguishable by characters of the mouth, buccal cavity, esophagus, shape of tail, length of sheath, etc., and also, as Lucker has shown in a series of papers (e.g., Lucker, 1938) by the number and arrangement of cells in the intestine.

The only important variation from this formula is the molting of some species within the egg, thus eliminating a period of feeding and growth outside the host; this, as already noted, occurs in *Syngamus* and it also occurs in *Nematodirus* spp. (Ransom, 1911; Maupas and Seurat, 1913) and in *Oswaldocruzia filiformis* (*Strongylus auricularis*, Zeder) (Maupas and Seurat, 1913). According to the latter authors, *Ostertagia marshalli* hatches as a second-stage larva and undergoes its second molt 2 or 3 days later without feeding. This is not true, however, of *O. circumcincta*. When both molts occur inside the egg the infective embryos may or may not hatch prior to being swallowed by a host, eggs containing infective third-stage larvae being infective as well as the free larvae.

Strongylacantha glycirrhiza, according to Seurat (1920b), hatches at the end of 48 hours but the larvae fail to feed, and at the end of a month have molted twice and are ensheathed in both shed cuticles, just as in the case of *Dictyocaulus* (see below).

A striking exception to the usual course of events occurs in the case of *Ollulanus tricuspis*, according to Cameron (1927). This parasite of the stomach of cats is viviparous. The eggs hatch in the uterus of the mother, and the larva undergoes its first molt *before* it is born, acquiring the typical tri-cuspid tail. Third-stage larvae are found free in the stomach of the cat, but it is not certain whether the second molt occurs before or after birth. This form is believed by Cameron to leave the stomach with the vomitus of the cat. When eaten by another cat with the vomitus the larvae change to fourth-stage larvae and finally adults. Some part of this development is believed to take place in the depths of the mucous membrane. No other method of exit from the cat has yet been found; no larvae were ever seen in the intestine, nor were mice infected when fed on cat stomach or infected vomitus. Continuous auto-infection is believed possible but improbable; Cameron suggests the possible production of a substance inhibiting complete larval development, as postulated by Fülleborn in the case of *Rhabdias bufonis* in the lungs of frogs.

The mode of access of the infective larvae to the final host varies in different species, even, sometimes, within the same genus. There are three possibilities: (1) penetration of the skin; (2) ingestion with food or water; (3) ingestion with a transport host. Skin penetration is characteristic of most of the hookworms (Family Ancylostomatidae) — *Ancylostoma, Necator, Uncinaria* and *Gaigeria*—but *Bunostomum* seems to be an exception in that, although the larvae, at least of *B. trigonocephalum*, seem to be capable of penetrating under certain conditions (Ortlepp, 1937, p. 207), they do not do so as readily as other hookworm larvae (Cameron, 1923; Schwartz, 1925), and normally infect by mouth. Although most of the hookworms are able to infect the host by mouth as well as through the skin, and may even be able to dispense with the parenteral migration (see above), Ortlepp (1937) was unable to cause infection in sheep by the oral route with larvae of *Gaigeria pachyscelis*. Most other members of the Strongyloidea and Trichostrongyloidea fail to penetrate the skin although a few (*Stephanurus dentatus, Nippostrongylus muris, Longistriata musculi, Trichostrongylus calcaratus*) are able to do so. Other species of *Trichostrongylus* apparently do not penetrate the skin. *Nippostrongylus muris* is almost wholly dependent upon skin penetration (Yokogawa, 1922), whereas for *Longistriata musculi* oral infection is probably more important in nature (Schwartz and Alicata, 1936).

The great majority of the worms belonging to the groups we are considering normally enter the host by mouth, with contaminated water or food. In most cases the larvae climb up on living vegetation and are more or less resistant to desiccation. This is true of all the Strongylidae so far as known (except *Stephanurus*), and all of the Trichostrongyloidea with the exception of the few mentioned in the preceding paragraph, and *Ollulanus*.

The development within the host involves varying degrees and types of migration. Skin-penetrating larvae usually follow the route described above for ancylostomes, but Schwartz and Alicata (1936) showed that the larvae of *Longistriata musculi* do not normally do so; they appear in the stomach within a few hours after skin penetration, and in the intestine soon after that, but they were not found in the liver, lungs or stomach walls. Their actual route was not determined. In the case of this worm, whether infection is by skin or mouth, the entire development takes place in the intestine, contrary to what happens in other skin-penetrating forms, even in the nearly related *Nippostrongylus*.

Nematodes infecting by mouth may or may not migrate via the blood stream. Most of the Trichostrongyloidea (e.g. *Cooperia, Ornithostrongylus, Ostertagia, Obeliscoides, Graphidium, Haemonchus, Hyostrongylus*, most species of *Trichostrongylus, Nematodirus*) perform no migration at all beyond a more or less temporary invasion of the glands or crypts of the stomach or duodenum. Some forms, e.g., *Ornithostrongylus quadriradiatus*, may reach the adult stage of development as early as the third or fourth day after infection (Cuvillier, 1937).

The Strongylidae show various gradations from invasion of the circulatory system and transportation with the blood, to mere temporary invasion of the glands. Of the three common species of *Strongylus* in horses each shows characteristic features in its migration, the larvae of *S. vulgaris* being found in aneurisms in the anterior mesenteric vein, those of *S. edentatus* under the peritoneal walls of the abdominal cavity, and those of *S. equinus* in liver and pancreas. According to the usually accepted view (see, for example, Neveu-Lemaire, 1936) *S. vulgaris* penetrates the walls of the intestine and migrates through the body via the circulatory system, passing through the capillaries of both liver and lungs to be distributed all over the body by the systemic arterial circulation. Ninety percent stop in the anterior mesenteric artery, to the walls of which they adhere by using the mouth as a sucker. The resulting irritation leads to the formation of an aneurysm and thromboses. Here they remain for 5 months, meanwhile growing and passing through two molts; one at a length of 3 to 4 mm, the other at a length of 7 to 10 mm. Having passed the final molt they release their holds and are carried by the blood stream to the walls of the cecum or colon. They remain imbedded in the walls in little nodules under the mucosa for about a month, and finally make their exit into the lumen. Olt (1932) thinks that the normal migration is via the lungs and trachea as in the case of hookworms, but that some larvae burrow through the intestinal walls and between the laminae of the mesenteries until they reach a large bloodvessel. If this is a large, heavy-walled vessel the slow passage through it leads to inflammation and the characteristic aneurysms. Wetzel and Enigk (1938a), on the other hand, believe they have convincing evidence that no

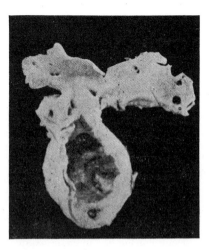

Fig. 183. *STRONGYLUS VULGARIS*

Verminous aneurysms affecting the anterior mesenteric artery. After Foster & Clark, 1937, Am. J. Trop. Med. v. 17 (1).

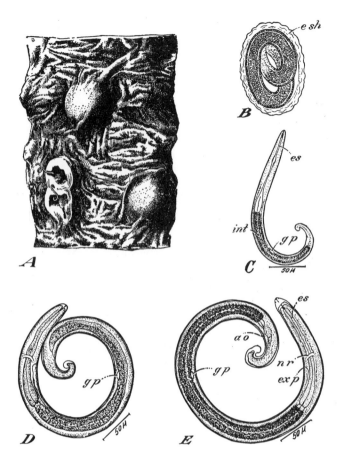

Fig. 184. *OESOPHAGOSTOMUM BIFURCUM* AND
METASTRONGYLUS SALMI

A—Nodules of *Oesophagostomum bifurcum* in the large intestine of an African (after Brumpt). B-E—*Metastrongylus salmi* (B—Egg with fully developed embryo; C—Newly hatched first stage larva; D—First stage larva undergoing first molt; E—Second stage larva undergoing second molt while still enclosed within the cuticle of first molt). A, after Chandler, 1940 (fig. 146) Int. to Parasit. B-E, after Alicata. 1935. U.S.D.A. Tech. Bull. 489.

Strongylus larvae migrate via the lungs and trachea, but undergo their whole development within the abdominal cavity.

S. edentatus larvae penetrate the walls of the intestine and the majority come to rest under the peritoneum, though the route followed in reaching this location has not been traced. Some, probably carried by the blood stream, reach the liver and lungs. After about 3 months, during which they grow much larger, the larvae migrate to the roots of the mesenteries and travel between the laminae to the walls of the cecum and colon. Here they become lodged for about a month in large subserous hemorrhagic nodules which eventually open into the lumen of the intestine.

S. equinus larvae penetrate the walls of the intestine and make their way to the liver and pancreas. It has generally been assumed that they arrive in these places via the blood stream, but Wetzel's observations (l.c.) throw doubt on this. After development to the fourth larval stage they return to the walls of colon and cecum, again by an undetermined route, and continue their growth in nodules in the walls of these organs. After reaching the final stage of development by a fourth molt they pass into the lumen.

The Trichoneminae of horses are believed not to migrate out of the intestine at all. Many of them, perhaps all, penetrate into the walls of the mucosa where they develop in nodules. They undergo the third molt when about 1 mm long, becoming what Ihle and Oordt (1923) call "Trichonema" larvae, provided with a provisional mouth capsule. The final molt occurs in the lumen of the intestine.

Triodontophorus tenuicollis is believed by Ortlepp (1925) to develop directly in the lumen of the cecum and colon, without even temporarily burying itself in the mucosa. He was never able to find larvae of this species in nodules. However, only fourth-stage larvae were found, and there is nothing in Ortlepp's observations to preclude a hookworm-like migration via lungs and trachea on the part of the third-stage larvae.

The Oesophagostominae have a life cycle in the host essentially the same as that of the Trichoneminae, the young worms tending to bury themselves in the mucosa, where they cause the formation of cysts or nodules. Here they undergo their development to the final stage, emerging into the lumen of the intestine at about the time of the final molt, or in some cases even later, when they have grown to a length of 4 or 5 mm.

According to Spindler (1933), *Oesophagostomum quadrispinulatum* (= *longicaudum*) of pigs produces inflamed liquefying cysts within 48 hours after infection, and the larvae begin escaping into the lumen after about 17 days. Similar inflamed cysts are produced by most other species of oesophagostomes, but Goodey (1924) failed to observe them in experimental infections with *O. dentatum* and Schwartz (1931) saw only small noninflamed nodules at the site of attachment of adult worms of this species in contrast to the inflamed lesions caused by *quadrispinulatum*. *Chabertia ovinus*, though nearly related to *Oesophagostomum*, also fails to develop in submucous nodules.

Stephanurus dentatus, (see Schwartz and Price, 1932; Ross and Kauzal, 1932) whether entering by skin or mouth, migrates to the liver via the blood stream. The third molt occurs about 70 hours after infection, and the larvae have a provisional mouth capsule. Normally such larvae escape from the capillaries in the liver and wander in the hepatic parenchyma until they reach the surface capsule. They wander under this for a time but eventually, 3 months or more after infection, break free into the body cavity and make their way to the perirenal fat tissue, perforating the walls of the ureters to establish connection with the outside world. They themselves become enclosed in capsules of host tissue.

II. METASTRONGYLOIDEA

In this superfamily of the Strongylina the early development follows somewhat different patterns from that of the other members of the suborder, except in a few instances (e.g., *Strongylacantha* resembles *Dictyocaulus* in hatching and then reaching the infective stage without feeding or growing, and the Syngamidae also resemble *Dictyocaulus* in having *optional* transport hosts).

Three principal types of development occur among the Metastrongyloidea: (1) the *Dictyocaulus* type, in which the larvae go through two molts and reach the infective stage, surrounded by one or both shed cuticles, without feeding or growing; (2) the *Metastrongylus* type, in which the first-stage larvae continue their development after ingestion by earthworms, and (3) the *Protostrongylus* type, in which the first-stage larvae, attracted by the mucus of snails or slugs, continue their development after entering the slime glands in the foot of these molluscs, and becoming encapsulated in the muscular connective tissue under the epithelium.

1. DICTYOCAULUS SPP.

The eggs of *D. filaria* and *D. viviparus* hatch in the bronchi, or at least in the intestine, as they are leaving the body of the definitive host, but those of *D. arnfieldi*, according to Wetzel and Enigk (1938) fail to hatch in the lungs, and usually do not hatch until a few hours after leaving the body. The first molt usually takes place at room temperature in from 1 to 2 days, and the second in from 3 days (in *D. arnfieldi*) to about 12 days (*D. filaria*) later. Usually both sheaths are present in early third-stage larvae, but the first cuticle is eventually lost. These infective larvae live a long time in moist soil or water, and are able to survive in earthworms if eaten by them, although they do not depend upon the earthworm as an intermediate host. The use of earthworms as transport hosts seems to be of less importance in the case of *Dictyocaulus* than in the case of *Syngamus* (see above). However, there is no evidence as yet that *Dictyocaulus* can use as large a variety of transport hosts as can *Syngamus*.

2. METASTRONGYLUS

Metastrongylus elongatus (= *apri*), *M. salmi*, and *Choerostrongylus pudendotectus*. The eggs of these worms contain fully developed embryos when deposited. Although usually stated to hatch in the bronchi or intestinal tract during passage out of the definitive host, Alicata (1935) found that they are usually passed in the feces unhatched, and remain unhatched until taken into the body of a susceptible intermediate host. The eggs or embryos may, however, remain viable for 3 months in moist soil.

When ingested by earthworms (species of *Helodrilus* and *Lumbricus*) the larvae burrow into the walls of the esophagus and proventriculus of these hosts. Alicata has found them there 16 hours after exposure to infection. They also enter the circulatory system and may be found in the hearts, but Schwartz and Alicata (1929) showed that migration via the blood stream was not an essential part of the life cycle of this worm in its intermediate host. In the earthworm the first molt occurs about 8 to 10 days or more after infection, and the second one a few days later, this molt beginning before the first cuticle has been shed. The second cuticle is retained by the third-stage larvae, which are now infective. The larvae do not spontaneously leave the host, and an earthworm may remain infective over winter, and probably at times for several years. Upon death of the earthworms the larvae are able to survive for 2 weeks in moist soil. Pigs become infected by eating infected earthworms or liberated infective larvae. After ingestion, according to Hobmaier and Hobmaier (1929), they migrate via the lymphatics or blood stream, undergoing the third molt in mesenteric lymph glands, and then proceed via the lymphatic and blood systems to the lungs, where they become mature after a fourth and final molt.

3. PROTOSTRONGYLINAE

All the members of the family Protostrongylinae resemble one another in requiring molluscs as intermediate hosts. In all cases the embryonated eggs hatch before leaving the body, or soon after, and the first-stage larvae may live in soil or water for several weeks, but without further development. The larvae are attracted by the slime of molluscs, and upon coming in contact with a mollusc they creep into furrows in the foot, whence they penetrate into mucous glands, burying themselves in the muscular connective tissue under the epithelum. Here they coil up and soon become enclosed in a tubercle resulting from encapsulation by the host. The first molt usually takes place after a week to 10 days at room temperature, the larvae having grown comparatively little in length, but having become thicker. The second molt usually takes place in from 10 or 12 days (*Aelurostrongylus, Muellerius, Crenosoma*) to 4 or 5 weeks (*Elaphostrongylus*), after which the larvae are infective when molluscs containing them are eaten. In most cases little specificity is shown with respect to the species of molluscs utilized as intermediate hosts, although, possibly because of the habits of the snails, certain species seem to be of prime importance. *Protostrongylus rufescens* develops primarily in *Helicella* (Hobmaier and Hobmaier, 1930); *Muellerius capillaris* can utilize a great variety of snails and slugs, although Pavlov (1937) found only *Helicella obvia* to be important in

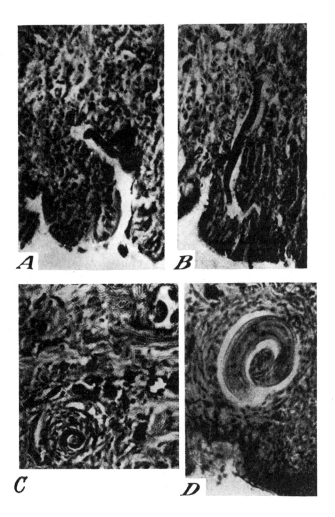

Fig. 185 DEVELOPMENT OF PROTOSTRONGYLINAE IN MOLLUSCS

Larvae of *Muellerius capillaris* in *Agriolimax agrestis*. A—Larvae in furrow of foot of mollusk a few hours after infection; B—On first day of infection (sagittal section); C—Coiled larva in foot on second day of infection (horizontal section); D—Larvae in sole of foot on 16th day of infection. After Hobmaier, 1934, Ztschr. f. Parasitenk. v. 6 (5).

Jugoslavia; *Aelurostrongylus abstrusus*, reported by Cameron (1927) to utilize mice as intermediate hosts, apparently erroneously, according to Hobmaier and Hobmaier (1935) develops in a variety of snails and slugs, but *Epiphragmophora* proved most suitable. Other forms in which a variety of molluscs have been shown to serve as hosts are *Aelurostrongylus falciformis* (Wetzel, 1938), *Crenosoma vulpis* (Wetzel and Müller, 1935), and *Elaphostrongylus odocoilei* (Hobmaier and Hobmaier, 1934).

Hobmaier (1934) believes that the utilization of molluscs as intermediate hosts by the Protostrongylinae grew out of the habit of the larvae of seeking protection from desiccation in the slime of the molluscs. This predilection for slime extends to the period of passage through the colon of the definitive host, for the larvae are commonly found burying themselves in the intestinal mucus and thus becoming located on the surface of fecal pellets instead of inside. In this position those larvae which were not protected from desiccation by the mucus, and subsequently the tissues, of snails would fail to survive. The larvae, as Hobmaier points out, differ widely in their habitat in the snail from the parthenitae of flukes, which probably develop in snails because these were ancestral hosts. Whereas fluke parthenitae are true internal parasites of molluscs, lungworm larvae are scarcely more than external parasites. Larvae ingested by snails usually pass all the way through the alimentary canal and fail to develop.

ASCARIDINA

OXYUROIDEA

1. Enterobius vermicularis

In spite of the fact that the oxyurid type of life cycle is the simplest and probably the most primitive of any found among nematodes parasitic in vertebrates, a search of the literature has failed to reveal a single instance in which a detailed molt by molt account of the life cycle has been described. The life cycle of *Enterobius vermicularis*, so far as it is known, will serve as an example of its type.

The adult female worms, with the uteri filled with developing eggs, live in the lower part of the large intestine and particularly in the rectum. They do not ordinarily deposit their eggs in the lumen of the intestine, but crawl out of the anus and deposit them in the perianal region, leaving trails of eggs as they creep about. Contact with air is apparently a stimulus to oviposition (Philpot, 1924). Although they frequently remain outside the anus and release the eggs in showers when the body ruptures, MacArthur (1930) and others state that they commonly retreat into the rectum, to repeat their egg-laying expeditions out of the anus over and over again, particularly at night.

The eggs when deposited by the females, or contained in the uterus of females which have voluntarily migrated out of the intestine, are fairly uniformly in the "tadpole" stage of development, apparently being unable to progress beyond this point without free oxygen. Within 6 hours after leaving the body they develop a coiled larva (ring-and-a-half embryo) which is infective. According to Brumpt (1922) the larva undergoes no molt before hatching nor, according to Philpot (1924), as a free larva in water. However, Alicata (1934) suggested that a molt within the egg shell might be general for the Ascaridina, and *Enterobius* might well be reëxamined. Chitwood (personal communication) believes he has seen a molt in the egg, and thinks there may be two.

Development of the larva in the egg will occur in oxygenated water, and in this medium the larvae commonly emerge in from 9 to 24 hours at 37° C., but they only live for a few days, so it is evident that water cannot be an important vehicle of infection. Exposed to air a considerable proportion of the eggs survive for at least 6 days at humidities above 62 percent (Jones and Jacobs, 1939).

When ingested the eggs hatch in the stomach or intestine, and the worms live during the early part of their development in the lower part of the small intestine, cecum and upper portions of the colon, not infrequently invading the appendix. Heller (1903) states that there definitely are two molts in the small intestine, and probably three. Chitwood, (personal communication) reports having seen a molt in the epithelium of the appendix. By analogy with other nematodes there is probably a total of four molts.

Although the worms have repeatedly been reported as burrowing into the mucous membranes, especially of the appendix (Penso, 1932), it seems probable that this is a habit only of the fourth-stage larvae. Chitwood (personal communication) reports having found the fourth-stage larvae in sections of the appendix. He has observed a definite period 6 to 9 days after infection when symptoms of invasion appeared, followed 4 to 7 days later by migration of the worms from the anus. Exposure to air after operation would account for the deposits of eggs which Penso reports and figures deep in the walls of the appendix.

There has been a large amount of discussion as to whether internal auto-infection by the worms can occur. The fact that infections persist even for many years in spite of the most careful efforts to prevent reinfection from the anus via the hands has lent support to this idea. However, the demonstration by Lentze (1935), and Nolan and Reardon (1939) of the ease with which airborne infections can occur seems sufficient to account for the persistence of infections. On the other hand, Zawadowsky and Schalimov (1929), Lentze (1935) and others have called attention to the failure of development and infection of eggs or embryos left under conditions such as exist in the lumen of the large intestine. It would be difficult to say that internal auto-infection could never occur, but the evidence is all in favor of the view that if it does occur it is an abnormal and exceptional condition.

Copulation of the young adult worms usually takes place in the upper parts of the colon or in the cecum, where the males live for some time. The females do not migrate to the rectum until they contain developing eggs. Ripe females begin to appear about 15 days after infection.

2. Other Oxyuroidea

The life cycles of other Oxyuridae are the same in essential features, but differ in details. *Oxyuris equi* differs in that the fourth-stage larva has a special structural development of the anterior portion or "corpus" of the esophagus which enables the larva to use it as a highly developed buccal capsule for adhering to the mucosa (Wetzel, 1931). The ripe females of this species creep out of the anus as do those of *Enterobius*, but this is probably not true of forms parasitic in rodents. The fourth-stage larva of *Dermatoxys veligera* is also provided with a special structure for adhering to the mucosa, but in this case the end is accomplished by the development of four conspicuous hooks on the head (Fig. 156, X, Y) (Dikmans, 1931), which is buried in the mucosa (Wetzel, 1932). These specializations for maintaining a position in the colon are of interest as indicative of a need for some sort of protection against expulsion from the body before maturity is reached, a need which may perhaps, as has already been suggested, have led to a deeper burrowing into the mucosa and ultimately to a parenteral migration.

According to Philpot (1924), *Aspiculuris tetraptera* has a life cycle strikingly like that of *Enterobius*, differing only in the earlier stage at which the eggs cease development before expulsion, and their failure to hatch outside the body. *Syphacia obvelata* differs in that the eggs have developed embryos when they leave the host. All stages of development from the youngest larva to adult can be found in the cecum of naturally infected mice, and are strikingly similar to those described and figured for *Aspiculuris*. *Tachygonetria longicollis* and *T. dentata* definitely undergo a molt before hatching from the egg. *Passalurus ambiguus*, according to Penso (1932), is capable of internal auto-infection; the gravid females burrow into the mucosa to deposit their eggs, and the larvae subsequently emerge to continue their development. Penso, however, postulates a similar behavior on the part of *Enterobius vermicularis*, and thinks that Wetzel's observations on *Dermatoxys veligera* were in error, the larvae with buried heads being emerging from, not entering, the mucosa. Although *Passalurus ambiguus* may sometimes deposit its eggs in the mucosa, Penso's observations need to be extended before this can be accepted as a normal or usual procedure.

Probstmayria vivipara (Atractidae) is, so far as known at present, unique among nematodes that are known to be obligatory parasites of vertebrates in reproducing continuously generation after generation in a single host. It is among the nematodes what the *Pupipara* are among the Diptera, or *Tunga* among fleas. Its larvae hatch in the uterus and grow almost to the size of the parents before being born (vide Ransom, 1907). They resemble the parents except for lack of development of the genital organs. No stage of development is known outside the body of the host. Transfer to new hosts is believed by Jerke (1902) to be accomplished by contamination of food or water by worms passed in the feces; such worms, he says, remain alive in feces for several days.

ASCARIDOIDEA

In the Ascaridoidea there are always one or more molts before the embryos leave the eggs and, with few if any exceptions, there is a phase of burrowing into the mucosa, and in many cases more extensive migration from the lumen of the intestine to the body cavity, liver, lungs or other tissues of the definitive or of an alternating host.

Heterakidae

The members of this family bridge the gap between the typical oxyurid life cycle and that of the ascaridids. At least one species, *Subulura brumpti* (see below), has become dependent upon an intermediate host.

The life cycle of *Heterakis gallinae*, according to Clapham (1933), is of the typical oxyurid type except that the females do not migrate out of the anus to deposit eggs, but oviposit in the ceca. Earlier writers have reported burrowing and encystment in the cecal walls, or penetration into cecal glands, but Clapham was unable to find any evidence of migration or burrowing, the larvae passing directly to the ceca within 48 hours and maturing in the lumen. The first molt occurs in the egg (Alicata, 1934), the third not until 10 days after infection.

Other species of *Heterakis* (*isolonche*, *beramporia*) burrow into the intestinal mucosa at some time during development and reach maturity in tumors which form around them. This possibly is the first step in the direction of the *Ascaris* type of life cycle.

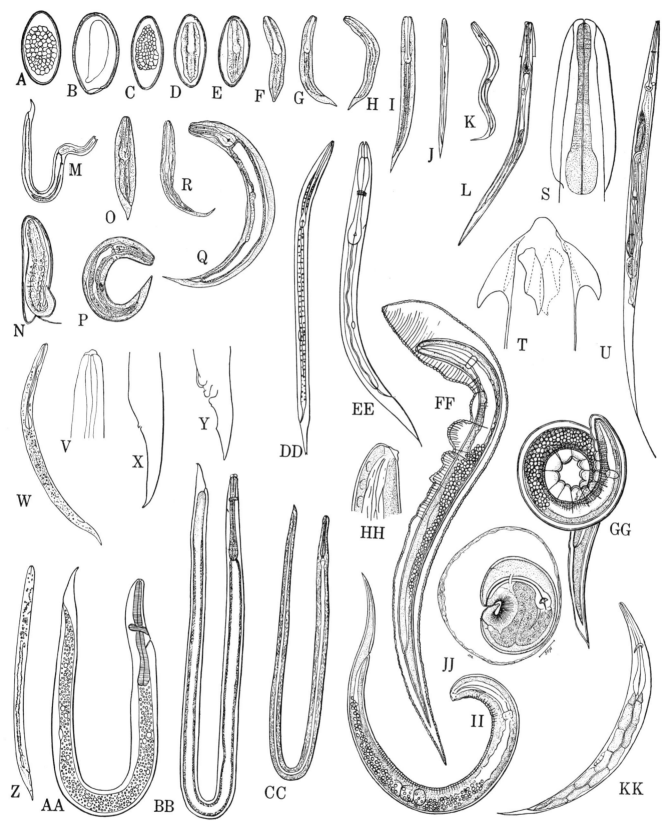

Fig. 186. DEVELOPMENT OF THE ASCARIDINA

A—*Passalurus ambiguus.* B—"*Oxyuris*" *brevicauda* showing emergence area and embryo in outline after incubation for 64 hours at 22° C. C-M—*Aspiculuris tetraptera* (C—Egg incubated in water 24 hours; D—43 hours; E—68 hours; F—Larva from intestine 4 hours after feeding; G—Larva from cecum 4 hours after feeding; H—After 18 hours; I—After 44 hours; J-K—10 days after feeding; J, male, K, female; L-M—18 days after feeding, L, female, M, male). N-Q—*Syphacia obvelata* (N—Uterine egg containing mature embryo; O—Hatched embryo; P—Youngest larva found in cecum; Q—Male measuring .81 mm.). R—*Enterobius vermicularis,* larva three days after hatching in Ringer's solution. S-T—*Dermatoxys veligera* (S—Head, T—Head, fourth stage). U—*Probstmayria vivipara,* lateral view of female containing a well developed embryo, a second less developed and two eggs. V-Y—*Ascaridia galli* (V—cephalic extremity of second stage larva showing oral prominence; W—Second stage, newly hatched; X—Tail of third stage female showing preanal swelling; Y—Tail of fourth stage

male). Z-CC—*Ascaris lumbricoides* (Z—Second stage (newly hatched); AA—Third stage; BB—Fourth stage (21 days); CC—Fifth stage (29 days old). DD-EE—*Cosmocercoides dukae* (DD—Newly hatched larva; EE—Infective larva). FF-II—*Contracaecum aduncum* (FF-GG—Hatched larvae; HH—Anterior end of larva armed with boring tooth; II—Larva from body cavity of *Ascartia bifilosa*). JJ—*Subulura brumpti,* encysted infective larva recovered from body cavity of the beetle *Alphitobius diaperinus.* KK—*Heterakis gallinae,* infective larva found newly hatched in the small intestine. A-R, after Philpot, J. Helminth., v. 2 (5), pp. 239-252. S-T, after Dikmans, 1931, Trans. Amer. Mic. Soc. v. 50 (4). U, after Ransom, 1907, Trans. Amer. Mic. Soc. v. 27. V-Y, after Roberts, 1937, Bull. 2 An. H. Sta. Queensland. Z-CC, after Roberts, 1934, Bull. 1, An. H. Sta. Queensland. DD-EE, after Harwood, 1930, J. Parasit. 17. FF-II, after Markowski, 1937, Bull. Acad. Polon. Ser. B, JJ, after Alicata, 1939, J. Parasit. 25. KK, after Clapham, 1933, J. Helminth. v. 11 (2).

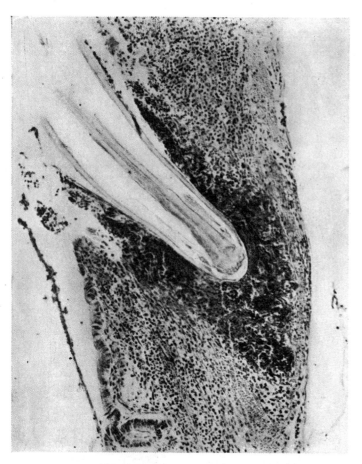

Fig. 187.

Dermatoxys veligera. Photomicrograph of fourth stage larva penetrating mucous membrane. After Wetzel, 1931, J. Parasit. v. 18.

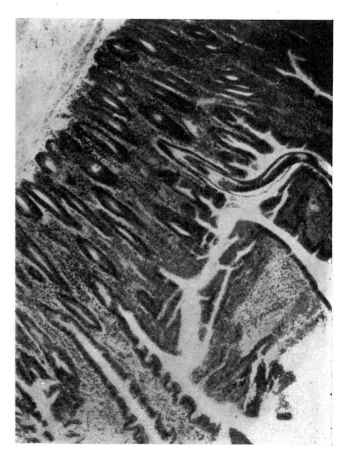

Fig. 188.

Ascaridia galli. Section of small intestine showing larva. After Roberts, 1937, Bull. No. 2. Animal Health Sta. Yeerongpilly, Queensland.

The life cycle of *Ascaridia galli* may well be a second step towards that of *Ascaris.* As elucidated by Ackert (1931), Alicata (1934) and Roberts (1937) this worm undergoes one molt in the egg and then normally remains enclosed in the egg until infection. There are three molts in the host, the first of these (second molt) occurring about 6 days after infection and the others at about 6-day intervals thereafter. After reaching the third stage, on about the ninth or tenth days, the larvae burrow down between the villi and penetrate into the glands of Lieberkuhn, the posterior ends of the bodies remaining free in the lumen. Itagaki (1927) observed that at certain seasons in Japan (midsummer and midwinter) the larvae habitually penetrated into the mucosa, about as described by Ackert and by Roberts, causing fibrous nodules, but that in spring and autumn they remain in the lumen. Roberts reported less tendency for the larvae to burrow into the mucosa in April and May than in November. Although on rare occasions the larvae penetrate too deeply and enter the peritoneal cavity, mesenteries, liver, or even the lungs (Ackert, 1923; Guberlet, 1924), it is clear that this is purely accidental.

Subulura brumpti, according to Alicata (1939), has departed from the usual heterakid life cycle pattern in requiring an intermediate host. This is the only member of the subfamily Subulurinae in which the life cycle has been investigated, and it is possible that the use of an intermediate host has become general in this group as it has in the Anisakinae.

Alicata was unable to infect chickens by feeding embryonated eggs, either just recovered from the uteri of gravid females, or incubated in water at about 24° C. for 1 week, but succeeded in producing infection by feeding naturally infected arthropods harboring the cysts in the body cavity. The cysts contain coiled nematodes having bulbed esophagi and conspicuous esophageal valves as in the adults (Fig. 186, JJ). High incidences of natural infection were found in the following arthropods collected on poultry farms in Hawaii: (Coleoptera) *Dermestes vulpinus, Gonocephalus seriatum, Ammophorus insularis, Alphitobius diaperinus;* and (Dermaptera) *Euborellia annu-*

lipes. Encysted larvae were also found in grasshoppers (*Conocephalus saltator*) 15 days after experimental infection.

COSMOCERCIDAE

At least some of the members of this family resemble the typical Ascarididae in that the larvae, burrowing into the mucosa, enter the circulatory system and reach the lungs, where they escape into the air spaces and eventually make their way back to the intestine via trachea and esophagus. They differ, however, in having a free-living phase outside the body. *Cosmocerca trispinosa* (= *Nematoxys longicauda*) has long been known to occur in the lungs of salamanders in an immature form, and in the intestine as an adult. Von Linstow considers its growth in the lungs as analogous to the growth of Anisakinae in an intermediate host. Harwood (1930) found that *Cosmocercoides dukae* (his *Oxysomatium variabilis*) undergoes a molt after 5 days of free life outside the body and, although his observations on development after infection are inconclusive, that the larvae are found in the lungs not only after subcutaneous inoculation but also after infection by mouth. They do not, however, penetrate the skin.

ASCARIDIDAE

The majority of the Ascarididae have a migratory phase before becoming adult in the intestine. The larvae, burrowing into the mucosa, enter the circulatory system and are carried via liver or lymphatic system to the heart, thence to the lungs where they become free in the air spaces, and thence via trachea and throat back to the alimentary canal. *Toxascaris leonina,* according to Wright (1935), does not perform this migration; the life cycle is similar to that of *Ascaridia* except that the second-stage larvae burrow into the mucous membranes almost immediately after hatching, and return to the lumen of the intestine after the third molt, on the 9th or 10th day. As shown by Fülleborn (1922) and others, some larvae penetrate all the way through into the body cavity and enter viscera by

Fig. 189.

Ascaris suum. Larva in section of mouse lung 1 week after infection. After Ransom, 1920, U.S.D.A. Yearbook.

this route, and some are probably picked up and carried by the circulatory system.

Ascaris lumbricoides. Long thought to have a direct development in the intestine, *Ascaris lumbricoides* was first shown to undergo a preliminary migration through the body by Stewart (1914); Stewart found that eggs fed to rats migrated to the lungs, and erroneously concluded that rats served as intermediate hosts. Shortly thereafter Ransom and Foster (1917) and Chandler (1918) called attention to the probability that the migration through the body was a part of the normal development in the definitive host; experimental proof, with details of the development, was supplied by Ransom and Foster in 1920. Details of the course of the migration were worked out and reported by Ransom and Cram in 1921, and further details were supplied by Roberts (1934).

The first-stage larva appears in the egg on about the eighth day at the optimum temperature of 30 to 33° C., and the first molt occurs in the egg on about the 18th day. Ransom and Foster (1920) first observed that the embryo underwent a molt in the egg. Later Alicata (1934) reported that the egg is not infective until after this molt; he also pointed out that the embryos of *Ascaridia lineata, Parascaris equorum, Toxocara canis, Toxascaris leonina, Heterakis gallinae* and the roach oxyurid *Blatticola blattae* also underwent a molt, a feature which may be common in the Ascaridina and which determines when the egg has reached the infective stage.

Normally the eggs of *Ascaris lumbricoides* hatch in the small intestine after being swallowed, but they will sometimes hatch when implanted subcutaneously or intraperitoneally (Ransom and Foster, 1920; Yoshida and Toyoda, 1938) or in artificial media containing glucose or various nitrogenous substances (Yoshida and Toyoda, *l.c.*).

The second-stage larva has a small, sclerotized, knob-like structure at the anterior end, called the "boring tooth." The larvae bore into the intestinal wall, mainly in the duodenum and upper part of the jejunum, after hatching; the majority have disappeared within 2 hours. The majority enter the blood stream after some hours and are found in the liver in from 18 hours to several days after infection. A few apparently enter lymphatics since they are sometimes found in mesenteric lymph glands, but from here they seem to go via mesenteric venules to the liver rather than directly to the lungs. Within 5 or 6 days all have left the liver and have gone to the lungs via the blood stream; some appear in the lungs within 18 hours, and they may continue to be found there for 10 or 12 days, although most numerous on about the third to fifth days. During the first 2 days of this migration the larvae grow considerably. About the fifth or sixth day the larvae in the lungs, measuring about 0.8 to 1 mm in length, undergo the second molt. The third-stage larva has three lips with papillae, lacks the boring tooth, has a highly developed muscular esophagus, has the intestinal cells packed with granules, has a distinct nerve ring and oval genital primordium, and a conical tail turned dorsad at the tip.

On the tenth to twelfth days the third molt occurs, also in the lungs. In the opinion of Roberts, although second- and third stage larvae may be found in the intestine prior to the tenth day (Ransom recovered larvae from the trachea as early as the third day), these larvae have not completed their development in the lungs and probably fail to establish themselves in the intestine. The suggestion is made that the occurrence of such larvae in the intestine may indicate unfavorable conditions in the lungs resulting from excessive infections. Roberts found some hundreds of fourth-stage larvae in the intestine on the 14th and 21st days, but no molting third-stage larvae were found between the 11th and 14th days. Fourth-stage larvae are 1.4 mm or more in length. The cuticle begins to show striations, fin-like lateral alae are present, the lips resemble those of the adult, the esophagus is less bulbous, and the sexes can be differentiated by a difference in length of tail. Rudimentary genital tubules are present in the body cavity.

After arrival in the intestine the larva grows enormously, reaching a length of 16 to 25 mm 29 days after infection. Larvae undergoing the fourth molt measure 17.5 to 22.5 mm (Roberts). The lateral alae have become inconspicuous, the genital tubules and body-wall muscles are comparatively well developed, and the characteristic features of the tail of the two sexes are present. Growth to maturity and beginning of reproduction takes several weeks.

It is obvious that the only striking difference between this life cycle and that of the heterakids is the entrance into the circulatory system when burrowing into the intestinal wall, the consequence of which is the migration through the body via liver, heart and lungs. The determining factor seems to be the age at which the larvae do their burrowing. *Enterobius* and *Dermatoxys*, as we saw, burrow as fourth-stage larvae, and some species of *Heterakis* do likewise and live as adults in the burrows; *Ascaridia* burrows while in the third stage; but *Ascaris* burrows immediately after hatching as a second-stage larva. The burrowing heterakid larvae are too large to enter or be sucked into blood vessels, whereas the *Ascaris* larvae can easily do so. The failure of *Toxascaris* larvae to enter the circulatory system except rarely may be found to be due to a difference in size, particularly in the diameter of the larvae.

Toxocara canis has essentially the same life cycle as *Ascaris lumbricoides*, and the same is true of *Neoascaris vitulorum* (*vide* Schwartz, 1922), of *Parascaris equorum* (*vide* Baylis, 1923), of *Ascaris columnaris* (*vide* Goodey and Cameron, 1923), and probably of all other Ascaridinae. According to Fülleborn (1921), *Toxocara canis* is frequently encapsulated in the tissues of mice or other abnormal hosts, which thereby become transport hosts.

ANISAKINAE

It has long been known that various members of this subfamily occur as immature worms in the body cavity, mesenteries and other organs of various vertebrates, and sometimes invertebrates, whereas the adults occur in vertebrates which prey upon these hosts. Although morphological characters often suggested affinities between larvae and adults there was little experimental evidence in support of them. Moreover the various larval forms were not clearly differentiated from each other. Baylis (1916) for instance, showed that a number of larval forms were confused under the name "*Ascaris capsularia*," which he believed on morphological and distributional evidence to be the larval form of "*Ascaris decipiens*" (now *Porrocaecum decipiens*). The same confusion probably holds for other species.

Thomas (1937a, 1937b) experimentally worked out the life cycle of *Contracaecum spiculigerum*. Eggs obtained from the proventriculus of a cormorant contained active molted larvae with a boring tooth after being incubated in water for 5 days, and on the sixth day they molted a second time and then hatched. Many attached themselves by the anterior end of the sheaths, which seemed adhesive, but they swam freely when detached. On the thirteenth day a third molt was in progress, with a cuticular tooth still present. When swallowed by tadpoles or guppies (*Labistes reticulatus*) the larvae shed their sheaths and were found free in the intestine or in the body cavity. About 3 months later larvae were found encysted in the mesenteries; they had grown to 1.3 mm in length (from less than 400 μ). In cysts developed by the host tissues they continue to grow until nearly adult size is reached. Unlike most nematodes the number of molts is not limited to four; as many as eight molted cuticles have been removed from encysted worms from a natural infection.

There is evidence that when an infected fish is eaten by another fish the larvae penetrate the intestinal wall and re-encyst in the mesentery. This was observed to occur when a parasitized guppy was fed to a black bass. In all cases the worms retain the cuticular "boring tooth" until the definitive host is reached, although three lips can be seen under the cuticles in older larvae. Natural infections with similar worms were found in several species of fish in Illinois. Sexual maturity is reached only in birds. Fledgling cormorants become infected when fed on infected guppies. The larvae at first penetrate into the glands of Lieberkuhn, and when fish are present in the ventriculus they leave the glands and penetrate into the food during its digestion.

Kahl, 1936, investigated the life cycle of *Contracaecum clavatum* and concluded that it can undergo partial development in a great variety of intermediate hosts, including *Sagitta*, Calanidae, amphipods and medusae among invertebrates, and in *Ammodytes* and *Merangus* among fishes. Wülker, 1929, thought there was a succession of three hosts,—plankton, plankton-eating fish, and piscivorous fish, but Kahl thinks that all three hosts are not necessary; development to the stage infective for the definitive hosts can take place directly in such fish as *Merlangus merlangus*. *Merlangus* can also serve as a definitive host, if infective larvae are swallowed with the flesh of smaller intermediate hosts.

Markowski (1937), influenced by Wülker's work, found that certain species of copepods served as first intermediate hosts for *C. aduncum*, and presented evidence for the view that a variety of plankton-eating or carnivorous fish might serve as second intermediate hosts, although he expressed doubt that the larvae developing in the parenteral organs of a fish would develop to maturity in the intestine of the same fish, even if it were a suitable host. Markowski did not consider the possibility of a plankton host being unnecessary. According to Kahl the larvae undergo their early development in the intestine of the intermediate hosts, and then, when about 5 mm long, acquire a boring tooth and penetrate into the body cavity where they molt again, but retain the sheath with tooth and posterior spine until eaten by the final host. Essentially then, the life cycle of this species is similar to that of *C. spiculigerum*, although according to Kahl the eggs develop embryos only after being swallowed by a host. For a species living in marine hosts this might be necessary. It is probable that all the species of *Contracaecum* conform very closely to the same pattern.

Thomas (1937c) worked out the life cycle of *Rhaphidascaris canadensis*. The eggs of the species may become embryonated after 8 hours outside the host and are infective within 24 hours, after one molt within the egg. When eaten by nymphs of dragonflies, these eggs hatch, the first cuticle is shed, and the larvae penetrate into the body cavity. Infected nymphs caused infection in guppies, which in turn caused infection in fingerling muskelunge. In Douglas Lake the livers of all yearling *Perca flavescens* are full of *Rhaphidascaris* cysts, whereas the plankton-feeding fingerlings are free of infection. Guppies can be infected directly by the embryonated eggs, the intervention of an invertebrate host apparently being unnecessary, as in the case of *Contracaecum aduncum*. In small bottom-feeding or nymph-eating fish, then, they become encapsulated in the mesenteries and liver and continue growth until eaten by species of *Esox*, in which the cycle is completed. *R. acus* of Europe presumably has a similar cycle, since the larvae are found in the inner organs of various cyprinoid, salmonid and percid fishes, whereas the adults are found in *Esox, Perca, Alosa* and *Anguilla*.

The observation of Baylis on the probable relation between *Porrocaecum decipiens* of seals and walruses and encapsulated larvae in various fishes have already been mentioned. A number of European writers have reported encysted larvae of *Porrocaecum* in insectivores (moles, shrews, desman) and Schwartz (1925) has reported them from under the skin of moles and shrews in the United States; he, and also Solonitzine, who has found the larvae of a *Porrocaecum* on the serous surface of the stomach of a desman (*Desmana moschata*), think the adult stage is probably reached in a bird of prey.

Walton (1936a) found evidence for a similar life cycle for *Multicaecum tenuicolle*. Encysted larvae were found in species of *Rana* and in *Siren;* 3 weeks after being fed to a young alligator, presumably parasite-free, several immature males and females were found. A similar cycle was found by Walton (1936b) for *Ophidascaris labiatopapillosa;* the larvae were encysted in mesenteries and muscles of *Rana* spp., the adults developing in *Natrix* spp. Similar larvae encysted in muscles of *Amphiuma*, however, failed to develop in *Natrix*. Ortlepp (1922) failed to get larvae of *O. filaria* to penetrate the mucous membranes when the ripe eggs were fed to a mouse, although those of *Polydelphis anoura* migrated to liver and lungs like typical Ascaridinae.

SPIRURINA

SPIRUROIDEA

The members of this superfamily, with a few exceptions, show a striking degree of uniformity in the general features of their life cycles. Although many species tend to live in the walls of the alimentary canal or in more distant locations in the body, the eggs, usually embryonated, escape with the feces, and usually hatch only after being eaten by an intermediate host. The embryos of *Habronema*, however, hatch before escaping from the body. In most cases there is some degree of specificity with respect to the intermediate host, but usually it is not very close. After ingestion by the intermediate host the first-stage larvae emerge from the egg, penetrate into the body cavity or tissues, undergo two molts, and become encapsulated as third-stage larvae. These larvae are usually not sheathed, as are the larvae of metastrongyles; the second cuticle is not needed as a protection, since this is provided by a capsule produced by the host, so is completely shed.

Infection of the definitive host is nearly always by ingestion of the infected intermediate host, although an alternative method occurs in the case of *Habronema* (see below). Not infrequently transport hosts may intervene between the true intermediate host and the definitive host, and it is possible that this can occur in all spiruroids. When the larvae are eaten by a host in which the worm is unable to reach maturity they burrow through the walls of the alimentary canal and become reëncysted. In most cases this seems to be an optional course of development which is frequently favorable to ultimate access to a definitive host (e.g., *Spirocerca, Habronema mansioni*) but in the case of at least one species, *Gnathostoma spinigerum*, a second intermediate host has apparently become indispensable in the life cycle. After reaching the final host the worms undergo two more molts before reaching maturity. Being too large to enter blood vessels in the intestinal wall, they usually reach their destination, if this is outside the alimentary canal, by direct migration through tissues or along natural passageways.

Gongylonema pulchrum will serve as an example of a typical spiruroid life cycle. *Gnathostoma spinigerum* and *Draschia megastoma* will serve to exemplify two important variations.

GONGYLONEMA PULCHRUM

The adult worms live imbedded in the mucous membranes of the esophagus, tongue and oral cavity. The eggs escape into the lumen and leave the body with the feces in a fully embryonated condition. No further development takes place until the eggs are ingested by a suitable intermediate host. This may be any of a large number of beetles, particularly scarabaeids, or cockroaches. Twenty-four hours after ingestion by *Blatella germanica*, according to Alicata (1935), empty egg shells are found in the crop and intestine. The absence of larvae in the lumen or wall of the intestine and the presence of a few still adhering to the wall of the crop, apparently ready to invade the body cavity, suggests that hatching takes place in the crop, and that the larvae find their way into the body cavity by piercing the wall of the crop. Forty-eight hours after ingestion of eggs, first-stage larvae are found in the body cavity, especially in the thoracic region.

The newly hatched first-stage larva is cylindrical with a spine and a small hook near the anterior end on the ventral side, behind which about 20 rings of minute spines encircle the anterior end of the body (Fig 190B); the tip of the blunt tail is encircled by 8 to 10 small refringent points, a character which is diagnostic of the first-stage larva. The filariform esophagus and intestine are about equal in length, both transparent. The larvae wander about in the body cavity and grow to double their original length in about 2 weeks, and at this time are preparing for the first molt (Ransom and Hall, 1916; Alicata, 1935). The actual molt, according to Alicata, does not occur until about the 19th day.

The second-stage larvae lose the cuticular armature at the anterior and posterior ends, which are bluntly rounded. The slender esophagus occupies about one-half the body length, and in older larvae becomes differentiated into an anterior muscular portion and a posterior glandular portion. These larvae increase in size to a length of 1.5 to 2 mm by the end of the fourth week, when they begin the second molt. At about this time they usually penetrate the muscles of the body wall, and sometimes, in heavy infections, other muscles, and they may become partially encysted prior to the second molt.

Third-stage larvae are found encysted at the end of about a month. This stage is distinguished by a raised lateral bor-

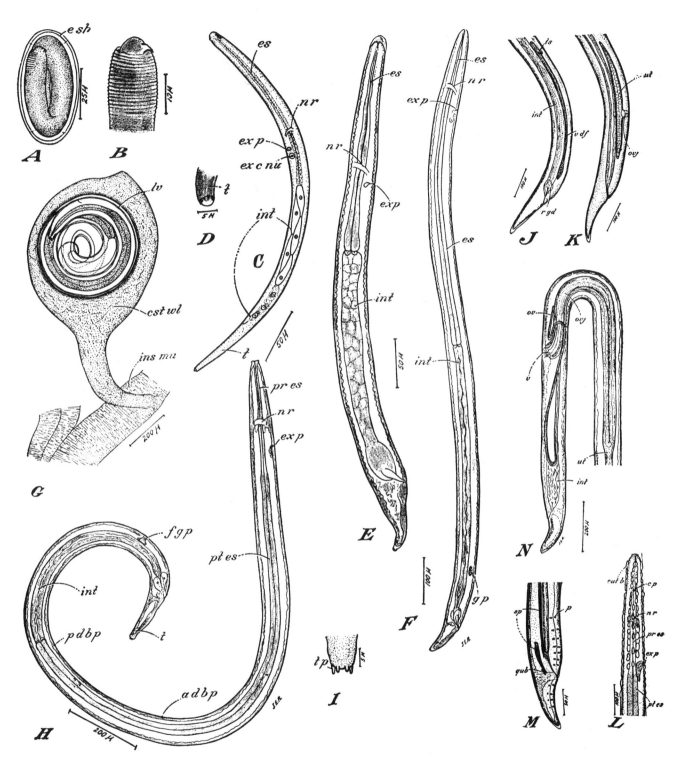

Fig. 190.

Development of *Gongylonema pulchrum.* A—Egg with fully developed embryo; B—First stage larva, anterior end; C—First stage larva from intermediate host, four days after experimental infection; D—Tail, lateral view; E—First stage larva undergoing first molt; F—Second stage larva; G—Third stage larva encysted in musculature of roach (*Blatella germanica*); H—Third stage, lateral view; I—Posterior end showing four digitiform processes; J—Posterior end of male undergoing third molt; K—Posterior end of female undergoing third molt. L—Fourth stage larva, anterior end; M—Posterior part of male in fourth molt; N—Region of vulva of larva undergoing fourth molt. After Alicata, 1935, U.S.D.A. Tech. Bull. 489.

der of the mouth and by four, occasionally only two, small digitiform processes on the tail. The larvae are found imbedded within the sarcoplasm surrounding a muscle fiber. As the cysts become well formed they are sometimes pushed out into the body cavity, remaining attached to the muscle by a thin strand, or eventually falling free. Baylis (1926) found that the larvae would escape from disintegrating cockroaches into water, and could be kept alive for a number of days, but since the larvae settle to the bottom he concluded that drinking water was not an important means of infection. Freed larvae were found to be incapable of skin penetration. The possibility exists, of course, that larvae, either in or out of their intermediate hosts, might reëncyst in some transport host; Alicata (l.c.) cites the finding of third stage larvae in the stomach wall of a mole.

Upon ingestion by a definitive host (Alicata used guinea pigs for experimental infections) the larvae are liberated in the stomach and may invade the esophagus within one-half hour after feeding, entering through the tissue at the gastroesophageal junction. They migrate upward through the epithelium of the esophagus and may reach the tongue as early as the third day. Larvae begin the third molt on the ninth day after ingestion, and many fourth-stage larvae are present by the twelfth day. These larvae are characterized by development of the reproductive organs, gradual development of the characteristic cuticular bosses at the anterior end, and loss of the caudal appendages. The final molt occurs about a month after infection; the minimum time required for growth to maturity seems not to have been determined definitely, but Ransom and Hall (l.c.) report the finding of egg-bearing females in a sheep about 3 months after infection, and Alicata (1935) obtained an adult male 70 days after infection.

GNATHOSTOMA SPINIGERUM

The adult worms live in tumors in the wall of the stomach of Felidae, or of the esophagus of mink, the eggs escaping into the alimentary canal through openings which eventually develop from the tumors into the lumen. The eggs escape from the body in an early stage of development (one to two-celled stage according to Prommas and Daengsvang, 1933; one to many-celled according to Refuerzo and Garcia (1938). In aerated water they become embryonated in a minimum of about 4 or 5 days, and in 2 days or more thereafter the embryos emerge from the egg in an ensheathed condition, being, therefore, in the second stage. These larvae have smooth cuticles devoid of spines or striations, and are armed with a spine at the anterior end.

The larvae usually live for only a few days in tapwater (Prommas and Doengsvang, 1933) although sometimes they may live for a month or more (Yoshida, 1934). Further development is known to occur only when the larvae are ingested by *Cyclops*. Attempts to infect mammals, fish, frogs, fleas and Cladocera have all been negative. The development of the larvae in *Cyclops* was independently discovered by Prommas and Daengsvang (l.c.) in Siam and by Yoshida (1934) in Japan. These workers showed that sheathless motile larvae were found in the stomachs of *Cyclops* soon after experimental exposure, and that by the following day they could be found in the body cavity. According to Refuerzo and Garcia (1938), the larvae in the body cavity 1 day after infection lose the sclerotized oral spine, and a fleshy enlargement representing the future lips develops at the anterior end. Three days later the cuticle becomes striated, its armature of spines develops, and a head bulb armed with four rows of spines, and connected with cervical sacs, is also present. The larvae seem to have completed their development to the infective stage by the sixth day.

Attempts to infect cats by feeding them infected *Cyclops* have been uniformly negative (Yoshida, 1934; Prommas and Daengsvang, 1936) but Prommas and Daengsvang succeeded in infecting a catfish, *Clarias batrachus*. The larvae were found in the muscles of the stomach or intestine of the fish 2 to 6 days after infection and after 6 days or more they were found, some free and some encysted, in body muscles. Chandler (1925a) had reported the presence of numerous gnathostome cysts in the mesenteries of Indian snakes, which he found to undergo further development in cats (1925a) until the adult morphology of *Gnathostoma spinigerum* was reached (1925b); Chandler also called attention to reports of probably identical larvae in pelicans and eagles. Subsequent to the work of Prommas and Daengsvang many other intermediate hosts, natural and experimental, have been added, including a considerable variety of fresh-water fishes, frogs, and snakes. In all of these the larva undergoes considerable growth, but does not develop more than 4 rows of spines on the head bulbs, in contrast to the 8 to 11 found in the adults

of *Gnathostoma spingerum*. It is probable that the larvae always become encysted ultimately.

Chandler (1925a) showed that when gnathostome cysts in snakes are fed to cats they penetrate through the alimentary canal and can be found parenterally within 2 days after infection. Some are found free in the abdominal cavity, under the parietal peritoneum, or in the capsules of the kidneys, but the majority, and nearly all later in the infection, are found burrowing in the liver. A single larva was also found in the liver of an experimentally infected guinea pig. The larvae in the livers of cats grow somewhat, and a vulva and rudimentary genital tubes develop within 6 days. No further development was observed in cats infected for as long as 4 weeks, although in the meantime there was extensive damage done to the liver. Subsequently (1925b) Chandler found, in naturally infected cats, all stages of development from (presumably) fourth-stage larvae burrowing in the liver, exactly like those obtained from experimental infections, to forms, still sexually immature, which had undergone the final transformation to the adult morphology, and had 8 to 11 rows of hooks on the head bulb, and complex spines on the body. Some of the worms which had undergone the final molt were found still in the liver, but others were evidently migrating out of the liver; a few were found in the mesentery or in the diaphragm, and several were in the stomach wall; one was free in the stomach. The worms in the stomach wall were not yet enclosed in hard-walled tumors, but occurred in submucous purulent cavities. It was evident from these observations that the worms, upon gaining access to a definitive host, migrate through the walls of the stomach or intestine to the abdominal cavity and enter the liver, where they burrow and feed actively for several weeks. They finally enter the wall of the stomach from the peritoneal side, and grow to maturity.

Africa et al (1936a) fed rats with encysted larvae and found the larvae in the liver and body muscles 8 to 25 days later. Infection of cats fed on gnathostoma cysts from cold-blooded hosts has been confirmed by Prommas and Daengsvang (1937), the prepatent period being 28 to 32 weeks, and by Africa et al (1936b), who found semi-mature worms in the diaphragm and in nodules in the stomach wall nearly 4 months after infection. It is clear that the formation of a tumor about the worms in the wall of stomach or esophagus, which finally opens into the lumen, is a late stage of development. It also seems evident from observations made by the writer (1925b) that these tumors, when in the stomach of cats, frequently become perforated into the peritoneum and are then fatal. Yoshida's observation that in mink the tumors form on the esophagus in the lower part of the thoracic cavity suggests that this may be the normal host and habitat, and that in these circumstances there is less danger of fatal parenteral perforation.

DRASCHIA MEGASTOMA

The life cycle of this worm is of particular interest since it represents an intermediate evolutionary step from that of the typical spiruroids to the filariae. It was first worked out in detail by Roubaud and Descazeaux (1921).

The female deposits embryonated eggs in the alimentary canal which, according to Roubaud and Descazeaux, hatch before leaving the body of the host. The first-stage larvae possess a hooklike structure similar to the hook of *Gongylonema*, but the larvae are in a very immature state, the contents of the body being granular in appearance, with no differentiation. These larvae are ingested by young maggots of flies, and there seems to be a fairly high degree of specificity. *Draschia megastoma* and *Habronema muscae* have been found to be capable of development in a number of species of *Musca* and also in *Muscina stabulans* and in *Fannia*, but actual transmission has been observed only in *Musca domestica*, and was definitely found by Roubaud and Descazeaux to fail in the case of *Muscina stabulans* because of inability of the larvae to escape from the proboscis of that species. *H. microstoma*, on the other hand, develops primarily in *Stomoxys*, but has been reported as developing in *Sarcophaga*, *Lyperosia* and *Musca* as well, though Roubaud and Descazeaux (1922b) state that it does not reach the infective stage in *Musca domestica*. Development of *Habronema* larvae has also been reported from *Drosophila*.

The ingested larvae bore through the walls of the alimentary canal of the maggots and enter the body cavity. They live free in the body cavity for only a brief time, and about the third day they penetrate into the Malpighian tubules. Here they become quiet, and undergo the first molt on the third or fourth day after ingestion. They lose the oral hook, become immobile, and grow very thick and sausage-

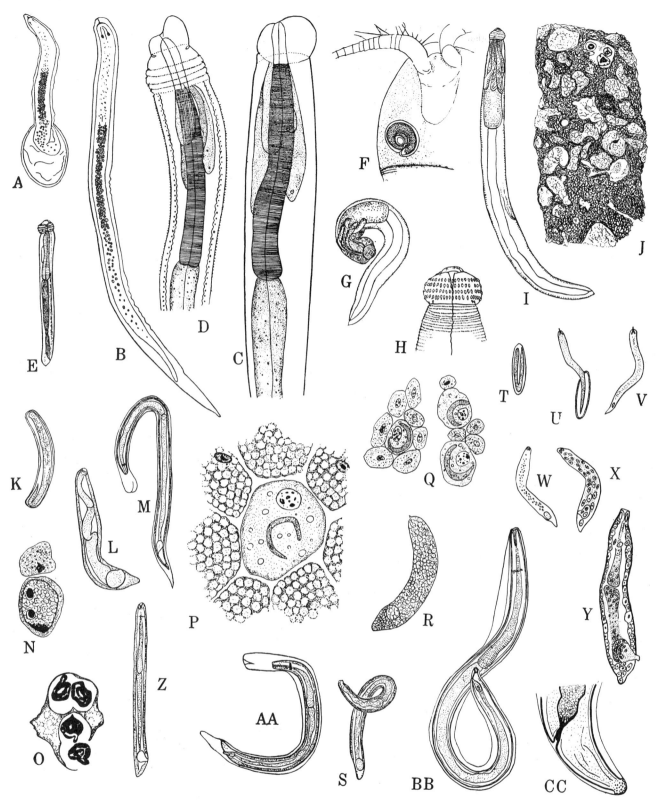

Fig. 191. DEVELOPMENT OF SPIRUROIDEA

A-J—*Gnathostoma spinigerum* (A—Larva emerging from egg through opercular end; B—Newly hatched larva with loose enveloping sheath, anterior spine; C—Anterior end of larva dissected from cyclops on first day of infection, no anterior spine but large fleshy lip, two pairs of contractile cervical sacs; D—Larva from Cyclops on fourth day of infection, cephalic bulb with four rows of minute spines, lip smaller; F—Larva in body cavity of cyclops; E—Same larva; G—Larva from cyst in mesentery of cobra; H—Head, bulb and lips of larva from liver of artificially infected cat; I—Gnathostome from liver of artificially infected cat; J—Section of liver of cat showing riddling of tissue by burrowing gnathostomes). K-M—Stages in the development of *Habronema muscae*. (K—Egg with embryo; L—Second stage larva; M—Third stage larva before the molt). N-O—Sections pointing out the histological reaction of the fat cells parasitized by *Habronema muscae* (N—Fat cells at the beginning of the infection showing peripherial thickening and hypertrophy of parasitized cell in relation to normal; O—Section of a fat sac enclosing many parasites). P—Fragment of fat tissue of larva of *Musca domestica* showing aciculate larva of *H. muscae* in hypertrophied and transformed fat cell. Q-S—Development of *H. microstomum* in Stomoxys. (Q—Group of adipose cells of the larva of *Stomoxys* of which three are infested with a young larva of *H. microstomum*; R—Sausage shaped larva; S—Older second stage larva). T-CC—*Draschia megastoma*. (T—Embryonated egg; U—Aciculate larva emerging from egg shell; V—Aciculate larva in intestine of fly; W—Second stage larva immediately after molt in malpighian tubules of fly larva; X-Y—Second stage larvae recovered from larva (X) and pupa (Y) of fly; Z—Full grown second stage larva; AA—Second stage larva about to molt; BB—Second stage larva, full grown and about to molt, removed from malpighian cyst; CC—Posterior end of same). A-D, after Refuerzo & Garcia, 1938, Philip. J. An. Ind. (5 (4)). E, F, after Prommas and Daengsvang, 1933, J. Parasit. G-J, after Chandler, 1925, Parasit. v. 17. KCC, after Roubaud and Descazeaux, Bull. Soc. Path. Exot., v. 14, 15.

like. An outline of the stoma appears at the head end, and a conspicuous caudal vesicle and outline of the pyriform rectum at the posterior end, but throughout the rest of the body the nuclei are still scattered without definite order. Gradually during the next few days the worm elongates, and the alimentary canal, nerve ring, and rectum become well developed. Meanwhile the tissue of the wall of the Malpighian tubule surrounding the larva degenerates and is finally reduced to a mere membrane, which serves as a sheath. On the eighth day, at about the time of emergence of the adult fly, the larvae begin to break loose into the abdominal cavity, still enclosed in the membrane, but they now molt a second time and their movements become very active, resulting in their soon freeing themselves.

These third-stage larvae, the infective forms, may appear as early as the ninth day. They migrate forward to the head of the fly, and collect in the interior of the labium. Attracted by warmth and moisture they move down into the labellum, and escape through the delicate membrane between the lobes of this structure when the fly is resting on a warm wet surface, e.g., the lips, nostrils or wounds of an animal. If on the lips the larvae have an opportunity to reach the stomach via the mouth, and grow to maturity in a normal manner, but from the nostrils they reach the lungs, and from the skin the subcutaneous tissues, and in either case fail to grow to maturity. There is no doubt but that animals could also be infected by swallowing flies harboring infective larvae, but in the case of habronemiasis of horses this would probably not be a common method in nature. On the other hand it would probably be the principal if not the exclusive method in the case of habronemiasis of insectivorous birds. Still another possibility—ingestion by a transport host—is suggested in the case of habronemiasis in birds of prey; this is supported by the finding of abundant larvae of *H. mansoni* encysted in the stomach walls of toads by Hsü and Chow (1938). This species had previously been recorded from the bearded vulture, *Gypaetus barbatus*, but several species of falcons were experimentally infected by feeding them larvae from toads.

Habronema muscae and *H. microstomum* have similar life cycles (*vide* Roubaud and Descazeaux, 1922a), but different in details. These two species, instead of undergoing development in the Malpighian tubes, develop in cells of the fat body, the thickened walls of the cells serving as temporary "cyst" walls. *H. microstomum*, which develops in the blood-sucking *Stomoxys*, might be expected to be introduced into the tissues when the insect pierces the skin, and be forced to find its way to the stomach by some roundabout parenteral route, but Roubaud and Descazeaux (1922b) point out an interesting biological adjustment which makes this unnecessary. They point out that, as a result of interference by the worms in its proboscis, the fly is unable to rasp a hole in the skin to suck blood, and is forced to revert to the habits of its ancestors and non-blood-sucking relatives, and obtain moisture and nourishment from the lips or other exposed moist surfaces.

The failure of the larvae of *Habronema* to become encysted in the intermediate host, there to remain until eaten by a definitive host, and the substitution of a voluntary exit from this host in response to warmth and moisture, are definite steps in the direction of a filarial life cycle. As remarked by Roubaud and Descazeaux (1922b), however, the habronemas are imperfectly adapted for parenteral parasitic life. Their larvae, in spite of the fact that they leave the body of the intermediate host on the surface of the body of the definitive host, are unable to penetrate the tissues, and are unable to reach maturity outside the alimentary canal. With (1) development of a parenteral adult habitat (already attempted by many spiruroids but always hampered by the necessity for the eggs to reach the alimentary canal), and (2) development of ability to enter the skin on the part of the infecting larvae, the only important change necessary to bring about a filarial life cycle is the substitution of the blood or skin for the alimentary canal as a means of exit for the larvae. Such a development could hardly fail to occur in the case of a parenteral parasite with a blood-sucking intermediate host.

OTHER SPIRUROIDEA

The life cycle of the majority of the Spiruroidea in which it has been determined conforms in general pattern to that of *Gongylonema*, except for the intermediate hosts involved. In some cases there seems to be far less specificity with respect to intermediate hosts than in others, but some instances of apparent specificity are probably due to incomplete data. Thus *Cheilospirura hamulosa* was not known to develop in anything but grasshoppers until Alicata (1937) showed that an amphipod and 10 species of beetles belonging to 7 different families, as well as several grasshoppers, could be utilized as intermediate hosts by this worm. On the other hand, Cram (1931) got negative results from feeding eggs of *C. spinosa* to cockroaches, ground beetles, sowbugs and crickets, but obtained development in two species of grasshoppers. Again, whereas *Tetrameres fissispina* is reported as capable of development in grasshoppers, roaches, *Daphnia*, *Gammarus* and earthworms, Swales (1936) found that the eggs of *T. crami* failed to hatch in various species of Cladocera, but developed readily in two species of amphipods. Members of the genera *Ascarops*, *Physocephalus* and *Spirocerca* seem to develop primarily in dung beetles; *Spirura*, *Protospirura* and *Gongylonema* in beetles or roaches; *Oxyspirura* in roaches; *Seurocyrnea* in roaches and grasshopper nymphs; *Acuaria* in grasshoppers; *Tetrameres* in various Orthoptera and Entomostraca; *Hartertia* in termites (workers); *Echinuria* in Cladocera, *Dispharynx* and *Hedruris* in isopods; *Cystidicola* in amphipods, and *Spiroxys* in copepods. Spiruroid larvae, possibly *Protospirura*, have been found in fleas also. Under experimental conditions *Physaloptera turgida*, according to Alicata (1938), is able to develop in cockroaches, but there is a possibility that other arthropods are utilized under natural conditions.

Spiroxys contorta, as reported by Hedrick (1935), differs from the majority of the Spiruroidea but resembles *Gnathostoma* in that the eggs become embryonated in water after leaving the body of the host. It differs from *Gnathostoma*, however, in that the definitive host can be infected directly by the third-stage larvae in *Cyclops*, without requiring a second intermediate host. In nature, however, transport hosts —fish, tadpoles, frogs, newts and dragonfly nymphs, and frequently turtles as well—are commonly made use of. The larvae of this worm are further peculiar in that they continue to grow after they reach the infective stage, both in *Cyclops* and in the various transport hosts. The development of a "sausage" form by the late first-stage larva of *Oxyspirura mansoni*, as figured by Kobayashi, is highly suggestive of *Habronema* or the filariae.

As far as known at present *Gnathostoma spinigerum* is the only spiruroid which *requires* a second intermediate host, but it is quite possible that this will be found to be true of other Gnathostomatidae as well, and perhaps of still other spiruroids. The larvae of *Echinocephalus* (family Gnathostomatidae) have been found encysted in the tissues of a bivalve, *Margaritifera vulgaris*, which is presumably the first intermediate host. Similar larvae have been found in a sea urchin. Since the adults occur in oyster-eating fishes no second intermediate host may be necessary.

The course of migration in the definitive host is usually, as noted above, by burrowing directly through tissues or natural cavities, or by migration along natural passageways. The path of *Oxyspirura mansoni* to the eye, according to Fielding (1926), is by way of esophagus, mouth and lachrymal duct, the larvae sometimes arriving in the eye 20 minutes after infected roaches are fed to chicks.

The migration route of *Spirocerca lupi* (=*sanguinolenta*) is not so clearly known. Faust (1927) thought that the larvae, after ingestion with the flesh of a transport host (hedgehog), reach the aorta via the portal system and lungs, but does not

---

Fig. 192.

Development of Ascaropsinae larvae. A-E—*Ascarops strongylina* (A—First stage larva, anterior end, lateral view; B—Larva recovered from an intermediate host three days after experimental infection; C—Larva undergoing first molt; D—Third stage larva, lateral view; E—Encysted larva, third stage). F-K—*Physocephalus sexalatus* (F—Anterior end, lateral view; G—Larva from intermediate host 2 days after experimental infection; H—Larva from intermediate host 12 days after experimental infection; I—Larva undergoing first molt; J—Encysted third stage larva (from Hobmaier, 1925); K—Third stage, tail). After Alicata, 1935, U.S.D.A., Tech. Bull. 489.

Fig. 193.

A-C & G—*Spiroxys contortus*; (A—Free-living larva with sheath; B—Five-day old larva from cyclops; C—*Cyclops leuckarti* with three larval nematodes; G—Fully developed larva from body cavity of cyclops, showing genital primordium). D-E—*Dispharynx spiralis* (D—Head; E—Tail). F—*Tetrameres americana*, tail of third stage larva. H-I—*Tetrameres crami* (H—Third stage larva from *Gammarus fasciatus* 32 days after infection; I—Diagrammatic illustration of papillae on tail of third stage larva). J—Larval spirurid larva from cat flea. K-M —*Protospirura muricola* (K—Lateral view of anterior extremity of infective larva; L—Lateral view of tail of 3.5 mm. specimen; M—Freehad sketch of rosette of papillae on tail of same). N-P—*Oxyspirura mansoni* (N—Larva just after hatching; O—Larva at end of first larval stage; P—Mature larva). Q—*Habronema mansoni*, larva. A-C, & G, after Hedrick, L. A., 1935, Tr. Am. Mic. Soc. v. 54(4). D-F, after Cram, E. B., 1931, U.S.D.A. Tech. Bull. 227. H, I, after Swales, 1936, Canad. J. Res. D. 14. J, after Alicata, J. E., 1935, J. Parasit. v. 21 (3). K-M, after Foster, A. O., and Johnson, C. M., 1939, Am. J. Trop. Med. v. 19 (3). N-P, after Kobayashi, H., 1928, Taiwan Igakk. Zasshi Formosa, No. 280. Q, after Hsü, H. F., and Chow, C. Y., 1938, China Med. J. Suppl. II.

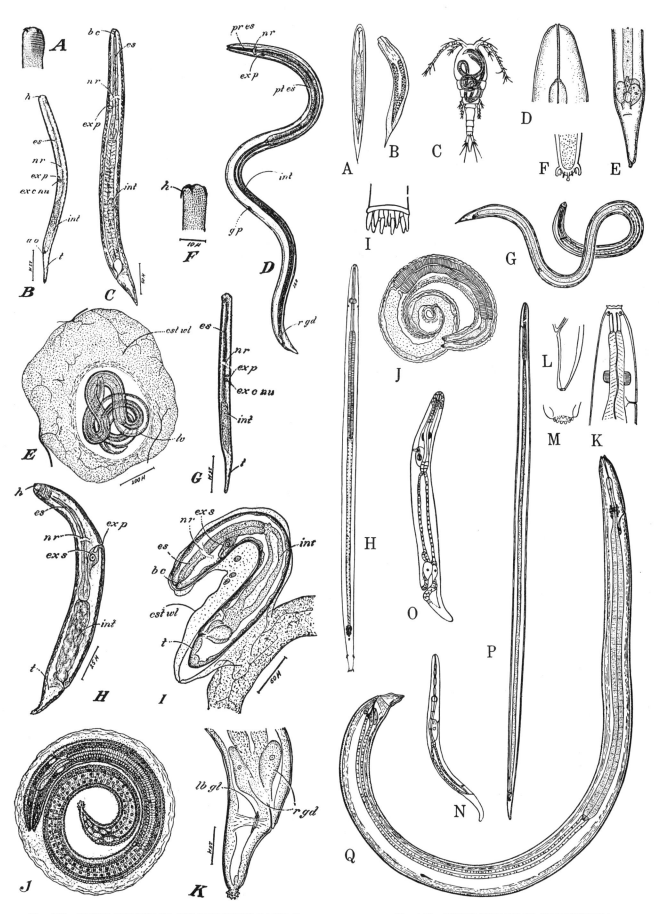

Fig. 192. DEVELOPMENT OF ASCAROPSINAE LARVAE

Fig. 193. DEVELOPMENT OF SPIRUROIDEA

make it clear how a worm 150 μ in diameter is able to pass through capillaries, or why the worms appear in the abdominal aorta before the thoracic, and never cause lesions in vessels anterior to the aortic arch. It seems far more likely that the larvae follow the route indicated by Hu and Hoeppli (1936); after penetrating the gastric wall they proceed to the coronary, gastroepiploic and coeliac arteries, and via these to the upper abdominal and lower thoracic portions of the aorta, eventually reaching the upper thoracic aorta from below.

In the aorta the worms attach themselves to the wall and cause the formation of characteristic nodules. Some worms remain in this position but many migrate outward through the aortic wall and through the intervening tissue until they reach the esophageal wall, in which they find a favorable habitat in which to reach maturity and reproduce. The eggs reach the lumen of the esophagus through a secondary opening from the tumor in its wall.

FILARIOIDEA

The Filarioidea are unique among nematodes, so far as is known at present, in having perfected a mechanism by which both exit from and entrance to a host takes place through the skin. The larvae of Dracunculoidea escape through the skin, though by a different mechanism, and the habronemas succeed in infecting a host when deposited on certain areas of skin (the lips) but in neither case is both exit and entrance accomplished by way of the skin. As noted under the discussion of Spiruroidea, the evolutionary process by which the life cycle of filariae developed is clearly foreshadowed by the course of events in the case of *Habronema*.

WUCHERERIA BANCROFTI

Manson's (1878) discovery of the ingestion of filarial embryos by mosquitoes and their development in these insects set a landmark in the history of medical entomology, since it was the first instance of a human blood infection being transmitted by an insect. Low (1900) first demonstrated the mechanism by which the larvae were returned from mosquitoes to man, and Annett, Dutton and Elliott (1901), Lebredo (1905), and Bahr (1912) added further details.

The adult worms live in the lymphatic system and liberate their larvae, known as microfilariae, into this system, whence they eventually, unless blocked, make their way into the blood stream. Their presence in the peripheral blood is periodic in most parts of the world, being present at night, but not in the daytime. Similar periodicity, though often less complete, is observed in many other filarial infections; in some species, however, e.g. *Loa loa*, there is a diurnal periodicity, and in others, e.g., *Dipetalonema perstans*, no periodicity has been observed. Two principal theories have been proposed to account for the periodicity: one, originally advanced by Manson, is that the larvae retire to internal organs during the day and enter the peripheral circulation only at night; the other, advanced by Lane (1929), is that the worms have cyclical parturition, producing their entire day's output of larvae at the same time each day, and that these worms are all destroyed in the host within 12 hours after they appear in the blood stream. Some support is given to this theory by O'Connor's (1931) observation at autopsies that at certain hours all the adult female filariae have their uteri crowded with embryos, while at other hours they are uniformly spent. On the other hand, the persistence for a year or more of microfilariae transferred to an uninfected host (Underwood and Harwood, 1939) is against this theory, though the fate of microfilariae in infected and non-infected hosts may not be at all comparable. As yet there is no unanimity of opinion as to the reason for microfilarial periodicity.

The microfilariae of *Wuchereria bancrofti* as seen in blood smears are covered by a sheath which has very generally been thought to be not a shed cuticle but a delicate, stretched vitelline membrane. Augustine (1937) questioned this, since he observed that developing microfilariae in the uterus of *Vagrifilaria columbigallinae* clearly show the vitelline membrane surrounding eggs containing coiled larvae, but none of the microfilariae from the vaginal region show any evidence of a sheath, and accumulations of crumpled hyaline objects interpreted as the remains of discarded vitelline membranes were found at a higher level in the uterus. Augustine was able to see no evidence of a sheath on the microfilariae of this species while they were in capillaries but was able to follow its formation on drying slides. He concludes, therefore, that the sheath is, as in other sheathed nematode larvae, the loosened but unshed cuticle from an incomplete ecdysis. This conclusion seems to us, however, to be very doubtful, since no other nematode larvae are known to molt at such an early stage in development, and

since two other molts have been observed during the course of development of the larvae in their mosquito hosts; this would bring them to the third stage, which is usual for infective larvae in intermediate hosts (see p. 237). Some species of filariae are not provided with sheaths.

The larvae are in a very immature state of development. They are covered by a layer of sub-cuticular cells, and within the body have a column of nuclei which subsequently develop into the esophagus and intestine.

This column of cells is broken at certain definite spots representing the future position of the nerve ring, the excretory pore and cell, and the anus. There are also a few large cells: an excretory cell just posterior to the excretory pore, a genital cell well behind the middle of the body, and a group of three cells previously reported as genital cells 2 to 4, but which Feng (1936) says give rise to the anus and rectum, and which Abe (1937) says belong to the sphincter between intestine and rectum, and are ultimately lost. There is a difference of opinion as to the existence of a stylet at the anterior end of the worm. The structure so called appears to be a rudimentary mouth cavity.

Upon ingestion by suitable species of mosquitoes the larvae become unsheathed in the stomach and penetrate into the body cavity, whence the majority migrate at once to the thoracic muscles, where development to the infective stage takes place. The factors which determine the suitability of particular mosquitoes have not been elucidated. Development takes place readily in mosquitoes of a variety of genera, including *Anopheles*, *Culex* and *Aëdes*, but sometimes nearly related species within these genera differ widely in their ability to serve as nurses. For example, *Culex quinquefasciatus* and *C. pipiens* are good hosts, whereas *C. vexans* is not; and *Aëdes variegatus* is a very good host whereas *A. aegypti* and *A. albopictus* are not. As yet nobody has succeeded in obtaining development in any arthropods other than mosquitoes.

Upon arrival in the thoracic muscles the larvae become quiescent, lying parallel with the muscle cells. Here in the course of 2 or 3 days they become considerably foreshortened, often to approximately half their original length, and at the same time grow considerably in girth, assuming what is known as the "sausage" stage. Only the caudal tip of the body fails to thicken, and is retained as an attenuated tail-like structure. Meanwhile a large excretory bladder develops and subsequently a large rectal cavity, and the outlines of the esophagus and intestine become defined. On the fifth day, according to Abe (1937), the larva undergoes its first molt, the cuticle developing an annular break near the anterior end. After this molt the larva reaches its maximum shortness and thickness and then, as the alimentary canal becomes well developed, begins to lengthen. As it approaches its maximum length it becomes active again and, according to Abe (l.c.), undergoes a second molt about the time it is ready to leave the thoracic muscles (in his experiments on the 13th day). The loosened cuticle breaks near the middle of the body and is shed. The larvae now become active and migrate out of the thorax. The majority go through the neck and head and move down into the interior of the labium, but a few get lost and can be found in the abdomen, legs, palpi, etc. Infective larvae commonly reach the labium about 2 weeks after infection in warm weather, but have been known to complete their development in 9½ days. In the labium they are stimulated by warmth, and when the mosquito is biting, escape through the delicate membrane where the labella join the shaft of the labium. The larvae do not, of course, interfere with skin-piercing as do the larvae of *Habronema* in the labium of *Stomoxys*, since in mosquitoes the labium itself is not a piercing or sucking organ. After leaving the proboscis and becoming free on the skin the larvae were believed by Fülleborn (1908), on the basis of experiments with *Dirofilaria immitis*, to penetrate into pores and enter through unbroken skin, but Yokogawa (1938) carried out a series of experiments which indicate that they can only enter broken skin, and presumably in nature use the wound made by the mosquito.

Nothing is known about the development of the larvae after they enter a human host until they reach maturity in the lymphatic system. *Dirofilaria immitis* requires about 9 months to reach maturity, and it is improbable that *Wuchereria bancrofti* takes any longer, if as long.

OTHER FILARIAE

The life cycles of comparatively few species of filariae are known, but among those that are known there is comparatively little variation. As already noted, some microfilariae are sheathed and some are not, but there is no evidence that the presence of a sheath has a "muzzling" effect in keeping the microfilariae from passing in or out of the capillaries, as Manson had thought. This was shown by O'Connor (1931) in the

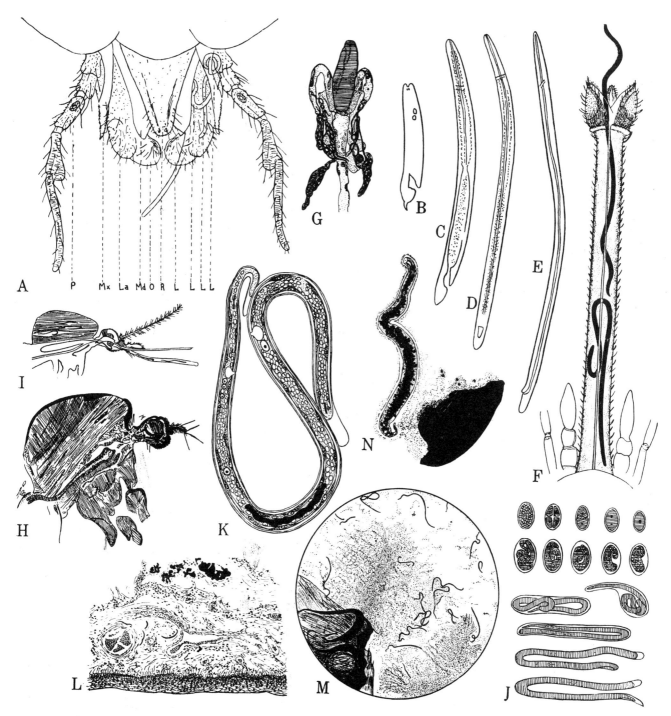

Fig. 194. DEVELOPMENT OF FILARIOIDEA

A—Mouth parts of *Simulium damnosum*, fixed in alcohol, cleared in warm clove oil showing position of larvae of *Onchocerca volvulus* emerging and in situ. B-E—*O. volvulus* (B—Early thoracic form, second day; C—Thoracic form, seventh day; D—Slightly later thoracic form; E—Proboscis form, ninth day). F—Mature larva of *Wuchereria bancrofti* escaping from proboscis of *Culex fatigans*. G—Larvae of *Dirofilaria repens* in *Anopheles* (11 days). H—*Wuchereria bancrofti* larvae two days after entering *Aedes variegatus*. I—Mature larvae of *W. bancrofti* in thoracic muscles and proboscis of mosquito. J—Embryonic development of *Loa loa*. K—Detailed drawing of *Wuchereria bancrofti* larva. L—Microfilaria in deeper layers of conjunctiva in case with disturbance of vision, keratitis, and iritis. M—Second larval stage of *Onchocerca* in the thoracic muscles of *Simulum metallicum*, approximately 48 hours after feeding upon infested patient. N—Third larval stage or so-called "sausage form" of *Onchocerca* on edge of thoracic muscles of *S. ochraceum* several days after feeding upon infested case. A-E, after Blacklock, D. B., 1926, Ann. Trop. Med. v. 20 (2). F, after Francis, E., 1919, Hyg. Lab. Bull. 117. G, H, & J, after Fuelleborn, F. Handb. path. Mikr. Jena v. 6. I & K, after Chandler, A. C., 1940 (Figs. 163 and 160). L-N, after Strong, et al., 1934, Contrib. Dept. Trop. Med., Harvard, VI.

case of *Wuchereria bancrofti* and by Harwood (1932) in the case of *Litomosoides carinii*.

The microfilariae of *Onchocerca*, which are unsheathed, differ from those of other filariae in that they live in the skin, and do not enter either the lymphatic or blood systems. The adult worms, living in subcutaneous tissues, are encapsulated by the host in hard nodules, through which the larvae are able to burrow and escape. The salivary secretion of the intermediate hosts (*Simulium*) seems to exert a definite chemotactic effect on the microfilariae, since they may be many times more numerous in the stomach of a fed fly than in a comparable quantity of tissue.

The intermediate hosts are usually Diptera. Fleas were stated by Breinl (1921) to serve as intermediate hosts for *Dirofilaria immitis* and Summers (1940) corroborated this, showing that development would occur in several species of fleas, and in a shorter time than in mosquitoes. Noë (1908) followed the development of *Dipetalonema grassii* in a tick, *Rhipicephalus sanguineus;* the microfilariae of this species are said to be too large to enter the blood circulation and are found in the lymph, which the ticks suck more than they do blood at the beginning of a meal. This work has not been confirmed and is open to suspicion in view of the fact that the embryos of related species (*Dirofilaria reconditum, Dipetalonema perstans*) live in the blood and develop in mosquitoes. Savani (1933) has also reported filariae in dog ticks in areas where *Dirofilaria immitis* is common. The intermediate hosts of *Wuchereria* and *Foleyella* are various mosquitoes; of *Dipetalonema perstans* and *Mansonella ozzardi, Culicoides;* of *Loa loa, Chrysops;* of *Onchocerca* spp., *Simulium* or *Culicoides;* and of *Dirofilaria*, fleas and mosquitoes.

There is some variation with respect to the site of development in the intermediate hosts. The majority of the species studied—*Wuchereria bancrofti, Microfilaria malayi* (adult perhaps unknown), *Dipetalonema, Mansonella, Dirofilaria reconditum*, and *Onchocerca* spp.—develop in the thoracic muscles of their dipteran hosts, but *Loa loa* develops principally in the muscles or fatty connective tissue of the abdomen of *Chrysops* (Connal and Connal, 1922), and *Dirofilaria immitis* develops in the haemocoele of fleas and in the Malpighian tubules of mosquitoes. These sites of development are of great interest in view of the similar sites utilized by *Draschia* and *Habronema* in muscoid flies.

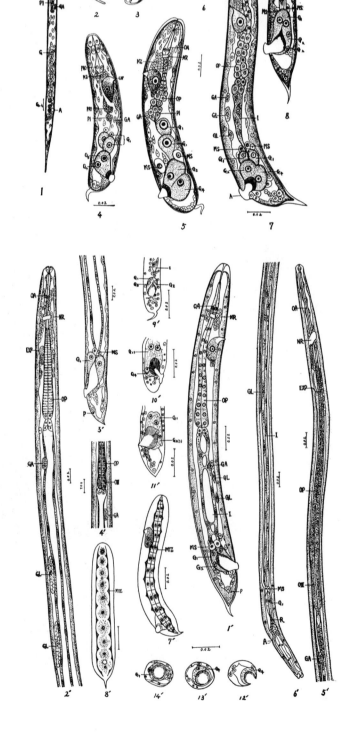

Fig. 195.

Development of *Wuchereria bancrofti.* 1—Larva 10 hours after infection. 2-4—Larva 2-3 days after infection. 5-6—Larva four days and three hours after infection. 7-8—Larva 5½ days after infection. 1'—Larva 5½ days after infection, just before first molt. 2'-3'—Larva 7½ days after infection. 4'—Larva 9½ days after infection of posterior end of esophagus. 5'-6'—Larva 11 days and 10 hours after infection.

CAMALLANINA

The members of both superfamilies of this suborder, Camallanoidea and Dracunculoidea, so far as known utilize copepods as intermediate hosts. There can be no doubt, from a consideration of the habitat and life cycle, that the Camallanoidea, dwelling as adults in the alimentary canal of aquatic hosts, are the more primitive, and that the tissue-dwelling Dracunculoidea, sometimes occurring in land animals, are a later evolutionary development. The relation of these two groups is comparable, in a broad way, with the relation of the Spiruroidea and the Filarioidea. In the case of the Filarioidea a habitat in the tissues is accompanied by evolution of a new method of exit and entrance of embryos via the skin, whereas in the Dracunculoidea it is accompanied by a new—but different—method of exit via the skin, suitable for an aquatic animal, but with retention of the primitive oral path of entry.

CAMALLANUS SWEETI

The life cycle of this worm was worked out by Moorthy (1938). The adult worms live in the intestine of a freshwater fish (*Ophicephalus gachua*) and produce free larvae which escape with the feces of the host. The embryos have a finely striated cuticle, a single dorsal denticle or boring cuticular tooth, and fairly well differentiated internal organs. On reaching water the larvae are swallowed by suitable species of *Cyclops* and reach the body cavity 2 or 3 hours after infection. These larvae undergo the first molt 24 to 36 hours later, and the second one after 5 to 7 days, in hot weather. The third-stage larvae are provided with ridged jaws suggestive of those of the adult, and have three unequal mucrones at the tip of the tail. No mention is made of these larvae becoming encysted in *Cyclops*. When infected *Cyclops* are eaten by small fish the larvae are activated by fish bile, escape from their copepod hosts and undergo further development, including possibly the third molt, in the intestines of these fish. The infection of the final host is thought to result from feeding on the second intermediate host, and the larvae undergo their fourth and final molt in the intestines of this host, acquiring the adult type of mouth. Whether the intervention of a second intermediate host is optional or obligatory was not determined, but in nature it would probably be the usual thing, since the final host does not ordinarily feed on *Cyclops* directly.

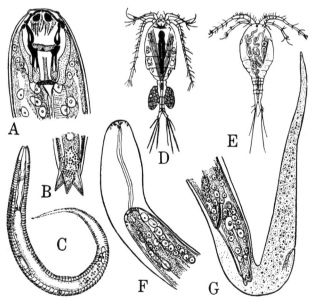

Fig. 196.

A-B—*Camallanus sweeti* (A—Head, fourth stage; B—Tail, same). C—*Procamallanus fulvidraconis*, mature embryo. D—Uninfected cyclops. E—Cyclops infected with *Dracunculus medinensis*. F-G—*Dracunculus medinensis* (F—Cephalic region undergoing second molt; G—Tail, same). A-B, after Moorthy, 1938, J. Parasit. v. 24 (4). C, after Li, 1935, J. Parasit., v. 21 (2). D, E, after Fuelleborn, 1913, Filariosen des Mensch. F, G, after Moorthy, 1938, Am. J. Hyg. v. 27 (2).

No encysted forms of *C. sweeti* were found in fish hosts, nor was any evidence found of their penetrating the walls of the intestine, but camallanid larvae of another type were found encysted in the body cavity, loosely attached to the intestines. These were observed to excyst when eaten by *Ophicephalus gachua*, but failed to undergo further development in that host.

An essentially similar development in *Cyclops* has been demonstrated for *Procamallanus fulvodraconis* by Li (1935), except that only one molt was observed. It seems probable that the first one was overlooked, since Li's figure of a 6-day-old larva corresponds with Moorthy's second stage larva of *Camallanus*, and his second-stage larva with Moorthy's third stage. However, Pereira et al (1936) state that *P. cearensis* develops only to the second stage in *Diaptomus*, the third and fourth stages being passed in the intestines of the fry of a fish other than the definitive host. Although they speak of this host as a "waiting host" (i.e., transport host) it would appear to be a true second intermediate host if their observation is correct that development to the third stage does not occur in *Cyclops*.

It will be seen that the camallanid life cycle is essentially the same as that of *Spiroxys* or of *Gnathostoma* except for the production of free embryos instead of eggs by the parent worms.

DRACUNCULUS MEDINENSIS

The adult female guinea worm, *Dracunculus medinensis*, when preparing for parturition, appears in the subcutaneous tissues of her host and produces a small ulcer on the surface of the skin. Upon stimulation by chilling of the skin, which happens in nature when the skin is plunged into water, she contracts violently in such a manner that a portion of the larva-filled uterus is prolapsed through a rupture in the cuticle, and the prolapsed portion of the uterus, bursting, liberates a small cloud of larvae. These larvae are unusually large (about 600 μ long), have a striated cuticle, a cuticular boring tooth or denticle, well-developed esophagus and intestine with dilated lumen, and a long filiform tail.

These larvae swim about in water and undergo further development only after being swallowed by certain species of copepods. The details of their development was worked out by Moorthy (1938). They reach the body cavity a few hours after being swallowed. They undergo two molts in the body cavity, the first one on the 5th to 7th day after infection, the second on the 8th to 12th day in hot weather. They start undergoing the second molt before casting off the exuviae of the first. The larvae grow very little in size, and actually decrease in length due to the loss of most of the filamentous tail. The third stage larvae increase slightly in size for about a week after the second molt, but after that undergo no further development; they are infective for the definitive host 4 to 8 days after the exuviae of the second molt are shed. They have a long esophagus of the adult type, and four mucrones at the tip of the tail. They remain active in the body cavity of the *Cyclops* for 4 or 5 weeks, but subsequently coil up and become quiet, but are not encysted. In addition to the usual type of larvae Moorthy also found a small proportion (1: 900) of "abnormal" larvae in which the tail is malformed. Moorthy suggested that these may have been males, but it is more likely that they should be regarded as abnormal individuals.

The early development of the larvae in the definitive host has not been followed. Sexually mature females 12 to 24 mm in length were found by Moorthy and Sweet (1938) in deep connective tissues of experimentally infected dogs 67 days after infection, and Moorthy believed that at this time fertilization had already taken place. Migration of the worms to the subcutaneous tissue and the formation of an ulcer for the egress of larvae occurs about a year after infection in man.

An essentially similar life cycle occurs in the case of *D. ophidensis* of garter snakes (Brackett, 1938). *Cyclops* infected with this species may be eaten by tadpoles and possibly other transport hosts; in tadpoles the larvae were found to remain free and viable in the body cavity for at least 2 weeks, but no further growth or development was observed.

The Philometridae, which have been found in a great variety of parenteral locations in aquatic hosts, have a life cycle essentially similar to that of *Dracunculus*. Thomas (1929) found that the first-stage larvae of *Philometra nodulosa* are devoured by *Cyclops* and invade its body cavity. Attempts at infection

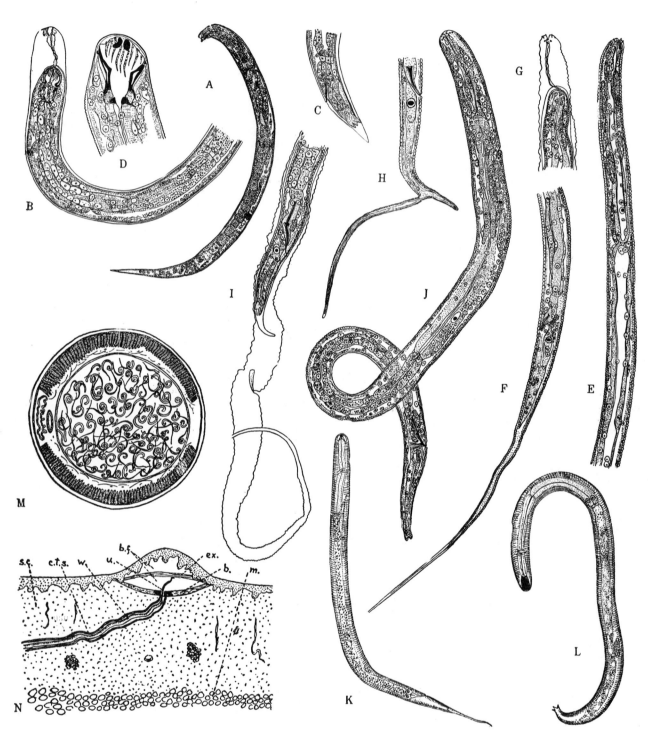

Fig. 197. DEVELOPMENT OF CAMALLANINA

A-D—*Camallanus sweeti* (A—First stage larva; B—Anterior end undergoing second molt; C—Posterior end undergoing second molt. D—Head, third stage). E-J—*Dracunculus medinensis* (E—First stage, anterior end; F—Same, posterior end; G—Anterior end moulting larva; H—Posterior end; I—Posterior end of normal larva undergoing first molt; J—Normal third stage larva). K-L—*Procamallanus fulvidraconis* (K—Larva, 6 days old; L—Larva, 14 days old). M—Cross-section of guinea worm showing uterus filled with embryos. x about 30 (after Leuckart). N—Diagram of guinea worm in the skin at the time of blister formation. A-D, after Moorthy, 1938, J. Parasit. v. 24 (4). E-J, after Moorthy, 1938, Am. J. Hyg., v. 27 (2). K-L, after Li, 1935. J. Parasit. v. 21 (2). M, N, after Chandler, 1940, Introduction to Parasit.

ot nsh from *Cyclops* a week after infection failed, presumably because of inadequate time for the larvae to reach the infective stage. Furuyama (1934) succeeded in completing the life cycle, in the case of *P. fujimotoi*, by feeding experimentally infected *Cyclops* to the definitive host. Young male and female worms were found in the body cavity, from whence the females subsequently migrated to their final habitat in the fins. In Philometridae the method of escape of the larvae is not as specialized as in the case of *Dracunculus;* the larvae of some species escape via the oviducts of the fish, while in the case of *P. fujimotoi* the ripe viviparous females leave the fins of their host, rupture, and liberate their larvae into the water. It is easy to see how the guinea-worm life cycle could have evolved from the camallanoid type by the substitution of escape of embryos through the skin for escape via the anus, which would be very simple in the case of parasites which reproduced in parenteral habitats in aquatic hosts.

TRICHUROIDEA

The members of this superfamily, with the exception of *Trichinella* and *Cystoöpsis* (see below), have a simple life cycle characterized by embryonation of eggs outside the body of the host; access to a new host by swallowing of eggs containing first-stage larvae provided with an oral spear; and direct migration, via the blood stream if outside the alimentary canal, to the site of development, without preliminary development elsewhere in the body. The life cycle of *Capillaria columbae*, recently worked out in detail by Wehr (1939), will serve as an example of the typical Trichuroidea.

CAPILLARIA COLUMBAE

The adults living in the small intestine are more or less imbedded in the mucosa, but the eggs make their way into the lumen and escape with the feces in an unsegmented state. Under favorable conditions of temperature, moisture, and oxygen segmentation occurs slowly, the first cleavage occurring in about 48 hours, the morula stage in about 3 days, and the infective first-stage larva in 6 to 8 days. No molting was observed to occur in the egg, and hatching does not normally take place before the egg is swallowed by a host. The entire development from newly hatched larvae to adult worms takes place in the small intestine of the definitive host.

The first-stage larva, like all other trichuroid larvae, has an oral spear. It has a long slender esophagus which posteriorly lies superficial to and only partly imbedded in the stichosome, which consist of two rows of opposing cells. The intestine is much shorter than the esophagus (ratio 1:3.5) and is terminated by a short rectum. The anus is subterminal.

The first molt occurs between 7 and 14 days after infection. The second-stage larvae are slenderer, and appear to have no oral spear; the stichosome consists of only a single row of cells, and the intestine is relatively longer. The second molt occurs about 14 days after infection. The third-stage larvae are still slenderer, with relatively longer intestine, and the genital primordium is long. The third molt occurs between 14 and 19 days after infection. The fourth-stage larvae are very slender, and sexually differentiated. The time of the final molt was not determined, but some sexually mature adults with eggs were found by the 19th day.

OTHER TRICHURIDAE

The available evidence indicates that the life cycle of *Trichuris* is essentially the same as that of *Capillaria columbae*, and it is probable that it is also the same for other species of *Capillaria* which inhabit the intestines of their hosts. The ability of *Capillaria* larvae to use transport hosts was shown by Wehr's (1936) demonstration that earthworms can serve as vectors for *C. annulata*, the crop-worm of chickens. Fülleborn's (1923b) figures of *Trichuris trichiura* larvae are strikingly similar to Wehr's figures of the first-stage larva of *Capillaria*. Although Neshi (1918, quoted by Yokogawa, 1920) reported the finding of four larvae of *Trichuris vulpis* in the lungs of a dog 21 hours after experimental infection, such migration on the part of *Trichuris* has not been observed by other workers either in normal or abnormal hosts (see Fülleborn, 1923a).

As Vogel (1930) pointed out, the entire group of Trichuroidea show a remarkable tendency to localization during their larval development in particular organs or tissues—what Vogel called ''organotropism.'' In all cases except *Trichinella* this organotropism continues throughout the adult life of the worms. Different species of Trichuridae are known to develop and live as adults in the esophagus, stomach, small intestine, cecum, colon, respiratory tree, liver, spleen, urinary bladder, and epithelium. The available evidence indicates that the newly hatched larvae of those species which do not grow to maturity in the intestine itself reach their destination by burrowing into the intestinal wall, entering the circulatory system, and escaping from the capillaries in the organ in which they are to develop Good evidence for this has been obtained in the case of *Capillaria hepatica* of the liver of rats. Vogel (1930) showed that if young larvae of *C. hepatica*, recovered from the liver a few days after infection, were planted in the spleen, lungs, or under the skin, a few would succeed in reaching the liver. Normally this worm penetrates the cecum, sometimes as early as 6 hours after infection (Luttermoser, 1938b), and is carried directly to the liver via the hepatic portal system (Fülleborn, 1924; Nishigori, 1925), only an exceptional few penetrating into the abdominal cavity, or being carried beyond the liver to the lungs and systemic circulation. In the case of *Trichosomoides crassicauda* of the urinary bladder of rats, Yokogawa (1921) fed embryonated eggs to rats and 1 to 4 days later found a few larvae in the abdominal and pleural cavities and the lungs; these he thought were *Trichosomoides* larvae from his feeding, but their size makes it evident that they were not.

An unusual situation with respect to transfer of infection to new hosts exists in the case of *Capillaria hepatica*, which is suggestive of a possible step in the evolution of the *Trichinella* life cycle. The eggs of this worm are deposited in the liver tissues of rats or mice, and remain there in an early stage of development (one to four cells), viable for at least 7 or 8 months (Luttermoser, 1938a). Only exceptionally do any of the eggs escape from the liver to be voided with the feces, and eating of an infected liver by a susceptible animal cannot result in infection because the non-embryonated eggs are not infective. Momma (1930) suggested flies as a factor in disseminating the eggs from decaying carcasses, and also showed that eggs in the feces of cats that have fed on infected rats are viable. Troisier and Deschiens (1930) and Shorb (1931) independently suggested that the usual method of transmission in nature is by ingestion of eggs that have become embryonated after being freed from the liver of an infected animal, either by decomposition or by being eaten by another animal, usually the latter.

TRICHINELLA SPIRALIS

The life cycle of this worm is unique among parasitic nematodes in that the period of waiting for a new host is passed in the parental host instead of in the open or in an intermediate host. The life cycle of *Capillaria hepatica*, described above, is a step in this direction, since in this case there are two periods of waiting, one in the liver tissue of the parental host, the other (the usual one) after embryonation in the open. In the case of Trichinella this double period is reduced to one by the complete elimination of the usual period of waiting outside the host, resulting from (1) precocious development to a burrowing larval stage in the uterus of the mother, and (2) consequent ability to infect the tissues of the parental host and to substitute development in this for the usual development in the open or in an alternate host.

The life cycle of this worm was one of the first to be worked out in its essential features, contributions having been made by Herbst, Küchenmeister, Leuckart, and Virchow from 1848 to 1860. The first entirely correct account of it was given by Leuckart (1860). The adult worms live in the small intestine. The females produce no egg shells, and the ova, unlike those of other Trichuroidea, develop precociously in the uterus, being born as active burrowing larvae, though in a very early stage of development, suggestive of microfilariae. There is an oral spear as in other members of the group, but the alimentary canal is rudimentary. This very immature larva enters the circulation, passing capillaries in both liver and lungs, and is distributed over the entire body. Presumably as the result of a special organotropism as suggested by Vogel (1930), the attraction in this particular case being the striated voluntary muscles, the larvae leave the capillaries and immediately penetrate through the sarcolemma into the interior of muscle cells, possibly by means of extra-corporeal digestion. As to whether the larvae actually penetrated into the muscle cells has long been a matter of dispute, but seems finally to have been settled by Jensen and Roth (1938). Immediately after penetration, accomplished by a boring movement of the spear-bearing head end, the larva is seen lying lengthwise just under the sarcolemma, or between the sarcolemma and adjacent muscle cells. Jensen and Roth think it likely that a histolytic enzyme is also involved in the penetration of the muscle cells and in dissolving the fibrillae inside.

Once inside the cells the larvae come to rest and begin their growth and differentiation, the muscle substance meanwhile undergoing degenerative changes. By the 17th day, according to Jensen and Roth, the larva has grown from 100 to 400 or 500 μ in length and has its esophagus and intestine clearly differentiated. According to Stäubli, however, it may have increased its length 10 times, to 800 to 1,000 μ, in from 10 to 14 days. After 11 days it begins to roll up spirally in a spindle-

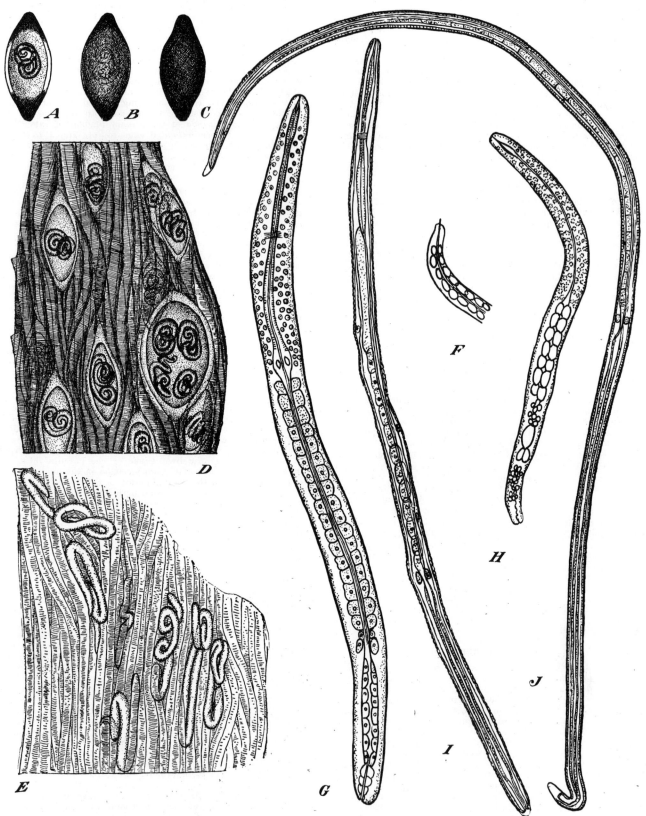

Fig. 198. DEVELOPMENT OF TRICHUROIDEA

A-C—Stages in calcification of *Trichinella* (A—Ends calcified; B—Thin layer of calcareous material over whole cyst; worm beginning to degenerate, C—Complete calcification). D—Larvae of *Trichinella spiralis*, encysted in striped muscle fibers in pork. Camera lucida drawing of cysts in infected sausage. E—Larvae of trichina worms burrowing in human flesh before encystment, from preparation from diaphragm of victim of trichiniasis. F-J—*Capillaria columbae* (F—Anterior end of unhatched first stage larva; G—Late first stage larva from intestine of pigeon 7 days after infection; H—Embryo or unhatched first stage larva; I—Second stage larva; J—Third stage larva in molt. A-E, after Chandler, 1940. Introduction to Parasitology. F-J, after Wehr, 1939, U.S.D.A. Tech Bull. 679

294

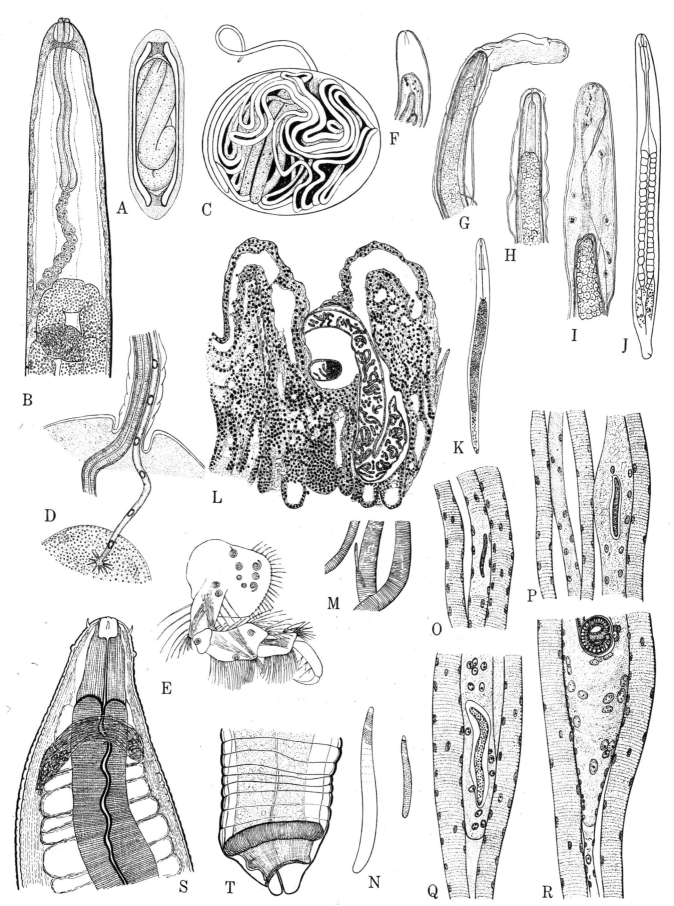

Fig. 199. TRICHUROIDEA AND DIOCTOPHYMATOIDEA

A-D—*Cystoopsis acipenseri* (A—Embryo; B—Head, male; C—Adult female; D—Connection of esophageal region and body proper of female.) E—*Cystoopsis* larva encysted in appendage of *Gammarus platycheir*. F-1— *Trichinella spiralis* (F. Molt at fourth hour; G—Molt after 14 hours; H—Third molt of female after 48 hours; I—Molt after 70 hours). J-K—*Trichuris trichiura* (J—From the cecum of a guinea pig; K—Larva pressed from egg). L-R—*Trichinella spiralis* (L—Section of intestine showing female in tunica propria; M—Young larva entering muscle; N—Young larvae; O-R—Stages in encystment and calcification). S-T—*Eustrongylides* (S—Head; T—Tail). A-D, and E, after Janicki, C., and Rasin, F., 1930, Ztschr. Wiss. Zool. v. 136. F-I, after Kreis, H. A., 1937, Zentralbl. f. Bakt. v. 138. J-K, after Fuelleborn, F., 1923, Arch. Schiffs- u. Tropen. v. 27. L-R, after Staubli, in Handb. path. Mikroorg. v. 8. Remainder original.

shaped enlargement of the muscle fiber, and after 4 to 6 weeks becomes encapsulated. If not ingested by a host suitable for their future development, the larvae ultimately die and there is fatty degeneration and finally calcification of the cysts. Trichinae are said to remain alive and infective for as long as 11 years in the muscles of swine, and to have lived for 12 to 24 years in man, according to Baylis. Prenatal infection with trichinae has been demonstrated in guinea pigs by Roth (1936); Mauss got negative results in rats, rabbits and hamsters. In spite of the fact that the larva undergoes so much growth and differentiation Stäubli was unable to detect any evidences of molts, and the writer has seen no reports of any being seen by later observers. By analogy with other nematodes, however, it seems probable that the infective larvae have undergone at least two molts. Infection has not been obtained with larvae less than 19 days old and only after 21 days can one obtain a high percentage of infections. This seems to indicate that the larvae undergo a change, such as a molt, prior to that time.

When the larva has undergone its full development, whether encapsulated or not, it is infective when eaten by another animal. Development in the intestine is extremely rapid, sexual maturity being reached and copulation occurring on the third day, and embryo production beginning on the fifth day. According to Kreis (1937) there are four molts in this brief period, at about 4, 12, 48 and 70 hours after ingestion. However, his evidence is not very convincing. According to recent investigations one molt was obtained after ingestion and the cuticle of the resultant nema passed uninterrupted over the vulva, indicating that at least one more molt would be necessary before maturity.

It is evident from this account that *Trichinella spiralis* is not only unique among nematodes in utilizing the parental host as a sole resting place while awaiting an opportunity to gain access to another host by cannibalism (insofar as it passes from individual to individual of one species) but it is also unique among the Trichuroidea in having different ''organotropisms'' for the larval development and for the adult development, the former being the striated voluntary muscles, particularly the most active ones (pectoral and tongue), the latter the mucous membrane of the small intestine.

CYSTOÖPSIS ACIPENSERI

This is an aberrant worm with respect to both its morphology and its life cycle. The females with their large spherical bodies and the small cylindrical males live in pairs in cyst-like cavities just under the skin of young sturgeons. According to Janicki and Rasín (1930), a well-developed vulva and muscular vagina are present, but they seem to be used only for the entrance of sperms and not for the exit of eggs. The spherical body is filled with numerous coils of the uterus filled with embryonated eggs. These, according to the authors quoted, escape only by a bursting of the thin wall of the cyst and rupture of the parasite.

Experimentally the embryonated eggs are eaten by certain species of amphipods, and the larvae, liberated in the stomach, penetrate into the body cavity. The young larvae possess a mouth spear like other Trichuroidea, and are in a very early stage of development. At the end of about 2 weeks they have reached their full size, and then migrate into the appendages or into muscle layers. Here they coil up after the manner of *Trichinella* larvae and soon become encapsulated. The capsule thickens with time, and by the end of 3 months cannot be broken under a coverglass. No experiments have been performed to prove the infectiousness of the larvae encysted in *Gammarus*, but there seems to be no reasonable doubt but that young sturgeons are infected by eating amphipods, and that the young worms migrate through the tissues of the host to their location in the skin as do some species of Trichuridae.

DIOCTOPHYMATINA

The life cycles of members of this group are very imperfectly known. The available information concerning the genus *Eustrongylides* has been summarized by Cram (1927). Larvae described by Rudolphi as *Filaria cystica* from under the peritoneum and in the abdominal muscles of certain Brazilian fish were regarded by Jägerskiöld as belonging to this genus. Ciurea (1924) found similar larvae in other fish from the Danube, and he also regarded them as belonging to *Eustrongylides*. Larvae found in Brazilian fishes by Schneider and by Leuckart are stated by Jägerskiöld to resemble *E. ignotus* of water birds.

Chapin (1926) found the preadult stage of this species in *Fundulus diaphanus* at Washington, D. C. From one to three specimens were found in each fish, and adult characters could be seen beneath the last cuticle, corresponding exactly to those of adult worms found by him in *Ardea herodias* from the same locality.

More recently Mueller (1934) reported similar larvae from *Fundulus*, in cysts attached to the mesenteries. They were 100 mm long by 0.685 mm in diameter, blood red in color, and the head was provided with 12 papillae, in 2 circles of 6 each, characteristic of the genus. Von Brand (1938) found a high percentage of *Fundulus* from Chesapeake Bay parasitized with this same larval form; individual fish harbored from 1 to 8 worms. Von Brand states that the encapsulated nematodes did not harm the host, but that after the host died they left their capsules and endeavored to escape from the dead host by burrowing through the tissues, eventually emerging through the gill region or body wall. He was able to keep the worms alive on sterile nutrient media for as long as 2 months, but there was no growth or development.

The larvae found by Ciurea are large, 28 to 70 mm long by 264 to 539 μ wide, and are rose-red or brown-red in color. On each side of the body near the anterior end is a row of small lateral papillae. The mouth aperture has the form of a cleft and has three small pointed papillae on each side of it, and three larger papillae just outside of these. The larvae have tails of two types, one enlarged near the end and regarded as that of the male, and the other rounded off without enlargement and regarded as that of the female. Whether the fish are first or second intermediate hosts is unknown. The thick-shelled eggs are undeveloped when they leave the body of the host.

Even less is known about the life cycle of *Dioctophyma renale*. The adults are usually found in the pelvis of the kidneys, particularly the right one, where they eventually destroy the entire parenchyma. Worms, often immature, are frequently found in other locations, particularly in the peritoneal cavity. The eggs, in an unsegmented condition, normally escape from the body with the urine. They develop slowly in water, requiring from 1 to 7 months for embryonation, according to the temperature, and then remain viable for at least 2 years, although they do not hatch. Beyond this point nothing is definitely known about the life cycle, but the frequency of infection in fish-eating animals makes its highly probable that fish serve as vectors for this worm as well as for *Eustrongylides*. Ciurea (1921) found a specimen 63 cm long in the peritoneal cavity of a dog fed, between 3 and 4 months previously, on 14 specimens of a cyprinid fish (*Idus idus*) from the Danube, but it is doubtful whether the worm was actually acquired from this feeding. Ciurea also found an active larva in the muscles of *Idus* which he thought might be that of *Dioctophyma*, but his figure and discription are more suggestive of an ascarid larva. Swales (1933) reported *D. renale* as a very common and important parasite of mink in Canada, and stated that on mink farms the infection was definitely associated with the feeding of fish to these animals.

It has generally been assumed that *Dioctophyma*, after entering the alimentary canal of a definitive host, is carried via the blood stream to the kidneys as a young larva, there to undergo its growth to maturity. Its occurrence in the peritoneal cavity was thought to be accidental and rare, and due to rupture of the cyst-like remnant of the kidneys after the complete atrophy of its parenchyma. As a matter of fact, however, the worms are very frequently found in the abdominal cavity of dogs, in the majority of cases without evidence of damage to the kidneys. Wislocki (1919) found them in that location in every one of 12 dogs which he examined, and in only two cases could a portal of entry through a partially destroyed kidney be surmised. Brown, Sheldon and Taylor (1940) found 13 of 20 infected dogs in North Carolina with worms in the body cavity only, 6 had worms in the right kidney as well, and 1 had them only in the right kidney. Lukasiak (1930) called attention to the fact that in spite of numerous searches, especially in the kidneys, larvae have never been found, and relatively young forms have been found not in the kidneys but in the abdominal cavity, by preference between the lobes of the liver. Stefanski and Strankowski (1936) found a case in which a worm in the abdominal cavity was clearly in process of entering the right kidney; its anterior end was lodged in the tissue of the right kidney, the substance of which had not been destroyed. From this and similar cases which they quote from the literature, and from the other evidence cited above, these authors conclude that the larvae of the worm, travelling via the blood stream, stop in the liver, grow, and finally continue their development in the abdominal cavity, probably penetrating the kidney only after the final molt, and hollowing out a canal in the substance of this organ. Since the larvae are probably too large to enter capillaries, it seems to us more probable that the worms reach the body cavity by directly burrowing into it, as do *Gnathostoma* larvae; we know of no evidence that the liver is involved at all.

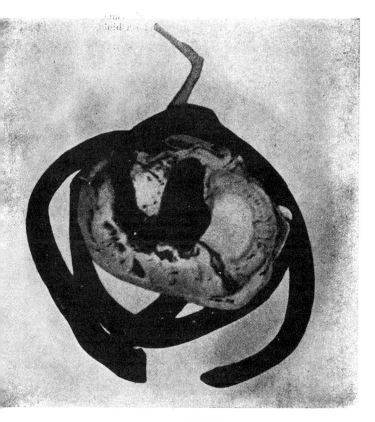

Fig. 200.
Dioctophyma renale female, anterior extremity of the parasite coiled in the pelvis of the right kidney. After Stefanski and Strankowski, 1936, An. de Parasit. Humaine et Comp. v. 14 (1).

Bibliography

ABE, S. 1937.—Development of the *Wuchereria bancrofti* in the body of the mosquito. [English summary.] J. Med. Assoc., Formosa, v. 36 (3): 516-519, 5 pls.

AFRICA, C. M., REFUERZO, P. G. and GARCIA, E. Y. 1936a.—Observations on the life cycle of *Gnathostoma spinigerum*. Philippine J. Sc., v. 59 (4): 513-521, 2 pls.
1936b.—Further observations on the life cycle of *Gnathostomum spinigerum*. Philippine J. Sc., v. 61 (2): 221-225.

ALICATA, J. C. 1934.—Observations on the period required for ascaris eggs to become infective. Proc. Helm. Soc. Wash., v. 1: 12.
1935.—Early developmental stages of nematodes occurring in swine. U. S. Dept. Agric., Tech. Bull., No. 489, 96 pp.
1937a.—The gizzard worm; and its transmission to chickens in Hawaii. Hawaii Agric. Exper. Sta. Cir. No. 11, 7 pp.
1937b.—Larval development of the spirurid nematode, *Physaloptera turgida*, in the cockroach, *Blattella germanica*. Papers on Helminthology, Jub. Skrjabin, pp. 11-14.
1939.—Preliminary note on the life history of *Subulura brumpti*, a common cecal nematode of poultry in Hawaii. J. Parasit., v. 25: 179-180.

ANNETT, H. E., DUTTON, J. E. and ELLIOTT, J. H. 1901.—Report of the malaria expedition to Nigeria. Pt. II, Filariasis. Liverpool Sch. Trop. Med., Mem. 4.

AUGUSTINE, D. L. 1923.—Investigations on the control of hookworm disease. XIX. Observations on the completion of the second ecdysis of *Necator americanus*. Am. J. Hyg., v. 3: 280-295.
1937.—Observations on living "sheathed" microfilariae in the capillary circulation. Tr. Roy. Soc. Trop. Med. & Hyg., v. 31 (1): 55-60.

BAHR, P. H. 1912.—Filariasis and elephantiasis in Fiji. Suppl. 1, J. Lond. Sch. Trop. Med., 192 pp., pls., maps.

BAYLIS, H. A. 1916.—Some ascarids in the British Museum (Natural History). Parasit., v. 8 (3): 360-378, 4 pls.
1926.—Further experiments with the *Gongylonema* of cattle; II. J. Trop. Med. & Hyg., v. 29 (20): 346-349.

BEACH, T. D. 1935.—The experimental propagation of *Strongyloides* in culture. Proc. Soc. Exper. Biol. & Med., v. 32 (9): 1484-1486.
1936.—Experimental studies on human and primate species of *Strongyloides*. V. The free-living phase of the life-cycle. Am. J. Hyg., v. 23 (2): 243-277.

BRACKETT, S. 1938.—Description and life history of the nematode *Dracunculus ophidensis* n. sp., with a re-description of the genus. J. Parasit., v. 24 (4): 353-361.

BRAUN, M. 1899.—Bemerkungen über den "sporadischen Fall von *Anguillula intestinalis* in Ostpreussen." Centralbl. Bakt. [etc.] Abt. I, Orig., v. 26 (20-21): 612-615, 1 pl.

BREINL, A. 1921.—Preliminary note on the development of the larvae of *Dirofilaria immitis* in dog fleas, *Ctenocephalus felis* and *canis*. Ann. Trop. Med. & Parasit., v. 14 (3): 389-392.

BROWN, H. W., SHELDON, A. J. and TAYLOR, W. W., JR. 1940 —The occurrence of *Dioctophyme renale* in dogs of North Carolina [Abstract]. J. Parasit. v. 26 (6) (Suppl.): 16.

BRUMPT, É. 1921.—Récherches sur la déterminisme des sexes et de l'evolution de anguillules parasites. Compt. Rend. Soc. Biol., Paris, v. 85 (23): 149-152.

BUCKLEY, J. J. C. 1934.—On *Syngamus ierei* sp. nov. from domestic cats, with some observations on its life cycle. J. Helm., v. 12 (2): 89-98.

CAMERON, T. W. M. 1923.—On the biology of the infective larva of *Monodontus trigonocephalus* (Rud.) of sheep. J. Helm., v. 1: 205-214.
1927a.—Observations on the life history of *Aelurostrongylus abstrusus* (Railliet), the lungworm of the cat. Ibid., v. 5: 55-66.
1927b.—Observations on the life history of *Ollulanus tricuspis* Leuck., the stomach worm of the cat., Ibid., v. 5 (2): 67-80.

CHANDLER, A. C. 1918.—Animal parasites and human disease, 1st ed., p. 275. New York.
1925.—The migration of hookworm larvae in soil. Indian Med. Gaz., v. 60 (3): 105-108.
1925b.—A contribution to the life history of a gnathostome. Parasit., v. 17 (3): 237-244.
1925c.—The helminthic parasites of cats in Calcutta and the relation of cats to human helminthic infections. Indian J. Med. Res., v. 13 (2): 213-227, 2 pls.
1938.—*Diploscapter coronata* as a facultative parasite of man, with a general review of vertebrate parasitism by rhabditoid worms. Parasit., v. 30 (1): 44-55.

CHAPIN, E. A. 1926.—*Eustrongylides ignotus* in the United States. J. Parasit., v. 13: 86-87.

CHITWOOD, B. G. 1933.—On some nematodes of the superfamily Rhabditoidea and their status as parasites of reptiles and amphibians. J. Wash. Acad. Sc., v. 23: 508-520.

CHITWOOD, B. G. and GRAHAM, G. L. 1940.—Absence of vitelline membranes on developing eggs in parasitic females of *Strongyloides ratti*. J. Parasit., v. 26 (3): 183-190.

CHU, T. 1936a.—A review of the status of the reptilian nematodes of the genus *Rhabdias* with a re-description of *Rhabdias fuscovenosa* var. *catenensis* (Rizzo, 1902) new rank. J. Parasit., v. 22: 130-139.
1936b.—Studies on the life history of *Rhabdias fuscovenosa* var. *catanensis* (Rizzo, 1902). Ibid., v. 22: 140-160.

CIUREA, J. 1921.—Sur la source d'infestation par l'eustrongyle géant (*Eustrongylus gigas* Rud.) Comp. Rend. Soc. Biol., Paris. v. 85 (27): 532-534.
1924.—Die *Eustrongylides*-Larven bei Donaufischen. Ztschr. Fleisch.- u. Milchhyg., v. 34 (13) 134-137.

CLAPHAM, P. A. 1934. Experimental studies on the transmission of gapeworm (*Syngamus trachea*) by earthworms. Proc. Roy. Soc. Lond., Ser. B, v. 115: 18-29.
1939a.—On flies as intermediate hosts of *S. trachea*. J. Helm., v. 17 (2): 61-64.
1939b.—Three new intermediary vectors for *Syngamus trachea*, J. Helm., v. 17 (4): 191-192.

CONNAL, A. and CONNAL, S. L. M. 1922.—The development of *Loa loa* (Guyot) in *Chrysops silacea* (Austen) and in *Chrysops dimidiata* (van der Wulp.). Tr. Roy. Soc. Trop. Med. & Hyg., v. 16: 64-89.

CORT, W. W. 1925.—Investigations on the control of hookworm disease. XXXIV. General summary of results. Am. J. Hyg., v. 5 (1): 49-89.

CRAM, E. B. 1927.—Bird parasites of the nematode suborders Strongylata, Ascaridata, and Spirurata. U. S. Natl. Mus., Bull. No. 140, 465 pp.
　　1931.—Developmental stages of some nematodes of the Spiruroidea parasitic in poultry and game birds. U. S. Dept. Agric. Tech. Bull. No. 227.

CUVILLIER, E. 1937.—The nematode, Ornithostrongylus quadriradiatus, a parasite of the domesticated pigeon. U. S. Dept. Agric. Tech. Bull. No. 569.

DARLING, S. T. 1911.—Strongyloides infections in man and animals in the Isthmian Canal Zone. J. Exper. Med., v. 14 (1): 1-24.

DE BLIECK, L. and BAUDET, E. A. R. F. 1926.—Contribution a l'étude du développement des strongylidés (sclérostomes) du gros intestin chez le cheval. Ann. Parasit., v. 4 (1): 87-96.

FAUST, E. C. 1927.—Migration route of Spirocerca sanguinolenta in its definitive host. Proc. Soc. Exper. Biol. & Med., v. 25: 192-195.
　　1931.—Human strongyloidiasis in Panama. Am. J. Hyg., v. 14: 203-211.
　　1933.—Experimental studies on human and primate species of Strongyloides. II. The development of Strongyloides in the experimental host. Am. J. Hyg., v. 18 (1), 114-132.

FAUST, E. C. and KAGY, E. S. 1933.—Experimental studies on human and primate species of Strongyloides. I. The variability and instability of types. Am. J. Trop. Med., v. 13: 47-65.

FENG, L. C. 1936.—The development of Microfilaria malayi in A. hyrcanus var. sinensis Wied. Chinese Med. J., Suppl. 1, pp. 345-367.

FIELDING, J. W. 1926.—Preliminary note on the transmission of the eyeworm of Australian poultry. Austral. J. Exper. Biol. & Med., v. 3: 225-232.

FOSTER, A. O. and CROSS, S. X. 1934.—The direct development of hookworms after oral infection. Am. J. Trop. Med., v. 14 (6): 565-573.

FREEBORN, S. B. 1923.—The control of the suckered roundworms of poultry. Cornell Vet., v. 13: 223-231.

FUELLEBORN, F. 1908.—Ueber Versuche an Hundefilarien und deren Uebertragung durch Mücken. Arch. Schliffs- u. Tropen-Hyg., Beihefte 8, 43 pp.
　　1914.—Untersuchungen ueber den Infektionsweg bei Strongyloides und Ankylostoma und die Biologie dieser Parasiten. Beihefte 5, Arch. Schiffs- u. Tropen-Hyg. v. 18: 26-80 (182-236), figs. 1-14, pls. 1-7.
　　1920a.—Perkutane Infektion bei Angiostomum nigrovenosum. Ibid. v. 24: 176.
　　1920b.—Ueber die Anpassung der Nematoden an den Parasitismus und den Infektionsweg bei Askaris und anderen Fadenwürmen des Menschen. Ibid. v. 24: 340-347. Fadenwürmen des Menschen. Ibid. v. 24: 340-347.
　　1921a.—Askarisinfeckion durch Verzehren eingekapselter Larven und über gelungene intrauterine Askarisinfektion. Ibid. v. 25: 367-375.
　　1921b.—Ueber die Wanderung von Askaris und anderen Nematoden-larven in Körper und intrauterine Askaris infektion. Ibid. v. 25: 146-149.
　　1923a.—Ueber die Entwicklung von Trichozephalus im Wirte. Ibid. v. 27: 413-420.
　　1923b.—Ueber den "Mundstachel" der Trichotracheliden-larven und Bemerkungen über die jungsten Stadien von Trichocephalus trichiurus. Ibid. v. 27: 421-425, 1 pl.
　　1924.—Ueber den Infektionsweg bei Hepaticola hepatica. Ibid. v. 28: 48-61.
　　1925.—Ueber die Durchlässigkeit der Blutcapillaren für Nematodenlarven. Ibid. v. 29, Beiheft 3, 100 pp. 4 pls.
　　1926.—Ueber das Verhalten der Hakenwurmlarven bei der Infektion per os. Ibid v. 30: 638-653.
　　1927.—Ueber das Verhalten der Larven in Körper des Wirtes. Ibid., Beihefte 2, v. 31, 56 pp., figs. 1-5.
　　1928.—Ueber den Infektionsweg bei Rhabdias bufonis (Rhabdonema nigrovenosum) des Frosches nebst Versuchen ueber die Lymphzirkulation des letzteren. Centralbl. Bakt. [etc.], Abt. I, Orig., v. 109: 444-462, 1 pl.

　　1929a.—On the larval migration of some parasitic nematodes in the body of the host and its biological significance. J. Helm., v. 7 15-26.
　　1929b.—Filariosen des Menschen. In Handb. Path. Mikroög., 6, Lief. 28, 1043-1224, 3 pls.

FURUYAMA, T. 1934.—On the morphology and life history of Philometra fujimotoi Furuyama, 1932. Keijo J. Med., v. 5 (3): 165-177.

GOODEY, T. 1922.—Observations on unsheathed larvae of nematodes. Ann Applied Biol., v. 9. 33.
　　1923.—Experiment on the feeding of embryonated eggs of Ascaris megalocephala to domesticated animals. Ann. Appl. Biol., v. 10: 116-121.
　　1924a.—The anatomy of Oesophagostomum dentatum (Rud.) a nematode parasite of the pig, with observations on the structure and biology of the free-living larvae. J. Helm., v. 2: 1-14.
　　1924b.—The anatomy and life history of the nematode Rhabdias fuscovenosa (Railliet) from the grass snake Tropidonotus natrix. Ibid., v. 2 (2): 51-64.

GOODEY, T. and CAMERON, T. W. M. 1923.—Observations on the morphology and life cycle of Ascaris columnaris Leidy, a nematode parasite of the skunk. J. Helm., v. 1: 1-8.

GRAHAM, G. L. 1936.—Studies on Strongyloides. I. S. ratti in parasitic series, each generation in the rat established with a single homogonic larva. Am. J. Hyg., v. 24 (1): 71-87.
　　1938.—Idem. II. Homogonic and heterogonic progeny of the single homogonically derived S. ratti parasite. Ibid., v. 27 (2): 221-234.
　　1939a.—Idem. IV. Seasonal variation in the production of heterogonic progeny by singly established S. ratti from a homogonically derived line. Ibid., v. 30 (Sect. D.), 15-27.
　　1939b.—Idem. V. Constitutional differences between a homogonic and a heterogonic line of S. ratti. J. Parasit., v. 25 (4): 365-375.

GUBERLET, J. E. 1924.—Note on the life history of Ascaridia perspicillum. Tr. Am. Micr. Soc., v. 43: 152-156.

HARWOOD, P. D. 1930.—A new species of Oxysomatium (Nematoda) with some remarks on the genera Oxysomatium and Aplectana and observations on the life history. J. Parasit., v. 17: 61-73.
　　1932.—A note on the tissue-penetrating abilities of sheathed microfilariae. Tr. Am. Micr. Soc., v. 51 (2): 153-154.

HELLER, A. 1903.—Ueber Oxyuris vermicularis. Deutsch Arch. Klin. Med., Berlin, v. 77, (1-2): 21-28, pls. 1-3.

HERRICK, C. A. 1928.—A quantitative study of infections with Ancyclostoma caninum in dogs. Am. J. Hyg., v. 8: 125-157.

HEDRICK, L. R. 1935.—The life history and morphology of Spiroxys contortus (Rudolphi); Nematoda: Spiruridae. Tr. Am. Micr. Soc., v. 54 (4): 307-335.

HOBMAIER, M. 1930.—Life history of Probostrongylus (Synthetocaulus rufescens. Proc. Soc. Exper. Biol. & Med., v. 28: 156-158.
　　1934a.—Lungenwurmlarven in Mollusken. Ztschr. Parasitenk., v. 6: 642-648.
　　1934b.—Elaphostrongylus odocoilei n. sp., a new lungworm in black-tail deer (Odocoileus columbianus). Description and life history. Proc. Soc. Exper. Biol. & Med., v. 31: 509-514.
　　1935.—Intermediate hosts of Aelurostrongylus abstrusus of the cat. Ibid., v. 32: 1641-1647.

HOBMAIER, A. and HOBMAIER, M. 1929.—Die Entwicklung der Larve des Lungenwurmes Metastrongylus elongatus (Strongylus paradoxus) des Schweines und ihr Invasionsweg, sowie vorlaufige Mitteilung über die Entwicklung von Choerostrongylus brevivaginatus. München. Tierärztl. Wochenschr., v. 80: 365-369, illus.

HU, C. H. and HOEPPLI, R. J. C. 1936.—The migration route of Spirocerca sanguinolenta in experimentally infected dogs. Chinese Med. J., Suppl. I., pp. 293-311.

HSÜ, H. F. and CHOW, C. Y.—On the intermediate host and larva of Habronema mansioni Seurat, 1914 (Nematoda). Chinese Med. J., Suppl. II, pp. 419-422.

IHLE, J. E. W. and v. OORDT, G. J. 1924.—On the development of the larva of the fourth stage of *Strongylus vulgaris* (Looss). Proc. K. Akad. Wetensch. Amsterdam, v. 27 (3-4): 194-200.

ITAGAKI, S., 1927.—On the life history of the chicken nematode, *Ascaridia perspicillum*. Rpt. Proc. 3rd World's Poultry Cong., Ottawa, pp. 339-344, Figs. 1-22.

JANICKI, C. and RASIN, K. 1930.—Bemerkungen über *Cystoöpsis acipenseri* des Wolga-Sterlets, sowie über die Entwicklung dieses Nematoden im. Zwischenwirt Ztschr. Wiss. Zool., v. 136: 1-37.

JENSEN, V. and ROTH, H. 1938.—Zur einwanderung der Trichinenlarve in die Quergestreifte Muskelfaser. Acta Path. & Microbiol. Scand., Suppl., v. 37: 259-268, 6 abb.

JERKE, H. W. M. 1902.—Eine parasitische Anguillula des Pferdes. Arch. Wiss. u. Prakt. Thierheilk., v. 29 (1-2): 113-127, 1 pl.

JONES, M. F. and JACOBS, L. 1939.—Studies on the survival of eggs of *Enterobius vermicularis* under known conditions of humidity and temperature [Abstract]. J. Parasit., v. 25 (6), Suppl., p. 32.

KAHL, W. 1936.—Beitrag zur Kenntnis des Nematoden *Contracaecum clavatum* Rud. Ztschr. Parasitenk., v. 8: 509-520.

KOBAYASHI H. 1928.—On the life history of *Oxyspirura mansoni* and the pathological changes in the conjunctiva and the ductus lacrymalis caused by this worm, with further observations on the structure of the adult worm. [In Japanese]. 50 pp., 8 figs., pls. 1-2, figs. 1-24; English summary, pp. 1-6. Reprinted from J. Formosa Med. Soc., Taiwan Igakukai Tasshi, (280), July, 1928.

KOCH, E. W., 1925.—Oxyurenfortpflanzung im Darm ohne Reinfektion und Magenpassage. Centralbl. Bakt. [etc.], Abt. I, Orig., v. 94: 208-236.

KOURI, P., BASNUEVO, J. G. and ARENAS, R. 1936.—Contribución al conocimiento del ciclo evolutivo del *Strongyloides stercoralis*. Nota previa. Rev. Parasit., Clin. & Lab., v. 2: 1-6.

KREIS, H. A. 1932.—Studies on the genus *Strongyloides* (Nematodes) Am. J. Hyg., v. 11 (2): 450-491.
1937.—Die Entwicklung der Trichinellen zum reifen geschlechtstier im Darme des Wirtes. Zentralbl. Bakt. [etc.], Abt. I, Orig., v. 138: 290-302.

LANE, C. 1929.—The mechanism of filarial periodicity. Lancet, Lond., (5521), v. 216, v. 1 (25): 1291-1293, 1 fig.

LEICHTENSTERN, O. 1886.—Fütterungsversuche mit *Ankylostoma-Larven*. Eine neue *Rhabditis*-Art in den Fäces vom Ziegelarbeitern. Centralbl. Klin. Med., v. 8 (39): 673-675.
1899.—Zur Lebensgeschichte der *Anguillula intestinalis*. Centralbl. Bakt. [etc.], Abt. I, Orig., v. 25: 226-231.
1905.—Studien über *Strongyloides stercoralis* (Bavay). Arb. K. Gsndhtsamte., Berlin, v. 22 (2): 309-350.

LENTZE, F. A., 1935.—Zur Biologie des *Oxyuris vermicularis*. Centrlbl. Bakt. [etc.], Abt. I, Orig., v. 135: 156-159.

LEUCKART, R. 1860.—Untersuchungen über *Trichina spiralis*. Zugleich ein Beitrag zur Kenntnis der Wurmkrankheiten. 57 pp., 2 pls., Leipz. & Heidelberg.
1866.—Zur Entwickelungsgeschichte der Nematoden [In his Helminthologische Mitteilungen]. Arch. Ver. Wiss. Heilkunde, Leipz., (n. F.), v. 2: 195-235.
[1883].—Ueber die Lebensgeschichte der sog. *Anguillula stercoralis* und deren Beziehungen zu der sog. *Ang. intestinalis*. Ber. Verhandl. K. Sächs. Gesellsch. Wiss. Leipz., Math.-Phys. Cl. (1882), v. 34: 85-107.

LI, H. C.—The taxonomy and early development of *Procamallanus fulvidraconis* n. sp., J. Parasit., v. 21 (2): 103-113.

LOOSS, A. 1898.—Zur Lebensgeschichte des *Ankylostoma duodenale*. Centralbl. Bakt. [etc.], Abt. I, Orig., v. 24: 483-488.
1905.—Die Wanderung der *Ancylostomum*- und *Strongyloides*-Larven von der Haut nach dem Darm. Compt. Rend. 6. Cong. Internatl. Zool., Geneva, pp. 225-233.
1911.—The anatomy and life history of *Anchylostoma duodenale* Dub. A monograph, Part 2. The development

in the free state. Rec. Egypt. Govt. Sch. Med., v. 4: 159-613.

LOW, G. C. 1900.—A recent observation on *Filaria nocturna* in *Culex*, probable mode of infection in man. Brit. Med. J., v. 1: 1456-1457.

LUCKER, J. T. 1934a.—Development of the swine nematode *Strongyloides ransomi* and the behavior of its infective larvae. U. S. Dept. Agric. Tech. Bull. No. 437, 30 pp.
1934b.—The morphology and development of the preparasitic larvae of *Posteriostomum ratzii*. J. Wash. Acad. Sc., v. 24: 302-310.
1936.—Preparasitic molts in *Nippostrongylus muris*, with remarks on the structure of the cuticula of Trichostrongyles. Parasit., v. 28 (2): 161-171.
1938.—Description and differentiation of infective larvae of three species of horse strongyles. Proc. Helm. Soc. Wash., v. 5 (1): 1-5.

LUKASIAK, J. 1930.—Anatomische und entwicklungsgeschichtliche Untersuchungen an *Dioctophyme renale* (Goeze, 1782). Arch. Nauk. Biol. Towarzyst. Nauk. Warszawsk., (1929), v. 3 (3): 100 pp., figs. 1-6, pls. 1-6.

LUTTERMOSER, G. W. 1938a.—Factors influencing the development and viability of the eggs of *Capillaria hepatica*. Am. J. Hyg., v. 27 (2): 275-289.
1938b.—An experimental study of *Capillaria hepatica* in the rat and the mouse. Ibid., v. 27 (2): 321-340.

MACARTHUR, W. P. 1930.—Threadworms and pruritis ani. J. Roy. Army Med. Corps, v. 55: 214-216.

McCoy, O. R. 1929.—The growth of hookworm larvae on pure cultures of bacteria. Science, v. 69 (1777): 74-75.

McINTOSH, A. and CHITWOOD, B. G. 1934.—A new nematode *Longibucca lasiura*, n. sp. (Rhabditoidea, Cylindrogastridae) from a bat. Parasit., v. 26: 138-140.

MANSON, P. 1878.—On the development of *Filaria sanguinis hominis*, and on the mosquito considered as a nurse. J. Linn. Soc. Lond., Zool., (75), v. 14: 304-311.

MARKOWSKI, S. 1937.—Ueber die Entwicklungsgeschichte und Biologie des Nematoden *Contracaecum aduncum* (Rudolphi, 1802). Bull. Internatl. Acad. Polon. Sc. e Lett., Ser. B, Sc. Nat. (II), pp. 227-242, 2 pls.

MAUPAS, E. and SEURAT, L. G. 1913.—La Mue et l'enkystement chez les Strongles du tube digestif. Compt. Rend. Soc. Biol., Paris, v. 74: 34-48.

MECINZKOW, E. 1865.—Ueber die Entwickelung von *Ascaris nigrovenosa*. Arch. Anat., Physiol. & Wiss. Med., Leipz., pp. 409-420, pl. 10, figs. 1-11.

MIYAGAWA, Y. 1916.—Ueber den Wanderungsweg der *Ankylostoma duodenale* innerhalb des Wirtes bei Oralinfektion und über ihren Hauptinfektionsmodus. Mitt. Med. Fakult. K. Univ., Tokyo., v. 15: 411-452.

MIYAGAWA, Y. and OKADA, R. 1930.—Biological significance of the lung journey of *Anchylostoma* larvae in the normal host. First Report. Jap. J. Exper. Med., v. 8: 285-308.
1931.—Idem, Second Report. Ibid., v. 9: 151-207.

MOMMA, K. 1930.—Notes on modes of rat infestation with *Hepaticola hepatica*. Ann. Trop. Med. & Parasit., v. 24 (1): 109-113.

MÖNNIG, H. O. 1930.—Studies on the bionomics of the free-living stages of *Trichostrongylus* spp. and other parasitic nematodes. Union So. Africa Dept. Agric., 16th Rpt. Dir. Vet. Service, pp. 175-198.

MOORTHY, V. N. 1938.—Observations on the development of *Dracunculus medinensis* larvae in Cyclops. Am. J. Hyg., v. 27 (2): 437-460.
1938b.—Observations on the life history of *Camallanus sweeti*. J. Parasit., v. 24 (4): 323-342.

MOORTHY, V. N. and SWEET, W. C. 1938.—Further notes on the experimental infection of dogs with dracontiasis. Am. J. Hyg., v. 27 (2): 301-310.

MORGAN, D. C. 1928.—*Parastrongyloides winchesi* gen. et sp. nov. A remarkable new nematode parasite of the mole and the shrew. J. Helm., v. 6: 79-86, 4 figs.

MUELLER, J. F. 1934.—Additional studies on parasites of Oneida Lake fishes including descriptions of new species. Bull. N.Y. State Coll. Forestry, v. 7: 335-373.

NESHI, G. 1918.—Ueber die Entwicklung des *Trichocephalus* innerhalb des Wirtes. (Japanese). Tokyo. Med. Wochenschr., No. 2080.

NÉVEU-LEMAIRE, M. 1936.—Traité d'helminthologie médical et vétérinaire, xxiii + 1514 pp., Vigot Frères, Paris.

NISHIGORI, M. 1925.—On the life history of *Hepaticola hepatica*. Second Report, J. Formosa Med. Soc., v. 247: 3-4.
    1928.—The factors which influence the external development of *Strongyloides stercoralis* and on auto-infection with this parasite. Ibid., v. 277: 1-56.

NOE, G. 1908.—Il ciclo evolutivo della *Filaria grassi*, mihi, 1907. Atti R. Accad. Lincei, Roma, Rendic. Cl. Sc. Fis., Math. & Nat., an. 305, 5. s., v. 17 (5), 1. semester: 282-293, figs. 1-4.

NOLAN, M. O. and REARDON, L. 1939.—Studies on oxyuriasis, XX. The distribution of the ova of *Enterobius vermicularis* in household dust. J. Parasit., v. 25: 173-177.

O'CONNOR, F. W. 1931.—Filarial periodicity with observations on the mechanism of the migration of the microfilariae from the parent worm to the blood stream. Puerto Rico J. Pub. Health & Trop. Med., v. 6: 263-272.

OKADA, R. 1931.—Experimental studies on the oral and percutaneous infection of *Anchylostoma caninum* (Four Reports), Jap. J. Exper. Med., v. 9: 209-280.

OLT, A. 1932.—Das Aneurysma verminosum des Pferdes und seine unbekannten Beziehungen zur Kolik. Deutsch. Tierärztl. Wochenschr., v. 40 (21): 326-332, figs. 1-3.

ORTLEPP, R. J. 1922.—On the hatching and migration in a mammalian host of larvae of ascarids normally parasitic in cold-blooded vertebrates. J. Trop. Med. & Hyg., v. 25: 97-100.
    1923.—The life history of *Syngamus trachea* (Montagu) v. Siebold, the gapeworm of chickens. J. Helm., v. 1 (3): 119-140.
    1925.—Observations on the life history of *Triodontophorus tenuicollis*, a nematode parasite of the horse. Ibid., v. 3: 1-14.
    1937.—Observations on the morphology and life history of *Gaigeria pachyscelis* Raill. and Henry, 1910: a hookworm parasite of sheep and goats. Onderstepoort J. Vet. Sc. & Anim. Indus., v. 8 (1): 183-212.

PAVLOV, P. 1937.—Récherches éxpèrimentales sur le cycle évolutif de *Synthctocaulus capillaris*. Ann. Parasit., v. 15: 500-503, pl. 14.

PEREIRA, C. V., VIANNA DIAS, M. and DE AZEVEDO, P. 1936.—Biologia do nematoide *Procamallanus cearensis* n. sp. (English summary, p. 225). Arch. Inst. Biol., Sao Paulo, v. 7: 209-226.

PHILPOT, F. 1924.—Notes on the eggs and early development of some species of Oxyurides. J. Helm., v. 2: 239-252.

PROMMAS, C. and DAENGSVANG, S. 1933.—Preliminary report of a study on the life cycle of *Gnathostoma spinigerum*. J. Parasit., v. 19 (4): 287-292.
    1936.—Further report of a study on the life cycle of *Gnathostoma spinigerum*. J. Parasit., v. 22 (2): 180-186.
    1937:—Feeding experiments on cats with *Gnathostoma spinigerum* larvae obtained from the second intermediate host. Ibid., v. 23 (1): 115-116.

RAILLIET, A. 1899.—Evolution sans hétérogonie d'un angiostome de la couleuvre à collier. Compt. Rend. Acad. Sc., Paris, v. 129 (26): 1271-1273.

RANSOM, B. H. 1906.—The life history of the twisted wire worm (*Haemonchus contortus*) of sheep and other ruminants. Bur. Anim. Indus., U. S. Dept. Agric., Cir. No. 93.
    1907.—*Probstmayria vivipara* (Probstmayr, 1865), Ransom, 1907, a nematode of horses heretofore unreported from the United States. Tr. Am. Micr. Soc., v. 27: 33-40.
    1911.—The nematode parasites in the alimentary tract of cattle, sheep and other ruminants. Bur. Anim. Indus., U. S. Dept. Agric., Bull. No. 127.

RANSOM, B. H., and CRAM, E. B. 1921.—The course of migration of *Ascaris* larvae. Am. J. Trop. Med., v. 1: 129-156, 2 pls.

RANSOM, B. H. and FOSTER, W. D. 1917.—Life history of *Ascaris lumbricoides* and related forms. J. Agric. Res., v. 11: 395-398.

    1920.—Observations on the life history of *Ascaris lumbricoides*. U. S. Dept. Agric., Bull. No. 817, 47 pp.

RANSOM, B. H. and HALL, M. C. 1916.—The life history of *Gongylonema scutatum*. J. Parasit., v. 2: 80-86.

REFUERZO, P. G. and GARCIA, E. Y. 1938.—The crustacean intermediate hosts of *Gnathostomum spinigerum* in the Philippines and its pre- and inter-crustacean development. Philippine J. Anim. Indus., v. 5 (4): 351-362, 5 pls.

ROUBAUD, E. and DESCAZEAUX, J. 1921.—Contributions à l'histoire de la mouche domestique comme agent vecteur des habronèmoses des équidés. Cycle évolutif et parasitisme de l'*Habronema megastoma* (Rudolphi, 1819) chez la mouche. Bull. Soc. Path. Exot., v. 14: 471-506.
    1922.—Evolution de l'*Habronema muscae* Carter chez la mouche domestique et de l'*H. microstomum* Schneider chez le stomoxe. Note préliminaire. Ibid., v. 15: 572-574.
    1922.—Évolution de l'*Habronema muscae* Carter chez dans leurs rapports avec l'évolution des Habronèmes d'équidés. Ibid., v. 15: 978-1001.

ROBERTS, F. H. S. 1934.—The large roundworm of pigs, *Ascaris lumbricoides* L. 1758. Anim. Health Sta., Queensland Dept. Agric. & Stock, Bull. No. 1, 81 pp., 11 figs., 2 pls.
    1937.—Studies on the biology and control of the large roundworm of fowls, *Ascaridia galli* (Schrank, 1788). Ibid., Bull. No. 2, 106 pp., 7 pls.

ROSS, I. C. and KAUZAL, G. 1932.—The life cycle of *Stephanurus dentatus* Diesing, 1839; the kidney-worm of pigs; with observations on its economic importance in Australia and suggestions for its control. Austral. Council Sc. & Indus. Res., Bull. No. 58, 80 pp., illus.

SANDGROUND, J. H. 1926.—Biological studies on the life cycle in the genus *Strongyloides*, Grassi, 1879. Am. J. Hyg., v. 6 (3): 337-388.
    1939.—*Cephalobus parasiticus* n. sp. and pseudostrongyloidiasis in *Macaca irus mordax*. Parasit., v. 31 (1): 132-137.

SCHWARTZ, B. 1922.—Observations on the life cycle of *Ascaris vitulorum*, a parasite of bovines in the Philippine Islands. Preliminary Paper. Philippine J. Sc., v. 20 (6): 661-669, 1 pl.
    1925a—Two new larval nematodes belonging to the genus *Porrocaecum* from mammals of the order Insectivora Proc. U. S. Natl. Mus., v. 67, Art. 17, 8 pp., 1 pl.
    1925b—Preparasitic stages in the life history of the cattle hookworm (*Bunostomum phlebotomum*). J. Agric. Res., v. 29: 451-458.
    1931.—Nodular worm infestation of domestic swine. Vet. Med., v. 26: 411-415.

SCHWARTZ, B. and ALICATA, J. E. 1929.—The development of *Metastrongylus elongatus* and *M. pudendotectus* in their intermediate hosts. [Abstract]. J. Parasit., v. 16: 105.
    1936.—Life history of *Longistriata musculi*, a nematode parasite of mice. J. Wash. Acad. Sc., v. 25 (3): 128-146.

SCHWARTZ, B. and PRICE, E. W. 1929.—The life history of the swine kidney-worm. Science, v. 70 (1825): 613-614.
    1931.—Infection of pigs through the skin with the larvae of the swine kidney worm, *Stephanurus dentatus*. J. Am. Vet. Med. Assoc., v. 79: 359-375.
    1932.—Infection of pigs and other animals with kidney worms, *Stephanurus dentatus*, following ingestion of larvae. J. Am. Vet. Med. Assoc., v. 81, n. s., v. 32 (3): 325-347.

SCOTT, J. A. 1928.—An experimental study of the development of *Ancylostoma caninum* in normal and abnormal hosts. Am. J. Hyg., v. 8 (2): 158-204.

SEURAT, L. G. 1920a.—Historie naturelle des nematodes de le Berbérie, Algiers. 221 + vi pp., 34 figs.
    1920b.—Développement embryonnaire et l'évolution du *Strongylacantha glycirrhiza* Beneden (Trichostrongylidae). Compt. Rend. Soc. Biol., Paris, v. 83: 1472-1474.

SHORB, D. A. 1931.—Experimental infestation of white rats with *Hepaticola hepatica*. J. Parasit., v. 17: 151-154.

SPINDLER, L. A. 1933.—Development of the nodular worm *Oesophagostomum longicaudum* in the pig. J. Agric. Res., v. 46: 531-542.

STAUBLI, C. 1913.—Trichinose. In Handb. Path. Mikroörg., Kolle u. Wassermann, Aufl. 2, Bd. 8, pp. 73-120, 3 pls.

STEFANSKI W. and STRANKOWSKI, M. 1936.—Sur un cas de pénétration du strongle géant dans la rein droit du chien. Ann. Parasit., v. 14: 55-60, 1 pl.

STILES, C. W. and HASSALL, A. 1899.—Internal parasites of the fur seal. In Jordan, D. S. et al. The fur seals and Fur-Seal Islands of the North Pacific Ocean. Pt. 3, pp. 99-177, figs. 1-100. Washington.

SUMMERS, W. A. 1940.—Fleas as acceptable intermediate hosts of the dog heartworm, Drofilaria immitis, v. 43: 448-450.

SWALES, W. E. 1933.—A review of Canadian helminthology. II. Additions to Part I, as determined from a study of parasitic helminths collected in Canada. Canad. J. Res., Sect. D, v. 8 (5): 478-482.
    1936.—Tetrameres crami Swales 1933, a nematode parasite of ducks in Canada. Morphological and biological studies. Ibid., v. 14: 151-164.

TAYLOR, E. L. 1928.—Syngamus trachea from the starling transferred to the chicken, and some physiological variations observed. Ann. Trop. Med. & Parasit., v. 22: 307-318.
    1935.—Syngamus trachea. The longevity of the infective larvae in the earthworm. Slugs and snails as intermediate hosts. J. Comp. Path. v. 48: 149-156.

THOMAS, L. J. 1929.—Philometra nodulosa nov. spec., with notes on the life history. J. Parasit., v. 15: 193-197.
    1937a.—On the life cycle of Contracaecum spiculigerum (Rud.), Ibid., v. 23: 429-431.
    1937b.—Further studies on the life cycle of Contracaecum spiculigerum. Ibid., v. 23: 572.
    1937c.—Life cycle of Rhaphidascaris canadensis Smedley, 1933, a nematode from the pike, Esox lucius. Ibid., v. 23: 572.

TRAVASSOS, L. 1926.—Entwicklung des Rhabdias fülleborni, n. sp. Arch. Schiffs- u. Tropen-Hyg., v. 30: 594-602.

TROISIER, J. and DESCHIENS, R. 1930.—L'Hepaticoliase. Ann. Méd., v. 27: 414-425.

UNDERWOOD, P. C. and HARWOOD, P. D. 1939.—Survival and location of the microfilariae of Dirofilaria immitis in the dog. J. Parasit., v. 25: 23-33.

VAN DURME, P. 1902.—Quelques notes sur les embryons de Strongyloides intestinalis et leur pénétration par le peau. Thompson Yates Lab. Rpt., Liverpool, v. 4 (2): 471-474, pl. 7.

VEGLIA, FRANK, 1916.—The anatomy and life-history of the Haemonchus contortus (Rud.), Dept. Agr. Union S. Africa, 3d & 4th Rpts. Director Vet. Res., pp. 349-500, figs. 1-60.

VOGEL, H. 1930.—Ueber die Organotropie von Hepaticola hepatica. Ztschr. Parasitenk., v. 2 (4): 502-505.

VON BRAND, T. 1938.—Physiological observations on a larval Eustrongylides (Nematoda). J. Parasit., v. 24: 445-451.

WHITE, R. H. 1920.—Earthworms—the important factor in the transmission of gapes of chickens. Md. State Coll. Agric., Bull. No. 234, pp. 103-118.

WALKER, H. D. 1886.—The gapeworm of fowls (Syngamus trachealis), the earthworm (Lumbricus terrestris) its original host, etc. Bull. Buffalo Soc. Nat. Sc., v. 5 (2): 47-71.

WALTON, A. C. 1937.—The Nematoda as parasites of Amphibia. III. Studies on life histories. J. Parasit., v. 23: 299-300.

WEHR, E. E. 1936.—Earthworms as transmitters of Capillaria annulata, the cropworm of chickens. N. Am. Vet., v. 17 (8): 18-20.
    1937.—Observations on the development of the poultry gapeworm, Syngamus trachea. Tr. Am. Micr. Soc., v. 56: 72-78.
    1939.—Studies on the development of the pigeon capillarid, Capillaria columbae. U. S. Dept. Agric., Tech. Bull. No. 679, 19 pp.

WETZEL, R. 1930.—On the biology of the fourth stage larva of Oxyuris equi (Schrank). J. Parasit., v. 17: 95-97.
    1931.—On the biology of the fourth stage larva of Dermatoxys veligera (Rudolphi, 1819) Schneider, 1866, an oxyurid parasitic in the hare. Ibid., v. 18: 40-43.
    1938.—Untersuchungen über die Entwicklung der Pferdestrongyliden. Sitzungsb. Gesellsch. Naturf. Freunde, Mar. 8, 1938, 18-19.
    1938.—Zur Biologie und systematischen Stellung des Dachslungenwurmes. Livro Jub. Travassos, pp. 531-535, 1 pl.

WETZEL R. and ENIGK, K. 1938a.—Wandern die Larvender Palisadenwürmer (Strongylus spec.) der Pferde durch die Lungen? Arch. Wiss. & Prakt. Tierheilk., v. 73 (2): 84-93.
    1938b.—Zur Biologie von Dictyocaulus arnfieldi, den Lungewurm der Einhufer. Ibid., v. 73 (2): 94-114.

WETZEL R. and MÜLLER, F. R. 1935.—Die Lebensgeschichte des schachtelhalmförmigen Fuchslungenwurmes Crenosoma vulpis und seine Bekämpfung. Deutsch. Peltztierzüchter, v. 19: 361-365.

WRIGHT, W. H. 1935.—Observations on the life history of Toxascaris leonina (Nematoda: Ascaridae). Proc. Helm. Soc. Wash., v. 2: 56.

WÜLKER, G. 1929.—Der Wirtwechsel der parasitischen Nematoden. Verhandl. Deutsch. Zool. Gellsch., v. 33: 147-157. (Zool. Anz., Suppl. 4).

YOKOGAWA, S. 1920.—On the migratory course of Trichosomoides crassicauda (Bellingham) in the body of the final host. J. Parasit., v. 7: 80-84.
    1922.—The development of Heligmosomum muris Yokogawa, a nematode from the intestine of the wild rat. Parasit., v. 14: 127-166.
    1926.—On the oral infection by the hookworm. Arch. Schiffs- u. Tropen-Hyg., v. 30: 663-679.
    1938.—Investigation on the mode of transmission of Wuchereria bancrofti (Preliminary Report). Tr. Soc. Path. Jap., v. 28: 619-624.

YOKOGAWA, S. and OISO, T.—Studies on oral infection with Ancylostoma. Am. J. Hyg., v. 6 (3): 484-497.

YOSHIDA, S. 1934.—Contribution to the study on Gnathostomum spinigerum Owen, 1836. Cause of esophageal tumor in the Japanese mink, with especial reference to its life history. Tr. 9th Cong. Far East. Assoc. Trop. Med., Nanking, v. 1: 625-630.

YOSHIDA, S. and TOYODA, K. 1938.—Artificial hatching of Ascaris eggs. Livro Jub. Travassos, Rio de Janeiro, pp. 569-577.

ZAWADOWSKY, M. M. and SCHALIMOV, L. G. 1929.—Die Eier von Oxyuris vermicularis und ihre Entwicklungsbedingungen sowie ueber die Bedingungen unter denen eine Autoinfektion von Oxyuriasis unmöglich ist. Ztschr. Parasitenk., v. 2: 12-43.

CHAPTER XX

FEEDING HABITS OF NEMATODE PARASITES OF VERTEBRATES

J. E. ACKERT and J. H. WHITLOCK, Kansas State College, Manhattan, Kansas

The obscured habitats of the parasitic nematodes preclude ready observations upon their feeding habits. Indications of their nutritive needs have been gained from chemical analyses of the worm bodies. Weinland (1901) found that glycogen made up one-fourth to one-third of the dry substance of the ascarid body. Flury (1912) was led to believe that ascarids had essentially the same chemical constitution as other animals. He found only such minor differences as a lack of uric acid, creatinine and the substitution of a high molecular alcohol (ascaryl alcohol) for glycerol in combination with fatty acids. From these studies it seemed probable that the nutritive needs of parasitic nematodes are fundamentally the same as those of other animals, although Ackert (1930) has shown that there is no evidence to indicate that *Ascaridia galli* needs Vitamin A, Vitamin B (complex), or Vitamin D.

As most of the research on the feeding habits of nematodes has been upon adult forms, they will be discussed first; then the larval forms will be compared with the adults, and the section will close with a brief review of digestion in the parasitic nematodes. Although there are many diverse groups of nematodes, few methods of parasitic feeding have been evolved. The similarity of these feeding habits in nematode groups which are widely separated morphologically would make a discussion of nutrition from a primarily taxonomic standpoint repetitious; hence the subject will be discussed from an ecological and physiological standpoint rather than from that of a morphological classification.

Ecologically, parasitic nematodes may be grouped as to whether they are associated with the physiological interior or exterior of the host body. The physiological exterior of the body, as here considered, is marked by any epithelial membrane lining a cavity which communicates with the exterior of the host body.

Most of the parasitic helminths are associated with the physiological exterior of the body, particularly the mucosae. This group will be subdivided upon the basis of being attached to the mucosae most of the time or usually unattached. Attached nematodes may hold their positions by the buccal capsule grasping the mucosa (*Ancylostoma*, *Necator*, *Strongylus*) or by penetration of the mucosa (*Physaloptera*, *Trichuris*). Nematodes unattached to the mucosa may be closely associated with it (*Haemonchus*, *Metastrongylus*) or not closely associated with it (*Ascaris*, *Ascaridia*, *Heterakis*, *Oxyuris*). Nematodes inhabiting the physiological interior of their hosts are best exemplified by *Dirofilaria*, *Spirocerca* and *Strongyloides*.

Nematodes in the Physiological Exterior of the Body

NEMATODES ATTACHED TO MUCOSA BY BUCCAL CAPSULE

The best examples of this group are the hookworms *Ancylostoma* and *Necator* which apparently remain attached to the intestinal mucosa much of the time. The sucker-like oral opening and the adjacent teeth or cutting plates afford effective means of attachment and blood-letting. Since the earliest recorded observations, blood has been considered as a probable food of hookworms. Grassi, according to Leichtenstern (1886), saw hookworms eject blood both from the mouth and the anus. Leichtenstern thought, as did Grassi, that much more blood is withdrawn by the parasite than is necessary for its food. As the faecally deposited red cells seemed to be practically unchanged, Leichtenstern inferred that the plasma must be the main source of nourishment. In 1888 Ernst noted the emission of blood from the mouth capsule and Whipple (1909) observed the oral and anal emission of blood in both *Necator* and *Ancylostoma*. Whipple believed that there was a rapid ingestion of blood by the parasite. Ackert and Payne (1923) who took *Necator suillus* repeatedly from the intestines of freshly killed swine frequently noted female specimens with bodies colored red from ingested blood. On the other hand some workers, notably Looss (1905) and Ashford and Igaravidez (1911), maintained that blood is not the normal food of hookworms. They observed worms lacking blood even when they were attached to the intestine. They found tissue elements and shreds of mucosa in both the esophagus and intestine and concluded that the parasites fed primarily on the mucous membranes; the blood in the tract was thought to be due to accidental hemorrhage from hookworm bites. Support for the view that portions of the mucosae serve as food for such worms was given by Hoeppli (1930), who found that *Ancylostoma duodenale* is more than a blood sucker. The piece of mucous membrane taken in by the mouth capsule is rasped by the teeth.

Blood from surface vessels pours into the buccal capsule where secretions from the esophageal glands partially digest the blood and loosened portions of the mucosae. "After this digestion has taken place, the liquefied masses are swallowed by the worm." Evidence that disintegration had taken place was furnished by staining the tissue at the bottom of the mouth capsule and in the lumen of the esophagus.

Wells (1931) in a series of ingenious experiments was able to observe living *Ancylostoma caninum* in the act of feeding. He was able to observe the attachment of the hookworms to the mucosa, study the details of the blood sucking process, the passage of the blood through the intestine of the worm, and its ejection through the anal orifice. From the volumes of blood withdrawn by the parasite and the rapid rate at which it passed through the intestine, Wells was of the opinion that the food of the hookworm consists of simple diffusable substances in the host blood.

In studies upon the food of the dog hookworm, *Ancylostoma caninum*, Hsü (1938) made serial sections of hookworms taken from living hosts and found red blood cells in all worms; he also found fragments of host tissue and white blood cells all of which were in stages of disintegration. As further evidence of blood as food for the hookworms, Hsü reported the finding of pigment granules in the cytoplasm of the worms' intestinal cells, which gave positive iron reaction. These granules, which were found in large quantities throughout the whole intestine, the author interpreted as owing their origin to the breaking down of red blood corpuscles. Hsü did not find intestinal contents of the host, worm eggs, or bacteria in the hookworms' intestines. He concluded that the food of *A. caninum* consisted of blood and mucosa cells.

Other nematodes that attach by means of buccal capsules include such forms as *Strongylus*, *Chabertia* and *Camallanus*, all of which are known to draw intestinal epithelium into their mouths. From the disintegrated condition of the tissue so drawn in, it is probable that epithelial tissue and blood form a portion of their food. Whitlock (unpublished), who has worked extensively with living equine strongyles, has noted in the worm intestines material resembling partially digested blood. Wetzel (1931b), studying *Chabertia ovina* (Fab.) in sheep colons, found that the nematodes feeds on the propria mucosae which it draws into its buccal capsule. The tissue fragments which are loosened by the gnawing of the nematode are partially digested, according to Wetzel, by secretions from the dorsal esophageal gland.

Support of the view that *C. ovina* attacks the mucous membrane is afforded by the work of Kauzal (1936) who found numerous small hemorrhages on the mucous membrane of the large intestine of sheep which he attributed to *C. ovina*. He examined 250 of these specimens quantitatively for iron which he assumed to be derived from haemoglobin. The presence of the haemoglobin and the reddish tint of the intestinal contents of the immature *C. ovina* led Kauzal to infer that this nema ingests considerable quantities of blood. That the attachment to the intestinal mucosae by the buccal capsule is a widespread feeding phenomenon among the Strongyloidea is further shown by Magath (1919) for *Camallanus americanus* in the turtle intestine and by Hoeppli and Hsü (1931) for *Kalicephalus* sp. in the enteric canal of snakes.

NEMATODES ATTACHED MUCH OF THE TIME BY PENETRATION OF THE MUCOSA

Typical examples of this group are *Trichuris* and *Physaloptera*. The food of *Trichuris* apparently is secured while the anterior extremity is imbedded in the mucous membrane of the large intestine. Christofferson (1914) who reviewed the literature on *Trichuris* (*Trichocephalus*) up to 1914, observed a peculiar cell transformation about the imbedded anterior portion of these nematodes. Hoeppli (1927) found such changes in human and baboon intestines in which the *Trichuris* made tunnels in the mucosa parallel with the surface of the intestinal lining. Surrounding the anterior ends of the trichurids in these tunnels, the epithelial cells, according to Hoeppli, were transformed into syncytial-like structures with eosinophilic homogeneous protoplasm and pycnotic nuclei as a result of the action of a liquifying secretion from the worm. Hoeppli's studies (1927, 1933) led him to believe that the liquified syncytial material was taken by the trichurids as food.

That *Trichuris* may take blood was indicated by the studies of Guiart (1908) who found blood-engorged trichurids. Garin, cited by Otto (1935), likewise found *Trichuris* filled with blood and reported that blood was found in the stools of 50 out of

54 trichuris-infected patients. In support of the view that these nematodes may feed on blood were the findings of Li (1933c) and Chitwood and Chitwood (1937) that the adult *Trichuris* bears a stylet capable of insertion sufficient for drawing blood. Chitwood and Chitwood (1937) showed that the anterior muscular part of the trichuroid esophagus possesses muscles capable of the dilation necessary for sucking. Moreover, they found by serial sections a large number of red corpuscles in the esophageal lumen of *Trichuris*. While Smirnov (1936), after a comprehensive study of the literature and of serial sections of worms, concluded that there was no convincing evidence that trichurids feed on blood, the fact remains that Whipple (1909) and Garin (1913) reported the occurrence of hemolytic enzymes in *Trichuris*. From the various studies made it appears that trichurids secure their food from the intestinal mucosa and that it may consist of liquified mucosal tissue and blood elements.

Studies on the food of *Trichinella spiralis* were made by Heller (1933) who introduced encysted trichina larvae enclosed in collodion sacs into the small intestines of cats and rats. While the meat around the larvae was digested in 6 to 8 hours, they made no growth in 1 to 3 days. *Trichinella* larvae enclosed in fine silk bags and thus kept away from the intestinal mucosa likewise did not develop. That these nearly adult trichinae do not feed on intestinal contents seems likely also from other tests by Heller who fed india ink along with meat containing encysted larvae, but failed to find any ink in the worms' intestines. Sections of intestinal tissue made after the encysted trichinae were fed showed that the freed larvae penetrated the intestinal mucosa, where the maturing trichinae doubtless secure their food.

Following the work of Heller, McCoy (1934) injected sterile *Trichinella* larvae from digested rat muscle into the amniotic sac of chick embryos 6 to 15 days old. Definite growth of the maturing larvae occurred in only about 1 or 2 percent of the trichinae. A single female developed to sexual maturity. Better success was attained in a second series of experiments in which the sterile larvae were injected into the amniotic sacs of rat embryos on approximately the 14th day of gestation. In 2 to 5 days, practically all worms were developing at nearly the normal rate and on the fifth day, numerous female trichinae were found with embryos developing in their uteri. These results give further evidence that *Trichinella* normally feeds upon host body fluids secured from the mucous membrane. Moreover, van Someren (1939) reported a functional buccal stylet in *T. spiralis* and indicated that it is used to lacerate the host tissue and release tissue fluids. From the examination of living specimens immediately after recovery, van Someren believed that the food, which is in a fluid state when ingested, consists of tissue fluids, cell contents, or perhaps predigested tissue acted on by a tissue lysate from the anterior esophageal glands.

Among other nematodes attached much of the time to the enteric mucosa is *Physaloptera*. Studies by Hoeppli and Feng (1931) showed that the mucosa about the anterior ends of these attached worms was liquified or partially digested, presumably from esophageal secretions from the nematodes. Studies of sectioned mucosa showed definite excavation of tissue immediately around the anterior ends of the worms, presumably from the taking of the liquified tissue as food.

UNATTACHED NEMATODES CLOSELY ASSOCIATED WITH THE MUCOSA

Many nematodes belonging to this group, while having poorly developed buccal capsules, are able to puncture the mucous membrane and draw blood. For example, Stadelmann (1891) found blood corpuscles in the nearly mature *Ostertagia ostertagi* (*Strongylus convolutus*) in nodules of the abomasum, and Dikmans and Andrews (1933) found such stages of *Ostertagia circumcincta* in the mucosa and partly free in the abomasal lumen of sheep. Unfortunately, however, much of the evidence is circumstantial. Thus Ransom (1911), writing of *Haemonchus contortus* and *Ostertagia ostertagi*, stated that they evidently suck blood for the heavily infected hosts are anaemic. Other writers simply state that they suck blood. Veglia's (1916) observations on living worms demonstrated that the oral lancet made definite cutting movements. Fallis (1938) placed *Nematodirus* among the blood suckers on the basis of a spectroscopical analysis of the body fluid which showed the absorption bands for oxyhaemoglobin. In the same year, Davey (1938), who studied the food of nematodes of the alimentary tract of sheep, questioned the spectroscopic demonstration of haemoglobin in nematodes as evidence of their being blood suckers. He found that haemoglobin was present in tissues other than the alimentary canal of *Nematodirus spathiger* and that its absorption bands had different positions from those in the blood of their hosts. These facts, together with his finding of haemoglobin in species of Trichostrongylidae long after any haemoglobin from the host would have decomposed, proved

that these nematodes could synthesize haemoglobin. Davey's (1938) culture tests with serum, blood digests, and defibrinated blood as food for *Ostertagia, Cooperia, Nematodirus* and *Trichostrongylus* were unsuccessful, as were also those on abomasum fluid for *Ostertagia circumcincta*. These negative results led him to the conclusion that these nematodes with rudimentary buccal capsules probably feed on tissue elements at or in the mucosa.

Other evidence of intimate association of trichostrongyles with the mucosa of the alimentary tract is available from rabbit nematodes. Alicata (1932) experimentally infected rabbits with *Obeliscoides cuniculi* by feeding infective larvae. Examination about 2 months later showed nematodes free on the mucous membrane of the stomach or under the membrane and into the submucosa. That such trichostrongyles feed from the enteric wall was the opinion of Wetzel and Enigk (1937) who, on infecting rabbits with *Graphidium strigosum*, found the stomach mucosa bloody. Enigk (1938) examined the intestinal contents of several sexually mature *G. strigosum* and found a colorless viscous mass containing nuclear remnants apparently from white blood cells, granules and bacteria. Other tests such as feeding the rabbits pulverized charcoal, trypan blue and carmine resulted in these substances being ingested by the nematodes. Also, injecting the hosts intravenously with trypan blue for several days resulted in the worms taking up several blue colored particles presumably desquamated mucosa cells. Enigk concluded that *G. strigosum's* food consists of gastric mucosa, gastric juice and stomach contents.

Other unattached nematodes closely associated with the mucosa include lungworms which inhabit the bronchi and bronchioles. Hung (1926) studying swine lungs infected with adult *Metastrongylus elongatus* frequently found eosinophiles and red blood corpuscles in the worms' intestines. The findings of Porter (1936), who made similar studies, indicated that the material in the worms' intestines consisted of elements identical with those found in the exudate surrounding the nematodes. In cross sections of the worms, Porter recognized large numbers of eosinophilic and neutrophilic polymorphonuclear leucocytes, lymphocytes and desquamated epithelial cells. Erythrocytes were seen in some instances. These and some of the leucocytes and epithelial cells appeared to have been digested in part by the worms.

From the findings of the investigators cited, the food of many of the unattached strongyles appears to consist mainly of substances derived from the mucosa, namely, leucocytes, erythrocytes, lymphocytes, plasma, exudates and desquamated mucosa cells, but also of some extra-mucosal material such as stomach contents.

UNATTACHED NEMATODES NOT CLOSELY ASSOCIATED WITH THE MUCOSA

Chief among the parasitic nematodes not closely associated with the intestinal mucosa are members of the Oxyuroidea and Ascaridoidea. Among the early observations of the food of such nematodes were those of Leuckart (1876) who found that the intestine of *Enterobius vermicularis* usually contained yellow fluid which on microscopical examination proved to be identical with the liquid host feces. Similar observations were made by Leuckart on *Oxyuris equi* whose intestinal contents contained small particles of vegetable material identical with the contents of the horse intestine.

Early in the present century, Weinberg (1907) examined the intestines of many *Ascaris* specimens but could not find red blood corpuscles in them and expressed the opinion that the horse *Parascaris* feeds on the contents of the host intestine.

To ascertain whether *Ascaris* feeds upon intestinal contents, Vogel, cited by Hoeppli (1927), fed powdered animal charcoal to a human patient infected with *Ascaris*. The results of the first test were negative, but in a second test carried out similarly, numerous charcoal particles were found in the intestine of the worm. On the other hand, a number of early workers held to the view that the ascarids are blood suckers. This view was derived, in part, from microscopical and chemical examinations which showed evidence of blood in the intestines of *Ascaris* and related forms. For example, Mueller (1929), on studying specimens of *Anisakis simplex* from the sperm whale stated that the intestine, in all cases, contained blood in considerable quantities with occasional fragments of muscle and other tissues. From the quantities of blood corpuscles present, Mueller was of the opinion that the nematode had a blood-sucking habit. Mueller was unable to determine the nature of the intestinal contents of any other genus of the Anisakinae that he studied.

If such ascarid forms are blood suckers some specimens should be found in contact with the mucosa. Hoeppli (1927) reported on the examination of 350 cadavers in which large numbers of ascarids were found. No evidence was available to show that any of these nematodes were attached to the mucosa. Hoeppli further stated that in Fülleborn's laboratory no cases had been found with the ascarid, *Toxocara canis*, attached to the dog intestine. Other workers on examining large numbers of horse intestines at slaughter houses always found *Parascaris equorum* free in the lumen of the gut.

Standard textbooks carry the statement that *Ascaris lumbricoides* feeds on intestinal contents but gnaws at the mucosa. This statement doubtless is due to the occasional finding of reddish spots in the intestinal epithelium in cases of ascarid infection. While such spots occur occasionally, those who have examined hundreds of mammalian and avian intestines which contained numerous ascarids can testify that in the great majority of cases, no evidences of the adult ascarids attaching the mucosa are available.

As to certain Ascaridoidea being attached to the intestinal wall presumably for feeding, Guiart, cited by Hoeppli (1927), found in the stomach of a dolphin the clear imprint of the worms' lips in pit-like depressions of the mucous membrane. Similar observations were made by Hoeppli (1927) on a *Contracaecum* sp. from a seal from northern waters. It is quite possible that instead of being attached, the dying worms pressed their anterior ends deeply into the mucous membrane of the dead host.

As to the food of ascarids, Archer and Peterson (1930), by giving patients infected with *Ascaris lumbricoides* a barium-cereal-meal, found that the enteric canals of the parasites showed string like shadows, indicating that the nematodes in the host intestine had swallowed the barium. These observations indicated that *Ascaris lumbricoides* feeds on the intestinal contents of man.

Following this work, Li (1933a) fed to six dogs, positive for ascarids, liquid chinese ink or powdered charcoal twice a day for several days. While most of the tests were negative, due presumably to a vermicidal action of the charcoal, one dog gave unquestioned positive evidence. The one female worm from the dog's intestine definitely showed charcoal and beef particles in its enteric tract. In a subsequent series of tests, Li (1933b) fed a mixture of powdered charcoal, clotted blood, striated beef muscle and starch granules to experimental animals harboring ascarids as follows: Dog, *Toxocara canis*; cat, *Toxascaris leonina*; and chicken, *Ascaridia galli*. The results from the dogs gave no positive evidence; that from the cat showed that the worm intestine contained charcoal, blood cells, and beef particles. These findings were confirmed by examination of paraffin sections of the worms. The results from four chickens showed charcoal and beef particles in the intestines of all worms including both male and female specimens. From similar experiments, in which starch granules were substituted for powdered charcoal, all worms recovered showed starch granules and some beef particles.

To ascertain the nature of food of the chicken cecal worm *Heterakis gallinae*, Li (1933b) fed infected chickens powdered charcoal and beef as before. On examination, most of the worms showed charcoal in the entire intestine. In further studies, Li opened the intestines of *Ascaris lumbricoides* from man and mounted the intestinal contents on slides for microscopic examination. While most of these contents could not be identified, Li found in one specimen two *Ascaris* eggs and a piece of striated muscle. The results of Li's experiments (1933a, 1933b) indicate that the intestinal contents of the host constitute part of the food of *Ascaris lumbricoides*, *Toxocara canis*, *Toxascaris leonina*, of mammals; and *Ascaridia galli* and *Heterakis gallinae* of fowls.

The findings of Li and of other workers cited, while showing that certain ascaroids take intestinal contents, do not preclude the possibility that these nematodes may also feed upon the intestinal epithelium. In a study of the food of the fowl nematode, *Ascaridia galli* (Schneider), Ackert and Whitlock (1935) deprived chickens infected with *Ascaridia galli* of food by mouth; the experimental chickens were nourished by intramuscular injections of glucose. The results of the first series of experiments on 141 chickens with worms of various ages indicated that little growth occurred in the worms after the host chickens were taken off the regular feed. In the second series in which 96 additional chickens were used, Ackert, Whitlock and Freeman (1940) used worm infections of one-week's duration in the tests. Experimental and control chicks under comparison were of the same age and the developing worms were from the same egg cultures. The results of this series of tests were very uniform, namely, that in the chickens given only water *per os* and intramuscular injections of glucose, the young *Ascaridia galli* ceased growing whereas the worms in the regularly fed control chickens made normal growth. These results indicate that the large nematode of chickens whose mouth parts are very similar to those of mammalian ascarids, did not secure nutriment from the intestinal epithelium of the host. These nematodes may have fed to some extent on duodenal mucus from the goblet cells but Ackert, Edgar and Frick (1939) have shown recently that such mucus may contain an inhibitory

growth factor for young *Ascaridia* that have been grown in the culture media developed by Ackert, Todd and Tanner (1938). This last group of workers prepared a salt-dextrose solution in which young *Ascaridia galli* will grow. On the introduction of mucus from growing chickens into the nutrient solution, the cultured *Ascaridia* ceased growing, whereas the control worms in the nutrient solution continued to increase in length. In the glucose-injected chickens, the duodenal mucus, while containing an inhibitory growth factor, may have afforded the *Ascaridia galli* food sufficient for maintaining life, but not for growth.

The literature cites cases in which blood has been found in the digestive tracts of ascaroids. For example, Mueller (1929) found blood in the intestine of *Anisakis simplex* and Guiart, cited by Lievre (1934), saw some *Ascaris* whose digestive tracts were full of blood. On the other hand, Lievre cited Brumpt as having performed numerous autopsies to see if *Ascaris* caused ecchymotic spots on the mucosa. But Brumpt was unable to find such spots, and the intestinal contents of the worms showed only chyme, never blood.

Indirect evidence of blood as a nutrient of ascarids is available from the finding of haemoglobin in the worms' bodies by such tests as the Benzidine blood test and spectroscopic analysis. That the former is an unreliable test for blood has been shown recently by Davey (1938). Using spectroscopic analysis, Lievre (1934) found traces of haemoglobin in the intestine of the dog ascarid, *Toxocara canis*. Even though the spectroscopic examination of blood was positive in 75 percent of the cases, the quantity of haemoglobin noticed was so small that Lievre was led to think that the haemoglobin had come from the flesh colored food of the animal. Lievre, on macerating the intestines of *Ascaris lumbricoides*, *Parascaris equorum* and *Ascaris suum* was unable to find any haemoglobin present in these worms by spectroscopic analysis. He concluded that there is no haemophagia in *Ascaris* and only in exceptional circumstances would there be ingestion of blood. Davey (1938) demonstrated haemoglobin in the dermo-muscular tube of *Toxocara canis* and in tissues other than the alimentary canal of *Ascaris*. He found, further, that the absorption bands of the haemoglobin in the tissues had different positions from those in the blood of their hosts, indicating that these nematodes were able to synthesize haemoglobin. Thus the presence of haemoglobin in the tissues of nematodes is not necessarily evidence that they feed on blood.

Further indicative evidence that ascarids may take blood is available from the work of Schwartz (1921) who found that the body fluid of *Ascaris lumbricoides* inhibited coagulation of rabbit blood to a moderate extent. Extracts of *Parascaris equorum* and of *Toxocara* sp. had a slight effect on the coagulation of sheep's blood. Whether or not this property of the ascarid body is utilized by the living nematodes is unknown. It is conceivable that ascarid nematodes living with hookworms which are known to draw blood in excess would swallow blood from time to time. But as other writers have indicated, this would be exceptional rather than normal.

In the light of our present knowledge, it appears that the oxyuroids and ascaridoids feed normally on the intestinal contents of the host including also any mucus, desquamated mucosa cells, and blood elements that may be free in the lumen of the intestine.

Nematodes in the Physiological Interior of the Body

The nematodes that live in the physiological interior of the body are exemplified by *Dirofilaria*, *Spirocerca* and intra-mucosal *Strongyloides*. In a recent study, Hsü (1938a) was led to believe that *Dirofilaria immitis* feeds exclusively on red and white blood cells. In the case of *Diplotriaena tricuspis*, Hsü (1938b), after studying the intestinal contents of this parasite of the crow, concluded that the worms' food consisted of the inflammatory exudate in the thoracic cavity. While there was evidence of blood being ingested, Hsü believed that it is not taken normally. As the adult *Wuchereria bancrofti* normally lives in lymph vessels and nodes, it doubtless normally feeds upon lymphocytes and other constitutents of lymph. When such encapsulation as shown by Faust (1939) occurs, the encapsulated worm dies, apparently from lack of food.

In a further study of the food of nematodes, Hsü (1938a) concluded that *Spirocerca lupi* feeds on inflammatory cells that pass through the nodule walls.

Observations on the food of another nematode of this group, *Strongyloides stercoralis*, were recorded by Askanazy (1900) who concluded that the mother worms in the intestinal mucosa fed on chyle; he found no indication that they take blood. Faust (1935), studying *Strongyloides* in the mucosa of the jejunum, found evidence of lytic action by the female worms, particularly around the head of the worm where disintegration of the tissue was observed. Considering the facts that the adult females spend much of their time in the intestinal mucosa and that they have not been observed to take blood, it is probable that they feed on chyle and the partially disintegrated tissues in their tunnels.

The Food of Larval Parasitic Nematodes

The nutrition of the larval parasitic nematodes is fundamentally like that of adult parasitic or free-living nematodes subject to the modifications imposed by the environment and the structure of the larvae in question. For example, the researches of McCoy (1929), Lepage (1933, 1937) and Glaser and Stoll (1938) on the free-living stages of Strongylina have revealed no essential differences between the mode of nutrition of these immature forms and that of the free-living stages of Rhabdiasidae (see Chu, 1936) and Strongyloididae (see Faust, 1932). The feeding mechanisms (buccal capsule and rhabditoid esophagus) and sources of food (bacteria or fluid organic matter) are essentially the same. The method of feeding of *Rhabditis* as described by Chitwood and Chitwood (1938, p. 76 of this series) is probably typical of this group.

Of the larvae of heteroxenous, parasitic nematodes in their intermediate host, no complete study of the feeding habits is available. But their locations in the intermediate hosts are fundamentally similar to those of various adult forms in primary hosts. Since many of such larvae increase in size without the presence of reserve food stuff, they must secure their nourishment from their host. From the foregoing it would be logical to conclude that larval nematodes in secondary hosts feed as do adult nematodes in analagous positions in primary hosts. Inactive encysted forms are at such low levels of metabolic activity in both types of host that simple diffusion is probably more than adequate to maintain the parasite.

The nutrition of immature nematodes in a primary host is, as far as known, like that of adult nematodes in similar positions except for (a) larval nematodes carrying reserve foodstuff and (b) larval nematodes which may be nourished by diffusion. For the rest it is possible to find a larval nematode feeding habit identical with each major type of adult nematode method of feeding.

Wetzel (1930) has shown fourth stage *Oxyuris equi* to feed like adult *Strongylus* sp. Ortlepp (1937) found the same true of larval *Gaigeria pachyscelis*. According to Ackert (1931), *Ascaridia galli* larvae penetrate the mucosa of the small intestine and feed much as do *Physaloptera* sp. or *Strongyloides* sp.

In a study of *Cooperia curticei*, Andrews (1939) noted the third stage larvae feeding in the lumen of the gut. The fourth stage larvae of this parasite had their anterior ends in the crypts of Lieberkühn and grew while in this position indicating the same type of nutrition as that observed in the adult Trichostrongylidae.

Immature *Probstmayria vivipara* are found free in the gut tube like ascarids indicating a similar mode of nutrition. According to Ransom (1911) immature *Oesophagostoma columbianum* feed on the cheesy material in the nodule making their mode of nutrition essentially similar to such forms as *Gnathostoma*. Ascarid larvae in the blood stream ingest and digest blood cells, according to Smirnov (1935), hence resembling adult *Dirofilaria*. Wetzel (1931a) has reported a case of what he considers to be extra-intestinal digestion by the fourth stage larva of *Dermatoxys veligera* which attaches to the intestinal mucosa by means of four cephalic hooks, a unique attachment mechanism in nematodes.

The recent development of culturing techniques for parasitic nematodes promises more information regarding their food. However, to date only one parasitic nematode has been cultured throughout its life cycle. This is *Neoplectana glaseri* which is parasitic in the Japanese beetle, *Popillia japonica* (Glaser, 1932). Attempts to grow *Haemonchus contortus* of sheep by Lapage (1933) and Glaser and Stoll (1938), and *Ascaridia galli* of chickens by Ackert, Todd and Tanner (1938) have been only partially successful. Hence, these are included with the discussion of the nutrition of the larval forms. No direct observations of the food of these parasitic nematodes have been made, but the fact that the nematodes have grown and developed in an artificial environment indicates that at least part of the environment is a source of food. Table 1 lists these attempts at culturing parasitic nematodes.

From these considerations it appears that the food of larval parasitic nematodes may include bacteria, enteric contents, vascular fluids and elements, and mucosal cells and tissues.

Digestion in Parasitic Nematodes

Most of the parasitic nematodes are placed in intimate contact with the host's physiological fluids which carry nutrient materials to its cells. Since these nutrients are in their simplest diffusable form it might be assumed that much of the nourishment of parasitic nematodes is derived from this source and that no true digestion is required. However, a number of

TABLE 1.—*Summary of attempts to culture parasitic nematodes.*

Author	Nematode	Degree of Growth	Successful Culture Media	Normal host
Glaser (1932)	*Neoplectana glaseri*	Complete	Dextrose - veal infusion agar with yeast	Japanese beetle
McCoy and Glaser (1936)	*Neoplectana glaseri*	Complete	Fermented potato medium	Japanese beetle
McCoy and Girth (1938)	*Neoplectana glaseri*	Complete	Veal infusion & preservatives	Japanese beetle
Glaser and Stoll (1938)	*Haemonchus contortus*	Last part of fourth larval stage	Agar, liver extract, sheep blood and kidney defibrinated blood	Sheep
Ackert, Todd and Tanner (1938)	*Ascaridia galli*	Measurable growth.	Incubating hens' eggs, starch, dextrose, commercial agar	Chicken

workers have demonstrated the existence of a true digestion in phylogenetically widely separated parasitic nematodes; hence, it is probable that they all carry on some form of digestion. According to the location of the digestive processes, various workers have distinguished between an intestinal and extra-intestinal digestion. Much of the evidence of extra-intestinal digestion rests upon the occurrence of necrotic or cytolyzed material around the anterior attached end or within the buccal capsule of parasitic worms. That such a condition of the host tissue is so often interpreted as extra-intestinal digestion is somewhat questionable since the effect of parasite excretions, simple trauma, mechanical pressure, and heterophilogenous proteolytic enzymes upon the host tissue would produce many of the conditions described as extra-intestinal digestion. This form of digestion may be possible in some nematodes, however, since Hoeppli (1927) has discovered an epitheliolytic material present in the anterior end but not in the posterior portion of *Strongylus*.

The intestinal digestion in nematodes has been the subject of work by a considerable number of investigators. Most of the studies have been confined to the demonstrations of enzymes within extracts of the parasites. Because of the early workers' limited knowledge of enzyme action, much of their results need confirmation before they can be accepted. Flury (1912), for example, made no attempt to critically evaluate the research of other workers. He simply listed the worker's name and the enzymes which he reported. In little of this early work was the action of bacterial enzymes adequately controlled. The demonstration of a peptolytic enzyme in the gut of *Toxocara canis* by Abderhalden and Heise (1909) is questionable for this reason. Nor was any particular attempt made in the early work to differentiate between intracellular and extracellular enzymes. Most of it was done with extracts of the parasite being studied, and peroxidases and proteases were reported as though the question of their respective origins was of little importance.

Recent researches have been more accurate. Enigk (1938) showed that *Graphidium strigosum* produces an amylase and a protease which are active in the gut tube of the parasite. He was unable to demonstrate a lipase. Chitwood (1938) demonstrated in an extract of the esophagus of *Ascaris lumbricoides*, a proteolytic enzyme which was inactive at its isoelectric point (ph 8.0) and most active in a weak acid solution. The fact that such workers as Wetzel (1928) and Hoeppli and his coworkers have demonstrated digested epithelial cells within the alimentary tracts of certain parasitic nematodes, gives evidence of the existence of proteolytic enzymes in parasitic nematodes. Enigk's (1938) finding of a varying reaction in the gut tube of *Graphidium strigosum* (pH 7.0 at the ends and 4.4-4.8 in the middle), and Van Someren's (1939) report of an acid reaction in the intestine and rectum of *Trichinella spiralis* are additional confirmation of the presence of enzymes because alteration of the reaction of the digestive tract of animals is universally coordinated with the optimum pH for the enzymes present.

Anticoagulants in blood sucking nematodes have been demonstrated by a number of workers. While such products are not primarily digestive, they doubtless prevent blocking of the parasite's alimentary tract with clotted blood; hence, they are an aid to digestion. Such products have been reported by Schwartz (1921) and Hoeppli and Feng (1933).

Careful consideration of the relative values of the researches demonstrating the presence of enzymes leads to the conclusion that at least one and probably more proteolytic enzymes are present in the intestine of many parasitic nematodes as well as at least one amylytic enzyme. Demonstrations of lipases have been impossible or questionable. Although little research has been done on digestion in parasitic nematodes, the demonstration of these enzymes lends weight to the hypothesis that digestion in parasitic nematodes is essentially like that of other animals possessing a digestive tract.

Bibliography

ABDERHALDEN, E., and HEISE, R. 1909.—Neber das Vorkommen peptolytischen Fermente bei den Wirbellosen. Ztschr. Physiol. Chem., v. 62:136-138.

ACKERT, J. E. 1930.—Vitamin requirements of intestinal nematodes. Anat. Rec., v. 47(3):363.
  1931.—The morphology of life history of the fowl nematode *Ascaridia lineata* (Schneider). Parasit., v. 23 (3):360-379, pls. 13-14, figs. 1-50.

ACKERT, J. E., EDGAR, S. A. and FRICK, L. P. 1939.—Goblet cells and age resistance of animals to parasitism. Tr. Am. Micr. Soc., v. 58:81-89.

ACKERT, J. E. and PAYNE, FLORENCE KING. 1923.—Investigations on the control of hookworm disease. XII. Studies on the occurrence, distribution and morphology of *Necator suillus*, including descriptions of the other species of *Necator*. Am. J. Hyg., v. 3(1):1-25, pls. 1-2, figs. 1-23.

ACKERT, J. E., TODD, A. C. and TANNER, W. A. 1938.—Growing larval *Ascaridia lineata* (Nematoda) *in vitro*. Tr. Am. Micr. Soc., v. 57:292-296.

ACKERT, J. E. and WHITLOCK, J. H. 1935.—Studies on ascarid nutrition. J. Parasit., v. 21:428.

ACKERT, J. E., WHITLOCK, J. H. and FREEMAN, A. E., JR. 1940. —The food of the fowl nematode, *Ascaridia lineata* (Schneider). J. Parasit., v. 26(1):17-32, fig. 1.

ALICATA, JOSEPH E. 1932.—Life history of the rabbit stomach worm *Obeliscoides cuniculi*. J. Agric. Res., v. 44(5):401-419, figs. 1-12.

ANDREWS, J. S. 1939.—Life history of the nematode *Cooperia curticei* and development of resistance in sheep. J. Agric. Res., v. 58:771-785, figs. 1-3.

ARCHER, V. W. and PETERSON, C. H. 1930.—Roentgen diagnosis of ascariasis. J. Am. Med. Assoc., v. 95:1819-1821.

ASHFORD, B. K. and IGARAVIDEZ, P. G. 1911.—Uncinariasis in Porto Rico. U. S. 61st Cong., 3d Sess., Senate Doc. 808, 335 pp.

ASKANAZY, M. 1900.—Ueber Art und Zweck der Invasion der Anguillula intestinalis in die Darmwand. Centrabl. Bakt. [etc.], Abt. 1, v. 27:569, figs. 1-4.

CHITWOOD, B. G. 1938.—Notes on the physiology of *Ascaris lumbricoides*. Proc. Helm. Soc. Wash., v. 5(1):18-19.

CHITWOOD, B. G. and CHITWOOD, M. B. 1937.—The histology of nemic esophagi. VIII. The esophagus of representatives of the Enoplida. J. Wash. Acad. Sc., v. 27(12):517-531, figs. 1-2.

CHRISTOFFERSON, N. R. 1914.—*Trichocephalus dispar* im Darmkanal des Menschen. Zieglers Beitr. Path. Anat. & Allg. Path., v. 57:474-515.

CHU, TSO-CHIH. 1936.—Studies on the life history of *Rhabdias fuscovenosa* var. *catanensis* (Rizzo, 1902). J. Parasit., v. 22(2):140-160, figs. 1-8.

DAVEY, D. G. 1938.—Studies on the physiology of the nematodes of the alimentary canal of sheep. Parasit., v. 30(3):278-295, fig. 1.

DIKMANS, G. and ANDREWS, J. S. 1933.—A note on the life history of *Ostertagia circumcincta*. J. Parasit., v. 20(2):107.

ENIGK, K. 1938.—Ein Beitrag zur Physiologie und zum Wirt-Parasit-verhältnis von *Graphidium strigosum* (Trichostrongylidae, Nematoda). Ztschr. Parasitenk., v. 10(3):386-414.

ERNST, J. 1888.—Einige Fälle von Ankylostomiasis, nebst Sectionsbefunden. Deutsch. Med. Wochenschr., v. 14:291-294.

FALLIS, A. MURRAY. 1938.—A study of the helminth parasites of lambs in Ontario. Tr. Roy. Canad. Inst., v. 22, pt. 1 (47):81-128, figs. 1-30.

FAUST, ERNEST CARROLL. 1932.—The symptomatology, diagnosis and treatment of *Strongyloides* infection. J. Am. Med. Assoc., v. 98:2276-2277.
  1935.—Experimental studies on human and primate species of *Strongyloides*. IV. The pathology of *Strongyloides* infection. Archiv. Path., v. 19:769-806.
  1939.—Human helminthology Philadelphia. Lea & Febiger. 780 pp.

FLURY, FERDINAND. 1912.—Sur Chemie und Toxokologie der Ascariden. Arch. Exper. Path. & Pharmakol., v. 67:275-392.

GARIN, C. 1913.—Recherches physiologiques sur la fixation et le mode de nutrition de quelques nematodes parasites du tube digestif de l'homme et des animaux. Univ. Lyon. n.s., I. Sc. Med., v. 34:160 pp., figs. 1-55.

GLASER, R. W. 1932.—Studies on *Neoaplectana glaseri*, a nematode parasite of the Japanese beetle (*Popillia japonica*). N. J. Dept. Agric. Circ. No. 211. 34 pp. pls. 1-3, figs. 1-17.

GLASER, R. W. and STOLL, NORMAN R. 1938.—Sterile culture of the free-living stages of the sheep stomach worm, *Haemonchus contortus*. Parasit., v. 30(3):324-332, figs. 1 3.

GUIART, J. 1908.—Le trichocephale vit aussi dans l'intestin grele et se nourrit de sang. Lyon Med., v. 110(6):325-326.

HELLER, M. 1933.—Entwickelt sich die *Trichinella spiralis* in der darmlichtung ihres wirtes? Ztschr. Parasitenk., v. 5:370-392.

HOEPPLI, R. 1927.—Ueber Beziehungen zwischen dem biologischen Verhalten parasitischer Nematoden und histologischen Reaktionen des Wirbeltierkoerpers. Arch. Schiffs-u. Tropen-Hyg., v. 31(3):207-290.
  1930.—Parasitic nematodes and the lesions they produce. Natl. Med. J. China., v. 16:103-110.
  1933.—On histolytic changes and extra-intestinal digestion in parasite infections. Lingnan Sc. J., v. 12 (Suppl.):1-11, pl. 1, figs. 1-3.

HOEPPLI, R. and FENG, L. C. 1931.—On the action of esophageal glands of parasitic nematodes. Chinese Med. J., v. 17:589-598, pls. 1-3, figs. 1-6.
  1933.—The presence of an anticoagulin in the esophagus of *Bunostomum trigonocephalum* from the intestine of sheep. Arch. Schiffs-u. Tropen-Hyg., v. 37(4):176-182.

HOEPPLI, R. and HSÜ, H. F. 1931.—Histological changes in the digestive tract of vertebrates due to parasitic worms. Chinese Med. J., v. 17:557-566, pls. 1-2, figs. 1-4.

HSÜ, H. F. 1938a.—Studies on the food and the digestive system of certain parasites. I. On the food of the dog hookworm *Ancylostoma caninum*. Bul. Fan Mem. Inst. Biol., Zool., Ser., v. 8(2):121-132, pls. 12-13, figs. 1 6.
  1938b.—Idem. II. On the food of *Schistosoma japonicum*, *Paragonimus ringeri*, *Dirofilaria immitis*, *Spirocerca sanguinolenta*, and *Rhabdias* sp. Ibid., v. 8:347-366.
  1938c.—Idem. IV. On the food of *Diplotriaena tricuspis* (Nematoda). Ibid., v. 8:403-406.

HUNG, SEE-LU. 1926.—The histological changes in lung tissue of swine produced by *Metastrongylus elongatus*. N. Am. Vet., v. 7(1):21-23, figs. 1-3.

KAUZAL, G. 1936.—Further studies on the pathogenic importance of *Chabertia ovina*. Austral. Vet. J., v. 12:107-110.

LAPAGE, GEOFFREY. 1933.—Cultivation of parasitic nematodes. Nature, v. 131:583-584.
  1937.—Nematodes parasitic in animals (Monograph). London. Methuen and Co., Ltd. 172 pp.

LEICHTENSTERN, O. 1886.—Weitere Beiträge zur Ankylostomafrage. Deut. Med. Wochenschr., v. 12:173-176.

LEUCKART, R. 1876.—Die Menschlichen Parasiten. pp. 301, 345.

LI, H. C. 1933a.—Parasitic nematodes: Studies on their intestinal contents. I. The feeding of dog ascaris, *Toxocara canus* (Werner, 1782). II. The presence of bacteria. Lingnan Sc. J., v. 12 (Suppl., May):33-41, pl. 1, figs. 1-3.
  1933b.—Feeding experiments on representatives of Ascaroidea and Oxyuroidea. Chinese Med. J., v. 47:1336-1342.
  1933c.—On the mouth-spear of *Trichocephalus trichu-*

rus and of a *Trichocephalus* sp. from monkey, *Macacus rhesus*. Ibid., v. 47:1343-1346.

LIEVRE, H. 1934.—A propos de l'Hematophagie des *Ascaris*. Comp. Rend. Soc. Biol., Paris, v. 116:1079.

LOOSS, A. 1905.—The anatomy and life history of *Agchylostoma duodenale* Dub. Rec. Egypt. Govt. Sch. Med., v. 3:1-158, pls. 1-10, figs. 1-100, photos 1-6.

McCOY, E. E. and GIRTH, H. B. 1938.—The culture of *Neoaplectana glaseri* on veal pulp. N. J. Dept. Agric. Circ. No. 285. 12 pp.

McCOY, E. E. and GLASER, R. W. 1936.—Nematode culture for Japanese beetle control. N. J. Dept. Agric. Circ. No. 265. 9 pp. figs. 1-5.

McCOY, OLIVER R. 1929.—The suitability of various bacteria as food for hookworm larvae. Am. J. Hyg., v. 10(1):140-156.
1934.—The development of adult trichinae in chick and rat embryos. J. Parasit., v. 20(6):333.

MAGATH, THOMAS BYRD. 1919.—*Camallanus americanus*, nov. spec. Tr. Am. Micr. Soc., v. 38(2):49-170, pls. 7-16, figs. 1-133.

MUELLER, JUSTUS F. 1929.—Studies on the microscopical anatomy and physiology of *Ascaris lumbricoides* and *Ascaris megalocephala*. Ztschr. Zellforsch., v. 8(3)361-403, pls. 1-18, figs. 1-80.

ORTLEPP, R. J. 1937.—Observations on the morphology and life-history of *Gaigeria pachyscelis* Raill. and Henry, 1910: A hookworm parasite of sheep and goats. Onderstepoort J. Vet. Sc. & Anim. Indus., v. 8(1):183-212, figs. 1-18.

OTTO, G. F. 1935.—Blood studies on trichuris-infested and worm-free children in Louisiana. Am. J. Trop. Med., v. 15(6):693-704.

PORTER, DALE A. 1936.—The ingestion of the inflammatory exudate by swine lungworms. J. Parasit., v. 22(4)411-412, fig. 1.

RANSOM, B. H. 1911.—The nematodes parasitic in the alimentary tract of cattle, sheep, and other ruminants. U. S. Dept. Agric., Bur. Anim. Indus. Bul. No. 127, 132 pp., figs. 1-152.

SCHWARTZ, BENJAMIN. 1921.—Effects of secretions of certain parasitic nematodes on coagulation of blood. J. Parasit., v. 7(3):144-150.

SMIRNOV, G. G. 1935.—Nutrition of the *Ascaris* larvae in the process of migration. Parazit., Perenosch. i Iadovit. Zhivotn. Slorn. Rabot. . . . Pavlovskii 1909-1935, pp. 298-306. Russian with English summary.
1936.—On the question of hematophagia in thread-worms and whipworms. Trudy Sec. Parasit., U.S.S.R. Inst. Exper. Med., v. 2:229-239. [Russian with English summary.]

VAN SOMEREN, VERNON D. 1939.—On the presence of a buccal stylet in adult Trichinella, and the mode of feeding of the adults. J. Helm., v. 17(2): 83-92, figs. 1-5.

STADELMANN, H. 1891.—Ueber den anatomischen Bau des *Strongylus convolutus* Ostertag, nebst einigen Bemerkungen zu seiner Biologie. Inaug.-Diss., Berlin.

VEGLIA, FRANK. 1916.—The anatomy and life history of the *Haemonchus contortus* (Rud.). Dept. Agric. Union S. Africa, 3d & 4th Rpts. Director Vet. Res., pp. 347-500, figs. 1-60.

WEINBERG, M. 1907.—Du Role des Helminthes. Ann. Inst. Pasteur, v. 21: 417-533.

WEINLAND, ERNST. 1901.—Ueber den Glycogengehalt einiger parasitischer würmer. Ztschr. Biol., v. 41: 69-74.

WELLS, HERBERT S. 1931.—Observations on the blood sucking activities of the hookworm, *Ancylostoma caninum*. J. Parasit., v. 17(4): 167-182, fig. 1.

WETZEL, R. 1928.—Pathogenic effects of *Strongylus equinus*, *edentatus* and *vulgaris* on mucosa of colon of horses. Deutsch. Tierärzte Wochenschr., v. 36: 719-722.
1930.—The biology of the fourth stage larvae of *Oxyuris equi*. J. Parasit., v. 17: 95-97, fig. 1.
1931a.—On the biology of the fourth stage larva of *Dermatoxys veligera* (Rudolphi 1819) Schneider 1866, an oxyurid parasitic of the hare. Ibid., v. 18:40-43.
1931b.—On the feeding habits and pathogenic action of *Chabertia ovina* (Fabricius, 1788). N. Am. Vet., v. 12(9): 25-28, fig. 1.

WETZEL, R. and ENIGK, K. 1937.—Zur Biologie von *Graphidium strigosum*, dem Magenwurm der Hasen und Kaninchen. Deutsch. Tierärzte. Wochenschr., v. 45(25): 401-405, figs. 1-3.

WHIPPLE, G. H. 1909.—The presence of a weak hemolysin in hookworm and its relation to the anemia of uncinariasis. J. Exper. Med., v. 11: 331-343.

CHAPTER XXI

CHEMICAL COMPOSITION AND METABOLISM OF NEMATODE PARASITES OF VERTEBRATES, AND THE CHEMISTRY OF THEIR ENVIRONMENT

THEODOR VON BRAND, Department of Biology, Catholic University of America and

THEODORE LOUIS JAHN, Department of Zoology, State University of Iowa

The metabolic processes of nematode parasites comprise a subject which has been under investigation for many years. Progress in the field, however, has been particularly rapid during the last decade, and it is the purpose of the present authors to present a summary of the known facts of metabolism together with the related subjects of the chemistry of the worms and of their environment. Recent reviews which deal with some of the subject matter here presented are those of Slater (1928), McCoy (1935), Lapage (1938) and v. Brand (1934, 1938).

Peculiarities of Environment Which May Influence Metabolism

The wide differences in the habitats of the various nematode parasites of vertebrates are undoubtedly correlated with wide differences in metabolic processes. The organisms which live in the digestive tract, blood stream, lungs, kidneys, subcutaneous tissue, etc., are subject to quite a variety of environmental conditions. In those cases where open contact with the blood stream or lymph is maintained the parasites are, of course, subjected to an environment very similar to that of the cells of the host body. Whenever a nematode is surrounded by a cyst wall which reduces the availability of oxygen, or is located in a region deprived of free blood circulation, metabolic processes are probably different from the processes in those species which live in the blood stream. Species that live in the digestive tract have an environment which is peculiar in many respects. The chemistry of blood is adequately described elsewhere, and the chemical environment within cysts and in chemically isolated tissues is practically unknown (except for cestode cysts, Schopfer, 1932). Therefore, the present discussion of environment is limited to the chemistry of the digestive tract.

From the viewpoint of nematology the chemistry of the intestinal contents is interesting for several reasons. A thorough knowledge of the chemistry of the environment may allow a better understanding of the physiology of the intestinal parasites, it may aid in the formulation of culture media suitable for growth *in vitro* (cf. Glaser and Stoll, 1938), and it may shed light on the problems of host specificity and on the possibility that experimental modifications of intestinal contents may be of use in controlling the activities of the nematodes. The effect on nematodes of many of the substances found in the intestine has not been studied. In the hope that the present discussion might serve as a partial outline of substances to be investigated, the authors have included a general discussion of the chemical compounds present.

THE SEQUENCE OF CHEMICAL EVENTS IN THE DIGESTIVE TRACT

In any discussion of the chemical composition of the contents of the digestive tract it is necessary to keep in mind the sequence of events which occurs as the ingesta pass through the alimentary canal. The chemical composition of the contents of the gut varies with diet, with species, and with the state of health. However, in any healthy animal on a constant diet there is a definite sequence to the chemical changes which occur.

In man, the stomach receives the mixture of food and saliva. To this is added mucus, pepsin, and hydrochloric acid. The material present in the duodenum is derived from four sources: chyme from the stomach, bile, pancreatic juice, and succus entericus. The stomach contents when emptied into the duodenum consist, among other things, of proteoses and peptones, starch, sugars, fat droplets, some fatty acids and glycerol, hydrochloric acid, plant fragments containing cellulose and undigested plant tissue, and water.

The bile contains mucin, the pigments biliverdin and bilirubin, the bile salts Na-taurocholate and Na-glycocholate, cholesterol, lecithin, fats, soaps, inorganic salts and water. The relative amounts of taurocholate and glycocholate vary with the species; the dog, for example, is entirely lacking in glycocholate. The pancreatic secretion contains Na_2CO_3 and the enzymes trypsin, lipase, and amylopsin. The succus entericus contributes the enzymes erepsin, lipase, maltase, invertase, lactase, and rennin, and a large amount of mucus and desquamated epithelial cells. Due to partial sterilization of food, or to the action of hydrochloric acid and bile salts, living bacteria are present only in small numbers in the duodenum and in

normal men may sometimes be absent (Kellogg, 1933). As these materials pass through the duodenum and jejunum digestion is completed, and the products of digestion and most of the bile, salts are absorbed. The bacteria increase in numbers, utilize some of the products of digestion and decompose others. As the material passes through the large intestine water is absorbed, and calcium, magnesium, iron, and phosphates are secreted by the intestinal wall.

The feces of an animal on a carnivorous diet are composed mostly of the intestinal secretions and bacteria. If vegetables make up a considerable part of the diet, the bulk of the feces is increased, and plant fragments appear in the feces, sometimes with the contained plant protoplasm only partially digested. The large bulk of undigested cellulose stimulates peristalsis, and consequently causes a more rapid passage of ingesta through the intestine, which results in the absorption of less water by the colon and a more liquid feces.

The materials which are present in the digestive tract and which may affect the metabolism of nemas are for convenience discussed under the following headings: (1) Composition of the intestinal gases, (2) Hydrogen ion concentration, (3) Dissolved materials (exclusive of gases), (4) Antienzymes. Nematodes, especially those which live in tissues, are known to secrete digestive enzymes, but these are more properly discussed under the subject of nutrition of the worms.

COMPOSITION OF THE INTESTINAL GASES

The composition of the gases in or in contact with the ingesta varies greatly in different parts of the digestive tract. The gas tension of the stomach contents varies at different periods following a meal and depends on the amount of air ingested with food. The action of HCl causes a release of bound CO_2, most of which is probably absorbed either in the stomach or upper intestine. The oxygen ingested with the food apparently undergoes a rapid decrease so that it is almost absent from the intestine below the duodenum. The analyses of von Brand and Weise (1932) show that very little oxygen is introduced into the intestine by the bile. These investigators also studied the oxygen content of fluid intestinal matter and of intestinal gases. They found that the oxygen content of the fluid of both the large and small intestines of almost all animals examined was practically nil. The only exception was one pig which contained quite appreciable amounts of oxygen. This might have been caused by the swallowing of large amounts of air, perhaps at the time of slaughtering. The values for all animals except the pig correspond to about 5 percent saturation. The data of several investigators on the oxygen content of intestinal fluids and gases are summarized in Tables 13 and 14. The data in Table 14 demonstrate the absence of oxygen in the gaseous content of the intestine of all animals except the pig. Long and Fenger (1917) found that oxygen was present in appreciable quantities in the intestinal gases of the pig, and this was confirmed by v. Brand and Weise. It has been assumed by Slater (1925) that the intestinal walls give off oxygen to the intestinal contents during digestion. This has not been proved experimentally, and Long and Fenger (1917) found that the oxygen content was lowest during active digestion. It seems probable (as indicated by the data of McIver, Redfield, and Benedict, 1926) that oxygen may diffuse inward from the intestinal wall, but it is also very likely that the bacteria present near the wall would consume this immediately so that very little oxygen from this source would ever reach the central portion of the lumen. The available evidence indicates that the environment of intestinal helminths is not devoid of oxygen but contains oxygen in only small quantities. Worms which live close to the intestinal wall may have access to larger amounts. In the case of the hookworm it is apparent from the observations of Wells (1931) that the blood sucking activities represent largely a respiratory function.

Analyses of intestinal gases other than O_2 are not numerous. The intestinal gases of man vary with diet. Ruge (1861) gives the following data for percentage composition:

Diet	CO_2	H_2	CH_4	N_2
Vegetables	21-34	1.5-4.0	44-55	10.19
Meat	8-13	0.7-3.0	26-37	45-64
Milk	9-16	43-54	0.9	36-38

The data of Fries (1906) show that the gases of man on a mixed diet are similar to those given above for a meat diet. Further analyses are given by Basch (1908). The absorption of intestinal gases is discussed by McIver, Redfield, and Benedict (1926), and the subject of human gastro-intestinal gases is reviewed by Ziegler and Hirsch (1925) and Lloyd-Jones and Liljedahl (1934).

The intestinal gases of the dog were analyzed by Planer (1860). His analyses demonstrated large amounts of CO_2 and N_2, and a smaller amount of H_2 throughout the digestive tract, a small amount of O_2 in the small intestine, and a small amount of H_2S in the large intestine when the dog was on bread or meat diets. On a vegetable diet H_2 largely replaced the N_2, while O_2 and H_2S were absent. In these analyses methane is conspicuously absent.

The intestinal gases of various herbivores have been analyzed by Tappeiner (1883), and the literature is summarized by Scheunert and Schieblich (1927). The data of Tappeiner (1883) on the percentage composition of the gases in cattle, sheep, and goats (all of which were quite similar) are as follows:

	Rumen	Small Intestine	Caecum and Colon
CO_2	65	62-92	about 30
O_2	.5	0	0
CH_4	30	.04-6.6	38.53
H_2	0.6-4.7	0-37	2-6
N_2	1-40	1	23-34

Data for the horse (Tappeiner, 1883) are:

	Stomach	Small Intestine	Caecum and Colon	Rectum
CO_2	75	15-43	55-85	29
O_2	0	0.57-76	0-.14	0
CH_4	0	0	11-33	57
H_2	14	20-24	1.7-2.2	0.8
N_2	10	37-60	.9-10.0	13

Data for the rabbit (Tappeiner, 1883) are:

	Stomach	Small Intestine	Caecum and Colon
CO_2	32	75	6
O_2	0	0	0.6
CH_4	0	2	21.0
H_2	0	18	0.6
N_2	68	6	72

Long and Fenger (1917) found a large amount of N_2 (74—92%) somewhat less CO_2 (5—28%), and about 5 percent O_2, but no methane or H_2 present in the small intestine of hogs.

The production of methane is probably caused mostly by bacterial decomposition of cellulose, although the data of Ruge indicate that it can also be produced by bacterial action when the animal is on a meat diet. The analyses of Tappeiner and others also indicate the presence of H_2S in some cases. H_2S and N_2 must be formed by the action of bacteria on protein. Ammonia is also formed by bacterial decomposition of protein, but it is usually bound by the acids of the intestine. Most of the CO_2 is probably of bacterial origin, although in the duodenum it may also be formed by the Na_2CO_3 of the pancreatic juice and the acid of the chyme. The NH_3 and the CO_2 of the succus entericus are discussed by Herrin (1937). Most of the intestinal gases are eliminated from the body by the lungs. Tacke (1884) found that 10 to 20 times as much of the intestinal gases of rabbits escape by means of the blood and lungs as by way of the anus.

The effect of the gases other than oxygen on intestinal nematodes is entirely unknown. Methane, H_2, and N_2 are probably without either beneficial or harmful effects. The utilization of oxygen will be discussed in the part dealing with metabolism of adult nematodes. The effect of CO_2 is unknown. Since it is incapable of further oxidation and since there is no evidence of chemosynthesis in the nematodes, the only apparent effect it could have would be the adjustment of intracellular pH. Since intestinal nematodes live in a medium usually saturated with CO_2 it is conceivable that they may depend on this substance as an intracellular buffer. Therefore, it may become important to maintain a high CO_2 content in in vitro cultures (Cf. possible role in growth of intestinal protozoa, Jahn, 1934, 1936). It should be noted that Weinland (1901) found that Ascaris survived longer in vitro when the medium was saturated with CO_2.

TABLE 13.—*Oxygen Content of Fluid Intestinal Masses*

Animal	Part of intestine	Oxygen in volume percent mean and ( ) extremes	Number of determinations	Investigator
Horse	Sm. intestine	0.024 (0.016 0.031)	2	Toryu (1934)
Dog	Sm. intestine	0.028	1	v. Brand & Weise (1932)
Cattle	Sm. intestine	0.013 (0.00 0.025)	2	v. Brand & Weise (1932)
Sheep	Sm. intestine	0.012 (0.00 0.025)	4	v. Brand & Weise (1932)
Pig	Sm. intestine	0.083 (0.00-0.358)	6	v. Brand & Weise (1932)
Cattle	Lg. intestine	0.010 (0.00 0.023)	3	v. Brand & Weise (1932)
Pig	Lg. intestine	0.00	1	v. Brand & Weise (1932)

TABLE 14.—*Oxygen Content of Intestinal Gases*

Animal	Part of intestine	Oxygen in volume percent mean and ( ) extremes	Number of determinations	Investigator
Horse	Sm. intestine	0.67 (0.57-0.76)	3	Tappeiner (1883)
Cattle	Sm. intestine	0.00	1	Tappeiner (1883)
Cattle	Sm. intestine	*	*	v. Brand & Weise (1932)
Goat	Sm. intestine	*	*	Tappeiner (1883)
Sheep	Sm. intestine	*	*	v. Brand & Weise (1932)
Pig	Sm. intestine	5.5 (1.2-14.2)	9	Long & Fenger (1917)
Pig	Sm. intestine	4.2 (0.4-8.2)	6	v. Brand & Weise (1932)
Dog	Sm. intestine	0.2 (0.0-0.7)	8	Planer (1860)
Horse	Lg. intestine	0.07 (0.00-0.14)	4	Tappeiner (1883)
Cattle	Lg. intestine	0.00	1	Tappeiner (1883)
Goat	Lg. intestine	0.03 (0.00-0.07)	3	Tappeiner (1883)
Sheep	Lg. intestine	0.00	1	Tappeiner (1883)
Rabbit	Lg. intestine	0.62	1	Tappeiner (1883)
Dog	Lg. intestine	0.00	6	Planer (1860)

*Not enough gas for analysis.

HYDROGEN ION CONCENTRATION

The pH of the stomach and intestine has been measured for a large number of animals, and some of the representative data are listed in Tables 15 and 16. Contents of the stomach of carnivores, omnivores, and herbivores with a simple stomach, and of the abomasum of ruminants are distinctly acid in character due to the secretion of hydrochloric acid. The rumen and omasum vary from neutral to distinctly alkaline. The pH of the duodenum is extremely variable but is usually acid because of the introduction of HCl from the stomach. The pH of the remainder of the small intestine is less acid than the duodenum, and there is usually a progressive rise toward neutrality or to a slight alkalinity; the pH seldom reaches a value higher than 8.0 or 8.2. The colon and caecum of most animals are neutral, slightly acid, or slightly alkaline. Some of the recent literature is reviewed by Lenkeit (1933).

The pH of the intestine may be lowered by the administration of large quantities of lactose, especially if the diet is low in protein. Robinson and Duncan (1931) found that the pH of the rat intestine could be lowered about one pH unit by the administration of 25 percent lactose with a low protein diet (other literature is cited by these authors). In man it is known that the acidity of the intestine may be considerably decreased if large amounts of lactose accompanied by *Lactobacillus acidophilus* are ingested (literature cited by Kopeloff, 1926, and Frost and Hankinson, 1931). Comparable results have been obtained with the domestic fowl (Ashcraft, 1933). The direct addition of mineral salts such as NaCl, $MgSO_4$, $CaCl_2$, $Ca(OH)_2$, $CaSO_4$, $NaHCO_3$, and NH_4Cl to the diet may have no effect on pH in experimental animals (McClendon et al, 1919; Heller, Owens, and Portwood, 1935; Mussehl, Blish, and Ackerson, 1933). However, positive results with mineral salts have been obtained by Robinson (1922), Shohl and Bing (1928) and others. A deficiency of vitamin D is also known to cause the intestinal contents to become alkaline due to lack of Ca absorption (Zucker and Matzner, 1923; Jephcott and Bacharach,

TABLE 15.—*The pH of Stomach Contents*

Animal	pH	Author
Man	minimum pH 1.0 to 2.5	McClendon and Medes (1925) Kahn and Stokes (1926)
Rat	3.2-4.6	Sun, Blumenthal, Slifer, Herber and Wang (1932)
Rat	1.8-5.6 (av. 3.6)	Eastman and Miller (1935)
Rat	3.3-3.9	Kofoid, McNeil and Cailleau (1932)
Horse	1.13-6.8 (50% between 1.1 and 3.3)	Schwarz, Steinmetzer, and Caithaml (1926)
Rabbit	1.8	McLaughlin (1931)
Cat	3.34	McLaughlin (1931)
Dog	1.5-2.0	Mann and Bollman (1930)
Dog	2.0-6.0	Schwarz and Danziger (1924)
Dog	1.37-5.7 (av. 3.47)	Nagl (1928)
Chick. gizzard	3.39	McLaughlin (1931)
Chick. gizzard	2.9-3.2	Ashcraft (1933)
Chicken proventriculus	5.9	McLaughlin (1931)
Chicken proventriculus	4.8-5.7	Ashcraft (1933)
Cattle abomasum	2.0-4.1	Schwarz and Kaplan (1926)
Cattle abomasum	3.8	Mangold (1925)
Sheep abomasum	3.15-5.25 (av. 4.0)	Davey (1938)
Cattle rumen	8.89 (8.61-9.68)	Schwarz and Gabriel (1926)
Cattle rumen	7.5-8.0	Kreipe (1927)
Sheep rumen	7.0-7.6	Ferber (1928)

1926; Redman, Willimott, and Wokes, 1927). The pH may be appreciably lowered by addition of cod liver oil to a rachitogenic diet. The effect of varying the proportions of protein, fat, and carbohydrate has been reported to cause no marked change in the pH of the intestine of rats (Redman, Willimott and Wokes, 1927), dogs (Grayzel and Miller, 1928; Graham and Emery, 1928), or man (Hume, Denis, Silverman, and Irwin, 1924). However, the data of Robinson and Duncan (1931) show consistently higher pH values for rats fed on grain and alfalfa than for rats on a high protein diet (Table 16). Eastman and Miller (1935) studied the effect of a number of diets on gastro-intestinal pH in rats.

It has been suspected for some time that the pH of the central portion of the lumen is not the same as that close to the intestinal wall. Evidence for this is found in the feces in that the surface of stools is more alkaline, apparently because of secretion of alkaline salts by the intestinal wall. Kofoid, McNeil, and Cailleau (1932) reported differences in the pH of contents and wall throughout the digestive tract of the rat. (Table 16). Robinson (1935) studied the effect of placing various salt solutions in the small intestine of dogs on the pH of the solution and decided that each portion of the digestive tract tended to produce a characteristic pH value in the solution, regardless of the initial pH. He concluded that the pH of the region close to the wall increases regularly from pH 6.5 to 7.5 or 8.0 throughout the length of the small intestine and pointed out that the pH close to the wall is probably largely independent of changes produced in the lumen by the action of bacteria. Ball (1939), by means of a capillary glass electrode, has measured the pH of the wall (data given in Table 16).

The possibility that pH may be a limiting factor in the distribution of sheep nematodes was investigated by Davey (1938). He found that *Ostertagia circumcincta* was able to live between pH 3.2 and pH 9.0. This range allows it to live in the abomasum of sheep (pH 3.2-5.25) but apparently may be one reason why it does not infest the stomach of the dog (pH 2 or less) or horse (pH 1.1-6.8) or the abomasum of cattle (pH 2.0 to 4.1). Two duodenal species from sheep, *Trichostrongylus colubriformis* and *T. vitrus*, were able to stand a continuous acidity as low as pH 3.6, but five other species (*Nematodirus filicollis*, *N. spathiger*, *Cooperia oncophora*, *Cooperia curticei*, *Strongyloides papillosus*) from the middle and lower small intestine were killed at acidities of pH 3.9 to 4.6. Since the duodenum is more acid than the ileum, the low resistance to acidity may be an important factor in preventing the five species from the middle and lower intestine from infesting the duodenum. It has been suggested by Lapage (1935a, 1938) that pH has an influence on the second ecdysis of trichostrongylid larvae (outside of the host) and that this may be of importance in allowing development of the parasite. The third ecdysis (in the intestine) might be similarly affected.

It seems possible that the presence of nemas in the digestive tract might cause a change in gastro-intestinal pH, either directly (perhaps because of lesions in the epithelium) or indirectly through the systemic reactions of the host. In cases of ancylostomiasis and intestinal schistosomiasis Eldin and Hassan (1933) found evidence of gastric disturbance which disappeared after removal of the worms. Fernandez (1934), however, found no correlation between gastric acidity and helminth parasites.

DISSOLVED SUBSTANCES (EXCLUSIVE OF GASES)

The dissolved materials of the digestive tract consist of the ingesta and various secretions listed above, the products of digestion, and the products of bacterial decomposition. Many of these, especially the carbohydrates, may serve as food for nematodes; many others may be toxic and may be effective in

TABLE 16.—*The pH of the digestive tract contents.*

Animal and Diet	Duodenum	Jejunum	Ileum	Caecum	Colon	Investigator
Man	2.27-7.8	----	----	----	----	Long and Fenger (1917)
Man	4.7-6.5	7.0	6.1-7.3	----	----	Karr and Abbott (1935)
Man	4.5-5.1	----	5.9-6.5	----	----	McClendon (1920)
Dog	2.0-7.6	7.0-7.6	6.0-8.0	----	7.4	Mann and Bollman (1930)
Dog	6.2-6.5	6.0-7.0	6.0-7.0	6.0-6.5	----	Graham and Emery (1927-28)
Dog	5.9	6.0-6.27	6.36	6.57	6.84	Grayzel and Miller (1928)
Dog	----	----	----	----	7.6-8.4	Heupke (1931)
Cat	6.5	6.9	6.8	----	5.25	McLaughlin (1931)
Rat	6.5	to	7.2	6.5-7.2	6.4-6.6	Sun, Blumenthal, Slifer, Herber, and Wang (1932)
Rat—grain & alfalfa	6.75[1]	7.7[1]	8.2[1]	7.0	7.2	Robinson and Duncan (1931)
Rat—high protein	6.4[1]	6.8[1]	7.3[1]	7.3	7.2	Robinson and Duncan (1931)
Rat—high base	6.4[1]	6.8[1]	7.25[1]	7.0	7.2	Robinson and Duncan (1931)
Rat	5.9[1] (4.2-6.9)	6.6[1] (5.0-7.3)	6.9[1] (5.6-7.7)	6.4 (5.1-7.4)	6.6 (5.4-7.5)	Eastman and Miller (1935)
Rat—lumen of gut	----	----	7.13	7.13	7.33	Kofoid, McNeil, and Cailleau (1932)
Rat—wall of gut	6.93	----	7.34	7.34	6.95	Kofoid, McNeil, and Cailleau (1932)
Rat—wall of gut	6.34	----	6.89	7.06	6.91	Ball (1939)
Rabbit	7.35	----	8.0	6.26	----	McLaughlin (1931)
Cattle	6.68	8.42	8.2	8.2	----	Danniger, Pfragner, and Schultes (1928)
Cattle	----	----	----	----	7.4-8.4	Heupke (1931)
Horse	6.72	----	7.09	8.12	----	Danniger, Pfragner, and Schultes (1928)
Hogs, calves, lambs	Indefinitely variable 6.48 to 7.76. More often acid than alk.			----	----	Long and Fenger (1917)
Fowl	6.3	----	6.22	1.9[2]	----	McLaughlin (1931)
Fowl—meat scrap	5.96	----	7.1	7.0	7.2	Ashcraft (1933)
Fowl—+ 20% lactose	6.51	----	7.16	5.1	6.3	Ashcraft (1933)

[1]Intestine divided into three approx. equal portions so that the measurements given may not correspond exactly to those of the duodenum, jejunum, and ileum.

[2]Possibly a misprint in the original paper.

causing the localization of nematodes in certain portions of the digestive tract.

The possibility that bile salts may affect the growth of intestinal parasites has been recognized for some time. According to Moorthy (1935) fresh bile from certain species of *Barbus* and from sheep and man is capable of killing *Cyclops* and of activating the enclosed larvae of *Dracunculus medinensis* to escape. De Waele (1934) claimed that the cestode, *Taenia hydatigena* (*Cysticercus pisiformis*), is able to infest dogs because of the absence of Na-glycocholate in dog bile and that since Na-glycocholate is toxic to the organism it can not develop in animals which secrete this substance.

Davey (1938) has investigated the effect of bile salts on sheep nematodes. He found that the species which infest the duodenum (*Trichostrongylus colubriformis* and *T. vitrinus*) of sheep were much more resistant to Na-taurocholate and Na-glycocholate than other species (*Nematodirus fillicollis*, *N. spathiger*, *Cooperia oncophora*, *Cooperia curticei*, and *Ostertagia circumcincta*) from the lower small intestine and abomasum. *Cooperia curticei*, which lives closer to the opening of the bile duct than the other species except *Trichostrongylus colubriformis* and *T. vitrinus*, has a resistance second only to *Trichostrongylus*. Since the bile salts are introduced by the bile duct and are largely reabsorbed in the small intestine, the concentration of bile salts decreases along the intestine. The high concentration in the upper small intestine probably prevents species other than *Trichostrongylus* from living in that region. In these experiments glycocholate seemed to be somewhat more toxic to *Trichostrongylus* than taurocholate. Davey mentioned the possibility that differential susceptibility to the two bile salts might be a factor in the determination of host specificity.

The products of bacterial decomposition are of several types:

1. Products of carbohydrate decomposition from:
   a. Hydrolysis of cellulose to glucose in the rumen and large intestine of herbivores.
   b. Fermentation of simple sugars to lower fatty acids in the small and large intestine of all vertebrates and in the rumen of ruminants.
2. Products of protein decomposition from:
   a. Hydrolysis of proteins to amino acids in the upper small intestine.
   b. Fermentation of amino acids to aporrhegmas and to lower products in the lower small intestine and large intestine of animals with simple stomachs and in the rumen of ruminants. Some of the products of fermentation are indol, skatol, paracresol, phenol, volatile fatty acids, H_2S, histamine, and tyramine. The relative amounts of these products depend on the type of protein and on the species of bacteria present.

At present there is little evidence that these substances are useful or harmful to intestinal nemas. Glucose is probably absorbed by nemas, and on this assumption changes in the diet or in the bacterial flora which would affect the distribution of glucose should affect the parasites. From the studies of Grove, Olmstead, and Koenig (1929) on the lower fatty acids in feces it seems as if the quantity and perhaps the distribution of these materials along the digestive tract is greatly affected by diet. It is also probable that the products of protein putrefaction may exert beneficial or harmful effects on the parasites. If so, then experiments in which the amount of protein putrefaction is controlled are in order. Such control is possible by the administration of large amounts of lactose and bacteria which ferment glucose to acid (review, Arnold, 1933). This treatment results in the replacement of the protein putrefying organisms of the coli-aerogenes group by those which ferment carbohydrate. The change in type of fermentation products is probably due to both the protein sparing action of carbohydrate and the change in flora produced by increased acidity of the intestine. Putrefaction could also be decreased by increasing the rate of passage of ingesta. It is possible to increase protein putrefaction at least in the large intestine by feeding such large quantities of protein that some of it escapes complete digestion and absorption in the small intestine. The putrefying organisms also increase under conditions of achlorhydria which result in an alkalinization of the intestine, and if the achlorhydria is severe they may even become implanted in the stomach. It seems probable that experimental modification of the intestinal contents through modification of the intestinal flora may bring about changes in the distribution of nemas along the intestine, and perhaps such experiments may result in methods of controlling or eliminating certain species. Any changes which may prevent ecdysis of larval nematodes might be extremely useful (Lapage, 1938).

It is known that H_2S is highly toxic to vertebrates and that it easily passes through most animal membranes. The studies of Enigk (1936) on the lethal effects of H_2S on the eggs of *Ascaris lumbricoides* and the studies of Lapage (1935) on the infective larvae of *Trichostrongylus* suggest that the outer covering of eggs and larvae may be permeable to H_2S and other sulfur compounds. Lapage (1935b) obtained considerable evidence that the permeability of the sheaths of larvae is changed by sulfur compounds. In these experiments the effect of pH was not carefully controlled, but the effect of 1 percent Na_2S on the ecdysis of infective larvae was more pronounced than that of 1 or 2 per cent NaOH. The sheaths became greatly distended due to intake of water. If this effect is really due to the sulfur compounds, this type of effect may give a chemical basis for the statements of Mudie (1934) and Johnston (1934) that the eating of garlic will cause the disappearance of threadworms from the human digestive tract. Lapage (1938) suggested that compounds which yield H_2S when subjected to the action of intestinal bacteria might eventually be used as anthelmintics.

Some of the products of protein putrefaction, especially H_2S, rapidly combine with molecular oxygen and when in solution produce very low oxidation-reduction potentials. Bergeim (1924) devised a chemical method of obtaining an index of the reducing power of intestinal contents, and he found that the amount of reduction varied with diet. Preliminary electrical measurements of the oxidation-reduction potential of the rat digestive tract (Jahn, 1933) have shown that the Eh value may be as low as —200 mv. in the caecum and somewhat higher in the lower small intestine. These measurements are well within the "anaerobic" range and support the conclusions mentioned above that oxygen is very scarce in the small intestine and absent in the caecum.

The osmotic pressure of the digestive tract is usually somewhat higher than that of the serum and tissues. Schopfer (1932) gives the following freezing point depressions for various animals: sheep, 0.70-0.83° C.; cow, 0.80° C.; horse, 0.74-0.77° C.; hog, 0.9-1.0° C.; and the elasmobranch *Scylliorhinus*, 2.4° C. With the exception of the elasmobranch the serum of the above animals has a molecular depression of about 0.55 to 0.65° C. Davey (1936b) gave a value of 0.55-0.63° C. for the abomasal contents of sheep. The osmotic pressure of the intestinal contents probably varies considerably with salt intake, but absorption and excretion are apparently rapid enough to prevent the osmotic pressure from ever becoming more than twice that of the blood. As will be discussed below (General Chemical Composition) the osmotic pressure of the medium determines that of the worms. However, the effect of this change in osmotic pressure on worm metabolism is unknown. Davey (1938) has shown that *Ostertagia circumcincta* is capable of living in NaCl which varied from .4 percent to 1.3 percent (0.9 percent is equivalent to a freezing point depression of 0.6° C.). In balanced salt solutions the range would probably be greater.

ANTIENZYMES

Since the nematodes of the vertebrate digestive tract live in a medium high in the concentration of proteolytic enzymes, the question of how they are able to resist digestion has often been mentioned in the literature. The mechanism seems to be at least dual: (1) the cuticle is relatively indigestible, and (2) the worms contain or secrete antienzymes, i.e., substances which inactivate the digestive enzymes. Evidence for this latter mechanism was first described by Weinland (1903) who described a substance with antitryptic powers in aqueous extracts of *Ascaris*. Dastre and Stassano (1904) believed that the action was antikinasic, but the experiments of Hamill (1906) confirmed the original conclusions of Weinland (1903). Hamill (1906) ascribed the following properties to the antienzyme: highly soluble in water and weak alcohol; insoluble in 85 percent alcohol; thermostable in neutral or acid solutions; thermolabile in weakly or strongly alkaline solutions; readily diffusible through membranes which retain colloids. Harned and Nash (1932) described an improved method for preparing high concentrations of antitrypsin by fractional precipitation with alcohol. They claimed that by varying the concentration of alcohol a preparation of antitrypsin could be obtained almost free of *Ascaris* protease. These investigators were able to demonstrate that their antitrypsin preparation also contained a weak antipepsin. A powerful trypsin inhibiting fraction was also recently isolated by Collier (1941) from *Ascaris*. An antitrypsin with chemical properties similar to those of *Ascaris* antitrypsin has been prepared from egg white by Balls and Swenson (1934).

Sang (1938) investigated the mechanism of the action of *Ascaris* antienzyme and confirmed the conclusion that the substance exerted both an antitryptic and an antipeptic activity. However, he could not confirm the result of Harned and Nash (1932) that the ratio of protease to antienzyme could be varied. Sang concluded that *Ascaris* protease and *Ascaris* antitrypsin and antipepsin are all one and the same substance, and he proposed that this substance be called "ascarase." His investigations showed that ascarase was readily diffusible and that it either is or is associated with a substance of the order of a primary albumose. It was precipitated by ammonium sulphate and 70 per cent alcohol, and was not destroyed by trypsin. Ascarase did not inhibit the action of papain. Von Bonsdorff (1939) was unable to confirm the existence of antitrypsin or antipepsin in *Ascaris* extracts, but he did find that the extracts inhibited proteolysis of casein by depepsinized gastric juice at pH 7.4.

Stewart and Shearer (1933) studied the digestion of protein by infected and noninfected sheep and concluded that the nematodes of the stomach inhibited the normal digestive processes. They then obtained an extract from the worms which was capable of producing a 40 to 75 per cent inhibition of the peptic digestion of casein. For this substance and for similar antienzymes of nematodes they suggested the term "nezyme." Andrews (1938) could not repeat the results of Stewart and Shearer on the lowered digestive action of infected sheep. He found that the digestibility coefficients were the same in infected and noninfected animals. Infected sheep did not gain weight as rapidly as controls, but Andrews concluded that this was probably caused by intestinal irritation.

The existence of antienzymes has also been reported for cestodes. However, de Waele (1933), on the basis of experiments on *Taenia saginata*, has questioned the existence of antienzymes and has assumed that protection of the worms from enzyme action is due entirely to the resistance of the cuticle. One basis for this assumption is found in the fact that pieces of worms but not whole worms may be digested by trypsin. This conclusion is subject to criticism in that when worm fragments are placed in an enzyme solution considerable dilution of any antienzyme may occur by diffusion and the antienzyme may thereby be rendered ineffective. In view of the chemical isolation of the antienzyme mentioned above (Hamill, 1906; Nash and Harned, 1932; Collier, 1941) de Waele's conclusion certainly can not be extended to the nematodes.

General Chemical Composition

DRY WEIGHT

There have been only a few determinations of the dry weight of parasitic nematodes, and the values recorded are fairly high. The average figures reported for *Ascaris lumbricoides* are 20.7 percent (Weinland, 1901) and 15 percent (Flury, 1912), for *Parascaris*, 21 percent (Schimmelpfennig, 1903) and 14.8 percent (Flury, 1912), and for a larval *Eustrongylides*, 25 percent (V. Brand, 1938). Flury (1912) measured the dry weight of various parts of the body and obtained the following results:

| | Dry weight in percent of fresh weight | |
	Ascaris lumbricoides	*Parascaris equorum*
Body wall	23.5-25.0	25.0
Alimentary tract	27.5	24.9
Body fluid	4.0- 6.7	5.0
Reproductive organs	25.0-33.3	24.0-27.4

It can be calculated from Flury's figures that these values represent the following fractions of the total dry weight: body wall 65 percent, alimentary tract 3 percent, body fluid 10 percent, and reproductive organs 20 per cent.

CARBOHYDRATES

Storage of carbohydrates in the form of polysaccharides seems to be quite common among the parasitic nematodes. Although chemical analyses have been made only for *Ascaris*, it seems likely that in this respect other species are very similar. Weinland (1901) and Flury (1912) found an optical rotation of +183° to +193° for the polysaccharide of *Ascaris*. Since these workers and Campbell (1936) identified the sugar resulting from hydrolysis as glucose, and since the solubility of the polysaccharide and its color reaction with iodine are typical of glycogen, it seems probable that the substance is true glycogen. Campbell (1936), however, observed antigenic properties of a polysaccharide fraction isolated from *Ascaris*. It does not seem likely that pure glycogen would be capable of inducing the formation of specific anti-bodies. One should therefore expect that another polysaccharide is associated, perhaps in very small amounts only, with the glycogen. However, in so far as metabolic processes are concerned, it is justifiable to speak of glycogen alone.

The occurrence of large amounts of glycogen in ascarids was established in a qualitative or semi-quantitative way by Claude Bernard (1859) and Foster (1865), but Weinland (1901) was the first to undertake a large series of quantitative determinations. The more recent data on the glycogen content are summarized in the following table:

Species	Sex	Glycogen in % of fresh substance	Country	Investigator
Ascaris lumbricoides	?	5.4	Germany	Weinland, 1901
Ascaris lumbricoides	?	6.6	Germany	Schulte, 1917
Ascaris lumbricoides	♀	7.2	Denmark	v. Brand, 1934
Ascaris lumbricoides	♀	8.7	Russia	Smorodincev and
	♂	6.1	Russia	Bebesin, 1936
Ascaris lumbricoides	♀	5.3	USA	v. Brand, 1937
Ascaris lumbricoides	♂	5.8	USA	v. Brand, 1937
Dog *Ascaris*	?	4.5	Germany	Weinland, 1901
Parascaris equorum	?	2.1	Germany	Schimmelpfennig, 1903
Parascaris equorum	♀	3.8	Japan	Toryu, 1933
Parascaris equorum	♂	2.9	Japan	Toryu, 1933
Ancylostoma caninum	mixed	1.6	USA	v. Brand and Otto, 1938
Strongylus vulgaris	?	3.5	Japan	Toryu, 1933
Filaria equina	?	2.2	Japan	Toryu, 1933
Larval *Eustrongylides*	....	6.9	USA	v. Brand, 1938

Apparently the glycogen content of parasitic nematodes is always high. The lowest value amongst the intestinal nematodes was found in *Ancylostoma*. This may be related to the fact that the hookworms have access to larger amounts of oxygen than the other intestinal helminths. It is curious that *Ascaris lumbricoides* analyzed in Denmark and Russia yielded higher average glycogen values than those in USA and Germany. It is unknown whether this is caused by a different diet of the host and therefore of the parasite in various countries, or merely to different handling of the pigs before slaughtering.

Sexual differences in glycogen content of parasitic nematodes do not seem to be pronounced. Smorodincev and Bebesin (1936) and Toryu (1933) found more glycogen in females than in males of *Ascaris* and *Parascaris*. Von Brand (1937), on the other hand, found slightly more polysaccharide in male ascarids.

So far, only adult nematodes of warm-blooded hosts have been analyzed, and contrary to what is known about many free-living invertebrates, no evidence of seasonal variation in the amount of stored glycogen has been found. The obvious explanation of this difference lies in the uniform conditions under which the parasitic organisms live throughout the year. From this viewpoint, it should prove interesting to survey parasites from poikilothermic and heterothermic hosts, in which such variations are more likely to occur.

The glycogen distribution in various organs and tissues has been investigated both by quantitative chemical methods and by differential staining. Toryu's (1933) analyses of various organs of *Parascaris equorum* are summarized in the following table:

Organ	Glycogen in percent of— fresh substance ♀	♂	total glycogen ♀	♂
Body wall (cuticle + subcuticle + muscles)	5.8	4.9	66	96
Intestine	0.6	0.6	2	2
Ovary	6.5	....	23	....
Uterus	1.6	....	9	....
Male reproductive system	....	0.5	....	2

The body wall is obviously the most important storage place for glycogen in worms of both sexes.

Differential glycogen staining has been used chiefly by v. Kemnitz (1912) and Martini (1916) working with *Ascaris* and *Oxyuris*, respectively. These workers extended the earlier investigations of Brault and Loeper (1904) and Busch (1905). It seems that in both cases the most intensive glycogen reactions are found in the plasmatic bulbs of the muscle cells of the body wall and in the hypodermis, especially in the region of the lateral chords, but it was also found in other organs, for example, the intestine (compare also Hirsch and Bretschneider, 1937) and the reproductive organs. Glycogen, however, was never found in the cuticle, the phagocytic organs and the nervous system. Additional data on the glycogen morphology of other parasitic nematodes (*Parascaris, Sclerostomum, Heterakis* and *Ancylostoma*) are found in the papers of Busch (1905), v. Kemnitz (1912), Fauré-Fremiet (1913), Toryu (1933) and Giovannola (1935). In these cases, the general pattern of glycogen storage seems to be similar to that of *Ascaris*. In accordance with the quantitative chemical observations much less glycogen was found by morphological methods in hookworms than in ascarids. In the former, however, the rays of the bursa are an important storage place, and probably represent an energy reserve for the male during the periods of copulation when it is detached from the intestinal wall (Giovannola, 1935).

Not much is known about the occurrence of carbohydrates of lower molecular weight in parasitic nematodes. Weinland (1901) found 1.6 percent, and Schulte (1917) found 0.9 percent glucose in *Ascaris lumbricoides*. It is, however, questionable whether these figures are not too high, due to a partial breakdown of glycogen during the analyses. According to Foster (1865) and v. Brand (1934) only very small amounts of reducing sugar occur in *Ascaris*. Fauré-Fremiet (1913) found 0.15 percent glucose in the body fluid of *Parascaris*.

ETHER EXTRACTABLE MATERIAL

The parasitic nematodes seem to contain only small amounts of material extractable with ether or petrol ether. The mean values for *Ascaris lumbricoides* vary from 1.2 to 1.6 percent (Weinland, 1901; Flury, 1913; Schulte, 1917; v. Brand, 1934; Smorodincev and Bebesin, 1936), and the value for a larval *Eustrongylides* is 1.1 percent (v. Brand, 1938).

The chemical compounds comprising the ether extract seem to be quite similar in *Ascaris* and *Parascaris* (Flury, 1912; Fauré-Fremiet, 1913; Schulz and Becker, 1933). According to Flury (1912) 100 gm of ether extractable material from *Ascaris* contains the following:

Volatile fatty acids	31.07 gm
Saturated fatty acids	30.89 gm
Unsaturated fatty acids	34.14 gm
Unsaponifiable matter	24.72 gm
Glycerol	2.40 gm
Lecithin	6.61 gm

The volatile fatty acids were represented chiefly by valeric and butyric acids, with small amounts of formic, propionic and acrylic acid. In *Parascaris* the whole series of volatile fatty acids has been reported (Schimmelpfennig, 1903). The saturated fatty acids of higher molecular weight were recognized as stearic acid with a small admixture of palmitic acid. Oleic acid was the chief representative of the unsaturated fatty acids. Flury's value for glycerol is probably too low. Schulz and Becker (1933), using newer methods, found glycerol values ranging up to 8.8 percent. It is, therefore, unnecessary to assume as seemed necessary to Flury (1912) that there is a combination of part of the fatty acids with the unsaponifiable matter. It is probable that all the fatty acids are present in form of glyceryl esters. The unsaponifiable material is of special interest because it contains a compound which so far has been found in no other animal. This substance was found independently by Flury (1912) and Fauré-Fremiet (1913), and it is known as ascaryl alcohol. It was recently reinvestigated by Schulz and Becker (1933), who assigned it the formula $C_{33}H_{68}O_4$. They state that its configuration is not yet sufficiently known, but that it may be an ethereal combination of glycerol with some higher alcohol. According to Fauré-Fremiet (1913) ascaryl alcohol occurs in the female reproductive cells only. Under these circumstances one wonders why neither Flury (1912) nor Schulz and Becker (1933) mention any other unsaponifiable substance, which should be expected in other parts of the body. Fauré-Fremiet (1913) found small amounts of cholesterol in the body fluid, the eggs, and the testes of *Ascaris*, but Bondouy (1910) found no cholesterol in *Strongylus equinus*. The ether extract of the latter species seems to be characterized by the presence of soaps.

Little is known about the distribution of the ether extractable material in different organs. Flury (1912) found it to comprise 1.00 percent of the body wall of *Ascaris* and 4.0 to 6.25 percent of the reproductive organs. The latter figure agrees with that given by Fauré-Fremiet (1913) for the testes. If allowances are made for the relative weights of body wall and reproductive systems, it seems probable that roughly the same amount of ether extractable material is stored in both these places. This is in marked contrast to the distribution of glycogen.

Microscopical examinations (v. Kemnitz, 1912; Fauré-Fremiet, 1913; Mueller, 1928/29; Hirsch and Bretschneider, 1937) have shown that fat droplets are deposited in the plasma bulbs of the muscles of *Ascaris*, in which the nuclei are usually surrounded by an accumulation of fat, in the four chords, and especially in the subcuticula. Stainable fat was also found in ganglion cells, the intestinal cells, and the reproductive organs. According to Mueller (1928/29) considerably more fat can be demonstrated with osmic acid in *Parascaris* than in *Ascaris*, although the pattern of fat deposition is the same in both species.

NITROGEN CONTAINING SUBSTANCES

Flury (1912) found 8.1 percent proteins in *Ascaris*. This is somewhat less than should be expected from Weinland's (1901) N figure of 1.80 percent. Flury (1912) ascertained the presence of albumin, globulin, albumoses and peptones, purinebases, amines and ammonia, and he identified a series of amino acids as degradation products of the worm protein. Recently Yoshimura (1930) performed a quantitative analysis of the amino acids resulting from the hydrolysis of ascarids with sulfuric and hydrochloric acid. His results are summarized in the following table:

Amino acids in percent of dry substance upon hydrolysis with			
hydrochloric acid		sulfuric acid	
Leucine	3.70	Leucine	15.54
Alanine	1.45	Histidine	0.45
Valine	0.79	Arginine	1.28
Proline	3.41	Lysine	2.58
Isoleucine	1.45	Tyrosine	2.09
Serine	0.72		
Glutaminic acid	3.93		
Aspartic acid	0.36		
Glycocoll	0.29		
Phenylalanine	0.02		

The N containing substances constituting the cuticle have already been discussed in another chapter (see page 32), and that characteristic of the eggs (chitin) is mentioned on page 177.

Fauré-Fremiet (1913) described under the name of ascaridine an intracellular protein of the spermatozoa of *Ascaris*. It contains 17.5 percent N, but no phosphorus or sulfur. The chemical constitution of this interesting compound is not yet sufficiently known. It is insoluble in cold distilled water, but dissolves rapidly in water of 50 to 51°C. This critical temperature varies greatly if the substance is dissolved in various salt solutions (Fauré-Fremiet and Filhol, 1937). According to Champetier and Fauré-Fermiet's (1937) roentgenographic studies ascaridine seems to be a semi-crystalline substance, but it can be changed experimentally into an amorphous state.

In recent years an increasing amount of attention has been given to the occurrence of respiratory pigments in parasitic nematodes. Haemoglobin seems to be widely distributed. It has been found in *Dioctophyma*, *Ascaris*, *Parascaris*, *Toxocara*, *Nematodirus*, species of *Trichostrongylus*, *Camallanus*, *Spirocerca*, a larval *Eustrongylides* and larvae of *Trichinella* (Aducco, 1889; Flury, 1912; Fauré-Fremiet, 1913; Keilin, 1925; Krüger, 1936; v. Brand, 1937; Davey, 1938; Stannard, McCoy and Latchford, 1938; Wharton, 1938, 1941; Hsü, 1938; Janicki, 1939). The best known case is that of *Ascaris* where it is found both in the body fluid and the body wall. The absorption bands of the haemoglobins occurring at these two places are slightly different, and this indicates the presence of two kinds of haemoglobin (Keilin, 1925). In all the above cases, where haemoglobin has been found beyond the intestinal wall, one can safely assume that it has been synthetized by the worm. Parts of the host haemoglobin molecule may, of course, be used in this process, but no definite data on this possibility have been obtained. Obviously, haemoglobin found in the intestinal tract of a worm will not fall in the same category, though in some instances it may play a physiologically similar role (hookworm, for example).

The only other respiratory pigments found so far are cytochrome, which is known to occur in *Ascaris*, *Parascaris* and *Camallanus* where the highest concentration is found in the eggs and sperm (Keilin, 1925; Wharton, 1941) and flavine found by Gourévitch (1937) in *Parascaris*.

INORGANIC SUBSTANCES

Ascaris lumbricoides according to Flury (1912) contains 0.76 percent inorganic substances, and a larval *Eustrongylides* according to v. Brand (1938) contains 1.1 percent.

A quantitative analysis of the inorganic substances of *Ascaris* by Flury (1912) gave the following results:

Na	1.104% of the dry weight
K	0.607
Ca	0.404
Mg	0.058
Al	0.131
Fe	0.019
Cl	1.272
PO₄	1.315
SO₄	0.114
SiO₂	0.029

Neither copper nor manganese was found, and it can be said that on the whole the composition of the ash of *Ascaris* seems to be quite similar to that of free living organisms.

The osmotic pressure of the tissues of several *Ascaris* species and that of the body fluid of *Parascaris* (Vialli, 1923, Schopfer, 1926, 1932) is similar, but not identical to that found in the host intestine. The osmotic pressure of the worms always seems to be a little lower, so that they live in a slightly hypertonic environment. It is noteworthy that chlorides seem to play only a minor role in producing the normal osmotic pressure of the body fluid of *Parascaris* (Marcet, 1865, Schopfer, 1932). The total osmotic pressure corresponds to a freezing point depression (Δ) of —0.62°C. whereas the osmotic pressure due to the chlorides is equivalent to a Δ value of —0.12°C. The osmotic pressure varies directly with that of the environment.[1] The osmotic pressure of *Proleptus obtusus* living in the marine elasmobranch *Scylliorhinus* is considerably higher than that of the other parasites mentioned and is slightly higher than that of *Scylliorhinus* blood ($\Delta = -2.40°$, Schopfer, 1932).

---

[1] Panikkar and Sproston (1941) give data for *Angusticaecum* sp. from the intestine of the tortoise. It is of interest that according to Stoll (1940) the first parasitic ecdysis of *Haemonchus contortus* is favored by hypotonic solutions.

Metabolism of Adult Nematodes

METABOLISM UNDER ANAEROBIC CONDITIONS

Most of the experiments on nematodes under anaerobic conditions have been performed with *Ascaris lumbricoides*. Bunge (1889) found that this species can be kept for several days in the absence of oxygen and that it produces during this time carbon dioxide and a volatile acid. Considerable progress was made by Weinland (1901) who performed quantitative determinations of the amounts of various substances consumed and produced and who recognized that carbohydrates were predominantly used. In starvation experiments of several days' duration he found that 100 gm of worms consumed 0.7 gm glycogen and 0.1 gm glucose in 24 hours. He found among the end products 0.4 gm carbon dioxide and 0.3 gm of a volatile fatty acid which he identified as valeric acid. Later Weinland (1904) found that caproic acid was also present in the ether soluble excreta of *Ascaris*. A quantitative study of fat and nitrogen in similar starvation experiments led Weinland (1901) to the conviction that both carbon dioxide and fatty acids were derived from the breakdown of glycogen, and he compared this process to the fermentations produced by microorganisms. This view concerning the anaerobic processes of *Ascaris* is still valid, although subsequent investigations necessitated certain changes in Weinland's conclusions. In the first place it was found that in addition to valeric and caproic acids, some formic, butyric (Flury, 1912) and lactic acid (v. Brand, 1934a) were also present in the excreta. At present it is certain that valeric acid is the chief end product, but there is some uncertainty as to the type of valeric acid excreted. It seems probable that it is normal valeric acid (Waechter, 1934), although Flury (1912) believed that he had identified iso-valeric acid. Krüger (1936) suggested the presence of methyl-ethyl-acetic acid, but Oesterlin (1937) pointed out that this identification was insufficiently supported by Krüger's data.

The second necessary modification of Weinland's conclusions concerns the intensity of the fermentation process. It was found that with increasing length of starvation a decreasing daily amount of glycogen was used and that less carbon dioxide was produced (Weinland, 1901; Schulte, 1917; v. Brand, 1934a, 1937; Krüger, 1936). In experiments conducted for only 24 hours with fresh worms about 1.4 gm of glycogen was used. This is twice as much as Weinland (1901) found for the average daily glycogen consumption (0.7 gm) in experiments which lasted as long as 6 days. It is, however, curious and not yet sufficiently understood, that despite the different lengths of their experimental periods, most of the above mentioned investigators found that between 0.2 and 0.3 gm of valeric acid was produced per day. Krüger (1936), however, found that about 0.5 gm fatty acid was excreted during the first 24 hours.

The last complete biochemical balance under anaerobic conditions was given by v. Brand (1934a) for females of *Ascaris lumbricoides*. He found that 100 gm of worms consumed, during 24 hours at 37°C., 1.39 gm glycogen and produced 0.71 gm carbon dioxide, 0.22 gm valeric acid, and 0.02 gm of lactic acid. No complete data are available for males. It has been found, however, that the glycogen consumption is identical in both sexes during the first 24 hours and that the more active males later consume more glycogen than the females (v. Brand, 1937a).

Parascaris equorum seems to have a quite similar carbohydrate metabolism. Fischer (1924) ascertained the production of small amounts of lactic acid. Toryu (1936a) found a small amount of lactic and propionic acid and a large amount of valeric acid, but no formic, acetic, butyric, caproic, malic, citric or succinic acids. His glycogen/acid balance for the first 24 hours of anaerobiosis for 100 gm of worms was as follows: Consumed: 1.39 gm glycogen. Produced: 0.65 gm valeric acid and 0.02 gm lactic acid. In addition carbon dioxide was produced and the amount of carbon dioxide differed markedly for females and males (Toryu, 1936b). It is not clear what animals were used for the glycogen/acid experiments, and therefore it is impossible to introduce reliable carbon dioxide values into the above balance.

The above data indicate that the end products of the anaerobic carbohydrate metabolism are chiefly lower fatty acids and therefore noticeably different from that of a vertebrate muscle. This concept has been criticized chiefly by Fischer (1924) and Slater (1925). The former investigator concedes that living *Parascaris* excrete only a small amount of lactic acid, and a larger amount of an unidentified acid. He found, however, that in minced material the production of lactic and the liberation of phosphoric acid was sufficient to account for the whole acidity observed in aerobically conducted experiments. Therefore, he concluded that there was no great difference between the glycogen breakdown in *Parascaris* and in vertebrates. In the opinion of the present writers, however, his observation indicates merely that through changes in the experimental conditions the course of the chemical reactions can be changed—a phenomenon well known in experiments with yeast and other lower plants. It should be remembered that Weinland (1902) found the same end products with extracts of *Ascaris* under anaerobic conditions as he had found in experiments with whole worms.

Slater (1925) demonstrated that bacteria capable of transforming sugar into volatile fatty acids could be isolated from a saline solution in which ascarids had been immersed. He failed, however, to show that they were present in sufficient numbers to account for all the organic acids produced in experiments with worms, and, furthermore, he did not demonstrate any substance which could have served as a substrate for such bacterial fermentation.

Several lines of evidence have been brought forward which seem to indicate a direct connection between nematodes and the production of lower fatty acids. The following two may be mentioned. The volatile acids are found not only in saline in which worms have been kept, but also in distillates of minced worms (Weinland, 1901) and in the ether extract of whole worms (Flury, 1912; Schimmelpfennig, 1903). Valeric acid has, furthermore, been found under both aerobic and anaerobic conditions, although one should expect that such a difference in the external conditions should have a deep influence on the development of a bacterial flora in the surroundings. For further information on this controversy compare the discussion of Slater (1928) with those of Weinland (1901) and v. Brand (1934b).

Several methods have been discussed in which valeric acid may originate from carbohydrate. Weinland (1901) favored the following equation: $4C_6H_{12}O_6 = 9CO_2 + 3C_5H_{10}O_2 + 9H_2$. It must, however, be emphasized, that the postulated hydrogen could not be found. Weinland (1901) had to assume that it was used at once in other reactions. He also discussed an equation proposed by Koenigs:

$$13C_6H_{12}O_6 = 12C_5H_{10}O_2 + 18CO_2 + 18H_2O.$$

Weinland rejected this equation because it did not predict nearly as much carbon dioxide as he found to be present. However, the excess might have originated either from bicarbonate or from protein decomposition. Jost (1928) has given the following chain of reactions which leads to Koenigs' equation. These equations are purely theoretical, but the series is interesting in that it shows a possible link between the production of lactic and valeric acids.

$$C_6H_{12}O_6 = 2\ C_3H_6O_3$$
Glucose   Lactic acid

$$\left.\begin{array}{l} CH_3.CHOH.COOH \\ CH_3.CHOH.COOH \end{array}\right\} \xrightarrow[\text{and dehydration}]{\text{Dismutation}} \left\{\begin{array}{l} CH_3.CO.COOH \\ CH_3.CH_2.COOH \end{array}\right. + H_2O$$
2 Lactic acid   Pyruvic acid
                        Propionic acid

$$CH_3.CO.COOH + CH_3.CH_2.COOH = \overset{\overset{\displaystyle CH_2.CH_2.COOH}{|}}{CH_3.COH.COOH} =$$
$$CH_3.CHOH.CH_2.CH_2.COOH + CO_2$$
γ Hydroxy-valeric acid

$$CH_3.CHOH.CH_2.CH_2.COOH = CH_3.CH_2.CH_2.CH_2.COOH + O$$
Normal valeric acid

$$1\ C_6H_{12}O_6 = 1\ CH_3.CH_2.CH_2.CH_2.COOH + H_2O + CO_2 + O$$

$$12\ C_6H_{12}O_6 = 12\ CH_3.CH_2.CH_2.CH_2.COOH + 12\ H_2O + 12\ CO_2 + 6\ O_2$$

$$1\ C_6H_{12}O_6 + 6\ O_2 = 6\ CO_2 + 6\ H_2O$$

$$13\ C_6H_{12}O_6 = 12\ CH_3.CH_2.CH_2.CH_2.COOH + 18\ CO_2 + 18\ H_2O.$$

In effect, then, 12 molecules of sugar would be transformed into 12 molecules of valeric acid, carbon dioxide and water, and the oxygen liberated during this process would be sufficient to oxidize completely a thirteenth molecule of sugar.

Toryu (1936a) proposed the equation: $4C_6H_{12}O_6 = 4CO_2 + 4C_5H_{10}O_2 + H_2O$. This equation needs no further consideration, since the O and H atoms on the two sides do not balance. Correctly written it would read: $4C_6H_{12}O_6 = 4CO_2 + 4C_5H_{10}O_2 + 4H_2O + 2O_2$. This obviously corresponds closely to an intermediate step of Koenigs' equation as formulated by Jost.

The amount of heat produced during the metabolism of *Ascaris lumbricoides* was first determined directly by Krummacher (1919). His experiments, however, were performed at a time at which oxygen was regarded as an inert gas for these worms. Krummacher's experiments were neither clearly aerobic nor anaerobic, and the data obtained are therefore difficult to interpret. Meier (1931), on Krummacher's suggestion, performed similar experiments under anaerobic conditions. He found a heat production of 0.300 gm cal per gm of worm per hour. On the basis of Weinland's chemical data and his own heat determinations he calculated that the fermentation process yields 22 percent of the energy obtainable by total oxidation of the carbohydrate. This is considerably more than usually found in bacterial fermentations. Undoubtedly, however, Meier's figure of 22 percent is far too high. His experimental periods lasted only from 4 to 12 hours, and he used presumably fresh worms. Therefore, the carbohydrate consumption must have been much higher than Weinland's figure. Furthermore, Schulte (1917) has demonstrated by direct comparisons of the heat of combustion with the glycogen content of fresh and starving ascarids that the carbohydrate metabolism accounts for only 80 percent of the total loss of calories from the body. Meier, however, assumed that the total heat production was due to carbohydrate fermentation. At present the data necessary for an exact balance sheet of the energies involved seems to be unavailable. A fair guess would place the energy yield of the fermentation between 6 and 12 percent. This is still more than that usually found in bacterial fermentations. Lactic acid fermentation, for example, yields only about 2.6 percent, and alcoholic fermentation yields 4 percent.

Changes under anaerobic conditions in the material extractable with ether have been studied less thoroughly than the changes in glycogen content. Weinland (1901) found that there was no change in the fat content of ascarids during starvation, and v. Brand (1934a) reached the same conclusion. Schulte (1917), on the other hand, observed a fat increase of 0.08 gm per 100 gm animals per day. He considered this fat to be a product of carbohydrate fermentation. It seems certain, at least, that no fat is consumed under anaerobic conditions. This is not astonishing, because it seems hardly possible that an anaerobic process could yield energy from an oxygen poor substance like fat (Weinland, 1901).

The nitrogen metabolism of *Ascaris* is not very great. For 100 gm of worms the amount of nitrogen excreted in 24 hours was found by Weinland (1904b) to be 15 to 20 mgm and by v. Brand (1934a) to be 29 mgm. One third of the excreted N is ammonia, and the greater part of the remainder can be precipitated by phosphotungstic acid (Weinland, 1904b). Flury (1912) found that the worms excreted not only ammonia but small amounts of amine bases, substances which gave the

biuret reaction, hydrogen sulfide (also Krüger, 1936), and mercaptan. According to v. Brand (1934a) about one fourth of the total excreted N is contained in discharged eggs. Chitwood (1938) found urea in a concentration of about 0.02 percent in the fluid from the excretory pore of freshly collected worms. After 24 hours of starvation the tests for urea were negative, and Chitwood doubts that the urea was formed by the worm. It may have been obtained from the host.

METABOLISM UNDER AEROBIC CONDITIONS

Weinland (1901) believed that *Ascaris* did not consume oxygen. However, he did observe that more carbon dioxide was evolved under aerobic than under anaerobic conditions. He explained this on the assumption that the extra carbon dioxide was due either to the metabolism of aerobically developing eggs or to that of an aerobic bacterial flora. His view was generally accepted until Adam (1932) proved that *Ascaris* was able to consume oxygen. The observations of Adam were soon confirmed and extended to other forms. The following table summarizes some of these data on oxygen consumption.

Species	Sex	O₂ consumption in gm per 100 gm worms in 24 hrs. at body temp.	Investigator
Ascaris lumbricoides	♂	0.38	Adam, 1932
Ascaris lumbricoides	♀	0.21	Adam, 1932
Ascaris lumbricoides	♀	0.21	v. Brand, 1934
Ascaris lumbricoides	♀	0.13	Harwood and
Ascaris lumbricoides	♂	0.21	Brown, 1934
Ascaris lumbricoides	?	0.55*	Krüger, 1936
Ascaris lumbricoides	?	0.27*	Krüger, 1936
Parascaris equorum	♀	0.08	Toryu, 1936
Parascaris equorum	♂	0.35	Toryu, 1936
Setaria equinum	?	0.89	Toryu, 1936
Ancylostoma caninum	?	more than ten times as much as female *Ascaris*	Harwood and Brown, 1934

The oxygen consumption of both *Setaria* and *Ancylostoma* is considerably higher than that of *Ascaris* or *Parascaris*. The former undoubtedly have easier access to oxygen and may therefore be better adapted to aerobic metabolism.

The amount of oxygen consumed by *Ascaris* is influenced by several factors. One factor is size, and small animals consume relatively more than large ones. However, it is doubtful if the difference in oxygen consumption of males and females can be explained merely on the basis of size. Krüger (1936) gave a formula which allows one to calculate approximately the increase of oxygen consumption with increasing weight. The formula is applicable only to worms which weigh over 1.4 gm. In smaller worms the increase is more rapid. Krüger stated nothing about the sex of his worms, but the deviation of his data from the formula begins near the average weight of males. In a recent paper Krüger (1940) shows that the O₂ consumption of ascarids of various sizes is fairly constant if referred to surface rather than weight.

The oxygen consumption of starving ascarids kept for long periods of time at the oxygen tension of air show a general tendency to increase (v. Brand, 1934a; Krüger, 1937). This might be an indication of adaptation to the abnormally high oxygen tension.

The oxygen consumption of *Ascaris* varies directly with the oxygen tension, regardless of whether whole worms, parts of worms or even minced material is used (Harnisch, 1933; Krüger, 1936). This is a striking contrast to what is known from massively built free-living organisms, like actinians. In these a similar dependence is observed in whole animals, but it disappears if minced material is used. The diffusion rate of oxygen is the limiting factor, and if the path through which oxygen has to diffuse is shortened by using minced animals, the oxygen consumption remains virtually unchanged over a wide range of tensions. This explanation can not hold for *Ascaris*. Harnisch, however, has found that the oxygen consumption of planarians and *Chironomus* larvae, which is normally independent of oxygen tension, may become dependent if the animals are subjected to anaerobic conditions prior to the experiments. In his opinion two kinds of aerobic processes must be distinguished: (1) a primary aerobic process which is considered to be independent of the oxygen tension, and (2) a secondary process which is considered to be dependent. In *Ascaris* only

*Krüger (1936) gives data of various sized worms. Those for worms of about the average size of males and females have been introduced in the table, the higher figure being for worms of 1.5 gm, the lower for worms of 4.5 gm.

315

the secondary aerobic process is present. Harnisch (1937) offered support of this view in the observation that washed minced *Ascaris* material has only a negligible oxygen consumption. The same material, suspended in *Ascaris* body fluid, has a very high oxygen consumption and surpasses even that of non-minced material. According to Harnisch this indicates the presence of a powerful oxidizing mechanism outside of the cells which may govern the entire aerobic processes of *Ascaris*. This, he claims, is in accordance with his explanation of experiments with artificially induced secondary aerobic processes in *Chironomus*. The cellular agents which govern the primary aerobic processes in *Chironomus*, however, could not be removed from the cells by washing (Harnisch, 1936).

The data of Kempner (1937) show that in a variety of biological materials the effect of oxygen tension on oxygen consumption varies with pH, CO_2 tension, salt content, and temperature. It is apparent that certain tissues heretofore considered to have a respiratory mechanism unaffected by oxygen tension really show an independence only in alkaline CO_2-free media in a certain temperature range. These observations of Kempner indicate that the whole question of oxygen tension versus oxygen consumption should be reëxamined, and that the respiration of no material can be said to be completely dependent or independent of O_2 tension unless the effects of the above factors have been investigated. It is possible that these factors may have some effect on the nematode data discussed above. A discussion of the theoretical relationship between oxygen tension and oxygen consumption is given by Marsh (1935).

It seems that all the different organs of *Ascaris* are able to consume oxygen. This has been shown for the body wall, intestine, ovaries, uterus and even the body fluid (Harnisch, 1935, 1937; Krüger, 1936). The largest absolute amount is consumed by the body wall, although the intestine shows the highest rate of oxygen consumption.

It is now generally believed that ascarids evolve larger amounts of carbon dioxide under aerobic than under anaerobic conditions (Weinland, 1901; v. Brand, 1934a; Krüger, 1936), and Harnisch (1937) has abandoned his previous contention to the contrary. The respiratory quotient in air is consistently very high. In fresh worms it may be about 4 or even higher, and in worms kept for several days in saline it is between 1.27 and 1.88 (Krüger, 1937). This indicates that the oxidation of metabolites is not complete and that even in the presence of oxygen the metabolism consists in part of anaerobic fermentations.

The excretion of organic acids under aerobic conditions, first seen by Weinland (1901), is definite proof of the presence of fermentations. The acids have been identified as small amounts of lactic acid (v. Brand, 1934a), formic, acetic, and probably butyric acid, a large amount of valeric acid and some unidentified higher acids (Oesterlin, 1937). Since these products are similar to those formed under anaerobic conditions (see above), it seems likely that the fermentations going on under aerobic and anaerobic conditions are identical. The amounts of acids excreted at the oxygen tension of air are definitely lower than under strictly anaerobic conditions (v. Brand, 1934a; Krüger, 1936, 1937), but at low oxygen tensions even more acids are excreted (Krüger, 1936).

It is customary in the nematode literature to refer to the oxidations which involve oxygen consumption and which lead to the production of carbon dioxide and water as oxidative metabolism and to refer to the molecular rearrangements and oxido-reductions which lead to the production of carbonic, lactic, valeric, and other acids and in which oxygen is not consumed as fermentative metabolism. Von Brand (1934a) and Krüger (1937), by basing calculations on the ratio of anaerobically evolved carbon dioxide to anaerobically excreted acids or similar data at low oxygen tensions, calculated the amounts of aerobically evolved carbon dioxide which originated in fermentative and in oxidative metabolism. This latter figure was used, in connection with the oxygen consumption, to calculate the true respiratory quotient which was found to be about 0.9 or 1.0. In some cases very low quotients were found, and these data are difficult to explain at the present time. The opinion of Harnisch (1933) that the aerobic processes do not lead to the production of CO_2 and that the respiratory quotient is zero has been generally abandoned.

Krüger (1936) found that the uncorrected respiratory quotient of ascarids kept in air instead of saline fell rapidly to about 1.0 and remained at this level for some time. This would indicate (Krüger, 1937) either that the fermentations cease altogether, or that the fermentations present do not lead to carbon dioxide production (e.g., lactic acid formation).

The question of what substances are oxidized has received some attention by v. Brand (1934a). He found that under aerobic conditions somewhat less glycogen is consumed than

under anaerobic ones. On an assumption similar to that made above for the carbon dioxide, he calculated the amounts of the consumed glycogen which had apparently been decomposed by fermentative and by oxidative metabolism. He arrived at the following balances:

Uncorrected balance for 100 gm worms starving at 37° C. under aerobic conditions:
Decomposed: 1.18 gm glycogen. Consumed: 0.21 gm oxygen. End products: 0.84 gm carbon dioxide + 0.16 gm valeric acid + 0.01 gm lactic acid.
Oxidative part of the metabolism:
Decomposed: 0.37 gm glycogen. Consumed: 0.21 gm oxygen. End products: 0.34 gm carbon dioxide + ?.
Fermentative part of the metabolism:
Decomposed: 0.86 gm glycogen. End products: 0.48 gm carbon dioxide + 0.16 gm valeric acid + 0.01 gm lactic acid.

The amount of glycogen which disappeared was so great that complete oxidation to carbon dioxide and water could not be assumed for all of that which was calculated to undergo oxidative metabolism. Probably only a partial oxidation takes place (formation of aldehydes?).

Harnisch (1935) thought that possibly iso-valeric acid would be oxidized to aceto acetic acid or β-hydroxy-butyric acid which in turn would be decomposed to acetone and carbon dioxide. However, chemical determinations on the excreta do not favor this view. This statement applies also to v. Brand's (1934a, b) original theory that fats may be changed into carbohydrate.

It seems as if *Ascaris*, in contrast to many free-living animals, does not contract a noticeable oxygen debt during a period of anaerobiosis (Adam, 1932; Harnisch, 1933). It was found (v. Brand, 1937b), however, that ascarids subjected to 20 hours anaerobiosis and then brought for 2 to 6 hours into aerobic conditions, resynthesized 1/20 to 1/10 of the glycogen consumed during the anaerobic period. This resynthesis is clearly an aerobic process, and it is apparently much less pronounced in *Ascaris* than in similarly treated vertebrate muscles. This may be due to the fact that in vertebrate muscle the end products accumulate, whereas in *Ascaris* they are excreted, and only those present in the body at the beginning of the aerobic period are available for resynthesis. It is unknown whether lactic acid or the lower fatty acids are resynthesized to glycogen.

There is still some controversy concerning the significance of the aerobic processes of *Ascaris*. Harnisch (1933) assumed that the aerobic processes would yield no energy, and he still thinks (Harnisch, 1935) that they play no role in the normal energy supply of the organism. This view is similar to that of Krüger (1937) who states that they are probably not linked to any specific organ function and that any derived energy is probably wasted. The present writers are of the opinion that at this time no definite statements regarding the possible utilization of this energy can be made.

The fact that the rate of the fermentative processes is reduced at the oxygen pressure of air, seems to indicate rather clearly that fermentations and oxidations are not entirely independent as Harnisch (1933) originally assumed. Whether Krüger's (1937) view is correct that the oxidations follow essentially the same course as in truly aerobic organisms, or whether Harnisch (1937) is right in assuming that they correspond only to the secondary aerobic processes occurring in free-living animals only under special conditions, must be decided by future investigations.

The aerobic metabolism of *Parascaris equorum* has been studied by Toryu (1934 to 1936b). He found an almost identical glycogen consumption under aerobic and anaerobic conditions, but since the worms excreted slightly less organic acids under aerobic conditions, he concluded that a small amount of glycogen was oxidized. Apparently the aerobic metabolism of *Parascaris* follows the same pattern as that of *Ascaris*.

The question of whether or not parasitic nematodes use fat under aerobic conditions is difficult to answer satisfactorily at the present time. In v. Brand (1934a) aerobic experiments no fat was used. In view, however, that his experiments lasted only 24 hours and that in general carbohydrate is consumed before the fat reserves are attacked, these experiments can not be accepted as conclusive evidence that no fat may be used during longer periods of starvation. Mueller (1928/29) observed that in explanted pieces of *Ascaris* a loss of morphologically demonstrable fat occurred after several days, and Hirsch and Bretschneider (1937) have shown that in starving ascarids much of the stainable fat disappeared from the intestinal cells after 6 days. These observations are suggestive that fat may be used, but they should be confirmed by quantitative chemical methods.[1] Bondouy (1910) detected a lipase in *Strongylus equinus*, and the possible significance of its presence warrants further study.

---

[1] In a recent paper v. Brand (1914) showed that *Ascaris* uses no fat for production of energy during an aerobic starvation period of 5 days.

The aerobic and anaerobic nitrogen metabolism of *Ascaris* has been compared by v. Brand (1934a). The amounts of nitrogen excreted both in soluble excreta and in eggs were very nearly identical in both cases. He assumed that at least a large part of the N metabolism was involved in the transformation of the protoplasm of the body into that of eggs. He also considered it likely that at least a large part of the nitrogen metabolism was always anaerobic. This view is supported by the fact that free-living animals, like the leeches, show, in contrast to *Ascaris*, a marked difference in the amount of nitrogen excreted under aerobic and under anaerobic conditions.

DEDUCTIONS CONCERNING THE METABOLISM
IN VIVO

Deductions concerning the nature of the metabolism of internal parasites can be drawn only from the chemical composition of their surroundings and their metabolism *in vitro*. Of special interest is the question of whether the nematodes parasitizing the intestine lead an anaerobic or an aerobic life. On the basis of the investigations of Bunge (1889) and Weinland (1901) the first possibility was accepted for many years as an undisputed fact. More recently certain investigators (Slater, 1925; Mueller, 1928/29; Adam, 1932; Davey, 1938a and b) have held the opposite view, i.e., that the worms can get enough oxygen in the intestine to allow an oxidative metabolism. Recently v. Brand (1938) has reviewed the question, and he believes that a general answer can not be given. Apparently the size or relative surface and the presence of respiratory pigments will have a great influence on whether a worm can or can not obtain sufficient oxygen at the low tensions prevailing in the intestine. Large parasites, like *Ascaris* or *Parascaris*, must be regarded as predominantly anaerobic organisms. As mentioned above, they show a marked fermentative metabolism even in air. Since their oxygen consumption is dependent on the oxygen pressure, one can be reasonably sure that fermentative metabolism will be relatively much greater in the intestine. Further signs of their adaptation to an anaerobic life are that they are remarkably resistant to the lack of oxygen *in vitro* and that they are able to excrete the end products of anaerobic metabolism. It seems, however, quite possible that the small amounts of oxygen available in the intestine are not entirely without significance. This may be indicated by the observations that the worms contain some haemoglobin, that stimulated *Ascaris* die much more rapidly in absence than in presence of oxygen, and finally that they are able to perform under suitable conditions such a clearly aerobic process as the resynthesis of glycogen.

Small nematodes, on the other hand, offer better opportunities for the diffusion of oxygen because of their relatively larger surface. This may explain why the sheep nematodes do not show (Davey, 1937, 1938a and b) the same resistance against lack of oxygen as *Ascaris*. The conclusion of Davey that these worms lead an aerobic life under natural conditions is, therefore, probably only in apparent contradiction with the statement made above in regard to large helminths.

An entirely different way of getting oxygen may be realized in worms sucking larger amounts of blood from their hosts. According to Wells (1931) the blood sucking activities of hookworms seem to serve largely as a respiratory function. His data allow the calculation that under optimal conditions 100 gm of worms could obtain 20 gm of oxygen from this source in 24 hours. This would be about ten times as much as Harwood and Brown (1934) found to be the actual oxygen consumption.

No data are known about the metabolism of adult parasitic nematodes which normally live outside the intestine. It is therefore unnecessary to enter into a similar discussion concerning their metabolism. On the whole one may assume that they will have frequently, though probably not in every case, better opportunities to get larger amounts of oxygen than the intestinal helminths.

SYNTHESIS OF RESERVE SUBSTANCES

There are only a few investigations which concern the question of the synthesis of reserve substances in parasitic nematodes. Hoffman (1934) and Krüger (1936) have shown that the heat production and the oxygen uptake of ascarids under both anaerobic and aerobic conditions are increased if sugar is present in the surrounding medium. Hirsch and Bretschneider (1937) fed ascarids iron saccharate and concluded from their histological investigation that it was absorbed as colloid and broken down only in a certain part of the intestinal cells into iron and sugar.

Quantitative determinations of the glycogen content of carbohydrate-fed ascarids have been performed by Weinland and Ritter (1902). They found no increase in the glycogen content of animals kept in solutions containing various carbohy-

drates, although glucose caused a lowering of the rate of utilization of body glycogen. More positive results were achieved by injecting the sugar solutions into the animals. In these experiments new glycogen was formed after injection of glucose and probably levulose. The consumption of body glycogen was decreased by injections of maltose and perhaps galactose, but not by injections of lactose.

Von Brand and Otto (1938) compared the glycogen content of hookworms from dogs which had been starved for 48 to 72 hours before death with those from dogs which had been given so much sugar during a similar period that the liver glycogen rose from 0.06 percent to 5.04 percent. No difference whatever in glycogen content of the worms was found. This may be related to the fact that hookworms obtain their food from the tissues rather than from the lumen of the intestine and therefore can gain their maximal food requirements even from a starving host.

So far no experiments have been performed on the deposition of fat in parasitic nematodes except the above-mentioned doubtful results of Schulte (1917) concerning the fat increase in ascarids under anaerobic conditions. The whole question of synthesis should prove interesting for future investigations.

Metabolism of Eggs and Larvae

The eggs of many parasitic nematodes show, like the adults, a surprising degree of resistance to lack of oxygen. The eggs of such forms as *Ancylostoma*, *Parascaris*, *Trichocephalus* or *Nematodirus* can be kept for days or even weeks in the absence of oxygen, but they do not complete their development (Looss, 1911; Bataillon, 1910; Zawadowski, 1916; Fauré-Fremiet, 1913; Zawadowski and Orlow, 1927; Zviaginzev, 1934; Dinnik and Dinnik, 1937). In *Parascaris* oxygen is unnecessary only during the early stages, i.e., maturation, fertilization and perhaps the first cleavage stages; for further development oxygen is indispensable (Fauré-Fremiet, 1913; Szweikowska, 1929; Dyrdowska, 1931). The need of oxygen for completion of development seems to be a general requirement, although the stage of development at which oxygen becomes necessary seems to vary somewhat with different species. Zawadowsky and Schalimow (1929), Schalimow (1931), and Wendt (1936) conclude that the necessity for oxygen begins in *Enterobius vermicularis* with the tadpole stage, and in *Oxyuris equi* with the gastrula stage. Relatively low oxygen pressures, however, are sufficient to insure normal development in *Ascaris* and *Ancylostoma* (Brown, 1928; McCoy, 1930).

The amount of oxygen consumed by one *Ascaris* egg in developing from the one-cell stage to the motile embryo is about 0.0025 cmm with only slight variations whether the development is completed in 21 days at 23°C or in 11 days at 30°C (Brown, 1928). Huff (1936) obtained a value of 0.0041 cmm for *Ascaris*, and Nolf (1932) obtained a value of 0.0027 for the eggs of *Trichuris*. It is surprising that an *Ancylostoma* egg requires for its development from the morula stage to the fully developed larva almost exactly the same amount of oxygen (0.0028 cmm at 23° C. according to McCoy, 1930) as an *Ascaris* egg, although development of *Ancylostoma* is completed in about 24 hours. Since these eggs are about the same size, it seems as if the difference in the rate of oxygen consumption mentioned above for the adults of these species is also present in the embryonic stages.

Huff (1936) observed that the oxygen consumption of *Ascaris* eggs increased more than five times after removal of the albuminous coating by antiformin. Friedheim (1933) found that the oxygen consumption of *Ascaris* eggs is considerably increased if they are immersed in a dilute solution of hallochrome (a pigment which is a reversible oxidation-reduction system isolated from the polychaete worm *Halla parthenopea* and which has an accelerative effect on respiration). The mechanism of the increase in respiration by either of these two methods is not known. Friedheim (1933) apparently used mixed stages of fertilized eggs, and there seems to be no reason for assuming that hallochrome could penetrate the egg shell. Therefore, one might expect the acceleration obtained to be due to an increase in the effective oxygen tension or to an increase of respiration in only those eggs on which an impermeable shell had not yet been formed. The experiments of Huff might also be explained as being caused by an increase in effective oxygen tension because of slow diffusion of oxygen through the albuminous coat, but no data concerning these possibilities are available. Since the R. Q. is always less than 1.0 (see below) the possible effect of oxygen tension could not be merely to change the ratio of oxidative and fermentative metabolism. The accelerations produced by Friedheim (1933) and Huff (1936) must, for the present, be accredited to changes in the rate of oxidative metabolism, and the reasons for the changes remain obscure.

The oxygen consumption of *Ascaris* or *Parascaris* eggs has also been reduced experimentally by ultracentrifuging and by exposure to cyanide (Zawadowsky, 1926; Huff and Boell, 1936). About 90 percent of the respiration was sensitive to cyanide, and it seemed that ultracentrifuging affected only the cyanide sensitive respiratory mechanism.

The respiratory quotient of *Parascaris* and *Ascaris* eggs has been found to be below 1, and this indicates that, in contrast to results on tissues of the adult worm, no fermentative processes are present in the eggs. The respiratory quotient determined at the beginning of development was about .80, and, with some variations in the case of *Parascaris*, it increased during the later stages to .92-.98 (Fauré-Fremiet, 1913a, 1913; Huff, 1936). The total energy liberated by one *Parascaris* during its development was 50 x 10⁻⁷ cal. (Fauré-Fremiet, 1913). Nolf (1932) found that the R. Q. of *Trichuris* decreased from a value of 1.0 for the first 5 days of development to a value of 0.73 for the 8th to 15th days.

In considering the chemical changes which occur in the eggs of parasitic nematodes during their development, one must distinguish clearly between processes which lead to the formation of the egg shells and processes which liberate energy. The shells, as far as they are formed from the ovum, consist essentially of the shell proper and the vitelline membrane. The shell is composed of chitin in such species as *Parascaris*, *Ascaris*, *Dioctophyma* and *Enterobius* (Fauré-Fremiet, 1913; Szwejkowska, 1929; Schmidt, 1936; Wottge, 1937; Chitwood, 1938; Jacobs and Jones, 1939). The investigations of Fauré-Fremiet (1913) and Szwejkowska (1929) have demonstrated that in *Ascaris* about half the glycogen stored in the oöcytes was used to form the glucosamine incorporated in the chitin. The latter has shown in addition that 26 percent of the total nitrogen of the egg was used during the chitin formation.

The vitelline membrane of the eggs of these and other species is of a lipoid nature (Fauré-Fremiet, 1913; Zawadowsky, 1928). Fauré Fremiet considered it to be mainly ascaryl alcohol, Wottge (1937) obtained a positive reaction for cholesterol, and Chitwood (1938) and Jacobs and Jones (1939) demonstrated that it gave sterol reactions. During the secretion of this layer certain changes in the chemical nature of the ether soluble substances, perhaps a saponification, seemed to occur. The necessity for further studies is indicated.

Chemical analyses of the egg indicate that both glycogen and fat are oxidized, and these data are in accordance with the above data on the respiratory quotient. Swejkowska (1929) found in *Parascaris* eggs just after fertilization about 0.46 percent volatile fatty acid and 0.53 percent higher fatty acids. After formation of the second polar body these substances had diminished to 0.34 and 0.36 percent respectively. For the same period it was calculated that in addition to the glycogen used in the formation of chitin an amount of glycogen corresponding to about 2.7 percent of the egg weight had disappeared. From Fauré-Fremiet's (1912, 1913) experiments it would appear that both fat and glycogen were used during the later developmental stages. All of these experiments were conducted under aerobic conditions. Dyrdowska (1931) found by the use of staining methods that the glycogen content of *Parascaris* eggs kept under anaerobic conditions underwent a slight diminution and that there was a marked decrease in the fat content. It seems desirable that this decrease in fat content should be verified with quantitative chemical methods since, as already stated above, it is difficult to understand how processes which liberate energy from fat could occur in the absence of oxygen. It should, furthermore, be remembered that Fauré-Fremiet (1913) gained the impression that the amount of fat in anaerobically kept eggs tended to increase.

With the exception of the above mentioned shifting of nitrogen from the ovum to the chitin shell, nothing is known about the nitrogen metabolism of eggs. Szwejkowska (1929) found no change in the total nitrogen content during the time of maturation, and Kosmin (1928) found the same nitrogen content (1.78 percent) in undeveloped and developed eggs. She points out that this may be caused by the impermeability of the vitelline membrane for protein degradation products which consequently might accumulate in the interior of the egg shells.

The fully developed embryo of *Ascaris* contains glycogen, even in eggs which have been stored for 6 months (Stepanow-Grigoriew and Hoeppli, 1926). This observation has a bearing on Pintner's theory (1922) concerning the physiological reason for the migration of parasitic worms through the host body prior to life in the intestine. Pintner was of the opinion that the chief function of the migration was to allow the worms to live for a time under aerobic conditions. This would allow them to accumulate a glycogen reserve which later on would enable them to begin life in the anaerobic intestine. The above mentioned observation of Stepanow-Grigoriew and Hoeppli (1926)

is not what one might expect on the basis of this theory. However, Stepanow-Griegoriew and Hoeppli (1926) and Giovannola (1936) found a definite accumulation of glycogen during the migration.

The fact that glycogen is still present in old embryos also indicates that the rate of metabolism in fully developed eggs is probably very much lower than in the developing eggs, and this problem seems worthy of quantitative consideration.

The young larvae of *Ascaris*, on the other hand, have a high level of metabolism, as evidenced by the investigation of Fenwick (1938). He found a preliminary phase of about half an hour during which the newly hatched larvae showed a low oxygen consumption. This he explained on the assumption that they had not yet become sufficiently adjusted to their new environment. Then followed an intermediate phase, lasting about an hour, in which 1,000 larvae consumed per hour 9.3 cmm oxygen at 37° C. After this the oxygen consumption decreased to a third level (0.928 cmm per 1,000) which was about 1/10 of the second level. This new rate of oxygen consumption was maintained throughout the rest of the experiments. Fenwick explained the high rate of the intermediate stage on the assumption that it was caused by the removal of an oxygen debt which the larvae had contracted while living within the egg shells. An investigation of the respiratory quotient of eggs containing infective embryos should prove helpful in answering this question.

The rate of metabolism of *Trichinella* larvae, according to the data of Stannard, McCoy and Latchford (1938), was about as high as that of *Ascaris* larvae in the third of Fenwick's stages. At body temperature in Tyrode solution the *Trichinella* larvae consumed 2.24 cmm oxygen per mgm dry weight per hour. In saline the value was 1.70, and in Tyrode without bicarbonate it was 1.78. The figures for 1,000 larvae in these solutions can be calculated to be about 1.12, 0.85 and 0.88 cmm oxygen per hour, respectively. The respiration was independent of the oxygen tension in the range of 1 to 100 percent oxygen. It was very sensitive to cyanide, but was stimulated by carbon monoxide and paraphenylene diamine. The respiratory quotient of the *Trichinella* larvae was always above 1, and the averages were from 1.13 to 1.17. It seems probable that under aerobic conditions some fermentations may take place, but most of the oxidative processes apparently proceed to completion. Fermentation alone was sufficient to keep the worms alive under anaerobic conditions, but apparently oxygen was necessary for enabling them to move.

The fermentation processes of the *Trichinella* larvae are very interesting, since they lead not only to the formation of carbon dioxide but to the formation of other as yet unidentified substances which are known to be non-acidic. In this respect they differ from all the other helminths. It is remarkable, furthermore, that substances like iodoacetate and others, which rapidly inhibit alcoholic fermentation or muscle glycolysis, were quite slow in their action on the anaerobic carbon dioxide production of these larvae (Stannard, McCoy and Latchford, 1938). McCoy, Downing and Van Voorhis (1941) showed that radioactive phosphorus fed to the host penetrates rapidly into the larvae. This observation indicates that the larvae may have an active metabolism inside the cyst.

The *Trichinella* larvae are clearly aerobic rather than anaerobic organisms. This is also true for the larvae of *Eustrongylides*, investigated by v. Brand (1938). He found that these worms survived much longer under aerobic than under anaerobic conditions. One hundred grams of worms in the presence of oxygen consumed 0.3 gm of glycogen in 24 hours at 37° C., and no organic acids could be found. Under anaerobic conditions 0.9 gm glycogen was consumed and organic acids equivalent to 30 cc n/10 acid were produced. The ratio between aerobically and anaerobically consumed glycogen was 1:3, a ratio which places these worms intermediate between most free-living worms which have ratios of about 1:5 and *Ascaris* with one of 1.0:1.3.

The experiments mentioned so far were performed with larvae which had been living under natural conditions in a host. From free-living stages of parasitic nematodes data are only available for *Ancylostoma caninum*. McCoy (1930) found that the oxygen consumption of infective larvae varied greatly with the temperature. At 7° C. it was imperceptible, but in the range of 17° C. to 42° C. the oxygen consumption increased about 9 percent for every degree rise in temperature, and followed an exponential curve, the b constant, of which was 1.0879. The actual oxygen consumption at 37° C. corresponded to 0.47 cmm per 1,000 larvae per hour, a figure somewhat lower, but of the same order of magnitude as those mentioned above for *Ascaris* and *Trichinella* larvae.

The free-living larvae of *Necator americanus*, and *Ancylostoma caninum* seem to derive their energy primarily from fatty substances stored in their body (Payne, 1923, Rogers, 1939), and the amount of fat demonstrable seems to be characteristic of the physiological age of the larvae (Payne, 1923; Cort, 1925). A decrease in the amount of fat granules was also observed by Giovannola (1936) in the filariform larvae of several species, especially if they were kept at 37° C.

It seems, however, that these larvae also consume glycogen. Giovannola (1936) found small amounts of glycogen in young rhabditiform larvae of *Necator*, *Ancylostoma* and *Nippostrongylus*, but none in the filariform stages. A comparable observation was made by Stepanow-Grigoriew and Hoeppli (1936) who found glycogen in one- or two-day old filariform larvae of *Strongyloides*, but never in three- to nine-day old larvae.

Bibliography

ENVIRONMENT AND ITS INFLUENCE ON METABOLISM

ANDREWS, J. S. 1938.—Effect of infestation with the nematode *Cooperia curticei* on the nutrition of lambs. J. Agric. Res., v. 57:349-360.

ARNOLD, L. 1933.—The bacterial flora within the stomach and small intestine. The effect of experimental alterations of acid base balance and of the age of the subject. Am. J. Med. Sc., v. 186:471-480.

ASHCRAFT, D. W. 1933.—Effect of milk products on pH of the intestinal contents of domestic fowl. Poultry Sc., v. 12:292-298.

BALL, B. H. 1939.—The pH of the digestive tract in the living albino rat as determined by the capillary glass electrode. Am. J. Physiol., v. 128:175-178.

BALLS, A. K. and SWENSON, T. L. 1934.—The antitrypsin of egg white. J. Biol. Chem., v. 106:409-419.

BASON, S. 1908.—The stomach and intestinal gases. N. Y. Med. J., v. 88:684-689, 738-741.

BERGEIM, O. 1924.—Intestinal chemistry. I. The estimation of intestinal reductions. II. Intestinal reductions as measures of intestinal putrefaction, with some observations on the influence of diet. J. Biol. Chem., v. 62:45-60.

VON BONSDORFF, B. 1939.—Influence of intestinal worms on proteolytic activity in vitro of trypsin, papain, and pepsin and human gastric juice at neutral reaction. Acta. Med. Scand., v. 100:459-482.

VON BRAND, T. 1934.—Das Leben ohne Sauerstoff bei wirbellosen Tieren. Ergeb. Biol., v. 10:37-100.
————. 1938.—The nature of the metabolic activities of intestinal helminths in their natural habitat: aerobiosis or anaerobiosis? Biodynamica (41), 13 pp.

VON BRAND, T. and WEISE, W. 1932.—Beobachtungen über den Sauerstoffgehalt der Umwelt einiger Entoparasiten. Ztschr. Vergleich. Physiol., v. 18:339-346.

COLLIER, H. B. 1941.—A trypsin inhibiting fraction of *Ascaris*. Canad. J. Res., v. 19B: 90-98.

DANNIGER, R., PFRAGNER, K. and SCHULTES, H. 1928.—Über die absolute Reaktion in dem Inhalt der einzelnen Darmabschnitte von Pferd und Rind. Pflüger's Arch., v. 220: 430-433.

DASTRE, A. and STASSANO, H. 1903.—Existence d'une Antikinase chez les parasites intestinaux. Comp. Rend. Soc. Biol. Paris, v. 55:131-132.

DAVEY, D. G. 1936.—Notes on the osmotic pressure of the contents of the stomach compartments of the sheep. J. Agric. Sc., v. 26:328-330.
————. 1938.—Studies on the physiology of the nematodes of the alimentary canal of sheep. Parasit., v. 30:278-295.

EASTMAN, I. M. and MILLER, E. G., JR. 1935.—Gastrointestinal pH in rats as determined by the glass electrode. J. Biol. Chem., v. 110:255-262.

ELDIN, M. S. and HASSAN, A. 1933.—Gastric functions in helminthic infections. J. Egypt. Med. Assoc., v. 16:735-752. [Abstract] Trop. Dis. Bull., 1934, v. 31:767.

ENIGK, C. 1936.—Untersuchungen über die Abtötung der Spulwurmeier und Coccidienoöcysten durch Chemikalien. Arch. Wiss. Prakt. Tierheilk. v. 70:439-448.

FERBER, K. E. 1928.—Die Zahl und Masse der Infusorien im Pansen und ihre Bedeutung für den Eiweissaufbau beim Wiederkäuer. Ztschr. Tierzücht. & Züchtungsbiol., v. 12: 31-63.

FERNANDEZ, F. 1934.—Parasitismo intestinal y jugo gastrico. Med. Paises Cálidos, Madrid, v. 7:336-338. [Abstract] Trop. Dis. Bull., 1934, v. 31:767.

FRIES, J. A. 1906.—Intestinal gases of man. Am. J. Physiol., v. 16:468-474.

FROST, W. D. and HANKINSON, H. 1931.—*Lactobacillus acidophilus*, an annotated bibliography to 1931. Milton, Wis.

GLASER, R. W. and STOLL, N. R. 1938.—Development under sterile conditions of the sheep stomach worm *Haemonchus contortus* (Nematode). Science, v. 87:259-260.

GRAHAM, W. R. and EMERY, E. S. 1927-28.—The reaction of the intestinal contents of dogs fed on different diets. J. Lab. & Clin. Med., v. 13:1097-1108.

GRAYZEL, D. M. and MILLER, E. G., JR. 1928.—The pH of the contents of the gastrointestinal tract in dogs, in relation to diet and rickets. J. Biol. Chem., v. 76:423-436.

GROVE, E. W., OLMSTED, W. H. and KOENIG, K. 1929.—The effect of diet and catharsis on the lower volatile fatty acids in the stools of normal men. J. Biol. Chem., v. 85: 127-136.

HAMILL, J. M. 1906.—On the mechanism of protection of intestinal worms, and its bearing on the relation of enterokinase to trypsin. J. Physiol., v. 33:479-492.

HARNED, B. K. and NASH, T. P., JR. 1932.—The protection of insulin by antiproteases, and its absorption from the intestine. J. Biol. Chem., v. 97:443-456.

HELLER, V. G., OWENS, J. R. and PORTWOOD, L. 1935.—The effect of the ingestion of saline waters upon the pH of the intestinal tract, the nitrogen balance and the coefficient of digestibility. J. Nutrition, v. 10:645-651.

HERRIN, R. C. 1937.—Ammonia content, pH, and carbon dioxide tension in the intestine of dogs. J. Biol. Chem., v. 118:459-470.

HEUPKE, W. 1931.—Über die Sekretion und Excretion des Dickdarms. Ztschr. Gesam. Exper. Med., v. 75:83-125.

HUME, H. V., DENIS, W., SILVERMAN, D. N. and IRWIN, E. L. 1924.—Hydrogen ion concentration in the human duodenum. J. Biol. Chem., v. 60:633-645.

JAHN, T. L. 1933.—Oxidation-reduction potential as a possible factor in the growth of intestinal parasites in vitro. J. Parasit., v. 20:129.
————. 1934.—Problems of population growth in the protozoa. Symposia in Quant. Biol. (Cold Spring Harbor), v. 2:167-180.
————. 1936.—Effect of aeration and lack of CO_2 on growth of bacteria-free cultures of protozoa. Proc. Soc. Exp. Biol. & Med., v. 33:494-498.

JEPHCOTT, H. and BACHARACH, A. L. 1926.—A rapid and reliable test for vitamin D. Biochem. J., v. 20:1351-1355.

JOHNSTON, F. 1934.—Threadworms. Brit. Med. J., v. 1:224.

KAHN, G. and STOKES, J. 1926.—The comparison of the electrometric and colorimetric methods for determination of the pH of gastric contents. J. Biol. Chem., v. 69:75-84.

KARR, W. G. and ABBOTT, W. O. 1935.—Intubation studies of the human small intestine. IV. Chemical characteristics of the intestinal contents in the fasting state and as influenced by the administration of acids, of alkalies, and of water. J. Clin. Invest., v. 14:893-900.

KELLOGG, E. L. 1933.—The duodenum. P. B. Hoeber, Inc., N. Y.

KOFOID, C. A., MCNEIL, E. and CAILLEAU, R. 1932.—Electrometric pH determinations of the walls and contents of the gastro-intestinal tracts of normal albino rats. Univ. Calif. Pub. Zool., v. 36:347-355.

KOPELOFF, N. 1926.—*Lactobacillus acidophilus*. Baltimore, 1926.

KREIPE, H. 1927.—Dissertation. Kiel. Cited by Lenkeit (1933).

LAPAGE, G. 1935a.—The second ecdysis of infective nematode larvae. Parasit., v. 27:186-206.

1935b.—The second ecdysis of the infective larvae of certain Trichostrongylidae in solutions of sodium sulphide and of organic compounds containing sulphur. J. Helm., v. 13:103-114.

1938.—Nematodes parasitic in animals. Chem. Pub. Co., N. Y.

LENKEIT, W. 1933.—Neuere Ergebnisse der vergleichenden Physiologie der Verdauung der Säugetiere. Ergeb. Physiol., v. 35:573-631.

LLOYD-JONES, O. and LILJEDAHL, E. M. 1934.—Alimentary Gas. Med. Rec., v. 139:320-323.

LONG, J. H. and FENGER, F. 1917.—On the normal reaction of the intestinal tract. J. Am. Chem. Soc., v. 39:1278-1286.

MANGOLD, E. 1929.—Handbuch der Ernährung und des Stoffwechsels der landwirtschaftlichen Nutztiere, v. 2:202.

MANN, F. C. and BOLLMAN, J. L. 1930.—The reaction of the content of the gastro-intestinal tract. J. Am. Med. Assoc., v. 95:1722-1724.

McCLENDON, J. F. 1920.—Hydrogen-ion concentration of the contents of the small intestine. Proc. Nat. Acad. Sc., v. 6:690-691.

McCLENDON, J. F. and MEDES, GRACE. 1925.—Physical chemistry in biology and medicine. Saunders Co.

McCLENDON, J. F., MYERS, F. J., CULLIGAN, L. C., and GYDESEN, C. S. 1919.—Factors influencing the hydrogen ion concentration of the ileum. J. Biol. Chem., v. 38:535-538.

McCOY, O. R. 1935.—The physiology of the helminth parasites. Physiol. Rev., v. 15:221-240.

McIVER, M. A., REDFIELD, A. C., and BENEDICT, E. B. 1926.—Gaseous exchange between blood and lumen of stomach and intestines. Am. J. Physiol., v. 76:92-111.

McLAUGHLIN, A. R. 1931.—Hydrogen ion concentration of the alimentary tracts of fowl, cat, and rabbit. Science, v. 73:191-192.

MOORTHY, V. N. 1935.—The influence of fresh bile on guinea-worm larvae encysted in Cyclops. Indian Med. Gaz., v. 70:21-23. [Abstract] Trop. Dis. Bull., 1935, v. 32:654.

MUDIE, E. C. 1934.—Threadworms. Brit. Med. J., v. 1:224.

MUSSEHL, F. E., BLISH, M. J. and ACKERSON, C. W. 1933.—Effect of dietary and environmental factors on the pH of the intestinal tract. Poultry Sc., v. 12:120-123.

NAGL, F. 1928.—Über die Titrations—und Ionenacidität im Mageninhalt des Hundes nach verschiedener Fütterung. Arch. Wiss. & Prak. Tierheilk., v. 58:198-203.

PLANER. 1860.—Die Gase des Verdauungsschlauches und ihre Beziehungen zum Blute. Sitzungsb. Akad. Wiss. Wien, Math.-Naturw. Kl., v. 42:307-354.

REDMAN, T., WILLIMOTT, S. G. and WOKES, F. 1927.—The pH of the gastrointestinal tract of certain rodents used in feeding experiments, and its possible significance in rickets. Biochem. J., v. 21:589-605.

ROBINSON, C. S. 1922.—Hydrogen ion concentration of the human feces. J. Biol. Chem., v. 52:445.

1935.—The hydrogen ion concentration of the contents of the small intestine. J. Biol. Chem., v. 108:403-408.

ROBINSON, C. S. and DUNCAN, C. W. 1931.—The effect of lactase and the acid-base value of the diet on the hydrogen-ion concentration of the intestinal contents of the rat and their possible influence on calcium absorption. J. Biol. Chem., v. 92:435-447.

RUGE, E. 1861.—Beiträge zur Kenntniss der Darmgase. Sitzungsb. Akad. Wiss. Wien, Math.-Naturw. Kl., v. 44:739-762.

SANG, J. H. 1938.—The antiproteolytic enzyme of Ascaris lumbricoides var. suis. Parasit., v. 30:141-155.

SCHEUNERT, A. and SCHIEBLICH, M. 1927.—Einfluss der Mikroorganismen auf die Vorgänge im Verdauungstraktus bei Herbivoren. Handb. Norm. & Path. Physiol., v. 3:967-1000.

SCHOPFER, W. H. 1932.—Recherches physico-chimiques sur le milieu interieur de quelques parasites. Rev. Suisse Zool., v. 39:59-194.

SCHWARZ, C. and DANZIGER, H. 1924.—Beiträge zur Physiologie der Verdauung. IV. Die H-Ionenkonzentrationen des aus dem Magen austretenden Mageninhaltes zugleich ein Beitrag zur Kenntnis der Magenentleerung. Pflüger's Arch., v. 202:478-487.

SCHWARZ, C. and KAPLAN, H. 1926.—Die H-Ionenkonzentration im Labmageninhalt des Rindes. Pflüger's Arch., v. 213:592-594.

SCHWARZ, C. and GABRIEL, F. 1926.—Die H-Ionenkonzentrationen im Panseninhalt des Rindes. Pflüger's Arch., v. 213:814-815.

SCHWARTZ, C., STEINMETZER, K. and CAITHAML, K. 1926.—Beiträge zur Physiologie der Verdauung. XVII. Die H-Jonenkonzentrationen im Mageninhalt des Pferdes. Pflüger's Arch., v. 213:595-601.

SHOHL, A. T. and BING, F. C. 1928.—Rickets in rats. IX. pH of the feces. J. Biol. Chem., v. 79:269-274.

SLATER, W. K. 1925.—The nature of the metabolic processes in Ascaris lumbricoides. Biochem. J., v. 19:604-610.

1928.—Anaerobic life in animals. Biol. Rev., v. 3:303-328.

STEWART, J. and SHEARER, G. D. 1933.—The effects of nematode infestations on the metabolism of the host. Third Report Univ. Cambridge, Inst. Anim. Path., pp. 58-129.

SUN, T., BLUMENTHAL, P. R., SLIFER, E. H., HERBER, E. C. and WANG, C. C. 1932.—The hydrogen-ion concentration of the alimentary tract of normal albino rats. Physiol. Zool., v. 5:191-197.

TACKE, B. 1884.—Über die Bedeutung der brennbaren Gase im thierischen Organismus. Inaug. Diss. Berlin.

TAPPEINER. 1883.—Die Gase des Verdauungsschlauches der Pflanzenfresser. Ztschr. Biol., v. 19:228-279.

TORYU, Y. 1934.—Contributions to the physiology of the Ascaris. II. The respiratory exchange in the Ascaris, Ascaris megalocephala Cloq. Sc. Rpt. Tohoku Imp. Univ., 4th Ser., v. 9:61-70.

DE WAELE, A. 1933.—Sur la Migration des Cestodes. Bull. Cl. Sc. Acad. Roy. Belg., Ser. 5, v. 19:649-660.

1934.—Etude de la fonction biliaire de l'invagination chez le cysticerque des cestodes. Ann. Parasit., v. 12:492-510.

WEINLAND, E. 1901.—Über Kohlehydratzersetzung ohne Sauerstoffaufnahme, einen tierischen Gärungsprozess. Ztschr. Biol., v. 42:55-90.

1903.—Über Antifermente. I and II. Ibid., v. 44:1-15, 45-60.

WELLS, H. S. 1931.—Observations on the blood sucking activities of the hookworm, Ancylostoma caninum. J. Parasit., v. 17:167-182.

ZIEGLER, J. and HIRSCH, W. 1925.—Über den Gasgehalt des Magendarmkanals und die Mittel zu seiner Beseitigung in Hinsicht auf die Röntgenuntersuchung. Fortschr. Geb. Röntgenstrahlen, v. 33:698-708.

ZUCKER, T. F. and MATZNER, M. J. 1924.—On the pharmacological action of the antirachitic active principle of cod liver oil. Proc. Soc. Exp. Biol. & Med., v. 21:186-187.

GENERAL CHEMICAL COMPOSITION

ADUCCO, V. 1889. — La substance colorante rouge de l'Eustrongylus gigas. Arch. Ital. Biol., v. 11:52-69.

BERNARD, CLAUDE. 1859.—De la matière glycogène chez les animaux dépourvus de foie. Compt. Rend. Soc. Biol. Paris, Ser. 3, v. 1:53-55.

BONDOUY, T. 1910.—Chimie biologique du Sclerostomum equinum. Thèse, Paris, p. 58.

v. BRAND, TH. 1934.—Der Stoffwechsel von Ascaris lumbricoides bei Oxybiose und Anoxybiose. Ztschr. Vergleich. Physiol., v. 21:220-235.

1937a.—The anaerobic glycogen consumption in Ascaris females and males. J. Parasit., v. 23:68-72.

1937b.—Haemoglobin in a larval *Eustrongylides*. J. Parasit., v. 23:316-317.

1938. — Physiological observations on a larval *Eustrongylides*. (Nematoda). J. Parasit., v. 24:445-451.

v. BRAND, TH. and OTTO, G. F. 1938.—Some aspects of the carbohydrate metabolism of the hookworm, *Ancylostomc caninum*, and its host. Am. J. Hyg., v. 27:683-689.

BRAULT, A. and LOEPER, M. 1904.—Le glycogène dans le developpement de certains parasites (Cestodes et Nematodes). J. Physiol. & Path. Gen., v. 6:503-512.

BUSCH, P. W. C. M. 1905.—Over de localisatie van het glycogeen bij enkele Darmparsieten. Diss. Utrecht. 109 pp.

CAMPBELL, D. H. 1936.—An antigenic polysaccharide fraction of *Ascaris lumbricoides* (from hog). J. Infect. Dis., v. 59:266-280.

CHAMPÉTIER, G. and FAURÉ-FREMIET, E. 1937.—Étude roentgenographique d'une protéine intracellulaire. Compt. Rend. Acad. Sc. Paris, v. 204:1901-1903.

DAVEY, D. G. 1938.—Studies on the physiology of the nematodes of the alimentary canal of sheep. Parasit., v. 30:278-295.

FAURÉ-FREMIET, E. 1913.—Le cycle germinatif chez l'*Ascaris megalocephala*. Arch. Anat. Micr., v. 15:435-757.

FAURÉ-FREMIET, E. and FILHOL, J. 1937.—La température de dispersion d'une protéine intracellulaire, l'ascaridine. J. Chim. Physique, v. 34:444-451.

FLURY, F. 1912.—Zur Chemie und Toxikologie der Ascariden. Arch. Exper. Path. & Pharmakol., v. 67:275-392.

FOSTER, M. 1865.—On the existence of glycogen in the tissues of certain Entozoa. Proc. Roy. Soc., v. 14:543-546.

GIOVANNOLA, A. 1935.—Osservazioni sulla nature delle coste nella borsa caudale degli Anchilostomi. Arch. Ital. Sc. Med., Colon., v. 16, Fasc. 6:1-7.

GOURÉVITCH, M. A. 1937.—Sur le dosage de la flavine; la flavine chez les invertébrés. Bull. Soc. Chim. Biol., v. 19:125-129.

HIRSCH, G. C. and BRETSCHNEIDER, L. H. 1937.—Die Arbeitsräume in den Darmzellen von *Ascaris*, die Einwirkung des Hungerns, die Sekretbildung. Cytologia, Tokyo, Fujii Jub., v. ?:424-436.

HSÜ, H. F. 1938.—Studies on the food and the digestive system of certain parasites. II. On the food of *Schistosoma japonicum, Paragonimus ringeri, Dirofilaria immitis, Spirocerca sanguinolenta* and *Rhabdias* sp. Bull. Fan Mem. Inst., Biol., Zool. Ser., v. 8:347-366.

JANICKI, M. J. 1939.—Untersuchungen zur Ernährungsfrage von *Dioctophyme renale* (Goeze, 1782). Zool. Poloniae, v. 3:189-223.

KEILIN, D. 1925.—On cytochrome, a respiratory pigment common to animals, yeasts and higher plants. Proc. Roy. Soc. Lond., Ser. B, v. 98:312-339.

v. KEMNITZ, G. 1912.—Die Morphologie des Stoffwechsels bei *Ascaris lumbricoides*. Arch. Zellforsch., v. 7:463-603.

KRÜGER, F. 1936.—Untersuchungen zur Kenntnis des aeroben und anaeroben Stoffwechsels des Schweinespulwurmes (*Ascaris suilla*). Zool. Jahrb. Abt. Allg. Zool. & Physiol., v. 57:1-56.

MARCET, W. H. 1865.—Chemical examination of the fluid from the peritoneal cavity of the Nematoda Entozoa. Proc. Roy. Soc. Lond., v. 14:69-70.

MARTINI, E. 1916.—Die Anatomie der *Oxyuris curvula*. Ztschr. Wiss. Zool., v. 116:142-543.

MUELLER, J. F. 1928-29.—Studies on the microscopical anatomy and physiology of *Ascaris lumbricoides* and *Ascaris megalocephala*. Ztschr. Zellforsch., v. 8:361-403.

PANIKKAR, N. K. and SPROSTON, N. G. 1941.—Osmotic relations of some metazoan parasites. Parasit., v. 33:214-223.

SCHIMMELPFENNIG, G. 1903.—Über *Ascaris megalocephala*. Beiträge zur Biologie und physiologischen Chemie derselben. Arch. Wiss. & Prakt. Tierheilk., v. 29:332-376.

SCHULTE, H. 1917.—Versuche über Stoffwechselvorgänge bei *Ascaris lumbricoides*. Pflügers Archiv., v. 166:1-44.

SCHOPFER, W. H. 1926.—Recherches physico-chimiques sur les liquides de parasites (*Ascaris*). II. Parasit., v. 18:277-282.

1932.—Recherches physico-chimiques sur le milieu interieur de quelques parasites. Rev. Suisse Zool., v. 39:59-194.

SCHULZ, FR. N. and BECKER, M. 1933.—Über Ascarylalkohol. Bioch. Ztschr., v. 265:253-259.

SMORODINCEV, I. and BEBESIN, K. 1936.—La teneur en glycogène des Ascarides. Compt. Rend. Acad. Sc. U. R. S. S., n. s., v. 2:189-191.

STANNARD, J. W., McCOY, O. R. and LATCHFORD, W. B. 1938.—Studies on the metabolism of *Trichinella spiralis* larvae. Am. J. Hyg., v. 27:666-682.

STOLL, N. R. 1940.—In vitro conditions favoring ecdysis at the end of the first parasitic stage of *Haemonchus contortus* (Nematoda). Growth, v. 4:383-406.

TORYU, Y. 1933.—Contributions to the physiology of the *Ascaris*. I. Glycogen content of the *Ascaris, Ascaris megalocephala* Cloq. Sc. Rpt. Tohoku Imp. Univ., 4th Ser., v. 8:65-74.

VIALLI, M. 1923.—Ricerche sulla pressione osmotica. II. nei Vermi. Rendic. Ist. Lombardo Sc. & Lett., v. 56: Rpts. 1-4.

WEINLAND, E. 1901a.—Über den Glykogengehalt einiger parasitischer Würmer. Ztschr. Biol., v. 41:69-74.

1901b.—Über Kohlehydratzersetzung ohne Sauerstoffaufnahme, einen tierischen Gärungsprozess. Ztschr. Biol., v. 42:55-90.

WHARTON, G. W. 1938.—Hemoglobin in turtle parasites. J. Parasit., v. 24. Suppl.: 21.

1941.—The function of respiratory pigments of certain turtle parasites. J. Parasit., v. 27:81-87.

YOSHIMURA, SH. 1930.—Beiträge zur Chemie der Askaris. J. Biochem., v. 12:27-34.

METABOLISM OF ADULT NEMATODES

ADAM, W. 1932.—Über die Stoffwechselprozesse von *Ascaris suilla* Duj. I. Teil. Die Aufnahme von Sauerstoff aus der Umgebung. Ztschr. Vergleich. Physiol., v. 16:229-251.

BONDOUY, T. 1910.—Chimie biologique du *Sclerostomum equinum*. Thèse, Paris. 58 pp.

v. BRAND, TH. 1934a.—Der Stoffwechsel von *Ascaris lumbricoides* bei Oxybiose und Anoxybiose. Ztschr. Vergleich. Physiol., v. 21: 220-235.

1934b.—Das Leben ohne Sauerstoff bei wirbellosen Tieren. Ergeb. Biol., v. 10:37-100.

1937a.—The anaerobic glycogen consumption in *Ascaris* females and males. J. Parasit., v. 23:68-72.

1937b.—The aerobic resynthesis of glycogen in *Ascaris*. J. Parasit., v. 23:316-317.

1938.—The nature of the metabolic activities of intestinal helminths in their natural habitat: aerobiosis or anaerobiosis? Biodynamica, No. 41:1-13.

1941.—Aerobic fat metabolism of *Ascaris lumbricoides*. Proc. Soc. Exp. Biol. Med., v. 46:417-418.

v. BRAND, TH. and OTTO, G. F. 1938.—Some aspects of the carbohydrate metabolism of the hookworm, *Ancylostoma caninum*, and its host. Am. J. Hyg., v. 27:683-689.

BUNGE, G. 1889.—Weitere Untersuchungen über die Athmung der Würmer. Ztschr. Physiol. Chem., v. 14:318-324.

CHITWOOD, B. G. 1938.—Notes on the physiology of *Ascaris lumbricoides*. Proc. Helm. Soc. Wash., v. 5:18-19.

DAVEY, D. G. 1937.—Physiology of nematodes. Nature, v. 140:645.

1938a.—The respiration of nematodes of the alimentary tract. J. Exper. Biol., v. 15:217-224.

1938b.—Studies on the physiology of the nematodes of the alimentary canal of sheep. Parasit., v. 30:278-295.

FISCHER, A. 1924.—Über den Kohlehydratstoffwechsel von *Ascaris megalocephala*. Bioch. Ztschr., v. 144:224-228.

FLURY, F. 1912.—Zur Chemie und Toxikologie der Ascariden. Arch. Exp. Path. & Pharm., v. 67:275-392.

HARNISCH, O. 1933.—Untersuchungen zur Kennzeichnung des

Sauerstoffverbrauchs von *Triaenophorus nodulosus* (Cest.). und *Ascaris lumbricoides* (Nemat.). Ztschr. Vergleich. Physiol., v. 19:310-348.

    1935.—Daten zur Beurteilung des Sauerstoffverbrauchs von *Ascaris lumbricoides* (nach Messungen an isolierten Organen). Ztschr. Vergleich. Physiol., v. 22: 50-66.

    1936.—Primäre und sekundäre Oxybiose der Larve von *Chironomus thummi.* Ztschr. Vergleich. Physiol., v. 23: 391-419.

    1937.—Zellfrei arbeitendes Oxydans im Gaswechsel von *Ascaris lumbricoides* und einigen Cestoden. Ztschr. Vergleich. Physiol., v. 24:667-686.

HARWOOD, P. D. and BROWN, H. W. 1934.—A preliminary report on the *in vitro* consumption of oxygen by parasitic nematodes. J. Parasit., v. 20:128.

HIRSCH, G. C. and BRETSCHNEIDER, L. H. 1937.—Der intraplasmatische Stoffwechsel in den Darmzellen von *Ascaris lumbricoides.* Teil II. Die Adsorption von Eisen und die Beteiligung der Golgikörper dabei. Protoplasma, v. 29: 9-30.

HOFFMANN, R. 1934.—Untersuchungen über die Wärmeentwicklung von *Ascaris lumbricoides* bei Fütterung mit Glykose, Fruktose und Galaktose. Ztschr. Biol. v. 95:390-400.

JOST, H. 1928.—Vergleichende Physiologie des Stoffwechsels. Bethes Handb. Physiol., v. 5:377-466.

KEMPNER, W. 1937.—The effect of oxygen tension on cellular metabolism. J. Cell. & Comp. Physiol., v. 10:339-364.

KRÜGER, F. 1936.—Untersuchungen zur Kenntnis des aeroben und anaeroben Stoffwechsels des Schweinespulwurmes (*Ascaris suilla*). Zool. Jahrb. Abt. Allg. Zool. & Physiol., v. 57:1-56.

    1937.—Bestimmungen über den aeroben und anaeroben Stoffumsatz beim Schweinespulwurm mit einem neuen Respirationsapparat. Ztschr. Vergleich. Physiol., v. 24: 687-719.

    1940.—Die Beziehung des Sauerstoffverbrauches zur Körperoberfläche beim Schweinespulwurm (*Ascaris lumbricoides*). Z. Zool., v. 152:547-570.

KRUMMACHER, O. 1919.—Untersuchungen über die Wärmeentwicklung der Spulwürmer. Ztschr. Biol., v. 69:293-321.

MARSH, G. 1935.—Kinetics of an intracellular system for respiration and bioelectric potential at flux equilibrium. Plant Physiol., v. 10:681-697.

MEIER, W. 1931.—Neuere Untersuchungen über die Wärmeentwicklung der Spulwürmer. Ztschr. Biol., v. 91:459-474.

MUELLER, J. F. 1928-29.—Studies on the microscopical anatomy and physiology of *Ascaris lumbricoides* and *Ascaris megalocephala.* Ztschr. Zellforsch., v. 8:361-403.

OESTERLIN, M. 1937.—Die von oxybiotisch gehaltenen Ascariden ausgeschiedenen Fettsäuren. Ztschr. Vergleich. Physiol., v. 25:88-91.

SCHIMMELPFENNIG, G. 1903.—Über *Ascaris megalocephala.* Beiträge zur Biologie und physiologischen Chemie derselben. Arch. Wiss. & Prakt. Tierheilk., v. 29:332-376.

SCHULTE, H. 1917.—Versuche über Stoffwechselvorgänge bei *Ascaris lumbricoides.* Pflüger's Archiv., v. 166:1-44.

SLATER, W. K. 1925.—The nature of the metabolic processes in *Ascaris lumbricoides.* Biochem. J., v. 19:604-610.

    1928.—Anaerobic life in animals. Biol. Rev., v. 3: 303-328.

TORYU, Y. 1934.—Contributions to the physiology of the *Ascaris.* II. The respiratory exchange in the *Ascaris, Ascaris megalocephala* Cloq. Sc. Rpt. Tohoku Imp. Univ., 4th. Ser., v. 9:61-70.

    1935.—Idem. III. Survival and glycogen content of the *Ascaris, Ascaris megalocephala* Cloq. in presence and absence of oxygen. Ibid., v. 10:361-375.

    1936a.—Idem. IV. Products from glycogen during anaerobic and aerobic existence of the *Ascaris, Ascaris megalocephala* Cloq. Ibid., v. 10:687-696.

    1936b.—Idem. V. Survival and respiratory exchange of the *Ascaris, Ascaris megalocephala* Cloq. intercepted from light in presence and absence of oxygen. Ibid., v. 11:1-17.

WAECHTER, J. 1934.—Über die Natur der beim Stoffwechsel der Spulwürmer ausgeschiedenen Fettsäuren. Ztschr. Biol., v. 95:497-501.

WEINLAND, E. 1901.—Über Kohlehydratzersetzung ohne Sauerstoffaufnahme bei *Ascaris,* einen tierischen Gärungsprozess. Ztschr. Biol., v. 42:55-90.

    1902.—Über ausgepresste Extrakte von *Ascaris lumbricoides* und ihre Wirkung. Ibid., v. 43:86-111.

    1904a.—Über die von *Ascaris lumbricoides* ausgeschiedenen Fettsäuren. Ibid., v. 45:113-116.

    1904b.—Über die Zersetzung stickstoffhaltiger Substanz bei *Ascaris.* Ibid., v. 45:517-531.

WEINLAND, E. and RITTER, A. 1902.—Über die Bildung von Glykogen aus Kohlehydraten bei *Ascaris.* Ztschr. Biol., v. 43:490-502.

WELLS, H. S. 1931.—Observations on the blood sucking activities of the hookworm, *Ancylostoma caninum.* J. Parasit., v. 17:167-182.

METABOLISM OF EGGS AND LARVAE

BATAILLON, E. 1910.—Contribution a l'analyse experimentale des phénomènes karyocinétiques chez *Ascaris megalocephala.* Arch. Entwicklungsmech., v. 30(1):24-44.

VON BRAND, TH. 1938.—Physiological observations on a larval *Eustrongylides* (Nematoda). J. Parasit., v. 24:445-451.

BROWN, H. W. 1928.—A quantitative study of the influence of oxygen and temperature on the embryonic development of the eggs of the pig ascarid (*Ascaris suum* Goeze). J. Parasit., v. 14:141-160.

CHITWOOD, B. G. 1938.—Further studies on nemic skeletoids and their significance in the chemical control of nemic pests. Proc. Helm. Soc. Wash., v. 5:68-75.

CORT, W. W. 1925.—Investigations on the control of hookworm disease. XXXIV. General summary of results. Am. J. Hyg., v. 5:49-89.

DINNIK, J. A. and DINNIK, N. N. 1937.—Influence de la température, de l'absence d'oxygène et du désséchement sur les oeufs de *Trichocephalus trichiurus* (L.). Med. Parasit. & Parasitic Dis., v. 5:603-618. [Russian with French summary.]

DYRDOWSKA, M. 1931.—Recherches sur le comportement du glycogène et des graisses dans les oeufs d'*Ascaris megalocephala* a l'état normal et dans une atmosphère d'azote. Comp. Rend. Soc. Biol., Paris, v. 108:593-596.

FAURÉ-FREMIET, E. 1912.—Graisse et glycogène dans le developpement de l'oeuf de l'*Ascaris megalocephala.* Bull. Soc. Zool. France, v. 37:233-234.

    1913a.—Le cycle germinatif chez l'*Ascaris megalocephala.* Arch. Anat. Micr., v. 15:435-757.

    1913b.—La segmentation de l'oeuf d'*Ascaris* au point de vue energétique. Comp. Rend. Soc. Biol., Paris, v. 75:90-92.

FENWICK, D. W. 1938.—The oxygen consumption of newly-hatched larvae of *Ascaris suum.* Proc. Zool. Soc. London, Ser. A, v. 108, Part 1:85-100.

FRIEDHEIM, E. A. H. 1933.—Das Pigment von *Halla parthenopea,* ein akzessorischer Atmungskatalysator. Biochem. Ztschr., v. 259:257-268.

GIOVANNOLA, A. 1936.—Energy and food reserve in the development of nematodes. J. Parasit., v. 22:207-218.

HUFF, G. C. 1936.—Experimental studies of factors influencing the development of the eggs of pig ascarid (*Ascaris suum* Goeze). J. Parasit., v. 22:455-463.

HUFF, G. C. and BOELL, E. J. 1936.—Effect of ultracentrifuging on oxygen consumption of the eggs of *Ascaris suum,* Goeze. Proc. Soc. Exp. Biol. & Med., v. 34:626-628.

JACOBS, L. and JONES, M. F. 1939.—Studies on oxyuriasis. XXI. The chemistry of the membranes of the pinworm egg. Proc. Helm. Soc. Wash., v. 6:57-60.

KOSMIN, N. 1928.—Zur Frage über den Stickstoffwechsel der Eier von *Ascaris megalocephala.* Tr. Lab. Exper. Biol. Zoopark, Moscow, v. 4:207-218. [Russian with German summary.]

LOOSS, A. 1911.—The anatomy and life history of *Anchylostoma duodenale* Duj. Part II. The development in the free stage. Rec. Egypt. Govt. Sch. Med., v. 4:163-613. [Not seen.]

McCOY, O. R. 1930.—The influence of temperature, hydrogen-ion concentration, and oxygen tension on the development of the eggs and larvae of the dog hookworm, *Ancylostoma caninum*. Am. J. Hyg., v. 11:413-448.

McCOY, Q. R., DOWNING, V. F. and VAN VOORHIS, S. N. 1941.—The penetration of radioactive phosphorus into encysted *Trichinella* larvae. J. Parasit., v. 27:53-58.

NOLF, L. O. 1932.—Experimental studies on certain factors influencing the development and viability of the ova of the human *Trichuris* as compared with those of the human *Ascaris*. Am. J. Hyg. v. 16:288-322.

PAYNE, F. K. 1923.—Investigations on the control of hookworm disease. XXX. Studies on factors involved in migration of hookworm larvae in soil. Am. J. Hyg., v. 3:547-583.

PINTNER, TH. 1922.—Die vermutliche Bedeutung der Helminthenwanderungen. Sitzungsb. Akad. Wiss. Wien, Math.-Naturw. Kl., Abt. I. v. 131:129-138.

ROGERS, W. P. 1939.—The physiological ageing of *Ancylostoma* larvae. J. Helm., v. 17:195-202.

SCHALIMOV, L. G. 1931.—A contribution to the biology of *Oxyuris equi*. Tr. Dynamics Develop., v. 6:181-196. [Russian with English summary.]

SCHMIDT, W. J. 1937.—Doppelbrechung und Feinbau der Eischale von *Ascaris* megalocephala. Ein Vergleich des Feinbaues faseriger und filmartiger Chitins. Ztschr. Zellforsch., v. 25:181-203.

STANNARD, J. N., McCOY, O. R., and LATCHFORD, W. B. 1938.—Studies on the metabolism of *Trichinella spiralis* larvae. Am. J. Hyg., v. 27:666-682.

STEPANOW-GRIGORIEW, J. and HOEPPLI, R. 1926.—Über Beziehungen zwischen Glykogengehalt parasitischer Nematodenlarven und ihrer Wanderung im Wirtskörper. Arch. Schiffs-u Tropen Hyg., v. 30:577-585.

SZWEJKOWSKA, G. 1929.—Recherches sur la physiologie de la maturation des oeufs d'*Ascaris*. Bull. Internatl. Acad. Polon Sc. & Lett., Ser. B, 1928:489-519.

WENDT, H. 1936.—Beiträge zum Entwicklungszyklus bei *Oxyuris vermicularis*. Ztschr. Kinderheilk., v. 58:375-387. [Not seen.]

WOTTGE, K. 1937.—Die stofflichen Veränderungen in der Eizelle von *Ascaris megalocephala* nach der Befruchtung. Protoplasma, v. 29:31-59.

ZAWADOWSKY, M. 1916.—Role de l'oxygène dans le processus de segmentation des oeufs de l'*Ascaris megalocephala*. (Note preliminaire.) Compt. Rend. Soc. Biol., Paris, v. 68:595-598.
  1926.—Zum Mechanismus der Wirkung von Zyankalium auf die lebende Zelle (Eier von *Ascaris megalocephala*). Biologia Generalis, v. 2:442-456.
  1928.—The nature of the egg-shell of various species of *Ascaris* eggs (*Toxascaris limbata* Reillet et Henry, *Belascaris mystax* Zeder, *Belascaris marginata* Rud., *Ascaris suilla* Duj.). Tr. Lab. Exper. Biol. Zoopark, Moscow, v. 4:201-206. [Russian with English summary.]

ZAWADOWSKY, M. and SCHALIMOV, L. G. 1929.—Is autoinvasion possible given the presence of *Enterobius (Oxyuris) vermicularis* in the intestine? Tr. Lab. Exper. Biol. Zoopark, Moscow, v. 5:1-42. [Russian with English summary.]

ZAWADOWSKY, M. and ORLOW, A. P. 1927.—Is there any possibility of autoinvasion during Ascariasis? Tr. Lab. Exper. Zoopark, Moscow, v. 3:99-116. [Russian with English summary.]

ZVIAGINZEV, S. N. 1934.—Contribution to the history of development of *Nematodirus helwetianus*. Tr. Dynamics Develop., v. 8:186-202. [Russian with English summary.]

ABBREVIATIONS

al, ala (bursa);
an, anus;
amph, external amphid;
amph gl, amphidial gland;
amph n, amphidial nerve;
amph p, amphidial pouch;
ar, arcade;
b, bulbar region of esophagus;
bac, bacillary cells (hypodermal glands);
bi, binding cell;
blb, swollen (bulbar) region of esophagus;
c, cephalic;
c, cuticle;
ca, cardiac region of intestine;
cc, coelomocyte;
ceph ppl, cephalic papilla;
ceph prob, cephalic (i.e., external) probola;
c gl, caudal gland;
ch, chord;
cl, cloaca;
cl b, stichosome (cell body);
clv, clavate cell;
c m, copulatory muscle;
c n, cephalic (papillary) nerve;
cr, cheilorhabdion;
cut, cuticle;
d, dorsal;
dd, dorsodorsal;
de, deirid (cervical papilla);
d gl, dorsal esophageal gland;
dl, dorsolateral;
dn, dorsal (somatic) nerve;
e, deirid (cervical papilla);
ed, externodorsal;
eg, fertilized egg;
ej 1, large ejaculatory gland;
ej 2, small ejaculatory gland;
el, externolateral;
es, esophagus;
es m, somato-esophageal muscle;
ex d, terminal excretory duct;
ex gl, excretory gland;
ex s, excretory sinus;
fa, fiber cell;
fi, fibril cell;
fu, (Fullzelle);
gng, ganglion;
g p, genital (caudal) papillae;
gub, gubernaculum;
h, depressor ani (H-shaped cell);
hu, sheath cell;
hy, hypodermis;
i, isthmus of esophagus;
id, internodorsal;
il, internolateral;
im, somato-intestinal muscles;
int, intestine;
l, lateral;
iv, internoventral;
lab prob, labial (i.e., internal) probolae;
lc, lateral (excretory) canal;
ld, laterodorsal;
l gl. (sub) lateral gland;
lu, lumbar;
lv, lateroventral;
m, somatic muscle;
mc, mesenterial cell;
mcn, medial caudal nerve;
me, mesentery;
ml, mediolateral;
mon gl, moniliform gland;
n, nerve;
nrv r, nerve ring;
oe, oesophagus;
on, onchium (tooth);
os, uvette;
or d, dorsal gland orifice;
or sv, subventral gland orifice;

ov, ovary;
ov d, oviduct;
ovij, ovejector;
ovj, ovejector;
p ex, excretory pore;
p gub, protractor of gubernaculum;
ph, external phasmid;
pl, posterolateral;
po c, metacorpus
ppl, papilla;
pr, prerectum;
pr c, procorpus;
ps, postero-subventral;
p sp, protractor of spicule;
p st, protractor of stylet;
ptr, protorhabdion;
pu, pulvillus;
r, rectum;
rc, latero-ventral commissure in Fig 8B, rectal commissure in Fig. 8C.
r gl, rectal gland;
r gub, retractor of gubernaculum;
r sp, retractor of spicule;
r st, retractor of stylet;
rt, retrovesicular;
rt c, retrovesicular commissure;
s, sensilla;
s c, sinus cell;
sd, subdorsal;
s m, sphincter muscles;
sp, spicule;
sp e, spermatozoan;
spn, spinneret;
s r, seminal receptacle;
st, stoma;
stc, stichocyte;
sto, stichosome;
sty, stylet;
sup or, supplementary organ;
s v, seminal vesicle;
sv, subventral;
sv gl, subventral esophageal gland;
t, testis;
td, terminus of dorsal (ray);
tr, telorhabdion;
tro, trophosome;
ut 1, distal part of uterus;
ut 2, proximal part of uterus;
ut f, uterine efferent;
v, vulva;
va, vagina;
v ch, ventral chord;
v d, vas deferens;
vl, ventrolateral;
vl c, ventrolateral commissure;
v m, vulvar muscle;
v n, ventral (somatic) nerve;
vul, vulva;
vv, ventroventral.

al, ala;
an, anus;
ap, apophysis;
cal, calomus;
cap, capitulum;
c gl, caudal gland;
ci, cirrus;
cl, cloaca;
co, corpus;
c m, copulatory muscle;
cru, crura;
cu, cun, cn, cuneus;
d r gl, dorsal rectal gland;
e, nucleus in esophago-intestinal valve;
ep c, epithelial cell;
g, gland cell nucleus;
gub, gubernaculum;

h, depressor ani;
i m, somato-intestinal muscle;
int, intestine;
lam, lamina;
l sp, left spicule;
m, marginal nucleus;
m c, mesenterial cell;
n, nerve cell nucleus;
pou, spicular pouch;
p gub, protractor muscle of gubernaculum;
p sp, protractor muscle of spicule;
r, radial nucleus;
r gl, rectal gland;
r sp, right spicule;
s, questionable nucleus;
sh, sheath;
s m, sphincter muscle;
sp, spicule;
sup or, supplementary organ;
sv r gl, subventral rectal gland;
tel, telamon;
v d, vas deferens;
x, questionable nucleus.

Chapters IX-XII

a, anterior cavity, Fig. 158;
ac, escort cell;
ad vc, anterior dorsoventral commissure;
a e l, anterior externolateral ganglion;
a il, anterior internolateral ganglion;
al, uc, anolumbar commissure;
amph gl n, amphidial gland nerve;
amph gn, amphidial ganglion;
an gn, anal ganglion;
ant ut, anterior uterus;
br gl, brown glands;
c gng, cephalic ganglion;
cl v, clavate cell;
da, dorsal antenna;
dc, dorsal cavity;
d ch, dorsal chord;
de n, deirid nerve;
d gn, dorsal ganglion;
dv c I-II, dorsoventral commissures;
ed, externodorsal ray;
ep c, epithelial cell;
ex bl, excretory bladder;
fa, fiber cell;
g, glia cell;
gc, coelomocyte;
ger c, germinal cell;
gon, gonad;
grm cl, germinal cell;

grm z, germinal zone;
grth z, growth zone;
h, hook;
l ant, lateral antenna;
l ch, lateral chord;
l dn, laterodorsal nerve;
l gn, lateral ganglion;
ln, lateral nerve;
l v c, lateroventral commissure;
l vn, lateroventral (subventral) somatic nerve;
lu gn, lumbar ganglion;
mas, mastax;
mc n, medial caudal nerve;
m el, medial externolateral ganglion;
m lu, metalumbar ganglion;
n r, nerve ring;
oog, oogonium;
or op, oral aperture;
ovj 1, ejectrix;
ovj 2, sphincter;
ovj 3, infundibulum;
p b, prebursal nerve;
p el, posterior externolateral ganglion;
ph gl, phasmidial gland;
pl gn, postlateral ganglion;
pl u, prolumbar ganglion;
post es, post-esophagus;
prb, prosboscis;
pu, post-uterine sac;
pv gn, postventral ganglion;
ric, ano-rectal commissure;
rchs, rachis;
rt gn, retrovesicular ganglion;
s, septum;
sd, 1-4, etc., subdorsal cephalic papillary neurones;
sd gn, subdorsal ganglion;
sem v, seminal vesicle;
s n, sinus nucleus;
s ncl, sinus nucleus;
sp gn, spicular ganglion;
spm, spermatozoön;
sv 1-5, etc., subventral cephalic papillary neurones;
sv c gn, subventral cephalic ganglion;
sv n, subventral ganglion;
sv nrv, subventral nerve;
sv p, subventral (lateroventral) papilla;
t cl, terminal cell;
t dn, terminal duct nucleus;
t p, caudal protuberance;
va gl, vaginal gland;
v def, vas deferens;
v gl, vaginal gland;
vlv, valve;
v u, vagina uterina;
v v, vagina vera.

SUBJECT INDEX

A

Acuariidae, p. 19.
Alaimidae, p. 24.
Allantonematidae, p. 15.
  life history, p. 256–264.
Alloionema appendiculatum
  life history, p. 249.
Anatomy,
  enumerative, nervous system, p. 165–172.
  minute, nervous system, p. 165–172.
  topographic, nervous system, p. 161–165.
Ancylostoma,
  female reproductive system, p. 145–147.
  life history, p. 272–273.
Ancylostomatidae, p. 16.
Angiostomatidae, p. 13.
Anguillicolidae, p. 19.
Anisakinae, life history, p. 281–282.
Antienzymes, p. 311–312.
Aphasmidia, p. 21–25.
  cephalic structures, p. 64–66.
  esophagus, p. 85–97.
  excretory system, p. 131–133.
  male reproductive system, p. 156.
  stoma, p. 71–75.
Aphasmidians, female reproductive
  system, p. 143–145.
Aphelenchoides, p. 3, p. 15.
Ascarid, excretory system, p. 131.
Ascarididae, p. 19.
  life history, p. 280–281.
Ascaridina, p. 18–19.
  cephalic structures, p. 59–60.
  esophagus, p. 80–83.
  female reproductive system, p. 147.
  intestine, p. 109.
  life history, p. 278–282.
  stoma, p. 69.
Ascaridoidea, p. 18–19.
  life history, p. 278–282.
Ascaris lumbricoides, nervous system,
  anatomy, p. 161–163.
Atractidae, p. 18.
Axonolaimidae, p. 22.
Axonolaimoidea, p. 22.

B

Belondiridae, p. 25.
Body cavity, p. 53–55.

C

Camallanidae, p. 19.
Camallanina, p. 19.
  cephalic structures, p. 61–64.
  esophagus, p. 83–85.
  female reproductive system, p. 147.
  life history, p. 291–293.
  stoma, p. 69.
Camallanoidea, p. 19.
Camallanus sweeti, life history, p. 291.
Capillaria columbae, life history, p. 293.
Cell, intestinal, p. 104–107.
Cephalic structures, p. 56–66.
Cephalobidae, p. 13.
  cephalic structure, p. 57.
Cephalobium microbivorum, life history,
  p. 250.
Chemical constitution, cuticular,
  p. 39–40.
Chromadorida, p. 21–22.
  esophagus, p. 85–88.
  intestine, p. 110.
Chromadoridae, p. 22.
Chromadorina, p. 22.
  cephalic structures, p. 65–66.
  stoma, p. 71–72.

Chromadoroidea, p. 22.
Cloaca, p. 118.
Cloacinidae, p. 16.
Coelomocytes, p. 53–55.
Comesomatidae, p. 22.
Cosmocercidae, p. 18.
  life history, p. 280.
Creagrocercidae, p. 13.
Criconematidae, p. 14–15.
Cucullanidae, p. 19.
Cucullanus, excretory system, p. 128.
Cuticle, external, p. 28–42.
  postembryonic development, p. 213–215.
Cyatholaimidae, p. 22.
Cylindrocorporidae, p. 13.
Cystoopsidae, p. 25.
Cystoopsis acipenseri, life history,
  p. 296.

D

Desmidocercidae, p. 21.
Desmodoridae, p. 22.
Desmodoroidea Steiner, p. 22.
Desmoscolecidae, p. 22.
Desmoscolecoidea, p. 22.
Diaphenocephalidae, p. 16.
Digestive tract, p. 9–10.
  chemical events, p. 308.
Dioctophymatidae, p. 25.
Dioctophymatina, p. 25.
  esophagus, p. 95–97.
  life history, p. 296.
  stoma, p. 75.
Dioctophymatoidea, ova, p. 185–186.
Dipetalonematidae, p. 19–21.
Diplogasteridae, p. 13.
Diplogasterids, cephalic structure, p. 57.
Ditylenchus, p. 2–3.
Dorylaimidae, p. 24.
Dorylaimina, p. 24–25.
  esophagus, p. 93–95.
  stoma, p. 74–75.
Dorylaimoidea, p. 24–25.
Dorylaimopsis, intestine, p. 110–112.
Draconematidae, p. 22.
Dracunculidae, p. 19.
Dracunculinae, p. 19.
Dracunculoidea, p. 19.
  ova, p. 182–183.
Dracunculus medinensis, life history,
  p. 291–293.
Draschia megastoma, life history,
  p. 284–286.
Drilonematidae, p. 13.
  life history, p. 251.
Drilonematoidea, p. 13.

E

Embryology, p. 202–212.
Embryonic sheath, p. 179–180.
Enoplida, p. 22–25.
  esophagus, p. 88–93.
  intestine, p. 112–113.
Enoplidae, p. 23.
Enoploidea, p. 23.
Enoplina, p. 23–24.
  cephalic structures, p. 66.
  stoma, p. 72–74.
Enterobius vermicularis, life history,
  p. 278.
Envelope, egg, p. 175–177.
Epsilonematidae, p. 22.
Esophago-sympathetic nervous system,
  p. 97.
Esophagus, p. 78–102, 97–100.
  postembryonic development, p. 217–219.
Excretory system, p. 11, 126–135.
  postembryonic development, p. 221.

F

Filariidae, p. 19.
Filarioidea, p. 19–21.
  life history, p. 288–291.
  ova, p. 183–185.

G

Gametogenesis, p. 191–201.
Gas, intestinal, composition, p. 308–309.
Gnathostoma spinigerum, life history,
  p. 284.
Gnathostomatidae, p. 19.
Gongylonema pulchrum, life history,
  p. 282–284.
Greefiellidae, p. 22.
Gubernaculum, p. 121–123.
Gut, posterior, p. 116–125.

H

Haemonchus contortus, life history,
  p. 273.
Heterakidae, p. 18.
  life history, p. 278–280.
Heterodera, p. 2.
Heteroderidae, p. 14.
Hydrogen ion concentration, p. 309–310.
Hypodermis, p. 8–9, 42–46.
  postembryonic development, p. 215.

I

Intestine, p. 102–115.
  postembryonic development, p. 219–221.
Ironidae, p. 24.

K

Kathlaniidae, p. 18.

L

Labial region, postembryonic develop-
  ment, p. 215–217.
Layering, cuticular, p. 37–39.
Leidynema appendiculatum, life history,
  p. 250–251.
Leptonchidae, p. 24–25.
Life history, p. 243–301.
Linhomoeidae, p. 22.

M

Membrane, p. 53–55.
  egg, p. 177–179.
  egg, chemistry, p. 186–187.
Mermithidae, p. 25.
Mermithoidea,
  female reproductive system, p. 149.
  ova, p. 185.
Mermithoidea Wülker, p. 25.
Mesenteron, see Intestine.
Metabolism,
  aerobic, p. 315–317.
  anaerobic, p. 314–315.
  of eggs and larvae, p. 317–319.
Metastrongylidae, p. 18.
Metastrongyloidea Cram, p. 18.
Metastrongyloidea, life history, p. 276–277.

Introduction to NEMATOLOGY

By the late **B. G. Chitwood** and **M. B. Chitwood**
with contributions by
R. O. Christenson, L. Jacobs, F. G. Wallace,
A. C. Walton, J. R. Christie, A. C. Chandler,
J. E. Alicata, J. E. Ackert, J. H. Whitlock,
T. von Brand, and T. L. Jahn

Few zoology texts have achieved the degree of universal
respect accorded the Chitwoods' classic *Introduction to
Nematology*. This remarkable treatise — originally
published in limited editions that have been
unobtainable for years — is considered today as one
of the foremost examples of truly comprehensive
scientific writing, prepared with such extraordinary care
and meticulous attention to detail that it remains
unchallenged as the best nematology text and reference
book available for beginning students, taxonomists,
morphologists, and those who use nematodes as
experimental animals.

This book presents a unified series of comparative
studies covering nematode anatomy, ontogeny, and
systematics, and aspects of physiology, with specific
historical outlines and detailed bibliographies for each
chapter. This new printing contains the complete,
unabridged text of the original (with the exception of
two non-essential chapters on epidemiology and
control). The hundreds of outstanding original drawings
have been reproduced without alteration, and the
massive bibliographies —"classics" in their own right —
are presented in full.

Publisher's Note: This new University Park Press edition of
Introduction to Nematology is the first ever issued by a
major scientific publisher, and the only one that combines
in a single volume the five separate parts of the earlier
editions. Any remaining copies of the original Chitwood
editions are now in restricted collections not generally
available to the public. University Park Press is glad to be
able to present this outstanding scientific treatise once again
to students and scholars throughout the world.

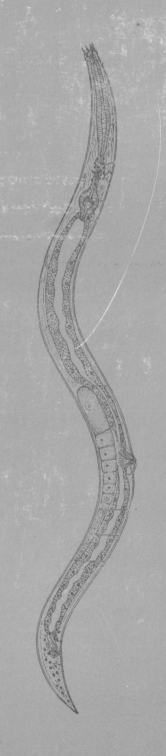

UNIVERSITY PARK PRESS

International Publishers in Science and Medicine
Chamber of Commerce Building
Baltimore, Maryland 21202